International Conference on Advanced Structural and Functional Materials Design 2008

AA002946

Journal of Physics: Conference Series Volume 165

Osaka, Japan
10-12 November 2008

Editors:

T. Kakeshita **Y. Fujiwara**

ISBN: 978-1-61738-369-4

Printed from e-media with permission by:

Curran Associates, Inc.
57 Morehouse Lane
Red Hook, NY 12571
www.proceedings.com

Some format issues inherent in the e-media version may also appear in this print version.

Printed by Curran Associates, Inc. (2010)

Institute of Physics
Dirac House, Temple Back
Bristol BS1 6BE UK

Phone: 44 1 17 929 7481
Fax: 44 1 17 920 0979

techtracking@iop.org

TABLE OF CONTENTS

ADVANCED MATERIALS FOR STRUCTURAL APPLICATIONS

ADVANCED MATERIALS FOR FUNCTIONAL APPLICATIONS

ADVANCED MATERIALS FOR STRUCTURAL AND FUNCTIONAL APPLICATIONS

The Ministry of Education, Culture, Sports, Science and Technology of Japan started the Priority Assistance for the Formation of Worldwide Renowned Centers of Research – Global COE Program. This program is based on the competitive principle where a third party evaluation decides which program to support and to give priority support to the formation of world-class centers of research. Our program *Center of Excellence for Advanced Structural and Functional Materials Design* was selected as one of 13 programs in the field of Chemistry and Materials Science. This center is composed of two materials-related Departments in the Graduate School of Engineering: Materials and Manufacturing Science and Adaptive Machine Systems, and 4 Research Institutes: Center for Atomic and Molecular Technologies, Welding and Joining Research Institute, Institute of Scientific and Industrial Research and Research Center for Ultra-High Voltage Electron Microscopy.

Recently, materials research, particularly that of metallic materials, has specialized only in individual elemental characteristics and narrow specialty fields, and there is a feeling that the original role of materials research has been forgotten. The 6 educational and research organizations which make up the COE program cooperatively try to develop new advanced structural and functional materials and achieve technological breakthrough for their fabrication processes from electronic, atomic, microstructural and morphological standpoints, focusing on their design and application: development of high performance structural materials such as space plane and turbine blades operating under a severe environment, new fabrication and assembling methods for electronic devices, development of evaluation technique for materials reliability, and development of new biomaterials for regeneration of biological hard tissues.

The aim of this international conference was to report the scientific progress in our Global COE program and also to discuss related research topics. The organizing committee gratefully thanks participants for presenting their recent results and for discussions with our COE members and international attendees.

November 2008

Professor Tomoyuki Kakeshita
Chairman of the Conference
Vice Dean, Graduate School of Engineering, Osaka University,
Division of Materials and Manufacturing Science, Graduate School of Engineering
Leader of Global COE Program, Osaka University,
"Center of Excellence for Advanced Structural and Functional Materials Design"

ORGANIZATION

Chairman:
T Kakeshita (Osaka University)

Advisory Board:
H Mehrer (University Münster, Germany), E K H Salje (University of Cambridge, United Kingdom), H-E Schaefer (University of Stuttgart, Germany), P Veyssiere (CNRS-ONERA, France)

Organizing Committee:
T Kakeshita, H Araki, H Fujii, S Fujimoto, Y Fujiwara, A Hirose, S Kirihara, M Mochizuki, H Mori, T Nagase, H Nakajima, T Nakano, R Nakatani, K Nogi, Y Setsuhara, Y Shiratsuchi, T Tanaka, T Terai, H Tsuchiya, N Tsuji, H Utsunomiya, H Yasuda, H Yasuda (Osaka University)

Executive Committee:
T Kakeshita, S Fujimoto, Y Fujiwara, A Hirose, T Tanaka, H Yasuda (Osaka University)

Conference Secretariat:
Y Fujiwara (Osaka University)

Proceedings Editors:
T Kakeshita and Y Fujiwara (Osaka University)

All papers published in this volume of *Journal of Physics: Conference Series* have been peer reviewed through processes administered by the proceedings Editors. Reviews were conducted by expert referees to the professional and scientific standards expected of a proceedings journal published by IOP Publishing.

Texture analysis and R-value variation after severe shear deformed and heat treated Al alloy sheets

Saidmurod Akramov [1], **Insoo Kim** [1*]

[1]School of Advanced Materials and Systems Engineering, Kumoh National Institute of Technology, Gumi, Gyung Buk, 730-701, Korea

[1]saidmurod23@yahoo.com, [1*]iskim@kumoh.ac.kr; corresponding author

Al alloy sheets were deformed by ECAP(equal channel angular pressing), asymmetry and frictional rolling without lubrication and subsequent heat treating, the textures and r-values of the samples measured by x-ray diffractometer and tensile tester, respectively. The $\bar{r}$ and $|\Delta r|$-values variations after ECAP, asymmetry rolling and frictional rolling without lubricant process and subsequent heat treating of Al alloy sheets are compared and analyzed by using the changes of inverse pole figures and f(g) values of ODFs.

1. Introduction

It has been found that the $\bar{r}$-values (average r-value) are increased and $|\Delta r|$-values are decreased with increasing <111> parallel to the ND of sheet plane in FCC metals [1]. To get the <111> parallel to the ND of sheet plane, many researchers have been studied in the various types of shear deformation processes. However, due to the high stored energy, the existence of structural heterogeneities and the critical annealing conditions in the deformed state, unwanted structures will be gotten during the heat treatment [2]. The $\bar{r}$-values of ECAPed, asymmetry rolled and frictionally rolled (severe shear deformation processes) and subsequent heat treated Al alloy sheets showed about 1.2-1.8 times higher than those of the initial Al sheets [3-8]. The $|\Delta r|$-values of ECAPed and subsequent heat treated AA1050 Al alloy sheets showed lower than those of the initial Al sheets. The $|\Delta r|$-values of asymmetry rolled and frictionally rolled and subsequent heat treated AA1050, AA3003 and AA5052 Al alloy sheets showed higher than those of the initial Al sheets [3-8]. In this paper, the effect of texture changes on the variations of $\bar{r}$-values and $|\Delta r|$-values after ECAPed, asymmetrically rolling and frictionally rolling without lubricant process is compared and analyzed by changes of inverse pole figures and f(g) values of orientation distribution functions (ODF) again.

2. Experimental procedure

AA1050 Al alloy sheets were used in present study for ECAP. For asymmetric and frictional rolling processes we used AA1050, AA3003 and AA5052 Al alloy sheets. The sheet samples, with dimensions of 35mm x 15mm x 2mm for ECAP and heat treated at 550°C during 2 hours (named initial Al sheet) and dimensions of 60mm x 40mm x 3.8mm for ECAP, asymmetric and frictional rolling processes heat treated at 500°C during 2 hours (named initial Al sheet) were prepared. The ECAP was performed in a die with an oblique angle (Φ) of 90° and a curvature angle (Ψ) of 20° at room temperature and at a constant speed of 2mm/sec by using route A(without any rotation between each pass) and C(180° rotation along rolling direction between each pass) types. To investigate texture and r-value development, samples were heat treated at 300°C for 1 hour in an air condition.

The initial Al sheets were asymmetrically rolled to 90% reductions on a laboratory asymmetrical rolling mill with roll radius ratio of 1.5. Samples were rotated 180° around transverse axis of Al sheets, rolled with more than 50% reduction in thickness and heat treated for 5 minutes at 200°C per each pass and the total number of rolling passes was 3. To obtain high friction coefficient, lubricant did not use in the rolling process. After the asymmetric rolling, samples were subsequent heat treated at temperatures of 250-300°C in a salt bath. The initial Al sheets were frictionally rolled to 90% reductions on a laboratory symmetrical rolling mill with of 150 mm in diameter. Samples were rotated 180° around transverse axis of Al sheets, rolled with more than 40-50% reduction in thickness and heat treated for 5 minutes at 200°C per each pass and the total number of rolling passes was 4. To obtain high friction coefficient, lubricant did not use in the rolling process. After the frictional rolling, samples were subsequent heat treated at temperatures of 275-300°C in a salt bath.

To know the texture changes of the deformed and heat treated samples, the incomplete pole figures of (111), (200), and (220) for each sample were measured by using X-ray goniometer. Texture measurements were performed at one tenth thickness layer (S=0.9) and at one fourth thickness layer (S=0.75) from the surface, which means 10% and 25% of sheet thickness were removed from upper surface, respectively. Also, observations of a half thickness layer (S=0) were performed.

To measure the r-value (ratio of true strain of width and true strain of thickness after tensile test), tensile test samples were prepared along the angles of 0°, 45°, 90°(180°), and 135° or 0°, 45°, 90° to ECAP or rolling direction, respectively. The $\bar{r}$-value and $|\Delta r|$-value were obtained from the measured r-value data. The $\bar{r}$-values were calculated by the use of $\bar{r} = (r_0 + 2r_{45} + 2r_{90} + 2r_{135} + r_{180}) / 8$ or $\bar{r} = (r_0 + 2r_{45} + r_{90}) / 4$, and $|\Delta r|$-values were calculated by the use of $|\Delta r| = (r_0 - r_{45}) + (r_{90} - r_{45}) + (r_{90} - r_{135}) + (r_{180} - r_{135}) / 4$ or $|\Delta r| = (r_0 - 2r_{45} + r_{90}) / 2$. Here, the $r_{0,\ 45,\ 90,\ 135,\ \text{and}\ 180}$ means the r-value of along the angles of 0°, 45°, 90°, 135°, and 180° to ECAP or rolling direction, respectively.

3. Results and discussion

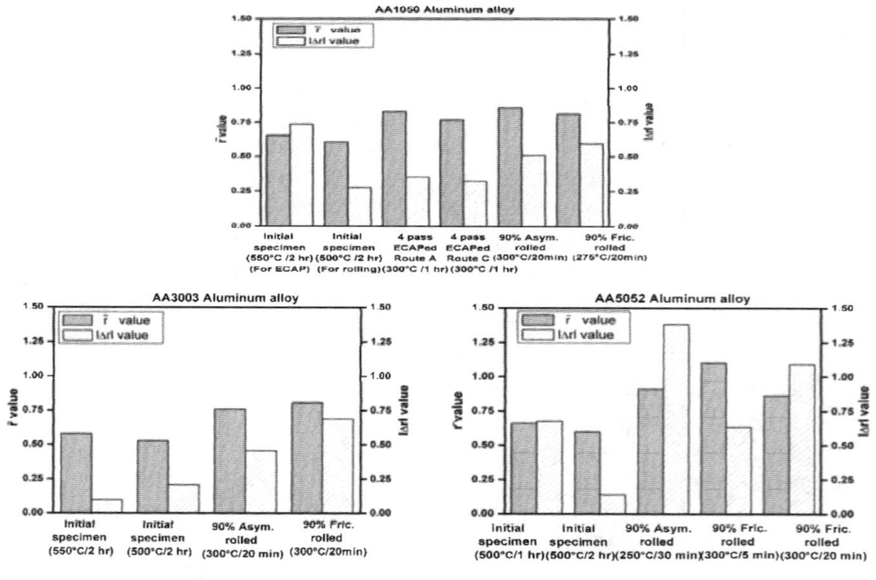

Figure 1 The $\bar{r}$-value and $|\Delta r|$-value variations after severe shear deformed and heat treated Al alloy sheets [3-8]

Figure 1 shows the $\bar{r}$-values and |Δr|-values variations after severe shear deformed (ECAP, asymmetry and frictional rolling) and subsequent heat treated Al alloy sheets [3-8].

The $\bar{r}$-values of ECAPed, asymmetrically rolled and frictionally rolled and subsequent heat treated Al alloy sheets showed about 1.2-1.5 times for AA1050 Al, 1.4-1.5 times for AA3003 Al and 1.4-1.8 times for AA5052 Al higher than those of the initial Al sheets in Figure 1.

Therefore, the increasing of $\bar{r}$-values in AA3003 and AA5052 Al alloys is higher than that of AA1050 Al alloy and the increasing of $\bar{r}$-values in AA5052 Al alloy is higher than that of AA3003 Al alloy after severe shear deformation and subsequent heat treatment in Figure 1.

The |Δr|-values of ECAPed, asymmetrically rolled and frictionally rolled and subsequent heat treated Al alloy sheets showed about 0.4-2.1 times for AA1050 Al, 2.2-3.3 times for AA3003 Al and 3.6-9.9 times for AA5052 Al higher than those of the initial Al sheets in Figure 1.

Therefore, the increasing of |Δr|-values in AA3003 and AA5052 Al alloys is higher than that of AA1050 Al alloy and the increasing of |Δr|-values in AA5052 Al alloy is higher than that of AA3003 Al alloy after severe shear deformation and subsequent heat treatment in Figure 1.

These changes of $\bar{r}$-values and |Δr|-values will be related to the change of textures as the next results.

Figure 2 shows inverse pole figures of AA1050 Al alloy, where S=0.9 data are obtained from one tenth surface thickness layer and S=0.75 data are obtained from one-fourth thickness layer and S=0 data are obtained from center layer of the Al sheet.

AA1050 initial Al samples for ECAP exhibit <001>//ND texture through the whole thickness layer due to heat treatment at 550°C for 2 hours in the Figure 2(a, b) and AA1050 initial Al samples for asymmetrical and frictional rolling exhibit also <001>//ND texture through the whole thickness layer due to heat treatment at 500°C for 2 hours in the Figure 2(c, d). After ECAPed AA1050 Al by using route A and heat treated at 300°C for 1 hour, <111>//ND and near <115>//ND textures are observed in the one-tenth thickness layer and <110>//ND texture are observed in one-fourth thickness layer in Figure 2(e, f). The increasing of r and $\bar{r}$-values is related to <111>//ND and <110>//ND textures in Figure 2 [1]. After ECAPed AA1050 Al by using route C and subsequent heat treated at 300°C for 1 hour, <110>//ND texture is observed in one-tenth thickness layer and <001>//ND and near <012>//ND textures are observed in on the one-fourth thickness layer in Figure 2(g, h). <110>//ND texture is related to increase r-value in Figure 2(g, h) [1]. Asymmetrically rolled and heat treated AA1050 Al at 300°C for 20 minutes shows random texture in one-tenth thickness and center layer in Figure 2(i, j). Frictionally rolled and heat treated AA1050 Al at 275°C for 20 minutes shows <001>//ND texture in one-tenth thickness and center layer in Figure 2(k, l).

Figure 3 shows inverse pole figures of AA3003 Al alloy, where S=0.9 data are obtained from one tenth surface thickness layer and S=0 data are obtained from center layer of the Al sheet.

AA3003 initial Al samples for asymmetric and frictional rolling show <001>//ND texture in the one-tenth thickness layer and random texture is observed in center layer due to heat treatment at 500°C for 2 hours in the Figure 3(a, b). After asymmetrically rolled AA3003 Al and heat treated at 300°C for 20 minutes, <111>//ND and <100>//ND textures are observed in the one-tenth thickness layer and <100>//ND texture is observed in center layer in Figure 3(c, d). <100>//ND texture was occurred after heat treatment in Figure 3(c, d). After frictionally rolled AA3003 Al and heat treated at 300°C for 20minutes, <113>//ND texture is observed in the one-tenth thickness layer and near <113>//ND, <100>//ND and <110>//ND textures are observed in center layer in Figure 3(e, f). <100>//ND texture is also occurred after heat treatment in Figure 3(e, f).

Figure 4 shows inverse pole figures of AA5052 Al alloy, where S=0.9 data are obtained from one tenth surface thickness layer and S=0 data are obtained from center layer of the Al sheet. AA5052 initial Al samples for asymmetry and frictional rolling show <001>//ND texture in the one-tenth thickness and center layer due to heat treatment at 500°C for 2 hours in the Figure 4(a, b). After asymmetrically rolled AA5052 Al and heat treated at 250°C for 30 minutes, <111>//ND and

<100>//ND textures are observed in the one-tenth thickness and center layer in Figure 4(c, d). After frictionally rolled AA5052 Al and heat treated at 300°C for 20minutes, strong <111>//ND and <100>//ND are observed in the one-tenth thickness layer and near <013>//ND texture is observed in center layer in Figure 4(e, f). The increasing of r and $\bar{r}$-values is related to <111>//ND texture in Figure 3 and 4 [1].

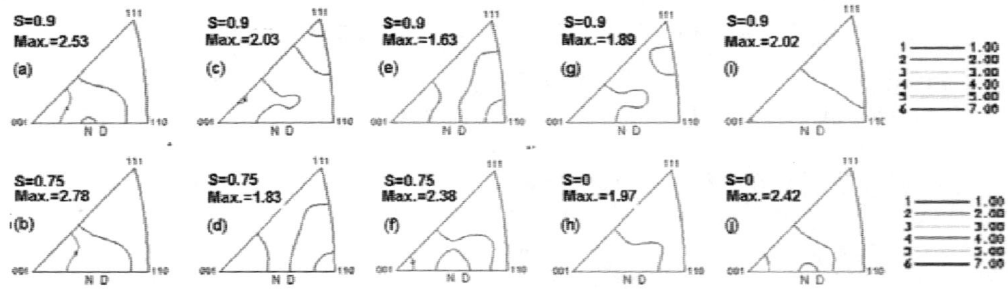

Figure 2 Inverse pole figures of one tenth thickness from the surface layer (S=0.9) and one-fourth and a half thickness layer (S=0.75 and S=0) of AA1050 Al alloy sheet: (a, b) initial Al sheet for ECAP, (c, d) initial Al sheet for asymmetric and frictional rolling, (e, f) 4 passes ECAPed and heat treated at 300°C for 1 hour (Route A), (g, h) 4 passes ECAPed and heat treated at 300°C for 1 hour (Route C), (i, j) 90% asymmetrically rolled and heat treated at 300°C for 20min., (k, l) 90% frictionally rolled and heat treated at 275°C for 20 min.

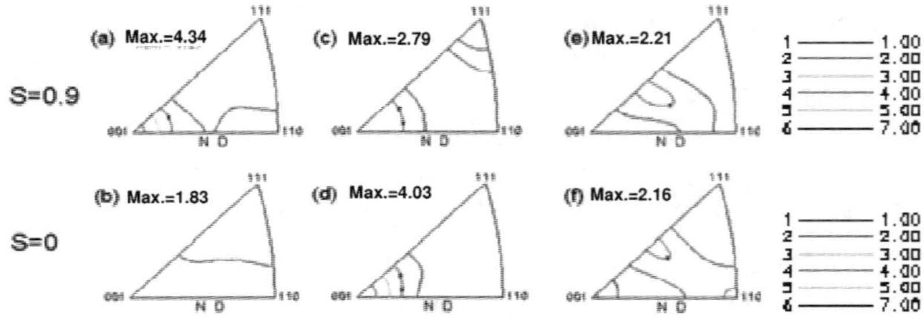

Figure 3 Inverse pole figures of one tenth thickness from the surface layer (S=0.9) and center layer (S=0) of AA3003 Al alloy sheet: (a, b) initial Al sheet, (c, d) 90% asymmetrically rolled and heat treated at 300C for 20 min., (e, f) 90% frictionally rolled and heat treated at 300°C for 20 min.

Figure 5 shows f(g) changes of AA3003 Al alloy, where S=0.9 data (a, b) and S=0 data (c, d) of the initial and asymmetric and frictional rolled and subsequent heat treated Al sheet, respectively.
Due to the full recrystalisation at 500°C for 2 hours, textures of initial AA3003 Al sheet show high {001}<001> and {013}<231> textures in both one-tenth thickness (S=0.9) and center layers (S=0). Textures of asymmetrically rolled and heat treated AA3003 Al sheet at 300° C for 20 minutes show high {001}<110> and {123}<634> textures in one-tenth thickness layer (S=0.9) and {001}<110>, {123}<634>, {012}<021>, {111}<112> and {111}<110> textures in center layer (S=0). Textures of frictionally rolled and heat treated AA3003 Al sheet at 300° C for 20 minutes show high {112}<111>, {111}<112>, {110}<112> and {110}<001> textures in one-tenth thickness layer (S=0.9) and {013}<231>, {112}<111>, {001}<100>, {110}<001> and {110}<111> textures in center layer (S=0).

The higher $\bar{r}$-values in Figure 1 are also related to increase the {111}<112>, {111}<110>, {123}<634>, {110}<001>, {112}<111> and {110}<111> texture components [1] after deformed and heat treated at 300° C for 20 minutes in Figure 5. Because, P. H. Lequeu et al [1] shown that the calculated $\bar{r}$-value of {111}<112>, {111}<110>, {110}<001> and {110}<112> ideal textures are higher than that of <100> // ND ideal texture. The |Δr|-value of the ECAPed and the subsequent heat treated Al sheet is lower than that of the initial Al sheet. This lower |Δr|-value is related to the decreasing of <100> // ND texture component, the increasing of other types of texture components at 300° C for 20 minutes heat treated Al sheet in Figure 5.

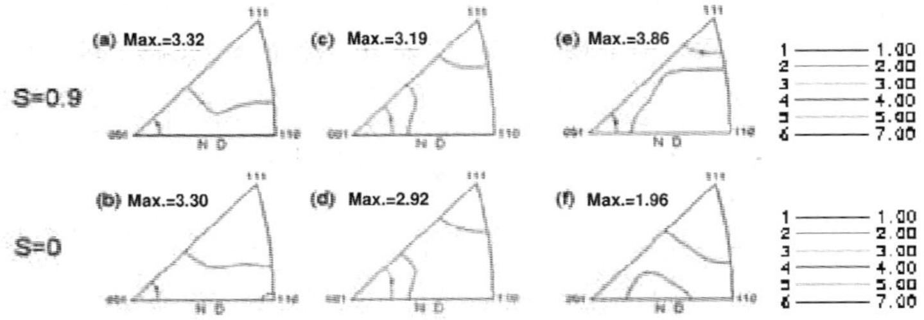

Figure 4 Inverse pole figures of one tenth thickness from the surface layer (S=0.9) and center layer (S=0) of AA5052 Al alloy sheet:(a, b) initial Al sheet, (c, d) 90% asymmetrically rolled and heat treated at 250°C for 30 min., (c) 90% frictionally rolled and heat treated at 300°C for 20 min.

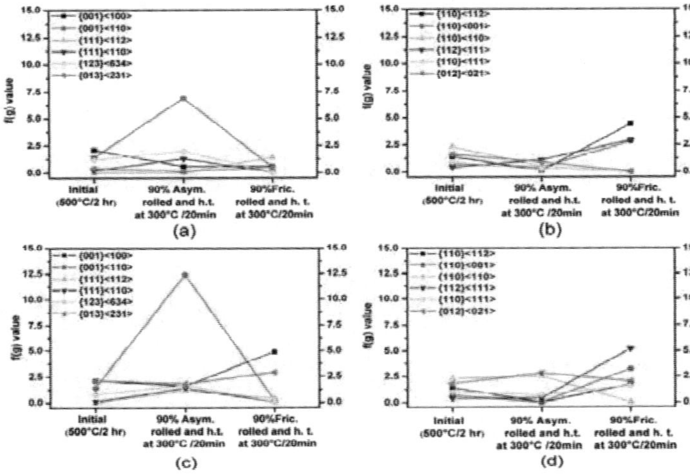

Figure 5 The f(g) value variations of various types of texture components obtained from the ODFs of the (a, b) one tenth surface layer (S=0.9) and (c, d) center layer (S=0) depth thickness of initial AA3003 Al sheet, and 90% asymmetrically rolled and heat treated at 250°C for 30 min. and 90% frictionally rolled and heat treated AA3003 Al sheets of at 300°C for 20 min.

Figure 6 shows f(g) value changes of AA5052 Al alloy, where S=0.9 data (a, b) and S=0 data (c, d) of the initial and asymmetric and frictional rolled and subsequent heat treated Al sheets, respectively.
Due to the full recrystalisation at 500°C for 2 hours, textures of initial AA5052 Al sheet show high {001}<001>, {110}<001> and {013}<231> textures in one-tenth thickness (S=0.9) and high {001}<001>, {110}<112>, {012}<021>, {001}<110> and {013}<231> textures center layers (S=0). Textures of asymmetrically rolled and heat treated AA5052 Al sheet at 250° C for 30 minutes show the increasing of {001}<110> and decreasing of {001}<100> and {013}<231> textures in one-tenth thickness layer (S=0.9) and increasing of {001}<110>, {123}<634>, {111}<110>, {111}<112> and {012}<021> and decreasing of {001}<100>, {013}<231>, {110}<112> and {110}<001> textures in center layer (S=0). Textures of frictionally rolled and heat treated AA3003 Al sheet at 300° C for 20 minutes show the increasing of {111}<112>, {111}<110>, {001}<110>, {012}<021> and {110}<110> textures and decreasing of {001}<100> and {013}<231> textures in one-tenth thickness layer (S=0.9) and increasing of {011}<110>, {112}<111> and {110}<112> textures and decreasing of {001}<100>, {013}<231> and {012}<021> textures in center layer (S=0). The higher $\bar{r}$-values in Fig. 1 are also related to increase the component of {111}<112>, {111}<110>, {123}<634>, {110}<001>, {112}<111> and {110}<111> textures [1] after deformed and heat treated at 250-300° C for 20-30 minutes in Figure 6. Because, P. H. Lequeu et al [1] have shown that the calculated $\bar{r}$-value of

{111}<112>, {111}<110>, {110}<001> and {110}<112> ideal texture omponents are higher than that of <100> // ND ideal texture. The $|\Delta r|$-values of the asymmetrically and frictionally rolled and the subsequent heat treated Al sheet are higher than those of the initial Al sheets. The higher $|\Delta r|$-value is related to the mixed effect of changes of texture components of the heat treated Al sheets at 250-300° C for 20-30 minutes in Figure 6.

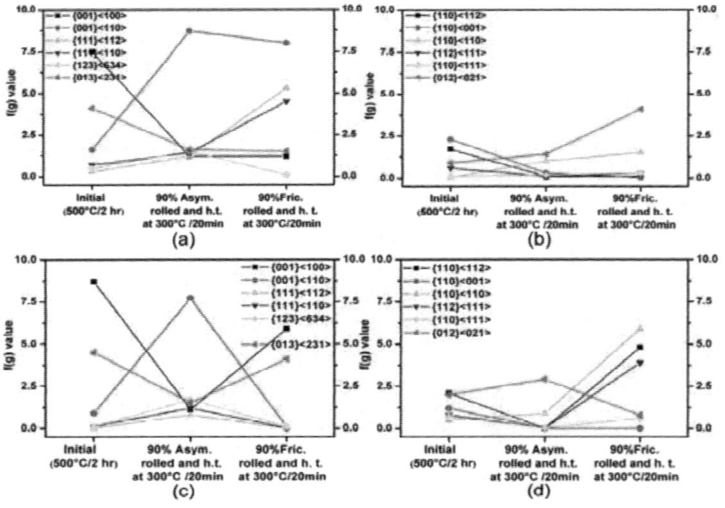

Figure 6 The f(g) value variations of various types of texture components obtained from the ODFs of the (a, b) one tenth surface layer (S=0.9) and (c, d) center layer (S=0) depth thickness of initial AA5052 Al sheet, and 90% asymmetrically rolled and heat treated at 250°C for 30 min. and 90% frictionally rolled and heat treated AA5052 Al sheets of at 300°C for 20 min.

4. Conclusions

1. The $\bar{r}$-values of ECAPed, asymmetry rolled and frictionally rolled and subsequent heat treated Al alloy sheets showed about 1.2-1.5 times for AA1050 Al, 1.4-1.5 times for AA3003 Al and 1.4-1.8 times for AA5052 Al higher than those of the initial Al sheets.
2. The $|\Delta r|$-values of ECAPed, asymmetry rolled and frictionally rolled and subsequent heat treated Al alloy sheets showed about 1.2-1.5 times for AA1050 Al, 1.4-1.5 times for AA3003 Al and 1.4-1.8 times for AA5052 Al higher than those of the initial Al sheets.

3. The changes of $\bar{r}$-values and $|\Delta r|$-values are related to the decreasing of {001}<100> and increasing of {111}<112>, {111}<110>, {123}<634>, {110}<001>, {110}<112>, {112}<111> and {110}<111> texture components after severe shear deformation and heat treatment of Al alloys.

Acknowledgment

This paper was supported by Research Fund, Korean Research Foundation Grant Funded by the Korean Government (MOEHRD) (KRF-2005-D00372) and Kumoh National Institute of Technology.

References

[1] Lequeu P H and Jonas J J 1988 Metal. Trans. A **19A** 105
[2] Hatherly M, Hutchinson W B 1979 An Intro. to Textures in Met., Inst. of Metallurgists, London
[3] Akramov S, Lee M G, Kim I, Sung D Y and Park B H 2005 Mater. Sci. Forum **475/479** 417
[4] Akramov S, Lee M G, Kim I, Sung D Y, Park B H 2005 Mater. Sci. Forum **495/497** 803
[5] Akramov S, Kim I, Lee M G and Park B H 2006 Sol. state phenon. **116/117** 324
[6] Akramov S, Kim I and Park N J 2007 Adv. Mater. research **26/28** 393
[7] Kim I, Akramov S and Jeong H B Mod. Phys. letter B in press
[8] Kim I, Akramov S, Jeong H B and No T K Mod. Phys. letter B in press

Secondary Reaction Zone Formations in coated Ni-base Single Crystal Superalloys

A Suzuki and C M F Rae

Rolls-Royce University Technology Partnership/University of Cambridge, Department of Materials Science and Metallurgy, Pembroke Street, Cambridge CB2 3QZ, UK

as643@cam.ac.uk

Abstract. Ruthenium (Ru) has been added to the latest 4th Generation Ni-base superalloys to improve phase stability and modify creep life. Various coatings are routinely applied to these advanced alloys to protect the turbine blade at elevated temperature, however, this creates several problems such as the precipitation of brittle Topologically Close-Packed (TCP) phases and the formation of Secondary Reaction Zones (SRZ). The SRZ forms under the plat-aluminized coating of turbine blades and consists of γ, γ' and TCP phases growing into substrate by the migration of high-angle grain boundaries. Surface residual stress and chemical super-saturation of alloying elements are associated to SRZ formation. In the thin sections of high-pressure turbine blades this is critical in determining blade performance and longevity. It is essential to know how Ru additions affect coating and SRZ morphologies during exposure. In this study, we focus on the effects of three variables on the SRZ formation: Ru concentration, alloy composition in Ru-containing alloys and surface finish. A series of Platinum-Aluminised superalloys containing 2-5wt% Ru and having various surface finishes was studied after isothermal exposure at 1100°C for up to 500h. The alloys were classified into two groups by their distinctive SRZ morphology. At the lowest Ru levels sporadic formation of SRZ was observed, whilst a continuous SRZ was formed in the higher Ru alloys. EBSD analysis revealed that the latter group have a higher nucleation rate of individual SRZ grains and also showed more rapid SRZ growth. The precipitation of TCPs in the substrate also inhibited the growth of the SRZ towards the end of the exposure further reducing the penetration of the SRZ into the substrate. It is concluded that Ru-additions to Ni-base superalloys are effective in impeding TCP phase formation in the substrate, but increase both the extent and the rate of SRZ formation beneath coating.

1. Introduction

Ni-base superalloys are used as single crystal castings to make high-pressure turbine blades which operate at up to 80% of their absolute melting temperature whilst retaining excellent mechanical properties and oxidation resistance. The latest 4th generation Ni-base superalloys, contain 2-5% Ruthenium (Ru), and this is very effective in improving the mechanical properties in part by suppressing the formation of deleterious intermetallic phases[1,2].

The improved mechanical properties come at the cost of the oxidation resistance and coatings, such as Platinum Aluminide coatings (PtAl) and MCrAlY, are necessary to protect the turbine blades from oxidation and corrosion during service, details of these coatings are given elsewhere [3-9].

Several major problems are associated with coated Ni-base superalloys leading to the loss of the coating, these include the precipitation of topologically close-packed (TCP) phases, the formation of the secondary reaction zone (SRZ) and the 'rumpling' of the coating caused by thermal cycling. These problems degrade the properties of coated Ni-base superalloys, and are potentially life-limiting to the turbine blades. In particular, TCP phase precipitation and Secondary Reaction Zone (SRZ) formation are regarded as the most serious problems in both coated and uncoated Ni-base superlloys.

Topologically Close-Packed (TCP) phases is a collective designation for several intermetallic compounds rich in the elements W, Mo, Re and Cr precipitated in Ni-base superalloys [10]. TCP phases form during service at elevated temperature in Ni-base superalloys that have high concentrations of these elements added to promote strength the loss of these elements from the matrix impairs the mechanical properties of the blade. Precipitation of the TCP phases deteriorates the ductility and creep resistance of Ni-base superalloys. Further details of TCP phases are given in references [11-13].

The Secondary Reaction Zone (SRZ) is an intermediate layer formed between an aluminized or Plat Aluminized coating and the substrate by a discontinuous precipitate reaction similar to recrystallization. Figure 1 shows a schematic image of the SRZ which is formed beneath the inter diffusion zone (IDZ) of the aluminide coating with the substrate. It transforms the metastable aluminium-enriched substrate microstructure into an equilibrium mix of γ, γ' and TCPs. The TCPs align perpendicular to the growth direction facilitated by the rapid diffusion path of the high angle boundary. Therefore, the SRZ is associated with a boundary having both high mobility and diffusivity.

Among several potential risk factors for SRZ formation [14,15], nucleation appears to be associated with surface damage prior to coating as the SRZ was not observed when the surface was electropolished [15]. The driving force is the difference in free energy associated with the phase changes plus any loss of dislocation density and the reduction in the interface area brought about the coarsening of the precipitates. Also, an alloy containing more than 4-5wt% Re is found to have a high probability of forming SRZ [14,15].

As well as the precipitation of TCP phases, SRZ formation results in coarsening the γ/γ' structure and causes the mechanical properties of the alloy to deteriorate. Also, SRZ formation beneath the coating is said to reduce the load bearing section of the blade [16]. The formation of SRZ indicates the inter-diffusion between coating and the substrate, so that phase instability is accompanied in the alloy [14].

There are many studies examining the effect of Ru on the structure and stability of 4th generation superalloys [17], however the effect of Ru coated alloy has not been fully clarified. This is the subject of this paper.

2. Experimental

Nominal compositions of the Ni-base single crystal superalloys use in this paper are listed in Table 1. All alloys were supplied in the fully heat treated condition. The alloys contain Ru in the range of 2-5wt%, replacing Ni all other elements remaining constant. The coatings were applied to surfaces cut normal to the bar axis, nominally [001]. Specimens are polished up to 600 SiC paper prior to Pt coating, and

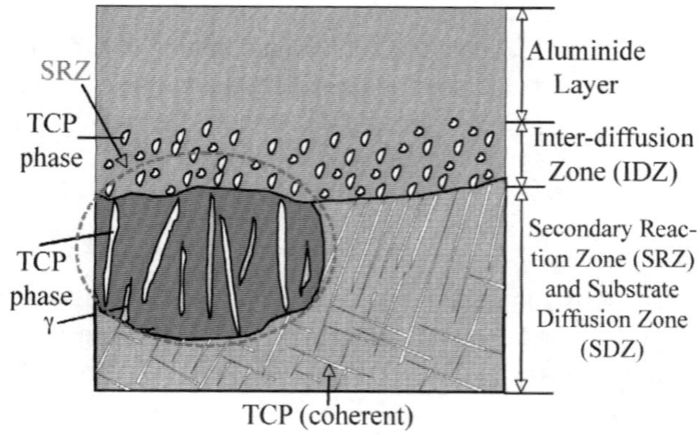

Figure 1. Schematic image of aluminized Ni-base superalloys.

Table 1. Alloy composition (wt%).

	Co	Cr	Mo	W	Re	Ru	Al	Ta	Hf	Ni
UCSX-2+2Ru	8.0	3.0	1.0	8.0	6.5	2.0	5.4	8.0	0.1	58.0
UCSX-2+3Ru	8.0	3.0	1.0	8.0	6.5	3.0	5.4	8.0	0.1	57.0
UCSX-2+5Ru	8.0	3.0	1.0	8.0	6.5	5.0	5.4	8.0	0.1	55.0

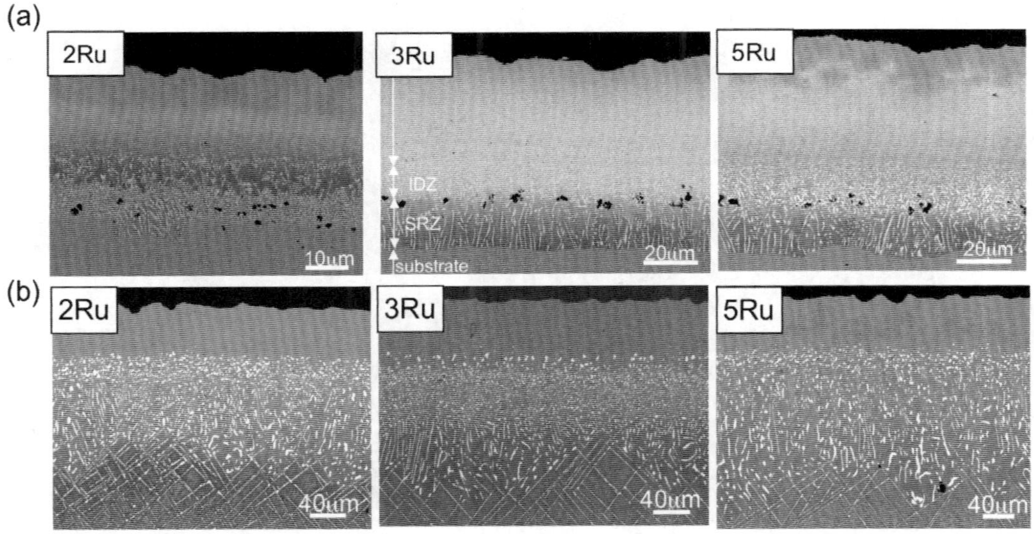

Figure 2. Summary of microstructures of UCSX alloys (a) as coated and (b) after 500h exposure.

vapour aluminised after Pt coating. Isothermal exposure tests were conducted at 1100°C to investigate the microstructure by means of SEM and EBSD (Electron Back-Scattered Diffraction).

The thickness of the SRZ was measured to determine the growth of the SRZ. For each specimen an average of 10 points was measured, depending the area fraction of SRZ. Further details of specimen preparations and SRZ thickness measurement should be referred elsewhere [18].

3. Results

3.1. Effect of Ru content on SRZ formation

Figure 2 (a) shows a montage of the microstructures of the as Pt-Aluminized specimens. All the specimens consist of β (Pt,Ni)Al surface layer, an Inter Diffusion Zone (IDZ), and the substrate in that order. The specimens had developed Secondary Reaction Zones (SRZ) during the 1 h heat treatment at 1100°C after aluminizing. Depending on the alloy composition the alloys showed a much greater effect. All the UCSX-2+2Ru samples showed sporadic SRZ formation, whereas the UCSX-2+3Ru and UCSX-2+5Ru specimens showed more continuous SRZ formation. This suggests that a higher amount of Ru in the substrate encourages SRZ nucleation at an early stage.

Figure 2(b) shows the microstructures after 500 h exposure. All the specimens have now developed a full SRZ layer and show profuse TCP precipitation in the SDZ. UCSX-2+2Ru in particular, showed the highest proportion of TCP phases in SDZ. Some of these TCP plates in the substrate are aligned with rows of incoherent precipitates in the SRZ indicating that the SRZ continues to grow after the TCPs form in the substrate. In other instances, TCP phases, precipitated in SDZ, prevent further growth of the SRZ. This seems to be the case where the TCP plate lies parallel to the boundary plane [18].

The higher Ru alloys, UCSX-2+3Ru and UCSX-2+5Ru, showed more SRZ growth than UCSX-2+2Ru. At the same time, these two alloys showed less TCP precipitation in the SDZ than UCSX-

2+2Ru. There is hence a negative correlation between TCP precipitation in the substrate and SRZ formation.

3.2. Microstructural Evolution of the SRZ and the SDZ

Figure 3 shows the growth kinetics of the SRZ in each specimen during the first 500 h exposure. Quantifying the SRZ thickness reveals a numbers of interesting trends. In the alloy UCSX-2+2Ru the SRZ ceases to grow after about 50 h exposure for the vapor aluminized specimens, whereas the SRZ in alloys UCSX-2+3Ru and UCSX-2+5Ru continues to grow substantially after 50 h.

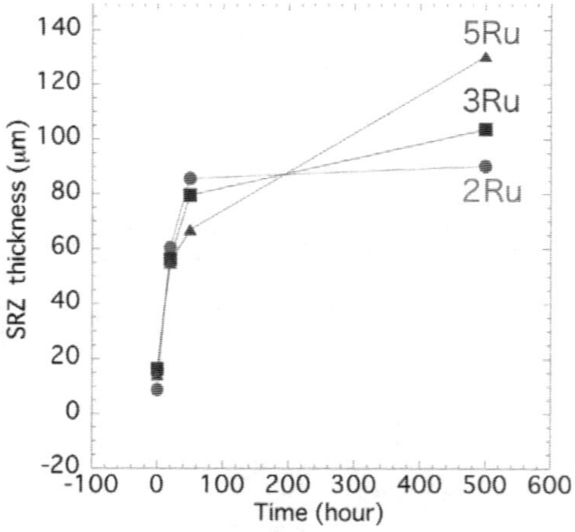

Figure 3. SRZ growth kinetics.

Figure 4 shows EBSD mappings and the corresponding pole figures for typical SRZ areas between the IDZ and the substrate. The normal to the coated surface for all alloys lies within 20° of the [001] direction, but the sections were cut at various angles giving a variety of substrate orientations perpendicular to [001] in Figure 4. Hence the [001] pole figures have in all cases been rotated to locate the substrate [001] orientation normal to the page and the [100] and [010] orientations along the x and y axes. The black areas correspond to the TCP phases which were generally not recognized by the software under the conditions used for mapping.

The number of SRZ grains nucleated during the early stages of growth varies greatly with alloy content. This has been quantified by counting the number of grains at the root of the SRZ adjacent to the coating and at the end of the SRZ adjacent to the substrate. A straight line was drawn at each location and all orientations where two or more adjacent points were indexed were included; the results are given in Table 2.

There is a clear effect of the higher Ru content resulting in a much higher nucleation rate, for example the number of grains increases by nearly 70% from the 2% Ru alloy to the 5% Ru alloy. The number of grains converges on a similar number as the SRZ grows with the 2% Ru alloy nucleating further grains and the 5% Ru alloy losing some orientations.

4. Discussion

The effect of Ru on the formation of SRZ is an issue of importance to the alloy developer. In this study the concentration of Ru was varied over the range found necessary to effectively suppress the formation of TCP phases in the substrate [15,19] whilst keeping the cost of the

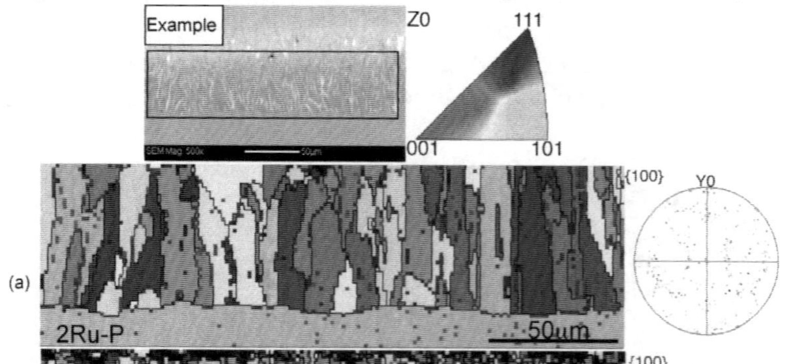

Figure 4. EBSD mapping and pole figures of (a) UCSX-2+2Ru, (b) UCSX-2+3Ru, and (c) UCSX-2+5Ru after the exposure of 20h.

Table 2. Number of grains in SRZ.

Alloy	2Ru	3Ru	5Ru
Number at root	31	47	52
Number at tip	36	40	44

Table 3. Percentage of twin boundaries in SRZ.

Alloy	2Ru	3Ru	5Ru
% of Twin boundaries	16.5	18.9	19.7

alloy within reasonable bounds and avoiding the formation of the delta phase [20].

The formation of the SRZ involves the nucleation of grains of a distinct orientation and their subsequent migration into the substrate by the passage of a high angle grain boundary. The boundary provides a fast diffusion path enabling the rationalization of the fine γ/γ' microstructure of the substrate to a lower-energy, coarser morphology and at the same time the precipitation of the TCP phases at approximately equilibrium volume fractions for the local composition. This is, of course, modified by the diffusion of the Al into the substrate which varies throughout the exposure. SRZ formation is thus a discontinuous precipitation process.

Ru increases the thickness of the SRZ for the specimens in Figure 3. In the first 50 hours SRZ grows in all three alloys at about the same rate but after that time growth continues in the 3% and the 5% alloy but not the 2% alloy. The continuous growth of the SRZ facilitates the diffusion of Al into the substrate thus reducing the Al levels in the SRZ and increasing the proportion of γ.

There is a significant increase in the nucleation rate as the Ru content increases as shown by the very large numbers of grains at the root of the SRZ in UCSX-2+3Ru and UCSX-2+5Ru. The lowest Ru alloy UCSX-2+2Ru was the only sample to show evidence of retained substrate between the coating and the SRZ, in all other cases the SRZ nucleated directly below the NiAl coating. It is not clear why this occurs but may reflect an increased driving force for the formation of the SRZ with Ru content.

It was expected that the number of grains in the SRZ would decrease as growth competition between the grains selected those with the boundary misorientation which had the greatest mobility. This does not appear to be the case as no preferred orientation for the grains of the SRZ was observed. The number of grains increased or decreased with growth towards a constant value. At the outer edge of the SRZ the average width of a grain is about 6 μm, very similar to the spacing of the TCPs. Growth of the TCPs is analogous to eutectoid precipitation with the TCPs lying approximately parallel to the growth direction, and it can be seen that the TCP phases frequently lie at the boundaries between the grains. It appears that the width of a grain is ultimately determined by the spacing of the TCP precipitates. This, in turn, should be a compromise between the diffusion distance along the moving boundary and the minimization of the interphase boundary energy between the γ' and the TCPs. As the volume fractions of the TCPs for the Polished specimens are all approximately 15% after 20 h it is understandable why the spacing of the grains adjusts to a similar value in all the specimens.

All the boundaries between the SRZ and the substrate were high angle boundaries with misorientation between 20°-60°, consistent with the report by Lavigne [21]. Hence some selectivity in either nucleation or growth in the very early stages of the SRZ formation has occurred to eliminate the less mobile lower angle boundaries.

We observe a clear increase in the number of grains nucleated as the Ru content increases. However this is associated with a *lower* growth rate of the SRZ in all alloys particularly in the first 50 h exposure. The same effect is seen with respect to the surface damage [18]. It is not clear that the grain size is *necessarily* the cause of the lower growth rate, merely that lower growth rate is associated with a smaller grain size resulting either from high surface damage or high Ru content. Further investigations are required to differentiate these effects.

The mechanism by which the number of grains reduces is clear: some orientations simply become occluded by others, particularly where these are separated by a TCP plate. But the mechanism by which the number of grains increases is less obvious; one possibility is simply the nucleation of new orientations by the formation of twins which form 20% of boundaries between SRZ grains in Table 3.

The percentage of twins increased with Ru content in substrate. Twin boundaries forms with lower energy than normal grain boundary. Thus, more formations of twin boundaries might be associated with deeper SRZ penetration in higher Ru alloys.

5. Conclusions

Secondary Reaction Zones in Pt-Al coated 4th generation Ni-base blade alloys were investigated, and following sequences are obtained.

- Ru is effective in inhibiting TCP formation in substrate, but promotes greater SRZ formation.
- TCP precipitation in the substrate inhibits SRZ growth.
- Higher Ru resulted in an increased nucleation rate of the SRZ.

Acknowledgements

The authors would like to thank Dr R.A. Hobbs, Dr N. Jones, and Dr B. Broomfield in Rolls-Royce plc for providing materials and fruitful discussion. Also, the authors will give great acknowledgement to Mrs M. Yoshida, Dr Y. Matsubara and Dr H. Murakami in NIMS Thermal Spraying Group to prepare specimens in this study.

References

[1] Koizumi Y et al, 2003, *Proc. of 31st Gas Turbine Fall Lecture Meeting* (Toyama, Japan: Gas Turbine Society of Japan), p259
[2] Yeh A C, Rae C M F and Tin S, 2004, *Superalloys 2004*, ed Green K A et. al. (Warrendale: TMS), p677
[3] Krishna G R, Das D K, Singh V and Joshi S V, 1998, *Mat. Sci. and Eng. A*, **251**, 40
[4] National Materials Advisory Board, Commision on Engineering and Technical Systems, National Research Council, 1996, *Coatings for High-Temperature Structural Materials – Trends and Opportunities* (Washington: National Academy Press)
[5] Angenete J and Stiller K, 2001, *Mat. Sci. and Eng. A*, **316**, 182
[6] Zhang Y, Haynes J A, Lee W Y, Wright I G, Pint B A, Cooley K M and Liaw P K, 2001 *Met. and Mat. Trans.* A, **32**, 1727
[7] Warnes B M, 2001, *Surf. and Coat. Tech.*, **146-147**, 7
[8] Stinger M J, Yanar N M, Topping M G, Pettit F S, Meier G H, 1999, *Z. Metalkd*, **12**, 1069
[9] Reed R C, 2006, *The Suprealloys – Fundamentals and Applications* (Cambridge: Cambridge University Press)
[10] Sinha A K, 1972, *Progress in Materials Science*, **15**, 81
[11] Rae C M F and Reed R C, 2001, *Acta Mater.*, **49**, 4113
[12] Pollock T M, 1995, *Mat. Sci. and Eng. B*, **32**, 255
[13] Karanaratne M S A, Rae C M F and Reed R C, 2001, *Met. and Mat. Trans. A*, **32**, 2409
[14] Walston W S, Schaeffer J C and Murphy W H, 1996, *Superalloys 1996*, ed Kissinger R D et. al. (Warrendale: TMS), p9
[15] Walston W S, Cetel A, MacKay R, O'Hara K, Duhl D and Dreshfield R, 2004, *Superalloys 2004*, ed Green K A et. al. (Warrendale: TMS), p15
[16] Zhao J C, Pfaendtner J A, Govern C and Jackson M R, 2005, *United States Patent*, 6921586
[17] Hobbs R A, Zhang L, Rae C M F and Tin S, 2008, *Met. and Mat. Trans. A*, **39,** 1014
[18] Suzuki A, Rae C M F, Yoshida M, Matsubara Y and Murakami H, 2008, *Superalloys 2008*, ed Reed R C et. al. (Warrendale: TMS), p777
[19] O'Hara K S, Walston W S, Ross E W and Darolia R, 1996, *United States Patent*, 5482789
[20] Tryon B and Pollock T M, 2006, *Mat. Sci. and Eng. A*, **430**, 266
[21] Lavigne O, Ramusat C, Drawin S, Caron P, Boivin D and Pouchou J L, 2004, *Superalloys 2004*, ed Green K A et. al. (Warrendale: TMS), p667

Microstructure homogeneity and mechanical properties of electrodeposited nanocrystalline Ni

D Terada [1], H Hase [1], M Sonobe [2], T Yamasaki [2], N Tsuji [1]

[1] Department of Adaptive Machine Systems, Graduate school of Engineering, Osaka University, 2-1 Yamadaoka, Suita, Osaka 565-0871, Japan

[2] Department of Materials Science & Chemistry, Graduate school of Engineering , University of Hyogo, 2167 Shosha, Himeji, Hyogo 671-2280, Japan

terada@ams.eng.osaka-u.ac.jp

Abstract. Microstructure and mechanical properties of an electrodeposited nanocrystalline Ni (nc-Ni) were investigated. Electrodeposition was done on two different kinds of Cu substrates. One was a Cu sheet having coarse grains, and the other was a sheet having a nanocrystalline Cu (nc-Cu) layer on the surface. The substrates were chemically removed after the electrodeposition of Ni. Then, sheets of fully dense nc-Ni with a thickness of approximately 100 μm were obtained. The surface of the sheets which had touched the substrate showed a mat surface appearance, while the other surface was mirror-like. In the nc-Ni electrodeposited on the coarse grained Cu substrate, the mirror-like surface showed homogeneous equiaxed grain structure, while the mat surface showed inhomogeneous structure. In the nc-Ni electrodeposited on the nc-Cu substrate, the microstructure of the mirror-like surface was similar to that in the nc-Ni deposited on the coarse grained Cu. However, the mat surface showed the homogenous structure, different from that in the nc-Ni deposited on the coarse grained Cu. This result indicates that the grain size and homogeneity of the microstructure of nc-Ni sheets were affected by the microstructure of the substrate. The electrodeposited nc-Ni performed strength four times higher than that of the conventionally coarse-grained pure Ni with grain size of 28 μm. In the tensile test, the total elongation of the nc-Ni electrodeposited on the coarse grained Cu was smaller than that of the nc-Ni electrodeposited on the nc-Cu. It is suggested that the homogeneity of microstructure significantly affects the mechanical properties of nc-Ni.

1. Introduction

Grain refinement is one of the ways to improve mechanical properties of metallic materials. It is well known that ultrafine-grained (UFG) materials fabricated by severe plastic deformation (SPD) processes [1-4] exhibit high strength and peculiar mechanical properties [5, 6]. However, it is difficult to fabricate nanocrystalline materials whose grain size is about several tens nanometers through the SPD process. Thus, it is still unclear whether the mechanical properties of the nanocrystalline materials are similar to that of the UFG materials or not.

Electrodeposition has been known as a technique for fabricating nanocrystalline materials having the mean grain size of several tens nanometers [6, 7]. In the present study, microstructure of an electrodeposited nanocrystalline Ni is observed in detail. In addition, mechanical properties of the nanocrystalline Ni are also evaluated by tensile test.

2. Experimental procedure

The material investigated in the present study is an electrodeposited nanocrystalline Ni (nc-Ni). The solution containing $NiSO_4$, $NiCl_2$, H_3BO_4, saccharine and sodium lauryl sulfate was used as the plating bath for the production of nc-Ni. Electrodepositions were done on two different kinds of Cu substrates at 65℃ and pH 5.0 with current density of 0.05A/cm^2 for 3hours. One of the substrates was a fully annealed Cu sheet having coarse grains, and the other was a sheet having a nanocrystalline Cu (nc-Cu) surface layer fabricated by plating. The substrates were chemically removed after the electrodeposition of Ni. Then, sheets of fully dense nc-Ni with a thickness of approximately 100 μm were obtained. The surface of the sheets which had touched the substrate showed a mat surface appearance, while the other surface which had touched the plating solution was mirror-like.

The structure of the nc sheets was characterized by TEM (transmission electron microscopy). In order to compare the structure of the mirror-like surface and the mat surface, the TEM specimens were prepared for both surfaces.

The mechanical properties of the nc-Ni were investigated by Vickers hardness test and tensile test. Tensile specimens 5mm in gage width and 10 mm in gage length (1/5 of JIS-5 specimen) were tested at ambient temperature at an initial strain rate of 8.3×10^{-4} s^{-1}. Vickers hardness was measured on the cross-sections of the nc-Ni sheets to clarify the difference throughout thickness of the sheets.

3. Results and discussion

3.1. Microstructure of the nc-Ni sheets

The TEM micrographs of the nc-Ni electrodeposited on the coarse grained Cu substrate were shown in Fig.1. The microstructure of mirror-like surface (solution side) and mat surface (substrate side) were shown in Fig. 1 (a) and (b), respectively. The mirror-like surface showed homogeneous equiaxed nanocrystalline grain structure. The mean grain size of the mirror-like surface was 31 nm. On the other hand, the mat surface showed inhomogeneous structure composed of small nanocrystalline grains and large grains with diameters larger than 100 nm. The mean grain size of the small grains and large grains in the mat surface region were 54 nm and 128 nm, respectively. The nc-Ni grains involve diffraction contrasts which might be caused by defects within the grains, such as dislocations. Some dislocations were observed in the nc-Ni grains by high resolution TEM observation (Fig.1 (c)). In Fig.1 (c), white lines show the lattice plane and an extra half plane was observed. The result confirms that the dislocations were introduced even in the electrodeposition process.

Figure 2 shows the TEM micrographs of the mirror-like surface (a) and the mat surface (b) in the nc-Ni electrodeposited on the nc-Cu substrate. The microstructure of the mirror-like surface was homogeneous, which was similar to that in the nc-Ni deposited on the coarse grained Cu (Fig.1 (a)). The mean grain size of the mirror-like surface was 28 nm. The mat surface also showed the relatively homogenous structure having the mean grain size of 40 nm, which was different from that in the nc-Ni deposited on the coarse grained Cu (Fig.1 (b)). These results indicate that the grain size and

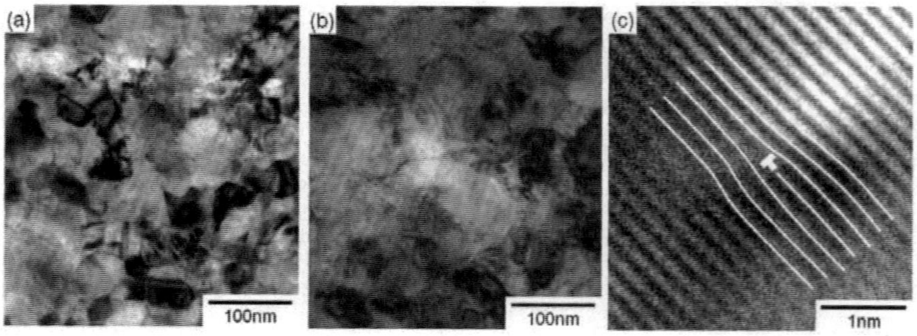

Figure 1. TEM micrographs of the nc-Ni electrodeposited on the coarse grained Cu substrate. (a) mirror-like surface (solution side), (b) mat surface (substrate side) and (c) high resolution TEM microstructure of mirror-like surface.

microstructural homogeneity of the electrodeposited Ni are affected by the microstructure of the substrate. Such an inhomogeneity in the electrodeposited materials has not been reported and focused. However, it is important when considering mechanical properties, as will be shown in the following sections.

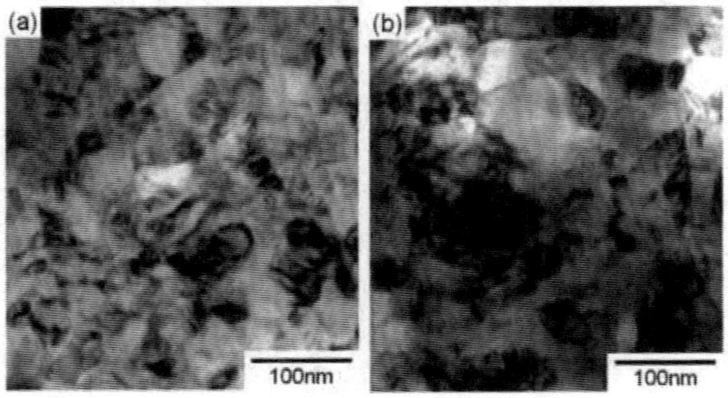

Figure 2. TEM micrographs of the nc-Ni electrodeposited on the nc-Cu substrate. (a) mirror-like surface (solution side) and (b) mat surface (substrate side).

3.2. Vickers hardness of the nc-Ni sheets

Vickers hardness of the nc-Ni is shown in Fig. 3 as a function of the normalized distance from the mat surface, t/t_0, where t is the distance from the mat surface and t_0 is the total thickness of the nc sheet. Therefore, $t/t_0 = 0$ indicates the mat surface, while $t/t_0 = 1$ the mirror-like surface. In Fig. 3, the circles and triangles indicate the hardness of the nc-Ni electrodeposited on the coarse grained Cu and the nc-Cu, respectively. In the nc-Ni electrodeposited on the coarse grained Cu substrate, the hardness near the mat surface was significantly smaller than that near the mirror-like surface. The hardness increased with increasing t/t_0 up to 0.2, and then the hardness showed nearly a constant value after that. On the other hand, in the nc-Ni electrodeposited on the nc-Cu substrate, the dependence of hardness on t/t_0 was not significant. The hardness distributions coincide well with the microstructural results shown in Figs. 1 and 2.

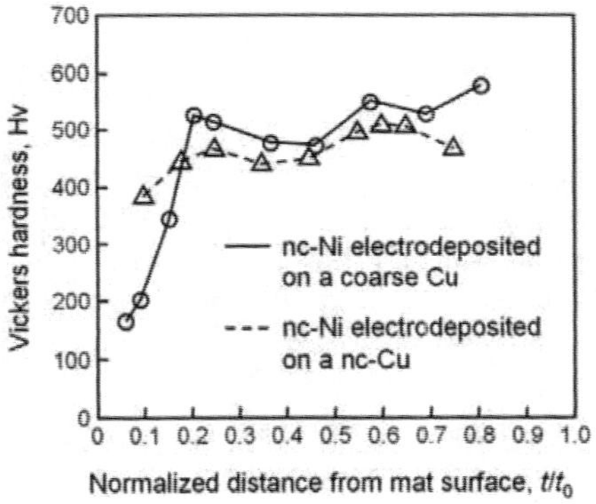

Figure 3. Vickers hardness as a function of the normalized distance from the mat surface, t/t_0.

3.3. Tensile test for the nc-Ni sheets

In most of reports for nanocrystalline metallic materials, Vickers hardness tests were carried out for investigating mechanical properties. Tensile test can give more information to understand deformation behaviors of the nanocrystalline materials. Thus, tensile test was done in the present study at ambient temperature at an initial strain rate of 8.3 x 10^{-4} s^{-1}. The stress-strain curves of the nc-Ni sheets obtained by the tensile test are shown in Fig. 4. For comparison, the curve of a coarse grained Ni (99.9% purity) having the mean grain size of 28 μm, which was prepared through conventional wrought and recrystallization processes, is also shown in the figure. For the coarse grained Ni, the yield stress was 43MPa, the ultimate tensile strength (UTS) of the coarse grained Ni was 331MPa and the total elongation was 65.5%. These are typical values of conventional polycrystalline pure Ni having coarse grains. It was found that the UTS of the nc-Ni electrodeposited on the coarse grained Cu was similar to that of the nc-Ni electrodeposited on the nc-Cu substrate. The 0.2% proof stress of the nc-Ni electrodeposited on the coarse grained Cu and the nc-Cu substrate were

0.98 GPa and 1.03 GPa, respectively. The UTS of the nc-Ni electrodeposited on the coarse grained Cu and the nc-Cu substrate were 1.42 GPa and 1.44 GPa, respectively. The UTS of nc sheets was more than four times higher that that of the coarse grained Ni.

The total elongation of the nc-Ni electrodeposited on the coarse grained Cu and the nc-Cu substrate were 3.2% and 4.6%, respectively. These are much smaller than the total elongation of the coarse grained Ni. The total elongation of the nc-Ni electrodeposited on the coarse grained Cu was smaller than that of the nc-Ni electrodeposited on the nc-Cu. From the stress-strain curves, it was found that the nc-Ni electrodeposited on the coarse grained Cu fractured around the UTS point. This indicates that early fracture occurred during uniform deformation. On the other hand, in nc-Ni electrodeposited on the nc-Cu substrate, the flow stress increased up to the UTS, decreased gradually after that, and then, the specimen was fractured. That is, this nc-Ni had certain amount of post-uniform elongation. These results suggest that the homogeneity of microstructure affects the mechanical properties of the electrodeposited nc-Ni.

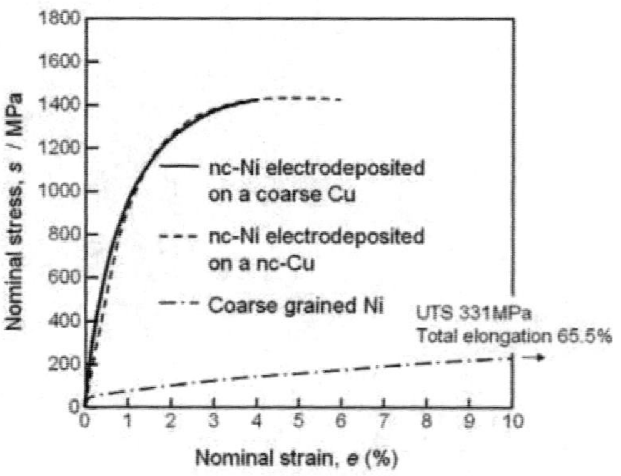

Figure 4. Stress-strain curves obtained by tensile test of the nc-Ni and coarse grained Ni sheets.

4. Summary

The microstructure and mechanical properties of the nc-Ni sheets electrodeposited on the Cu substrates having different microstructures were investigated. From TEM observation and Vickers hardness test, the nc-Ni electrodeposited on the nc-Cu substrate showed relatively homogeneous nanocrystalline microstructure, while that electrodeposited on the coarse Cu substrate showed inhomogeneous microstructure. It was found that the microstructure of the nc-Ni could be affected by the microstructure of the substrate.

The tensile test results showed that the total elongation of the nc-Ni electrodeposited on the nc-Cu was larger than that of the nc-Ni electrodeposited on the coarse grained Cu. The difference in the total elongation corresponded with the difference in the post-uniform deformation behavior. The homogeneity of microstructure was suggested to affect the mechanical properties, especially post-uniform elongation, of the nc-Ni sheets.

Acknowledgements

This study was supported by Priority Assistance for the Formation of Worldwide Renowned Centers of Research - The Global COE Program (Project: Center of Excellence for Advanced Structural and Functional Materials Design)in Osaka University through the Ministry of Education, Culture, Sports, Science and Technology (MEXT), Japan. The authors are grateful to Professor H. Mori and Dr. T. Sakata of the Research Center for Ultra-High Voltage Electron Microscopy at Osaka University for their help in using the high resolution TEM.

References

[1] Langdon T G 2007 *Mater Sci Eng.* **A 462** 3
[2] Islamgaliev R K, Yunusova N F, Sabirov I N, Sergueeva A V, Valiev R Z 2001 *Mater Sci Eng* **A319** 877
[3] Tsuji N, Saito Y, Lee S H, Minamino Y 2003 *Adv Eng Mater* **5** 338
[4] Song R, Ponge D, Raabe D, Speer J G, Matlock D K 2006 *Mater Sci Eng* **A441** 1
[5] Tsuji N, Ito Y, Sato Y and Minamino Y 2002 *Scripta Mater* **47** 893.
[6] Huang X, Hansen N and Tsuji N 2006 *Science***32** 249
[7] Suryanarayana S 1995 *Int Mater Reviews* **40** 41
[8] Yamasaki T, Schlobmecher P, Ehrilich K and Ogino Y 1998 *NanoSturucuted Materials* **10** 375

Influence of heat-treatment on microstructure and plastic deformation behavior in Ni_3V single crystals with the $D0_{22}$ structure

K Hagihara[1], M Mori[1], T Kishimoto[1], Y Umakoshi[2]

[1]Division of Materials and Manufacturing Science, Graduate School of Engineering, Osaka University, 2-1 Yamadaoka, Suita, Osaka 565-0871, Japan
[2] National Institute for Materials Science, 1-2-1, Sengen, Tsukuba, 305-0047, Japan
E-mail: hagihara@mat.eng.osaka-u.ac.jp

Abstract. The control of microstructure in Ni_3V single crystals such as variant and anti-phase boundary (APD) was attempted by quenching from the disordered state followed by annealing at several temperatures. In the heat-treatments, the microstructure strongly varied depending on the quenching speed from the disordered state. In slow-quenching, the lamellar structure composed of two variants was developed after annealing, as reported in many polycrystalline samples. However, only one of three variants was preferentially grown in the specimen rapidly quenched from the disordered state followed by annealing. The yield stress of slow-quenched specimen showed more than twice the value of the fast-quenched specimen.

1. Introduction

Recently, multi-phase alloys composed of Ni_3X-type compounds, for example see [1-7], are expected to be a new high-temperature structural material, since they show superior mechanical properties including the occurrence of the yield stress anomaly [8-12]. One of the promising components of them, $D0_{22}$-Ni_3V is recently focused [3-7]. Ni_3V is the Kurnakov-type compound with an order-disorder transformation temperature of 1318K, where the phase transformation occurs from A1(fcc) to $D0_{22}$ structure. In the $D0_{22}$ structure, two $L1_2$ unit cells put together accompanied by translation of a/2<110>. Therefore the phase transformation involves the loss of three-fold axis. As a result, the Ni_3V crystal always contains three variants whose [001] c-axis is mutually perpendicular. In addition, the anti-phase domains (APD) are also sometimes introduced in alloy under some heat-treatment conditions due to the order-disorder transition. It is expected that such variations in microstructures strongly affect to the plastic behavior of Ni_3V. In this study, the change in microstructure depending on a heat-treatment and its influence to the plastic deformation behavior are investigated using the single crystals.

2. Experimental procedure

Mother ingot with a stoichiometric composition of Ni-25.0at.%V was prepared by arc-melting under high purity Ar gas. Using them, single crystal was grown by the floating zone (FZ) method at a growth rate of 2.5 mmh^{-1} under Ar gas flow. The gained single crystals were encapsulated into silica-tube under Ar atmosphere, and the control of microstructure was conducted by annealing at 1473K under disordered state for 10.8ks followed by quenching in iced water, and subsequent annealing was done at 1073K or 1173K for 10ks. In these heat-treatments the microstructure was strongly varied whether the silica-tube which encapsulated the specimen was immediately broken in water or not, in quenching from 1473K. In this study the former is called fast-quenching and the latter is called slow-quenching. The features of microstructure were examined in a transmission electron microscope (TEM, JEOL JEM-3010) operated at 300kV. In the heat-treatment containing fast-quenching, some plate-like shapes of specimens with the thickness of about 1mm were prepared, and the influence of the shape of specimen on the change in microstructure was examined. The crystal orientation was examined by X-ray back-Laue reflection method, and the normal directions of the plate-like specimens were controlled to be parallel to the [100], [110] or [111]. In this paper the Miller indices for crystallographic planes and directions in Ni_3V are given using fcc notation for simplicity. Rectangle specimens with the size of $2x2mm^2x5mm$ and having the loading axis parallel to [111] orientation were cut, and the mechanical properties of the specimens were examined by compression tests. The tests were done at room temperature at a nominal strain rate of $1.7x10^{-4}$ s^{-1} in a vacuum.

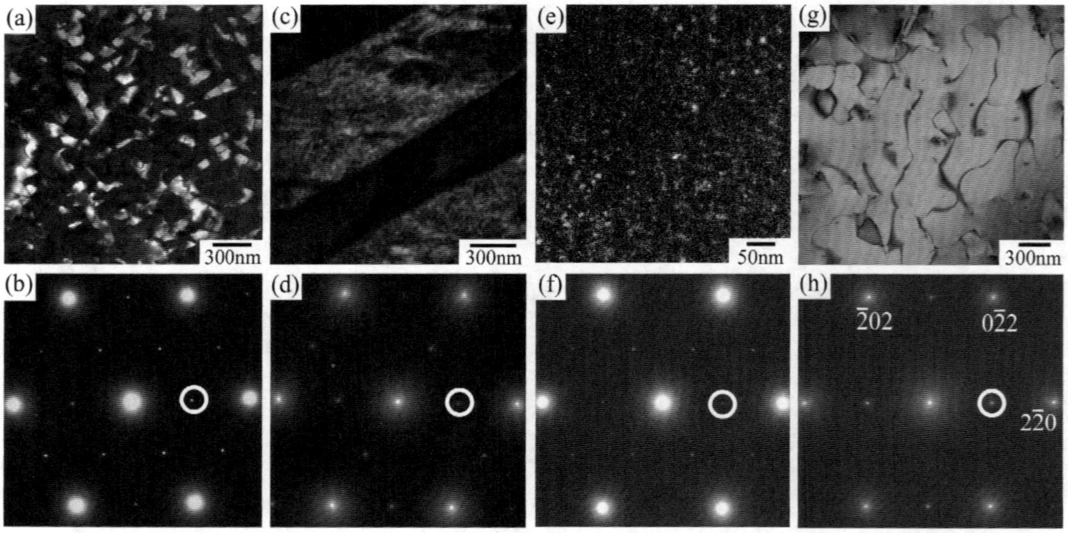

Figure 1 (a, c, e, g) Dark-field images of microstructure and (b, d, f, h) corresponding diffraction patterns (DPs) observed along [111] direction. (a, b) As slowly-quenched specimen from 1473K and (c, d) further annealed at 1173K for 10ks. (e, f) As fast-quenched specimen from 1473K and (g, h) further annealed at 1073K for 10ks, respectively. The $1\,\bar{1}\,0$ ordered spots used for the dark-field observation are circled in (b, d, f, h).

3. Results and discussion

3.1 Variation in microstructure depending on cooling rate in quenching from the disordered state

In this study, the control of microstructure was attempted by quenching from 1473K under disordered state into iced water, and subsequent annealing. In these heat-treatments the microstructure such as an arrangement of variant could be controlled as keeping the crystal orientation (as single crystalline form). As one of the notable controlling factors of microstructure in these heat-treatments, the features of microstructure showed a large variation depending on the cooling rate in quenching.

In case that the silica-tube which encapsulated the specimen was not broken in water in quenching (called slow-quenching), the specimen was covered by three variants with diameter of around 100nm, as shown in fig.1(a, b). By the further annealing, the sizes of variants were gradually increased, and finally two of three variants formed the lamellar structure with the interface parallel to {101} as shown in fig.(c, d). The features of the change in microstructure are similar to the previous many reports investigated with polycrystals [13-15]. Especially, the microstructure gained by as slow-quenched shown in fig.(a) was similar to the one called multi-variant structure (MVS) [15]. The controlling factors on the evolution of microstructure were discussed by them, and suggested that the minimization of elastic strain energy due to the transformation strain and reduction in the interfacial energy between the variants exhibit the important roles [13-15].

While in case that the silica-tube was immediately broken in water in quenching (called fast-quenching), the three variants were formed as the same as in the slow-quenched specimen, but their sizes became very small to be around 10nm in diameter as shown in fig.(e, f). In addition, in the dark field observation it seemed that a small amount of disordered A1(fcc)-phase was remained around those small variants. When the further annealing was conducted to the specimen, only one of three variants was preferentially grown largely to the millimeter size, contrary to the case in slow-quenched samples. As shown in fig. 1(g), many APDs were contained in the grown single variant in addition to some stacking faults and dislocations. Singh et al. previously reported the similar results in the polycrystalline specimen prepared by rapid-quenching using the melt-spun method [16].

These results demonstrate that the controlling mechanisms of the growth of microstructure vary depending on the initial size of the variant and/or the existence or nonexistence of the disordered phase in the quenched specimen.

In the fast-quenching followed by subsequent annealing process, the morphology of the variants could be controlled by changing the shape of specimens. When the size of specimen was large to be about 2x2mm^{2}x5mm, that used for compression test, two or three large plate-like shapes of variants were often formed as to divide the specimen on center. However, in case that the thickness of specimen was reduced to about 1mm, only one large single variant were frequently grown and covered the specimen. It is to note that in the plate-like shape of specimen, the selectivity of growing variant was not equivalent among the three variants, but it could be controlled by changing the crystal orientation of the normal direction in the plate-like specimen. Three kinds of plate-like specimens whose surfaces were parallel to (100), (110) or (111) were prepared and then heat-treated. It was confirmed that large single variant were grown in all cases. However, in case of (100) specimen, at the high frequency the grown single variant showed the feature that its a-axis was parallel to the plate normal as shown in fig.2; the variant whose c-axes was in-plane direction was preferentially grown in the specimen. Also in case of (110) specimen, the variant whose c-axis is in-plane direction was preferentially grown compared to other two variants. The results imply the frequency of nucleation of three variants in the plate-like specimen might be varied depending on whether their c-axes are close to the plate normal direction or not, although such difference in volume fraction of three variant could not be clearly detected in the TEM observation in the as-quenched specimen. The thermal flow direction and/or surface energy are considered to affect this phenomenon, and further studies are needed to clarify this.

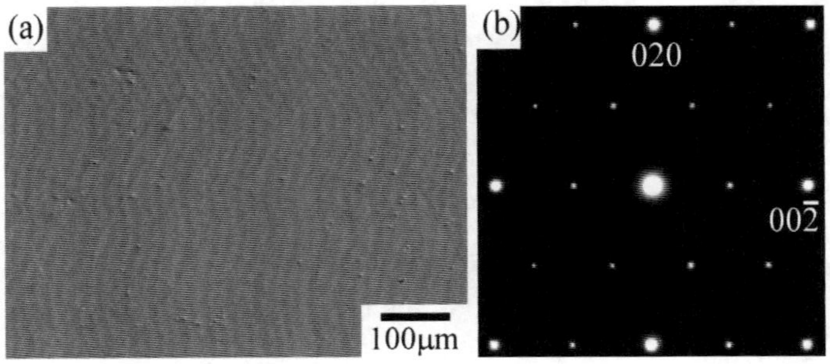

Figure 2 (a) Optical micrograph of specimen fast-quenched from 1473K followed by annealing at 1073K for 10ks, whose surface normal is parallel to (100). (b) Corresponding DP in TEM. Observed beam direction is almost parallel to the foil normal. The DP corresponds to [100] zone pattern (along a-axis) in the $D0_{22}$ structure.

3.2 Mechanical properties of the Ni_3V single crystals after quenching

Figure 3 shows the room temperature yield stress in as fast-quenched and slow-quenched specimens compressed at [111] orientation. For comparison, the yield stress of as FZ-grown crystal, in which two variants form lamellar structure [17], is also shown in the graph. At the [111] orientation, the Schmid factors (SF) for possible deformation mode in Ni_3V exhibit the equivalent values in three variants; SF for {111}<110] slip is 0.272 and SF for {111}1/6<112] twinning is 0.157^{c}, respectively. The yield stresses show significant variations in the three specimens, indicating the difference of microstructure strongly affects the plastic deformation behavior of Ni_3V. The slow-quenched specimen and as-FZ-grown specimen consist only of the ordered variants, but the yield stress shows much higher value in former than in latter. In the slow-quenched specimen three variants existed with the diameter of around 100nm, but in the as FZ-grown crystal the thickness of the variants was much larger to be about 700nm in the crystal. This clearly indicates that the variant boundaries acted as strong obstacle for deformation and the refinement of them effectively increased the yield stress. On the other hand, the yield stress in fast-quenched specimen showed much lower value than others. This may indicate the coexistence of disordered phase and the difference of degree of order in the variants also strongly influence on the yield stress in Ni_3V. On these points, the effect of annealing after quenching from disordered state to the yield stress and plastic deformation behavior is now under survey.

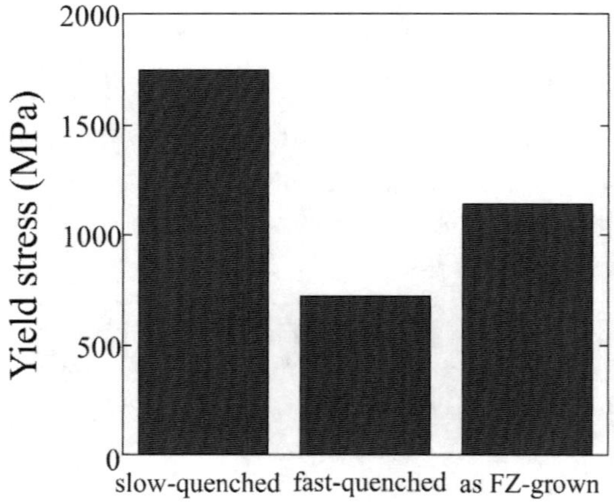

Figure 3 Room temperature yield stress in as slow-quenched, as fast-quenched and as FZ-grown specimens in compression at [111] orientation.

4. Conclusion

In Ni_3V single crystal, the microstructure such as the variant and the APD could be controlled by the heat-treatment as keeping the crystal orientation (single crystalline form). The microstructure is strongly varied depending on the quenching speed from disordered state. In slow-quenched specimen, three variants with the diameter of about 100nm are formed. The lamellar structure composed of two of three variants is developed by the further annealing. On the other hand, three variants are also formed in fast-quenched specimen, but their sizes become very small to be about 10nm. In this case, only one of three variants is preferentially grown largely to millimetre size by the further annealing. The yield stress of slow-quenched specimen shows much higher value than those in fast-quenched specimen and as FZ-grown specimen. The result demonstrates that the variant boundaries act as strong obstacle for deformation, and hence the refinement of them effectively increases the yield stress.

Acknowledgement: This work was supported by Priority Assistance of the Formation of Worldwide Renowned Centers of Research – The Global COE Program (Project: Center of Excellence for Advanced Structural and Functional Materials Design) from the MEXT, Japan.

References

[1] Tomihisa K, Kaneno Y and Takasugi T 2002 *Intermetallics* **10** 247
[2] Hagihara K, Nakano T and Umakoshi Y 2002 *Sci.Tec. Adv. Mater.* **3** 193
[3] Nunomura Y, Kaneno Y, Tsuda H and Takasugi T 2004 *Intermetallics* **12** 389
[4] Nunomura Y, Kaneno Y, Tsuda H and Takasugi T 2006 *Acta Mater.* **54** 851
[5] Shibuya S, Kaneno Y, Yoshida M, Shishido T and Takasugi T 2007 *Intermetallics* **15** 119
[6] Soga W, Kaneno Y, Yoshida M and Takasugi T 2008 *Mat. Sci. Eng. A* **473** 180
[7] Kaneno Y, Soga W, Tsuda H and Takasugi T 2008 *J. Mater. Sci.* **43** 748
[8] For example, Yamaguchi M and Umakoshi Y 1990 *Prog. Mat. Sci.* **34** 1
[9] Hagihara K, Nakano T and Umakoshi Y 2000 *Acta Mater.* **48** 1469
[10] Hagihara K, Nakano T and Umakoshi Y 2003 *Acta Mater.* **51** 2623
[11] Hagihara K, Nakano T and Umakoshi Y 2003 *Scripta Mater.* **48** 577
[12] Hagihara K, Tanaka T, Nakano T and Umakoshi Y 2005 *Acta Mater.* **53** 5051
[13] Tanner LE *Phys. Stat. Sol.* 1968 **30** 685
[14] Singh JB, Sundararaman M, Mukhopadhyay P and Prabhu N 2003 *Intermetallics* **11** 83
[15] Suzuki A, Kojima H, Matsuo T and Takeyama M 2004 *Intermetallics* **12** 969
[16] Singh JB, Sundararaman M, Banerjee S and Mukhopadhyay P 2005 *Acta Mater* **53** 1135
[17] Hagihara K, Mori M and Umakoshi Y 2007 *Mater. Sci. Forum* **561-565** 407

Journal of Physics: Conference Series **165** (2009) 012005 doi:10.1088/1742-6596/165/1/012005

Energetics of formation process of a <001> prismatic dislocation loop via the collision between two 1/2<111> loops in α-iron

K Arakawa and H Mori

Research Center for Ultra-High Voltage Electron Microscopy, Osaka University, 7-1 Mihogaoka, Ibaraki, Osaka 567-0047, Japan

E-mail: arakawak@uhvem.osaka-u.ac.jp

Abstract. It has been proposed by Marian *et al.* [1] that a [001] interstitial-type dislocation loop can be formed in body-centered cubic iron via the collision between a 1/2[111] loop and $1/2[\bar{1}\bar{1}1]$ loop, which undergo one-dimensional glide diffusion, and the subsequent shear reaction. However, the formation of [001] loops through this reaction has not been reproduced by other works even though the two 1/2<111> loops collided with each other. In the present paper, the origin of the difficulty in this reaction is discussed within the framework of isotropic elasticity theory. The sign of the driving force for the reaction is heavily dependent on the reaction path. The two 1/2<111> loops colliding to form a [110] junction can transform to a single [001] loop when a shear loop generated within the 1/2[111] loop propagates in sync with the other shear loop within the $1/2[\bar{1}\bar{1}1]$ loop. However, unsynchronized motion of the two shear loops significantly suppresses the propagation of the shear loops, which might be caused by the thermal fluctuation at finite temperatures. This will be one of the origins of the difficulty in the formation of [001] loops through the collision between the two 1/2<111> loops.

1. Introduction

Body-centered cubic (bcc) iron-based steels are major candidate materials for the future nuclear-fission and nuclear-fusion reactors. It is known that prismatic perfect dislocation loops—agglomerations of self-interstitial atoms—with the Burgers vector, **b**, of <001> are formed in bcc iron upon high-energy particle irradiation at elevated temperatures, as well as dislocation loops with **b**=1/2<111> [2, 3].

The detailed process of the formation of <001> loops, however, has not been fully understood. Eyre and Bullough [3] proposed that an extremely small 1/2[011] imperfect dislocation loop with a stacking fault, which is the agglomeration of [011] dumbbells on the (011) plane, is unfaulted and changes into a 1/2[111] loop or [001] loop through the following shear reactions on the loop plane, when it grows:

$$\frac{1}{2}[011]+\frac{1}{2}[100]\rightarrow\frac{1}{2}[111] \quad (1)$$

$$\frac{1}{2}[011]+\frac{1}{2}[0\bar{1}1]\rightarrow[001]\,. \quad (2)$$

In accordance with the result obtained by Dudarev *et al.* [4], the latter reaction is expected to more frequently occur than the former reaction at high temperatures; this will yield the formation of <001> loops.

On the other hand, Masters [2] proposed that the coalescence between a 1/2[111] loop and $1/2[\bar{1}\bar{1}1]$ loop leads to the formation of a single [001] loop, via the following reaction:

$$\frac{1}{2}[111]+\frac{1}{2}[\bar{1}\bar{1}1]\rightarrow[001]. \tag{3}$$

This kind of loop-loop interaction as well as the behavior of individual loops must be influential in the microstructural evolution under irradiation. In 2002, Marian *et al.* [1] showed a picture of the process expressed by equation (3), based on the idea that 1/2<111> loops can undergo one-dimensional glide diffusion in their **b** direction [1, 5, 6]. They confirmed by using molecular dynamics calculations that when a 1/2[111] loop lying on a {011} plane (loop A) accidentally collides with a $1/2[\bar{1}\bar{1}1]$ loop lying on another {011} plane (loop B) side by side via their one-dimensional motion, a [001] junction can be formed. They proposed that when the sizes of the two loops are comparable and the loop A plane involves the **b** of loop B and the loop B plane involves the **b** of loop A, propagation of a shear loop within loop A generated due to a shear stress applied by loop B and that within loop B generated by loop A expand the [001] junction and finally form a single [001] loop. In contrast, other molecular dynamics calculations conducted by Osetsky *et al.* [7] and Dmitry *et al.* [8] have not reproduced the formation of a [001] loop even when loop A collided with loop B. In addition, in a recent experiment by using *in-situ* transmission electron microscopy [9], the formation of a [001] loop was not observed even when the collision occurred, and rather it was observed that a larger loop absorbed smaller one after the collision. It should be noted that these results do not necessarily mean that the formation of a [001] loop via the mechanism proposed by Marian *et al.* [1] can not be realized. However, these results seem to imply that the Marian's mechanism [1] rarely works if any. In the present paper, the origin of the difficulty in the formation of [001] loops via Marian's mechanism [1] will be touched on, by evaluating the sign of the driving force for the reaction based on non-singular isotropic elasticity theory [10].

2. Verification of the Marian's mechanism [1]

When loop A collides with loop B side by side, there are two manners of the collision between the two loops, depending on the spatial configuration of them. A [001] junction is formed in one case, and a [110] junction is formed in the other case. The necessary condition for the formation of the [001] loop via the shear reaction is not the formation of the [001] junction, as claimed by Marian *et al.* [1], but that of the [110] junction via the collision between loop A and loop B, if we adopt the traditional terminology in dislocation theory. Hereafter, the loop shape approximates the rectangular, for simplicity.

2.1. Driving force for the formation of the [001] loop

In order to examine whether the driving force for the [001] loop formation is positive or negative and to clarify the origin of the promotion or suppression of the reaction, we evaluate the driving force for the propagation of the shear loops for [011] junction case [9] and extract the origin of the sign of the driving force. Figure 1 shows the dislocation structure for the calculation. Here, the dislocation directions point into the paper or the right. The loop A plane is (011), and the loop B plane is $(0\bar{1}1)$. We assumed that the segments transformed to $[00\pm1]$ ones could immediately glide and cancel the 90-degree bend to reduce the total energy, as shown by the segment named η and ζ. The total elastic energy, W, is calculated by the summation of the self-energies of all the segments and the interaction energies between all different segments. The self- and interaction energies were calculated within the framework of nonsingular isotropic elasticity theory, proposed by Cai *et al* [10]. The total driving force for the [001] loop formation, F, was evaluated by the following equation:

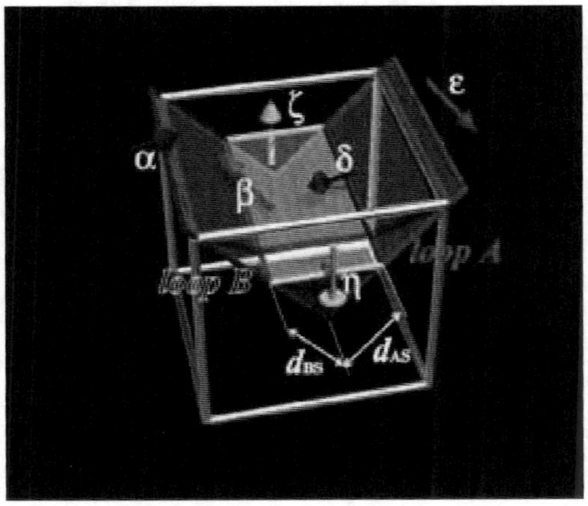

Figure 1. Dislocation structure for the calculation of the driving force for the formation of a [001] loop. This structure is formed via the shear loop generation after the [110] junction formation due to the collision between loop A and loop B. The dislocation directions point into the paper or the right. The loop A plane is (011), and the loop B plane is $(0\bar{1}1)$. The Burgers vectors for typical segments are: $1/2[\bar{1}\bar{1}1]$ for α, 1/2[111] for β, $1/2[11\bar{1}]$ for δ, $1/2[\bar{1}\bar{1}\bar{1}]$ for ε, $[00\bar{1}]$ for η, and [001] for ζ. The length d_{AS} and d_{BS} are also shown.

$$F = -\frac{\partial W}{\partial(d_{AS} + d_{BS})}. \tag{4}$$

Figure 2 shows one of examples for the calculated results of the driving force for the [001] loop formation as a function of $d_{AS}+d_{BS}$. Initially, the shape of loop A and loop B was square, and all the side length values were 10 nm. The length d_{AS} and d_{BS} were alternately incremented by a length of $a/3$, where a is the lattice constant (= 0.28665 nm). In this figure, in addition to the driving force calculated from the total energy, the components originated from the typical interaction energies and the total self-energies are shown. From this figure, it is clarified that the driving force is always positive mainly due to the repulsive interaction between β and δ and the attractive interaction between δ and ε and that of α and β when the distances between the noted pair of segments are short, although the driving force from the total self-energies becomes negative after the reaction progresses to some extent. Thus, if a shear loop propagates in sync with the other shear loop, the [001] loop can be easily formed.

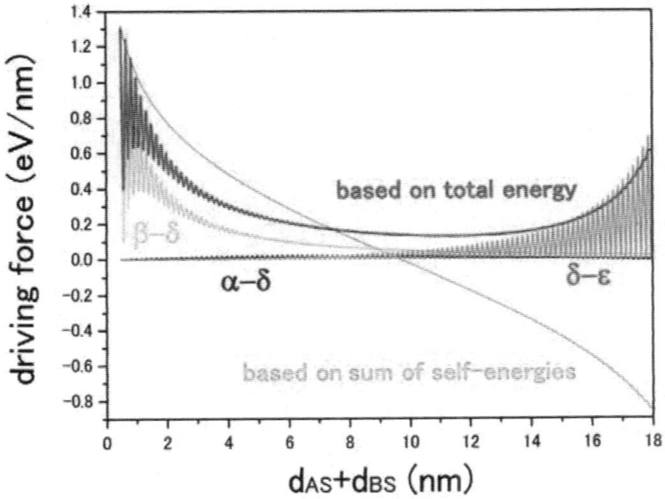

Figure 2. Driving force for the formation of a [001] loop as a function of $d_{AS}+d_{BS}$. Initially, the shape of loop A and loop B is square, and all the side length values are 10 nm. The length d_{AS} and d_{BS} are alternately incremented by a length of $a/3$, where a is the lattice constant.

Figure 3 shows another example of the calculated result of the driving force. In this case, only d_{BS} was firstly increased from a to $6a$ while d_{AS} was fixed. After it, d_{AS} was increased while d_{BS} was fixed. As shown in this figure, while δ progresses after β precedes, it undergoes significant attractive force from β and it yields the negative driving force. The origin of this attractive force only locally appearing is due to the spatial variation in the sign of the shear stress around edge dislocations [11]. Thus, unsynchronized propagation of the two shear loops suppresses the formation of the [001] loop. At finite temperatures, the thermal fluctuation will enhance this kind of unsynchronized motion of the shear dislocations. This will be one of the origins of the difficulty in the formation of [001] loops via the shear reaction following the collision between loop A and loop B.

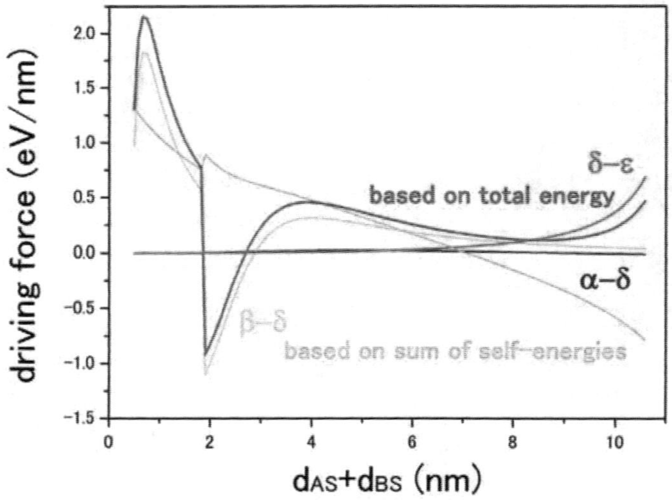

Figure 3. Driving force for the formation of a [001] loop as a function of $d_{AS}+d_{BS}$. Initially, the shape of loop A and loop B is square, and all the side length values are 10 nm. Only d_{BS} is firstly increased from a to $6a$ while d_{AS} is fixed. After it, d_{AS} is increased while d_{BS} is fixed.

3. Concluding remarks

A 1/2[111] loop (loop A) and a $1/2[\bar{1}\bar{1}1]$ loop (loop B) must collide with each other so that they form a [110] junction to form a single [001] loop through the shear reaction following the collision. By the calculations based on isotropic elasticity theory, we have found that the driving force for the shear loop propagation within loop A and loop B (a [001] loop formation) after the [110] junction formation is heavily dependent on the reaction path. Synchronized propagation of the two shear loops within loop A and loop B yielded the promoted formation of the [001] loop; whilst, unsynchronized propagation of the two shear loops significantly suppressed the [001] loop formation. At finite temperatures, such an unsynchronized motion of the shear dislocations will be enhanced due to the thermal fluctuation; it will be one of origins of the difficulty in the formation of [001] loops via the shear reaction following the collision between loop A and loop B.

Acknowledgements

This work was financially supported by Priority Assistance for the Formation of Worldwide Renowned Centers of Research—The Global COE Program (Project: Center of Excellence for Advanced Structural and Functional Materials Design) from the Ministry of Education, Culture, Sports, Science and Technology (MEXT), Japan.

References

[1] Marian J, Wirth B D and Perlado J M 2002 *Phys. Rev. Lett.* **88** 255507.
[2] Masters B C 1965 *Philos. Mag.* **11** 881.
[3] Eyre B L and Bullough R 1965 *Philos. Mag.* **12** 31.
[4] Dudarev S L, Bullough R and Derlet P M 2008 *Phys. Rev. Lett.* **100** 135503.
[5] Osetsky Y N, Bacon D J, Serra A, Singh B N and Golubov S I 2003 *Philos. Mag.* A **83** 61.
[6] Arakawa K, Ono K, Isshiki M, Mimura K, Uchikoshi M and Mori H 2007 *Science* **318** 956.
[7] Osetsky Y N, Serra A and Priego V 2000 *J. Nucl. Mater.* **276** 202.
[8] Terentyev D, Malerba L, Klaver P and Olsson P 2008 *J. Nucl. Mater.* **382** 126..
[9] Arakawa K and Mori H 2008 *in preparation.*
[10] Cai W, Arsenlis A, Weinberger C R and Bulatov V V 2006 *J. Mech. Phys. Solids* **54** 561.
[11] Hirth J P and Lothe J 1982 *Theory of Dislocations* (New York: Wiley).

Formation and microstructure of Al_2O_3-YAG eutectic ceramics by phase transformation from metastable system to equilibrium system

Tomoya Nagira[1], Hideyuki Yasuda[1], Masato Yoshiya[1]

[1]Department of Adaptive Machine Systems, Osaka University, 2-1 Yamadaoka, Suita, Osaka 565-0871, Japan

E-mail:nagira@ams.eng.osaka-u.ac.jp

Abstract. Unidirectionally solidified Al_2O_3-YAG($Y_3Al_5O_{12}$: yttrium-aluminum-garnet) eutectic ceramic composites have been recognized as encouraging heat-resistance materials because of the superior mechanical properties at high temperatures. In addition to the excellent mechanical properties at high temperatures, some interesting solidification phenomena have been reported in the Al_2O_3-Y_2O_3 system. The Al_2O_3-YAG equilibrium eutectic at 2099 K and the Al_2O_3-YAP metastable eutectic at 1975 K exist in the Al_2O_3-Y_2O_3 system. The heating the metastable eutectic up to temperatures above the metastable eutectic temperature produced the undercooled melt. Solidification in the equilibrium path accompanied the melting of the metastable eutectic. The solidification process using undercooled melt resulted in the fine and uniform eutectic structure. In this study, the effect of the initial Al_2O_3-YAP particles size on the undercooled melt formation was examined. The Al_2O_3-YAP particles with diameters more than several μm resulted in the transformation through the undercooled melt. EBSD analysis showed that the domains of Al_2O_3 grains with same crystallographic orientation were observed and that their domain size depended on the Al_2O_3-YAP particles size. On the other hand, for the Al_2O_3-YAP particles with a diameter of 500 nm, the each Al_2O_3 grain with diameter of about 1 μm had the different crystallographic orientations, which suggested that the transformation from metastable eutectic to equilibrium eutectic occurred in the solid state. The increase in the Al_2O_3-YAP free surface area suppressed the undercooled melt formation.

1. Introduction

The melt growth composite (MGC) of ceramics are of considerate practical concern due to the excellent mechanical properties at high temperatures. Especially, unidirectionally solidified Al_2O_3-YAG eutectic ceramics consisting of single crystal Al_2O_3 and single crystal YAG is attractive for high-temperature materials [1-2]. The composites showed the flexural strength of 360-400 MPa from room temperature to 2073 K [2]. In addition to the excellent mechanical properties at high temperatures, some interesting solidification phenomena have been reported in the Al_2O_3-Y_2O_3 system [3-4]. There are two eutectic systems in the phase diagram of Al_2O_3-Y_2O_3 system in the Al_2O_3 rich part, one is Al_2O_3-YAG equilibrium eutectic system and the other is Al_2O_3-YAP($YAlO_3$: yttrium aluminum-perovskite) metastable eutectic system. In the composition range of 13.5-28.5 mol%Y_2O_3, the selection of the two eutectic metastable systems is determined because of the suppression in the nucleation of

YAG [5-6]. When heating the melt with the metastable eutectic composition above 2273K, Al_2O_3-YAP metastable system is indispensably selected.

Our previous papers [7] showed that heating the Al_2O_3-YAP metastable eutectic particles with diamters 20-45 μm up to temperatures above the metastable eutectic temperature of 1975 K but below equilibrium eutectic temperature of 2099 K produced the undercooled melt. Solidification in the equilibrium path accompanied melting of the metastable eutectic system. The coupling of the melting and the solidification enabled a nearly adiabatic transformation, resulting in the fine and uniform eutectic structure. This unique solidification process is attractive for fabricating the dense Al_2O_3-YAG composites with a fine microstructure.

The effect of the Al_2O_3-YAP particle size on the microstructure was also examined [8-9]. The Al_2O_3-YAG compact produced by the Al_2O_3-YAP metastable eutectic particles with diameters 20-45 μm had the fine and uniform eutectic structure with a lamellar spacing of submicron order. In the Al_2O_3-YAG compact produced by the Al_2O_3-YAP particles with diameters less than 20μm, the Al_2O_3 phase with a diameter of 2.5 μm were dispersed in the YAG matrix. In order to increase the surface area of Al_2O_3-YAP particle, the Al_2O_3 and YAP powder mixtures with a diameter of 500 nm were prepared. Transformation to Al_2O_3-YAG eutectic occurred in the solid state reaction [10]. The microstructure of Al_2O_3-YAG compact showed that the Al_2O_3 phases with the diameter of 1-2 μm were uniformly dispersed in the YAG matrix.

It is reported that YAG phase is likely to nucleate at the Al_2O_3/YAP phase boundaries when the Al_2O_3-YAP compact was heated at low heating rate of 0.16 K/s and that 11% volume expansion in the phase transformation (Al_2O_3+YAP→YAG) suppresses the nucleation of the YAG phase[4]. Thus, the increase in the free surface area of the Al_2O_3-YAP particles for volume expansion can influence on undercooled melt formation. In this study, the effect on the Al_2O_3-YAP particle size on the undercooled melt formation was examined from the view point of the crystallographic orientation. The preferential site for YAG nucleation in the Al_2O_3-YAP bulk was also examined by AFM analysis.

2. Experimental Procedure

The α-Al_2O_3 powder and Y_2O_3 powder were weighed according to the Al_2O_3-23.5mol%Y_2O_3 which is the Al_2O_3-YAP metastable eutectic composition. Their powders were ball milled with ethanol for 24 h to obtain homogeneous slurry and was dried in a vacuum. The mixed powder was melted in a Mo crucible by an induction furnace in an Ar atmosphere. The Al_2O_3-YAP metastable eutectic structure was obtained by solidifying the melt kept the above 2300 K. The solidified specimens were crushed into particles and sieved through a 45-63 μm, 32-45 μm and 20 μm in decreasing order. The Al_2O_3-YAP metastable eutectic particles of 20-45 μm and less than 20 μm in size were prepared.

The polymerized complex method [11], which provides the fine and uniform ceramics particle, was utilized in order to prepare the nano-sized YAP particles. Citric acid and ethylene glycol were added in the proportion of 4 moles and 180 moles, respectively, for each mole of metal cation. Citric acid, $Al(NO_3)_3 \cdot 9H_2O$ and $Y(NO_3)_3 \cdot 6H_2O$, according to the nominal composition of $YAlO_3$, were dissolved in ethylene glycol at 473 K. The formation of the polymer between ethylene glycol and the metal citrate complexes was promoted at 473-573 K. When the colloidal solution was condensed, it became highly viscous. Viscous polymeric product was decomposed to a fine powder at 723 K. The powder precursor was calcined at 1573 K for 18 ks in air to enhance the crystallization and eliminate the organic contents. The obtained powders and α -Al_2O_3 powders with a diameter of 500 nm were completely mixed with a planetary ball milling apparatus for 72 ks at 300 rpm according to the metastable eutectic composition.

The desired Al_2O_3-YAP metastable eutectic particles were inserted into the Mo mold, which has outer diameter of 10 mm and inner diameter of 6 mm. The Mo mold was inserted into the carbon die and then the Al_2O_3 powder was put into carbon die to avoid the energization of the Mo die. The graphite die was heated to 2023 K at the heating rate of 1.7 K/s under the pressure of 40 MPa. The specimen was kept for 60 s at the desired temperature in vacuum.

3. Results and Discussion

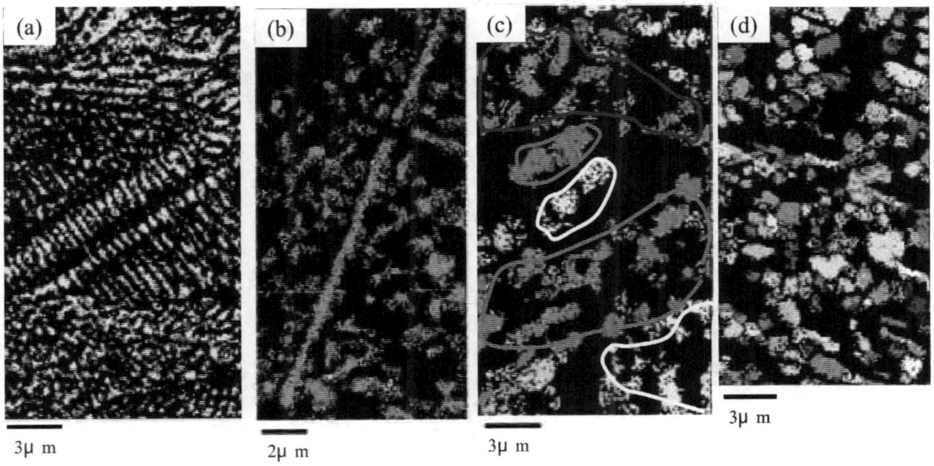

Figure 1. Al_2O_3 crystallographic orientation mappings of the (a)Al_2O_3-YAP particles and the Al_2O_3-YAG compacts produced by metastable eutectic particles with diameters (b) 20-45 μm, (c) less than 20 μm and (d) 500 nm. Each circle in Figure 1(c) shows the domains of Al_2O_3 grains with the same crystallographic orientations.

The Al_2O_3 crystallographic orientation mappings of Al_2O_3-YAG compacts produced by the Al_2O_3-YAP particles with diameters 20-45 μm, less than 20 μm and 500 nm are shown in Figure 1(b), (c) and (d) respectively. For comparison, the Al_2O_3 crystallographic orientation mapping of Al_2O_3-YAP particles is shown in Figure 1(a). Al_2O_3-YAP particles had the lamellar structure with a lamellar spacing of about 0.5 μm. The Al_2O_3 grains with same crystallographic orientation in the range of at least 50 μm were observed. For Al_2O_3-YAG compacts produced by the Al_2O_3-YAP grains with diameters 20-45 μm and less than 20 μm, the domains of the Al_2O_3 grains with the same crystallographic orientation were observed, and their domains size were approximately 20 μm and 4-15 μm, respectively. These results suggested that the Al_2O_3 grains in each domain connected with each other. The volume of undercooled melt during the transformation from the metastable system to equilibrium system depended on initial Al_2O_3-YAP particle size. As to YAG phase, the domains of YAG grain with a same crystallographic orientation were not observed regardless of the Al_2O_3-YAP particles size, which indicated that the eutectic structure formed through undercooled melt was composed of discontinuously grown YAG grains with continuously grown Al_2O_3 grains[10]. On the other hand, for the Al_2O_3-YAG compact formed by the Al_2O_3-YAP particles with a diameter of 500 nm, the most of Al_2O_3 grains with 1-2 μm had the different crystallographic orientation, which suggests that the transformation from the metastable system to the equilibrium system occurred in the solid state.

YAG nucleation in the free surface area of the Al_2O_3-YAP bulk with diameter of 10 mm was observed by AFM analysis. Since the Al_2O_3 and YAP phases were distinguished by using the difference in their hardness, final polishing was performed with Al_2O_3 suspension. The concave and convex portions indicate YAP and Al_2O_3, as shown in Figure 2(a). After polishing, the Al_2O_3-YAP bulk was heated at the heating rate of 1.7 K/s up to 1773 K which is under metastable eutectic temperature and was then quenched. AFM observation showed that the prominent parts were observed in the free surface area of Al_2O_3-YAP bulk as shown in Figure 2(b). The prominent areas were formed by the volume expansion of 11 % when YAG phase was formed in the solid state. Thus, YAG can be preferentially formed at the free surface area of Al_2O_3-YAP bulk even under the condition of the lower temperature than metastable eutectic temperature and the high heating rate. Based on these results, the increase in the free surface area of Al_2O_3-YAP particle promoted the YAG nucleation.

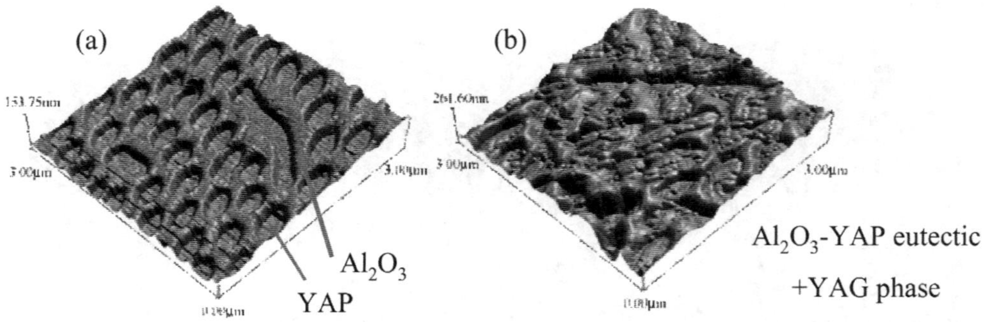

Figure 2. AFM surface images of the Al_2O_3-YAP bulk (a) before annealing and (b) after annealing at 1773K.

4. Conclusions

The effect on the initial Al_2O_3-YAP particle size on the undercooled melt formation and the preferential site for YAG nucleation in the Al_2O_3-YAP bulk were examined. EBSD analysis showed that the domain size of the Al_2O_3 grains with the same crystallographic orientation for the Al_2O_3-YAP particles with diameters 20-45 μm and less than 20 μm was approximately 20 μm and 4-15 μm, respectively. The volume of undercooled melt during the transformation from the Al_2O_3-YAP metastable system to Al_2O_3-YAG equilibrium system depended on Al_2O_3-YAP particle size. On the other hand, for the Al_2O_3-YAP particle with a diameter of 500 nm, Al_2O_3 grains had the different crystallographic orientation, which indicate the solid-state transformation from metastabale eutectic to equilibrium eutectic. AFM analysis showed that YAG nucleation preferably occurred on the free surface area of the Al_2O_3-YAP bulk. Based on these results, the increase in the Al_2O_3-YAP free surface area promoted YAG nucleation and suppressed the undercooled melt formation.

Acknowledgements

This study was supported by Priority Assistance for the Formation of Worldwide Renowned Centers of Research - The Global COE Program (Project: Center of Excellence for Advanced Structural and Functional Materials Design) from the Ministry of Education, Culture, Sports, Science and Technology (MEXT), Japan. This works was also supported by a Giant-in-Aid for Young Scientists (B)(No.20760506) and Nippon Sheet Glass Foundation For Materials Science and Engineering.

References

[1] Mah T, and Parthasarathy T A 1990 *Ceram. Eng. Sci. Proc.* **11,** 1617.
[2] Waku Y, Nakagawa N, Wakamoto T, Ohtsubo H, Shimizu K 1997 *Nature* **389** 49.
[3] Parthasarathy T A, Mah T 1993 *J. Am. Ceram. Soc.* **76** 29.
[4] Randall S Hay 1994 *J. Am. Ceram. Soc.* **77** 1473.
[5] Caslavsky J L, Viechinickl D J 1980 *J. Mater. Sci.* **15** 1709.
[6] Yasuda H, Ohnaka I, Mizutani Y and Waku Y 2001 *Sci. Tech. Adv. Mater.* **2** 67.
[7] Yasuda H, Ohnaka I, Mizutani Y, Sugiyama A, Morikawa T, Takeshima S, Sakimura T and Waku Y 2004 *Sci. Tech. Adv. Mater.* **5** 207.
[8] Nagira T, Yasuda H, Sakimura T and Kawaguchi A 2007 *Materials Transactions* **48** 2312.
[9] Nagira T, Yasuda H, Yoshiya M and Uesugi K 2007 *SP07 5th decennial international conference on Solidification Processing* 211.
[10] Nagira T and Yasuda H 2005 *Collected abstract of the 2005 Autumn Meeting of the Japan Institute of Metals* 289.
[11] Kakihana M, Yoshimura M, Mazaki H, Yasuoka H and Borjesson L 1992 *J. Appl. Phys.* **71** 3904.

Surface modification of β-Type titanium alloy by electrochemical potential pulse polarization

Shinji Fujimoto, Vedarajan Raman and Hiroaki Tsuchiya

Division of Materials and Manufacturing Science, Graduate School of Engineering, Osaka University, 2-1 Yamadaoka, Suita, Osaka 565-0871, Japan

fujimoto@mat.eng.osaka-u.ac.jp

Abstract. In the present work, we report the formation of a porous oxide / hydroxide surface layer on the Ti-29Nb-13Ta-4.6Zr (TNTZ) alloy achieved by the combination of an alkali immersion and a potential pulse polarisation process. The alkali treatment has been employed for pure titanium to produce amorphous and porous layer prior to hydroxyapatite (HAp) growth. But, in the case of TNTZ, immersion in 5M NaOH at the open circuit potential (OCP) at 60°C for 24 hours, did not yield any uniform layer, instead a thick deposited layer with highly cracked one. The cracks were attributed to the growth of a tantalum enriched particulate. In order to avoid the crack formation, the electrochemical behaviour of the alloy and the pure alloying elements (Ti, Nb, Ta and Zr) was investigated to produce a uniform surface with the application of a square wave modulated potential pulse polarization, leading to the formation of a relatively uniform porous layer on the alloy.

1. Introduction

Titanium and its alloys are widely used for orthopaedic implant applications due to their favourable mechanical properties and biocompatibility. The alloys such as Ti-6Al-4V and Ti-6Al-7Nb are suitable for implant materials, but, their elastic modulus is higher compared with that of bone. β type titanium alloy, such asTi-29Nb-13Ta-4.6Zr (TNTZ) alloy, has been developed to achieve elastic modulus close to bone [1-4]. Similar to other biomedical titanium alloys, TNTZ alloy also might be difficult to attain favourable bonding with bone and tissues directly. Therefore, some surface modification is required to improve its biocompatibility.

Various surface modification methods have been proposed to establish biocompatibility [5]. One such treatment includes the alkali treatment where titanium and titanium alloys are immersed in a high concentration of NaOH solution, which leads to the formation of a thin-amorphous sodium titanate layer on the surface [6-8]. The sodium titanate layer has a sponge-like homogeneous porous structure with the pore size of approximately 1 μm as shown in Fig 1. A subsequent heat treatment of the alkali treated specimens modifies the layer. The heat-treated layer when immersed in a simulated body fluid (SBF), results in the

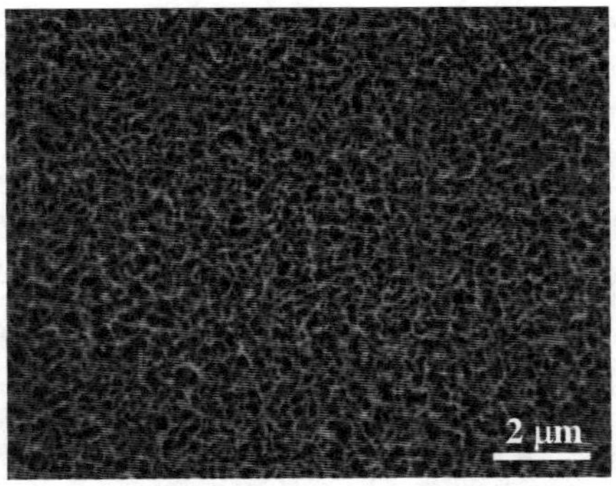

Fig.1 SEM morphology of pure Ti after alkli treatment.

formation of Ca-P layer [9]. This alkali treatment is much simpler, more economical and capable of producing uniform coating on complex shape implant. On the other hand, such homogeneous porous structure is hard to form on Ti alloys. In the present work, we report the electrochemical behaviour of each alloying elements contained, then propose the process to produce a porous layer on TNTZ alloy with the combination of the alkali treatment and an electrochemical pulse polarization.

2. Experimental

Specimens examined were pure Ti, pure Nb, pure Ta, pure Zr (the purity level of all the metals are in the range of 99.95% up) and the TNTZ alloy with the dimensions of 10mm x 10mm x 2mm. The specimens were mechanically ground with emery paper up to 2000 grit, and further polished to mirror finish with 3 μm alumina. The specimens were ultrasonically degreased in acetone and distilled water, successively, followed by rinsing with distilled water and finally drying in air.

For electrochemical measurements the conventional three-electrode electrochemical cell was used with Pt wire and Ag/AgCl as the counter and reference electrode, respectively. The specimens were pressed against an O-ring with Cu plate in the electrochemical cell, leaving approx. 60mm^2 exposed to an electrolyte. All experiments were carried out in 5M NaOH at 60°C. The electrolyte was prepared from analytical grade chemicals and demonised water.

For immersion tests, the specimens were immersed with a 1mm platinum wire in a 5M NaOH solution for a period of 1 day at 60°C. The specimens were prepared by holding potential at an open-circuit potential (OCP) and desired applied potentials. The morphology of the specimens were observed using a JEOL 6500 series Field Emission Scanning Electron Microscope (FE-SEM) with the Energy Dispersed X-ray (EDX) unit.

3. Results and discussion

2.1.1 Alkali treatment.

The conventional surface treatment for a bioactive surface was prepared for the TNTZ alloy by immersing in 5M NaOH at 60°C for 24 hours. Morphology of the surface was examined with SEM. The TNTZ alloy produced a thick, highly cracked surface. In order to understand the crack growth, the morphology was studied at different exposure time.

After 1 hour of exposure, the surface exhibited particulates incorporated in a continuous film as seen in Fig. 2(a). Fig.2(b) shows the morphology after 3 hours of exposure time, the surface exhibited some cracks, which initiate around the particle. After 24 hours, the surface was highly cracked as shown in Fig.2 (c). The layer and white particulates grew in thickness and in size and amount, respectively, with time. Therefore, the excessively broken film with the growing thickness can be attributed to the continuous growth of the white particulate. EDX analysis was carried out to characterize the chemical composition of the white particulates. The position analysed is marked as A in Fig.3. For comparison, the surrounding area indicated as B in Fig.3 is also examined. The EDX analysis revealed that tantalum is enriched in the particulate (Table 1), compared with the surrounding layer with trace amounts of oxygen, sodium and the other elements. Fig.4(a) shows an as-polished specimen of the TNTZ alloy exhibiting few tantalum enriched spots. After 30 minutes of immersion in

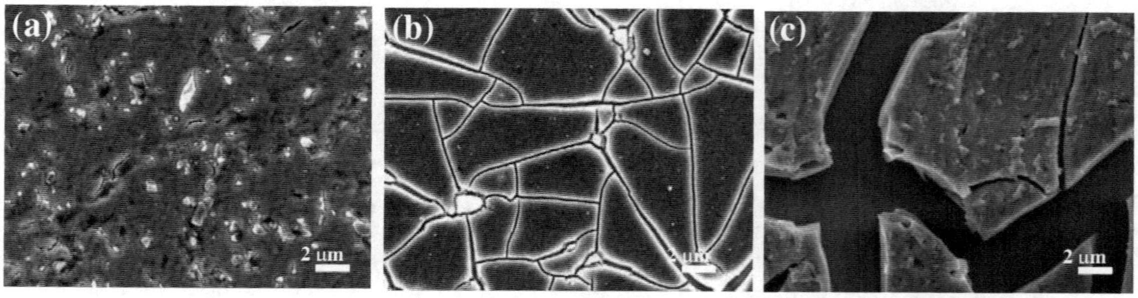

Fig.2 TNTZ alloy treated at open circuit potential for (a) 1 hr, (b)3 hrs and (c) 24 hrs

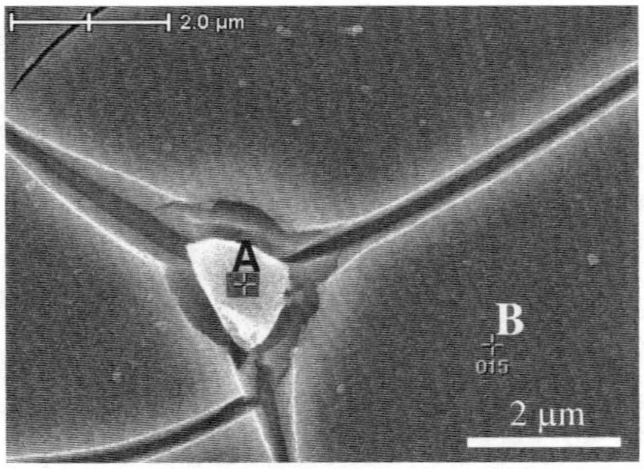

Fig.3 SEM micrograph of TNTZ alloy treated at OCP for 12 hrs, markers A and B point the position of EDX analysis.

Table 1 EDX data of TNTZ alloy treated at OCP for 12 hours. (mass%)

Elements	A	B
O (K)	**3.78**	**21.36**
Na (K)	**0.54**	**8.78**
Ti (K)	**10.73**	**43.69**
Zr (L)	**1.52**	**6.49**
Nb (L)	**4.46**	**7.21**
Ta (M)	**78.97**	**12.47**
Total	**100.00**	**100.00**

Fig. 4 (a) Mapping for Ta on untreated TNTZ alloy, (b) Mapping for Ta on a TNTZ alloy treated in 5 M NaOH for 30 minutes at 60°C.

5M NaOH at 60°C, on the other hand, many localised tantalum enriched spots were identified as shown in Fig.4 (b). In other words, the localized enrichment of tantalum occurred during immersion in the alkali medium. Kokubo et al., reported the formation of a crystalline $Na_2Ta_4O_{11}$, sodium tantalate, over the tantalum substrate after immersion in a 0.5M NaOH at 60°C for 24 hours[10]. The tantalum enriched particulates observed in the present work might be the localised formation of a crystalline sodium tantalite. Considering that these spots were not observed when specimen is immersed within 30 minutes, the tantalum enriched particulates grow into the surface layer from the bottom of the layer, resulting in destructing the uniform nature of the film. These small cracks propagate and the film fragments into many pieces with the growing thickness of the layer as shown in Fig. 2. In order to overcome this problem, electrochemical methods can be adopted. Hence, the electrochemical behaviour of the alloy and its constituent elements was measured.

2.1.2 Open Circuit Potential

The changes in the open circuit potentials for the TNTZ alloy and its elements with time measured in 5M NaOH at 60°C for 24 hours are shown in Fig.5. After 24 hours, all the metals attained a noble potential compared to the initial potential. The OCP of TNTZ alloy was comparable with that of Ta and Nb rather than that of Ti or Zr. During the first one-hour immersion, OCP of all the metals dropped in the less noble direction and then stabilized with exposure time. The decrease of OCP with time indicates the dissolution of the air formed films. It was also noted that there was no distinct change in the order of the nobility compared to 24 hours immersion. The TNTZ alloy exhibited OCP as low as –1.41V, a potential lower than that of pure Nb and Ta. The potentials of pure Ta and the TNTZ alloy shifted less noble with exposure time until 24 hours. The behaviour of TNTZ alloy was more closely comparable to pure Ta or Nb rather than Ti although the mass % of Ti in the alloy

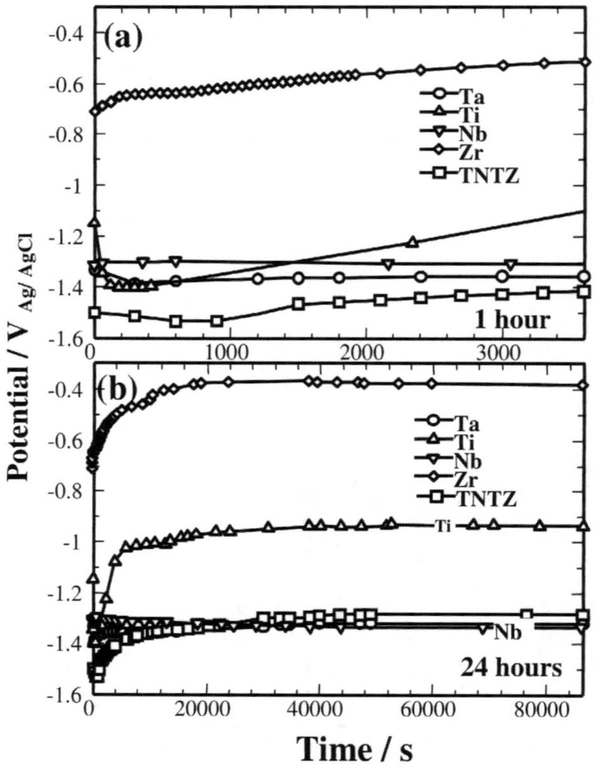

Fig. 5 (a) OCP during 1 hour of immersion, (b) OCP during 24 hours of immersion.

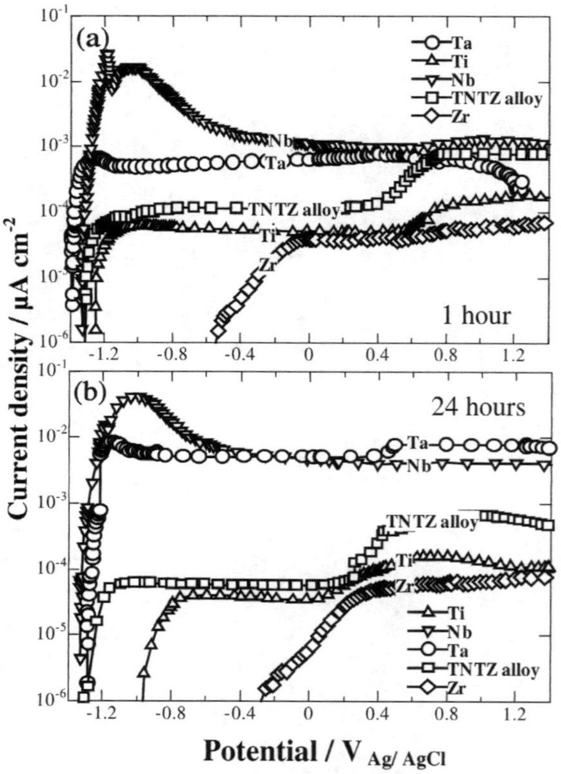

Fig.6 (a) Anodic polarisaiton after 1 hour of immersion, (b) anodic polarisaiton after 24 hours of immersion

composition was higher. This behaviour of the alloy can be attributed to the high alloying content in the alloy. Furthermore, Nb and Ta are more reactive to aggressive alkali medium than Ti.

2.1.3 Polarization studies

Anodic polarization curves were also measured for the TNTZ alloy and its constituent elements after 1 hour and 24 hours of immersion in 5M NaOH at 60°C with a scan rate of $60 mVs^{-1}$.

Fig.6(a) shows the anodic polarization curves obtained after 1 hour of immersion. Nb and Ta exhibited relatively high current densities in both active and passive regions, while Zr showed the lowest current density with a highly stable passive region. Ti exhibited essentially no active dissolution peak, while showing a small current increase at around 0.4 V, which might be due to a change in the stability of the oxide film. The TNTZ alloy had a current density pattern similar to that of Ti with a pronounced increasing current at more noble than 0.4 V.

Fig.6(b) shows the anodic polarisation curves measured after 24 hours of immersion. Nb, Ta and TNTZ alloy had a very low initial OCP, while Ti had a nobler one compared to the TNTZ alloy and Zr had the initial OCP as high as –0.4V. The polarization behaviour of these metals was similar to that after 1 hour immersion. A significant observation after 24 hours was that the active dissolution peak of Nb which had two peaks during 1 hour immersion merged into a single broad peak and the current density of Ta was increased by a decade. The passive current density of Ta was higher than that of Nb in the passive region. On the other hand, there was no remarkable change in the current density behaviour of the TNTZ alloy and Ti.

The OCP studies together with the polarisation studies reveal the complex oxide growth on the TNTZ alloy. The alloy has exhibited a potential change resembling Nb and Ta, and also the polarization curve resembling Ti. As the content of Zr was small, there was no significant participation of Zr in the electrochemical behaviour of the TNTZ alloy.

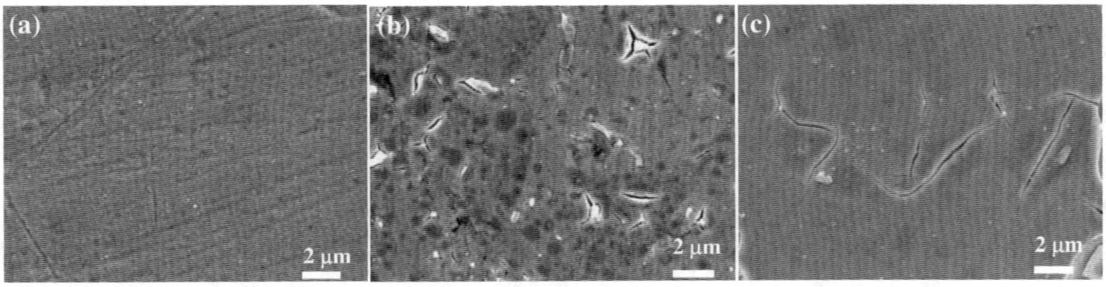

Fig.7 SEM images of a uniform layer formed on TNTZ alloy by pulse polarisation after (a) 1 hr, (b) 3 hrs and (c) 24 hrs.

2.1.4 Electrochemical – alkali treatment

In order to produce a uniform layer over the TNTZ alloy, a potential pulse polarization is introduced. The alloy is immersed in 5 M NaOH at 60 °C, then polarized with the applied potential modulated as square wave. The higher and lower potential of square wave pulses, E_H and E_L, respectively, and the duration of each pulse, t_H, t_L, are varied. The film was formed in the following condition, E_L = -1.2V, E_H = -0.94V; t_L = 2 s, t_H = 5 s. The applied potentials for E_L and E_H were chosen from the electrochemical behaviour of the TNTZ alloy and the alloying elements. The anodic polarisation behaviour of tantalum, after 1 hour and 24 hours of immersion, reveal an active dissolution peak at -1.2V (E_L). While, the E_H (-0.94V) was determined from the OCP of pure titanium after 24 hours of immersion. After the treatment. the morphology of the surfaces produced by pulse polarization technique appeared as shown in Fig.7. There was no particulate (a) after 1 hour of treatment. A few spots with the tantalum rich particles were seen after (b) 3 hours of treatment. At the end of 24 hours, a uniform film with a few cracks was observed.

The variation in current with the applied square wave modulated potential was cathodic during the application of E_L and anodic during E_H. The cathodic behaviour of the alloy shifts the equilibrium of reaction of tantalum with the medium. Hence, during the 1 hour treatment, no particulate was observed. But with time the equilibrium could have been attained and the particles emerge, cracking the uniform layer. The application of E_H, might enhance the growth of the film. The thickness of the film after 24

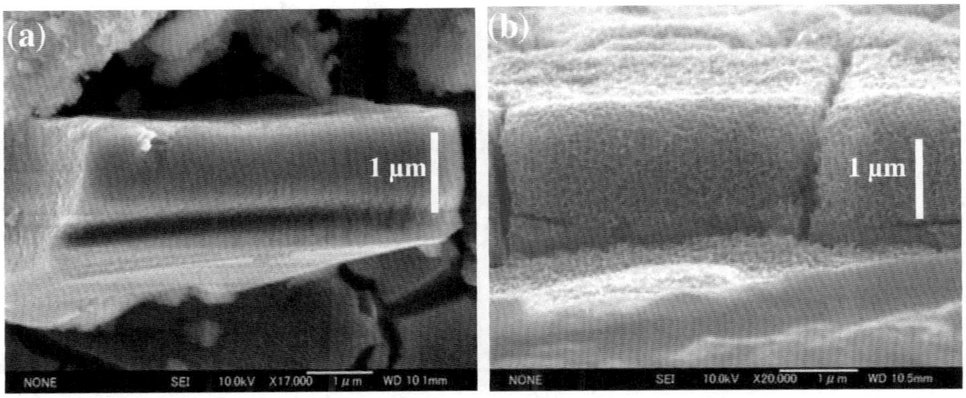

Fig.8 (a) Cross sectional view of the layer formed over TNTZ alloy by alkali treatment at OCP, (b) Cross sectional view of the layer formed over TNTZ alloy by pulse polarisation.

hours of treatment was approx. 2 μm. This was comparable with the thickness of the film observed after the conventional alkali treatment as apparent from Figs.8 (a) and (b).

4. Conclusion

In the present work, the electrochemical behaviour of the β-type TNTZ alloy in 5 M NaOH at 60 °C is examined in order to establish a surface modification process using alkali treatment. The fragmented nature of the layer, which is obtained by the simple natural immersion, is associated to the localised precipitation of a tantalum enriched particulate oxide during the initial stage of the immersion. Application of a square wave modulated potential over the TNTZ alloy apparently hindered the growth of those tantalum enriched particulates resulting in the formation of the uniform layer of sodium titanate / titanium hydroxide layer with a porous network structure.

Acknowledgements

This research is supported by the Global COE program (Project: Advanced Structural and Functional Materials Design, Osaka University) sponsored by the Ministry of Education, Sports, Culture, Science and Technology, Japan.

References

[1] M. Niinomi, D. Kuroda, K. Fukunga, M. Morinaga, Y. Kato, T. Yashiro, A. Suzuki, Mat. Sci. Eng. A263 (1999) 193.
[2] N. Sakaguchi, M. Niinomi, T.Akahori, J. Takeda, H. Toda, Mat. Sci. Eng. C 25(2005) 363.
[3] T. Akahori, M. Niinomi, H. Fukui, A. Suzuki, Mat. Transactions, 45(5) (2004)1540.
[4] T. Naganawa, Y. Ishihara, T. Iwata, M. Koide, M. Ohguchi, Y. Ohguchi, Y. Murase, H. Kamei, N. Sato, M. Mizuno, T. Noguchi, J Preriodontol (2004) 1701.
[5] S.-J. Li, R. Yang, M. Niinomi, Y.-L.Hao, Y.-Y. Cui, Biomat. 25(2004) 2525.
[6] H.M.Kim, F. Miyaji, T.Kokubo, T.Nakamura, J.Biomed. Mater. Res. 32 (1996) 409.
[7] H.M.Kim, F. Miyaji, T.Kokubo, T.Nakamura, J. Mater Sci Mater Med 8 (1997), 341.
[8] B.-H. Lee, J.-K. Kim, Y.-D. Kim, K. Choi, K.-H. Lee, J. Biomed.Mater. Res. 69A (2004) 279.
[9] H. M. Kim, H. Kaneko, T. Kokubo, T. Miyazaki, T. Nakamura, Key Eng. Mat. 240/242 (2003)11.
[10] T. Miyazaki, H.-M. Kim, F. Miyaji, T. Kokubo, and T. Nakamura, J. Biomed. Mater. Res. **50**, 35 (2000).

Electrolytic processing of MgO coatings

R Hashaikeh and J A Szpunar

Department of Mining and Metallurgical Engineering, McGill University, Montreal, QC H3A 2B2, Canada

E-mail: jerzy.szpunar@mcgill.ca

Abstract. MgO coating was formed on nickel superalloy substrate. The coating process consisted of two steps: cathodic electrolytic deposition of MgOH, and heat treatment. The heat treatment was necessary to calcinate and to sinter the deposit. The deposit morphology and composition, before and after heat treatment, were studied by X-ray diffraction and the scanning electron microscopy. The influence of deposition parameters on deposition yields and deposit morphology is studied and discussed.

1. Introduction

Metal ceramic coatings have played an increasing role in protecting metals from stresses, wear, chemical attack and high temperatures. The application of these coatings is growing continuously in the automotive, aeronautic and electro domestic industries. The most common methods used to fabricate metal ceramic coatings are: plasma spraying, CVD, sol gel and PVD.

Recently, the use of the electrolytic method to fabricate metal/ceramic coatings has been investigated. This is a low temperature method which provides coatings with a homogeneous microstructure and high sinterability.[1] Gal-Or et al.[2][3] produced a ceramic composite coating of titanium oxide and zirconium oxide. The zirconia coat was deposited on a titanium substrate from an aqueous solution with zirconyl salt ($ZrO(NO_3)_2$). The deposited coat was amorphous and cracked. Sintering of the coatings in air at 780 °C caused crystallization of the tetragonal and monoclinic ZrO_2 polymorphs phases having nanosize crystallites. Oxidation of the titanium substrate during heat treatment resulted in the formation of TiO_2 at the site of cracks between the zirconia islands.

Aries[4] used the electrolytic method to fabricate an alumina barrier on stainless steel. The alumina barrier was obtained by thermal dehydration of aluminium hydroxide deposited from an aqueous solution of aluminium sulphate. Before deposition, the surface of the metal substrate was pretreated to form a functional conversion coating characterized by a strong adhesion to the substrate. The pretreatment involved a chemical treatment of the substrate or what is known as "anodizing process." An anodic oxidation of the metal in an electrochemical cell with a suitable electrolyte leads to the formation of a homogeneous, rough-surface oxide layer with a thickness of about 200 nm. Alumina was then deposited by the electrochemical method to form a two-layer coat.

A great deal of research has been devoted to demonstrate the feasibility of electrolytic deposition of important ceramic materials like: Al_2O_3, ZrO_2, CeO, TiO_2, RuO_2, PbO_2, ZnO, Nb_2O_5, $BaTiO_3$, PZT and $YBa_2Cu_3O_7$.[1]

The present work investigates the possibility of magnesia electrolytic deposition and describes the formation of magnesia coating on nickel superalloy. Magnesia has the highest thermal expansion coefficient among ceramics, which is an important property for the application of metal ceramic coating.

2. Mechanism of Deposition

If two electrodes are immersed in an aqueous solution of magnesium nitrate ($Mg(NO_3)_2$) and connected to a source of current of a sufficiently high potential, there will be a passage of electric charge through the solution and at the same time various chemical reactions will take place at the electrodes. In an aqueous solution, the magnesium nitrate dissolves to give magnesium ions.

$$Mg(NO_3)_2 \Rightarrow Mg^{2+} + 2NO_3^-$$

During cathodic synthesis, the cathodic reaction is used to generate hydroxide ions at the electrode surface; created mainly by the electrolysis reaction of water.

$$2H_2O + 2e^- \Rightarrow H_2 + 2OH^-$$

The hydroxide ions may also be produced by other cathodic reactions that include the reaction of reduction of oxygen

$$O_2 + 2H_2O + 4e^- \Rightarrow 4OH^-$$

In the case of nitrate salts, the reduction reaction of nitrate produces hydroxide ions:

$$NO_3^- + H_2O + 2e^- \Rightarrow NO_2^- + 2OH^-$$

$$NO_3^- + 6H_2O + 8e^- \Rightarrow NH_3 + 9OH^-$$

Mg^{2+} ions are then hydrolyzed by the electro-generated hydroxide ions to form a cathodic deposit.

The dehydration of magnesium hydroxide to form MgO occurs during heat treatment.

$$Mg(OH)_2 \Rightarrow MgO + H_2O$$

In the following sections the effects of electrochemical deposition parameters on the deposit morphology and the microstructure evolution in the as-deposited and sintered coatings will be analyzed.

3. Experimental Procedure

Nickel superalloy (Inconel 600) was used as a substrate. The substrate specimen has dimensions of 2x1.5x0.6 cm^3. The electrochemical bath consisted of a 250ml beaker. The electrolyte was composed of 0.5M Magnesium Nitrate ($Mg(NO_3)_2$), dissolved in 50% water and 50% ethanol. The substrate was the cathode and the anode material was graphite with a distance of 2 cm between them. The source of current was (EG&G Instruments Potentiostat) rectifier. Deposits were obtained on the cathode with current densities ranging from 2.5-20 mA/cm^2, and duration from 5 to 60 minutes. All experiments were performed without stirring. The cell Voltage and current density were measured with AVOmeters. The deposits were dried in air at room temperature. The coating weights were measured before and after the deposition. Specimens were fired at 700 °C and 1100 °C in air; TGA was used to study the weight change during firing. The microstructure of the deposit was characterized after different stages of the process, using the optical and scanning electron microscopy (SEM). The phase composition and the crystallization state were determined by X-ray diffraction (XRD).

4. Results and Discussion

Figure 1 shows a SEM image of the deposited coat; the obtained deposit was a gel-like transparent solid deposit with a homogeneous coverage of the substrate. The color of the deposit depended on the current density. The deposit color was yellow at low current density, yellow-orange at intermediate current density and brown at high current density. No visible change was observed when the deposit was removed from the bath, which indicates that the $Mg(OH)_2$ deposit did not dehydrate with drying.

Figure 1. SEM image of as deposited $Mg(OH)_2$ coating on Nickel superalloy:
a) Homogeneous covering of the substrate.
b) A cracked deposit.

Both the deposit quality and quantity depended on the current density. The amount deposited in the same period of time increases with increasing current density. High deposition rates can be obtained at high current densities; but the deposit quality is best at low current density. At a current density greater than 5 mA/cm^2, the deposit forms cracks after drying. However, at low current density, the deposit is homogeneous without any cracks.

The dependence of the deposited weight on deposition time at constant current density of 2.5 mA/cm^2 is linear. During deposition, the thickness of the coat increases. As this is a non-conductive deposit, the resistance to current increases, and since we are working at a constant current density the voltage across the cell increases with time.

The weight of the deposited $Mg(OH)_2$ decreases rapidly with time as temperature increases. This decrease in weight is related to the evaporation of the co-deposited solvent, which was about 20% of the deposit weight. At a temperature of 200 °C the nickel superalloy substrate started to oxidize and the sample weight started to increase, but at around a temperature of around 300 °C the weight of the sample decreased. This decrease is expected to be the result of the calcination reaction of $Mg(OH)_2$ to form MgO.

Figure 2 shows the change in morphology of the deposit with heat treatment. The calcination process used to form MgO, the crystallization process of MgO and the sintering of the MgO all these processes occur during the heat treatment.

Figure 2. Morphology of the deposit after different heat treatment:
1. Sintered at 700 °C for ½ hr.
2. Sintered at 1100 °C for 1 hr.

X-ray diffraction (XRD) spectra from the as-deposited coat, after being separated from the substrate, showed a weak level of crystallization with two small peaks identified as MgO and $Mg(OH)_2$. Annealing of the coat together with the substrate in air for 1 hour at 1100 °C caused the oxidation of aluminium and the crystallization of MgO, see Figure 3.

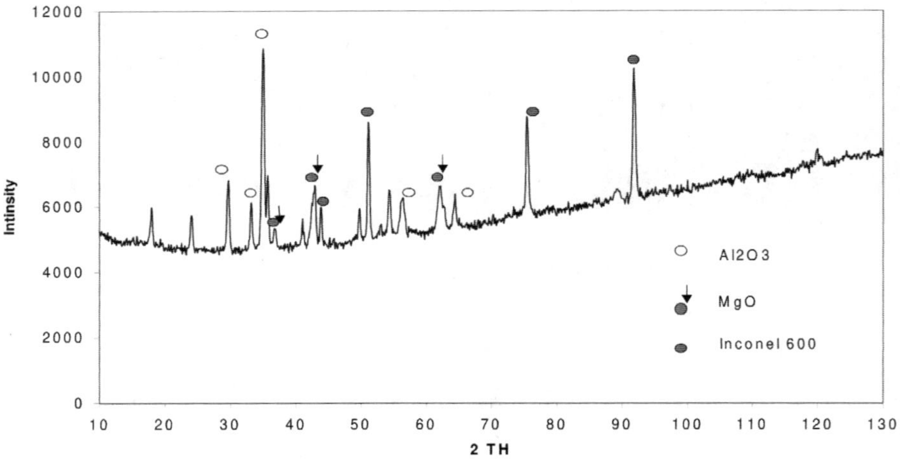

Figure 3. X-ray diffraction of sintered deposit at 1100 °C.

5. Summary

Electrolytic coating of ceramics is a simple process that can be used to produce ceramics and ceramic coatings. In this work MgO coating was produced using the electrolytic technique. The main problem that has to be overcome when using this technique is the cracking of the coating, which results from the shrinkage that takes place during the drying process. A crack-free coating of magnesium oxide on nickel superalloy substrate was produced in this work by depositing magnesium hydroxide from magnesium nitrate aqueous solution. The magnesium hydroxide was then calcinated to produce MgO.

References

[1] N. B. Dahotre and T. S. Sudarshan, "Intermetallic and Ceramic Coatings", *Marcel Dekker, Inc.*, NewYork, pp83-146 (1999).
[2] L. Gal-Or, I. Silberman, and R. Chaim, "Electrolytic ZrO_2 Coatings; I. Electrochemical aspects", *J. Electrochem. Soc.*, **138**, No. 7, 1939-1942 (1991).
[3] L. Gal-Or, I. Silberman, and R. Chaim, "Electrolytic ZrO_2 Coatings; II. Microstructural Aspects", *J. Electrochem. Soc.*, **138**, No. 7, 1942-1946 (1991).
[4] L. Aries, "Preparation of Electrolytic Ceramic Films on Stainless Steel Conversion Coatings", *Surface Engineering*, **114**, n.3, 235-240 (1998).
[5] I. Zhitomirsky, A. Petric, "Electolytic and Electrophoretic Deposition of CeO_2 Films", *Material letters*, **40**, 263-268 (1999).
[6] R.Chaim, G. Stark, L. Gal-Or, "Electrochemical ZrO_2 and Al_2O_3 Coatings on SiC Substrates", *Journal of Materials Science*, **29**, n.23, 6241-6249 (1994).
[7] B. E. Russ and J. B. Talbot, " A Study of the Adhesion of Electrophoretically Deposited Phosphors", *J. Electrochem. Soc.*, **145**, No.4, 1245-1256 (1998).b

Journal of Physics: Conference Series **165** (2009) 012009 doi:10.1088/1742-6596/165/1/012009

Crystallographic characterization of stress corrosion cracking initiation on type316L stainless steel in high temperature and high pressure water

Tomo Saito[1], **Hiroya Masaki, Fahmi Azhari Mukhlis, Hiroaki Tsuchiya and Shinji Fujimoto**
Division of Materials and Manufacturing Science, Graduate School of Engineering, Osaka University, 2-1 Yamada-oka, Suita, Osaka 565-0871, Japan

tomo.saito@mat.eng.osaka-u.ac.jp

Abstract. Intergranular stress corrosion cracking (IGSCC) and transgranular stress corrosion cracking (TGSCC) of Type316L stainless steel are examined by a slow strain rate test (SSRT) in a dilute sodium sulphate solution at 288 °C. Crystallographical orientation measurements on Type316L are performed by electron backscatter diffraction (EBSD) in order to grasp crack initiation during SSRT. The relationship between crystal orientation and crack initiation of IGSCC and TGSCC can be characterized using EBSD data.

1. Introduction

Stress corrosion cracking (SCC) of austenitic stainless steel is classified into two types – intergranular stress corrosion cracking (IGSCC) and transgranular stress corrosion cracking (TGSCC). IGSCC occurs as a result of the sensitization of stainless steel, that is, the formation of Cr depleted zone induced by Cr carbide precipitation at grain boundaries. TGSCC has been observed in concentrated chloride solution. Since the middle 1990's it has been reported that even low carbon stainless steels such as Type316L stainless steel are subjected to SCC in high temperature and high pressure water environments [1-3]. TGSCC is initiated at surface where severe plastic deformation is introduced by surface machining and then shifted to IGSCC. Although extensive studies have been achieved on stress corrosion cracking of low carbon austenitic stainless steels in such environments, neither the crystallographic features nor their relationship to crack initiation have been clarified.

In the present work, we establish procedures to characterize the crystallographic features of crack initiation observed for cold worked Type316L stainless steel in high temperature and high pressure water environments using the slow strain rate test (SSRT) and electron backscatter diffraction (EBSD). SSRT is known to provide the SCC susceptibility of materials in any corrosive environment within a relatively short experimental time [4-6]. EBSD is a technique that is useful for obtaining crystallographic characteristics of metals. For example, plastic deformed metals or alloys were investigated using EBSD [7-8].

2. Experimental procedures

2.1. Material

The material examined was Type316L stainless steel (0.015% C, 0.52% Si, 0.96% Mn, 0.020% P, 0.004% S, 12.39% Ni, 16.28% Cr, 2.12% Mo and the balance Fe). The steel was subjected to cold-rolling with a reduction ratio of 20 %. Tensile specimens, cut from the material, were mechanically ground with SiC abrasive paper and diamond paste, and then mirror-finished by mechano-chemical polishing. The shape and dimensions of a tensile specimen is shown in Fig. 1.

[1] To whom any correspondence should be addressed.

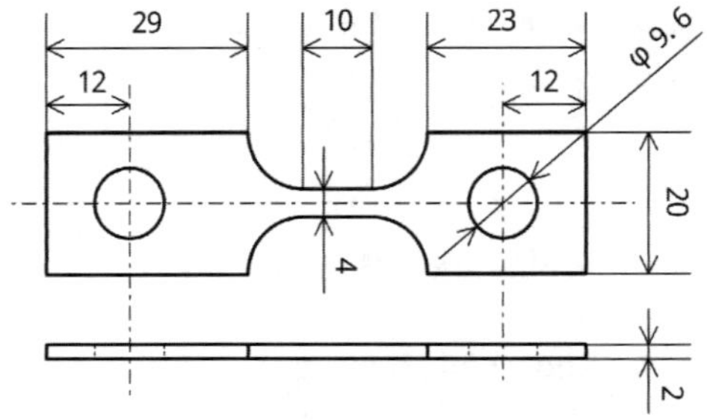

Figure 1. The shape and dimensions of a tensile specimen.

2.2. *Slow strain rate test (SSRT)*

The slow strain rate tests were performed in dilute sodium sulphate solutions controlling dissolved oxygen contents (8 ppm or less than 2 ppb) at 288 °C. The specimens were elongated with a strain rate of $5x10^{-7}$ s^{-1} by 10% and then the strain was maintained for 10 hours.

2.3. *Electron backscatter diffraction (EBSD)*

Crystallographic characteristics of Type316L were determined by EBSD in a scanning electron microscope (SEM). The EBSD measurements were conducted before and after SSRT. In the former case, SEM observations were required after SSRT on the same locations as EBSD analyses were carried out. The positions of analyzed areas were identified by micro-indent marks as illustrated in Fig. 2. In the latter case, on the other hand, EBSD analysis was conducted after SSRT on selected areas to obtain crystallographic characteristics of crack initiation sites. However, in this case, granular hydrous oxides formed during SSRT were removed by Ar ion sputtering. From the EBSD analysis and SEM observations we characterized the crack initiation sites in terms of crystallographic features such as the grain boundary misorientation.

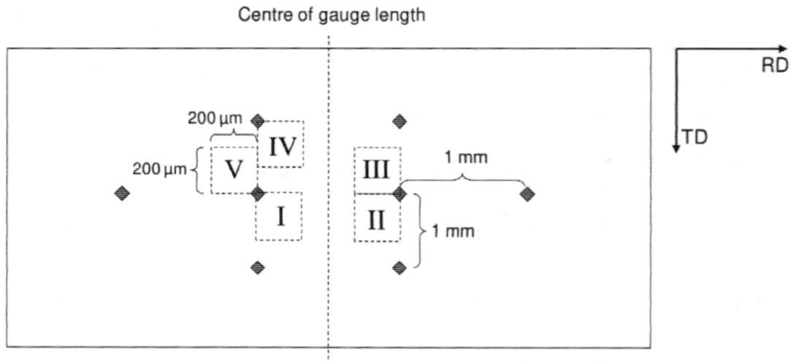

Figure 2. Positions of the micro-indent mark and EBSP analysis areas on the gauge area of the tensile specimen surface of the type316L stainless steel

3. Results and discussion

3.1. *SEM and EBSD observations*

Figure 3(a) and 3(b) show the Image Quality (IQ) map and the Inverse Pole Figure (IPF) map, respectively, obtained from the EBSD measurement. The maps were acquired before SSRT. Corresponding SEM top-view image is also included in Fig. 3(c). The SEM image was taken at the identical position as the EBSD maps. Black lines in Fig. 3(b) indicate the positions of cracks confirmed by the SEM observation. The maps and image shown in Fig. 3 were obtained from area I as presented in Fig. 2. The clear IQ map reflects the fact that crystallographic data were obtained with high reliability. Figure 3(b) clearly indicates that both intergranular and transgranular cracks were initiated on 20% cold-worked Type 316L stainless steel in high temperature and high pressure water.

IQ map, IPF map and corresponding SEM top-view image obtained after SSRT are shown in Fig. 4. The SSRT was conducted in high temperature and high pressure water containing dissolved oxygen less than 2 ppb. The environment resulted in the formation of granular hydrous ferric oxides on the stainless surface. Prior to the EBSD measurement, therefore, the oxides formed during SSRT were removed by Ar ion sputtering. Remarkable is that clear IPF map could be obtained even after SSRT in a high temperature and high pressure water though IQ map shown in Fig. 4(a) is less clear compared to Fig. 3(a). These results clearly demonstrate that this procedure is effective to get crystallographic information.

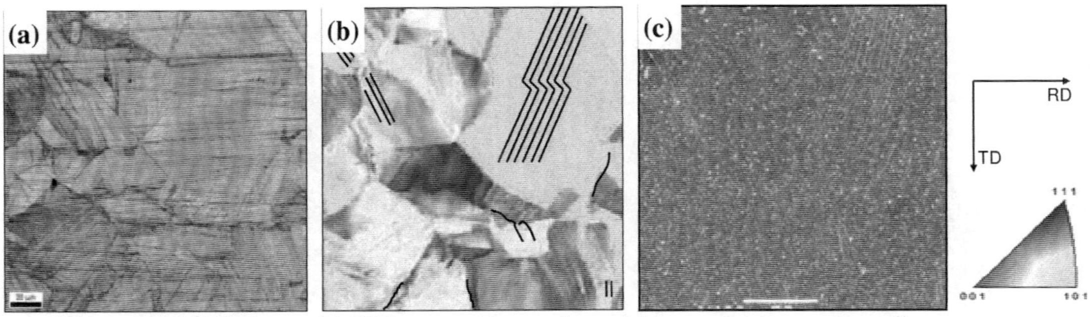

Figure 3. (a) IQ map, (b) IPF map with crack positions and (c) corresponding SEM top-view image. The maps and image were obtained from area I as indicated in Fig. 2. The EBSD measurement was carried out before SSRT and the crack positions were confirmed from the SEM image taken after SSRT.

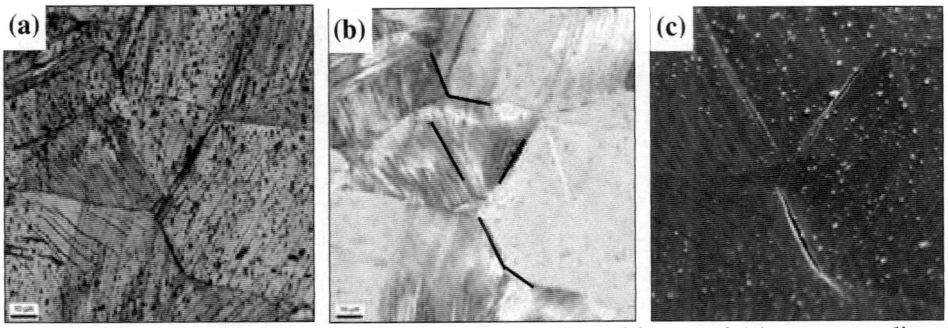

Figure 4. (a) IQ map, (b) IPF map with crack positions and (c) corresponding SEM top-view image. The maps and image were obtained after SSRT. Prior to the measurement, the granular oxides formed during SSRT were removed by Ar ion sputtering.

3.2. The relationship between crack and crystallographic characteristics

IGSCC and TGSCC are discussed using EBSP data obtained before SSRT. EBSD data not only shows crystal orientation at the metal surface but also allow us to characterize several crystallographic parameters such as misorientation angle. Misorientation angle is defined as angular difference between adjacent crystals. Misorientation angles at grain boundaries can be related to crack initiations at the boundaries. Figure 5(a) shows the relationship between crack probability and misorientation angle. Crack probability is defined as the ratio by the number of grain boundaries with specific misorientation angle to the number of cracks initiated at the grain boundaries. It is apparent from Fig. 5(a) that most cracks are initiated at the misorientation angle of about 30° and 50°.

For the TGSCC, we measured the crack angle. The crack angle is defined as the angle between the load direction and the crack direction on the surface. Figure 5(b) shows the number of cracks for each crack angle. It is clear that there is preferential crack angle at 60° for TGSCC. One can deduce from this result that slip deformation is one of main factors for TGSCC.

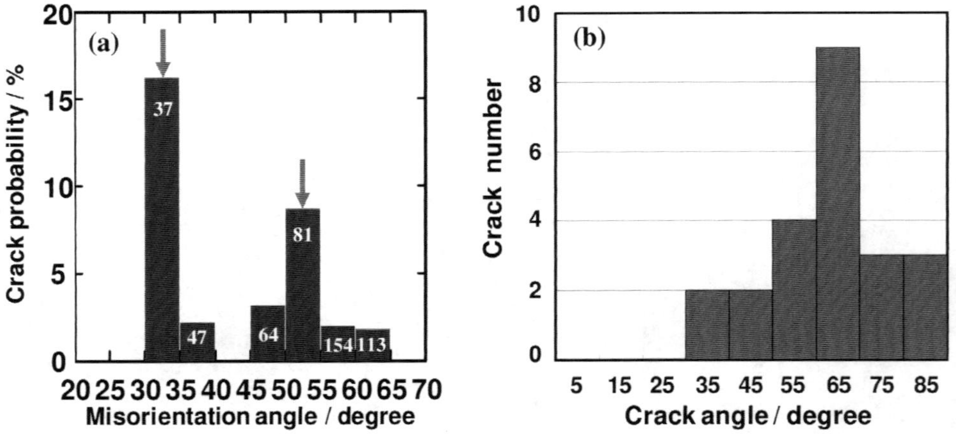

Figure 5. Relationship (a) between misorientation angle and crack probability and (b) between crack angle and number of cracks. The numbers in (a) indicate the number of grain boundary with each misorientation angle.

4. Conclusion

In the present work, we established the procedures to obtain crystallographic characteristics on cold-worked Type 316L stainless steel even after SSRT in high temperature and high pressure water environment and clearly demonstrated that samples can be analyzed using EBSD data.

Most IGSCC were initiated at grain boundaries with specific misorientation angles (30° and 50°). For TGSCC, most cracks were initiated at an angle of 60° against load direction.

Acknowledgments

This work was supported by the Global COE program (Project: Advanced Structural and Functional Materials Design) from the Ministry of Education, Sports, Culture, Science and Technology, Japan.

References

[1] J. Congleton and W. Yang 1995 *Corros. Sci.* **37** 429
[2] T. Magnin, A. Chambreuil and B. Bayle 1996 *Act. Mater.* **44** 1457
[3] J. Deleume, D. Poquillon, V. Garat, J. M. Cloue and E. Andrieu 2008 *Corros. Sci.* **50** 737
[4] R. Nishimura and Y. Maeda 2004 *Corros. Sci.* **46** 769
[5] E. Villalba and A. Atrens 2008 *Eng. Fail. Anal.* **15** 617
[6] Z. Lu, T. Shoji, Y. Takeda, Y. Ito and S. Yamazaki 2008 *Corros. Sci.* **50** 698
[7] M. Kamaya, A. J. Wilkinson and J. M. Titchmarsh 2005 *Nucl. Eng. Des* **235** 713
[8] L. N. Brewer, M. A. Othon, L. M. Young, and T. M. Angeliu 2006 *Microsc. Microanal.* **12** 85

Ultrafine grained steels managing both high strength and ductility

N Tsuji
Dept. Adaptive Machine Systems, Osaka University, Suita 565-0871, Japan

E-mail: tsuji@ams.eng.osaka-u.ac.jp

Abstract. Mechanical properties of the ultrafine grained (UFG) steels having mean grain sizes much smaller than 1 μm were systematically shown from the original experimental data. The UFG steels performed very high strength that reached to 2-4 times of the starting materials having conventionally coarse-grained structures. On the other hand, the uniform tensile elongation was commonly limited within a few percents when the materials have single-phased UFG structures. The limited uniform elongation was explained in terms of early plastic instability. On the basis of the understanding, the ways to manage both strength and ductility in the UFG steels were indicated. The effectiveness of making the UFG structures multi-phased was confirmed.

1. Introduction

Grain refinement has been always one of the most important subjects in microstructure control of structural metallic materials. However, the minimum mean grain size that has been conventionally achieved in bulky materials is around 10 μm. In case of commercial steels, the champion value has been 5 μm in the steel plate thermomechanically processed by so-called controlled rolling [1]. On the other hand, it has become recently possible to fabricate the steels with ultrafine grained (UFG) structures of which mean grain size is smaller than 1 μm, at least in laboratory scales [2]. One of the ways to obtain the UFG steel is thermomechanical processing under critical conditions, where phase transformation from heavily deformed austenite to ferrite is basically used. For example, grain sizes around 1 μm have been achieved in carbon steels through heavy one-pass deformation at low temperatures around 500°C in undercooled austenite region followed by rapid cooling [3]. The other route to realize the UFG structures is severe plastic deformation above logarithmic equivalent strain of about 4 [4]. The SPD can be applied not only to steels but also to most of workable metals and alloys, and the UFGs in 100 nm dimensions or even nanocrystals have been obtained in various kinds of metals and alloys [4].

The UFG materials perform the strength 2-4 times higher than that of the conventionally grain-sized materials [4,5]. On the other hand, however, the UFG materials usually exhibit limited ductility (especially limited uniform elongation) [4,5]. This imbalance between strength and ductility is the biggest issue of the UFG materials for practical application in the future. If the issue can be overcome, the UFG materials are quite attractive, because high strength can be achieved in simple chemical compositions without special alloying elements. The mechanical properties of the UFG materials have not yet been systematically clarified. In the present paper, strength and ductility of the UFG ferritic steels are systematically shown from the experimental evidences, and the way to manage both strength and ductility in the UFG steels is discussed.

2. Experimental

A Ti-added ultralow-carbon interstitial free (IF) steel was highly deformed by the accumulative roll-bonding (ARB) process, in order to obtain an UFG structure. The ARB is a kind of SPD process using rolling deformation, which was originally developed in the present author's group [4,6]. **Figure 1** illustrates the principle of the ARB process. Two pieces of the plain IF steel sheet 1 mm thick, 30 mm wide and 300 mm long were stacked after degreasing and wire-brushing the contact surfaces. The stacked sheets were spot-welded at the top and tail ends, held in an electric furnace set at 500°C for 600 s, and then roll-bonded by 50% one-pass rolling using a two-high rolling mill with a roll diameter of 310 mm without lubrication. The roll-bonded sheet was cut into two and the above mentioned procedures were repeated up to 7 cycles. Because von Mises equivalent strain for 50% rolling is 0.8, the total strain applied to the material by 7 cycles of the ARB is 5.6. The ARB processed sheets were annealed at various temperatures ranging from 400 °C to 800 °C for 1.8 ks to change the grain size.

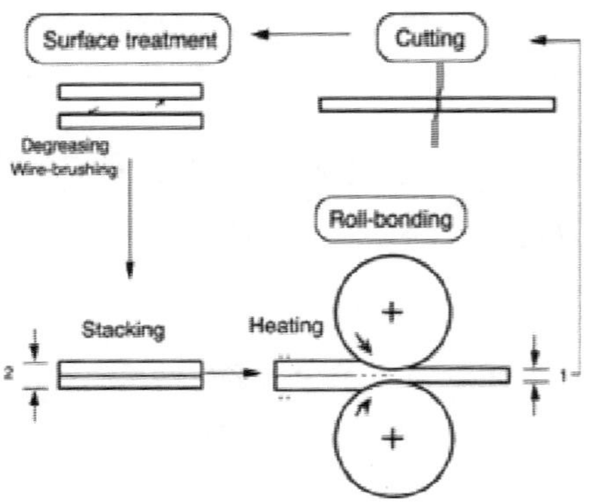

Fig.1 Schematic illustration showing the principle of the ARB process.

Another thermomechanical processing [7,8] was applied on a plain low-carbon steel (JIS-SS400) to fabricate another kind of UFG structure. The detailed procedures of the process were shown in the previous article [7]. The sheet of the SS400 having fully martensitic structure was obtained by austenitization and water-quenching. The martensite sheet was conventionally cold-rolled by 50% total reduction and annealed at various temperatures from 200 °C to 700 °C for 1.8 ks. By annealing under appropriate conditions, the UFG structures composed of UFG ferrite with mean grain size of 100 nm and nano-carbide dispersing within the UFG ferrite matrix uniformly can be formed [7,8]. This is a simple but novel process to obtain multi-phased UFG structure in carbon steels [7].

Microstructures of the obtained UFG steels were observed on the sections perpendicular to the transverse direction (TD) of the sheets by means of scanning electron microscopy (SEM), electron back-scattering pattern analysis in a field-emission type SEM (FE-SEM/EBSP) and transmission electron microscopy (TEM). Mechanical properties of the specimens were measured in tensile test at room temperature and initial strain rate of 8.4 x 10^{-4} s^{-1}. Tensile specimen with gage width of 5 mm and gage length of 10 mm (1/5 miniature of JIS-5 specimen) was used for the tensile test. An extensometer was attached to the specimen in the tensile test, in order to measure exact displacement.

2 Mechanical Properties of UFG Steel with Ferrite Single Phase

Figure 2 shows the stress-strain curves of the IF steel ARB processed by various cycles. The strength of the sheet greatly increases by just 1 cycle of the ARB, while the tensile elongation significantly decreases. The strength rises up with increasing the ARB cycle and the tensile strength reaches to 900 MPa after 7 ARB cycles which is 3.2 times higher than that of the starting sheet (280 MPa). On the other hand, the elongation of the sheets shows nearly constant value after 1 ARB cycle. Especially, the uniform elongation is limited within a few %. These are typical mechanical properties of the SPD materials having single phase [6,9,10].

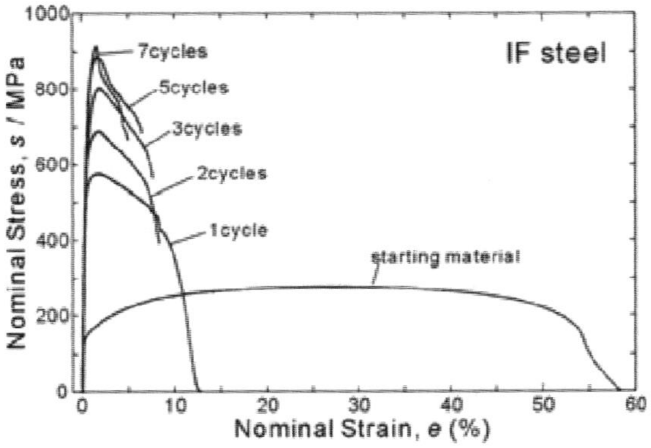

Fig.2 Stress-strain curves of the IF steel ARB processed by various cycles at 500°C.

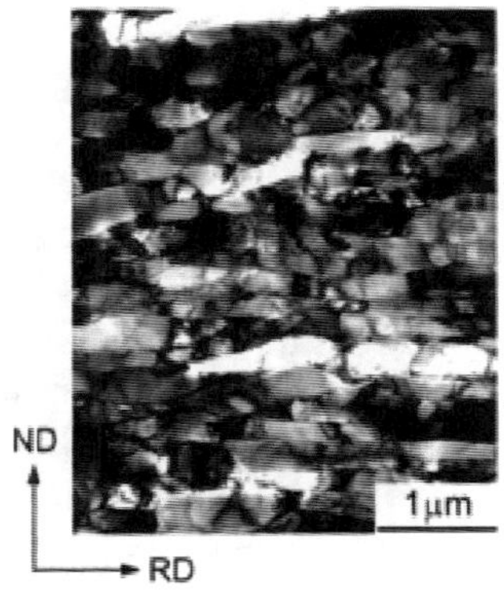

Fig.3 TEM microstructure of the IF steel ARB processed by 6 cycles at 500°C. Observed from TD.

A TEM microstructure of the IF steel ARB processed by 6 cycles is shown in **Fig.3**. The ARB processed materials typically reveal the UFG microstructure elongated along the rolling direction (RD) of the sheets. The mean grain thickness in Fig.3 was 210 nm. It has been already confirmed that the elongated UFGs are mostly surrounded by high-angle grain boundaries and the ARB processed sheets are uniformly filled with the elongated UFGs throughout thickness [11-14]. The material having such a microstructure performs very high strength but limited uniform elongation (Fig.2).

When the ARB processed IF steel is annealed at various temperatures, a continuous change of the microstructures happens, which is called *continuous recrystallization* [5,15,16]. **Figure 4** shows TEM microstructures of the IF steel ARB processed by 5 cycles and then annealed at various temperatures for 1.8 ks. Through these processes, the bulky sheets of IF steel having various mean grain sizes ranging from 0.2 μm to 20 μm can be obtained, which are useful to test the mechanical properties of the UFG materials systematically.

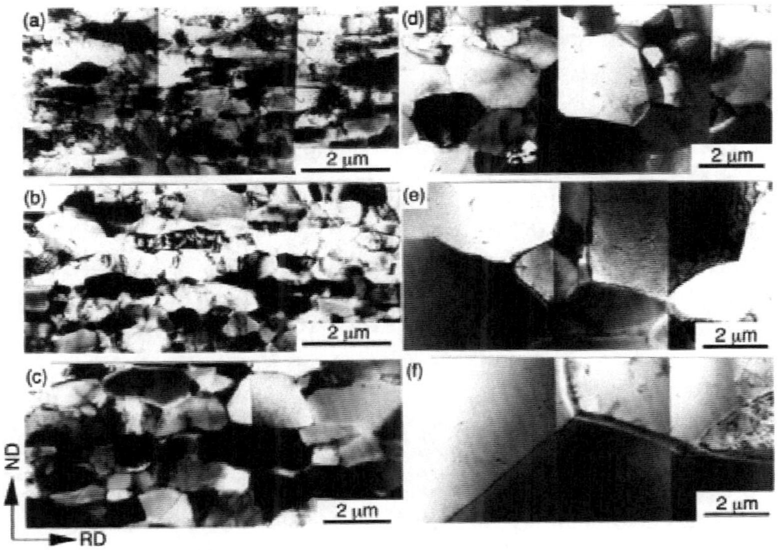

Fig.4 TEM microstructures of the IF steel ARB processed by 5 cycles at 500°C and then annealed for 1.8 ks at (a) 400°C, (b) 500°C, (c) 600°C, (d) 625°C, (e) 650°C and (f) 700°C. Observed from TD.

Stress-strain curves of the IF steel with various mean grain sizes, which were fabricated by the ARB and annealing process, are shown in **Fig.5**. The stress-strain behaviors changes as the mean grain size decreases. It is obvious that the flow stress, especially the yield stress, significantly increases with decreasing the grain size. On the other hand, the elongation, especially the uniform elongation, suddenly drops when the grain size becomes smaller than 1 μm. Uniform elongation in tensile test is determined as the strain at which macroscopic necking initiates. Macroscopic necking is usually explained by plastic instability. Equation (1) indicates the well-known Considére criterion for plastic instability.

$$\sigma \geq \left(\frac{d\sigma}{d\varepsilon}\right) \quad (1)$$

According to this equation, necking initiates when the flow stress (left-hand term) becomes equal to the strain-hardening rate (right-hand term). The plastic instability is verified in the ultrafine grained IF steel with various mean grain sizes in **Fig.6**. In Fig.6, the dotted lines are the true stress-strain curves while the solid lines strain-hardening rate obtained by differentiating each stress-strain curve. The point at which two curves meet, i.e., the plastic instability point, coincided well with the uniform elongation of these specimens actually measured. That is, the limited uniform elongation in the UFG ferritic steel is understood in terms of plastic instability. As is shown in the stress-strain curves (Fig.5), grain refinement significantly increases the yield strength of the material. On the other hand, strain-hardening is not enhanced by grain refinement, but it rather decreases with decreasing the grain size. As a result, plastic instability happens at very early stage of tensile test in the UFG ferritic material, resulting in limited uniform elongation. This is an inevitable nature of the UFG materials having single phase.

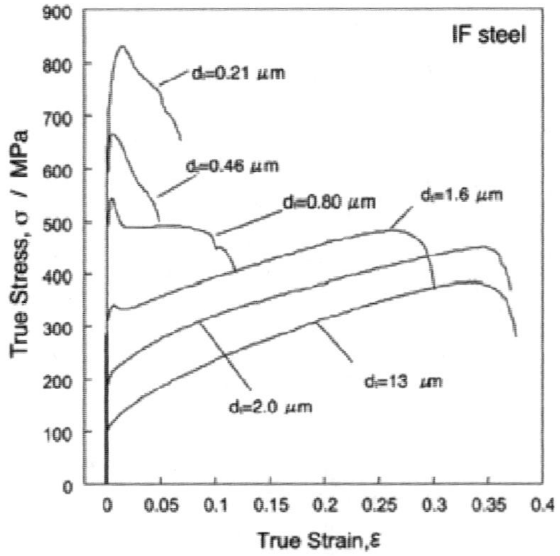

Fig.5 True stress-strain curves of the IF steel with various mean grain sizes fabricated by the ARB and annealing process.

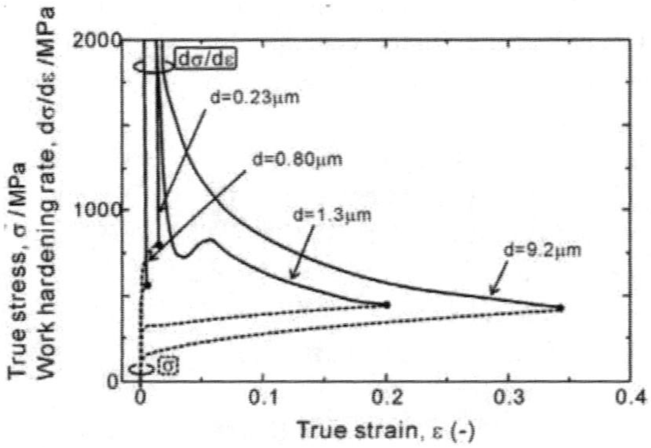

Fig.6 True stress and strain-hardening rate as a function of true strain in the IF steel having various grain sizes fabricated by the ARB and annealing process.

3 Mechanical Properties of Multi-Phased UFG Steel

Understanding on the basis of plastic instability also tells us how to improve the ductility of the UFG materials. If the strain-hardening (right-hand term in eq.(1)) is enhanced by some means, it is expected that the plastic instability is delayed even in the UFG material. One of the ways to enhance strain-hardening is to disperse fine second phase(s) within the UFG ferrite matrix uniformly. **Figure 7** is a TEM microstructure of the low-C steel (SS400) processed in the latter thermomechanical treatment (Martensite-method) [7]. The starting sheet having as-quenched martensite structure was conventionally cold-rolled by 50% and then annealed at 500°C for 1.8 ks. The specimen shows a kind of UFG structure. It is noteworthy that a number of carbides with nano-meter sizes uniformly precipitate within the UFG ferrite matrix. This is because the as-quenched martensite was a supersaturated solid solution of carbon. Consequently, the structure resulted from the Martensite-method is a multi-phased UFG structure composed of nearly equiaxed UFG ferrite having mean grain size of 100 nm and nano-carbides uniformly disperse within the UFG ferrite.

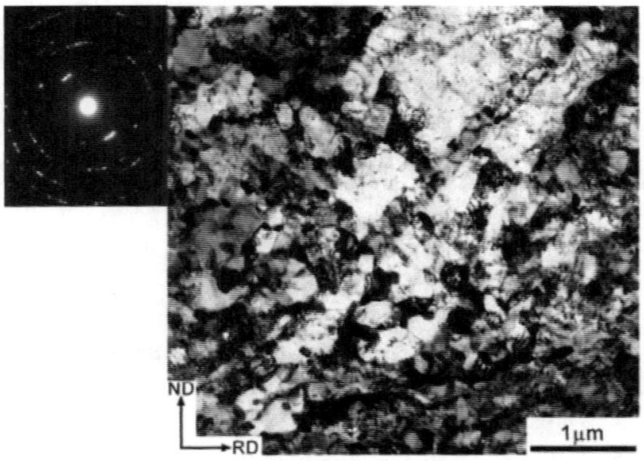

Fig.7 TEM microstructure of the SS400 steel 50% cold-rolled and annealed at 773K for 1.8ks. Observed from TD. The starting microstructure was martensite.

The multi-phased UFG steel fabricated by the Martensite-method actually managed both high strength and adequate ductility. Stress-strain curves of the low-C steel Martensite-processed are shown in **Fig.8**. Strain-hardening is clearly enhanced and the specimen annealed at 550°C showed tensile strength of 850 MPa, uniform elongation of 10% and total elongation of 10% at the same time. The standard tensile strength of this material is 400 MPa. The result clearly shows that dispersing fine precipitates is effective to manage both strength and ductility in the UFG steels.

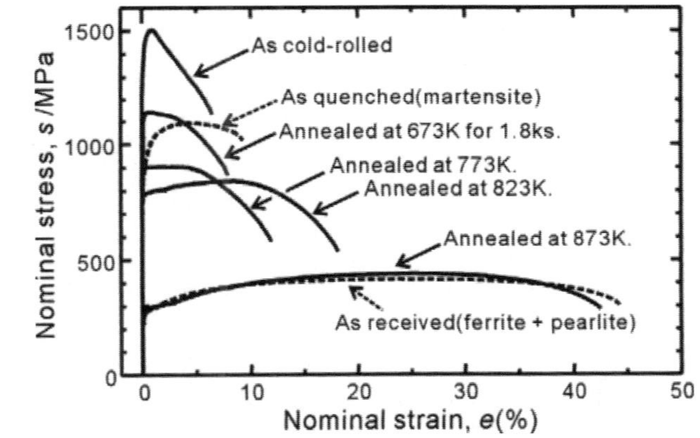

Fig.8 Nominal stress-strain curves of the SS400 steel 50% cold-rolled and annealed at various temperatures for 1.8ks. Starting microstructure was martensite.

4 Summary

Mechanical properties of UFG steels were shown based on the author's original experimental data. UFG steels perform very high strength but limited uniform tensile ductility when they have single phase structure. The limited uniform elongation was understood in terms of early plastic instability, which is an inevitable feature of the UFG single-phased materials. It was also indicated at the same time that dispersing nano-carbides within the UFG ferrite matrix is an effective way to improve the uniform elongation of the UFG steels. Steels have various options of multi-phased structure, such as, ferrite + cementite, ferrite + martensite, ferrite + bainite, and so on. It was clearly shown in the present paper that the studies of the UFG steels should be focused on "multi-phase" in future.

References

[1] Maki T 1989 *Netsusyori* **29** 278-284.
[2] Tsuji N 2002 *Tetsu-to Hagané* **88** 359-369.
[3] Adachi Y, Tomita T and Hinotani S 1999 *Tetsu-to Hagané* **85** 620.
[4] Tsuji N, Saito Y, Lee S-H, and Minamino Y 2003 *Adv. Eng. Mater* **5** 338-344.

[5] Tsuji N, Ito Y, Saito Y and Minamino Y 2002 *Scripta Mater.* **47** 893-899.
[6]Saito Y, Tsuji N, Utsunomiya H, Sakai T and Hong R-G 1998 *Scripta Mater* **39** 1221-1227.
[7] Tsuji N, Ueji R, Minamino Y and Saito Y 2002 *Scripta Mater* **46** 305-310.
[8] Ueji R, Tsuji N, Minamino Y and Koizumi Y 2002 *Acta Mater.* **50** 4177-4189.
[9] Tsuji N, Saito Y, Utsunomiya H and Tanigawa S 1999 *Scripta Mater* **40** 795-800.
[10] Tsuji N, Ueji R, Ito Y and Kim H-W 2006 *Ultrafine Grained Materials IV, Eds. Zhu YT, Langdon TG, Horita Z, Zehtbauer MJ, Semiatin SL and Lowe TC. TMS* 81-88.
[11] Tsuji N, Ueji R and Saito Y 2000 *Materia Japan* **39** 961.
[12] Tsuji N, Ueji R and Minamino Y 2002 *Scripta Mater.* **47** 69-76.
[13] Kamikawa N, Tsuji N and Minamino Y 2004 *Sci. Tech. Adv. Mater* **5** 163-172.
[14] Li BL, Tsuji N and Kamikawa N 2006 *Mater. Sci. Eng. A* **423** 331-342.
[15] Tsuji N, Ueji R, Ito Y and Saito Y 2000 *Proc. of the 21st Risø Int. Symp.* 607-616.
[16] Tsuji N, Kamikawa N and Minamino Y 2004 *Mater. Sci. Forum* **467-470** 341-346.

Grain refinement of magnesium alloy sheets by ARB using high-speed rolling mill

Hiroshi Utsunomiya[14], Ken-ichi Izumi[1,2], Tetsuo Sakai[1] and Toshiji Mukai[3]

[1]Division of Materials and Manufacturing Science, Graduate School of Engineering, Osaka University, 2-1 Yamadaoka, Suita, Osaka 565-0871, Japan
[2]Graduate Student. Now at Daikin Industries, Ltd., Settsu Osaka 566-8585, Japan
[3] Structural Metals Center, National Institute for Materials Science, 1-2-1 Sengen, Tsukuba, Ibaraki 305-0047, Japan

E-mail: uts@mat.eng.osaka-u.ac.jp

Abstract. Applications of ARB to magnesium alloys were limited due to low deformability. The authors recently found that the rollability of the alloys is significantly improved in high-speed rolling. It is supposed that the severe plastic deformation of magnesium alloy sheets is feasible if rolling in ARB processes is conducted at high speed. In this study, AZ31B and ZK60A sheets are processed by ARB up to five cycles at 423K with a speed of 1000m/min. Vickers hardness increases with increasing number of ARB cycles, while the tensile strength shows the maximum after the second cycle. The grain size is reduced significantly at the first cycle and decreases gradually from the second cycle. The mean grain sizes after five cycles are 1.6μm for AZ31B and 1.8μm for ZK60A. It is concluded that ARB using high-speed rolling is effective for grain refinement of magnesium alloys.

1. Introduction

Magnesium alloys are lightweight structural materials, however wrought products are not used so widely as cast products. Although strengthening of magnesium alloys is demanded, cold/warm working processes have not been applied widely because of the low ductility. This is due to the lack of slip systems below 500K. Rolled sheets are used for laptop PCs, automobiles, etc. The sheets are mostly manufactured by multi-pass hot rolling in industries. Applicable reduction in thickness in cold/warm rolling is limited by edge-cracking and fracture.

Accumulative roll bonding (ARB) is the severe plastic deformation process applicable to sheets [1]. Ultra-grain refinement by ARB process was reported for most metals and alloys. However, applications of ARB to magnesium alloys are limited due to the low deformability. Some researchers conducted hot (>500K) ARB where deformability is sufficient but strain hardening is negligible [2-5]. The authors applied cold ARB to Mg-Li alloy sheet which shows higher deformability due to dual-phase (*hcp* + *bcc*) microstructure [6]. Formation of ultrafine grains by cold ARB was reported. However, effects of normal ARB (ARB below recrystallization temperature) to practical magnesium alloys have not been investigated.

[4] To whom any correspondence should be addressed.

The authors recently found that high-speed rolling is very effective to improve the deformability of magnesium alloys in rolling [7-9]. For instance, AZ31 sheets can be reduced 60% without fracture in one-pass cold rolling at 2000m/min. Major reason for the improvement is the temperature rise during the rolling. In the high-speed heavy rolling, large plastic work and frictional work convert to heat while the heat is less transferred to cold rolls due to short contacting time. Accordingly, the sheet temperature increases so that non-basal slip or dynamic recrystallization takes place [9,10].

It is supposed that the SPD of magnesium alloy sheets is feasible if the rolling operation in ARB process is conducted at high speed. In this study, AZ31B and ZK60A sheets are processed by ARB at 423K with a speed of 1000m/min. AZ31 is the most popular magnesium alloy, while ZK60 is a high-strength wrought alloy.

2. Experiment

Prior to ARB, AZ31B (Mg-3%Al-1%Zn-0.4%Mn) sheets were annealed after rolling and ZK60A (Mg-5.4%Mn-0.56%Zr) sheets were solution-treated after extrusion. The both sheets were 2.5mm thick, 30mm wide and 300mm long. A high-speed rolling mill of which roll diameter was 530mm was used. One sheet was held for 900s in an electric furnace at 423K. The thickness of the sheet was reduced 50% by one pass rolling operation at 1000m/min under unlubricated conditions. Bonding of two sheets was not performed at the first cycle. After passing the roll gap, the sheet was immediately quenched into water shower. At the second cycle, the 1.25mm thick sheet was cut into two halves in length. The sheet surfaces were degreased but not scratch brushed. The sheets were stacked and processed under the same ARB conditions as shown in Figure 1. ARB was repeated up to five cycles (equivalent strain=4.0) for both AZ31B and ZK60A sheets.

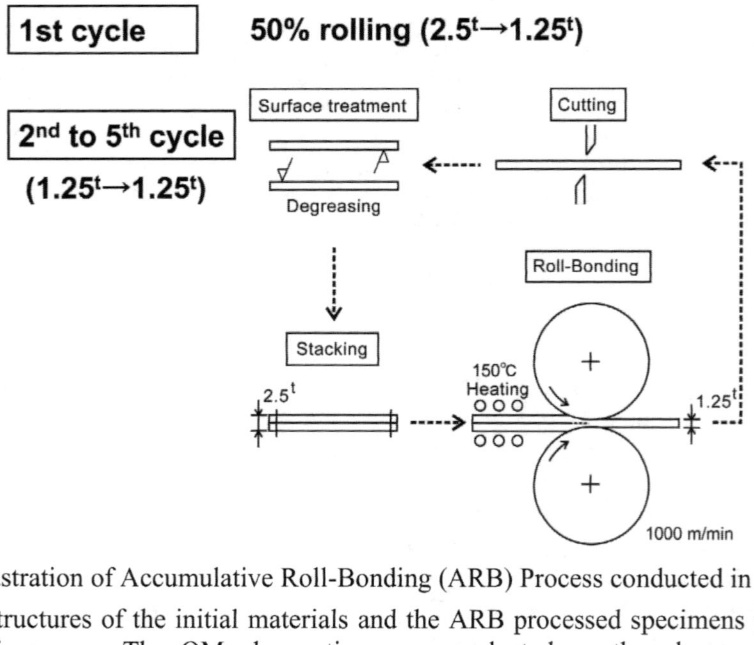

Figure 1. Illustration of Accumulative Roll-Bonding (ARB) Process conducted in this study.

The microstructures of the initial materials and the ARB processed specimens were observed with an optical microscope. The OM observation was conducted on the plane perpendicular to the transverse direction (TD plane, or longitudinal section). The mean intercept length was measured as the grain size. After chemical polishing, the pole figures on the mid plane were measured by the Schulz reflection method with Cu-Kα X-ray radiation. The intensity was normalized by that of randomly oriented magnesium sample. Mechanical properties of the sheets were measured by tensile test at room temperature on an Instron-type machine. The tensile-test specimens were machined so that the tensile axis was parallel to the rolling direction. The gauge length was 10 mm and its width was 5 mm. The crosshead speed was 0.5 mm/min, so that the initial strain rate was $8.3 \times 10^{-4}\ s^{-1}$.

3. Results and Discussion

After the first ARB cycle, both AZ31B and ZK60A alloy sheets showed small edge cracks. As the propagation of edge cracks is relatively slow, the width was not trimmed much after each cycle. The both alloys were subjected to five cycles without fracture. However it was not possible to take sufficient tensile specimen from the ZK60A sheets after five cycles.

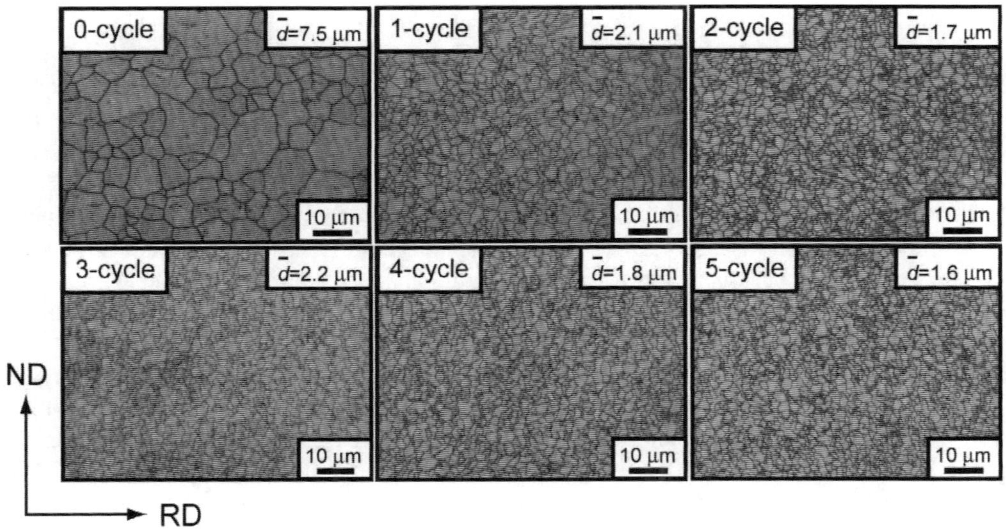

Figure 2. Optical micrographs of AZ31B sheets as a function of number of ARB cycles.

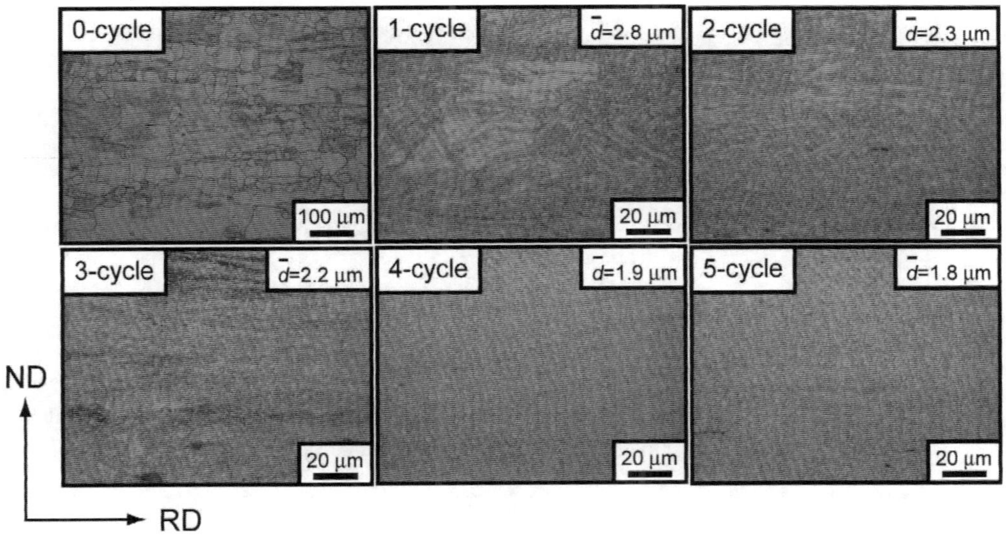

Figure 3. Optical micrographs of ZK60A sheets as a function of number of ARB cycles.

Change in microstructure of AZ31B is shown in Figure 2. In the figure, '0-cycle' means the initial microstructure. The initial sheet is covered with equiaxed grains of which mean grain size is 7.5μm. The grain size is significantly reduced to 2.1 μm at the first cycle may be due to dynamic recrystallization under high Z parameter [9,10]. Grain size does not change much after the second cycle. Microstructural evolution of ZK60A is shown in Figure 3. The initial microstructure of ZK60A is not homogeneous containing equiaxed and elongated grains due to the previous extrusion process. With increasing number of ARB cycles, the fraction of fine equiaxed grains increases, the microstructure becomes more homogeneous and the mean grain size decreases gradually. Changes in the mean grain size are compared in Figure 4. The grain size is significantly reduced at the first cycle and decreases gradually in general from the second cycle. Difference between the two alloys is not so large. Enlarged micrographs of the 5-cycle processed sheets are shown in Figure 5. The mean grain sizes are 1.6μm for AZ31B and 1.8μm for ZK60A.

Changes in (0002) pole figures are shown in Figure 6. All the ARB-processed sheets have basal texture. After the first ARB cycle, AZ31B shows the basal texture with two peaks on RD. It means that *c* axis is tilted 15-20 deg. back and forth to the rolling direction. The angle between the two peaks θ is shown on the figure. Basal texture with double peaks is a typical rolling texture of magnesium alloys. ZK60A accompanies two other peaks on the TD at the early stage. The two peaks are retained components of the extrusion texture (0002)$\perp$ED. With increasing number of ARB cycles, θ decreases and the basal texture turns into single-peak basal texture. This is due to the sheet-stacking in ARB process [11]. It is known that single-peak basal texture is a shear texture or surface texture component of magnesium alloys [12] so that the texture forms beneath the surfaces. As the two sheets are roll-bonded together in ARB process, the surface locates at the center of the thickness at the next cycle. Therefore, the texture on the midplane changes from the rolling texture to the shear texture with increasing number of ARB cycles.

Changes in tensile strength and elongation are shown in Figure 7. The tensile strength increases at the first cycle. The increase of ZK60A is higher than that of AZ31B. After the second cycle, the elongations of the both alloys is low and the tensile strength decreases gradually with increasing the number of ARB cycles. AZ31B and ZK60A show similar trend, though the tensile strength of ZK60A is approximately 30MPa higher than that of AZ31B. It seems that the change in tensile strength does

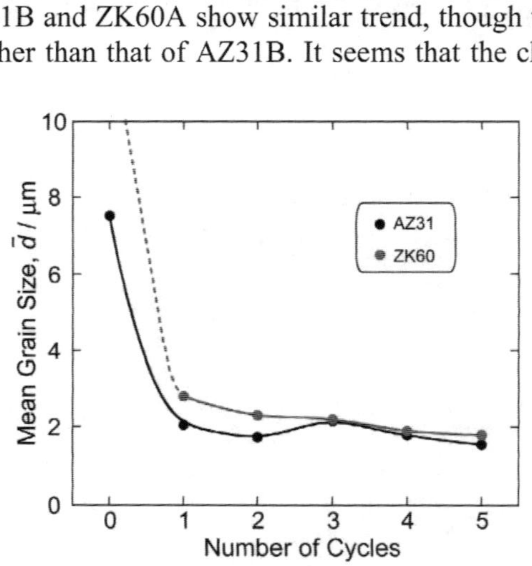

Figure 4. Mean grain size as a function of number of ARB cycles.

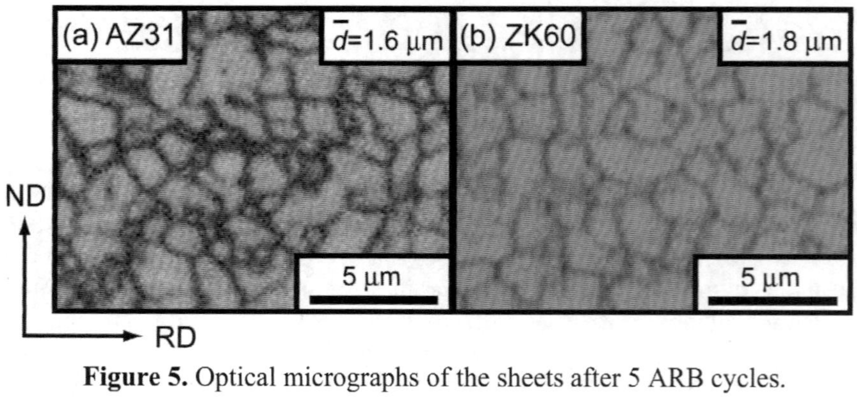

Figure 5. Optical micrographs of the sheets after 5 ARB cycles.

not agree with that in grain size. Change in Vickers hardness is shown in Figure 8. The hardness

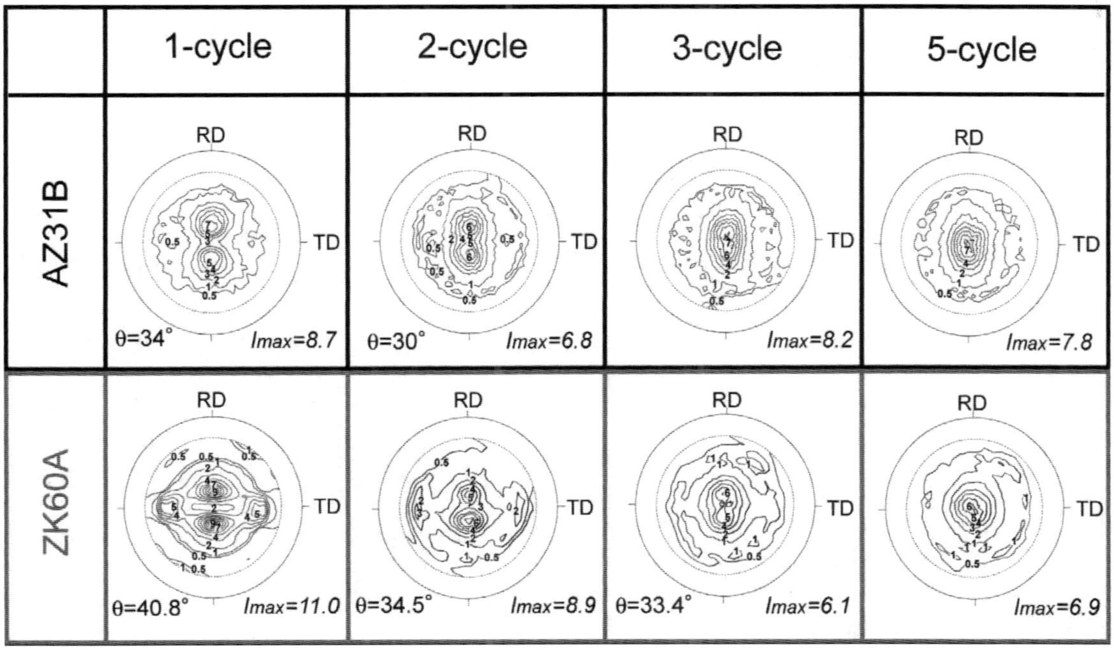

Figure 6. (0001) pole figures of AZ31B and ZK60A as a function of number of ARB cycles.

increases until the second cycle and does not change much after the second cycle. These trends correspond with that in the grain size shown in Figure 4. The hardness increases with decreasing grain size. It is notable that the impression formed by the indentation on longitudinal section is not square but diamond (rhombus) shape elongated in the thickness direction. This is due to the plastic anisotropy on longitudinal section. In order to investigate the anisotropy, two diagonal lengths of the impression were measured separately. HV^{ND} is the hardness calculated with the diagonal length in the thickness direction only, while HV^{RD} is the hardness with the diagonal length in the rolling direction. They are also shown on Figure 8. HV^{RD} is always higher than HV^{ND} due to the evolution of basal texture as expected. The flow stress of magnesium alloys is asymmetric and tensile flow stress along *a*-axis is higher than that in *c*-axis [13]. As *c*-axis is aligned in the thickness direction (ND) in the textured sheet, the impression is longer in the direction. The anisotropy generally increases with number of ARB cycles. It means that textured magnesium sheets are easier to be compressed in the thickness direction than in the rolling direction.

In most metals and alloys, tensile strength shows good correlation with Vickers hardness [14]. However, it is found that the hardness and the strength do not show similar trend with an increase in the number of ARB cycles. The fact implies that unmatured fracture occurs during tensile test of the ARB-processed sheets. A possible explanation for the fracture is the evolution of single-peak basal texture. Number of slip systems is insufficient for magnesium alloys so that more shear banding and twinning and fracture is expected at room temperature, in particular in well-textured sheets.

4. Conclusion

AZ31B and ZK60A sheets are processed up to five ARB cycles (strain of 4.0) at 423K by ARB with a speed of 1000m/min. Vickers hardness increases with the number of ARB cycles, while the tensile strength shows the maximum after the second cycle. The grain size is reduced significantly at the first cycle and decreases gradually from the second cycle. The mean grain sizes after five cycles are 1.6μm

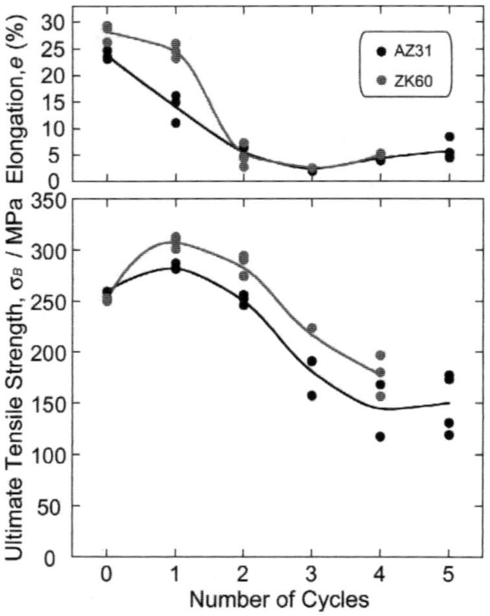

Figure 7. Tensile properties as a function of number of ARB cycles.

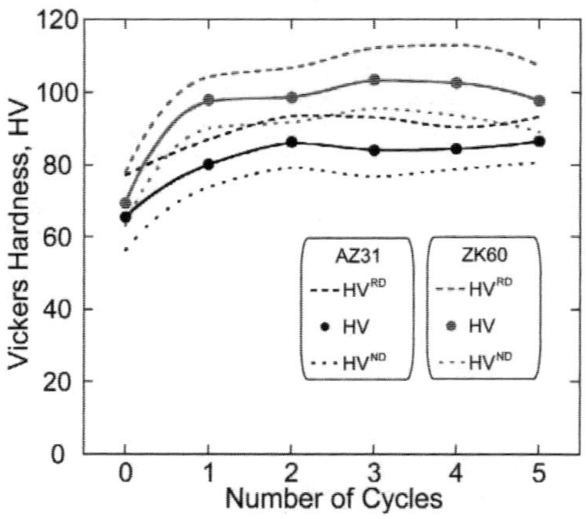

Figure 8. Vickers hardness as a function of number of ARB cycles (dotted lines show Vickers hardness calculated from one diagonal length of the impression).

for AZ31B and 1.8μm for ZK60A. It is concluded that ARB process is effective for grain refinement and hardening of magnesium alloys.

Acknowledgement
This work was supported by the National Institute for Materials Science, Tsukuba, Japan and by "21st Century COE program: Center of Excellence for Advanced Structural and Functional Materials Design" from MEXT of Japan.

References

[1] Saito Y, Utsunomiya H, Tsuji N, and Sakai T 1999 Acta Mater. 47 579
[2] Perez-Prado M T, del Valle J A and Ruano O A 2004 Scr. Mater. 51 1093.
[3] Bowles A and Horton J A 2006 Magnesium Technology in the Global Age, ed. M O Pekguleryuz and L W F Mackenzie, (CIM), pp183-192
[4] Fatemi-Varzaneh S M, Zare-Hanzaki A and Haghshenas M 2008 Int. J. of Modern Physics B. 22 2833.
[5] Zhan M Y, Li Y Y and Chen W P 2008 Trans Nonferrous Met. Soc. China 18, 309
[6] Izumi K, Utsunomiya H, Sakai T and Mukai T 2006 Magnesium Technology in the Global Age, ed. M O Pekguleryuz and L W F Mackenzie, (CIM), pp 229-241
[7] Utsunomiya H, Sakai T, Minamiguchi S and Koh H 2006 Magnesium Technology 2006, ed A A Luo. N R Neelameggham and R S Beals, (Warrendale, TMS), pp 251-254
[8] Sakai T, Utsunomiya H, Koh H and Minamiguchi S 2007 Materials Science Forum 539-543 3359.
[9] Koh H, Sakai T, Utsunomiya H and Minamiguchi S 2007 Mat. Trans. 48 2023
[10] Li H, Hsu E, Szpunar J, Utsunomiya H, Sakai T 2008 J. Mater. Sci. 43 7148
[11] Lee S H, Saito Y, Tsuji N, Utsunomiya H, Sakai T 2002 Scr. Mater. 46 281
[12] Hargreaves A 1945 J Inst Metals 71 73
[13] McDonald J C and Bakarian P W 1965 Trans. Metall. Soc. AIME 233 95
[14] Tabor D 1951 The Hardness of Metals.(Oxford:Oxford University Press) p 107

Journal of Physics: Conference Series **165** (2009) 012012 doi:10.1088/1742-6596/165/1/012012

Recent progress in micro and nano-joining

Y Zhou, A Hu, M I Khan, W Wu, B Tam, M Yavuz
Centre for Advanced Materials Joining
Department of Mechanical and Mechatronics Engineering, University of Waterloo
200 University Avenue West, Waterloo, N2L 3G1

E-mail: nzhou@uwaterloo.ca

Abstract. Micro and nano-joining has been identified as a key enabling technology in the construction of micromechanical and microelectronic devices. The current article reviews recent progress in micro and nano-joining. In particular, laser micro-welding (LMW) of crossed 316 LVM stainless steel (SS) wire was compared to conventional resistance micro-welding (RMW) and was successfully employed in welding a Pt-Ir /SS dissimilar combination. Welding of Au nanoparticles was realized using femtosecond laser irradiation and its application in the surface enhanced Raman spectroscopy was investigated. Brazing between carbon nanotube (CNT) bundles and Ni electrodes was attained in vacuum, resulting in the development of a novel CNT filament of incandescent lamps.

1. Introduction

Welding and joining is an essential step in the fabrication of various devices at all scales including macro, micro and nano. As devices become increasingly smaller, challenges faced in micro and nano-joining must be overcome before these joints are safely implemented. For examples, researchers must determine how to robustly join these miniature building blocks while avoiding excessive damage. The current article reviews recent challenges faced in micro and non-joining while detailing recent progress in these areas.

Crossed wire welding [1,2,3,4] is a critical fabrication technique in the manufacture of implantable medical devices. Difficulties associated with resistance micro-welding (RMW) of new materials and combinations has resulted in an initiative to explore alternative joining methods. Recently, laser micro-welding (LMW) has been shown to be a potential candidate to replace RMW [4]. In this study a comparison between RMW and LMW of crossed 316LVM SS wire was conducted. In addition, benefits associated with dissimilar welding using the LMW process are discussed.

Various nanojoining approaches have been developed, including e-beam exposure [5], focusing ion beam deposition [6], ultrasonic welding [7], sputtering/evaporating Au and post-annealing [8,9] and nanosoldering with low melting point solders [10,11]. This article will review current progress in nanojoining using femtosecond laser welding and vacuum brazing processes. Both processes have potential for large scaling and batch-nanojoining, which are promising for practical applications.

2. Experimental

2.1. Resistance and Laser Micro-welding

Laser welding was conducted using a Miyachi Unitek LW-50A laser welding system capable of 50J pulse energy and 5 kW peak power. A MacGregor DC400P direct-current (DC) controller and Unitek 80A/115 weld head are used for RMW. Fixturing was used to bond the wires at right angles (90°) for both the RMW and LMW process. 316 LVM SS (0.38 mm diameter) and Pt-Ir (0.4 mm diameter) wire was used in this study. Tensile testing was used to determine the joint breaking force (JBF). Stainless steel metallographic samples were etched using 5 ml HNO_3, 25 ml HCl, and 30 ml H_2O at an elevated temperature (80°C). Etching time varied between 3 to 5 seconds.

2.2. Nano-Welding

Femtosecond laser welding was carried out with a 100 fs 800 nm Ti:sapphire laser system with a pulse energy of 250 μJ/pulse and repeatition rate of 1 kHz [12]. Au nanoparticles were synthesized by the citrate redution of $HAuCl_4$. Welding was done in liquid by focusing the laser into a quatz cell or in solid by depositing Au nanoparticles onto a TEM Cu-grid.

Vaccum brazing was conducted using a high vacuum tube furnace under 10^{-6}Torr. The brazing material has a nominal composition of Ag–63.0%, Cu–35.25%,Ti – 1.75% with a melting point of 815°C, while solidifies at 775°C. A detailed brazing procedure was reported elsewhere [13]. The CNT filament made of CNT bundles and welded to two Ni electrodes was tested in a vacuum chamber (10^{-4} Torr).

3. Results and Discussion

3.1. Micro-welding of Crossed Stainless Wires

Figure 1 shows the joint breaking force (JBF) as a function of peak power and current for the resistance and laser micro welding processes, respectively [3,4]. Each process requires a minimum parameter input to provide enough energy to initiate bonding. The modification of the process parameters results in an increase in JBF up to a maximum. Similar peak joint breaking force near 80 N was attained for both processes. Excess energy inputs resulted in over welded joints having insufficient strength.

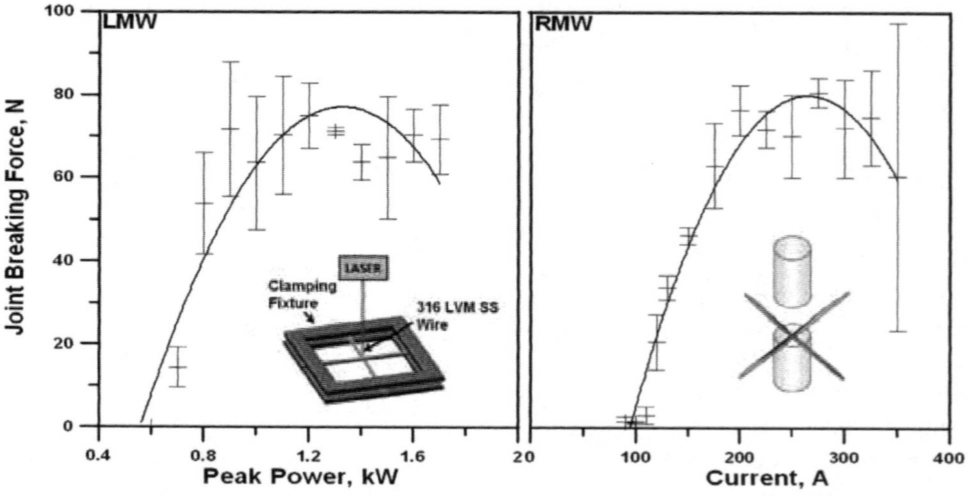

Figure 1: JBF of crossed SS wire for LMW and RMW [3,4]

Observed weld cross-section for laser and resistance welded 316LVM SS for conditions producing optimal JBF are shown in Figure 2. A dendritic solidification structure was observed for both welding processes, indicating peak temperatures surpassed the liquidus temperature. RMW exhibited a less directional structure compared to LMW, which is likely due to the contact of the copper electrodes. Thus, fusion welding mode and similar joint strengths can be achieved for both processes.

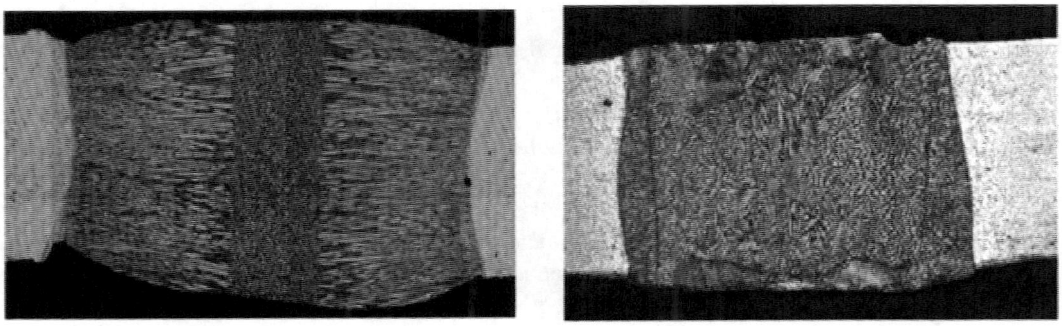

Laser Micro-Weld **Resistance Micro-Weld**

Figure 2: Fusion weld cross-section for 316LVM SS [3,4]

Limitations observed with RMW of dissimilar material have shown to be resolved using LMW. For example, differences in resistivity between Pt-Ir alloys and stainless steel make it difficult to attain acceptable dissimilar RMW joints. Figure 3a shows excessive melting of the higher resistivity SS in the dissimilar RMW joint. However, as shown in Figure 3b, the LMW joint results in an acceptable weld with a smoother surface. Hence, current challenges with micro-welding of new materials and combinations can be overcome by implementing alternative micro-welding processes.

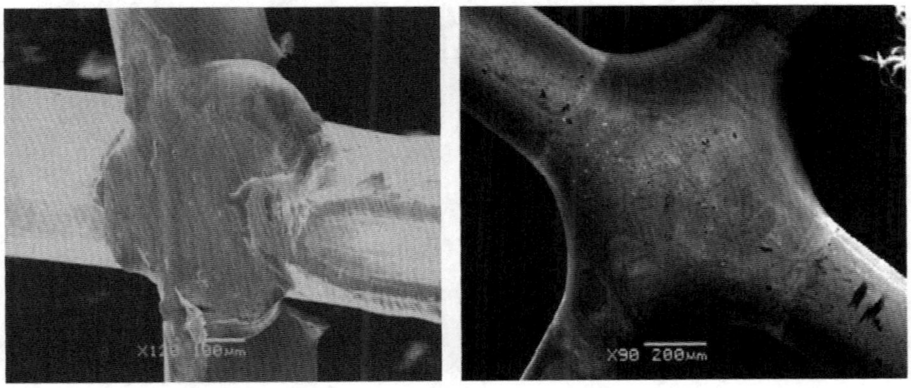

a) RMW b) LMW

Figure 3: Crossed SS and Pt/IR dissimilar weld

3.2. Femtosecond Laser Welding of Au Nanoparticles
Figure 4 shows as-grown (before welding) and welded Au nanoparticles irradiated in water at a laser intensity of $4x10^{14}$ W/cm^2 for 10 min. As-grown Au nanoparticles (before welding) have a mean size of 15 nm. After femtosecond laser irradiation, some 1-3 nm Au particles were generated. These very tiny Au nanoparticles were formed through the fragmentation of Au nanoparticles by laser irradiation. A previous study has identified the electron ejection as the first step of photofragmentation [14]. This

electron emission causes nanoparticles to become positively charged and the repulsion among the charges leads to the fragmentation. Meanwhile, we obtain about 70%-80% nanoparticle welded together. In some parts 2-3 particles join together. In some cases several particles (including both large particles around 15 nm and small particles around 1-3 nm) form a curved chain or a network. The dynamic procedure of welding between nanoparticles is still unclear. Our study shows that femtosecond laser can induce the melt of surface nanoscopic layer if the pulse energy is properly chosen [15]. As a result of electron ejection (8-10% of total bonding electrons), the lattice bond gets loose, leading to the lattice melt [16]. This melt occurs during the ultrafast pulse width and is thus ultrafast and nonthermal. The lattice beneath the surface melting layer keeps unchanged. Probably this ultrafast melt results in a nanojoining of Au nanoparticles. Our recent results show that the photofragmentation can be suppressed by lowering the focused light intensity.

Welded Au nanoparticles can work as the probe for surface enhanced Raman effect (SERS) for single molecular detection because the surface plasmon near the vicinity of welded neck region is enhanced. A 5 fold improvement of SERS signals was achieved using welded Au nanoparticles in comparison with as-grown Au nanoparticles for the polyyne molecules diluted in acetone at the same concentration.

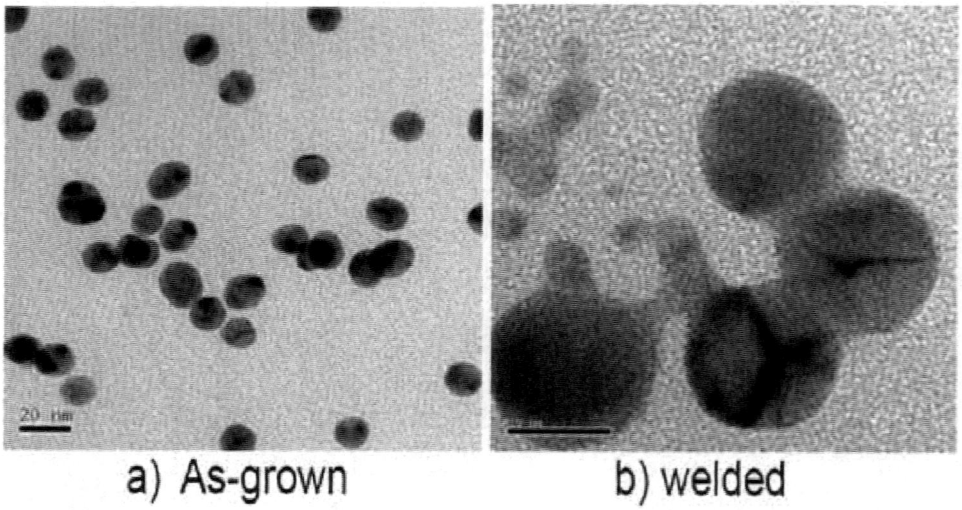

Figure 4: TEM images of (a) as-grown (before welding) and (b) welded Au nanoparticles. Note that the sacle bars are different, (a) 20 nm and 10 nm in (b).

3.3. Vaccum brazing of CNT

Cu-Ag-Ti brazing alloy has successfully wetted CNT surfaces at high temperatures by formation of C-Ti bonds. This is confirmed by the local composition analysis by X-ray energy dispersive spectroscopy, element mapping and X-ray photoelectron energy spectra [13]. This technique was employed in the fabrication of a filament with a 2 cm long CNT bundle. Figure 5 presents the transporting measurement in a vacuum chamber under 10^{-4} Torr. It is evident that the CNT-Ni electrode connection displays good Ohmic behaviour and brazing leads to a dramatic decrease of contacting resistance comparing to both mechanical connection and joining using Ag paste. Based on SEM characterization, the CNT bundle diameter was about 100 μm. At 15 V, the filament current density reaches 3000 A/cm^2 for brazing, which is near double of a reported value of 1670 A/cm^2 using the similar CNT bundle material while joining with Ag-paste [17].

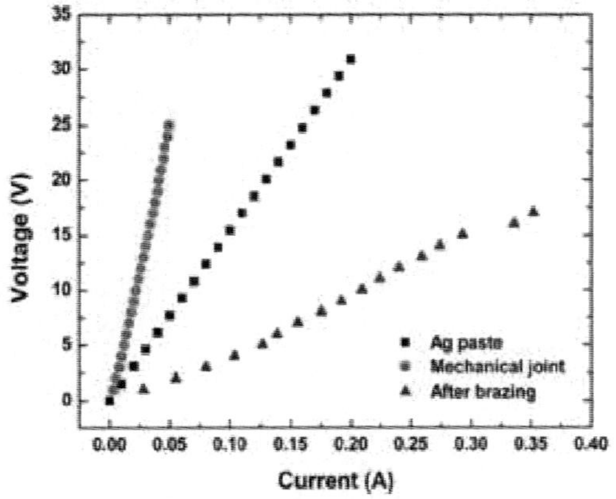

Figure 5: I-V curves of three kinds of jointed CNT filaments: (upper to lower) mechanical joint, Ag-pasted and vacuum brazing.

Figure 6 shows the emission spectra for the CNT and tungsten filament at 1570 °C. A stronger infrared luminescence (IR>750 nm) was observed for the CNT incandescent lamp when compared to the tungsten filament at the same temperature. This indicates that CNT filament is capable of working at higher temperatures; hence, higher emission effects are expected.

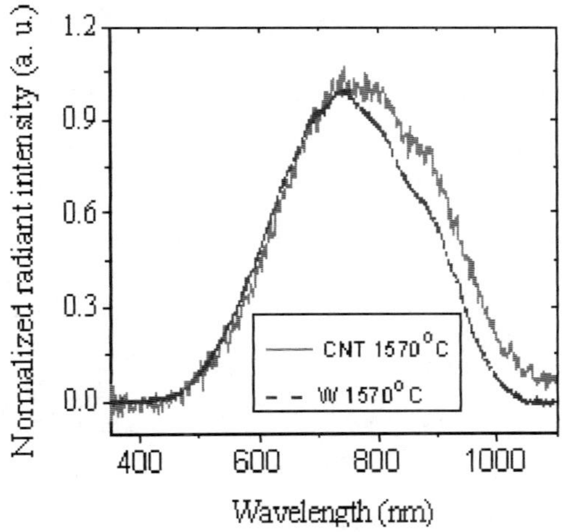

Figure 6: Emission spectra of a CNT filament and commercial tungsten lamp at a filament temperature of 1570°C.

4. Conclusions and Outlook

The current article detailed recent progress in micro and nano-joining. Key finding include:

1) Similar welds strengths and fusion welding modes can be achieved for crossed wire welding using the RMW and LMW process. The latter has versatile capacities to join various materials, especially for less conducting/insulating biomaterials, like bio-electrodes.

2) Nanojoining Au particles and brazed CNTs have been successfully implemented as probes for the surface enhanced Raman spectroscopy and energy saving filaments for incandescent lamp, respectively.

3) Developed nano-joining techniques can enable large scale batch processing and may open new frontiers in the construction of complicated nanodevices with 3D architecture and different building blocks.

5. Acknowledgements

This work is partially supported by the Natural Science and Engineering Research Council (NSERC) of Canada and the Canada Research Chair (CRC) in Microjoining. The authors are also grateful to the Department of Mechanical Engineering, Tsinghua University, China for providing the carbon nanotubes.

6. References

[1] S. Fukumoto and Y. Zhou: Metall. Mater. A, 2004, vol. 35A, pp. 3165-76.

[2] S. Fukumoto, Z. Chen, and Y. Zhou: Metall. Mater. A, 2005, vol. 36A, pp. 2717-24.

[3] M. I. Khan, J. M. Kim, M. L. Kuntz, Y. Zhou: Accepted Metall. Mater. A, September 2008

[4] M. I. Khan and Y. Zhou, Processing's in Materials and Processes for Medical Devices Conference (MPMD) September 2007, California

[5] T. Terrones, F. Banhart, N. Grobert, J. -C. Charlier, H. Terrones, and P. M. Ajayan: Phys. Rev. Lett., 2002, vol. 89, pp. 075501-1-4.

[6] B. Wei, R. Spolenak, P. Kohler-Redlich, M. Ruehle, and E. Arzt: Appl. Phys. Lett., 1999, vol. 74, pp. 3149-51.

[7] C. Chen, L. Yan, E. S. Kong, and Y. Zhang: Nanotechnology, 2006, vol. 17, pp. 2192-7.

[8] P. J. de Pablo, E. Graugnard, B. Walsh, R. P. Andres, S. Datta, R. Reifenberger, Appl. Phys. Lett., 1999, vol. 74, pp. 323-5.

[9] W. Il Park, J. S. Kim, G. Yi, M. H. Bae, and H. –J. Lee: Appl. Phys. Lett. 2004, vol. 85, pp. 5052-4.

[10] C. O. Girit, and A. Zettl: Appl. Phys. Lett. 2007, vol 91, pp. 193512-4.

[11] E. Ide, S. A. Hirose, K. F. Kobayashi: Acta Mater. 2005, vol. 53, pp. 2383-93.

[12] A. Hu, J. Sanderson, A. a. Zaidi, C. Wang, T. Zhang, Y. Zhou, W. W. Duley: Carbon 2008, vol 46, pp. 1792-828.

[13] W. Wu, A. Hu, X. Li, J. Q. Wei, Q. Shu, K. L. Wang, M. Yavuz, Y. Zhou: Mater. Lett. 2008, vol 62, pp. 4486-8.

[14] P. V. Kamat, M. Flumiani, and G. V. Hartland: J. Phys. Chem. B 1998, vol. 102, pp. 3123-8.

[15] A. Hu, M. Rybachuk, Q. -B. Lu, and W. W. Duley: Appl. Phys. Lett. 2007, vol. 91, pp. 131906-8.

[16] P. Stampfli, and K. H. Bennemann: Phys. Rev. B 1992, vol. 46, 10686-92.

[17] Q. K. Shu, J. Q. Wei, K. L. Wang, C. G. Li, Y. Jia, and D. H. Wu: J. Appl. Phys. 2007, vol. 101, pp. 084306-1-4

Journal of Physics: Conference Series **165** (2009) 012013 doi:10.1088/1742-6596/165/1/012013

Surface hardening of two cast irons by friction stir processing

Hidetoshi Fujii[1], Yasufumi Yamaguchi[2], Toshifumi Kikuchi[1], Shoji Kiguchi[2], Kiyoshi Nogi[1]
[1]Joining and Welding Research Institute, Osaka University, 11-1 Mihogaoka, Ibraki, Osaka 567-0047, Japan
[2] Faculty of Science and Engineering, Kinki University, 3-4-1 Kowakae, Higashi-Osaka, Osaka, Japan

E-mail: fujii@jwri.osaka-u.ac.jp

Abstract. The Friction Stir Processing (FSP) was applied to the surface hardening of cast irons. Flake graphite cast iron (FC300) and nodular graphite cast iron (FCD700) were used to investigate the validity of this method. The matrices of the FC300 and FC700 cast irons are pearlite. The rotary tool is a 25mm diameter cylindrical tool, and the travelling speed was varied between 50 and 150mm/min in order to control the heat input at the constant rotation speed of 900rpm. As a result, it has been clarified that a Vickers hardness of about 700HV is obtained for both cast irons. It is considered that a very fine martensite structure is formed because the FSP generates the heat very locally, and a very high cooling rate is constantly obtained. When a tool without an umbo (probe) is used, the domain in which graphite is crushed and striated is minimized. This leads to obtaining a much harder sample. The hardness change depends on the size of the martensite, which can be controlled by the process conditions, such as the tool traveling speed and the load. Based on these results, it was clarified that the FSP has many advantages for cast irons, such as a higher hardness and lower distortion. As a result, no post surface heat treatment and no post machining are required to obtain the required hardness, while these processes are generally required when using the traditional methods.

1. Introduction

Cast iron has excellent material characteristics, such as abrasion resistance, corrosion resistance, machinability, and vibration absorptivity, and accordingly, it has been used in various industrial fields, such as automobile parts, industrial machine parts, and machine tool parts. Currently, a weight saving and advanced features are being requested, thus the material of choice is changing from cast irons to aluminum alloys and magnesium alloys, but in contrast to these developments, some high added values, such as a higher performance [1-3] and thinner product [4,5] are requested for the cast iron components.

In this study, riction stir processing (FSP), whose principle is the same as friction stir welding [6], was used as one of the surface treatment methods for cast irons. Many such studies of FSP have already been intensively conduced for aluminum alloys [7-13]. This method solves many problems, such as gas inclusion and grain growth in the weld of aluminum alloys [14]. For ferroalloys, on the other hand, there is no report concerning the FSP because some issues, such as the durability of the tool, have to be solved, although some research studies on friction stir welding [15-17] have begun.

2. Experimental

Five-mm-thick flake graphite cast iron (FC300) and nodular graphite cast iron (FCD700) plates were used as the test materials. Table 1 shows their chemical compositions. The matrix for both materials is perlite, and therefore, the Vickers hardness of the nodular graphite cast iron is 200-230HV, while for the flake graphite cast iron, it is 170-210HV.

A cylindrical tool with a 25mm diameter was used and the tilt angle was 3 degrees. The tool material was a tungsten carbide-based alloy [15-17]. The travelling speed was varied between 50 and 150mm/min in order to control the heat input, while the rotation speed of the tool was constant at 900rpm. After performing the FSP, the hardness of the matrix was measured on a cross section (0-1.5mm depth) using a Vickers testing machine. The microstructure was observed using an optical microscope.

Table 1 Chemical composition of specimens (mass%).

	C	Si	Mn	P	S	Mg	Cu
FC300	3.07	1.65	0.75	0.07	0.050	–	–
FCD700	3.60	2.25	0.35	0.02	0.008	0.045	0.70

3. Results and discussion

3.1 Hardness and microstructure

Figure 1 shows the surface appearances of the FCD700 cast iron friction stir processed under various conditions. All experiments were performed while increasing the load during the process. The optimum conditions of the FSP were determined by observing the groove defect formed at the center of the sample and the peeling off of the material due to the adhesion to the tool as defects.

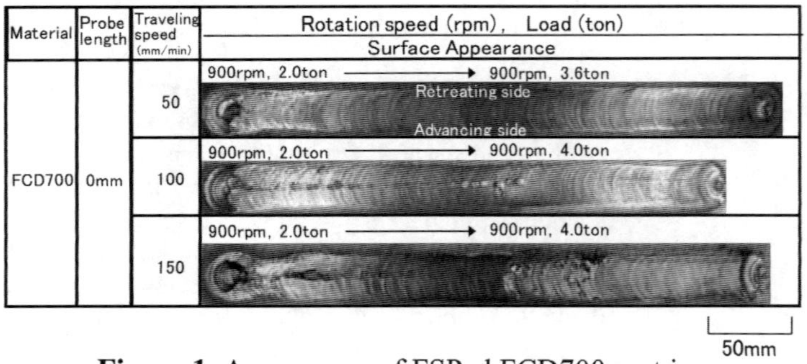

Figure 1. Appearance of FSPed FCD700 cast iron.

For the FCD700 cast iron, an excellent surface without any defects was obtained at 50mm/min, when the load exceeded about 3.0×10^3 kgf. On the other hand, at 100 or 150 mm/min, a load of 3.5×10^3kgf or more was necessary, and the required minimum load increased with the increasing tool traveling speed. Moreover, the modified region becomes narrower when compared to the 50mm/min case.

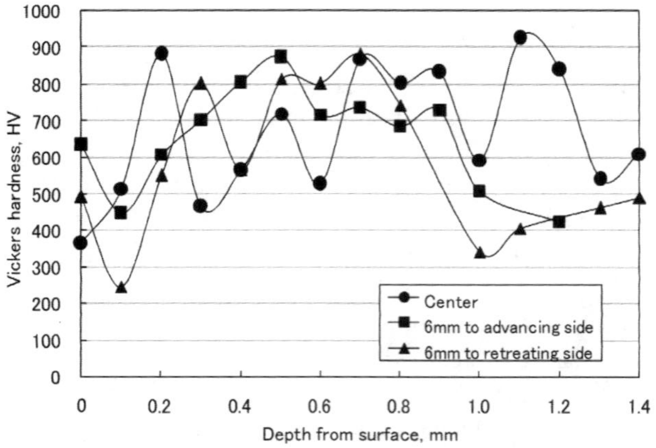

Figure 2. Vickers hardness distribution of FSPed FCD700 from the surface.

The formation of the defects is significantly related to the heat input. When the heat input is insufficient, for example, under the conditions of 900 rpm and 2×10^3 kgf, a defect is formed during the early stage by scooping out the material, as shown in Fig.1. This is because the cast iron does not soften due to the insufficient heat input, and then a flash is formed by a cutting-like phenomenon. When the heat input is excessive, the surface material is peeled off as if the material is melted because it softens too much. A good microstructure formed by the plastic flows without the surface being peeled off can be obtained by controlling the heat input properly. Therefore, it is necessary to adjust the process conditions, such as the rotation speed and the traveling speed, in order to optimize the heat input. In this case, there was neither no distortion nor a change in the dimensions of the material, which is completely different from that obtained by other surface hardening methods.

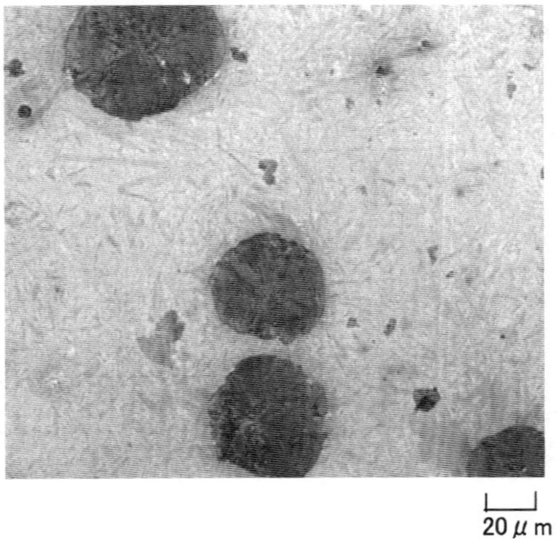

Figure 3. Microstructures of FSPed FCD700.

Figure 2 shows the Vickers hardness distribution in the depth direction at the central part and at 6mm from the center on the advancing side and the retreating side on a cross section vertical to the welding direction after the FSP. It was found that a high and comparatively steady hardness is obtained in the area from 0.2 to 1.0 mm in depth while a low value was observed at 0.1 mm from the surface. The hardness between 0.2 to 1.0 mm exceeds 700 HV.

Figure 3 shows the microstructure at a high magnification. A very fine needlelike martensite structure is observed. It is considered that this structure was generated because the material was locally heated and rapidly cooled during the FSP. On the other hand, the size and density of the martensite are changed even in the same martensite structure. The martensite structure is fine, and its density is high in a very hard microstructure which exceeds 700 HV.

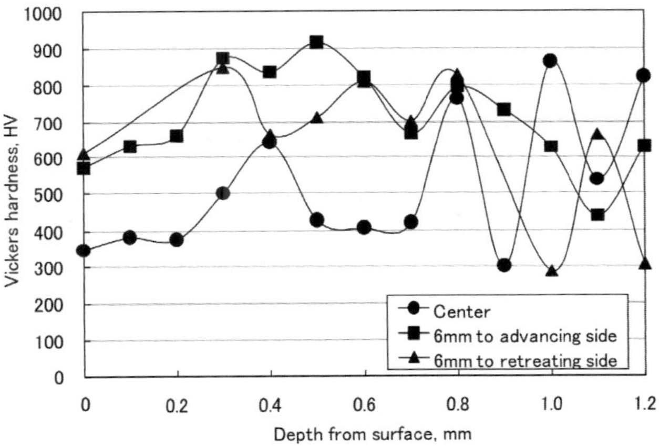

Figure 4. Vickers hardness distribution of FSPed FC300 from the surface.

3.2 Effect of graphite shape

Figures 4 show the Vickers hardness distribution of the flake graphite cast iron friction stir processed in a way similar to the nodular graphite cast iron. An average value higher than 700 HV is obtained to about 1 mm depth, and values higher than 800 HV were also measured at many points. Figure 5 shows an example of the microstructure. A

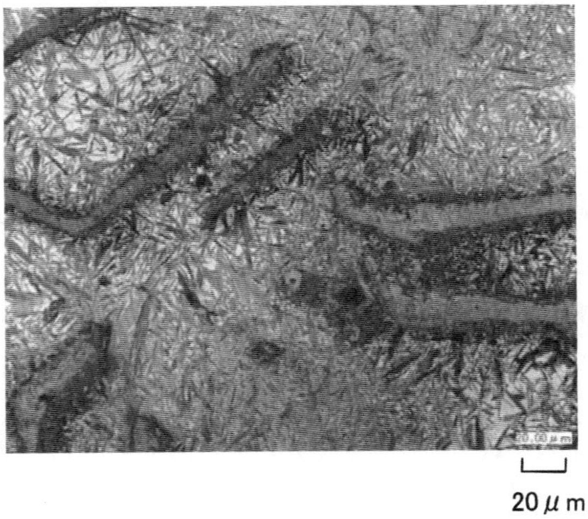

Figure 5. Microstructure of FSPed FC300.

martensitic structure is formed throughout the region.

It can be seen in Fig.4 that the hardness of the central part decreases near the surface. Since the hardness of the mother material is lower, the bottom of the tool enters more deeply compared to the FCD 700. Accordingly, the domain is expanded, where the graphite was crushed and striated by plastic flow, and the hardness then decreased.

4. Conclusions

The friction stir processing was successfully performed on the surface of cast iron using a rotating φ25mm WC based cylindrical tool at the speed of 900 rpm (friction stir processing), and the following points have been clarified;

1. The average hardness of about 700 HV for the matrix is obtained for both the flake graphite cast iron and the spheroidal graphite cast iron. This is because an extremely fine martensite phase is formed by this process.
2. It is considered that a very fine martensite structure is formed because the FSP generates the heat very locally, and a very high cooling is constantly obtained.
3. The hardness changes depending on the size of the martensite, which can be controlled by the process conditions, such as the tool traveling speed and the load.

Based on these results, it was clarified that the FSP has many advantages when used for cast irons, such as higher hardness and lower distortion.

Acknowledgments
The authors wish to acknowledge the financial support of the "Priority Assistance of the Formation of Worldwide Renowned Centers of Research-The global COE Programs (Project: Center of Excellence for Advanced Structural and Functional Materials Design) and a Grant-in-Aid for the Cooperative Research Project of Nationwide Joint-Use Research Institutes on Development Base of Joining Technology for New Metallic Glasses and Inorganic Materials from the Ministry of Education, Sports, Culture, Science and Grant-in-Aid for Science Research from the Japan Society for Promotion of Science and Technology of Japan. The authors also wish to acknowledge the financial support for young researchers from the Japan Foundry Engineering Society.

References

[1] Yamaguchi Y, Kiguchi S, Sumimoto H and Sato T 2002 J.JFS, **74** 512
[2] Aso S, Goto M, Ike H, Syobuzawa Y and Konishi N 2002 J.JFS **74** 83
[3] Horikawa Y, Noguchi T, Kimura Y and Miyakoshi Y 2006 J.JFS **78** 152
[4] Igarashi Y and Nakae H 2002 J.JFS **74** 30
[5] Kowata T, Horie H, Nakamura M, Hiratsuka S and Senda A 1993 J.JFS **65** 209
[6] ThomasW M 1991 International Patent Appl.No.PCT/GB92/02203 and GB Patent Appl. No.9125978.8(U.S.Patent No.5,460,317
[7] Mishra R S, Mahoney M W, Mcfadden S X, Mara N A and Mukherjee A K 1999 Scripta Mater., **42** 163
[8] Berbon P B, Bingel W H, Mishra R S, Bampton C C and Mahoney M W 2001 Scripta Mater. **44** 61
[9] Ma Z Y, Mishra R S and Mahoney M W 2003 Acta Mater. **50** 307
[10] Kwon Y J, Shigematsu I and Saito N 2003 J.Jpn.Inst.Met. **67** 547
[11] Charit I and Mishra R S 2005 Acta Mater., **53** 4211
[12] Su J Q, Nelson T W and Sterling C J 2005 Mater.Sci.Eng.A **405** 277
[13] Liu H J, Fujii H and Nogi K 2004 Mater. Sci. Tech. **20** 399
[14] Japan Welding Soceity 2006 Friction Stir Welding, Sanpo Shuppan 16
[15] Fujii H, Ueji R, Takada Y, Kitahara H, Tsuji N, Nakata K and Nogi K 2006 Mater. Trans., **47** 239
[16] Fujii H, Cui L, Tsuji N, Maeda M, Nakata K and Nogi K 2006 Mater. Sci. Eng. A, **429** 50
[17] Cui L., Fujii H, Tsuji N and Nogi K, 2007 Scripta Mater., **56** 637

Computational simulation of weld microstructure and distortion by considering process mechanics

M Mochizuki[1], Y Mikami[1], S Okano[1] and S Itoh[1]

Division of Materials and Manufacturing Science, Graduate School of Engineering, Osaka University, 2-1 Yamada-oka, Suita, Osaka, Japan

E-mail: mmochi@mapse.eng.osaka-u.ac.jp

Abstract. Highly precise fabrication of welded materials is in great demand, and so microstructure and distortion controls are essential. Furthermore, consideration of process mechanics is important for intelligent fabrication. In this study, the microstructure and hardness distribution in multi-pass weld metal are evaluated by computational simulations under the conditions of multiple heat cycles and phase transformation. Because conventional CCT diagrams of weld metal are not available even for single-pass weld metal, new diagrams for multi-pass weld metals are created. The weld microstructure and hardness distribution are precisely predicted when using the created CCT diagram for multi-pass weld metal and calculating the weld thermal cycle. Weld distortion is also investigated by using numerical simulation with a thermal elastic-plastic analysis. In conventional evaluations of weld distortion, the average heat input has been used as the dominant parameter; however, it is difficult to consider the effect of molten pool configurations on weld distortion based only on the heat input. Thus, the effect of welding process conditions on weld distortion is studied by considering molten pool configurations, determined by temperature distribution and history.

1. Introduction

Highly precise fabrication is being demanded, and so microstructure and distortion control are very important. It is important to determine welding process conditions by considering combinations of the various aspects of "process mechanics," including the appearance of a bead, penetration shape through a cross-section, fabrication efficiency, microstructure, weld defects during and after welding, environment, thermal distortion, residual stress, and so on. In this study, computational simulations are performed to study the effects of welding process conditions and molten pool configurations on weld microstructures and distortion from the viewpoint of process mechanics.

For multi-pass weld metal, it is widely known that mechanical properties such as yield stress, tensile strength and fracture toughness are affected by the interpass temperature and heat input [1], [2]. Heterogeneity caused by multiple heat cycles makes it difficult to understand the results of mechanical testing. It is therefore very important to investigate weld thermal cycles and microstructure in relation to mechanical properties. Multiple welding heat cycles particularly affect joint performance. Many experimental and practical approaches have been applied to correlate welding conditions and joint performance [3]-[5], but the theoretical procedures of evaluation are very effective for complicated weld phenomena and the prediction of mechanical properties of welded joints [6]-[10]. Computational simulation is a convenient method of evaluation [11]-[15]. Numerical simulations using modified

CCT diagrams can be used to evaluate the relationship between the hardness of multi-pass weld metal and welding conditions with respect to weld thermal cycles and microstructure.

A widely used equation for evaluating weld distortion is average heat input $Q = \eta IE/v$, where η: heat efficiency, I: welding current, E: arc voltage, v: welding speed. For example, many prediction equations and analytical methods of weld distortion have been developed based on the empirical data of average heat input Q in conventional welding [4], [16]-[18]. In conventional welding, the recommended welding conditions are more or less fixed when the boundary conditions of the welding are given [19]-[21]. However, it is difficult to use only Q to consider the effect of multiple heat sources or waveforms, such as the pulse condition, on weld distortion [22]-[25]. The evaluation of weld distortion of complicated controlled welding processes should consider the molten pool configuration. Consequently, in this study, precise weld distortions depending on precise welding conditions are evaluated both by numerical analysis using the thermal elastic-plastic finite-element method and actual experiments with different molten pool configurations. The molten pool configuration is determined from the temperature distribution and history during the weld process, and weld distortion is determined from the temperature field. The effect of the energy density distribution on weld circularity distortion is also investigated. An appropriate molten pool configuration to reduce weld distortion is discussed by using process mechanics.

2. Prediction of weld microstructure and hardness under different interpass temperatures

Welded joints with root gap 8 mm and bevel angle 35° were fabricated. The materials used were JIS G 3106 SN490B for the base metal and JIS Z 3312 YGW11 for the welding wire. The temperature during welding was measured at a location on the surface of the base metal, as shown in Fig. 1. Interpass temperature is defined as the temperature measured at this location just before the start of each welding pass. Joints were welded under two conditions of interpass temperatures: "continuous welding" and "interpass temperature-controlled welding." In the continuous welding condition, the interpass temperature is not controlled. In the interpass temperature-controlled welding condition, the interpass temperature is maintained at a temperature lower than 350°C during the entire welding process.

The fraction of each type of microstructure was calculated from weld thermal cycles based on CCT diagrams. In this study, the hardness of each microstructure was estimated from the cooling rate of the weld metal [14], [25]. However, CCT diagrams of the weld metal are not available, even for single-pass weld metal, due to the maximum temperature distribution; therefore, modified CCT diagrams for multi-pass weld metals were created. To create CCT diagrams of the weld metal in multiple heat cycles, a simulated weld thermal cycle test was conducted to observe the microstructural changes. The microstructures formed in multi-pass weld metal can differ depending on the maximum temperature, and the differences are found by comparing the microstructures of the heat-affected zone in multi-pass welds. The maximum temperature of the simulated weld thermal cycle for creating CCT diagrams was selected as follows: 1350°C for the coarse-grained region and 900°C for the fine-

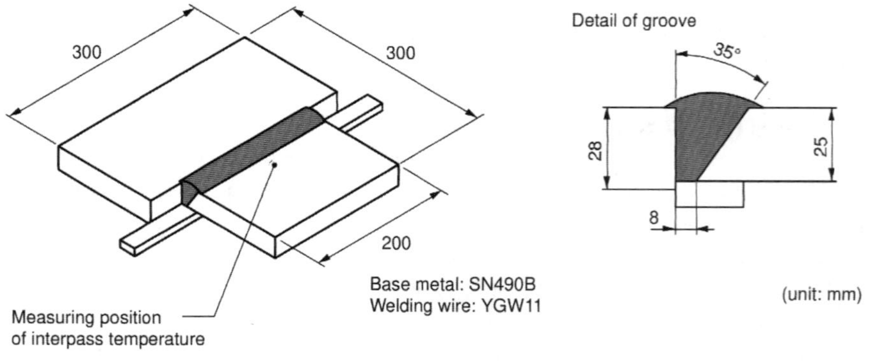

Figure 1. Configuration of a multi-pass welded joint.

grained region.

Observations of the microstructure and the results of the simulated weld thermal cycle test show that the microstructure of multi-pass weld metals is more dependent on the maximum temperature in reheating than in welding conditions such as the interpass temperature. Therefore, CCT diagrams for two maximum temperatures were created. In the CCT diagrams, ferrite + pearlite, bainite and austenite are considered to be the microstructural fractions [4], [6]-[10], [22]. The hardness distribution of the multi-pass weld metals is estimated with the linear rule of mixture. The fraction of the microstructure corresponding to weld thermal cycles is calculated from the newly created CCT diagrams.

The calculated distribution of the fraction of the ferrite + pearlite microstructure is shown in Fig. 2 together with macro cross sections of the welds. The distribution of ferrite + pearlite approximately corresponds to the reheated zone in the macro cross sections. The effect of the last weld pass by continuous welding is larger than that by the interpass temperature-controlled welding. The microstructural distribution by interpass temperature-controlled welding thus becomes more complicated than that by continuous welding.

The hardness distribution in the weld metal across the thickness direction is shown in Fig. 3. Hardness curves are estimated from the measured hardness values. As seen in these figures, the distribution of the microstructure corresponds to the distribution of hardness in both depositing sequences. Vickers hardness distributions are predicted precisely by using numerical analysis that considers the difference of welding conditions. The microstructure and hardness distribution are predicted when the appropriate CCT diagram for the multi-pass weld is used and when a detailed calculation of the weld thermal cycle is carried out.

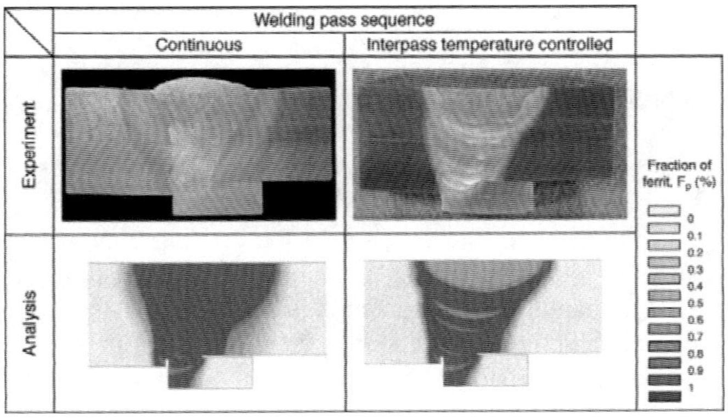

Figure 2. Comparison of macro cross-section and calculated distribution of fraction of ferrite + pearlite.

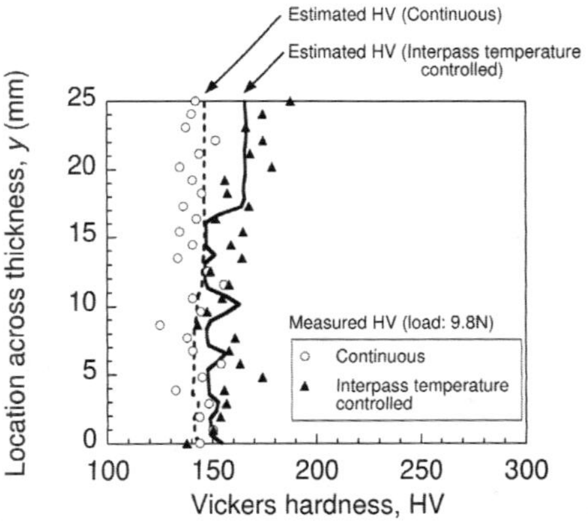

Figure 3. Estimated and measured hardness distribution in weld metals.

3. Effect of molten pool configuration and energy density distribution on weld distortion

The energy density distribution of the heat input affects weld distortion, because the temperature distribution and history change when a heat source with a different energy density is used, even for the same total heat input. A profiler for laser beams is used to measure the energy density distribution, and the effect of the different distributions on weld distortion are studied.

The profiler for measuring energy density distributions scans the surface of a small-diameter opening drilled in the chip of the laser beam-receiving unit by moving in a circular and axial direction. The laser beam is introduced via the chip to a photodiode to sense the intensity of penetrating energy passing through the opening. The distribution of energy on the measuring plane is sent to a computer to process the position of the opening and the intensity of the energy. The three-dimensional energy density distribution is plotted by feeding the chip in the height direction to measure and compute the energy distribution at each level.

Two types of laser beams are measured: Gaussian-like and flattop. Figure 4 shows the results of the distribution of the laser energy density. The light-focusing diameter of the Gaussian-like distribution laser with a laser output of 300 W is approximately 450 μm, and the light-condensing diameter of the flat-top distribution laser is approximately 230 μm for the same laser output. The Gaussian-like distribution laser is designed so that an area with peripheral output lower than the center has a relatively wide range. In the flattop distribution laser, the area with output lower than the center steeply attenuates, in contrast. Thus, two different kinds of profiles were identified.

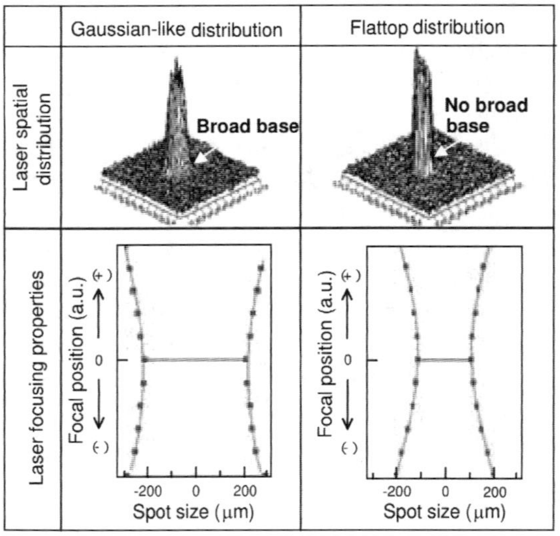

Figure 4. Results of monitoring laser beam profile.

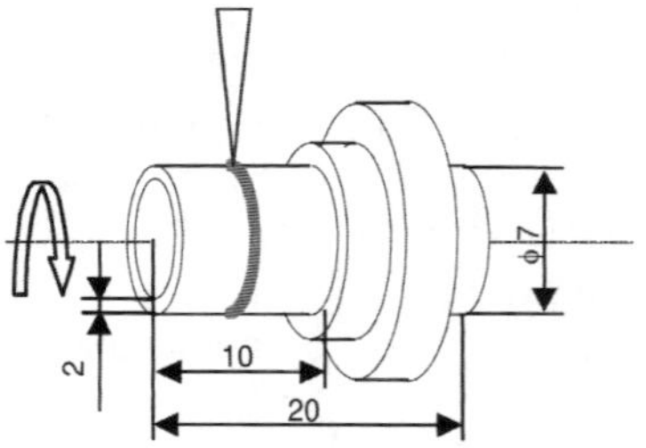

Figure 5. Configuration of bead-on-pipe component specimen.

The penetration geometry using these two laser profiles was evaluated by measuring the cross-sectional shape in bead-on-pipe irradiation. The bead width and penetration depth were measured as functions of the laser output and processing rate in an experiment to verify the melting conditions resulting from the beam profile. The laser beam is irradiated in a bead-on manner onto the test piece, which is a pipe made of type 304 stainless steel, as shown in Fig. 5. A comparison of the cross-section at the same bead penetration depth is shown in Fig. 6, and the cross-sectional geometry under the same laser welding conditions is shown in Fig. 7. The penetration depth tends to increase as the output increases, and the flattop distribution tends to have a larger penetration depth.

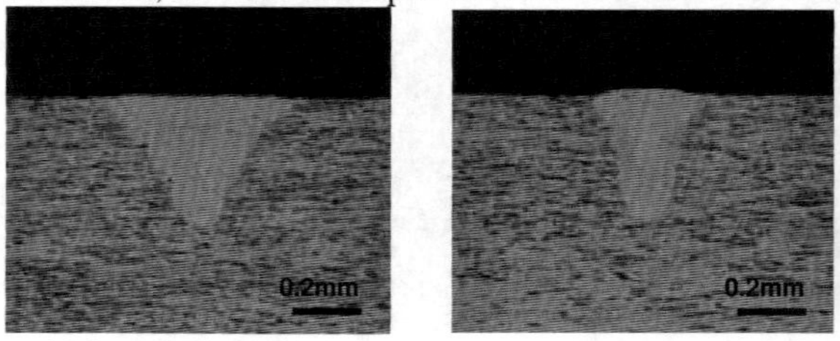

(a) Gaussian-like distribution (200rpm, 300W)

(b) Flattop distribution (300rpm, 300W)

Figure 6. Comparison of cross-section at the same penetration depth.

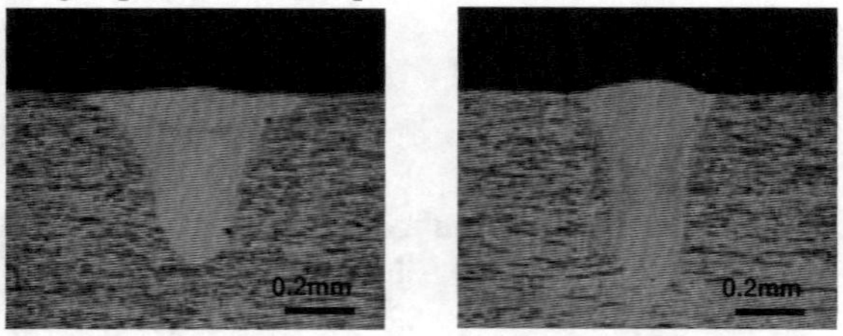

(a) Gaussian-like distribution (200rpm, 350W)

(b) Flattop distribution (200rpm, 350W)

Figure 7. Comparison of cross-section with the same laser welding condition.

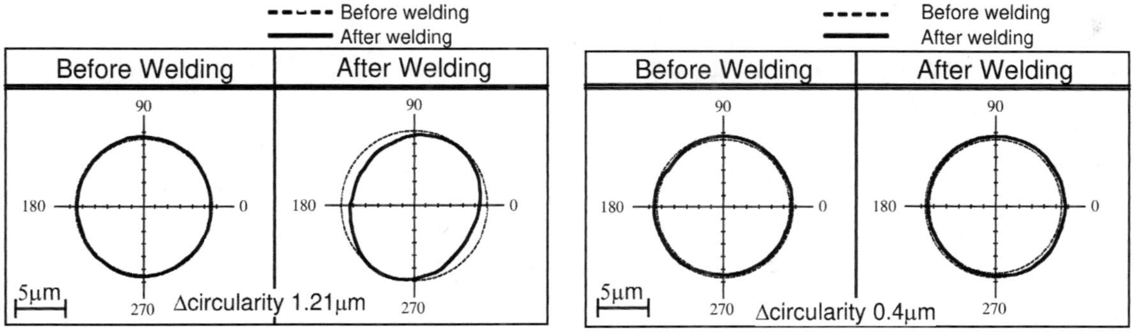

(a) Gaussian-like distribution (b) Flat-top distribution

Figure 8. Measurements of circularity in bead-on-pipe specimen.

A comparison between the cross-sectional shapes in Figs. 6 and 7 indicates that the Gaussian-like distribution beam allows the bead to expand because the upper portion, which does not contribute to the welding depth, absorbs much of the heat and spreads. The Gaussian-like distribution facilitates the melting of the low-energy portion that then expands the bead width, and the keyhole geometry caused by the energy density efficiently guides energy in the thickness direction due to multiplex reflection. In contrast, the flattop laser allows deeper penetration with the same bead width. The beam profiles for these two types of lasers are very different in light focus, as is confirmed by the cross-sectional geometry.

The difference in effects between the flattop distribution beam and the Gaussian-like distribution beam also is due to the variation in circularity. Circularity measurements are taken by the minimum area center method. In this method, the difference between the radii of two circles representing the minimum difference between the inscribed circle and the circumscribed circle is defined as the circularity value. Circularity is measured before and after welding to identify the difference between the pre-welding and post-welding circularities.

Figure 8 shows the results of the circularity measurement after bead-on irradiation of the pipes with the flattop distribution laser and the Gaussian-like distribution laser. The processing conditions fulfilling a penetration depth of 0.4 mm are determined from the results of the experiment in the previous section, and the actual circularity variation is measure.

To achieve penetration at the same level as that with the flattop distribution beam, the Gaussian-like distribution beam requires larger welding deformation due to the irradiation of energy that does not contribute to penetration. The Gaussian-like distribution beam, which traditionally is used, is effective in improving circularity as the processing rate increases, even when compared with the flat-top distribution beam at the same rate. However, processing rate limitations such as insufficient penetration remain higher than those of the flattop.

It is thus understood that the energy density distribution greatly affects weld distortion. This means not only the total heat input but also the heat input distribution, even for the same total heat input, is important for weld distortion reduction and control.

4. Summary

Highly precise fabrication is in demand, and therefore microstructure and distortion controls are essential. Thus, the process mechanics are important for intelligent fabrication. The microstructure and hardness distribution of multipass weld metals resulting from metallurgical and mechanical heterogeneity are estimated by a numerical simulation with weld thermal cycles and microstructural phase transformation. Modified CCT diagrams for the multi-pass weld metal are created to consider the effect of complicated welding heat cycles and to calculate phase transformation. The estimated hardness distribution shows good agreement with the experimental values by using an appropriate CCT diagram for the multi-pass weld and precise numerical simulation. The effect of energy density distributions at the same total heat input on weld distortion is also investigated in this paper. By understanding the relation between weld distortion and molten pool configuration and determining the appropriate welding conditions, more precise fabrication can be obtained.

Acknowledgements
The authors thank Dr. Hideaki Shirai of Denso Corporation for his cooperation in this study. This study was supported by Priority Assistance for the Formation of Worldwide Renowned Centers of Research - The Global COE Program (Project: Center of Excellence for Advanced Structural and Functional Materials Design), and also Grant-in-Aid for Scientific Research (B) – No. 20360393, from the Ministry of Education, Culture, Sports, Science and Technology (MEXT), Japan.

References
[1] Architectural Institute of Japan Ed. 1996 *Japanese Architectural Standard Specification JASS 6 Steel Work* 7th Ed. Architectural Institute of Japan
[2] Japanese Standards Association Ed. 1999 *JIS Z 3312 - MAG Welding Solid Wires for Mild Steel and High Strength Steel* Japanese Standards Association
[3] Mukai A, Nakano T, Okamoto H and Morita K 2000 *J. Steel Const. Eng.* **7** 13-25
[4] Okazawa T, Sakamoto S, Kuramochi M, Fukai S and Yamada K 2001 *J. Steel Const. Eng.* **8** 1-8
[5] Mochizuki M and Toyoda M 2004 *Pipeline Technology* Denys R Ed. Scientific Surveys Ltd Beaconsfield 897-915.
[6] Johnson WA and Merl RF 1939 *Trans. AIME* **135** 416-458
[7] Avrami M 1939 *J. Chem. Phys.* **7** 103-112
[8] Avrami M 1940 *J. Chem. Phys.* **8** 212-224
[9] Avrami M 1941 *J. Chem. Phys.* **9** 177-184
[10] Koistinen DP and Marbuerger RE 1959 *Acta Metal.* **7** 50-60
[11] Mochizuki M, Hayashi M and Hattori T 2000 *Trans. ASME J. Pres. Ves. Tech.* **122** 27-32
[12] Mochizuki M, Hattori T and Nakakado K 2000 *Trans. ASME J. Eng. Mater. Tech.* **122** 108-112
[13] Mochizuki M, Mikami Y, Yamasaki H and Toyoda M 2006 *IIW Doc. X-1607-06* International Institute of Welding
[14] Mikami Y, Mochizuki M and Toyoda M 2006 *Mater. Sci. Forum* **512** 379-382
[15] Mochizuki M, Inuzuka M, Nishida H, Nakata K and Toyoda M 2006 *Sci. Tech. Weld. Join.* **11** 366-370
[16] Okada T 1977 *J. Japan Weld. Soc.* **46** 202-207
[17] Watanabe M 1983 *Prep. National Meeting Japan Weld. Soc.* **33** 308-309
[18] Toyoda M, Mochizuki M and Mikami Y 2006 *Mater. Sci. Forum* **512** 19-24
[19] Mochizuki M 2006 *Sci. Tech. Weld. Join.* **11** 496-501
[20] Mochizuki M 2007 *Nucl. Eng. Des.* **237** 107-123
[21] Shirai H, Okano S, Mochizuki M and Toyoda M 2008 *Mater. Sci. Forum* **580-582** 667-670
[22] Mochizuki M and Toyoda M 2004 *J. de Phys. IV* **120** 635-648
[23] Mochizuki M, An GB and Toyoda M 2005 Key Eng. Mater. **297-300** 2784-2789
[24] Katsuyama J, Mochizuki M, Mori H, Asano W, An GB and Toyoda M 2008 *Mater. Sci. Forum* **580-582** 573-576
[25] Mochizuki M and Toyoda M 2007 *Trans. ASME J. Pres. Ves. Tech.* **129** 619-629

Journal of Physics: Conference Series **165** (2009) 012015 doi:10.1088/1742-6596/165/1/012015

Friction stir welding of Zr-based bulk metallic glass

Y S Ji[1], H Fujii[1], M Maeda[1], K Nakata[1], H Kimura[2], A Inoue[2], K Nogi[1]
[1]Joining and Welding Research Institute, Osaka University, 11-1 Mihogaoka, Ibraki, Osaka 567-0047, Japan
[2]Institite for Materials Research, Tohoku University, Sendai 980-8577, Japan

E-mail: fujii@jwri.osaka-u.ac.jp

Abstract. A $Zr_{55}Cu_{30}Al_{10}Ni_5$ bulk metallic glass plate was successfully welded below its crystallization temperature by friction stir welding. The flash formation and heat concentration at the shoulder edge was minimized using a wider tool and the angle of the recessed shoulder surface was 3°. To analyze the crystallization of the base material and stir zone, the microstructure and mechanical properties were analyzed using DSC, XRD, TEM, and micro-hardness. As a result, it was found that the amorphous structure and original mechanical properties were maintained in the whole joints.

1. Introduction

Bulk metallic glasses (BMGs) have been produced in various kinds of alloy systems, since the $Au_{75}Si_{25}$ alloy was the first amorphous alloy using a rapid solidification process in 1960. Bulk metallic glasses (BMGs) have excellent properties that include a high strength, good wear resistance, good magnetic properties, corrosion resistance, good forming ability, etc. [1-5]. These properties can be attributed to the atomic structure of the amorphous phase. Accordingly, BMGs are currently of interest as next generation materials and active research of BMGs are in progress. However, size limitation is a long-standing problem for structural applications because of a high cooling rate (1~100 K/s) requirement for the production of BMGs. To solve this problem, several recent studies on the joining of BMGs have been done [6-12]. In this study, the $Zr_{55}Cu_{30}Al_{10}Ni_5$ bulk metallic glass was joined using friction stir welding. Mechanical properties and the microstructure of welded joints were then investigated.

2. Experimental

A $Zr_{55}Cu_{30}Al_{10}Ni_5$ bulk metallic glass master alloy ingot was made by arc melting using high purity Zr (99.9%), Cu (99.9%), Ni (99.9%) and Al (99.9%). Fully amorphous specimens of 75×50×1 mm and 75×50×2 mm metallic glass square plates were prepared by copper mold casting under an Ar atmosphere. The stir-in-plate welding was performed using a position-controlled FSW machine. The tool, which was made of SKD61, had a 25 mm shoulder diameter, 5 mm probe diameter and a 1 mm or a 2 mm probe length with a 3° recessed shoulder surface. The probe was a screw-type and the tool was not tilted. The tool rotation speed was varied between 60 and 170 rpm to control the heat input while the travelling speed was constant at 100 mm/min.

Thermal properties of $Zr_{55}Cu_{30}Ni_5Al_{10}$ bulk metallic glass joints were measured by differential scanning calorimetry (DSC, Bruker, DSC 3300SA) at a heating rate of 0.33 K/s under flowing Ar gas. Microstructures of specimens were examined by X-ray diffraction (XRD) and transmission electron microscope (TEM). An XRD analyses were conducted (Bruker, AXS D8 DISCOVER) using Co-Kα radiation. TEM was done at 200 kV using a JEOL2030. Vickers micro-hardness (HV) values were measured using a (Akashi, AAV 501) digital micro-hardness tester with a testing load of 100 gf (=0.98 N).

3. Results and discussion

3.1 Tool shape

To minimize flash formation in this study a tool with a wide shoulder of 25 mm was used although a shoulder of approximately 12 mm is generally used.

On the other hand, the angle of the recessed shoulder surface of the tool is 3°. When the angle of the recessed shoulder surface is 10°, the heat is concentrated at the shoulder edge, causing the surface to become pitted at the shoulder. When the angle of the recessed shoulder surface is 0°, a lot of flash is formed. However, when the angle of the recessed shoulder surface is 3°, a comparatively good surface is obtained.

Figure 1 shows surface appearances of the stir-in-plate specimens of $Zr_{55}Cu_{30}Al_{10}Ni_5$ bulk metallic glass obtained at different rotation speeds. When the rotation speed is 60 rpm defects are formed on the surface, as shown in Figure 1(a), because the heat input is not sufficient. When the rotation speed is between 80 rpm and 170 rpm, large defects and cracks are not formed in the stir zone, as shown in Fig. 1(b), (c) and (d).

3.2 Microstructure and mechanical properties

Figure 2 shows XRD patterns of the stir zone of stir-in-plate specimens. When the rotation speed is between 80 and 150 rpm, XRD patterns of the stir zone shows typical broad patterns. This result confirms that the temperature of the stir zone during the stir-in-plate welding was below the crystallization temperature.

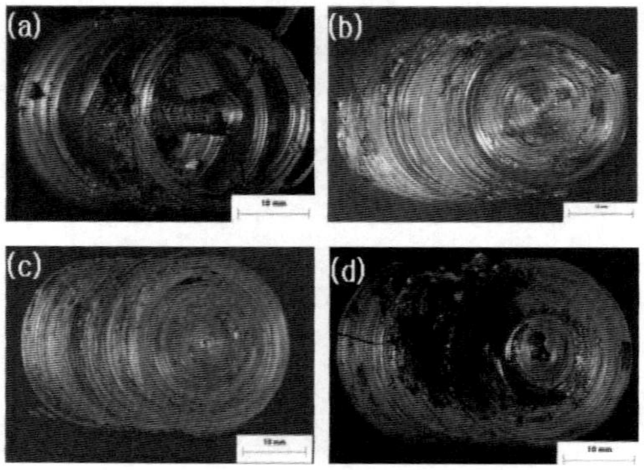

Figure 1. A microscopic overview of the stir zone in $Zr_{55}Cu_{30}Al_{10}Ni_5$ bulk metallic glass: (a) 60 rpm, 100 mm/min, (b) 80 rpm, 100 mm/min, (c) 150 rpm, 100 mm/min, and (d) 170 rpm, 100 mm/min.

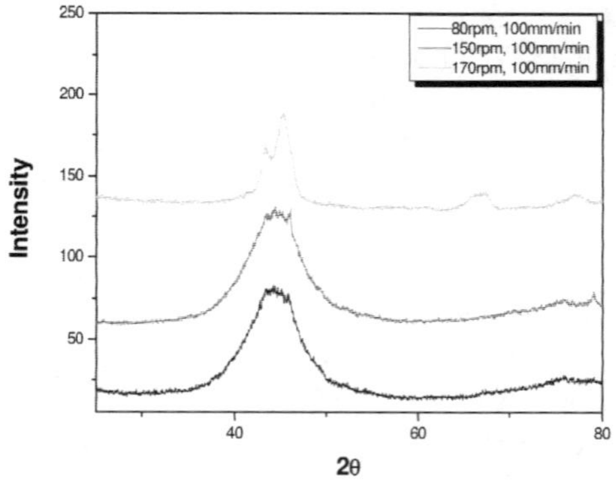

Figure 2. X-ray diffraction patterns of the stir zones obtained at different welding speeds.

However, when the rotation speed is 170 rpm crystallization behavior is observed in the XRD spectrum of stir zone. In this case, the welding was performed in a short period. Accordingly, the temperature of the stir zone during the stir-in-plate welding was estimated to be about above the crystallization temperature.

Figure 3 shows the hardness distribution on a cross section perpendicular to the welding direction of the stir-in-plate specimens. There is no significant change the in hardness between the base material and the stir zone when the rotation speed is between 80 and 150 rpm, as shown in Figure 3. The average value of the hardness in the base material and stir zone is 523 HV.

Figure 4 shows DSC results of the base material and stir zone of the stir-in-plate specimens obtained at a 0.33 K/s heating rate. The glass transition temperature (T_g), the crystallization temperature (T_x) and the super-cooled liquid region range ($\Delta T=T_x-T_g$) are about 680, 750 and 70 K, respectively. As shown in Figure 4, the exothermic peaks are clearly observed without any shift in T_g and T_x. These values are similar to that of the base material.

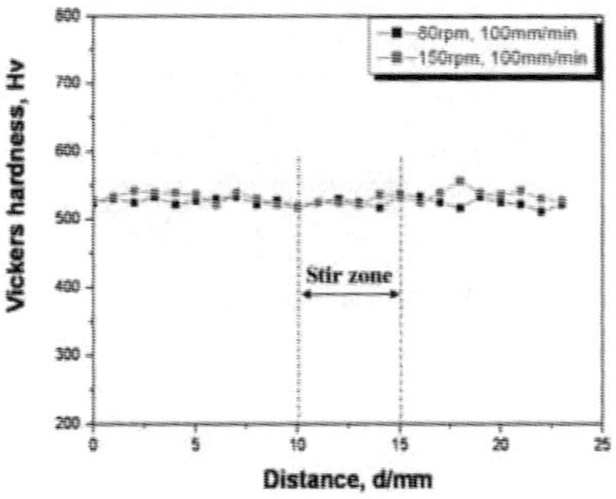

Figure 3. The hardness distribution on a cross section perpendicular to the welding direction.

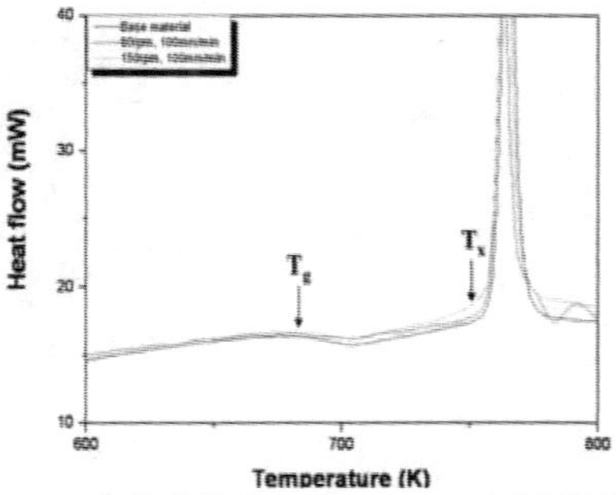

Figure 4. DSC (heating rate=0.33 K/s) of the base material and the stir zone.

Figure 5 shows bright-field TEM images and corresponding selected-area-diffraction (SAD) patterns of the base material and the stir zone in the stir-in-plate specimens at 150 rpm. As shown in Figure 5 (a) and (b), no nanocrystalline particles were observed in the bright-field images of the base material or the stir zone. It is recognized from this TEM result that the base material and stir zone are homogeneous and amorphous at 150 rpm.

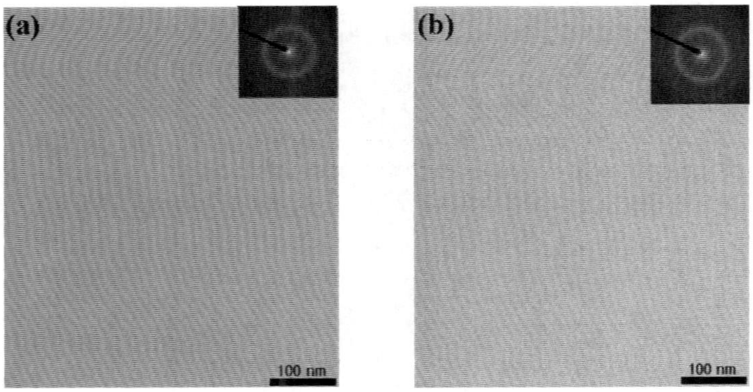

Figure 5. TEM images of stir-in-plate specimens at 150 rpm: (a) base material and (b) stir zone.

3.3 Effect of sample thickness

Figure 6 shows surface appearances and the possible range of the stir-in-plate of $Zr_{55}Cu_{30}Al_{10}Ni_5$ bulk metallic glass specimens of 2 mm thickness. At a rotation speed of 80 rpm, defects are formed on the surface because the heat input is not sufficient. At rotation speeds between 100 rpm and 150 rpm, no large defects or cracks are formed in the stir zone as shown in Figure 6. However, the possible welding range for the 2 mm plates is narrower than that for the 1 mm plates because a higher heat input is necessary for thicker plates.

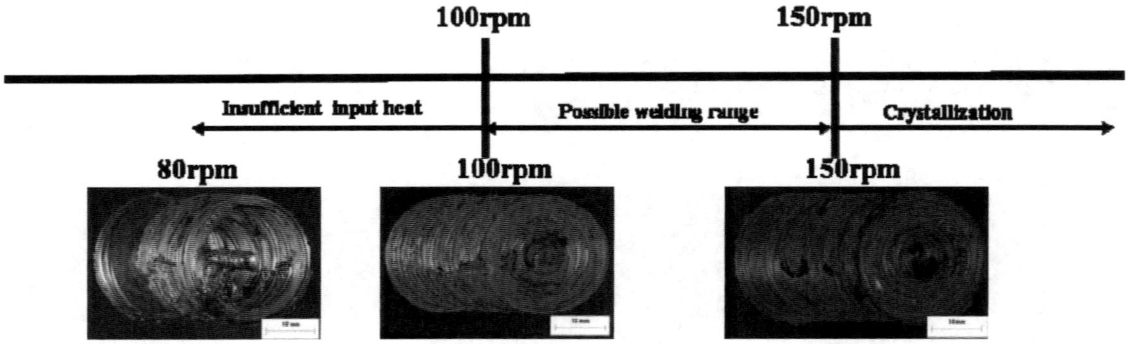

Figure 5. Possible welding range and surface appearance for 2 mm thick $Zr_{55}Cu_{30}Al_{10}Ni_5$ bulk metallic glass.

4. Conclusions

$Zr_{55}Cu_{30}Al_{10}Ni_5$ bulk metallic glass was welded by friction stir welding. The obtained results are summarized as follows:

(1) $Zr_{55}Cu_{30}Al_{10}Ni_5$ bulk metallic glass was successfully welded below its crystallization temperature using friction stir welding.

(2) Flash formation was minimized using a tool with a wider shoulder of 25 mm and heat concentration at the shoulder edge was prevented by the 3°angle of the recessed shoulder surface.

(3) The microstructure and mechanical properties of welded joints were analyzed. It was found that the amorphous structure and original mechanical properties were maintained.

Acknowledgments

The authors wish to acknowledge the financial support of a Grant-in-Aid for the Cooperative Research Project of Nationwide Joint-Use Research Institutes on Development Base of Joining Technology for New Metallic Glasses and Inorganic Materials, "Priority Assistance of the Formation of Worldwide Renowned Centers of Research-The global COE Programs (Project: Center of Excellence for Advanced Structural and Functional Materials Design) from the Ministry of Education, Sports, Culture, Science and Grant-in-Aid for Science Research from the Japan Society for Promotion of Science and Technology of Japan.

References

[1] Inoue A 2000 Acta. Mater. **48** 279
[2] Johnson W L 2002 JOM. **54** 40
[3] Pang S J, T Zhang K. Asami and A Inoue 2002 Corrosion Sci. **44** 1847
[4] Busch R, Bakke E and Johnson W L 1998 Acta. Mater. **46** 4725
[5] Waniuk T A, Busch_R, Masuhr_A and Johnson W L 1998 Acta. Mater. **46** 5229
[6] Wong C H, Shek C H 2003 Scripta. Mater. **49** 393
[7] Kawamura Y, Shoji Ohno T and Ohno Y 2003 J. Non-Cryst Solid. **317** 152
[8] Kawamura Y, Ohno Y, Chiba A 2002 Mater. Sci. Forum. **553** 386.
[9] Kawamura Y, Ohno Y 2001 Scripta. Mater. **45** 279
[10] Kawamura Y, Ohno Y 2001 Mater. Trans. JIM. **42** 717
[11] Kawamura Y, Ohno Y 2001 Scripta. Mater. **45** 127
[12] Kim J H, Kawamura Y 2008 Mater. Proce. Tech. **207** 112

Journal of Physics: Conference Series **165** (2009) 012016 doi:10.1088/1742-6596/165/1/012016

Evaluation of interfacial microstructures in dissimilar joints of aluminum alloys to steel using nanoindentation technique

Tomo Ogura[1], Yuichi Saito[2], Keisuke Ueda[2], Akio Hirose[1]
[1] Division of Materials and Manufacturing Science, Graduate School of Engineering, Osaka University
[2] Graduate student, Division of Materials and Manufacturing Science, Graduate School of Engineering, Osaka University
2-1 Yamadaoka, Suita, Osaka 565-0871, Japan

E-mail: tomo.ogura@mapse.eng.osaka-u.ac.jp

Abstract. The characteristics of interfacial microstructures with additional elements in dissimilar 6000 system aluminum/steel joints were basically evaluated using tensile test, EPMA, TEM and nanoindentation. For Si (and Cu)-added alloy (S1 and SC), EPMA analysis showed that Si (and Cu) was enrichment in the reaction layers, which were formed during diffusion bonding. SAED pattern clarified that the reaction compounds at the interface changed from AlFe intermetalic compounds to AlFeSi intermetalic compounds by Si addition. Nanoindentation technique was successfully applied to the interfacial microstructures to understand directly the nanoscopic mechanical properties in the interfacial microstructures. The hardness and Young's modulus of Al_3Fe intermetalic compounds was lower than those of Al_2Fe_5 intermetalic compounds. Moreover, the hardness and Young's modulus of AlFeSi(Cu) compounds were lower than those of Al_3Fe, indicating that the crystal system changed from orthorhombic structure to cubic structure. Joint strength of SC/steel joints was higher than that of the aluminum alloy with no additional element (Base)/ steel joint, indicating that interfacial microstructure was modified by the addition of Si and Cu to the 6000 system aluminum alloy. These results suggest that the nanoscopic mechanical properties at the interface microstructures affect greatly the macroscopic deformation behavior of the aluminum /steel dissimilar joints.

1. Introduction

Development of high reliable dissimilar joining process is important because interfacial microstructures formed during joining process cause greatly the tensile strength of the joints [1]. The dissimilar joining of aluminum alloys steels for a light- weight auto body has been considerably researched, however, the formation process of interfacial microstructures based on the addition of alloying elements, in particular, Si and/or Cu which are industrially utilized, are still controversial. As for the method to examine nanoscopic mechanical properties of the interfacial microstructures directly, nanoindentation techniques have been quite useful because nanoindentation becomes quite effective in investigating local deformation and hardness of alloys in a nanometer-scale [2]. In the present study, the characteristics of interfacial microstructures with additional elements in dissimilar aluminum/steel joints were basically evaluated using tensile test, EPMA, TEM and nanoindentation.

2. Experimental procedure

Four types of 6000 system aluminum alloys (Base (Si: 0.6, Cu: 0(in mass%)), S1 (Si: 1.5, Cu: 0), C1 (Si: 0.6, Cu: 1.0) and SC (Si: 1.5, Cu: 1.0)) and a low carbon steels (SPCE) including 0.01C, 0.15Mn and 0.01Si (in mass%) were used. After polishing the surface of the samples using emery paper (#2000), the surface was cleaned by acetone. Diffusion bonding was carried out in a vacuum of 1.0×10^{-2} Pa at heating temperatures (T) ranging from 773 to 848K for a holding time (t) from 150 to 1800s with a pressure of 2.5MPa. The heating rate was 3K/s.

After bonding, the interfacial microstructure was observed with scanning electron microscopy (SEM). The composition analysis was also performed using electron probe microanalysis (EPMA). The thickness of the reaction layer at the bonded interface was measured using SEM images. Tensile test was carries out with an Instron type tensile testing machine. The cross-head speed is 3.0×10^{-2}mm/s. The nanoindentaion measurement was performed toward the interface of joints using the instrument of ENT-1100a. The applied load was 100mgf and the load and release times were 10 and 1s, respectively.

3. Results and discussion

3.1Interfacial microstructures of aluminum alloy/SPCE joints

Figure 1 shows EPMA analysis of the interface of the SC/SPCE joint bonded at 785K for 1800s. Si and Cu were found to be enrichment in the reaction layer, in particular, near SC side. In the commercial 6000 alloys/SPCE joints, two kinds of Al-rich Fe-Al type IMC layer were formed during diffusion bonding [3]. Therefore, it is considered that IMC layer was changed by the additional elements. The effect was also confirmed by TEM observation shown in figure 2. SAED (Selected area electron diffraction) pattern identified the IMC layer of the C1/SPCE joint as $Al_{12}Fe_3Si$. Thus, as for the effects of additional elements, Si(and Cu) was considered to modify the reaction IMC layer at the interface from AlFe binary IMC to AlFeSi(Cu) ternary(quaternary) IMC.

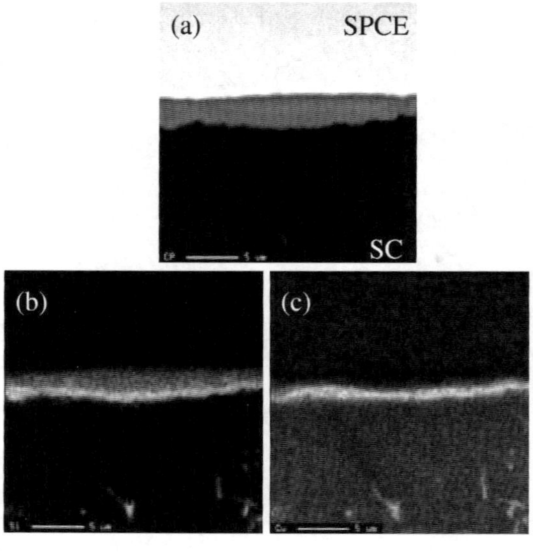

Figure 1. EPMA analysis of the interface of the SC/SPCE joint. (a) BE image, (b) Si mapping and (c) Cu mapping.

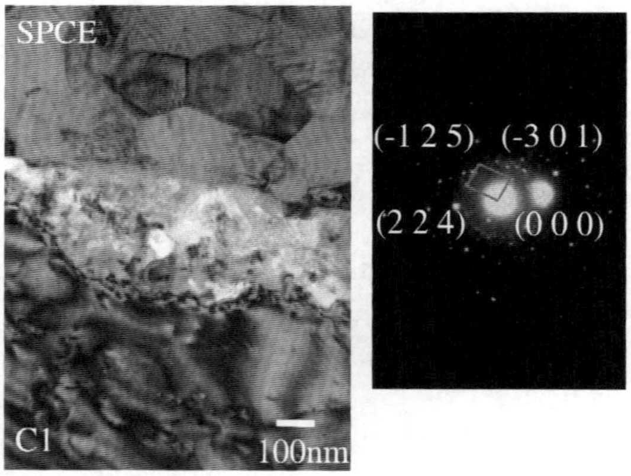

Figure 2. TEM image of the interface of C1/SPCE joint and the corresponding SATD pattern (the analysis area is in the circle).

3.2 Nanoscopic mechanical properties of interfacial microstructures

Nanoindentation technique was successfully applied to the interfacial microstructures in the investigated joints in order to understand directly the nanoscopic mechanical properties in the interfacial microstructures. Figure 3 shows a typical backscattered electron image of the interfacial microstructure after a nanoindentation test. It is clearly seen that some of triangle indents were loaded inside interfacial layer along an interface of the joint shown as circles in figure 3. Figure 4 and figure 5 show hardness and Young's modulus of interfacial reaction layers of the investigated joints. The interfacial reaction layer at SPCE side was identified as Al_5Fe_2 by EDX analysis regardless of additional elements. The reaction layer at Al side was Al_3Fe (Base), $Al_{12}Fe_3Si$ (S1 and C1) and AlFeSi containing Cu (SC).The hardness and Young's modulus of Al_3Fe IMC was lower than those of Al_2Fe_5 IMC. Moreover, the hardness and Young's modulus of AlFeSi(Cu) compounds were lower than those of Al_3Fe. The difference of hardness and Young's modulus value was considered to be the different of crystal system, because the crystal system of Al_3Fe and $Al_{12}Fe_3Si$ are orthorhombic structure and cubic structure, respectively.

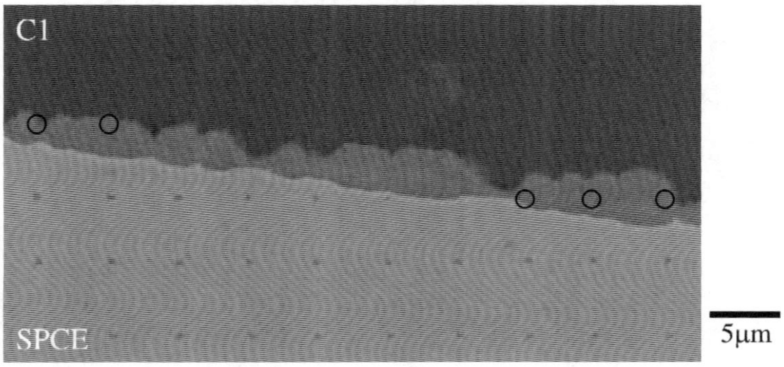

Figure 3. A backscattered electron image of nanoindentation marks around the interface of C1/SPCE joint. Some of nanoindentation marks were loaded inside the interfacial reaction layer (in the circle).

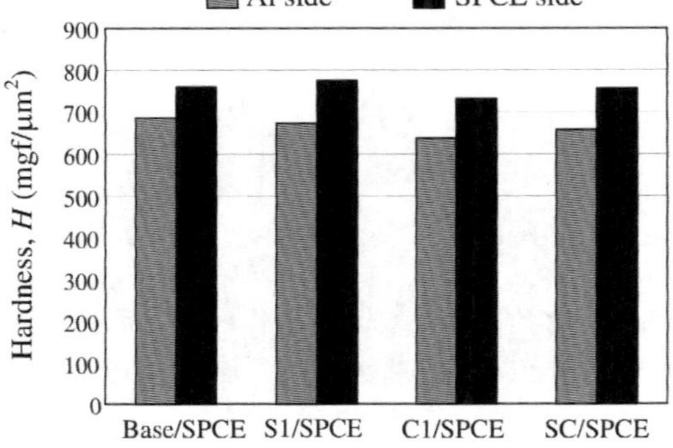

Figure 4. Hardness of the interfacial reaction layer of Al alloy/SPCE joints.

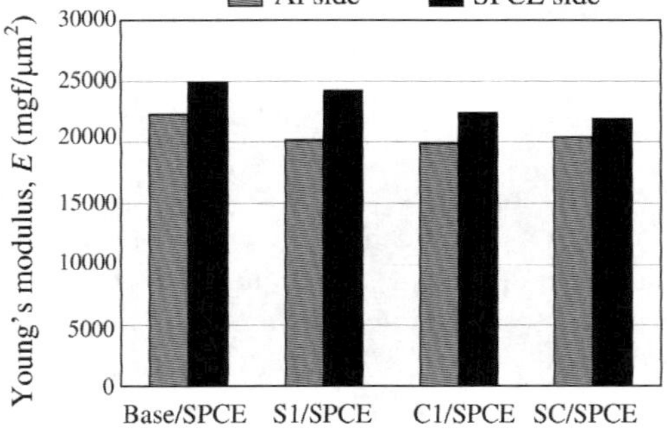

Figure 5. Young's modulus of the interfacial reaction layer of Al alloy/SPCE joints.

3.3 Joint strength

Form the subsection *3.1* and *3.2*, it was revealed that the interfacial reaction layer was to be multinary by additional elements and the nanoscopic mechanical properties were also changed. To clarify the effects of the interfacial microstructure changes on the macroscopic mechanical properties, tensile test was carried out. Figure 6 shows the maximum strength of the investigated joints. Tensile property was improved by the addition of Si and Cu to the aluminum alloy. Moreover, the thickness of the interfacial reaction layer in maximum strength of SC/SPCE joint became thinner than that of Base/SPCE joint. Based on the obtained results in the present work, therefore, it was indicated that the nanoscopic mechanical properties at the interface microstructures affect greatly the macroscopic deformation behavior of the aluminum /steel dissimilar joints and Si and Cu addition to the aluminum alloy is quite effective to modify the interfacial microstructures of aluminum alloy/SPCE joints.

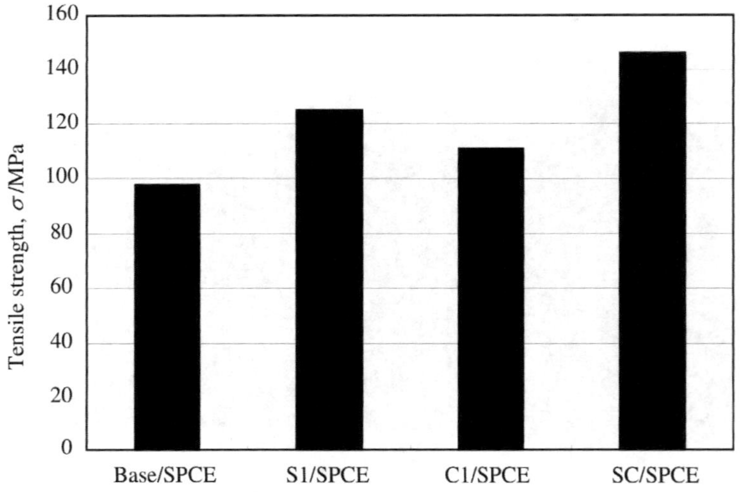

Figure 6. Maximum strength of Al alloy/SPCE joints.

4 Conclusions

1. High bondability was achieved through the reaction layer changes from AlFe intermetalic compounds to AlFeSi(Cu) intermetalic compounds by Si (and Cu) addition.
2. The hardness and Young's modulus of AlFeSi(Cu) intermetalic compounds was lower than those of AlFe intermetalic compounds because of the crystal system changes.
3. The nanoscopic mechanical properties at the interface microstructures affect greatly the macroscopic deformation behavior of the aluminum/steel dissimilar joints.

Acknowledgement

The authors would like to thank Prof. T. Nakano and Dr. T. Ishimoto for their significant help with nanoindentation measurement. This study was supported by Priority Assistance for the Formation of Worldwide Renowned Centers of Research - The Global COE Program (Project: Center of Excellence for Advanced Structural and Functional Materials Design) from the Ministry of Education, Culture, Sports, Science and Technology (MEXT), Japan.

Reference

[1] Hirose A 2007 *Journal of JSAE*, **61** 18.
[2] Ogura T, Hirosawa S and Sato T 2004 *Sci. Tech. Adv. Mater.*, **5** 491.
[3] Umeshita H *et al.*2007 *Smart Proc. Technol.*, **2** 199.

Enantiomorph identification in organic crystals by electron diffraction

K Tanaka[1], S Fujio[1], H Inui[1], R Ueji[2] and N Sumida[2]

[1] Department of Materials Science and Engineering, Kyoto University, Sakyo-ku, Kyoto 606-8501, Japan

[2] Department of Advanced Materials Science, Kagawa University 2217-20 Hayashi-cho, Takamatsu 761-0396, Japan

haruyuki.inui@materials.mbox.media.kyoto-u.ac.jp

Abstract. The convergent-beam electron diffraction (CBED) method we proposed recently for enantiomorph identification has successfully been applied to some amino acid crystals such as glutamic acid and threonin. Enantiomorph identification (either left-handed or right-handed form) can readily be made within the framework of the proposed method by noting the asymmetric intensity distribution of Bijvoet pairs of reflections in the CBED pattern taken along an appropriate zone-axis orientation. Although the proposed method usually requires only a single CBED pattern, some effort to eliminate the ambiguity of 180°-rotation of the CBED pattern about the incident beam is needed for enantiomorph identification for these organic crystals because of the lack of HOLZ (higher-order Laue zone) reflection disks.

1. Introduction

Objects may have both left-handed and right-handed forms in the absence of improper rotations (rotoinversions and rotoreflections) in the crystallographic description [1]. This is usually called enantiomorphism or chirality. The two members of the pair of enantiomorphic crystals or molecules are mirror-related. Since one of the two enantiomorphic crystals exhibit bioactivities different from another in most cases, distinction of enantiomorph (chirality) of these enantiomorphic crystals is sometimes very important especially when they are used in foods and pharmaceutical drugs [2,3]. Thus, many organic crystals and molecules are sometimes the subject of enantiomorphic investigation and some asymmetric synthesis methods are often adopted to promote the preferential nucleation for one of the two enantiomorphic crystals and molecules [2]. Enantiomorph is often distinguished by polarimetry for many organic molecules and, to the lesser content, organic molecular crystals [2,4], whereas for inorganic crystals and many organic molecular crystals including proteins, this is usually made by X-ray diffraction utilizing anomalous scattering that breaks the Friedel's law. However, these X-ray diffraction methods generally requires a relatively large-sized single crystals of high quality [5], which are sometimes difficult to prepare. We have recently proposed a new electron diffraction method for enantiomorph identification, in which the intensity asymmetry of Bijvoet pairs of ZOLZ (zero-th order Laue zone) and/or FOLZ (first order Laue zone) reflection disks in a CBED (convergent-beam electron diffraction) pattern taken along an appropriate zone-axis orientation is monitored and compared with the asymmetry in the corresponding computer simulated CBED pattern

[6]. This new method is usually made with a conventional transmission electron microscope (TEM) and allows in principle to identify all possible enantiomorphic crystals that are allowed to exist in crystallography [6] and has indeed been experimentally applied successfully to inorganic crystals belonging to the point groups of 622 [6-11], 6 [11], 312 and 321 [12] and so on [13].

Although the proposed method should find out the most fruitful applications in organic crystals, there is an inherent difficulty in observing organic crystals with TEM due mainly to the sublimation of these crystals under electron illumination, which arises from their low thermal and electrical conductivity [14]. However, we have recently applied the proposed method successfully to an organic superconductor κ-(BEDT-TTF)$_2$Cu(NCS)$_2$ by reducing electron illumination damage with the real-time video-recording of CBED patterns [15]. In view of the fact that the organic superconductor contains heavy metal atoms such as copper, the extent of the difficulty in observations is considered to be less significant when compared to other organic crystals consisting only of light elements. Of these organic crystals, those referred to as amino acids (containing at most sulfur) would be a nice example to check the applicability of the proposed electron diffraction method, since they are often the subject of enantiomorph identification for their bioactivity [3]. In the present study, we investigate how easily the electron diffraction method we recently proposed can be applied to organic crystals of the most difficult class, taking examples of some amino acids such as glutamic acid and threonin.

2. Amino acids and principles for enantiomorph identification

Of more than 700 amino acids so far identified, 26 have been known to be a constituent of proteins. These 26 amino acids are usually called L-α amino acids, since they are in general of the L-form in vivo, exhibiting a dextro-rotatory in polarimetry. Many of them crystallize into a form belonging to the space group either $P2_1$ or $P2_12_12_1$, with some exceptions occurring, for example, in cystine with the space group of $P6_122$ [2,3]. This indicates that we need to consider only three different point groups for enantiomorph identification within the framework of the proposed method; the point groups 2, 222 and 622. In the present study, we investigate the point groups 222 and 622, taking glutamic acid ($P2_12_12_1$), threonin ($P2_12_12_1$) and cystine ($P6_122$) as specific examples for these cases.

The proposed method requires to take a CBED pattern along the 'appropriate' zone-axis orientation in order to observe Bijvoet pairs of reflections symmetrically with respect to the symmetry line m-m', as shown in figure 1. Since Bijvoet pairs of reflections are Bragg reflections of space group symmetry equivalents to the two members of a Friedel pair, the appropriate zone-axis orientations to symmetrically observe them vary with crystal point group. The appropriate zone-axis orientations are analyzed to be of the $\langle uv0\rangle$-, $\langle 0vw\rangle$- and $\langle uow\rangle$-types for the point groups of 2 and 222, and $\langle u\bar{u}0w\rangle$-, $\langle uu2\bar{u}w\rangle$- and $\langle uvt0\rangle$-types for the point group of 622 [6]. Once Bijvoet pairs of reflections are observed symmetrically with respect to the symmetry line m-m', the phase distribution of these Bijvoet pairs of reflections is not symmetrical with respect to the symmetry line m-m', since the two members of Bijvoet pair reflections have equal amplitudes ($|F|$) but opposite phases

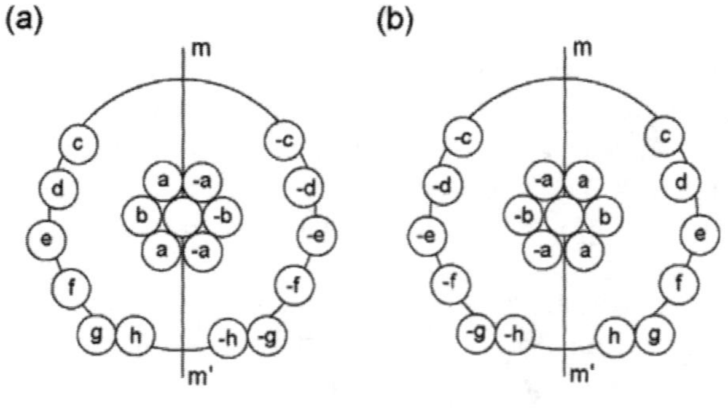

Figure 1. Schematic illustration of zone-axis CBED patterns of (a) right-handed and (b) left-handed enantiomorphic crystals, in which Bijvoet pairs of reflections are observed symmetrically with respect to the symmetry line m-m'. The nomenclatures a, b, c, … indicated at reflection disk positions schematically depict phase angles of the corresponding reflection disks.

(φ) in the expression of structure factors for any hkl reflection as in the case of a Friedel pair. Then, the intensity distribution of Bijvoet pairs of reflections in the CBED pattern is expected to be asymmetrical with respect to the symmetry line m-m', as a result of multiple scattering among reflections in the zone-axis CBED pattern. Since the right-handed crystal can be converted to the left-handed one by changing the coordinates of atom positions from (x,y,z) to $(-x,-y,-z)$, the asymmetrical phase distribution with respect to the symmetry line m-m' in the zone-axis CBED pattern for one of the two members of the enantiomorphic crystal is antisymmetric with that for the other (Figures 1(a) and (b)). Then, the asymmetric intensity distribution in the zone-axis CBED pattern with respect to m-m' is also expected to be antisymmetric for the two members of the enantiomorphic crystal, by which enantiomorph identification is made.

3. Experimental Procedures

L-α glutamic acid and L-threonin were purchased in the form of powders from Osaka Asahi Metal. Co. Ltd. These powder samples were crushed in liquid nitrogen in order to obtain very thin samples so as to minimize the temperature rise of specimen during TEM observations. They are then mounted on a microgrid for TEM observations. CBED experiments were made with the JEM-2000FX TEM operated at 100 kV at ambient temperature and with the JEM-3010 TEM operated at 100 kV in the temperature range from 10 K to ambient temperature. CBED patterns were recorded in real-time video-recording to minimize the recording time. Calculation of CBED patterns was made with a Win HREM software [16] with the atomic coordinates tabulated in [17] for the L-forms of glutamic acid, threonin and cystine, respectively.

4. Results

L-α cystine is formulated to be $C_6H_{12}N_2O_4S_2$ (molecular weight: 240.30) and is reported to belong to the space group of $P6_122$ in the L-form. Since enantiomorphism in this case is related to the screw axis parallel to the hexagonal c-axis, the space group of the D-form is changed to $P6_522$. Of the appropriate

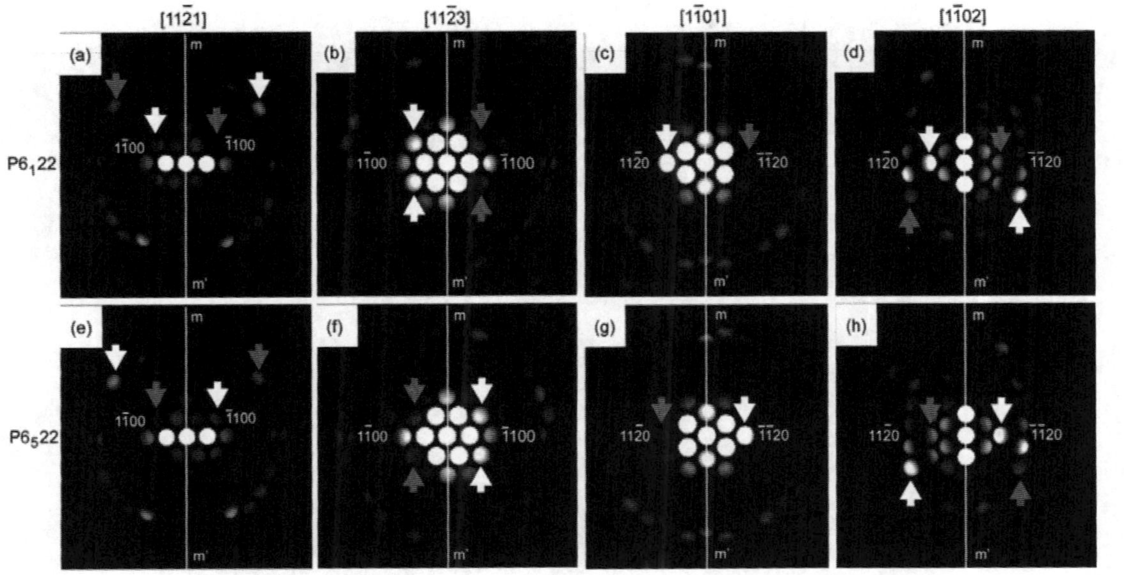

Figure 2. Calculated CBED patterns of L-α cystine taken along (a),(e) $[11\bar{2}1]$, (b)(f) $[11\bar{2}3]$, (c)(g) $[1\bar{1}01]$ and (d)(h) $[1\bar{1}02]$ zone-axis orientations. The space groups assumed in the calculation are $P6_122$ for (a)-(d) and $P6_522$ for (e)-(h). The accelerating voltage and crystal thickness used in the calculation are 100 kV and 22.4 nm, respectively.

zone-axis orientations allowed for the point group of 622, we choose $[11\bar{2}1]$, $[11\bar{2}3]$, $[1\bar{1}01]$ and $[1\bar{1}02]$ to calculate zone-axis CBED patterns to see whether the expected intensity asymmetry for Bijvoet pairs of reflection disks appears or not. As seen in figure 2, the intensity distribution is asymmetric with respect to the m-m' line for each of these CBED patterns. The intensity asymmetry for Bijvoet pairs of reflection disks occurs for both ZOLZ and FOLZ reflection disks for all these CBED patterns. When the L-form is changed to the D form (and vise versa), the asymmetric intensity distribution with respect to the m-m' line is reversed, by which enantiomorph identification is readily made. This indicates that enantiomorph identification can readily be made with only such a CBED pattern even for organic crystals consisting only of light elements, once the CBED pattern is taken along an appropriate zone-axis orientation.

This was indeed confirmed by experiments made for glutamic acid and threonin. L-β glutamic acid is formulated to be $C_5H_9NO_4$ (molecular weight: 147.13) and belongs to the space group of $P2_12_12_1$. The L-form is converted to the D-form by changing the sign of the atomic coordinates within the same space group. An experimental CBED pattern taken along [101], which belongs to the appropriate zone-axis orientations of the <*u*0*w*>-type, is shown in figure 3(a). The intensity asymmetry with respect to the m-m' line is clearly observed not only for a Bijvoet pair of 020-$0\bar{2}0$ but also for those of $15\bar{1}$-$1\bar{5}\bar{1}$ and $\bar{1}51$-$\bar{1}\bar{5}1$. However, unlike in figure 1, the CBED pattern of figure 3(a) exhibits only Bijvoet pairs of ZOLZ reflections and lacks the intensity for Bijvoet pairs of FOLZ reflections. The absence of the intensity for Bijvoet pairs of FOLZ reflections was similarly confirmed even at 10 K. This causes an ambiguity of 180°-rotation of the CBED pattern of figure 3(a) about the incident beam. In the present case, the existence of the [100] zone-axis upward and [001] zone-axis downward was confirmed upon tilting the specimen about [010]. Then, Bijvoet pairs of ZOLZ reflections are all unambiguously indexed as in figure 3(b). Figures 3(b) and (c) are calculated [101] CBED patterns of

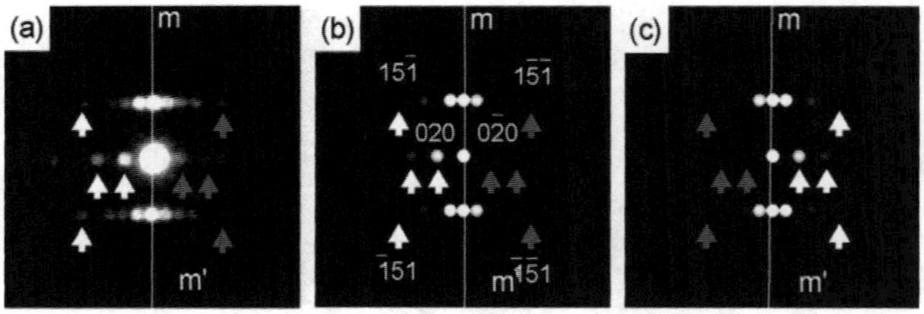

Figure 3. (a) Experimental and calculated [101] zone-axis CBED patterns for β glutamic acid of (b) the L- and (c) D-forms. In the calculation, the value of crystal thickness was assumed to be 20.2 nm.

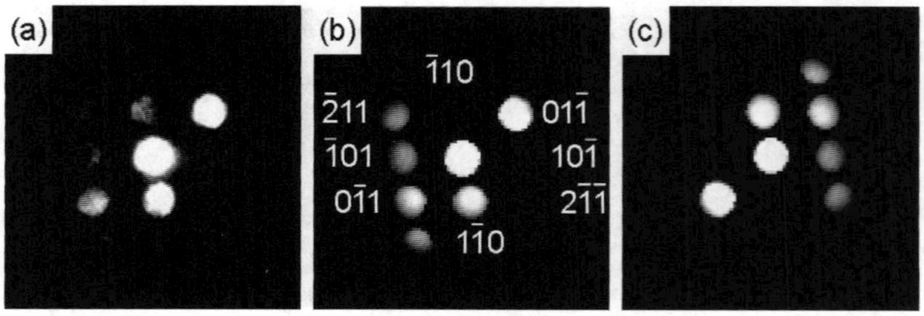

Figure 4. (a) Experimental and calculated [111] zone-axis CBED patterns for threonin of (b) the L- and (c) D-forms. In the calculation, the value of crystal thickness was assumed to be 33.0 nm.

the L- and D-forms of glutamic acid with the assumption of the value of crystal thickness of 20.2 nm. In the experimental pattern (figure 3(a)), the intensity of the 020 ZOLZ reflection disk is stronger than that of the $0\bar{2}0$ disk, indicating that glutamic acid investigated corresponds to the L-form.

L-threonin is formulated to be $C_4H_9NO_3$ (molecular weight: 119.12) and belongs to the space group of $P2_12_12_1$ and the L-form is converted to the D-form by changing the sign of the atomic coordinates within the same space group. Enantiomorph identification is thus made for threonin as in the case of glutamic acid. A CBED pattern experimentally taken along the [111] zone-axis orientation and the corresponding calculated patterns assuming the L- and D-forms are shown in Figures 4(a)-(c), respectively. Since the [111] zone-axis orientation does not belong to any of the appropriate zone-axis orientations, the symmetry line m-m' cannot be defined and the intensity distribution of all reflection disks in the experimental pattern should be compared with the calculated patterns. As in the case of glutamic acid, the experimental [111] CBED pattern of threonin also lacks the intensity of FOLZ reflections, giving rise to the ambiguity of 180°-rotation of the CBED pattern about the incident beam. To eliminate this ambiguity, the specimen was tilted about $[1\bar{1}0]$ so as to determine which direction (upper-right or lower-left) of the CBED pattern corresponds to the way to reach the [110] or [001] zone-axis orientations. Then, Bijvoet pairs of ZOLZ reflections are all unambiguously indexed as in figure 4(b). When the intensity for the Friedel pair of $01\bar{1}$ - $0\bar{1}1$ is compared with each other in the calculated CBED patterns, the intensity of the former disk is stronger than that of the latter disk for the right-handed form (figure 4(b)), whereas the opposite is true for the left-handed form (figure 4(c)). The intensity asymmetry in the experimental pattern (figure 4(a)) clearly indicates that threonin investigated corresponds to the L-form.

5. Discussion

Enantiomorph identification by electron diffraction was successfully made for amino acids consisting only of light elements for the first time. This was made at ambient temperature without cooling the specimen by noting the asymmetric intensity distribution for Bijvoet pairs of ZOLZ reflection disks arranged symmetrically with respect to the m-m' line in a CBED pattern taken along the appropriate zone-axis orientation (figure 1). In spite of the difficulty that the intensity for Bijvoet pairs of FOLZ reflections is too low to be detected in the CBED pattern, enantiomorph identification could be made successfully within the framework of the method we proposed recently [6] only by taking an additional SAED pattern to eliminate the ambiguity of 180°-rotation of the CBED pattern about the incident beam. The difficulty in observing FOLZ reflections in organic crystals may arise from the following factors [14]; (i) the atomic numbers of constituent atoms are in general low, (ii) thermal vibration of constituent atoms is high at ambient temperature and (iii) electron illumination damage quickly occurs introducing short-range disorder. Since the visibility for the intensity of Bijvoet pairs of FOLZ reflections remained almost identical even at 10 K, the factor (ii) may not make the main contribution. The factor (iii) can also be effectively avoided if real-time video recording is employed. Indeed, the intensity of FOLZ reflections was clearly visible for a while (a few seconds) in some CBED patterns with a smaller radius for the FOLZ ring. We thus believe that enantiomorph identification for organic crystals is also possible only with a single CBED pattern (as in many inorganic crystals) if an appropriate zone-axis orientation, where Bijvoet pairs of ZOLZ and FOLZ reflection disks are arranged symmetrically with respect to the m-m' line and the radius for the FOLZ ring is small in the CBED pattern, is properly chosen.

6. Conclusions

The CBED method we proposed recently for enantiomorph identification has successfully been applied to some amino acids such as glutamic acid and threonin. Enantiomorph identification is made at ambient temperature by making real-time video recording without cooling the specimens, which is generally believed to be necessary for reducing electron illumination damage for organic crystals. Although the proposed method usually requires only a single CBED pattern, an additional diffraction pattern is needed for enantiomorph identification of amino acids in order to exactly determine the crystallographic geometry of the relevant CBED pattern. This is due to the fact that the intensity of high-angle scattering from the organic crystal, which appears as HOLZ reflection disks, is too low to be imaged in the relevant CBED pattern even the specimen is cooled down to 10 K.

Acknowledgements
This work was supported by Grant-in-Aid for Scientific Research (A) (No. 18206074) from the Ministry of Education, Culture, Sports, Science and Technology (MEXT), Japan and in part by the Global COE (Center of Excellence) Program of International Center for Integrated Research and Advanced Education in Materials Science from the MEXT, Japan.

References
[1] Hahn T (ed) 1996 *International Tables for Crystallography, Volume A: Space-Group Symmetry, 4th revised ed.* (Dordrecht: The International Union of Crystallography by Kluwer Academic Press)
[2] Jacques J, Collet A, Wilen S H 1994 *Enantiomers, Racemates, and Resolution* (Malabar: Krieger)
[3] Crossley R 1995 *Chirality and Biological Activity of Drugs* (London: CRC)
[4] Wade L G JR 1987 *Organic Chemistry* (Englewood Cliffs: Prentice-Hall)
[5] Drenth J 1994 *Principles of Protain X-Ray Crystallography* (Berlin: Springer-Verlag)
[6] Inui H, Fujii A, Tanaka K, Sakamoto H, Ishizuka K 2003 *Acta Cryst.***B59** 802
[7] Inui H, Fujii A, Hashimoto H, Tanaka K, Yamaguchi M, Ishizuka K 2003 *Acta Mater.* **51** 2285
[8] Sakamoto H, Fujii A, Inui H, Tanaka K, Yamaguchi M, Ishizuka K 2003 *Mater. Sci. Forum* **426-432** 1783
[9] Sakamoto H, Fujii A, Tanaka K, Inui H 2005 *Acta Mater.* **53** 41
[10] Fujii A, Sakamoto H, Tanaka K, Inui H 2006 *Intermetallics* **15** 154
[11] Fujio S, Tanaka K, Inui H 2007 *Intermetallics* **15** 245
[12] Inui H, Fujii A, Sakamoto H, Fujio S, Tanaka K 2006 *J. Appl. Crystallogr.* **40** 241
[13] Inui H, Tanaka K, Kishida K, Fujio S 2007 *Mater. Sci. Forum* **539-543** 1457
[14] Reimer L 1984 *Transmission Erectron Microscpoy* (Berlin: Springer-Verlag)
[15] Fujio S, Tanaka K, Inui H, Ueji R, Sumida N, Yamochi H, Saito G, J. Appl. Crystallogr. (submitted)
[16] Ishizuka K 1998 in: M. Shiojiri and N. Nishio (Eds.), *Proc. Int. Symp. on Hybrid Analyses for Functio nal Nanostructure*, (Tokyo: Japanese Society of Electron Microscopy) p 69
[17] The Cambridge Structural Database: www.ccdc.cam.ac.uk

Excitation efficiency and saturation dynamics of near-infrared emission from Si nanocrystals embedded in a SiO_2-matrix

D Timmerman, W D A M de Boer, and T Gregorkiewicz

Van der Waals – Zeeman Institute, University of Amsterdam, Valckenierstraat 65, NL-1018 XE Amsterdam, The Netherlands

d.timmerman@uva.nl

Abstract. We have investigated photoluminescence (PL) and linear and induced absorption of Si nanocrystals (NCs) in a SiO_2 matrix. By measuring PL intensity dependence on the excitation photon fluence, we conclude that (i) the excitation wavelength independent saturation level is reached when on average a single photon is absorbed per NC, and (ii) excitation cross-section is proportional to the linear absorption only in the low-energy range; for higher energy it increases faster than absorption, thus indicating more efficient excitation which appears due to space-separated quantum cutting (SSQC). In order to shed more light on the mechanism behind the SSQC, we have compared spectral and temporal characteristics of induced absorption for excitation wavelengths below and above the SSQC threshold. These were found to be very similar, thus indicating that the SSQC process is most likely very fast, possibly taking place already during photon absorption by NCs.

1. Introduction

1.1. Si nanocrystals

Light generation in bulk Si is very inefficient. Its indirect bandgap leads to very long radiative lifetimes of excited electron-hole pairs and consequent domination of nonradiative recombination channels. In Si nanocrystals (NCs), radiative recombination is enhanced due to quantum confinement effect which relaxes the $\boldsymbol{k}$-conservation rule. Furthermore, the energy structure of Si NCs can be altered by changing their size; decreasing the size of NCs will increase both the bandgap and the spacing between the discrete energy levels. Therefore a solid-state dispersion of Si-NCs in SiO_2 matrix is an interesting material system whose optical properties can be tuned over a large energy range.

1.2. Cooperative carrier generation processes

One more consequence which can be expected due to the quantum confinement-induced level discretization is suppression of phonon cooling of excited carriers – the effect known as the *phonon bottleneck*. Competing cooling processes can in this case play a more prominent role. One of these is the inverse Auger recombination, where a high-energy electron/hole generates an electron-hole pair. This effect of impact excitation is well-known in bulk semiconductors [1], where it has low probability. It has been reported to take place also in different types of semiconductor NCs – PbS, PbSe, InAs and CdSe [2–4], and, recently, in a colloidal suspension of Si NCs [5]. In case of NCs, such a process results in generation of two (or more) electron-hole pairs in a NC upon absorption of a single photon of sufficient energy and is termed *multiple exciton generation* (MEG) or *carrier multiplication* (CM). The lifetime of multiple excitons in a single NC is very short (~ps), as wavefunction overlap in confined environment leads to strong carrier-carrier interactions and efficient Auger recombination. Consequently, regardless of how many excitons are generated within a NC, only one of them – the "last" remaining electron-hole pair – will have a sufficiently long lifetime to allow for its radiative recombination.

Recently, we have shown that in closely packed ensembles of Si NCs, a variation of the MEG process appears, in which a hot carrier cools by generating an electron-hole pair in a neighboring NC – the process termed *space-separated quantum cutting* (SSQC) [6]. In this case, the two electron-hole pairs formed by absorption of a single high-energy photon are localized in different NCs. Therefore their mutual interaction is strongly reduced, thus increasing their lifetimes and enhancing chance for radiative recombination and emission of two photons.

1.3. Carrier dynamics in Si nanocrystals

Presently occurrence of MEG is indirectly inferred from dynamics of free carrier absorption under (extremely) low power excitation, so as to exclude the possibility of multi-photon absorption within a single NC. In order to understand physical mechanisms of cooperative carrier generation processes, detailed knowledge of carrier dynamics in nanocrystalline environment is imperative. This is presently not the case; while in bulk Si these processes have been thoroughly investigated and satisfactory understanding has been obtained, carrier relaxation and recombination in NCs is still poorly known. Very specifically, this concerns the afore-mentioned suppression of the phonon-bottleneck effect. As a result, the debate on the MEG process continues [7]. Here the existing reports claim that the generation of the multiple carriers is very fast and either proceeds instantaneously during the absorption process itself [8] or during the first ~100 fs after photon absorption [9]. Consequently, different scenarios are considered, in which the high-energy photon either transfers its energy directly to two separate carriers by formation of a virtual state or creates a "hot" exciton whose cooling takes place via efficient impact excitation of another carrier. In case of Si NCs, yet additional complication to this process is introduced by indirect nature of the bandgap. It is fair to say that the actual mechanism of the generation of the multiple carriers by a single photon is still not understood. Elucidation of this process constitutes the major motivation of this study.

2. Experimental

2.1. Samples

The reported investigations have been performed on thin films (~micrometer) of Si NCs embedded in a matrix of SiO_2. These have been prepared by a radio-frequency co-sputtering method. SiO_2 and Si targets were simultaneously sputtered on a fused quartz substrate. Post annealing in N_2 ambient at 1100°C was subsequently conducted in order to segregate this substoichiometric mixture into Si NCs and SiO_2. In this method, size of Si NCs can be tuned by varying the amount of silicon during sputtering and, to certain extent, by the annealing temperature. The actual size of the produced samples was determined by high resolution transmission electron microscopy. The results presented in this report have been obtained for Si NCs of 3.1 nm diameter with a size distribution of ~14%.

2.2. Experimental setups

In this study photoluminescence (PL) and linear and (time-resolved) induced absorption (IA) have been investigated.

The setup used for PL measurements consisted of a tunable optical parametric oscillator pumped by the third harmonic of a pulsed Nd:YAG laser (354 nm) operating at a repetition rate of 10 Hz and with a pulse length of 5 ns. The PL spectra were resolved with a Jobin-Yvon TRIAX 320 monochromator and detected with a Hamamatsu R5509-72 photomultiplier. The illuminated area of the sample was kept constant by use of a pinhole, in order to get a flat profile of the laser beam and to ensure identical excitation spot in all measurements.

The induced absorption experiment was performed using a femtosecond pump-probe configuration, implemented with two Ti:Sapphire pumped OPAs for a tunable pumping wavelength. The excitation pulse was split in a primary pump pulse and a secondary probe exploring properties of the excited state. The probe pulse was passed though a white light convertor and mechanically delayed in respect to the primary pulse by means of an optical delay line thus allowing for time-resolved investigations. The detected induced absorption signal $I_{detection}$ was then corrected as follows: $I_{detection} = (I_{pump+probe} - I_{.probe})/I_{.probe}$, with $I_{pump+probe}$ and I_{probe} corresponding to absorptions of the probe beam under and without the influence of the pump, respectively. Experimental resolution was ~200 fs.

The linear absorption of the samples was determined with a Varian-50 UV-VIS spectrophotometer, for wavelengths between 200-1100 nm.

All the measurements discussed in this report were obtained at room temperature.

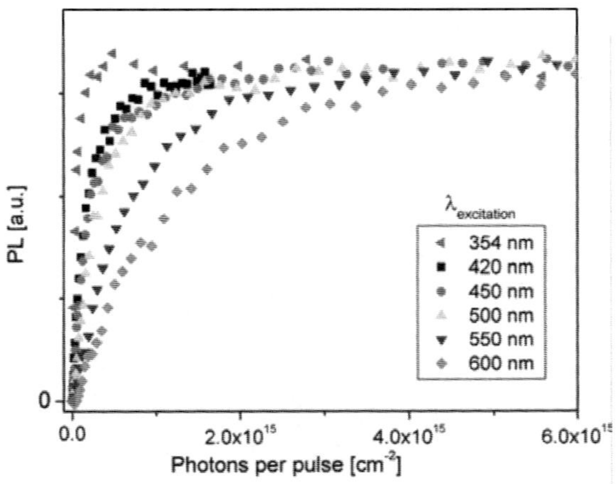

Figure 1. Intensity dependence of NC-related PL on the number of pump photons per pulse (fluence) for different excitation wavelengths. The slope of the initial part, related to the excitation cross-section, becomes steeper for shorter wavelengths, while the saturation level remains identical.

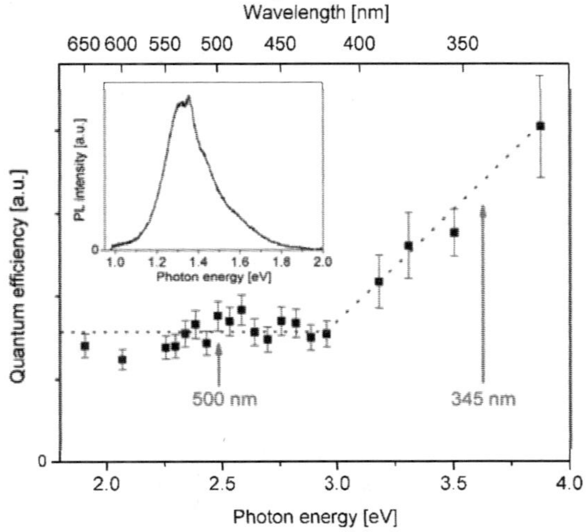

Figure 2. Relative quantum efficiency (see text for explanation) as function of the excitation photon energy. The quantum efficiency is constant up to photon energy of ~ 3 eV and the increases, indicating a second excitation mechanism. The dashed line acts as a guide to the eye. The arrows mark excitation wavelengths used in IA experiments. In the inset, PL spectrum of the NCs under study.

3. Experimental results

3.1. Excitonic photoluminescence

In the initial experiments, we have investigated intensity of PL due to exciton recombination. Figure 1 shows the results as obtained for different excitation wavelengths ranging from 354 to 650 nm. It can be concluded that (i) the saturation level is independent of excitation wavelength while (ii) the actual photon fluence at which PL intensity saturates decreases for shorter excitation wavelengths. The latter effect indicates increase of the excitation cross-section of exciton-related PL, which can be determined from the initial slope of intensity dependence, in the linear regime. In order to check whether this effect can be attributed to the wavelength dependence of absorption cross-section of Si NCs, the absorbance of the sample has been independently measured in a separate experiment. In figure 2 we plot the ratio of the excitation cross-section and absorbance, termed the *relative quantum efficiency*, as a function of quantum energy of the incoming photons. As can be seen, this parameter remains constant for low quantum energies until a threshold value of $E_X \approx 3$ eV, and shows strong increase for more energetic photons.

3.2. Induced absorption

Induced absorption measurements have been performed at two pump wavelengths of $\lambda_{pump} = 500$ nm (2.5 eV) and $\lambda_{pump} = 345$ nm (3.6 eV), i.e. below and above the energy threshold value E_X for the quantum efficiency increase, as indicated with arrows in figure 2. Figure 3 illustrates the spectral dependence of the induced absorption as obtained with a fixed delay of $\Delta t = 20$ ps for the visible and the infrared probe range at both pump settings. At both pumping wavelengths, the photon fluxes have been chosen such as to assure low and equal average number of electron-hole pair per NC: $\langle N_{e\text{-}h}\rangle_{345\ nm} \approx \langle N_{e\text{-}h}\rangle_{500\ nm} < 1$.

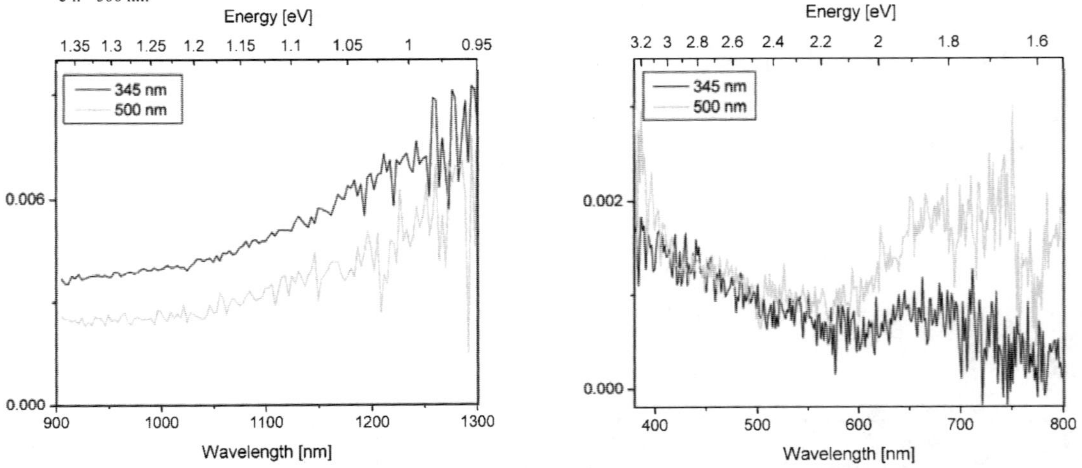

Figure 3. IA results for 345 nm and 500 nm pumping for a time delay of 20 ps between pump and probe pulses. Two probe ranges are shown: (a) the visible (380-800 nm) and (b) NIR (900-1300 nm). Excitation conditions are such that $\langle N_{e\text{-}h}\rangle_{345\ nm} \approx \langle N_{e\text{-}h}\rangle_{500\ nm} < 1$. Only minor differences appear.

Also dynamics of IA have been measured under the same experimental conditions, i.e. for $\lambda_{pump} = 345$ nm and $\lambda_{pump} = 500$ nm and the visible and near-IR probe ranges. The results are depicted in figure 4 for the time window of the first 200 ps after the pump pulse. As can be seen, apart for the intensity difference which can be inferred from figure 3, the measured IA dynamics are quite similar for both pump values.

4. Discussion

The investigated PL emission – see the spectrum depicted in the inset to figure 2 – originates from exciton recombination in Si NCs. Therefore saturation of this PL band points to the limited number of excitons participating in that emission. Since there appear to be no fundamental limits to generation of electron-hole pairs in NCs as the incoming photon flux increases, the fact that PL intensity saturates clearly indicates that only a limited number of generated excitons can recombine radiatively. Moreover, that number seems to be independent of the photon energy, while it is well known that absorption probability of photons increases with their quantum energy. (We also note that the number of illuminated NCs is kept constant by a pinhole). Such a situation is indeed to be expected since, as discussed in the introduction, multiple excitons localized in the same NC undergo very efficient nonradiative recombination. Therefore, when excitation exposure is shorter than the radiative re-combination, only a single photon might be emitted by excited NC regardless of the actual number of electron-hole pairs which have been generated in it. Since in the present case laser pulses of ~5 ns are used while the exciton radiative recombination time is in the microsecond range, this PL intensity limitation indeed can be expected. In order to investigate that, we have estimated the amount of NCs that have been excited after a laser pulse over the excitation power range using a Poissonian distribution

$$N_x = N_{NC} \times \frac{< NC^* >^x}{x!} e^{-<NC^*>},$$

where x – indicates the number of photons absorbed per NC, $<NC^*>$ is the average number of photons absorbed per NC, and N_{NC} is the amount of NCs in the illuminated volume. We have also verified that kinetics of PL did not depend on the excitation wavelength. Taken together, these results show that the saturation of the PL signal sets in at the moment that almost all of the NCs have absorbed at least one photon. This implies that each NC contributes a maximum of one photon to the PL signal, confirming validity of the earlier presented reasoning.

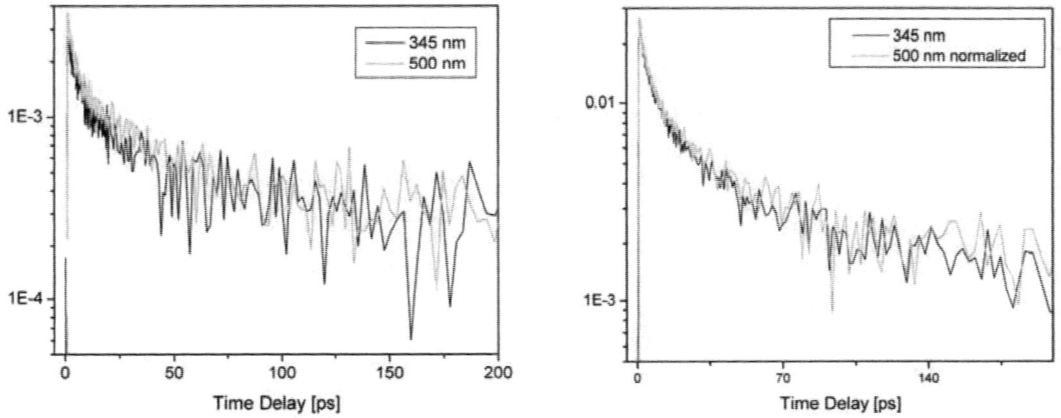

Figure 4. Normalized IA dynamics for 345 nm and 500 nm pumping for two different probing wavelengths of 530 nm (a) and 1200 nm (b). Temporal dependence appears to be independent of pump energy.

In contrast to the saturation level, the excitation cross-section of excitonic PL clearly increases for shorter wavelengths. This is to be expected, since absorption cross-section of bulk Si for higher energy photons is larger and the same can be expected also for Si NCs. However, the results depicted in figure 2 clearly show that this effect accounts for the increase of excitation cross-section in the low photon energy range: the relative quantum efficiency, which simply reflects the number-of-photons-in to the number-of-photons-out ratio, is indeed constant but only until the threshold value of $E_X \approx 3$ eV, thus reflecting linear relation between the two. For photon energy larger than ~ 3 eV the quantum efficiency increases. This effect indicates onset of a different excitation mechanism which starts to operate in parallel to the standard one. It has been assigned to the quantum cutting process in which an additional exciton can be generated at the expense of kinetic energy of a hot carrier [6]. As can be seen, the efficiency of this new process increases steeply with photon energy: for the shortest wavelength used in the current study, the total quantum efficiency is almost tripled in comparison with that attained by the "standard" excitation process. This illustrates the potential of the SSQC process of multiple carrier generation and gives good prospects for its photovoltaic, and possibly also photonic, applications.

As discussed in the introductory section, one can expect that IA could shed light on the microscopic mechanism behind the SSQC effect. This is based on the past experience in other materials, where practically all the information available on MEG has been obtained from IA investigations [2–5, 7]. We therefore compare results of IA experiments performed under conditions which do not allow for the SSQC process ($\lambda_{pump} = 500$ nm, $h\nu < E_X$) with these for which the SSQC process can be expected ($\lambda_{pump} = 345$ nm, $h\nu > E_X$). We note that in both cases the excitation fluence has been chosen so as to assure equal and small number of electron-hole pairs to be excited per NC $\langle NC^* \rangle < 1$, in order to assure that absorption of multiple photons by a single NC, which could preclude observation of SSQC related effects, can be excluded.

Figure 3 presents the IA spectra as obtained with the two excitation wavelengths for the visible (figure 3a) and the infrared (figure 3b) probe ranges. In general, bands in an absorption spectrum are usually related to specific transitions. In the current case, these could be intra-band transitions and, for sufficiently large pump quantum energies, transitions from NC- to SiO_2-related levels. Different excitation photon energies should enable absorption in different excited state levels. Consequently, the IA spectrum is likely to change depending on pump wavelength. As can be concluded, some differences can only be seen in the visible, while the spectra obtained for the infrared are identical. On the other hand, some structure in absorption spectrum can appear due to interference caused by internal reflections within the optically active layer. Such interference "fringes" have indeed been observed in the linear absorption spectrum of the investigated samples in the visible. Therefore we conclude that IA spectra are quite similar under sub- and sup-threshold pumping which indicates ultrafast cooling of carriers within their bands: regardless of their original energies defined by different pumping wavelengths, within the rise time of the signal the carriers cool to the (lowest) state from which similar absorptions are enabled by the probe.

From the above observation, an important conclusion can be drawn. The similarity of IA spectra measured for different excitation photon energies implies cooling on a time scale similar to (or faster than) the setup resolution (~200 fs). This in spite of the theoretically predicted retardation of phonon-assisted energy dissipation in the quantum confined regime. This conclusion is also confirmed when IA transients taken under the two pumping conditions are compared – see figure 4: transients for different pump photon energies reveal identical dynamical behavior and also similar amplitude. This implies that the same relaxation and recombination processes are observed, leading to an identical initial situation for absorption probing.

5. Conclusions

In the presented study, we conclude on the very fast (< 200 fs) thermalization of carriers excited by photon absorption in Si nanocrystals. This implies that the SSQC process inferred from excitation cross-section of excitonic PL is likely to occur on a similar time scale, or faster. This gives preference to theoretical models which suggest the MEG process to take place during photon absorption. Nevertheless, it is only fair to mention that the above interpretation of experimental results should be considered very critically; current conclusions on the ultrafast carrier cooling should be verified using other experimental approaches.

References

[1] Bude J, and Hess K 1992 *J. Appl. Phys.* **79** 2473

[2] Ellingson R, Beard M C, Johnson J C, Yu P, Micic O I, Nozik A J, Shabaev A and Efros A L 2005 *Nanolett.* **5** 865-871

[3] Pijpers J J H, Hendry E, Milder M T W, Fanciulli R, Savolainen J, Herek J L, Vanmaekelbergh D, Ruhman S, Mocatta D, Oron D, Aharoni A, Banin U and Bonn M 2007 *J. Phys. Chem. C* **111** 4146–4152

[4] Schaller R D, Petruska M A and Klimov V I 2005 *Appl. Phys. Lett.* **87** 253102

[5] Beard M C, Knutsen K P, Yu P, Luther J M, Song Q, Metzger W K, Ellingson R J and Nozik A J 2007 *Nanolett.* **7** 2506-2512

[6] Timmerman D, Izeddin I, Stallinga P, Yassievich I N and Gregorkiewicz T 2008 *Nature Photonics* **2** 105 - 109

[7] Trinh T M, Houtepen A J, Schins J M, Hanrath T, Piris J, Knulst W, Goossens A P L M and Siebbeles L D A 2008 *Nano Lett.* **8** 1713–1718

[8] Schaller R D, Agranovich V M and Klimov V I 2005 *Nature Physics* **1** 189-194

[9] Schaller R D, Sykora M, Pietryga J M and Klimov V I 2006 *Nano Lett.* **6** 424-429

Femtosecond laser-driven shock synthesis of hexagonal diamond from highly oriented pyrolytic graphite

Tomokazu Sano[1], Kengo Takahashi[1], Osami Sakata[2], Masayuki Okoshi[3], Narumi Inoue[3], Kojiro F. Kobayashi[4], and Akio Hirose[1]

[1] Divison of Materials and Manufacturing Science, Graduate School Engineering, Osaka University, 2-1 Yamada-Oka, Suita, 565-0871 Osaka, Japan
[2] Japan Synchrotron Radiation Research Institute / SPring-8, 1-1-1 Kouto, Sayo-cho, Sayo-gun, Hyogo 679-5198, Japan
[3] National Defence Academy of Japan, 1-10-20 Hashirimizu, Yokosuka, Kanagawa 239-8686, Japan
[4] Fukui University of Technology, 3-6-1 Gakuen, Fukui 910-8505, Japan

Email: sano@mapse.eng.osaka-u.ac.jp

Abstract. We synthesized hexagonal diamond directly from highly oriented pyrolytic graphite (HOPG) using a femtosecond laser pulse without catalyst. A femtosecond laser pulse with wavelength of 800 nm, pulse width of 130 fs, the intensity of 2×10^{15} W/cm^2 was irradiated onto the HOPG surface in air. Crystalline structures of the fs laser-affected region in the HOPG were analyzed using grazing-incidence XRD method. We found that the hexagonal diamond which is the metastable high-pressure phase of carbon appeared in the HOPG which was irradiated by the femtosecond laser normal to the basal plane. We suggest that the femtosecond laser-driven shock wave induces the graphite - hexagonal diamond transformation and that the hexagonal diamond is synthesized due to the rapidly cooling in the shock heated region.

1. Introduction

A stable phase of carbon is graphite in the lower pressure regime and cubic diamond in the higher pressure regime [1]. The cubic diamond is one of the most important materials in the field of the manufacturing industry due to its high hardness and thermal conductivity. Hexagonal diamond was discovered in the Canyon Diablo iron meteorite [2] and is considered the metastable high-pressure phase of carbon. The hexagonal diamond is synthesized artificially using the shock compression method from graphite with catalyst [3] and from cubic diamond without catalyst [4], the static compression method [5], and the chemical vapor deposition method [6]. There still exist many unresolved subjects on the synthesis mechanism of cubic and hexagonal diamonds. Metallization of carbon is predicted theoretically under more than 1.1 TPa [7]. The metallic carbon is the lightest metal on earth if we can synthesize and retain it under the atmospheric pressure. Therefore, exploration of the high-pressure polymorphic carbon is significantly important for not only the industry but the materials and planetary sciences.

A solid material transits directly to the plasma, which expands to the vacuum explosively, after an intense femtosecond laser pulse is irradiated to the solid. The shock wave is driven by the recoil pressure during the plasma expansion and propagates into the solid. There is no interaction of the femtosecond laser pulse with the expanding plasma because the plasma expansion occurs after the whole femtosecond laser pulse injection, which is the definitively different point from the nanosecond laser pulse irradiation. The expansion velocity of the plasma is faster than that of the nanosecond laser induced plasma. Therefore, the disturbance due to heating by the radiation from the plasma is negligible. The femtosecond laser-driven shock wave is used for the investigation of the materials dynamics under pressure [8]. A high pressure ε phase of iron, which has not been synthesized using conventional compression methods, was synthesized using the femtosecond laser-driven shock wave [9,10]. The femtosecond laser-driven shock wave is the potential tool to synthesize high-pressure materials.

The purpose of this study is to investigate synthesized structures of carbon after the femtosecond laser irradiation to the graphite. We expect that diamond will be synthesized if the strong shock wave is driven. But we do not expect whether the structure of the diamond is cubic or hexagonal because the rapidly cooling effect is significant for the femtosecond laser-driven shock wave compression compared to the conventional compression methods such as the flyer impact method and the nanosecond laser shock method.

2. Experimental method

The target used in this study was the highly oriented pyrolytic graphite (HOPG, NT-MDT Co., Russia, 10 mm × 10 mm × *t* 1.5 mm) with the ZYA quality, the mosaic spread of 0.4 – 0.7 deg, and the mass density of 2.267 g/cm^3. A femtosecond laser pulse (Spitfire, Spectra-Physics Inc.) with wavelength of 800 nm, pulse width of 130 fs, and spatial energy distribution of the near Gaussian firstly passed through an aperture of 6 mm diameter, and secondly was focused onto a target surface in air using a plano-convex lens with the focal length of 70 mm. The femtosecond laser pulse was irradiated normally to the basal plane of the HOPG target. A single pulse was shot at a point, and the shot to shot interval was 100 μm on the surface. Two kinds of the laser pulse energy, 0.7 and 5.5 mJ, was used. The spatially averaged laser intensity was approximately 3×10^{14} and 2×10^{15} W/cm^2 for the laser pulse energy of 0.7 and 5.5mJ, respectively, because the laser spot size was approximately 50 μm. Crystalline structures were analyzed using the grazing incident XRD method at the BL13XU in the SPring-8 [11]. The synchrotron X-ray with the wavelength of 1.000 Å passed through a slit with the horizontal width of 410 μm and the vertical width of 50 μm, and was irradiated to the sample. The incident angle α was fixed and the detector angle δ was varied. A solar slit was located in front of the detector, which is the YAP scintillation counter.

3. Results and discussion

The laser irradiated area was explosively removed and the deep crater was created for the laser pulse energy of 5.5 mJ, while the significant crater was not created for the laser pulse energy of 0.7 mJ. The deep crater created explosively indicates that the explosive ablation occurs and that the strong shock wave was driven.

Grazing incidence XRD patterns are shown in Fig. 1 for the laser pulse energy of (a) 0.7 and (b) 5.5 mJ. The incident angle of the X-ray to the sample surface was 0.1 deg, the angular resolution of the detector was 0.02 deg/step, and the exposure time was 5.0 s/step. The critical angle of carbon for the light wavelength of 1.000 Å is 0.13 deg. Therefore, there is no scattered X-ray if the HOPG surface is perfectly planar. However, the laser irradiated region is roughened so that the X-ray is scattered at the region. Peaks of (100) and (101) planes of the graphite appear after the 0.7 mJ laser pulse irradiation as shown in Fig. 2 (a). This means that the plastic deformation induced by the femtosecond laser-driven shock wave collapsed the orientation in the laser irradiated region. We found peaks of not only (100), (101), and (102) planes of graphite but (002) and (101) planes of hexagonal diamond appeared after the 5.5 mJ laser pulse irradiation. This means that the laser pulse of 5.5 mJ induced the shock wave, which is strong enough to induce the graphite - diamond transition.

Grazing incidence XRD patterns of the HOPG which was irradiated by the laser pulse of 5.5 mJ are shown in Fig. 2 for the incident angle 0.1, 0.2, and 0.3 deg. The X-ray penetration depth is 10 and 18 μm for the incident angle of 0.2 and 0.3 deg, respectively. The smaller incident angle shows the larger peak of the hexagonal diamond. This means that the induced hexagonal diamond is distributed mainly near the surface.

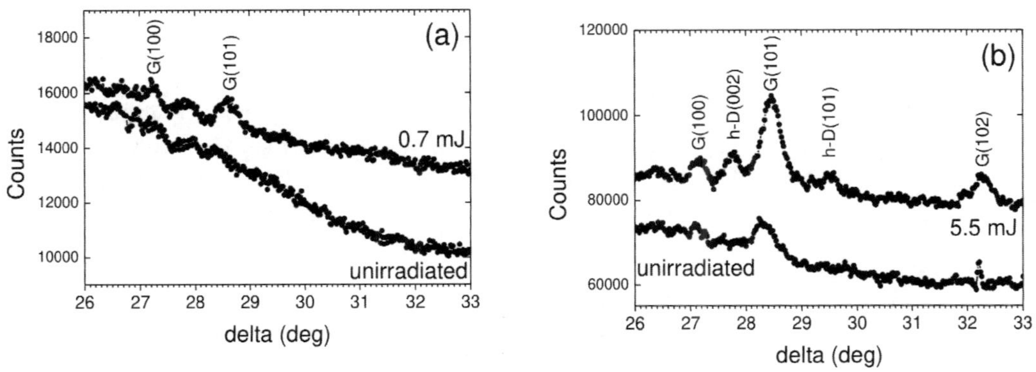

Figure 1. Grazing incidence XRD patterns of the femtosecond laser irradiated HOPG. The laser pulse energy: (a) 0.7 and (b) 5.5 mJ. G: Graphite, h-D: hexagonal diamond.

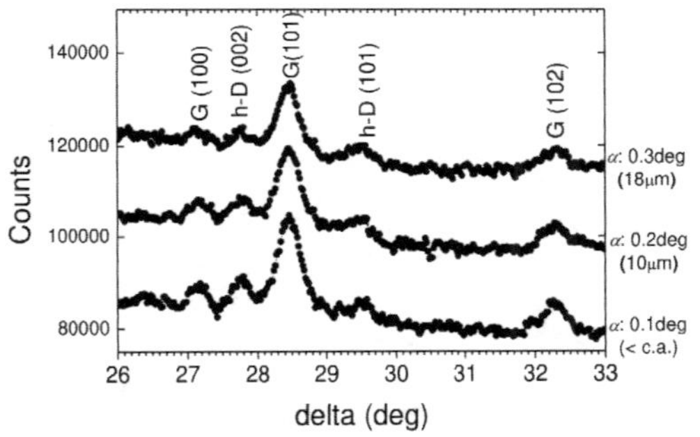

Figure 2. Grazing incidence XRD patterns for the incident angles of 0.1, 0.2, and 0.3 deg.

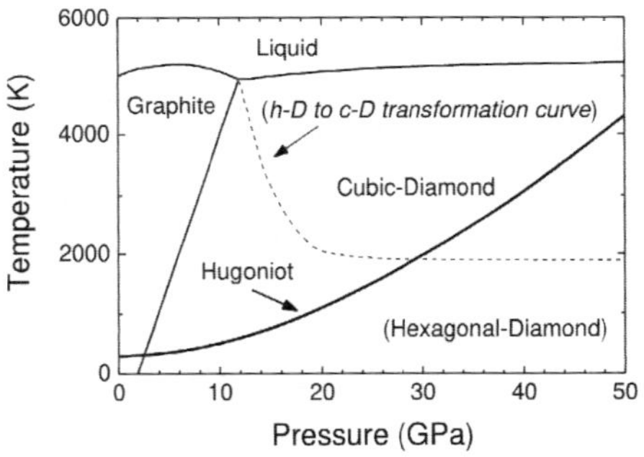

Figure 3. Calculated temperature-Hugoniot described on the carbon phase diagram.

We calculated the pressure-temperature curve under shock compression (temperature-Hugoniot) using the thermodynamic equation expressed below [12,13] in order to investigate the synthesis mechanism of hexagonal diamond;

$$C_{V\mathrm{H}}\frac{dT_{\mathrm{H}}}{dV_{\mathrm{H}}}+\frac{\gamma_0}{V_0}C_{V\mathrm{H}}T_{\mathrm{H}}=\frac{1}{2}\frac{dp_{\mathrm{H}}}{dV_{\mathrm{H}}}\left(V_0-V_{\mathrm{H}}\right)+\frac{1}{2}\left(p_{\mathrm{H}}-p_0\right), \tag{1}$$

where T is the temperature, p is the pressure, V is the specific volume, C_V is the constant-volume specific heat, γ is the Grüneisen parameter, subscript 0 is an ambient state, and subscript H is a shocked state. Constant values such as C_V and γ are referred to Nellis's data [14]. The calculated result is shown on the phase diagram of carbon [1] in Fig. 3. The Hugoniot curve passes through the hexagonal diamond to cubic diamond transformation curve at approximately 30 GPa. The femtosecond laser-driven shock pressure, which was estimated 100 - 300 GPa for the pulse intensity of around 10^{14} W/cm^2 [8], is strong enough to induce the graphite - hexagonal diamond - cubic diamond transformation. Mechanism of the transformation from graphite to hexagonal diamond is diffusionless and from hexagonal diamond to cubic diamond is diffusive. More than microseconds is required to induce the diffusive hexagonal to cubic diamond transformation in contrast to the relaxation time of picoseconds in case of the diffusionless graphite - hexagonal diamond transformation. Therefore, the femtosecond laser-driven shock compression with the duration of sub-nanoseconds [15] can only induce the graphite → hexagonal diamond transformation. We suggest that the recovery of the hexagonal diamond induced by the femtosecond laser shot is due to the rapidly cooling in the shock heated region [16].

4. Conclusions

We found that the hexagonal diamond which is the metastable high-pressure phase of carbon is synthesized in the HOPG, which was irradiated by the femtosecond laser normal to the basal plane, due to the shock induced graphite - hexagonal transformation. Measurements of both shock and particle velocities to estimate the shock pressure and duration are necessary to solve the detail mechanism.

Acknowledgements

This study was supported by Priority Assistance for the Formation of Worldwide Renowned Centers of Research - The Global COE Program (Project: Center of Excellence for Advanced Structural and Functional Materials Design). The synchrotron radiation experiments were performed at the BL13XU in the SPring-8 with the approval of the Japan Synchrotron Radiation Research Institute (Proposal Nos. 2006A1613, 2006B1553, and 2008B1160).

References

[1] F. P. Bundy, W. A. Bassett, M. S. Weathers, R. J. Hemley, H. K. Mao, and A. F. Voncharov, *Carbon* **34**, 141 (1996).
[2] C. Frondel and U. B. Marvin, *Nature* **214**, 587 (1967).
[3] L. F. Trueb, *J. Appl. Phys.* **39**, 4707 (1968).
[4] H. He, T. Sekine, and T. Kobayashi, *Appl. Phys. Lett.* **81**, 610 (2002).
[5] F. P. Bundy and J. S. Kasper, *J. Chem. Phys.* **46**, 3437 (1967).
[6] A. Misra, P. K. Tyagi, B. S. Yadav, P. Rai, D. S. Misra, V. Pancholi, and I. D. Samajdar, *Appl. Phys. Lett.* **89**, 071911 (2006).
[7] M. T. Yin and M. L. Cohen, *Phys. Rev. Lett.* **50**, 2006 (1983).
[8] R. Evans, A.D. Badger, F. Falliès, M. Mahdieh, T.A. Hall, P. Audeberg, J.-P. Geindre, J.-C. Gauthier, A. Mysyrowicz, G. Grillon, and A. Antonetti, *Phys. Rev. Lett.* **77**, 3359 (1996).
[9] T. Sano, H. Mori, E. Ohmura, and I. Miyamoto, *Appl. Phys. Lett.* **83**, 3498 (2003).
[10] T. Sano, H. Mori, O. Sakata, E. Ohmura, I. Miyamoto, A. Hirose, and K. F. Kobayashi, *Appl. Surf. Sci.* **247**, 571 (2005).
[11] O. Sakata, *et al.*, *Surf. Rev. Lett.* **10**, 543 (2003).
[12] J. M. Walsh and R.H. Christian, *Phys. Rev.* **97**, 1544 (1955).
[13] T. Sano and Y. Sano, *J. Appl. Phys.* **90**, 3754 (2001); *J. Appl. Phys.* **90**, 5576 (2001).
[14] W. J. Nellis, A. C. Mitchell, and A. K. McMahon, *J. Appl. Phys.* **90**, 696 (2001).
[15] S. D. McGrane, D. S. Moore, D. J. Funk, and R. L. Rabie, *Appl. Phys. Lett.* **80**, 3919 (2002).
[16] J. P. Colombier, P. Combis, R. Stoian, and E. Audouard, *Phys. Rev. B* **75**, 104105 (2007).

Journal of Physics: Conference Series **165** (2009) 012020 doi:10.1088/1742-6596/165/1/012020

Effects of static magnetic field and gas atmosphere on solidification of silicon by electromagnetic levitation

S Ueno[1], H Kobatake[2], H Fukuyama[2], S Awaji[3] and H Nakajima[1]
[1]The Institute of Scientific and Industrial Research, Osaka University, 8-1 Mihogaoka, Ibaraki, Osaka 567-0047, Japan
[2]Institute of Multidisciplinary Research for Advanced Materials, Tohoku University, 2-1-1 Katahira, Aoba-ku, Sendai, 980-8577, Japan
[3]Institute for Materials Research, Tohoku University, 2-1-1 Katahira, Aoba-ku, Sendai, 980-8577, Japan

E-mail: ueno23@sanken.osaka-u.ac.jp

Abstract. The effects of static magnetic field and hydrogen gas atmosphere on the solidification of silicon with hydrogen atom were examined using electromagnetic levitation method in hydrogen atmosphere. Since the convection in the levitated silicon melt which is caused by the electromagnetic field is restrained by a high static magnetic field, it is possible to examine a solidification of silicon from an equilibrium silicon melt with hydrogen without an inflluence of convection. The cooling rate of the melt increases with increasing hydrogen partial pressure. Since the undercooling of the melt does not change with the static magnetic field, the grain size on the surface does not change with the static magnetic field. However, the morphology of the grain surface drastically changed from flat and smooth to undulation.

1. Introduction

A containerless process such as electromagnetic levitation is useful to study of rapid solidification from the melt [1,2]. When a high static magnetic field is applied for the levitation process and the convection in the levitated melt which is caused by electromagnetic field is restrained by strong Lorentz force from the static magnetic field, the convection in the melt can be ignored, namely, the melt can deal with a rigid body [3]. A rigid body consisted of liquid phase is convenient to measure the thermophysical properties of the melt [4,5] and the solidification from a rigid body is suitable to examine an effect of convection on the solidified structure. Yasuda et al. suggest that the microstructure of carbon steel changes from equiaxed grain to columnar grain by static magnetic field application from 0 to 10 T because of inhibition the fragmentation of dendrite arm by the convection during the solidification [6]. When silicon is solidified by electromagnetic levitation method in high static magnetic field in hydrogen atmosphere, it is expected that an equilibrium silicon melt with hydrogen except an effect of convection in the melt is obtained.

In this work, the solidification of silicon was performed by electromagnetic levitation method at high static magnetic field of 0-5 T in 25%H_2-75%Ar and 50%H_2-50%Ar mixed gas in order to examine the effects of static magnetic field and hydrogen atmosphere on the solidification of silicon with hydrogen atoms.

2. Experimental procedure

0.8 g of silicon with 99.999% purity was used for the solidification. Silicon bulk was put on a quartz dish in spiral induction heating coil and was heated by a semiconductor laser with wave length of 808 nm. Then, the silicon melt was levitated in the spiral induction coil as shown in Fig.1. When the silicon was completely melted, the temperature correction for the monochrome pyrometer was performed by adjusting the melting point of silicon 1685 K. Then, H_2-Ar mixed gas or Ar gas was introduced in the chamber up to 0.1 MPa. The melting condition kept at 1773 K for 0.6 ks. Then, the power of the spiral induction coil turns up to a constant value of 38 A to move the silicon melt sphere at a fixed position of upside the spiral induction coil for cooling the melt. After solidifying of the surface of the sphere, the sample was caught by the quartz dish and the power of the induction coil was switched off. The solidification was monitored by a CCD camera from top of the chamber.

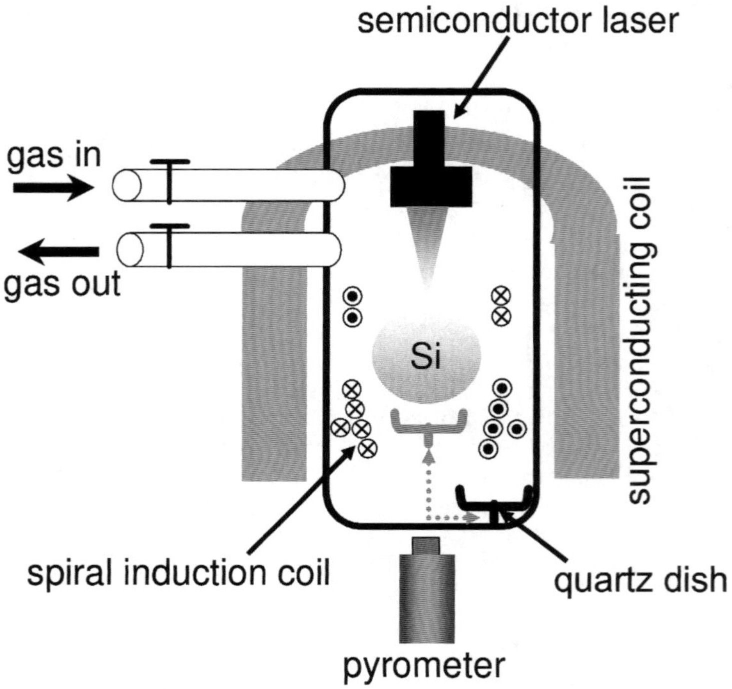

Figure 1 Apparatus for electromagnetic levitation furnace.

3. Result and discussion

Kobatake et al. suggested that the thermal conductivity of liquid silicon in static magnetic field of 2-4 T shows a constant value [4]; the convection in the levitated silicon melt due to the electromagnetic field is restrained by strong Lorentz force from the static magnetic field. Hence, in the present study, the solidification was performed under a static magnetic field of 5 T which is enough to restrain the convection in the melt. Since the hydrogen diffusion coefficient in solid silicon at the melting point is in order of 10^{-4} cm^2/s [7] and the diameter of melt sphere is about 0.8 cm, it is considered that the holding time of 0.6 ks is enough to make an equilibrium state of silicon melt with hydrogen. The temperature measurement error was within 10 K.

Figure 2 shows the cooling curves of the melts where, the static magnetic field is controlled to 5 T. The cooling rate of the melt for the case of the melting in 50%H_2-50%Ar is larger than that for the case of the melting in 25%H_2-75%Ar. Since the thermal conductivity of hydrogen gas is larger than that of argon gas, the cooling rate of the surface increased with increasing hydrogen partial pressure.

Figure 3 shows the dependence undercooling of the melts. The undercooling did not change with the static magnetic field. Yasuda et al. reported that the undercooling of copper alloy varies from 0K to 300K and does not depend on the static magnetic field [6]. On the other hand, the undercooling of pure copper melt varies from 0K to 100K [6]. In the present study, the undercooling of the melts varies less than 50 K and has a constant value with the static magnetic field. Hence, it can be seen that silicon melts with hydrogen atoms acts like a melt of pure metals.

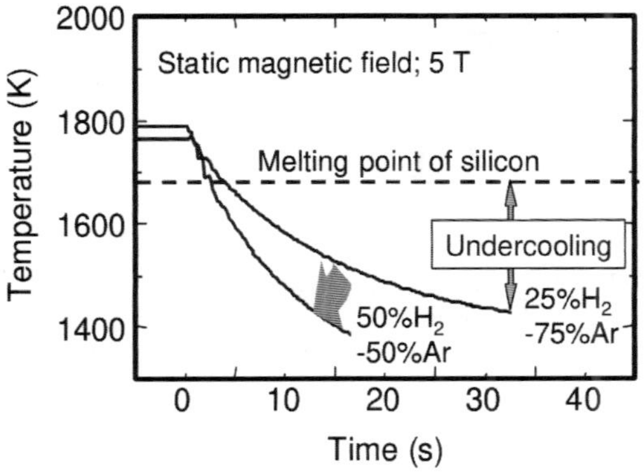

Fig.2 The cooling curves of the melts where, the static magnetic field is controlled to 5 T.

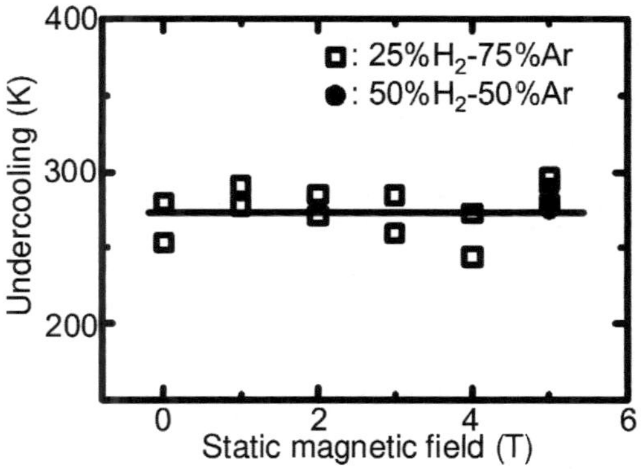

Fig.3 The dependence undercooling of the melts on the static magnetic field.

Since the undercooling of the melt does not change with the static magnetic field, size of the surface grains as shown in Fig.4 does not change with the static magnetic field as shown in Fig.5. However, the morphology of the grains surface changes with the static magnetic field from flat and smooth to undulation as shown in Fig.5. Small undulation is formed on individual grain surface for the sample solidified at 3 T. On the other hand, a facet crystal is appeared on the grain surface for the sample solidified at 5 T. This result suggests that the static magnetic field effects on the solidification of silicon, especially, on the nucleation of crystal.

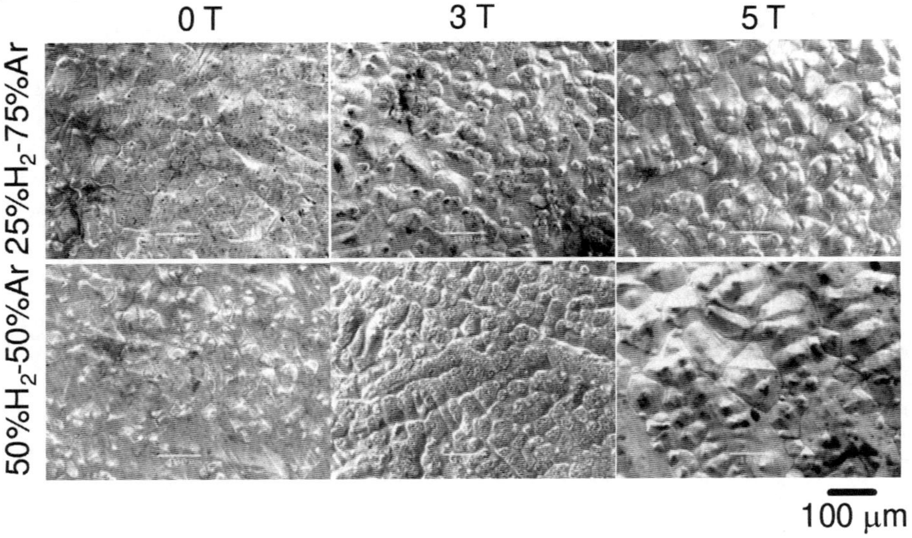

Fig.4 The morphology changes of the grains surface with the static magnetic field.

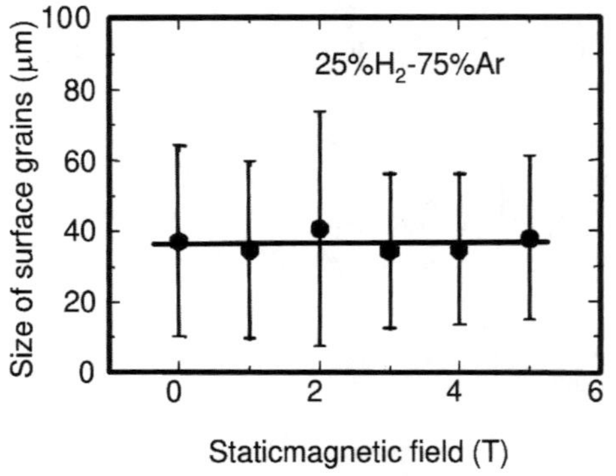

Fig.5 The dependence of size of surface grain on the static magnetic field.

4. Conclusion

Solidification from an equilibrium silicon melt with hydrogen was carried out by an electromagnetic levitation method under high static magnetic field. The cooling rate of the melt for the case of the melting in 50%H_2-50%Ar is larger than that for the case of the melting in 25%H_2-75%Ar. The static magnetic field affects on the solidification of silicon, especially, on the nucleation of crystal.

Acknowledgment

The present work was supported by Materials Science and Technology Research Center for Industrial Creation in Institute of Science and Collaborative Research Project with Institute of Materials Research, Tohoku University.

References

[1] Li D and Herlach DM 1997 *J. Mater. Sci.* **32** 1437
[2] Battersby SE, Cochrane RF and Mullis AM 1999 *J. Mater. Sci.* **34** 2049
[3] Yasuda H, Ohnaka I, Ninomiya Y, Ishii R, Fujita S and Kishio K 2004 *J. Cryst. Growth* **260** 475
[4] Kobatake H, Fukuyama H, Minato I, Tsukada T and Awaji S 2007 *Appl. Phys. Lett.* 90094102
[5] Tsukada T, Fukuyama H and Kobatake H 2007 *Int J. Heat Mass Tran.* **50** 3054
[6] Yasuda H, Ohnaka I, Ishii R, Fujita S and Tamura Y 2005 *ISIJ Int.* **45** 991
[7] Wieringen AV 1956 *Physica* **22** 849

Three-dimensional alignment of $FeSi_2$ with orthorhombic symmetry by an anisotropic magnetic field

Noriaki Nakatsuka, Hideyuki Yasuda, Tomoya Nagira, Masato Yoshiya

Department of Adaptive Machine Systems, Graduate School of Engineering, Osaka University, Suita, Osaka 565-0871, Japan

E-mail: yasuda@ams.eng.osaka-u.ac.jp

Abstract. This paper presents the three-dimensional alignment of β-$FeSi_2$ particles in a resin under an anisotropic magnetic field. The alignments obtained under a static magnetic field and a rotating magnetic field proved that the magnetic susceptibilities of β-$FeSi_2$ are $\chi_c>\chi_b>\chi_a$. The magnetic anisotropy was sufficiently large under a magnetic field of 10T for the crystallographic alignment. Under an anisotropic magnetic field (oscillating magnetic field), a-, b- and c-axes of β-$FeSi_2$ particles suspended in a resin were aligned each other. A pseudo single crystal β-$FeSi_2$ with the orthorhombic structure was fabricated by the oscillating magnetic field.

1. Introduction

Development of cryogen-free superconducting magnets enables us to use a relatively high magnetic field exceeding 2T for materials processing. One of the applications in materials processing is fabrication of the crystallographically aligned materials, in which a certain crystallographic orientation aligned each other. Para- and diamagnetic materials of which crystal symmetry is lower than the cubic symmetry have the magnetic anisotropy, while the ferromagnetic materials even with the cubic symmetry have the anisotropy. The uni-axial alignment under a magnetic field has been extensively studied [1-16]. Even in the coarsening process, the aligned structure was achieved [17,18]. Thus, the uni-axial alignment has been achieved for various materials.

Recently, the three-dimensional alignment, in which a-, b- and c-axes of grains were aligned each other, has been realized [19,20]. The magnetic susceptibilities of a-, b- and c-axes are different each other for a material with low symmetry like orthorhombic structure. An elliptic magnetic field, in which the direction of the magnetic field rotates with the modulated angular velocity, is used for the three-dimensional alignment. The time-average intensity of the elliptic magnetic field becomes anisotropic. When a crystal cannot follow the rotation, the crystal tends to rotate in accordance with the time-averaged magnetic field. As a result, the crystallographically aligned structure is formed in accordance with the magnetic anisotropy. Until now, the three dimensional alignment has been obtained only for L-alanine [20]. Although the mechanism was essentially explained in the previous study [19], the fundamental of the alignment is explained in the next section for describing the experimental procedures.

The β-$FeSi_2$ intermetailic compound with the orthorhombic symmetry has received considerable attention because of its high potential as optoelectronic and thermoelectric materials. Since β-$FeSi_2$ intermetailic compound is produced through the peritectoid reaction, it is rather difficult to produce a single crystal. The single crystals are desired for evaluating intrinsic physical properties and for improving material performance. In this paper, the three-dimensional alignment of β-$FeSi_2$ particles in a resin under an anisotropic magnetic field is presented as a preliminary study for fabricating a bulk single crystal by the magnetic alignment. In addition, the magnetic anisotropy of β-$FeSi_2$ is evaluated by using a static magnetic field and a rotating magnetic field.

2. Motion of magnetized particles under an anisotropic magnetic field

2.1. Crystallographic alignment under a static magnetic field

Magnetic susceptibilities of a single crystal are generally given by equation (1),

$$\chi = \begin{pmatrix} \chi_{11} & 0 & 0 \\ 0 & \chi_{22} & 0 \\ 0 & 0 & \chi_{33} \end{pmatrix} \tag{1}$$

The magnetic energy is also given by equation (2).

$$U = -\int_0^H \mathbf{M} \cdot d\mathbf{H} \tag{2}$$

The crystallographic alignment is described by considering motion and rotation of a crystal under a magnetic field. The crystals move and/or rotate in a favourable direction to minimize the magnetic energy.

In the case of a paramagnetic crystal with the cubic symmetry, the magnetic susceptibilities are $\chi_{11}= \chi_{22}= \chi_{33}$. Since the magnetic susceptibility is isotropic, the magnetization does not depend on crystal orientation. Thus, the crystallographical alignment is not obtained.

The magnetic susceptibilities becomes $\chi_{11}= \chi_{22}= \chi_a$, $\chi_{33}= \chi_c$ for a crystal with the hexagonal or the tetragonal symmetry. Under a static magnetic field, the magnetization for the crystal is expressed by

$${}^t\mathbf{M} = \begin{pmatrix} \chi_a & 0 & 0 \\ 0 & \chi_a & 0 \\ 0 & 0 & \chi_c \end{pmatrix} {}^t\mathbf{H}. \tag{3}$$

As shown in equation (3), the magnetization, **M**, is not always parallel to the magnetic field, **H**. The imposed magnetic field and the induced magnetization cause a torque on the crystal, given by equation (4).

$$\mathbf{T} = \mathbf{M} \times \mathbf{H} \tag{4}$$

In the case that χ_c is larger than χ_a, c-axis tends to be parallel to the magnetic field and a- and b-axes distribute in a plane perpendicular to the magnetic field. The imposition of a static magnetic field results in the uni-axial alignment in which the crystallographic orientation with the maximum magnetic susceptibility aligns each other.

2.2. Crystallographic alignment under a rotating magnetic field

In the case that χ_a is larger than χ_c for the hexagonal and the tetragonal symmetry, c-axis randomly distributes on a plane perpendicular to the magnetic field, because the magnetic susceptibility in any crystal orientation on the a-b plane is the same.

The uni-axial alignment can be achieved for the hexagonal or the tetragonal crystal with $\chi_a > \chi_c$ by using a rotating magnetic field. Here, one considers that a crystal with $\chi_a > \chi_c$ is suspended in a fluid under a rotating magnetic field with a constant angular velocity. When the angular velocity is sufficiently high, the crystal cannot follow the change of magnetic field direction. In this situation, the crystal rotates in accordance with the time-averaged magnetic field. Since the average intensity of the rotating magnetic field is constant on a plane perpendicular to the rotation axis, a- and b-axes randomly lie on the plane and c-axis is aligned along the rotation axis. Namely, the rotating magnetic field achieves the uni-axial alignment in which the crystallographic orientation with the minimum magnetic susceptibility aligned along the rotation axis.

2.3. Crystallographic alignment under a oscillating magnetic field

The magnetic susceptibilities of a crystal with the orthorhombic symmetry are

$$\chi = \begin{pmatrix} \chi_a & 0 & 0 \\ 0 & \chi_b & 0 \\ 0 & 0 & \chi_c \end{pmatrix}. \qquad (5)$$

Here, one consider the case of $\chi_a > \chi_b > \chi_c$. When the angular velocity of the rotating magnetic field is periodically modulated, the time-average magnetic field becomes elliptic. For example, the intensities of the time-averaged magnetic field in x-, y- and z- directions become $H_x > H_y > H_z$. If the angular velocity is sufficiently high, the crystal rotates in accordance with the time-average magnetic field. As a result, the three-dimensional alignment in which a-, b- and c-axes of crystals are aligned each other can be achieved by the elliptic magnetic field. The detail of the three-dimensional alignment has been studied [19].

3. Experimental procedure

3.1. Preparation of β-$FeSi_2$ particles

Mother alloy of 33.3at%Fe-66.7at%Si was made by arc melting. The ingot was annealed at 1173K for 24h to increase volume fraction of β-$FeSi_2$ intermetailic compound. After annealing, the other phases such as FeSi and $FeSi_2$ (high temperature phase) were observed only in the grain boundary region. The ingot was crashed mechanically into the particles (average diameter: 2.2 μm). The particle size was sufficiently smaller than the grain size in the ingot. Thus, most of the β-$FeSi_2$ particles consisted of a single grain of β-$FeSi_2$. The β-$FeSi_2$ particles were suspended in a resin. Volume fraction of β-$FeSi_2$ particles was 0.2%. The crystallographic alignment was examined by X-ray diffraction.

Since the magnetic susceptibility of the FeSi compound is much higher than that of β-$FeSi_2$, even small amount of the FeSi compound remaining in the crashed particles remarkably increases the apparent magnetic susceptibility of the specimen. However, most of the crashed particles did not contain the FeSi compound. The motion of the β-$FeSi_2$ particles is dominantly controlled by the magnetic property of β-$FeSi_2$. Thus, the overall trend in the alignment of the β-$FeSi_2$ particles is not influenced by the remaining FeSi compound.

3.2. Magnetic fields

Since the magnetic properties of β-$FeSi_2$ were not known, the magnetocrystalline anisotropy was examined by using a static magnetic field and a rotating magnetic field. As mentioned in the previous section, the orientation of the maximum magnetic susceptibility can be identified by imposing a static magnetic field. The static magnetic field of 10T was horizontally imposed. The resin was cured under the static magnetic field by adding a curing agent. The curing specimen was placed for more than 6h under a magnetic field.

The orientation of the minimum magnetic susceptibility can be identified by using a rotating magnetic field. In this study, the specimen was rotated at the constant angular velocity (9rpm). As the resin cured, the viscosity of resin remarkably increased. The β-$FeSi_2$ particles did not follow the change of the magnetic field direction and consequently the rotation was controlled by the time-averaged magnetic field.

Figure 1 shows a schematic illustration of the elliptic magnetic field used in this study (It is referred to as the oscillating magnetic field). The rotating direction was periodically changed. The rotating speed was 9 rpm and the oscillation angle was 30 degrees.

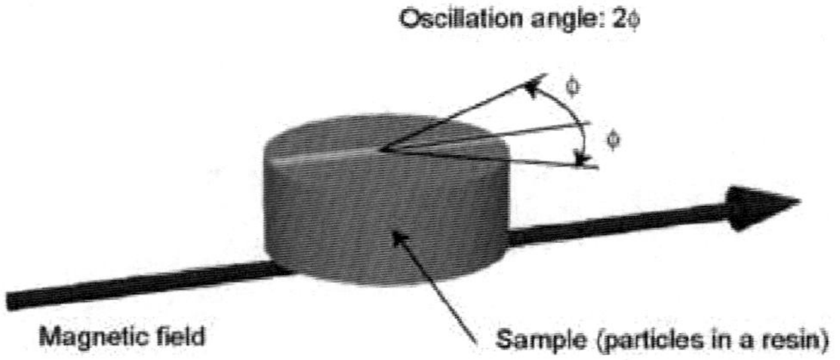

Figure 1. Oscillating magnetic by changing the rotation direction periodically.

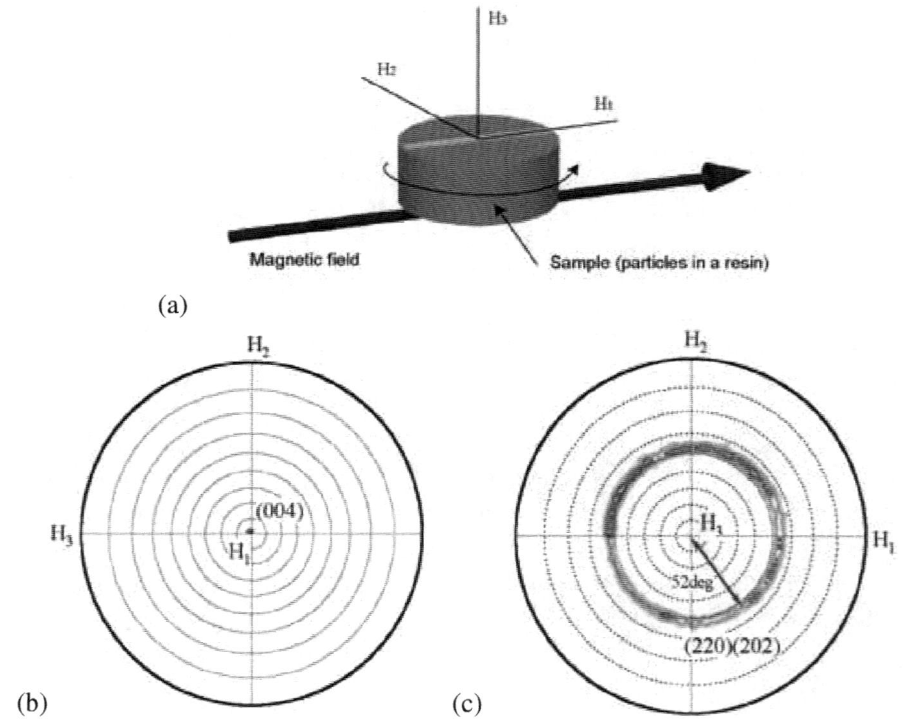

Figure 2. (a) Configuration of the specimen in the magnet, (b) stereoprojection of (004) plane in the specimen cured under a static magnetic field of 10T and the stereoprojection of (220) and (202) planes in the specimen cured under the rotation magnetic field (10T).

4. Crystallographic alignment under a magnetic field

4.1. Alignments under the static magnetic field (10T) and the rotating magnetic field

Figure 2(a) shows the configuration of the specimen in the magnet bore. The axis H_1 is the direction of the static magnetic field. As shown in figure 2(b), the diffraction of (004) plane was clearly observed around the H_1 axis. Thus, the [001] direction (c-axis) is parallel to the static magnetic field. The clear alignment indicates that the magnetic susceptibility of the c-axis is maximum among those of a-, b- and c-axes. In addition, the difference in the magnetic energy, $(1/2)(\chi_c - \chi_{a,b})H^2V$, is sufficiently large, comparing to the thermal energy, kT. Here, V is the volume of β-$FeSi_2$ particle.

The alignment of the β-$FeSi_2$ particles under the rotating magnetic field is shown in figure 2(c). Since the angular velocity was constant, the time-averaged intensity of the magnetic field is the same on the H_1-H_2 plane. The diffractions of (220) and (202) planes lie around the axis H_3 (the rotation axis). The tilt angle between the diffraction and the rotation axis was approximately 52 degrees. According to the crystal structure of β-$FeSi_2$ intermetallic compound, the diffraction pattern indicates that the [100] direction (a-axis) was parallel to the rotation axis. Thus, the magnetic susceptibility of the a-axis is minimum.

The alignments under the static magnetic field and the rotation magnetic field proved that the magnetic susceptibilities of the β-$FeSi_2$ are $\chi_c > \chi_b > \chi_a$. Furthermore, the differences in the magnetic energy, $(1/2)(\chi_c - \chi_b) H^2V$ and $(1/2)(\chi_b - \chi_a) H^2V$, are sufficiently large under the magnetic fields. Thus, the anisotropy in the magnetic susceptibilities of the β-$FeSi_2$ satisfied the requirement for the three-dimensional alignment.

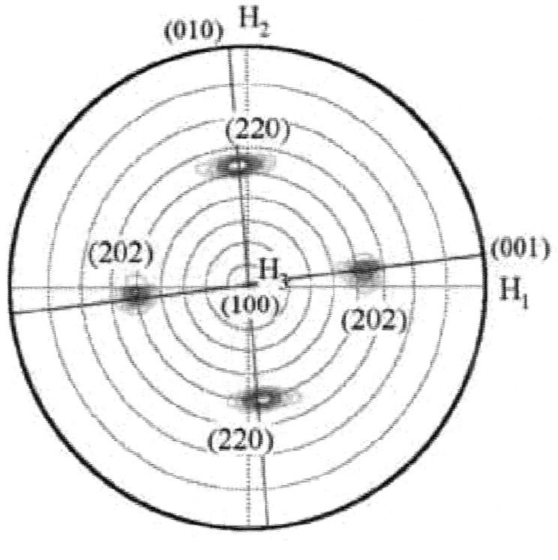

Figure 3. Stereoprojection of (220) and (202) planes in the specimen cured under the oscillating magnetic field. The oscillating angle was 30 degrees.

4.2. Alignment under the oscillating magnetic field

To achieve the three-dimensional alignment, the oscillating magnetic field was imposed on the β-$FeSi_2$ particles suspended in the resin. The stereographic projection of (220) and (202) planes is shown in figure 3. The distributions of (220) and (202) planes showed that the c-axis is parallel to the axis H_1, the b-axis parallel to the axis H_2 and the a-axis parallel to the axis H_3. The result indicated that a pseudo single crystal of β-$FeSi_2$ was fabricated by imposing the oscillating magnetic field.

Degree of the alignment was evaluated by measuring the half maximum full-width (HMFW) of the diffraction peaks of (220) and (202) planes. The values of HMFW ranged from 4 degrees to 10 degrees. The modification of oscillating magnetic field (oscillating angle, angular velocity, modulation of angular velocity etc.) is expected to improve the degree of the three-dimensional alignment.

5. Summary

The magnetic anisotropy of β-$FeSi_2$ was identified by using the static magnetic field (10T) and the rotating magnetic field (10T, 9rpm). The alignments obtained under the static magnetic field and the rotating magnetic field proved that the magnetic susceptibilities of $FeSi_2$ are $\chi_c>\chi_b>\chi_a$. The oscillating magnetic field achieved the three-dimensional alignment of β-$FeSi_2$ particles in the resin. Namely, a pseudo single crystal of β-$FeSi_2$ was fabricated by the oscillating magnetic field.

Acknowledgement

This study was supported by Priority Assistance for the Formation of Worldwide Renowned Centers of Research - The Global COE Program (Project: Center of Excellence for Advanced Structural and Functional Materials Design) from the Ministry of Education, Culture, Sports, Science and Technology (MEXT), Japan. This work is also supported in part by a Grant-in-Aid for Scientific Research and the JSPS Asian Core Program ''Construction of the World Centre on Electromagnetic Processing of Materials''

References

[1] Farrell D E, Chandrasekhar B S, DeGuire M R, Fang M M, Kogan V G, Klem R J and Finnemore D K 1987 *Phys. Rev. B* **36** 4025
[2] Lusnikov A, Miller L L, McCallum R W, Mitra S, Lee W C and Johnson D C 1989 *J. Appl. Phys.* **65** 3136
[3] Tkaczyk J E, Lay K W 1990 *J. Mater. Res.* **7** 1368
[4] Rango P de, Lees M, Lejay P, Sulpice A, Tournier R, Ingold M, Germi P and Pernet M 1991 *Nature* **349** 770
[5] Sarkar S and Nicholson P S 1992 *Appl. Phys. Lett.* **61** 494
[6] Arendt R H, Garbauskas M F, Lay K W and Tkaczyk J E 1991 *Physca C* **176** 131
[7] Holloway A, McCallum R W and Arrasmith S R 1993 *J. Mater. Res.* **8** 727
[8] Stassen S, Rulmont A, Vanderbemden Ph, Vanderschueren A, Gabelica Z, Cloots R and Ausloos M 1996 *J. Appl. Phys.* **79** 553
[9] Yasuda H, Ohnaka I, Shimamura K, Fukuda T and Watanabe K 2000 *Proc. 3rd Int. Symp. on Electromagnetic Processing of Materials*, (Nagoya, Japan, April 2000), pp 647-652
[10] Yamamoto Y, Yasuda H, Ohnaka I and Kishio K 2001 *Proc. 4th Pacific Rim Int. Conf. on Advanced Materials and Processing*, (Sendai, Japan, 2001), pp 293-296
[11] Roberts R W 1956 *Phys. Rev.* **104** 607
[12] Yasuda H, Ohnaka I, Kawakami O, Ueno K and Kishio K 2003 *ISIJ International* **43** 942
[13] Suzuki T S, Sakka Y and Kitazawa K 2001 *Adv. Eng. Mater.* **3** 490
[14] Suzuki T S and Sakka Y 2002 *Jpn. J. Appl. Phys.* **41** L1272
[15] Sakka Y, Suzuki T S, Tanabe N, Asai A and Kitazawa K 2001 *Jpn. J. Appl. Phys.* **41** L1416
[16] Nakahira A, Konishi S, Honda Y, Yasuda H and Ohnaka I 2003 *Trnas. Mater. Res. Soc. Jpn.* **28** 287
[17] Yasuda H, Nakahira A, Ohnaka I, Yamamoto Y and Kishio K 2003 *Materials Transaction* **43** 2555
[18] Yasuda H, Ohnaka I, Yamamoto Y, Wismogroho A S, Takezawa T and Kishio K 2003 *Materials Transaction* **43** 2550
[19] Kimura T and Yoshino M 2005 *Langmuir* **21** 4805
[20] Kimura T, Kimura F and Yoshino M 2006 *Langmuir* **22** 3464

Journal of Physics: Conference Series **165** (2009) 012022 doi:10.1088/1742-6596/165/1/012022

Fundamental study for solvent growth of silicon carbide utilizing Fe-Si melt

T Yoshikawa, S Kawanishi and T Tanaka
Division of Materials and Manufacturing Science, Osaka University, 2-1 Yamadaoka, Suita, Osaka, 565-0871, Japan.

E-mail: yoshikawa@mat.eng.osaka-u.ac.jp

Abstract. Fe-Si melt is a candidate for use as an alloy solvent for rapid liquid phase growth of SiC because of the high solubility of carbon in molten iron. In this work, the equilibrium phase relationship between SiC and the liquid phase of the Fe-Si-C system was investigated at 1523 K by the phase equilibration technique, and was further studied with the thermodynamic calculation. It was found that Fe- 36 mol% Si melt equilibrates with SiC at the temperature and possesses the higher carbon solubility than silicon-based melt. The first trial of the SiC crystal growth experiment was then carried out with Fe- 36 mol% Si melt by means of temperature difference method, and formation of SiC layer was obtained on the graphite substrate. Accordingly, it was found possibility for rapid growth of SiC by the solvent growth method with Fe- 36 mol% Si melt.

1. Introduction

Silicon carbide (SiC) has attracted attention as a basic semiconductor material for high temperature electronics and high power switching devices due to its wide bandgap, high electrical breakdown field and high thermal conductivity. Physical vapor deposition (PVD) is the standard method for the production of a bulk SiC single crystal, but it requires the controlled operation for SiC sublimation above 2200 K. The development of a low cost method for SiC material production is thus essential before wide-scale use of SiC semiconductor. We have focused on the liquid phase growth technique for the low temperature production of SiC.

Regarding to the liquid phase growth method of SiC, several research have been conducted using silicon-based alloy solvents such as silicon, Si-Ge or Si-Ti melts at above 1780 K, aiming for epitaxial growth on a SiC substrate [1 - 3]. Although epitaxial growth was achieved, the growth rate was not as high as that in the PVD method. This may be caused by the low solubility of carbon in the silicon-based solvent which accounts 80 ppmw at 1800 K in molten silicon [4]. We have applied the iron-based solvent containing silicon because of the high solubility of carbon in molten iron. The low eutectic temperatures of the Fe-Si system of around 1500 K can be effectively used for liquid phase growth at lower temperatures than has been done previously [1 – 3].

For the liquid phase growth of SiC using Fe-Si solvent, the phase diagram for the Fe-Si-C system is indispensable for determining the solvent composition. Investigation on the phase equilibrium in the Fe-Si-C system, however, is limited in the iron-rich corner and only a literature dealing with the liquid phase region saturated with SiC is available [5].

Therefore, the equilibrium phase relationship between SiC and the liquid phase in the Fe-Si-C ternary system was investigated in the present work. It was measured by an equilibration technique and was also estimated using thermodynamic calculation at 1523 K. Fe- 36 mol% Si alloy was found to be in equilibrium with SiC at the temperature. Furthermore, the growth experiment of SiC using the Fe-36 mol% Si solvent was carried out at 1523 K by means of temperature difference method. Formation of the SiC layer was obtained on the graphite substrate and the applicability of the alloy was clarified as the solvent for the liquid phase growth of SiC.

2. Equilibrium phase relationship between SiC and Fe-Si melt

Phase relationship between SiC and Fe-Si melt was only reported by Chipman et al. [5]. It was shown that the liquid composition in equilibrium with both SiC and graphite was Fe- 33.2 mol% Si at 1473 K and increased the Si content to Fe-37.7 mol% Si by increasing temperature to 1963 K. Here, the phase equilibrium between SiC and Fe-Si melt and the solubility of carbon in the melt was investigated around the liquid composition of Fe- 35 mol% Si.

Experiments were carried out with a vertical SiC resistance furnace. Fe-30, 33 and 36 mol% Si alloys were prepared by melting high purity electrolytic iron (99.992 %) and semiconductor-grade polycrystalline silicon (12N) in an alumina crucible at 1733 K under an argon atmosphere and then quenched in water. Five grams of Fe-Si alloy together with a dense graphite tablet or a SiC compact were inserted into an alumina Tammann tube (ϕ12mm) and kept in a furnace at 1523 K for 16 or 24 hours under an argon atmosphere. SiC compacts were prepared by the following methods: Equimolar mixture of 5N silicon and 3N carbon powder was compressed to form a disk at 50 MPa and then subjected to reactive liquid phase sintering by induction heating or spark plasma sintering (SPS) at temperatures higher than 1723 K under vacuum. Carbon analysis of the alloy was carried out by the combustion-IR absorption method. Silicon content was determined by SiO_2 gravimetric analysis or ICP atomic emission spectrometry.

Experimental results for the equilibrium composition of Fe-Si alloys saturated with SiC or graphite are shown in Fig. 1. The Fe-36 mol% Si alloy was in contact with SiC when the alloy was kept with either a graphite disk or a SiC compact, whereas no SiC layers were seen in the Fe-30 and 33 mol% Si alloys. The liquid composition in equilibrium with both graphite and SiC thus exists between 33 – 36 mol% Si at 1523 K. Chipman et al. [5] also reported comparable silicon content for that composition even though the carbon content seemed to be slightly lower than the content here.

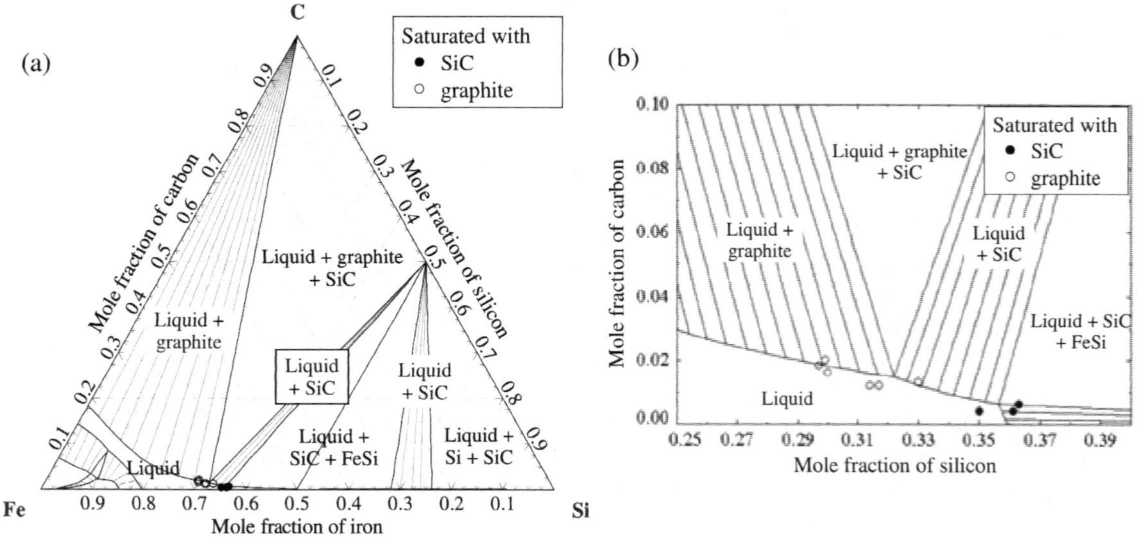

Figure 1 Experimental results for the equilibrium phase relation of Fe-30 – 36 mol% Si with the corresponding solid phase at 1523 K together with calculated results. (a) entire isothermal section of the Fe-Si-C system and (b) partial section.

The equilibrium phase relationship between SiC and the liquid phase in the Fe-Si-C system is further evaluated by thermodynamic calculations. The excess molar Gibbs energy in the liquid phase is described by a Redlich-Kister type polynomial given by Eqs. (1) and (2). Contributions of the binary Fe-Si, Fe-C and Si-C systems are taken into account here.

$$\Delta G^{Ex}_{m,liq} = x_{Fe} x_{Si} L_{Fe-Si,liq} + x_{Fe} x_C L_{Fe-C,liq} + x_{Si} x_C L_{Si-C,liq} \quad (1)$$

$$L_{i-j,liq} = \sum_{m=0}^{n} \left\{ {}^m L_{i-j,liq}(T)(x_i - x_j)^m \right\} \quad (2)$$

where $L_{i\text{-}j,\,liq}$ represents an interaction parameter between components i and j in the liquid phase and was refered to the literature [6-8]. The equilibrium calculation was carried out with thermodynamic calculation software (FactSage 5. 5). The calculated results of the isothermal section of the Fe-Si-C system are shown in Fig. 1 (a) and (b). A good agreement was obtained between experiments and calculations. The validity of the estimated equilibrium relationship of the iron-rich liquid phase saturated with SiC in the Fe-Si-C system was thus confirmed.

To optimize the Fe-Si solvent for rapid crystal growth of SiC, composition dependence of carbon solubility in Fe-Si melt was calculated as shown in Fig. 2. The solubility of carbon in iron-rich Fe-Si alloy was found to be 100 times higher than that in silicon-rich alloy. Hence, the Fe-Si melt of Fe-36 mol% Si composition which possesses high carbon solubility is expected to be a suitable solvent for rapid liquid phase growth of SiC.

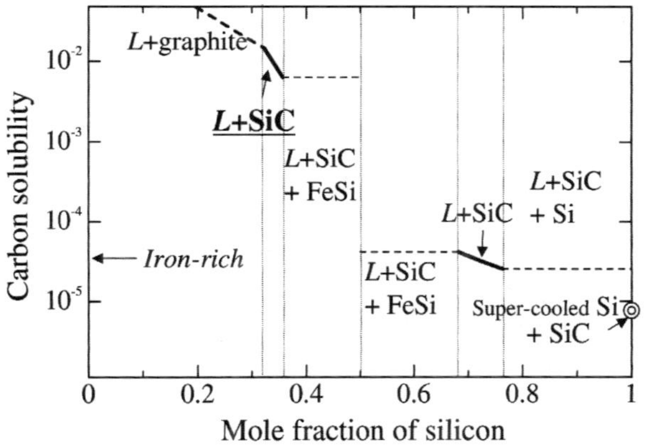

Figure 2 Calculated composition dependence of carbon solubility of the Fe-Si melt at 1523 K.

3. Trial of the liquid phase growth of SiC with Fe- Si solvent

First trial of the liquid phase growth of SiC with Fe- Si solvent was conducted by means of temperature difference method. The Fe-Si alloy is put between two graphite substrates and is placed in the temperature gradient. The difference in the carbon content at two interfaces between the alloy and graphite substrates brings mass transfer of carbon from high to low temperature interface, resulting in continuous dissolution of graphite and precipitation of SiC at high and low temperature interfaces, respectively.

The Fe- 36 mol% Si alloy prepared by the above-mentioned method was cut and polished to the square plate (5 × 5 × about 1 mm). The alloy was sandwiched between two graphite substrates and was suspended in the temperature gradient region in a vertical SiC resistance furnace at 1523 K under an argon atmosphere and kept for 16 h. Experiment was carried out with the different temperature gradients of 14 and 21 K / cm.

The cross sections around the graphite substrate at low temperature interface are shown in Fig. 3. Since the grown layers of SiC were observed on graphite substrates, the applicability of the Fe-Si alloy as a solvent for the liquid phase growth of silicon carbide was clarified. Further, temperature gradient affects the thickness of the grown layer, which indicates that its growth kinetics is at least partly limited by the mass transfer in the solvent. In future, the authors will try rapid growth of SiC by enhancing the forced convection in the solvent.

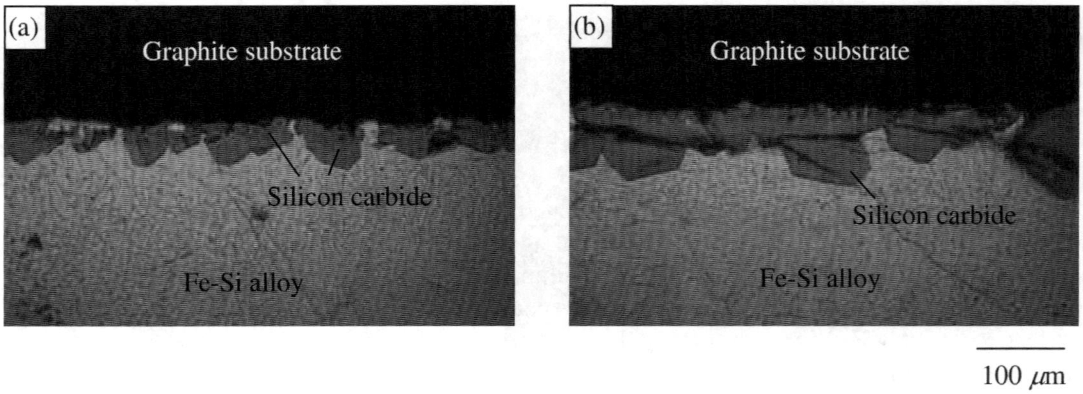

Figure 3 Microstructures around the low temperature interface between Fe-Si alloy and a graphite substrate after holding at 1523 K for 16 h. Temperature gradient; (a) 14 K / cm and (b) 21 K / cm.

4. Conclusion

In order to use Fe-Si melt as a solvent for rapid liquid phase growth of SiC, the fundamental study was carried out in the present work. The equilibrium composition of the Fe-Si melt with SiC was investigated at 1523 K by the equilibration of the Fe-Si melt with SiC or graphite. It was also studied with the thermodynamic calculation. Fe- 36 mol% Si melt was found to equilibrate with SiC at the temperature and possess the higher carbon solubility than silicon-based melt. The SiC crystal growth experiment was then carried out with Fe- 36mol%Si solvent by temperature difference method, and the formation of SiC layer was observed on the growth substrate of graphite. Accordingly, it was found possibility for rapid growth of SiC by the solvent growth method with Fe- 36 mol% Si melt.

References

[1] Suzuki A, Ikeda M, Nagano N, Matsunami H and Tanaka T 1976 *J. Appl. Phys.* **47** 4546.
[2] Filip O, Epelbaum B, Bickermann M and Winnacker A 2004 *J. Crys. Growth* **271** 142.
[3] Ujihara T, Munetoh S, Kusunoki K, Kamei K, Usami N, Fujiwara K, Sazaki G and Nakajima K 2005 *Thin Solid Films* **476** 206.
[4] Scace RI and Slack GA 1959 *J. Chem. Phys.* **30** 1551.
[5] Chipman J, Fulton JC, Gokcen N and Caskey, GR 1954 *Acta Metall.* **2** 439.
[6] Hultgren R, Desai PD, Hawkins DT, Gleiser M and Kelley KK 1973 *Selected Values of the Thermodynamic Properties of Binary Alloys* (Ohio: American Society for Metals) pp 871-883.
[7] Gustafsson P 1985 *Scand. J. Metall.* **4** 259.
[8] Kawanishi S, Yoshikawa T and Tanaka T, *Materials Transactions*, submitted.

Acknowledgement

This research is partly supported by a Grant-in-Aid for Young Scientists (Start-up) and Priority Assistance for the Formation of Worldwide Renowned Centers of Research - The Global COE Program (Project: Center of Excellence for Advanced Structural and Functional Materials Design) from the Ministry of Education, Culture, Sports, Science and Technology (MEXT), Japan.

Photoreflectance study of β-$FeSi_2$ epitaxial films grown by molecular beam epitaxy

Yoshikazu Terai[1], Keiichi Noda, Syoutaro Hashimoto, Yasufumi Fujiwara

Division of Materials and Manufacturing Science, Graduate School of Engineering, Osaka University, 2-1 Yamadaoka, Suita, Osaka 565-0871, Japan

E-mail : y-terai@mat.eng.osaka-u.ac.jp

Abstract. Band structure of β-$FeSi_2$ epitaxial and polycrystalline films was investigated by photoreflectance (PR) measurements. In the PR spectra, two critical points were obtained at E_g = 0.918 eV and E_i = 0.895 eV in the films. The E_g consistent with the direct transition energy at *Y* point in the Brillouin zone of β-$FeSi_2$. The E_i was assigned as a transition-energy between conduction band and the acceptor-level originating from Si vacancies in β-$FeSi_2$. The energy from top of valence band to the acceptor-level was determined to be 23 meV.

1. Introduction

Semiconducting β-$FeSi_2$ has attracted much interest as silicon-based optoelectronics materials. β-$FeSi_2$ thin films on Si substrate show photoluminescence (PL) at 1.54 μm, which is a wavelength for fiber optics communications [1-3], but the PL almost disappears in bulk samples [4]. The fact indicates that there is a difference in optical transition between bulk and thin film. Since β-$FeSi_2$ can be viewed as a moderate distortion of the CaF_2 structure by way of the Jahn-Teller effect [5], band structure and optical properties are strongly influenced by slight lattice distortions [6]. In addition, impurities and defects also affect the optical properties by perturbations of the band structure. However, band structure including impurities and defects levels is still uncertain in β-$FeSi_2$ thin films on Si substrates. Recently, we have suggested that point defects at Si sites in β-$FeSi_2$ act as non-radiative recombination centers in a relaxation processes of photoexcited carriers. The results of Al-doping [7] and long-time annealing [8] in β-$FeSi_2$thin films supported the existence of Si point defects, but the defect level related to the Si vacancy has not been reported.

In this contribution, we have investigated photoreflectance (PR) spectra of β-$FeSi_2$ thin films in order to clarify the band structure and revealed an impurity band transition related to the Si vacancy.

2. Experiments

In this paper, we have prepared three kinds of β-$FeSi_2$ thin films on Si substrate. The first was grown by reactive deposition epitaxy (RDE) method. In the RDE-grown epitaxial film with the thickness of 100 nm, Fe of 99.9985% purity was deposited on Si(111) substrate with a resistivity of 1-3 kΩ·cm at 670 °C. The density of Si point defect in the RDE film was considered to be rich, because the supply of Si atoms depended on thermal diffusion from the Si substrate. The second was grown by molecular beam epitaxy (MBE) methods, in which Fe and Si were co-evaporated on Si(111) substrate at 700 °C using template technique. The total film thickness was 200 nm. In the MBE-grown epitaxial film, the

[1] Corresponding author : Y. Terai, (Tel.: +81-6-6879-7548; fax: +81-6-6879-7536).

full width at half maximum (FWHM) of XRD rocking curve corresponding to β-$FeSi_2$ (202)/(220) diffraction peak was 0.24°. The FWHM was smallest in the three films, which shows that the MBE epitaxial film was highest-oriented crystalline on the Si substrate. The third was β-$FeSi_2$ polycrystal films of 50 nm produced by ion beam synthesis (IBS). In the IBS growth, mass-separated $^{56}Fe^+$ ion were implanted into Si(100) substrates at room temperature. Therefore, the purity of the iron source was the highest in the three films. Details of the growth method have been published [9].

PR measurements at 77 K were performed at near-normal incident in the 0.75-1.2 eV range. A 150 W halogen lamp was used as a probe source in conjunction with a single grating monochromator. The pump source was 532 nm laser mechanically chopped at a frequency of 140 Hz. The averaged power density at the sample surface was 326 mW/cm^2. The detector was an InGaAs photodiode. The experimental results were fitted by a least-squares fit algorithm to theoretical equations. In the low field limit, PR signal (ΔR/R) are given by a generalized Lorentzian functional form [10]

$$\Delta R / R = \mathrm{Re}\left[\sum_{j=1}^{n} C_j e^{i\theta_j}(E - E_j + i\Gamma_j)^{-m}\right], \qquad (1)$$

where, C_j and θ_j are the amplitude and phase of the line shape, and E_j and Γ_j are the energy and broadening parameter of the interband transitions. The exponent, m, depends on the type of critical point transition. In this paper, the exponent m is assumed to be $m = 2.5$ for three dimensional band-to-band transition.

3. Results and Discussion

Figure 1 shows PR spectra in three kinds of β-$FeSi_2$ thin films. In all the samples, two major features are observed at 0.89 - 0.92 eV. The primary feature at ~0.92 eV shows a third derivative spectral line shape, on the other hand, the weak feature at ~0.89 eV shows a first derivative one. We obtained two transition energies of E_g (~0.92 eV) and E_i (~0.89 eV) by the fitting using eq. (1). The fitting parameters are listed in Table I. Though the crystal quality and the purity of iron source are different, the transition energies of E_g and E_i are almost the same in the three films. In the model of PR [10], the third derivative line shape is due to a direct band to band transitions, while the first derivative one is indicative of an impurity- or defects-related transition. The E_g was assigned to the direct transition energy at *Y* point in the Brillouin zone of β-$FeSi_2$. The energy was consistent with that reported before [11]. On the other hand, E_i corresponds to an transition energy from band to impurity- or defect-level formed in the band-gap. There is a possibility that transition metals (Mn, Co, Ni) contaminating Fe source form the impurity-levels in the band-gap. However, the

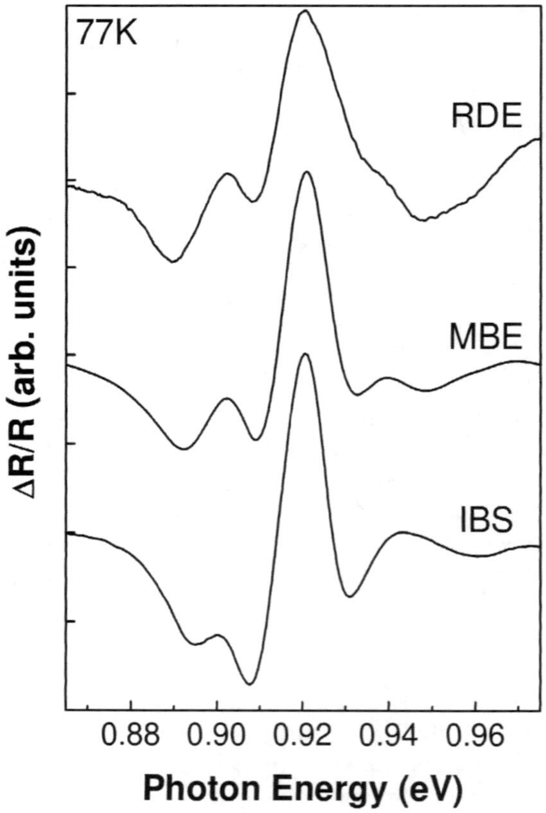

Figure 1. PR spectra of β-$FeSi_2$ epitaxial (RDE, MBE) and polycrystal (IBS) films at 77 K.

TABLE I. Lists of the fitting parameter using eq.(1). E_g and E_i denote the transition energy near 0.92 eV and 0.89 eV, respectively

	Transition	Energy (eV)	C_j ($\times 10^6$)	Γ_j (meV)
RDE	E_g	0.917	30	15
	E_i	0.896	10	15
MBE	E_g	0.918	10	11
	E_i	0.895	10	15
IBS	E_g	0.918	10	9
	E_i	0.895	10	15
MBE annealed	E_g	0.911	9	9
	E_i	0.887	7	15

possibility is excluded because the IBS-grown film which is free of transition-metal contamination shows the transition at E_i. As seen in Fig. 1, the E_i-transition is dominant in the RDE sample. The result suggests that Si vacancies in β-$FeSi_2$ are related to the E_i-transition. In β-$FeSi_2$, the existence of point defects at Si sites has been reported in a study of first principle calculations [12] and EPR measurements [13]. The Si vacancy forms a defect-level and acts as acceptor for β-$FeSi_2$. Therefore, the E_i-transition is assigned to the transition from conduction-band to the acceptor level formed by the Si vacancy. In the schematic band structure of β-$FeSi_2$ (Fig. 2), these optical transitions are shown.

In order to confirm the identification of E_i-transition, the MBE-grown sample was annealed at 900°C for 16 h. The annealing reduces the density of Si vacancy in β-$FeSi_2$ by a thermal diffusion of Si atoms from the Si substrate to β-$FeSi_2$ layer. Figure 3 shows the comparison of PR spectra between the as-grown and the annealed films. Fitting parameters to eq. (1) in the annealed film is also listed in Table I. After the annealing, the amplitude of E_i clearly decreases and both transition energies shift to

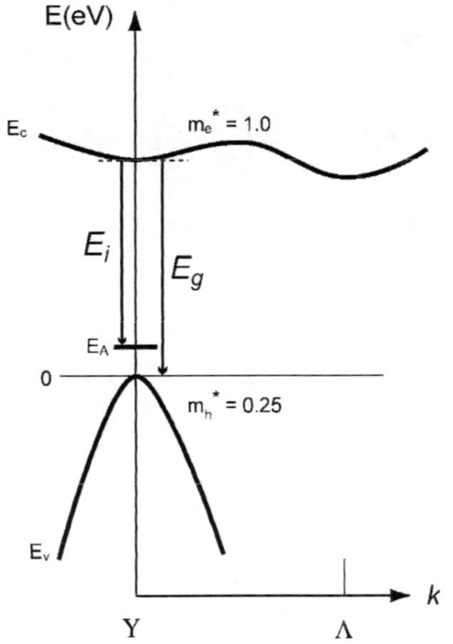

Figure 2. Schematic band structure of β-$FeSi_2$.

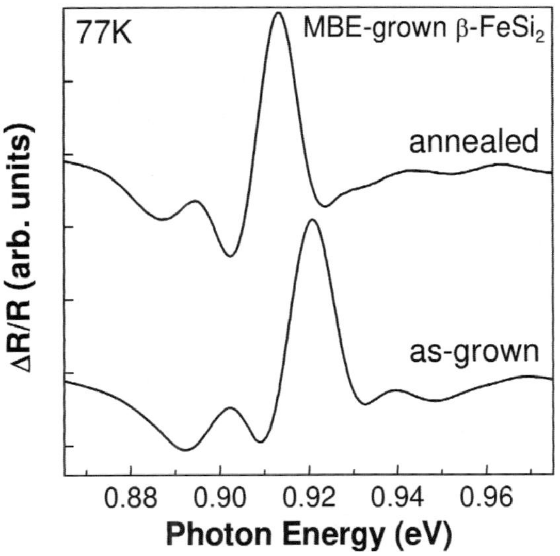

Figure 3. PR spectra of as-grown and annealed samples grown by MBE.

lower photon energy. The origin of the energy shift is not certain, but the decrease of the E_i-amplitude is understood by the thermal diffusion of Si atoms. These results convince us that the E_i corresponds to the transition energy from conduction-band to the acceptor level formed by the Si vacancy and the acceptor level is 23 meV from top of the valence band at *Y* point in the Brillouin zone.

4. Conclusions

Band structure of β-$FeSi_2$ thin films was investigated by PR measurements. The critical point in PR spectra at E_g = 0.918 eV was due to the direct band-to-band transition at *Y* point in the Brillouin zone of β-$FeSi_2$. The other critical point at E_i = 0.895 eV was revealed as a transition-energy between conduction band and the acceptor-level originating from Si vacancies in β-$FeSi_2$. The energy from top of valence band to the acceptor-level was determined to be 23 meV.

Acknowledgments

This work was supported in part by Grant-in-Aid for Young Scientists (B) No. 20760199 from the Ministry of Education, Culture, Sports, Science and Technology. This work was also supported in part by The Global COE Program (Project: Center of Excellence for Advanced Structural and Functional Materials Design) from the Ministry of Education, Culture, Sports, Science and Technology, Japan and by Iketani Science and Technology Fundation, Japan.

References

[1] Dimitriadis C A, Werner J H, Logothetidis S, Stutzmann M, Weber J, and Nesper R 1991 *J. Appl. Phys.* **68** 1726
[2] Maeda Y, Homewood K P, Suemasu T, Sadoh T, Udono H, and Yamaghchi K 2004 *Thin Solid Films* **461** 1
[3] Maeda Y, Homewood K P, Sadoh T, Terai Y, Yamaghchi K, and Akiyama K 2007 *Thin Solid Films* **515** 8101
[4] Udono H, Kikuma I, Okuno T, Masumoto Y, Tajima H and Komuro S 2004 *Thin Solid Films* **461** 182
[5] Christensen N E 1990 *Phys. Rev. B* **42** 71
[6] Miglio L and Meregalli V 1998 *J. Vac. Sci. Technol. B* **16** 1604
[7] Terai Y and Maeda Y 2004 *Appl. Phys. Lett.* **84** 903
[8] Maeda Y, Terai Y, Itakura and M, Kuwano N 2004 *Thin Solid Films* **461** 160
[9] Maeda Y, Umezawa K, Hayashi Y, Miyake and K, Ohashi K 2001 *Thin Solid Films* **381** 256
[10] Aspnes D E 1980 *Handbook on Semiconductors* (Amsterdam : North-Holland)
[11] Birdwell A G, Shaffner T J, Chandler-Horowitz D, Buh G H, Rebien M, Henrion W, Stauß P, Behar G, Malikova L, Pollak F H, Litter C L, Glosser R and Collins S 2004 *J. Appl. Phys.* **95** 2441
[12] Tani J, Kido H 2003 *J. Alloys Comp.* **352** 153
[13] Miki T, Matsui Y, Matsubara K, Kishimoto K 1994 *J. Appl. Phys.*.**75** 1693

Low-temperature growth of GaAs with high quality by metalorganic vapor phase epitaxy

Harunori Sakaguchi[1], Tomoyoshi Mishima[1], Takeshi Meguro[2] and Yasufumi Fujiwara[3]

[1] Research and Development Laboratory, Hitachi Cable, Ltd.
5-1-1 Hitaka-cho, Hitachi, Ibaraki 319-14, JAPAN
[2] Takasago Works, Hitachi Cable, Ltd.
880 Isagozawa-cho, Hitachi, Ibaraki 319-18, JAPAN
[3]Division of Materials and Manufacturing Science, Graduate School of Engineering, Osaka University
2-1 Yamadaoka, Suita, Osaka 565-0871, Japan

E-mail: sakaguchi.harunori@hitachi-cable.co.jp

Abstract. Growth of GaAs by MOVPE (Metal-Organic Vapor-Phase Epitaxy) at a much lower temperature than conventional conditions of 650°C required to produce highly resistive buffer layers and molecular-layer-level abrupt heterojunctions and doped-layer interfaces. It has been widely recognized that growth at 550~600°C causes poor morphological surfaces on the grown layers. This paper describes a fundamental improvement of the low-temperature growth of GaAs, resulting in smooth surfaces with very low impurity concentrations. Excellent GaAs layers can be grown at 500°C by increasing partial pressure of AsH_3 during growth. The method has been applied to high quality epitaxial layers for electronic device production.

1. Introduction

High purity and highly resistive buffer layers and molecular-layer-level abrupt interfaces are key to recent improvement of compound semiconductor electronic devices. Although MOVPE is now recognized as the most suitable mass-production method of epitaxial layers for electronic and optical devices, its growth temperature has usually been 650~850°C, which enhances thermal diffusion of lattice atoms at interfaces. When the growth temperature is reduced to 550~600°C, which is widely used in MBE, a result is poor morphological surfaces of the MOVPE GaAs layers. Therefore, there have been few reports on development of low temperature growth techniques using MOVPE.

It has been reported that the growth of GaAs by MOVPE is governed by mass transport of molecules at the range of 600~850°C and by kinetic movement of molecules on the surface at the range of 400~600°C[1]. This suggests that an unstable growth takes place at the transition temperature of the two ranges, which leads to the rough surface. Thus, if the growth is performed at fully kinetic conditions (below 500°C), mirror-like stable surfaces can be obtained again. This paper unveils our fundamental technology on the low temperature growth of GaAs with a smooth surface and high purity. The method has been applied to mass-production of the epitaxial layers for various GaAs-based devices.

2. Experimental

GaAs epitaxial layers were grown by MOVPE under an atmospheric pressure of H_2 using the material gases of TMG (Trimethyl gallium) and AsH_3. The flow rate of H_2 was 20 l/min. The substrate crystals were undoped LEC (Liquid Encapsulated Czochralski)-grown semi-insulating GaAs wafers with (100) surface. The growth temperature was changed in the range between 400 and 650°C. The measured items were (1) surface observation by an optical microscope, (2) electrical properties by the van der Pauw method, and (3) photoluminescence properties at liquid He temperature.

3. Results and Discussion

3.1. Low temperature growth of GaAs

The growth rate dependency on the growth temperature is shown in Figure 1. From the slope in the figure, the growth below 500°C was in the kinetic controlled region while above 600°C was in the mass transport limited region. The transition temperature was 550°C. The activation energy of the growth in the kinetic controlled region was 19.7 kcal/mol. This result agrees with that of D.H.Reep[1]. The layer grown below 400°C could not be observed. It is estimated that the growth rate at 400°C was 1/10 of that at 600°C. The reason must be little decomposition of TMG at that low temperature.

The photographs of the surface of the layers grown at the temperature range 450~600°C are shown in Figure 2. The thickness of the layers was 0.5~2μm. The layers grown at 600°C

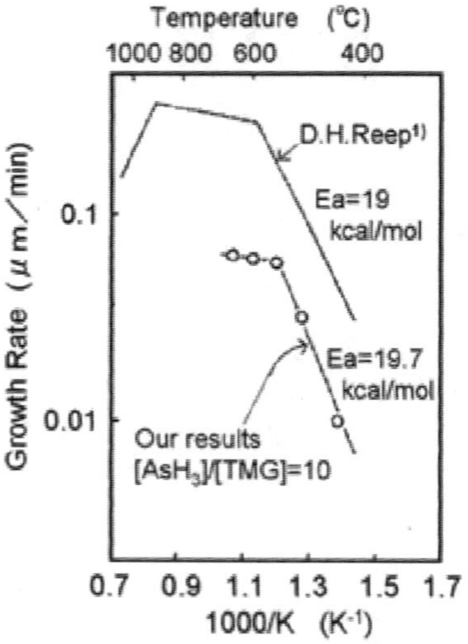

Figure 1. Dependence of growth rate on reciprocal growth temperature for epitaxial layers on (100) GaAs with P(TMG)=10.1 Pa and $P(AsH_3)$= 1.01×10^2 Pa ($[AsH_3]$ / [TMG] =10)

(mass transport limited region) had smooth surfaces, while the layers grown at 550°C (transition region) had milky surfaces. At the lower temperature of 450~500°C (kinetic controlled region), the grown layers successfully had smooth surfaces, as expected. A few hillocks like oval defects on MBE-grown wafers were observed on the surface of the layers grown at 500°C. More and smaller hillocks were observed on the surface of layers grown at 450°C.

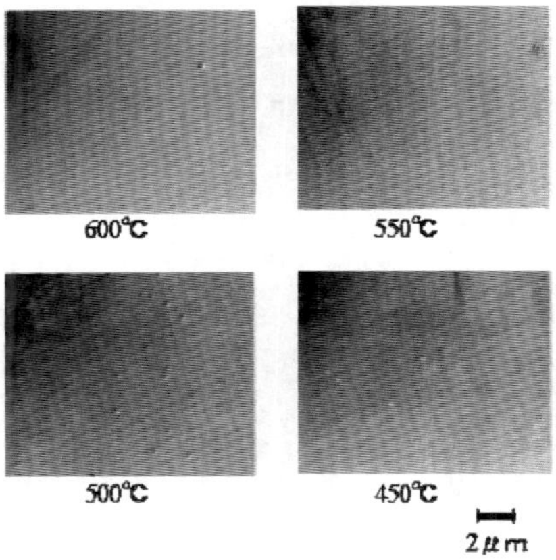

Figure2. Surface of low-temperature growth GaAs.

3.2. The characteristics of GaAs grown at low temperature

The carrier concentration of undoped layers grown between 450°C and 700°C are shown in Figure 3. Above 600°C, the layers had high resistivity. The layers grown between 500°C and 550°C became p-type and the hole concentration increased with decreasing growth temperature. These p-type GaAs layers have the high values of mobility 270~360 cm^2/Vs. The layers grown at 450°C became highly resistive (or fully depleted). The photoluminescence spectra from undoped GaAs layers grown at various values of temperature are shown in Figure 4. The acceptor in the p-type GaAs layers grown at low temperature is identified as carbon from the photoluminescence spectra. The intensity of photoluminescence from the layer grown at 450°C is very weak. This means that the properties of the grown crystal layer must be poor.

3.3. The effect of AsH_3 in low temperature growth

From photoluminescence spectra, carbon was determined to be the acceptor of the p-type layer grown at 450-500°C. It is well known that in the higher temperature growth range the incorporation of carbon into

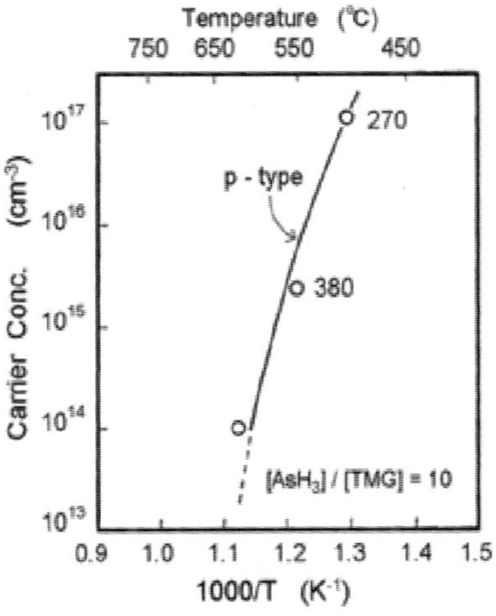

Figure 3. Growth temperature dependence of carrier concentration for epitaxial layers on(100) GaAs with P(TMG) =10.1Pa and $P(AsH_3)$ $=1.01\times10^2$ Pa ($[AsH_3]$ /[TMG] =10). The values near the circle show the mobility of the layers [cm^2/Vs].

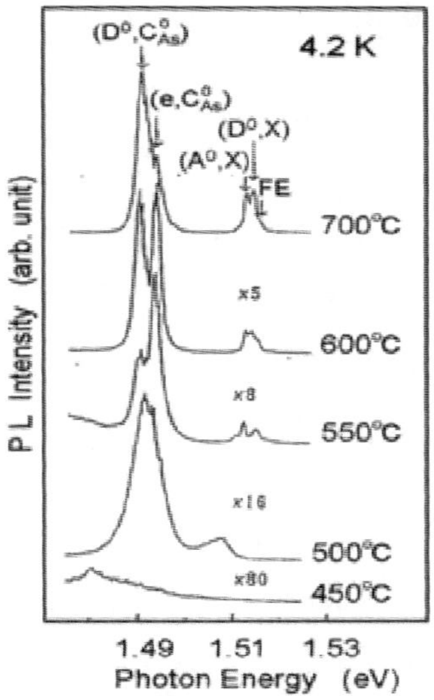

Figure 4. Low-temperature photoluminescence spectra of undoped GaAs epitaxial layers grown at various temperatures on (100) GaAs. P(TMG)=10.1 Pa and $P(AsH_3)$= 1.01×10^2 Pa ($[AsH_3]$/[TMG]=10)

MOVPE-grown GaAs crystals can be suppressed by increasing AsH_3 partial pressure P(AsH_3). An experimental study was added for the confirmation of the effect of P(AsH_3) at low temperature. The growth temperature was 500°C and P(AsH_3) was varied from 1.01×10^2 to 5.05×10^2 Pa. The carrier concentration of the undoped GaAs layers grown under various P(AsH_3) is shown in Figure 5. The hole concentration of GaAs decreases with increasing P(AsH_3) or molar ratio of AsH_3 to TMG ([AsH_3]/[TMG]), and the layers become semi-insulating at [AsH_3]/[TMG]=40 and 50. The photoluminescence spectra from these layers are shown in Figure 6. The intensity of the peak originated in carbon acceptor relatively decreases with increase of P(AsH_3) ([AsH_3]/[TMG]). The fine structures of the exiton peak can be observed in the photo- luminescence spectra from layers grown at [AsH_3]/[TMG]=50 . This means that high purity GaAs crystals can be grown at high P(AsH_3) ([AsH_3]/[TMG]) through low temperature growth.

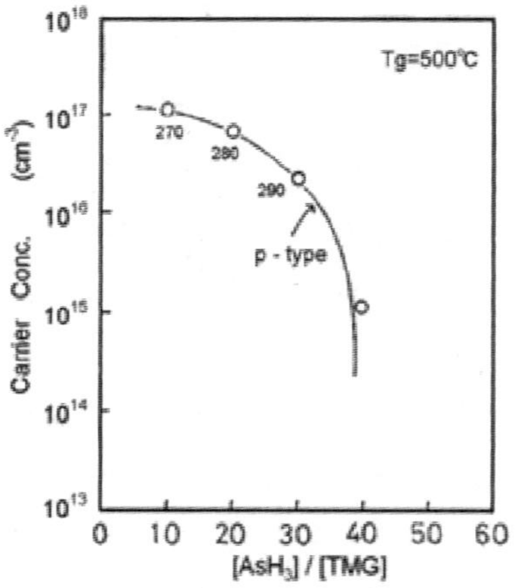

Figure 5. Effect of AsH_3 pressure ([AsH_3]/[TMG]) on the carrier concentration of undoped GaAs epitaxial layers grown at 500°C with *P*(TMG)=10.1Pa. The values near the circle show the mobility of the layers.

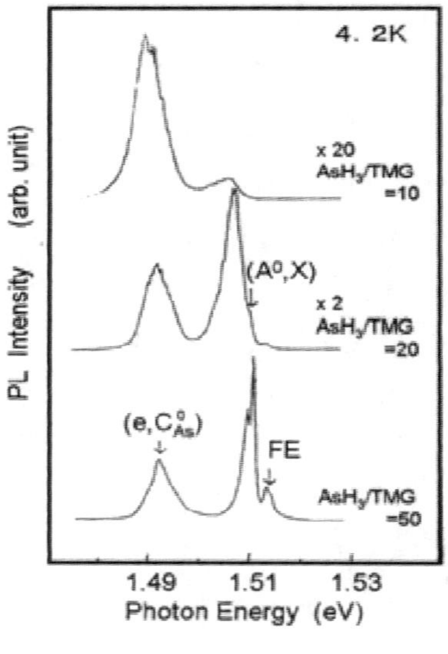

Figure 6. Low-temperature photo-luminescence spectra from undoped GaAs epitaxial layers grown at 500°C with various AsH_3 pressures ([AsH_3]/[TMG]).

4. Summary

Low temperature growth of GaAs was studied, and very smooth surfaces were successfully found on the layers grown at 450~500°C by MOVPE. Undoped GaAs layers grown at 500~550°C are p-type and their hole concentration increased with decreasing growth temperature. The acceptor in these layers was identified as carbon from photoluminescence spectra at 4.2K. GaAs crystal layers with high purity were grown at 500°C by increasing the partial pressure of AsH_3 during growth.

References

[1] Reep D H and Ghandhi S K 1983 *J.Electrochem.Soc.* **130** 675

[2] Tsuchiya T, Nagai H, Meguro T and Sakaguchi H 1994 *J. Cryst. Growth* **145**, 403

[3] Otoki Y,Sahara M,Nagai H,Sakaguchi H,Takahashi S and Kuma S 1997 *Mater.Sci.Eng.*B **44**, 8

[4] Otoki Y,Kamogawa H,Meguro T,Kashiwa M and Sakaguchi H 1998 *Digest of MANTEC98*, 137

Journal of Physics: Conference Series **165** (2009) 012025 doi:10.1088/1742-6596/165/1/012025

Development of new-type 1.5 μm light-emitting devices based on Er,O-codoped GaAs

Yasufumi Fujiwara, Yoshikazu Terai and Atsushi Nishikawa

Division of Materials and Manufacturing Science, Graduate School of Engineering, Osaka University, 2-1 Yamadaoka, Suita, Osaka 565-0871, Japan

E-mail: fujiwara@mat.eng.osaka-u.ac.jp

Abstract. We have fabricated two types of 1.5 μm light-emitting devices based on Er,O-codoped GaAs (GaAs:Er,O) using organometallic vapor phase epitaxy (OMVPE). In a device exhibiting a room-temperature lasing at the GaAs band-edge, the threshold current density increased with Er concentration. The Er intensity rose up steeply and then decreased gradually in the spontaneous emission region. In the stimulated emission region, it remained almost constant, reflecting ultrafast energy transfer to Er ions. In a device exhibiting a room-temperature lasing from strained GaInAs quantum wells (QWs) embedded in GaAs:Er,O, on the other hands, the lasing wavelength was designed to tune to the energy separation between the second excited states $^4I_{11/2}$ and the ground state $^4I_{15/2}$ of Er^{3+} ions. The Er intensity revealed initially steep increase with injected current density in the region for spontaneous emission from the GaInAs QWs. In the stimulated QW emission region, however, the intensity continued to increase with the current density.

1. Introduction

Rare-earth (RE)-doped semiconductors are a new class of materials with various promising potentials. RE ions doped in semiconductors exhibit a characteristic emission due to intra-4f shell transitions of RE ions. The emission shows sharp spectrum, temperature-stable wavelength and host-independent wavelength. Significant attention has been paid to Er-doped semiconductors because it can emit 1.5μm light. The wavelength is quite important for present silica-based optical communication.

We have succeeded in atomically-controlled doping of Er to various III-V semiconductors such as GaAs and InP by organometallic vapor phase epitaxy (OMVPE) [1]. In Er,O-codoped GaAs (GaAs:Er,O), one kind of Er center was formed selectively as an Er atom located at the Ga sublattice with two adjacent O atoms (hereafter referred as Er-2O) together with two As atoms, resulting in drastic enhancement in intensity of Er-related luminescence due to intra-4f shell transitions of Er^{3+} ions [2,3]. We have fabricated GaInP/GaAs:Er,O/GaInP double-heterostructure (DH) light-emitting diodes (LEDs), aiming at realizing the stimulated emission at the Er center under current injection. Radiant Er-2O electroluminescence (EL) has been successfully observed at around 1.54 μm at room temperature under forward bias [4]. Extremely large Er excitation cross section of 10^{-15} cm^2 was obtained under current injection, which is five orders larger than optical excitation cross section in the conventional Er-doped fiber amplifiers (10^{-20}~10^{-21} cm^2) [5]. For the improvement in the device performance, we have investigated quantitatively the mechanism of excitation and relaxation of Er ions by means of a pump-probe reflection technique, and found that there is a characteristic ultrafast

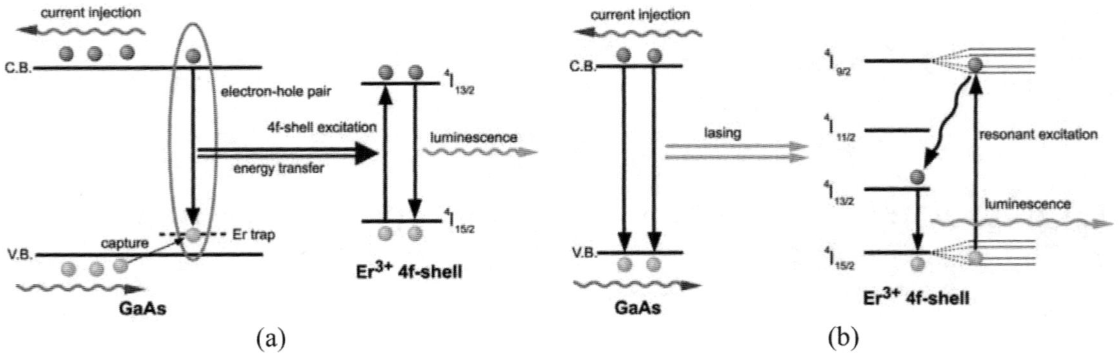

Figure 1. Schematic drawing for excitation mechanism of Er^{3+} ions under current injection; (a) indirect excitation and (b) direct excitation.

trapping process for nonequilibrium carriers in GaAs:Er,O [6,7].

The excitation mechanism of Er ions doped in the LEDs is shown in figure 1(a). Current injection introduces a free electron and a free hole. If we assume a trap as a donor-like hole trap, the injected hole is captured by the trap. The positively charged trap attracts the injected electron to create an electron-hole pair. The electron-hole pair recombines and part of the recombination energy is transferred to Er ions for 4f-shell excitation. Therefore, this excitation mechanism can be called indirect excitation. Now, we consider direct excitation with a lasing in the host material as shown in figure 1(b). If the wavelength is well matched to the energy separation of 4f-shell transitions, we can expect resonant direct excitation. In this case, the indirect excitation can be also expected. Therefore, the intensity of the Er luminescence should be enhanced under double excitation.

In this article, we report on fabrication and EL properties of two types of GaAs:Er,O-based laser diodes (LDs). One is a GaInP/GaAs:Er,O/GaInP LD exhibiting a lasing at the GaAs band-edge. The other is a LD with strained GaInAs quantum wells (QWs) embedded in the GaAs:Er,O, exhibiting a lasing from the QWs. The lasing wavelength was designed to tune to the energy separation between the second excited states $^4I_{11/2}$ and the ground state $^4I_{15/2}$ of Er^{3+} ions. The two LDs showed a quite different behavior of the Er intensity against the injected current density in the region for the stimulated emission from the GaAs band-edge and the GaInAs QWs.

2. OMVPE growth

The LDs were grown on (001)-oriented Si-doped *n*-GaAs and Zn-doped *p*-GaAs substrates at 545 °C by a low-pressure OMVPE system with a horizontal reactor. The reactor pressure was 76 Torr. Triethylgallium (TEGa), trimethylindium (TMIn), tertiarybutylarsine (TBAs) and tertiarybutylphosphine (TBP) were used as sources. H_2S and diethylzinc (DEZn) were also used for *p*- and *n*-type doping sources, respectively. Er was doped using $Er(DPM)_3$ [$(C_{11}H_{19}O_2)_3Er$] which contains six O atoms bonded to one Er atom in one molecule. The Er source was maintained at 80 °C and introduced into the reactor by a H_2 flow through the Er source cylinder (hereafter referred as Er flow). Er concentration of GaAs:Er,O layers was below the detection limit of secondary ion mass spectroscopy (SIMS) measurements (< 10^{17} cm^{-3}). $^{18}O_2$ of 49.9 ppm in Ar was used as an O_2 source in the growth of GaAs:Er,O. The O_2 content in growth ambient was set at 0.6 ppm.

3. LD exhibiting a lasing at the GaAs band-edge

The schematic drawing of the device structure is shown in figure 2. The growth sequence was initiated on a *n*-GaAs substrate by a 0.2-μm-thick, *n*-GaAs (1 x 10^{18} cm^{-3}) buffer layer, which was followed by a 1.0-μm-thick *n*-GaInP (3 x 10^{17} cm^{-3}) cladding layer and a 0.2 μm GaAs:Er,O active layer. A *p*-GaInP (6 x 10^{17} cm^{-3}) cladding layer was also 1.0 μm thick and a 0.1-μm-thick *p*-GaAs (1 x

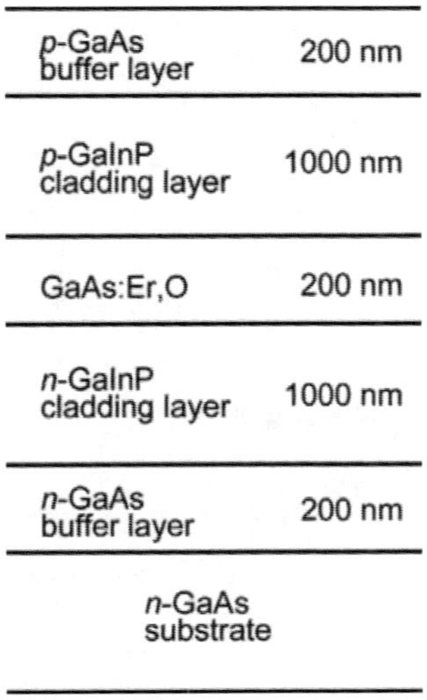

Figure 2. Device structure of a LD with GaAs:Er,O as an active layer.

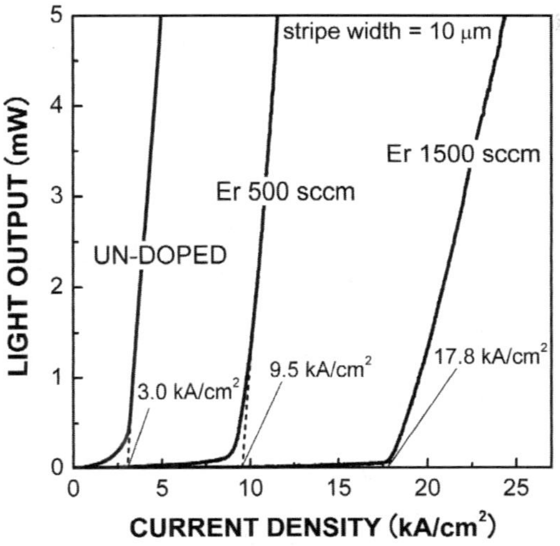

Figure 3. *I-L* characteristics for GaAs band-edge light (λ = 889 nm) in the LDs prepared with different Er flow rates, which are compared with those in a LD with undoped GaAs.

10^{19} cm^{-3}) contact layer was grown on top of the *p*-type cladding layer. Broad-stripe laser tips of the Fabry-Perot type were fabricated by a standard photolithography process.

In figure 3, the output of GaAs band-edge light (λ = 889 nm) for the GaInP/GaAs:Er,O/GaInP DH LDs is shown as a function of forward current density (*I-L*). The result in a LD with an undoped GaAs active layer is also shown for comparison. All the Er-doped LDs show a lasing at the GaAs band-edge at room temperature. The threshold current density (J_{th}) is also found to increase with the Er flow rate, suggesting that an energy-transfer from GaAs host to the 4f-shell of Er ions increases with the Er concentration.

The J_{th} has been analyzed as a function of reciprocal cavity length (*1/L*) [8]. The LDs showed a linear relationship between the J_{th} and *1/L*. Furthermore, the coefficient of the J_{th} against *1/L* also increased with the Er flow. It is well known that the J_{th} is simply expressed by the following equation [9],

$$J_{th} = \frac{d}{\beta\Gamma}(\alpha_{fc} + \frac{ln(1/R)}{L}) + dJ_0 . \qquad (1)$$

Here, d is the active layer thickness, β is gain factor, α_{fc} is internal loss due to free carrier absorption, Γ is the optical confinement factor, R is the reflectance and J_0 is the nominal current density at which the gain vanishes. As seen in the equation, the coefficient of the J_{th} against *1/L* depends only on $1/\beta$ because d, Γ and R are the same in all the LDs. Therefore, the large coefficient in the Er-doped LDs indicates that a gain factor becomes small by the Er doping.

Pump and probe reflectance spectroscopy for GaAs:Er,O with Er concentration of 10^{18}~10^{19} cm^{-3} revealed that the hole capturing and the electron-hole pair formation take place within 1 ps and 3-30 ps, respectively. Furthermore, the formation time is shorter in the sample with higher Er concentration [6,7]. In GaAs:Er,O with extremely low Er concentration (the Er flow rate: 1500 sccm), the fast decrease in $\Delta R/R$ was also observed, indicating that the fast hole capturing and the electron-hole pair formation are achieved at such low Er concentration. In the Er-doped LDs, the reduced gain factor was observed. The behaviour is quantitatively understood by the ultrafast carrier-capturing as follows. In high-gain region of $g > 30$ cm^{-1}, the gain factor β is approximately written by $\beta = a\tau_n/q$, where a is constant, τ_n is lifetime of photoexcited carriers and q is elementary charge [10]. Therefore, the reduced gain factor indicates that the lifetime of photoexcited carriers becomes short with increasing Er flow.

In order to confirm the ultrafast carrier-capturing in the Er-doped LD prepared with the Er flow rate of 1500 sccm, the intra-4f shell emission due to Er^{3+} ions at 1.54 μm was measured at 4.2 K under forward bias. Figure 4 shows EL intensity of the GaAs band-edge emission (λ = 831 nm) (a) and the Er-related emission (λ = 1539 nm) (b) as a function of current density. The LD on an *n*-GaAs substrate shows the stimulated emission at the GaAs band-edge with J_{th} = 9.5 kA/cm^2 at 4.2 K. The EL intensity of the Er-related emission saturates at 2 kA/cm^2 and decreases gradually at higher current density. The decrease is due to the Auger effect, energy transfer from excited Er ions to injected carriers. In the stimulated emission region, on the other hand, the EL intensity becomes almost constant. The constant EL intensity is closely related to saturated concentration of injected carriers during laser action. In GaAs, carrier lifetime varies from τ ≈ 1 ns at spontaneous recombination to τ ≈ 200 ps at stimulated recombination [11,12]. The observation of both the band-edge stimulated emission and the Er-related spontaneous emission in the stimulated emission region is a clear evidence that the carrier-capturing time at the bound state produced by Er-doping is quite shorter than the lifetime of the band-edge stimulated emission. Similar results were also obtained on a *p*-GaAs substrate. Furthermore, the lasing wavelength of the LDs was 831 nm, whose energy is smaller than the reported energy separation between the third excited states $^4I_{9/2}$ and the ground states $^4I_{9/2}$ of Er^{3+} ions. Therefore, the resonant direct excitation by the band-edge lasing could never be expected in the LDs.

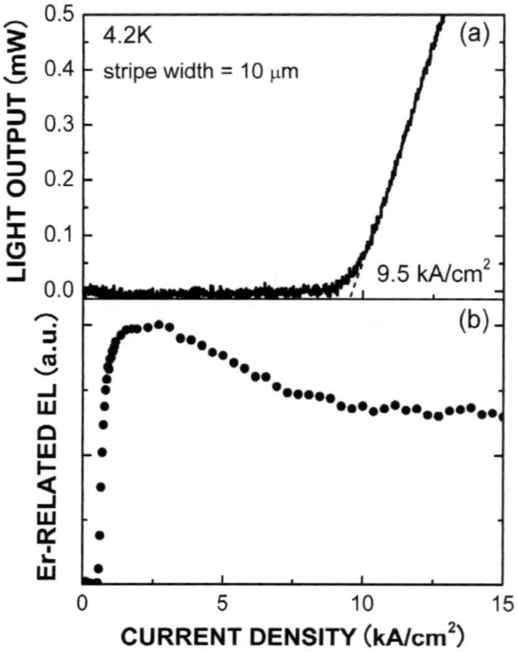

Figure 4. EL intensity of the (a) GaAs band-edge emission (λ = 831 nm) and (b) Er-related emission (λ = 1539 nm) as a function of current density in the LD prepared with 1500 sccm.

4. LD exhibiting a lasing from GaInAs QWs

The device structure is schematically shown in figure 5, which is a well-known separated confinement heterostucture (SCH-structure) [13]. A 0.2-μm-thick, *p*-GaAs (2 x 10^{19} cm^{-3}) buffer layer and a 1.0-μm-thick *p*-GaInP (1 x 10^{18} cm^{-3}) cladding layer were subsequently grown on a *p*-GaAs substrate. Two 8-nm-thick $Ga_{0.76}In_{0.24}As$ QWs with a 20-nm-thick GaAs:Er,O layer as a barrier were embedded in the center of a GaAs:Er,O waveguide layer with thickness of 200 nm. A *n*-GaInP (6 x 10^{17} cm^{-3}) cladding layer

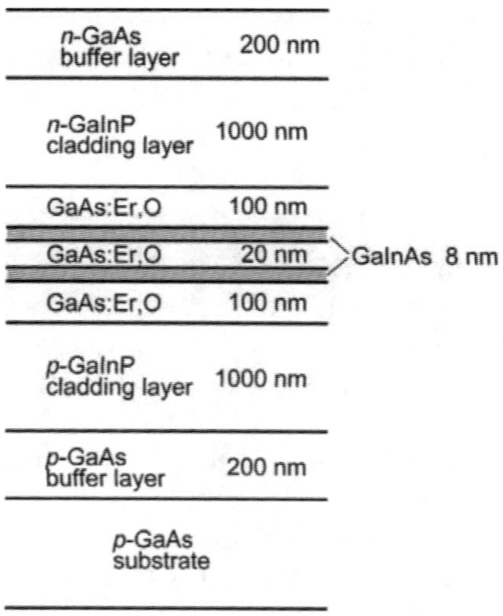

Figure 5. Device structure of a LD with two GaInAs QWs embedded in GaAs:Er,O.

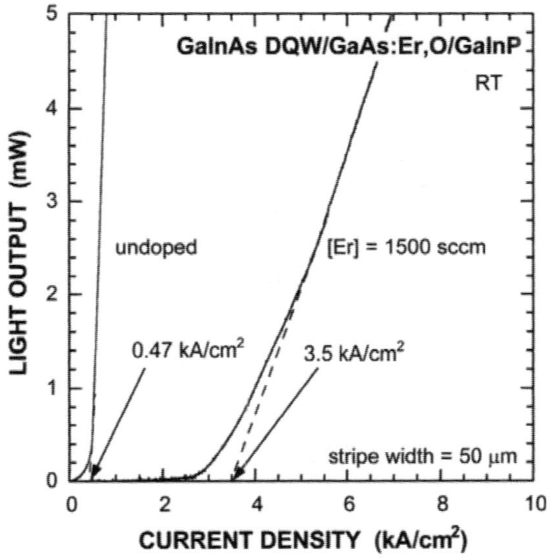

Figure 6. *I-L* characteristics for the QW emission (λ = 1033 nm) in the LD prepared with GaAs:Er,O as a waveguide layer, which is compared with those in the LD with an undoped GaAs waveguide layer.

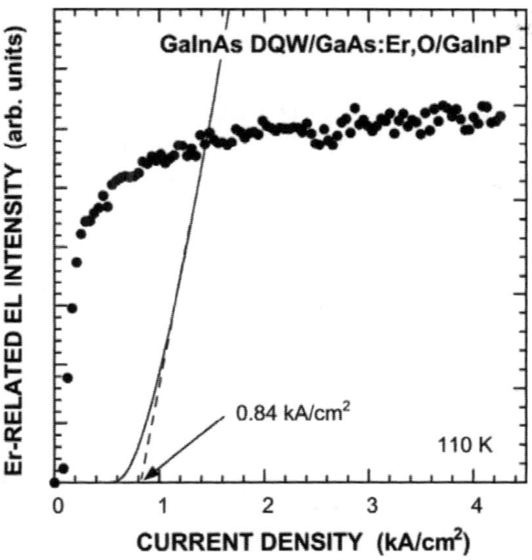

Figure 7. Er intensity at 110 K as a function of the current density. *I-L* characteristics for the QW emission (λ = 981 nm) at the temperature are also shown for comparison.

was also 1.0 μm thick and a 0.2-μm-thick *n*-GaAs (8 x 10^{17} cm^{-3}) contact layer was grown on top of the *n*-type cladding layer. In the growth of GaAs:Er,O, the Er flow rate was fixed at 1500 sccm.

Figure 6 shows the output of GaInAs QW emission (λ = 1033 nm) in the LD as a function of forward current density (*I-L*) at room temperature, which is compared with the result for a GaInAs QW-LD with undoped GaAs instead of GaAs:Er,O. Clear laser action can be observed in both the LDs. The J_{th} in the LD with GaAs:Er,O as a waveguide layer is seven times larger than that in the LD with an undoped GaAs waveguide layer. The increase in the J_{th} also reflects the ultrafast carrier capture by an Er-related trap in GaAs:Er,O.

What wavelength is required for the direct excitation of Er^{3+} ions? Hogg *et al.* reported a spectroscopic study on GaAs:Er,O with excitation in the near-band-edge region [14]. They observed two extremely sharp signals, attributed to $^4I_{15/2}$ - $^4I_{11/2}$ intra-4f-shell transitions of Er^{3+} ions in the Er-2O center, at 981.4 nm and 984.1 nm on a broad background due to indirect excitation by way of defect levels. In this study, therefore, the lasing wavelength of the LD was changed by the device temperature to tune the objective wavelengths. The Er intensity (λ = 1539 nm) at 110 K is shown in figure 7 as a function of the current density. The *I-L* characteristics for the GaInAs QW emission are also shown in the figure. The J_{th} at the temperature is 0.84 kA/cm^2 and the lasing wavelength is 981 nm. In the region for the spontaneous emission from the GaInAs QWs, the Er intensity increases steeply with the current density and then starts to saturate. In the region of the stimulated QW emission, on the other hand, the intensity continues to increase. This behavior is quite different from the behavior in the GaInP/GaAs:Er,O/GaInP DH LDs exhibiting a lasing at the band-edge of GaAs. The disappearance of the Auger effect in the GaInAs QW-LD is attributed to spatial separation between QWs and Er ions. The gradual increase in Er intensity under the QW lasing might be due to the resonant direct excitation of Er^{3+} ions. Further study on the increase of the intensity is now in progress.

5. Conclusions

Two types of GaAs:Er,O-based LD were fabricated by OMVPE and effects of the lasing on the Er luminescence was investigated. In the LD exhibiting a lasing at the GaAs band-edge at room temperature, the J_{th} increased with the Er flow rate during the growth and its reciprocal cavity length dependence indicated the effective gain factor of the LDs decreased by the codoping of Er and O, which is due to ultrafast capturing of injected carriers at an Er-related trap. The fast capturing process allowed us to observe both the spontaneous intra-4*f*-shell emission due to Er^{3+} ions at 1.54 μm and the stimulated emission at the GaAs band-edge in the Er-doped LDs. In the LD with GaInAs QWs embedded in the GaAs:Er,O, the lasing wavelength was designed to tune to the energy separation between $^4I_{15/2}$ - $^4I_{11/2}$ intra-4f-shell transitions of Er^{3+} ions. The J_{th} for the lasing at room temperature was seven times larger than that of a GaInAs QW-LD with an undoped GaAs waveguide layer. The increase in the J_{th} also reflected the ultrafast carrier capture by an Er-related trap in GaAs:Er,O. In the region for spontaneous emission from the GaInAs QWs, the Er intensity revealed initially steep increase with injected current density, which was followed by a saturation. In the stimulated QW emission region, the intensity continued to increase with the current density, indicating a feasibility of the resonant direct excitation of Er^{3+} ions.

Acknowledgements

This work was supported in part by Grant-in-Aid for Creative Scientific Research No. 19GS1209 from the Japan Society for the Promotion of Science. This work was also supported in part by Grant-in-Aids for Exploratory Research No. 19656082, for Scientific Research of Priority Areas "Panoscopic Assembling and Highly Ordered Functions for Rare Earth Materials" No. 17042016 and 19018014, and the Global COE Program "Advanced Structural and Functional Materials Design" from the Ministry of Education, Culture, Sports, Science and Technology (MEXT).

References

[1] Fujiwara Y, Ofuchi H, Tabuchi M, Takeda Y 2000 *InP and Related Compounds -Materials, Applications and Devices-, Optoelectronic Properties of Semiconductors and Superlattices* vol. 9, ed M O Manasreh (The Netherlands: Gordon and Breach Science Pub.) p 251
[2] Takahei K and Taguchi A 1993 *J. Appl. Phys.* **74** 1979
[3] Takahei K, Taguchi A, Horikoshi Y and Nakata J 1994 *J. Appl. Phys.* **76** 4332
[4] Koizumi A, Fujiwara Y, Inoue K, Urakami A, Yoshikane T and Takeda Y 2003 *Jpn. J. Appl. Phys.* **42** 2223
[5] Koizumi A, Fujiwara Y, Urakami A, Inoue K, Yoshikane T and Takeda Y 2003 *Physica B* **340-342** 309
[6] Fujiwara Y, Takemoto S, Nakamura K, Shimada K, Suzuki M, Hidaka K, Terai Y and Tonouchi M 2007 *Physica B* **401-402** 234
[7] Shimada K, Terai Y, Takemoto S, Hidaka K, Fujiwara Y, Suzuki M and Tonouch M 2008 *Appl. Phys. Lett.* **92** 111115
[8] Terai Y, Hidaka K, Fujii K, Takemoto S, Tonouchi M and Fujiwara Y 2008 *Appl. Phys. Lett.* **93** 231117
[9] Pilkuhn M, Rupprecht H and Blum S 1964 *Solid-State Electron.* **7** 905
[10] Stern F 1973 *IEEE J. Quantum Electron.* **9** 290
[11] Dapkus P D, Holonyak N Jr, Burnham R D, Keune D L, Burd J W, Lawley K L and Walline R E 1970 *J. Appl. Phys.* **41** 4194
[12] Lee M H, Holonyak N Jr, Campbell J C, Groves W O, Craford M G and Keune D L 1974 *Appl. Phys. Lett.* **24** 310
[13] Panish M B, Casey H C Jr, Sumski S and Foy P W 1973 *Appl. Phys. Lett.* **22** 590
[14] Hogg R A, Takahei K and Taguchi A 1996 *J. Appl. Phys.* **79** 8682

Luminescence properties of Eu-implanted GaN-based semiconductors

H Kasai, A Nishikawa, Y Terai, and Y Fujiwara

Division of Materials and Manufacturing Science, Graduate School of Engineering, Osaka University, 2-1 Yamadaoka, Suita, Osaka 565-0871, Japan

Hitoshi.Kasai@mat.eng.osaka-u.ac.jp

Abstract. We investigated the relationship between optical properties and crystal quality of Eu-implanted GaN-based semiconductors. Emission peaks that are due to intra-4f shell transitions in Eu^{3+} ions were observed in Eu-implanted GaN, InGaN, and AlGaN. Emission wavelengths from the Eu ions were constant regardless of the matrix materials. The emission peak intensity in the Eu-implanted AlGaN was higher than that in the Eu-implanted InGaN or GaN because of less matrix damage by ion implantation.

1. Introduction

Rare-earth (RE) doped wide-gap semiconductors such as GaN show sharp emission peaks in the visible range and this emission originates from the RE 4f shell by optical excitation. Red, green and blue emissions were obtained from Eu-, Er- and Tm-ions, respectively [1,2]. These optical properties have a potential to be applied to novel visible light-emitting diodes, laser diodes and displays [3,4]. Thus it is important to investigate the luminescence properties of the RE doped GaN-based semiconductor materials in detail. In this paper, we investigated luminescence properties and crystal quality of various Eu-implanted GaN-based semiconductors such as GaN, InGaN, and AlGaN systematically and report the influence of the matrix materials on Eu emission.

2. Experimental

Eu ion was implanted into various GaN-based semiconductors by ion implantation with an accelerating voltage of 200 keV. To investigate the Eu dose dependence on luminescence properties, Eu was doped into free-standing GaN substrates with doses ranging from 1×10^{13} cm^{-2} to 1×10^{15} cm^{-2}. The implantation depth in GaN was calculated to be about 40 nm at 200 keV. To investigate effects of matrix materials, Eu was also doped into GaN, InGaN and AlGaN epilayers that were grown on sapphire substrates by organometallic vapour phase epitaxy (OMVPE) at a dose of $1x10^{14}$ cm^{-2} and 200 keV. In compositions were 8, 13 and 17% in the InGaN epilayers and Al compositions were 5, 12 and 18% in AlGaN epilayers. Thicknesses of the InGaN and AlGaN epilayers ranged from 50 nm to 120 nm. A schematic of the structures is shown in figure 1. After the ion implantation all samples were annealed at 1000 °C with N_2 ambient for 1 min in a lamp heating furnace. Luminescence properties were examined by room temperature photoluminescence (PL) measurements using a He-Cd laser (325nm). The crystal quality was evaluated by double-crystal-x-ray diffraction measurements.

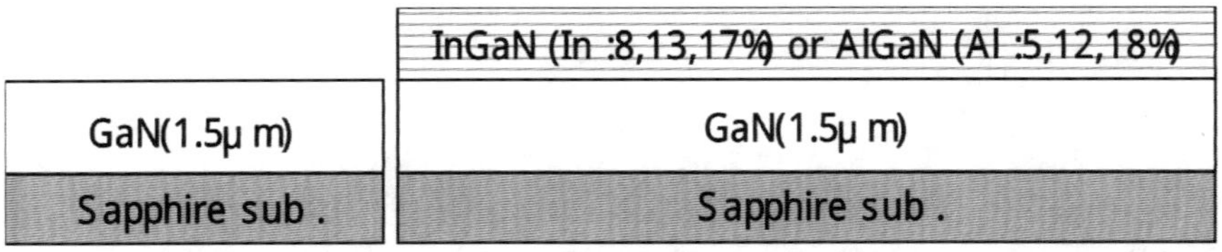

Figure 1. A schematic of GaN, InGaN and AlGaN epilayer structures on sapphire.

3. Results and discussion

Figure 2 shows PL spectra for the Eu-implanted freestanding GaN substrates after annealing. Before Eu implantation, a strong emission peak was observed at 367 nm that was due to the band-edge emission of GaN. This band-edge emission peak disappeared after Eu implantation because of implantation damage and no luminescence peak was observed in the spectral range from 350 nm to 700 nm. After annealing, emission peaks were observed at 544 nm, 601 nm, 623 nm and 666 nm and were due to intra-4f shell transitions of Eu^{3+} ions. The assigned transitions are $^5D_1$-$^7F_1$, $^5D_0$-$^7F_1$, $^5D_0$-$^7F_2$ and $^5D_0$-$^7F_3$, respectively [5,6]. As seen in the inset of Figure 2 the intensity of the emission line at 623 nm increases with an increase in the Eu dose. However, the rate of increase decreases and is almost saturated in the high Eu-implanted region.

Figure 3 shows XRD (ω-2θ) profiles for the Eu-implanted free-standing GaN substrates (a) after Eu implantation and (b) after subsequent annealing. In addition to the GaN(0002) reflection peak a Eu-related peak emerges at a lower angle because of Eu-induced strain after Eu implantation. The Eu-related peak disappears after annealing because of the relaxation of induced strain by reconstruction and diffusion. The GaN(0002) reflection peak is still present. However, the shape of the GaN(0002) peak becomes asymmetrical and a billow is present at the lower angle side because of the residual Eu. The height of the billow increases as the Eu dose increases. The FWHM of the rocking curve in the Eu-implanted GaN substrate increases after Eu implantation because of implantation damage. However, after annealing it recovers slightly owing to relaxation of the strain and improved crystal quality.

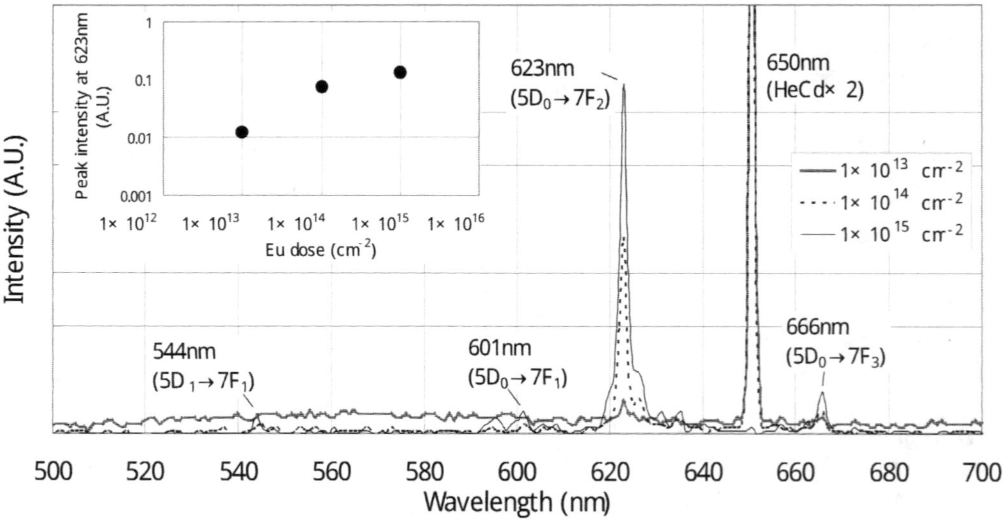

Figure 2. PL spectra for Eu-implanted GaN substrates after annealing. The inset shows the Eu dose dependence of the 623 nm emission peak intensity in Eu-implanted GaN substrates.

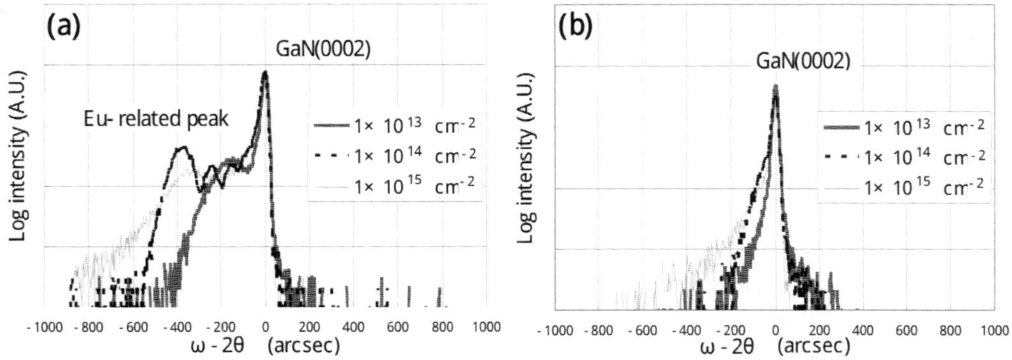

Figure 3. XRD (ω-2θ) profiles in GaN substrates (a) after Eu implantation and (b) after annealing.

We will now discuss the effects of matrix materials in Eu-implanted GaN-based semiconductors. Figure 4 shows the In and Al composition dependence of the PL emission peak intensity at 623 nm in the Eu-implanted InGaN and AlGaN epilayers after annealing. The emission peaks that are due to a band-edge emission from InGaN and AlGaN epilayers disappeared after Eu implantation as observed in the Eu-implanted GaN. After annealing, the emission peak due to Eu ions was observed at 623 nm for all samples. Wavelengths of Eu emission peaks remain constant regardless of the matrix materials because the origin of the emission is the intra-4f shell transitions of Eu ions. The emission peak intensity in the Eu-implanted InGaN is lower than that in the Eu-implanted GaN while the peak intensity of the Eu-implanted AlGaN is higher than that in the Eu-implanted GaN and increase with Al composition.

For Eu-implanted InGaN samples, the XRD (ω-2θ) peak of InGaN was absent because of implantation damage and not recovered after annealing, as shown in figure 5(b), (c). The recovery of this peak, as seen for the Eu-implanted GaN, was not evident after annealing. On the other hand, the XRD peak of AlGaN (Al 18%) is still present even after Eu implantation but shifts to a lower angle because of the Eu-induced strain, as shown in figure. 6(b). After annealing, XRD peaks of all the AlGaN samples were recovered. In particular, the XRD peaks of AlGaN (Al 12%) and AlGaN (Al 18%) regained their initial levels before implantation, as shown in figure 6(c). This means that AlGaN had less implantation damage compared with GaN or InGaN because atomic bonds in AlGaN are stronger than those in GaN and InGaN. According to these results the Eu emission peak intensity is dependent on damage to the matrix materials or crystal quality after Eu implantation and annealing, since the Eu ions in GaN-based semiconductor are not excited by He-Cd laser directly, but are excited through the energy transfer from the matrix materials. For more emission peak intensity in Eu-doped InGaN it is necessary to reduce the matrix damage and to improve the crystal quality by optimizing the implantation or annealing conditions.

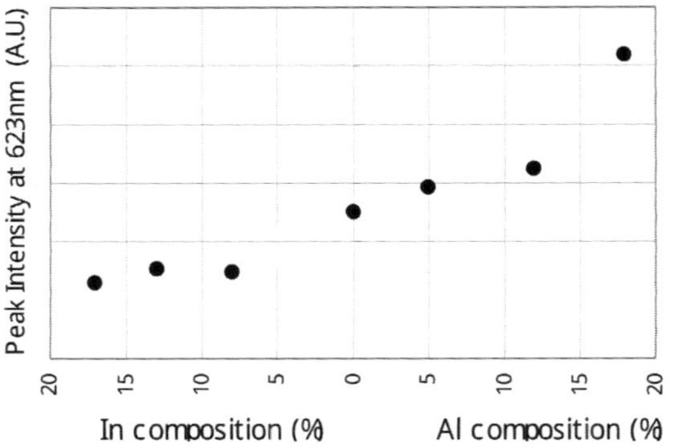

Figure 4. In or Al composition dependence of the 623 nm emission peak intensity in InGaN or AlGaN epilayers.

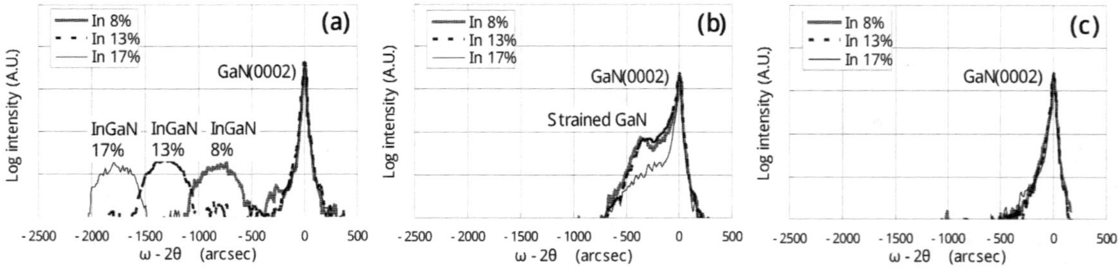

Figure 5. XRD (ω-2θ) profiles in InGaN epilayers (a) before and (b) after Eu implantation and (c) after annealing.

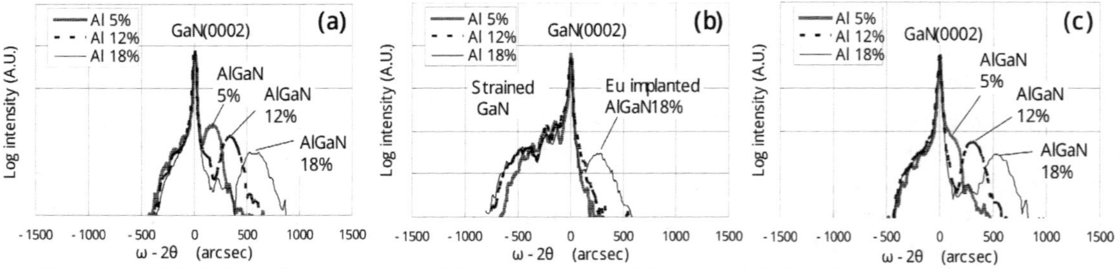

Figure 6. XRD (ω-2θ) profiles in AlGaN epilayers (a) before and (b) after Eu implantation and (c) after annealing.

4. Conclusion

We observed emission peaks due to intra-4f shell transitions of Eu^{3+} ions in Eu-implanted GaN at room temperature. The peak intensity increased as the Eu dose increased and became saturated in the high dose region. The wavelength of the Eu emission peaks remained constant regardless of the matrix materials. However, the peak intensity in the Eu-implanted AlGaN was higher than that in Eu-implanted GaN or InGaN and it increased with Al composition. This behaviour of Eu emission peak intensity was due to the difference in implantation damage in the matrix materials. AlGaN was less affected by implantation damage because atomic bonds in AlGaN are stronger than those in GaN and InGaN.

The authors thank Mr. Sato (Japan Atomic Energy Agency), Prof. A. Wakahara and Mr. Shimojo (Toyohashi University of Technology) for ion implantations.

References

[1] Steckl A J and Zavada J M 1999 *MRS Bul.* **24,** 16
[2] Steckl A J, Heikenfeld J C, Lee S, Garter M J, Baker C C, Wang Y and Jones R 2002 *IEEE J.Sel. Top. Quantum Electron.* **8,** 749
[3] Park J H and Steckl A J 2004 *Appl. Phys. Lett.* **85,** 20
Strite S and Morkoc H 1992 *J. Vac. Sci. Technol.* B **10** 1237
[4] Wang Y Q and Steckl 2004 *A J Phys Lett* **84,** 15
[5] Nakanishi Y, Wakahara A, Okada H, Yoshida A, Ohshima T, Itoh H, Nakao S Saito K and Kim Y T 2003 *Nucl. Instrum. Methods, Phys. Res.* B **206,** 1033
[6] Shirakata S, Sasaki R and Kataoka T 2004 *Appl. Phys. Lett.* **84**, 12

Metaloraganic chemical vapor deposition of Er-doped ZnO thin films with 1.54 μm photoluminescence

Keisuke Yamaoka[1], Yoshikazu Terai[1], Takashi Yamaguchi[1], Ha Ngoc Ngo[2], Tom Gregorkiewicz[2], Yasufumi Fujiwara[1]
[1] Division of Materials and Manufacturing Science, Graduate School of Engineering, Osaka University, Yamadaoka 2-1, Suita, Osaka 565-0871, Japan
[2] Van der Waals-Zeeman Institute, University of Amsterdam, Valckenierstraat 65, NL-1018XE Amsterdam, The Netherlands

E-mail: keisuke.yamaoka@mat.eng.osaka-u.ac.jp

Abstract. Er-doped ZnO (ZnO:Er) thin films were grown by metalorganic chemical vapor deposition. Subsequent annealing improved the crystallinity of the ZnO:Er by suppressing nonradiative centers in the films. The annealed ZnO:Er showed clear 1.54 μm PL originating from the $^4I_{13/2} \rightarrow {}^4I_{15/2}$ transitions of Er^{3+} ions. This was in addition to luminescence from the ZnO host. Er-related PL intensity increased when prepared with annealing temperatures of 800–900 °C, while annealing at 1000 °C reduced the intensity. PL of 1.54 μm with high intensity was obtained in the ZnO:Er film that exhibited band-edge emission of free- or bound-excitons.

1. Introduction

Interest in ZnO has been growing steadily over the past decade because of its wide applicability in many types of devices such as light emitting devices and transparent field effect transistors. For application in electroluminescent (EL) devices, ZnO thin films are potentially useful as light-emitting materials. They possess a large exciton binding energy (60 meV) for ultraviolet (UV) luminescence and also for visible and infrared (IR) emission by impurity doping [1,2]. Rare-earth (RE) doped ZnO, in particular, enables us to tune the luminescent wavelength and this is dependent on the intra-4*f* shell transition of RE ions [3]. Luminescence spectra from RE ions have very sharp emission lines at their temperature-stable wavelengths since the 4*f* shell is well shielded by the outer 5*s* and 5*p* electrons.

Er-doped ZnO (ZnO:Er) has attracted much attention because of its effective 1.54 μm emission which is due to the intra-4*f* shell transition of Er^{3+} ions [4,5]. The 1.54 μm wavelength is important for optical communication systems as it is compatible with the low-loss window of silica-based optical fibers. Recently, ZnO:Er thin films prepared by several deposition methods such as laser ablation [4], sputtering [6], wet-processing techniques [7] and ion implantation [8] have been widely studied. Metalorganic chemical vapor deposition (MOCVD) has advantages for device fabrication with a requirement for heterojunctions and large-area deposition. This technique has however not been used for the development of ZnO:Er films.

In this paper, we report growth of ZnO:Er films by MOCVD for the first time and the effects of post-growth annealing on structural and luminescent properties of the films.

2. Experimental

ZnO:Er films were grown on *c*-plane sapphire substrates using a dedicated MOCVD system. In the bell-jar chamber, the substrates were kept on a horizontal substrate holder. Gas-feed tubes were attached vertically to substrates and positioned 10 cm from the substrates. Diethylzinc (DEZn), O_2 and trisisopropylcyclopentadienylerbium [$(i\text{-}C_3H_7C_5H_4)_3Er$], abbreviated as $Er(i\text{-}PrCp)_3$, were used as source materials. DEZn that was evaporated in a cylinder at 60 °C and O_2 gas were separately introduced into the chamber at flow rates of 1.0 sccm and 10.0 sccm, respectively. A cylinder of $Er(i\text{-}PrCp)_3$ was maintained at 90 °C and Ar flow through the cylinder was fixed at 250 sccm. During growth, the reactor pressure was 3.7 Torr. The substrate temperature was kept at 400 °C using a resistive heater. Samples were grown for 10 min at a growth rate of about 100 nm/min. Subsequently, the grown films were annealed at 800–1000 °C in ambient O_2 for 30 min.

Crystalline structures of films were characterized by X-ray diffraction (XRD) measurements using $K\alpha$ radiation. Ultraviolet and visible photoluminescence (PL) measurements were carried out at 77 K. A 325 nm He-Cd laser and a charge coupled device were used for sample excitation and signal detection, respectively. For near infrared (IR) PL measurements at room temperature (RT) and at 4.2 K, samples were illuminated by the 355 nm third harmonic of a *Q*-switched Nd:YAG laser. PL spectra and decay signals were collected using a monochromator and a liquid-nitrogen-cooled Ge photodiode.

3. Results and discussion

Figure 1 shows XRD patterns of as-grown and annealed ZnO:Er films. In the as-grown film, peaks of ZnO(002) and (101) planes are observed at 34.5° and 36.3°, respectively. The higher intensity of the ZnO(002) peak indicates a *c*-axis preference for the wurtzite ZnO:Er crystal. Both the (002) and (101) peaks clearly shift to low diffraction angles by post-growth annealing. The film annealed at 800 °C shows a (002) peak with a diffraction angle of 34.43°, which is almost the same as that of bulk ZnO [9]. The full width at half maximum (FWHM) of the peaks decreased as the annealing temperature increased. These results indicate that annealing relaxes strain induced during the MOCVD growth and improves the crystalline quality of the ZnO:Er films. No peaks of the Er_2O_3 phase were observed in as-grown or annealed films.

Figure 2 shows 77-K PL spectra of the ZnO:Er films. Near band-edge PL peaks at ~3.4 eV and a broad band at ~2.4 eV are only observed in the annealed samples indicating that annealing decreases nonradiative centers present in the as-grown ZnO:Er. Shapes of the spectra change drastically as the annealing temperature changes. The intensity of near band-edge PL peaks increased at annealing

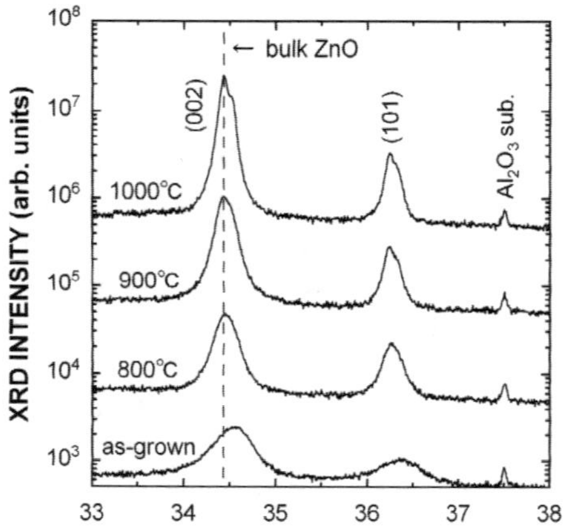

Figure 1. XRD patterns of as-grown and annealed ZnO:Er films. The dashed line corresponds to the (002) diffraction angle of bulk ZnO.

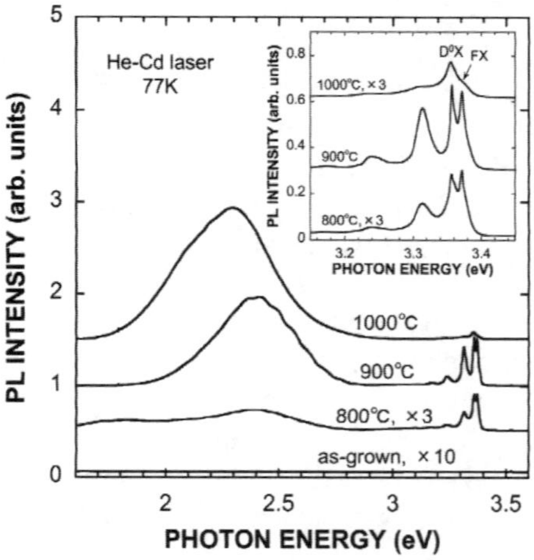

Figure 2. 77-K PL spectra of ZnO:Er films in the UV and visible region. The inset shows PL spectra around the band-edge emission.

temperatures from 800 to 900 °C and decreased at annealing temperatures from 900 to 1000 °C. On the other hand, the green-yellow band becomes large as the annealing temperature increases, accompanied by a redshift of the peak. The visible luminescence band of ZnO is known to include several peaks which are typically attributed to native defects such as oxygen vacancies (V_O) at ~2.4 eV, zinc vacancies (V_{Zn}) and oxygen interstitials (O_i) at 2.0–2.3 eV [10]. Therefore, the red shift of the band results from an increase in defects, presumably V_{Zn} or O_i, that are formed by high-temperature annealing in an O_2 atmosphere.

PL spectra around the band-edge emission are shown in the inset of Figure 2. Peaks at 3.357 and 3.371 eV correspond to neutral-donor-bound exciton (D^0X) transitions and free-exciton (FX) transitions, respectively [11]. Spectra at 800 and 900 °C show a clear FX peak, while there is no obvious FX peak at 1000 °C. These spectral changes also suggest that optical processes in the ZnO host vary significantly because of high-temperature annealing at 1000 °C. The origin of the peak at 3.313 eV with LO-phonon replicas is uncertain, but the emission might be attributed to acceptor-bound-exciton transitions [12], donor-acceptor pair recombinations [13] or two electron satellites of the D^0X [14].

The inset of Figure 3 shows 4.2-K PL spectra of ZnO:Er films annealed at 900 °C in the near-IR region. PL at around 1.54 μm (0.805 eV) that corresponds to the intra-4*f* shell transition from the first excited state ($^4I_{13/2}$) to the ground state ($^4I_{15/2}$) of Er^{3+} ions is observed in the annealed samples. The 1.54 μm emission confirms that Er^{3+} ions are successfully incorporated into ZnO films by the MOCVD method. In addition, the Er-related emission under the bandgap excitation indicates an energy transfer between the ZnO host and Er^{3+} ions. The PL peak intensity at 1.538 μm (0.806 eV) as a function of the annealing temperature is plotted in Figure 3. The intensity increases at 800–900 °C and decreases at 900–1000 °C. Figure 4 shows PL decay transients at 1.538 μm for the ZnO:Er film annealed at 900 °C. The decay curve was fitted with a double exponential form and lifetimes of the fast and slow components are ~2 ms and ~7 ms, respectively. Similar nonexponential decay transients containing fast and slow components have been observed for various Er-doped semiconductors [15]. The lifetime of ~7 ms is comparable to that of Er-doped SiO_2 [16] and longer than that of Er-doped GaN (~2 ms) [17].

Crystallinity and luminescent properties of ZnO:Er films change significantly by post-growth annealing. XRD and PL results show that lattice strain and a large amount of defects remain in as-grown films. Annealing at 800–900 °C induces relaxation of strain and a reduction of defects that act as nonradiative recombination centers. As a result, the high intensity Er-related 1.54-μm emission is

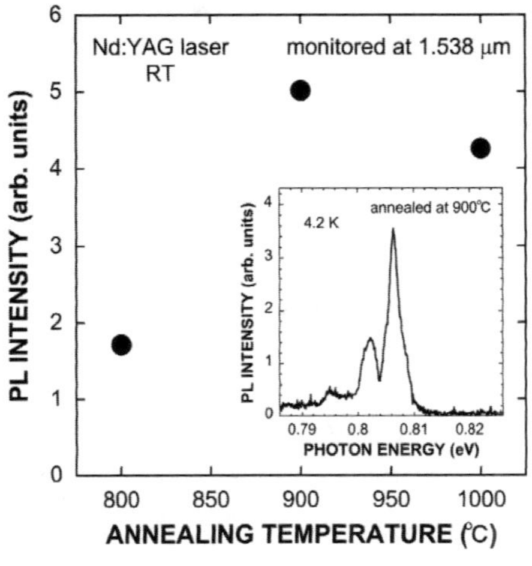

Figure 3. Dependence of 1.538-μm PL intensity at RT on annealing temperature. The inset is 4.2-K PL spectra of the ZnO:Er annealed at 900°C.

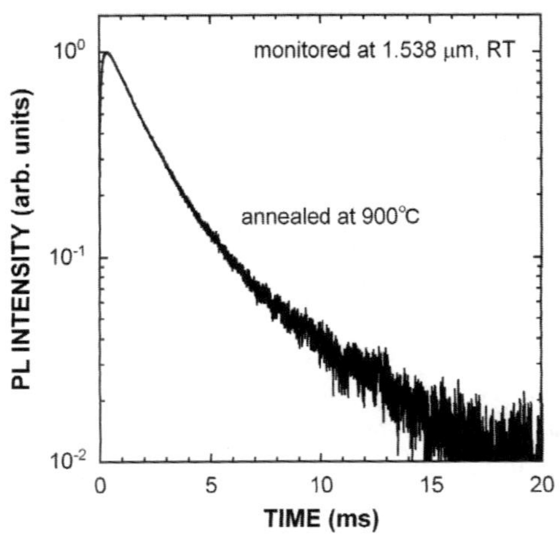

Figure 4. Time-resolved PL intensity at 1.538 μm in ZnO:Er annealed at 900°C. The data was taken at RT.

observed in the ZnO:Er film that was annealed at 900 °C. Although the ZnO:Er film that was annealed at 1000 °C shows the highest crystal quality from XRD, both the 1.54 μm PL and the band-edge emission were degraded. High-temperature annealing in an O_2-rich environment could introduce additional defects such as V_{Zn} or O_i into the ZnO host of high crystallinity. This would lead to a weak band-edge emission and a redshift of the visible PL band. These results indicate that carrier-recombination processes in the ZnO host directly affect the efficiency of the Er-related 1.54 μm luminescence.

4. Conclusion

Structural and luminescent properties of ZnO:Er thin films grown by MOCVD were investigated as a function of post-growth annealing temperature. Annealing at 800–900 °C in O_2 improved crystalline quality and reduced nonradiative centers, resulting in an enhancement of 1.54 μm PL because of the $^4I_{13/2} \rightarrow {}^4I_{15/2}$ transitions of Er^{3+} ions. High-temperature annealing at 1000 °C changes luminescent processes in the ZnO host by introducing defects that results in quenching of 1.54 μm PL.

Acknowledgements

This study was supported by the Priority Assistance for the Formation of Worldwide Renowned Centers of Research – The Global COE Program (Project: Center of Excellence for Advanced Structural and Functional Materials Design), and by a Grant-in-Aid for Creative Scientific Research (No. 19GS1209) from the Japan Society for the Promotion of Science (JSPS). One of the authors (K. Y.) was also supported by a Grant-in-Aid for JSPS Fellows (No. 1910120).

References

[1] Meyer B K, Polity A, Farangis B, He Y, Hasselkamp D, Krämer T and Wang C 2004 *Appl. Phys. Lett.* **85** 4929
[2] Ishihara J, Nakamura A, Shigemori A, Aoki T and Temmyo J 2006 *Appl. Phys. Lett.* **89** 091914
[3] Bachir S, Azuma K, Kossanyi J, Valat P and Ronfard-Haret J C 1997 *J. Lumin.* **75** 35
[4] Komuro S, Katsumata T, Morikawa T, Zhao X, Isshiki H and Aoyagi Y 2000 *Appl. Phys. Lett.* **76** 3935
[5] Zhou Z, Komori T, Yoshino M, Morinaga M, Matsunami N, Koizumi A and Takeda Y 2005 *Appl. Phys. Lett.* **86** 041107
[6] Song H and Kim Y J 2007 *J. Eur. Ceram. Soc.* **27** 3745
[7] Sun Y, Chen Y, Tian L, Yu Y, Kong X, Zeng Q, Zhang Y and Zhang H 2008 *J. Lumin.* **128** 15
[8] Fukudome T, Kaminaka A, Isshiki H, Saito R, Yugo S and Kimura T 2003 *Nucl. Instrum. Meth.* B **206** 287
[9] Zeuner A, Alves H, Hofmann D M, Meyer B K, Heuken M, Bläsing J and Krost A 2002 *Appl. Phys. Lett.* **80** 2078
[10] Reshchikov M A, Xie J Q, Hertog B and Osinsky A 2008 J. Appl. Phys. **103** 103514
[11] Özgür Ü, Alivov Y I, Liu C, Teke A, Reshchikov M A, Doğan S, Avrutin V, Cho S J and Morkoç H 2005 *J. Appl. Phys.* **98** 041301
[12] Fonoberov V A, Alim K A, Balandin A A, Xiu F and Liu J 2006 *Phys. Rev. B* **73** 165317
[13] Zhang B P, Binh N T, Segawa Y, Wakatsuki K and Usami N 2003 *Appl. Phys. Lett.* **83** 1635
[14] Meyer B K, Alves H, Hofmann D M, Kriegseis W, Forster D, Bertram F, Christen J, Hoffmann A, Straßburg M, Dworzak M, Haboeck U and Rodina A V 2004 *Phys. Stat. Sol.* (b) **241** 231
[15] Wu X, Hömmerich U, Mackenzie J D, Abernathy C R, Pearton S J, Schwartz R N, Wilson R G and J. M. Zavada 1997 *Appl. Phys. Lett.* **70** 2126
[16] Polman A 1997 *J. Appl. Phys.* **82** 1
[17] Hansen D M, Zhang R, Perkins N R, Safvi S, Zhang L, Bray K L and Kuech T F 1998 *Appl. Phys. Lett.* **72** 1244

Journal of Physics: Conference Series **165** (2009) 012028 doi:10.1088/1742-6596/165/1/012028

Photoluminescence of $CuInS_2$-based semiconductor quantum dots; Its origin and the effect of ZnS coating

Katsuhiro Nose[1], Naoya Fujita[1], Takahisa Omata[1], Shinya Otsuka-Yao-Matsuo[1], Wataru Kato[2], Masato Uehara[3], Hiroyuki Nakamura[3], Hideaki Maeda[2,3], Hayato Kamioka[4], and Hideo Hosono[4,5]

[1]Division of Material and Manufacturing Science, Graduate School of Engineering, Osaka University, 2-1 Yamada-oka, Suita, Osaka 565-0871, Japan
[2]Department of Molecular and Material Sciences, Interdisciplinary Graduate School of Engineering Science, Kyusyu University, 6-1 Kasugakoen, Kasuga 816-8580, Japan
[3]Nanotechnology Research Institute, National Institute of Advanced Industrial Science and Technology (AIST), 807-1, Shuku-machi, Tosu, Saga 841-0052, Japan
[4]Transparent Electro-Active Materials Project, Exploratory Research for Advanced Technology-Solution Oriented Research for Science and Technology (ERATO-SORST), Japan Science and Technology Agency, in frontier Collaborative Research Center, Tokyo Institute of Technology, 4259 Nagatsuta, Midori-ku, Yokohama 226-8503, Japan
[5]Materials and Structures Laboratory and Frontier Collaborative Research Centre, Tokyo Institute of Technology, 4259 Nagatsuta, Midori-ku, Yokohama 226-8503, Japan

E-mail: omata@mat.eng.osaka-u.ac.jp

Abstract. The origin of the size-dependent photoluminescence (PL) of $CuInS_2$-ZnS alloyed quantum dots and the effect of a ZnS thin layer coating on PL properties were studied. The PL emission band shifted to the shorter wavelength region with a decrease in the crystal size. A time-resolved measurement indicated that the PL lifetime for the lower energy component of the emission band was longer than that of the higher energy component. Although this is a characteristic feature of the donor-acceptor pair (DAP) recombination, the size dependent shift of the band was too large to be attributed to DAP recombination. The emission was attributed to an electronic transition from the quantum confined conduction band to an acceptor level. By coating with a ZnS thin layer the PL quantum yield increased up to 31%.

1. Introduction

The synthesis and optical properties of colloidal $CuInS_2$-ZnS alloyed quantum dots (QDs), designated as ZCIS QDs hereafter, have been recently reported [1]. The size-dependent optical band gap and the photoluminescence (PL) emission due to the quantum confinement effect were clearly observed the QDs. They are, therefore, expected to be used as cadmium-free and practical semiconductor QDs phosphors. However, the origin of the PL emission remains elusive. An understanding of the origin of their emission is important for the improvement of their optical properties. For application of QDs to electroluminescent devices they need to be surface-coated with inorganic materials, i.e., fabrication of a core/shell structure to form a type-I quantum well structure. This is necessary to maintain the quantum confinement of electrons and holes in the core-QDs even when the QDs are piled up and in contact with adjacent QDs in deposited films. We thus studied the origin of the PL emission band in ZCIS QDs and the effect of a ZnS thin layer coating on the PL property.

2. Experimental

2.1. Synthesis of ZCIS QDs

0.25 mmol of zinc diethyldithiocarbamate (DDCZ) was dissolved in 3.0 mL of tri-*n*-octylphosphine (TOP) at room temperature. Then, 1.0 mL of this solution and 4.0 mL of 1-octadecene (ODE) were mixed; mixed solution was designated the Zn-S solution. 0.1 mmol of CuI was dissolved in 3.0 mL of oleylamine at 60 °C. 0.1 mmol of InI_3 was dissolved in 3.0 mL of oleylamine at room temperature. 2.5 mL of Zn-S solution and 1.25 mL of copper and indium solutions were mixed. This mixed solution in a glass vial was heated in an oil-bath and maintained at 200 °C for 60~300 s with argon gas bubbling.

2.2. Synthesis of ZCIS/ZnS core/shell QDs

The surface coating of the ZCIS QDs by a ZnS thin layer, i.e., the fabrication of core/shell-type ZCIS/ZnS QDs, was carried out under two different sets of conditions to prepare core/shell QDs with different ZnS thickness. Approximately 2.26 mg of ZCIS QDs were extracted form 1.0 mL of colloidal solution. For a very thin ZnS coating, the ZCIS QDs extracted were re-dispersed in 1.5 mL of 5.5 mM DDCZ solution in TOP and ODE containing of 32 μL oleic acid as surfactant. For a relatively thicker ZnS coating, the ZCIS QDs extracted were re-dispersed into 2.0 mL of 35.3 mM DDCZ solution in TOP and ODE containing 210 μL oleic acid. These were reacted at 140 °C for 5 min.

2.3. Characterization

Optical absorption and PL spectra of colloidal solutions were measured for solutions in which QDs extracted from the as-synthesized colloidal solutions were re-dispersed into chloroform. PL spectra of powdered QDs were measured for the dried QDs extracted from the as-synthesized colloidal solution. Time-resolved PL was measured by using a streak camera and a YAG-OPO laser as an excitation source. The size of QDs and their distribution was evaluated by high-resolution transmission electron microscopy (HR-TEM). PL quantum yeilds (PLQY) were evaluated by comparison of the PL intensity with that of the standard sample of rhodamine B [2].

3. Result and discussion

Figure 1 shows optical absorption and PL spectra of colloidal ZCIS QDs synthesized using various reaction times. In the optical absorption spectra, shoulders that indicate the optical band gap shifted from 519 nm for the 300 s reaction to 437 nm for the 60 s reaction. The optical band gap increased as the reaction time decreased. In the PL spectra the broad PL emission band shifted from 657 nm for the 300 s reaction to 562 nm for the 60 s reaction which is corresponding to the optical band gap shift. The Stokes' shift, which corresponds to the energy difference between the optical band gap and the emission energy was calculated to be 500~600 meV. This value is too large to be attributed to a direct excitonic recombination because the Stokes' shift is at most 100 meV for excitonic recombination [3]. Such a large Stokes' shift indicates that the emission was attributable to the defect-related.

A streak camera image of time-resolved PL of ZCIS QDs is shown in Figure 2 (a). It is clearly observed that the PL lifetime for the longer wavelength component is longer than that for the shorter wavelength component. Figure 2 (b) shows PL decay curves at 610 nm, 630 nm and 650 nm emissions. The decay curves were well reproduced assuming a biexponential curves similar to the CdSe QDs cases [4]. PL lifetimes were 9 ns and 38 ns for the 610 nm emission, 19 ns and 53 ns for the 630 nm emission and 25 ns and 79 ns for the 650 nm emission. The observation of the longer emission lifetime for the longer emission wavelength is not corresponding to the direct exciton recombination dynamics. The longer emission lifetime for the longer wavelength emission component is a well-known feature

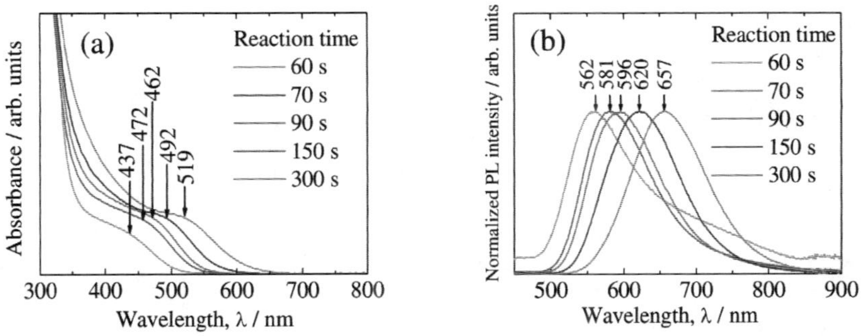

Figure 1. The optical absorption (a) and PL spectra (b) of a ZCIS QDs colloidal solution synthesized using various reaction times.

of a donor-acceptor pair (DAP) recombination [5]. However, the size dependent energy shift of the PL emission of the present samples, which was 320 meV between the 300 and 60 s reactions, was too large assuming that the energy shift came from the increase in the Coulomb interaction depending on the DAP distance. The large size-dependent shift of the emission energy (320 meV), which corresponds to approximately 70% of the size-dependent shift of the optical band gap (450 meV), indicates that the emission is related to the quantum confined level. According to the effective mass approximation calculation of the size-dependent optical band gap, an increase in the energies of the conduction band, the 1S(e) state, and the valence band, the 1S(h) state, are proportional to the reciprocal effective masses for electrons and holes, respectively [6]. This means that a smaller effective mass results in a larger increase in energy of the quantum confined level. The effective mass of an electron is generally smaller than that of a hole and therefore the size-dependent increase in energy of the conduction band should be greater than that of the valence band. Based on this situation, the size-dependent shift in emission which reaches up to 70% of that in the optical band gap indicates that the quantum confined conduction band should relate to the emission. Thus the electronic transition responsible for the emission is concluded to be the transition from the quantum confined conduction band to an acceptor level. Currently, the acceptor species relating to this emission is not clarified. Further study is needed to determine the defect species.

HR-TEM images of ZCIS QDs, obtained from the 300 s reaction, and ZCIS/ZnS core/shell QDs synthesized from the very thin ZnS coating condition are shown in Figure 3 (a) and (b), respectively. The insets of the respective images are the crystal size distributions obtained from HR-TEM observations. Average crystal sizes of ZCIS QDs and ZCIS/ZnS core/shell QDs were 2.7 nm and 3.5 nm, respectively. The increase in crystal size from 2.7 nm to 3.5 nm after the ZnS coating indicated that the ZnS layer with 0.4 nm thickness deposited on the surface of ZCIS QDs.

PL spectra of colloidal and powdered ZCIS and ZCIS/ZnS core/shell QDs, which were obtained under conditions used for the relatively thick ZnS coating, are shown in Figure 4 (a) and (b),

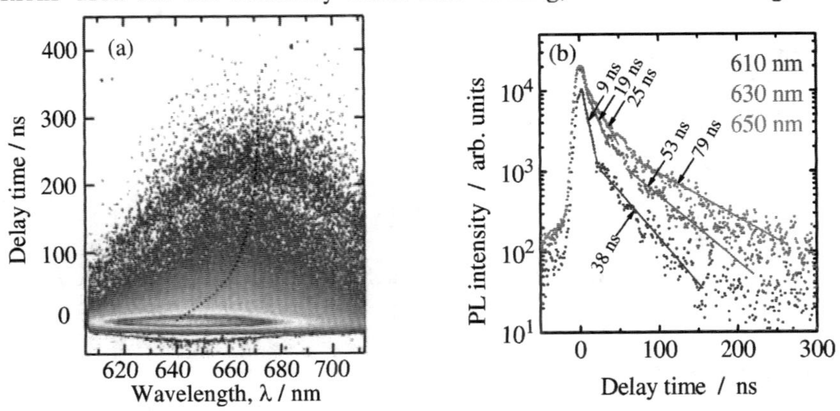

Figure 2. A streak camera image of time-resolved PL (a) and decay curves at 610, 630 and 650 nm emissions (b). The sample was synthesized at 200 °C for 300 s.

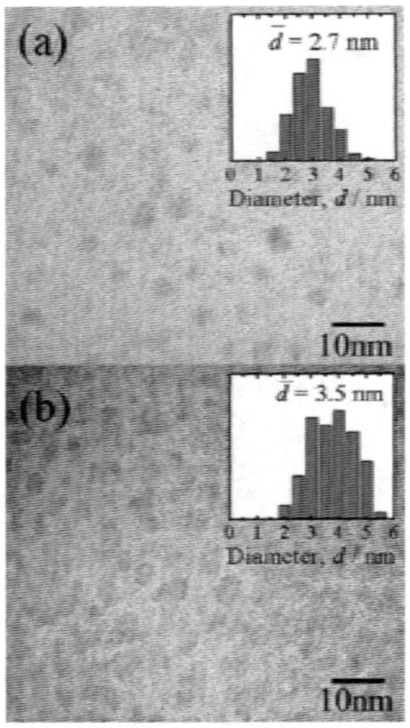

Figure 3. HR-TEM images of ZCIS QDs (a) and ZCIS/ZnS core/shell QDs (b).

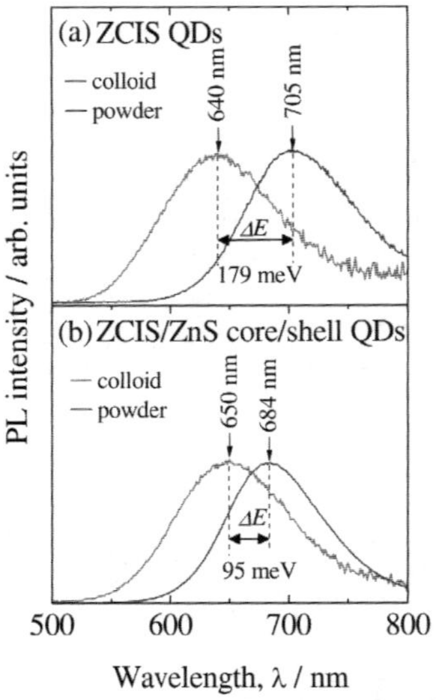

Figure 4. PL spectra of ZCIS QDs (a) and ZCIS/ZnS core/shell QDs (b).

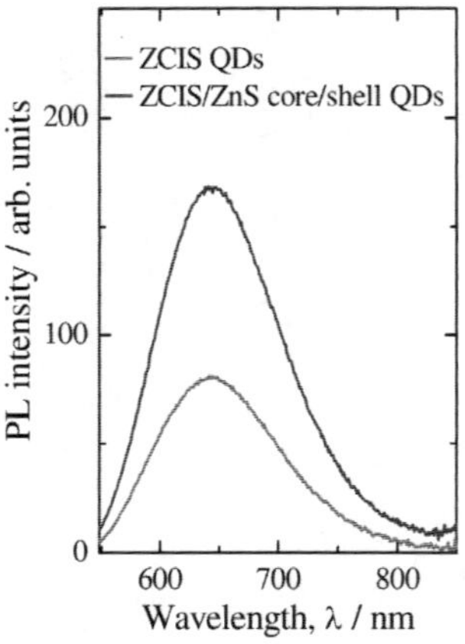

Figure 5. PL spectra of colloidal ZCIS QDs and ZCIS/ZnS core/shell QDs.

respectively. In the PL spectra of ZCIS QDs, the PL emission energy of the powdered form was 179 meV smaller than that of the colloidal form. Because the surfactant molecules, which confined electrons and holes within the QDs, were partially removed in powdered form, the each QDs should directly contact with adjacent ones. Thus the red-shift of the PL emission was attributed to the leakage of electrons and holes because of a partial breakdown of the quantum confinement effect. In PL spectra of ZCIS/ZnS core/shell QDs the energy difference between colloidal and powdered QDs was 95 meV and was smaller than that of bare ZCIS QDs. This result indicated that the ZnS coating was effectively maintaining the quantum confinement effect even in powdered form. Figure 5 shows PL spectra of colloidal ZCIS QDs and ZCIS/ZnS core/shell QDs with the relatively thick ZnS shell. The ZnS coating doubled the PL intensity and the PL quantum yield was 31%.

4. Conclusion

The origin of PL emission from ZCIS QDs was attributed to electron and hole recombination between the quantum confined conduction band and an acceptor level although the acceptor species is still unidentified. The ZnS coating was effective in maintaining the quantum confinement effect in powdered QDs and increased the PL quantum yield.

Acknowledgement

This work was supported by Priority Assistance for the Formation of Worldwide Renowned Centers of Research-The Global COE Program "Center of Excellence for Advanced Structural and Functional Materials Design" from the Ministry of Education, Culture, Sports, Science and Technology, Japan.

References

[1] Nakamura H, Kato W, Uehara M, Nose K, Omata T, Otsuka-Yao-Matsuo S, Miyazaki M, and Maeda H, 2006 *Chem. Mater.* **18** 3330
[2] Donegá C M, Hickey S G, Wuister S F, Vanmaekelbergh D, and Meijerink A, 2003 *J. Phys. Chem. B* **107** 489.
[3] Murray C B, Norris D J, and Bawendi M G, 1993 *J. Am. Chem. Soc.* **115** 8706.
[4] Javier A, Magana D, Jennings T, and Strouse G F, 2003 *Appl. Phys. Let.* **83** 1423.
[5] Thomas D G, Hopfield J J, and Augustyniak W M, 1965 *Phys. Rev.* **140** A202.
[6] Brus L E, 1984 *J. Chem. Phys.* **80**, 4403.

Magnetic behavior of single nanostructured magnet

O Kitakami[1], S Okamoto[1], N Kikuchi[1], T Shimatsu[2], H Sato[1] and H Aoi [2]

[1] IMRAM, Tohoku University, Sendai 980-8577, Japan
[2] RIEC, Tohoku University , Sendai 980-8577, Japan

E mail: kitakami@tagen.tohoku.ac.jp

Abstract. For future advancement in magnetic and spin electronic devices, it is essential to understand the magnetic behaviour and spin dynamics of a single nanomagnet. However, no effective method has been available for detection of such extremely small magnetic moment. To overcome this difficulty, we have newly developed a highly sensitive magnetic detection technique utilizing the anomalous Hall effect (AHE). This technique has systematically revealed the magnetic behaviors of various magnetic materials in the nanoscale regime. In this article, we demonstrate how single nanomagnets of $L1_0$ FePt and Co/Pt multilayers behave depending on their size, especially focusing on the magnetization reversal process and the bistability condition necessary for digital memory use.

1. Introduction

Rapid advances in recent high density magnetic memories and various spin electronic devices need deep understanding of the magnetic behavior and spin dynamics of their individual constituent nanomagnets. For instance, the areal data density of the most advanced hard disk drive (HDD) technology is expected to reach 1 tera bits per square inch (bpsi) within several years. For this density, the bit size reduces as small as 10 nm, and this size is comparable to the characteristic length of magnetic materials, such as exchange length or domain wall width. It is interesting to know how such a tiny magnet behaves in response to a magnetic field, temperature, and so on. This problem is very important both from scientific and technological points of view. In addition to magnetic recording mentioned above, spin electronic devices also consist of nanomagnets. For example, there are spin-valve magnetic sensor,[1] a magnetic random access memory consisting of very small magnetic tunnel junctions,[2] a spin torque oscillator being able to generate a microwave,[3] and a spin field effect transistor.[4] All these spin electronic devices consist of tiny nanomagnets. Therefore, for development of future high density memories and spin electronic devices, it is indispensable to quantitatively understand the magnetic behavior and spin dynamics of these nanomagnets. According to the micromagnetic theory, the exchange correlation length is a good measure to discuss the magnetic behavior of a nanomagnet.[5] When the particle size is smaller than the exchange length, all spins in a particle rotate in a coherent manner. This is the so-called Stoner-Wohlfarth model. On the other hand, when the particle is larger than the exchange length, the magnetization behave incoherently and switches its polarity via very complicated paths, such as vortex-like curing mode or nucleation of a reversed embryo. For typical magnetic materials, such as $L1_0$ FePt, this exchange length is estimated to be around 30 nm.[6] Now let us consider the possibility to detect magnetic information from such a small size magnet. A Co particle with the diameter of 30 nm has the total magnetic moment of 10^{-14} emu. This value is much smaller than the sensitivity limit of conventional magnetometers, so that no effective means are available for detection of such extremely small

magnetic moment with the only exception of a micro SQUID [7] and a micro-Hall device based on a semiconductor heterojunction with a two dimensional electron gas.[8] So the challenge we have to tackle is to establish a highly sensitive detection method for nanomagnets. To overcome this difficulty, we have newly developed a highly sensitive magnetic detection technique using the anomalous Hall effect (AHE), which gives the sensitivity better than 10^{-15} emu. [9]
In the present paper, we elucidate the static magnetic properties of single $L1_0$ type FePt [6, 9, 10] and Co/Pt multilayer dots,[11] especially focusing on their reversal process and the bistability condition necessary for digital memory use. In order to extend our knowledge on the nanomagnet behavior to the nanosecond (ns) regime, we have developed an ultrafast pulse field generator being able to produce a large amplitude pulse field with the rise time of sub-ns and studied the switching process of a single Co/Pt dot in the nanosecond (ns) regime using the pulse field generator.[12]

2. Experimental

$L1_0$ FePt films with nominal thickness of 10 nm was deposited on MgO (100) at T_s=1173 K by sputtering. Co/Pt multilayers (ML) with the structure of Si/Ta(2nm)/Pt(5nm)/[Co(0.2~1.4nm)/Pt(1.0nm)]$_3$ were prepared at room temperature by the same method. The magnetic anisotropy constants were evaluated by analyzing the anomalous Hall effect (AHE) curves based on the generalized Sucksmith-Thompson method.[13] The deposited continuous films were patterned into the dots with the diameter of 50~1000 nm by means of electron beam lithography and Ar ion etching. Subsequently a Pt/Al film of 6 nm in thickness was deposited onto the dots as an electrode material, and subsequently it was patterned into a cross shape. Figures 1(a) and 1(b), respectively, show the SEM micrograph of the single Co/Pt dot covered with the cross-shaped Hall electrode and its AHE curve (red line) along with the continuous film (black line). The equilibrium domain state was checked by measuring the remanent AHE value (V_{AHE}) after saturation with the in-plane field of 3 T. If the single-domain state (SD) is energetically favorable, V_{AHE} goes back to the +/ - saturation. In this case, the dot is judged to be bistable, being suitable for digital memory use. If it is not so, the dot will take multi-domain state (MD). The reversal mechanism was examined by the angular dependence of the remanence coercivity H_r [11,14] and the energy barrier for the magnetization reversal was evaluated by analyzing the sweep rate and the pulse duration dependences of the dynamic coercivity H_r based on the Neel-Arrhenius law. The ns switching experiment was performed using the home-built pulse field generator described in detail in ref.12. Figure 2 demonstrates the pulse profiles for various coaxial cable length L for the charging voltage V_0 = 100 – 400 V. Note that the pulse width can be controlled by varying L and its maximum field reaches H_p = 2.7 kOe at the center of the microcoil with the inner diameter of 10 μm.[12]

3. Results and discussion

3.1. Magnetic properties of $L1_0$-FePt nanomagnet

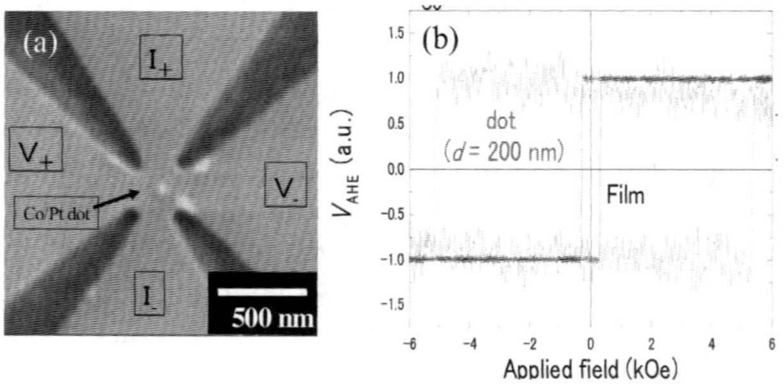

Figure 1 (a) SEM micrograph of a Co/Pt dot covered with a cross-shaped Hall electrode, (b) AHE curves for the continuous film (black) and the single dot (red).

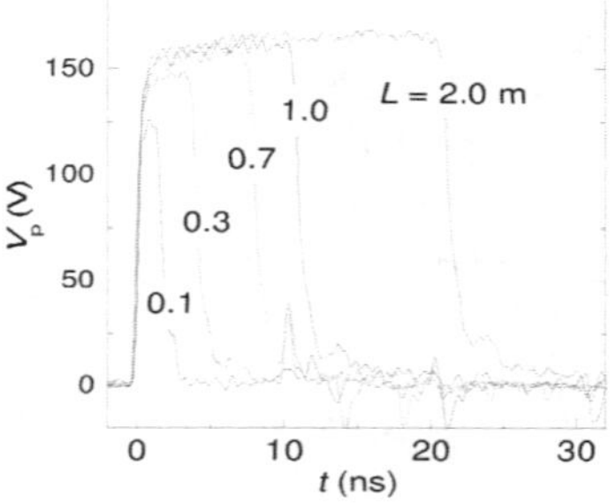

Figure 2 Pulse profiles after transmission through a Cu coil for charge voltage V_0 = 400V and coaxial cable length L= 0.1-2.0 m.

By means of the AHE method, we have measured the magnetization curves of single crystal FePt dots with various diameters. The result is summarized in fig.3.[6,9,10] As shown in this figure, the dot can be fully demagnetized for the size larger than 60 nm, indicating that the multi-domain structure is an equilibrium state. For D = 2000 nm, the initially demagnetized dot is continuously magnetized to the positive saturation. For D = 70 nm, the magnetization is discretized into two steps, suggesting the existence of two strong pinning sites in the dot. As D decreases below 60 nm, it is any more unable to demagnetize the dot, indicating that the dot is intrinsically bistable, that is, the dot can take either positive or negative saturation. This bistability is suitable for digital memory devices. As D reduces further down to 20 nm, the switching mode changes from incoherent mode to coherent one.[6,9,10] Thus, for the first time, we have clarified that a FePt dot behaves in a coherent manner as the dot size becomes smaller than nearly half the exchange length. Detailed analyses of the angular dependence of H_r and energy barrier E_b showed that the magnetization reversal in the single FePt dot larger than $D \sim$ 20 nm below which the coherent rotation is dominant proceeds via the nucleation of a reversed embryo followed by its rapid expansion, and the barrier E_b responsible for thermal stability is found to nearly coincide with the nucleation energy for the reversed embryo with the dimension of the wall width.[6,9,10] This picture is consistent with the intuitive nucleation model for permanent magnets proposed by Givord et al.[15] To a rough approximation, the bit size becomes smaller than 20 nm for recording density of above 1 Tbpsi, so that the constituent particles should be less than 20nm. Our present results assure that $L1_0$ FePt nanomagnets in this size region satisfy the bistability condition and behave in accordance with the coherent rotation mode, in addition to excellent thermal stability due to its high magnetic anisotropy energy (MAE). Now let us discuss the potential of $L1_0$ FePt as a ultra high density recording material. Our recent study on the size effect on the order/disorder transformation of equiatomic FePt verified full $L1_0$ ordering and the resulting huge MAE for the particles of above $D \sim$2 nm,[16] so that thermal stability condition $E_b \sim K_u V \geqq 60\ k_B T$ can be satisfied even for D = 4 nm. This result suggests the possibility for the data density of $1/(4\times4\ \mathrm{nm}^2) \sim 40$ Tbpsi under the assumption of 1 bit/particle recording.

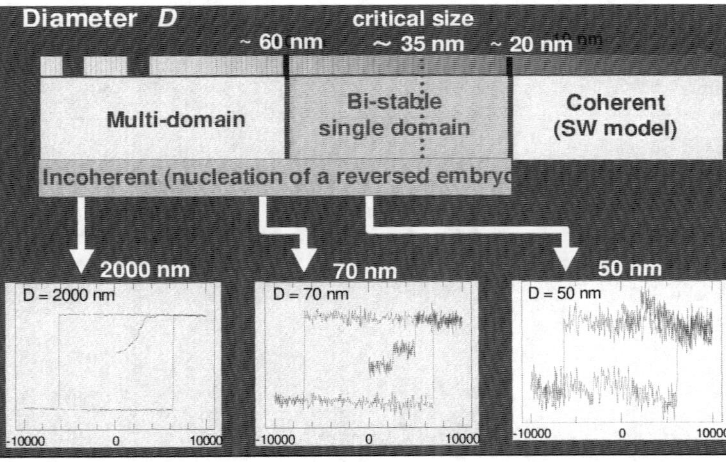

Figure 3 Magnetic behavior and AHE magnetization curves for $L1_0$ FePt nanodots with various diameters D.[9,10] The exchange length is calculated to be around 35 nm for a $L1_0$ FePt particle with the aspect ratio of 5, according to the micromagnetics theory.

3.2. Magnetic properties of Co/Pt multilayer nanomagnet

Figures 4(a) and 4(b), respectively, show the dot size and angular dependences of H_r for the [Co (t_{Co})/Pt (2 nm)]$_3$ dots measured by using the Hall device shown in fig.1. As the dot size reduces, H_r rapidly increases and its angular dependence varies from the 1/cosθ shape to the concave shape, suggesting the change in the reversal process from the pinning type to the nucleation type. From the

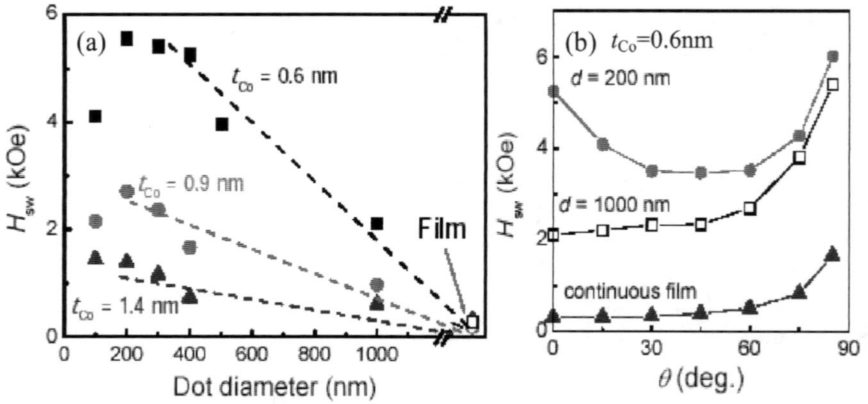

Figure 4 (a) Switching field H_{sw} *vs* dot diameter *d* and (b) angular dependence of H_{sw} at $T = 300$ K for $[Co(t_{Co})/Pt(2\ nm)]_3$ dots.

AHE measurements, we obtained the phase diagram for the magnetic behaviors of the Co/Pt dots, as shown in fig.5.[11] The Co/Pt dot favors a single-domain state (SD) when the dot size *d* is smaller than the equilibrium domain width of the continuous film as marked by the asterisks, otherwise the dot favors a multi-domain state (MD). Therefore the bistability necessary for digital memory use can be realized by downsizing the dot below the equilibrium domain size of the continuous film. As for magnetization reversals, the process changes from the nucleation type (N) to the pinning type (W) around $d \sim 600$ nm. We also note in fig.5 that the region of the bistability (SD) and nucleation mode (N) enlarges with reducing the dot size *d*. In order to investigate the reversal process in more detail, we measured and analyzed the temperature & field-sweep-rate dependences of the remanence coercivity. The results are shown in figs. 6 and 7. Note in Fig.7 that the energy barrier E_b monotonically increases with K_u, and both E_b measured at $T = 100$ and 300 K lie on the same line, indicating that the nucleation process remains invariant irrespective of the temperature. The K_u dependence of E_b shown in fig.7 can be explained by the nucleation model assuming a bubble-shaped reversed embryo with the dimension of the domain wall width.[11]

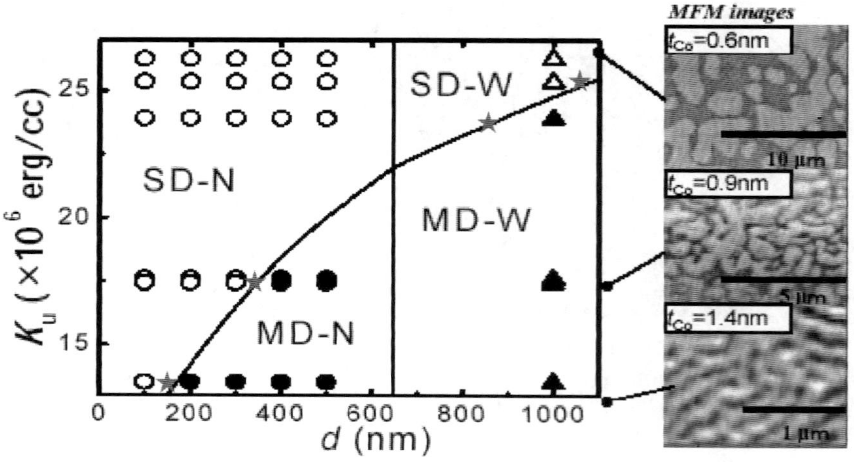

Figure 5 Bistability and reversal process of $[Co(t_{Co})/Pt(2\ nm)]_3$ dots as functions of the perpendicular magnetic anisotropy K_u and dot size *d*. The open and solid symbols, respectively, denote the single-domain (SD) and multi-domain (MD) states. The circles and triangles indicate the nucleation (N) and pinning (W) type reversals, respectively. The asterisks represent roughly averaged domain width of the Co/Pt continuous films determined by the MFM images shown in right panels.

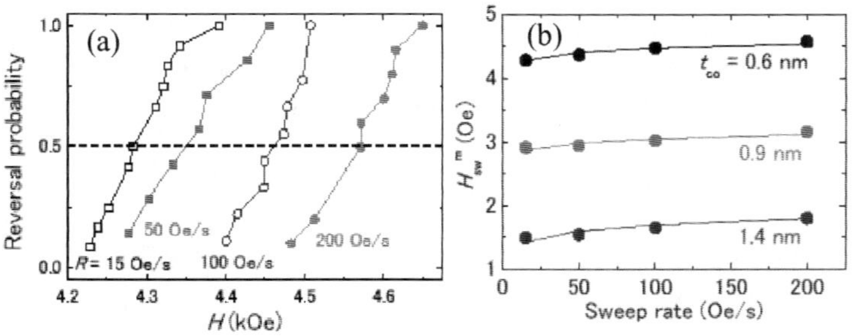

Figure 6 (a) Reversal probability vs reverse field H at various sweep rate R and (b) median switching field *vs* R for the dots with the structure $[Co(t_{Co})/Pt(2nm)]_3$. The solid lines in (b) are the fittings based on the Neel-Arrhenius model.

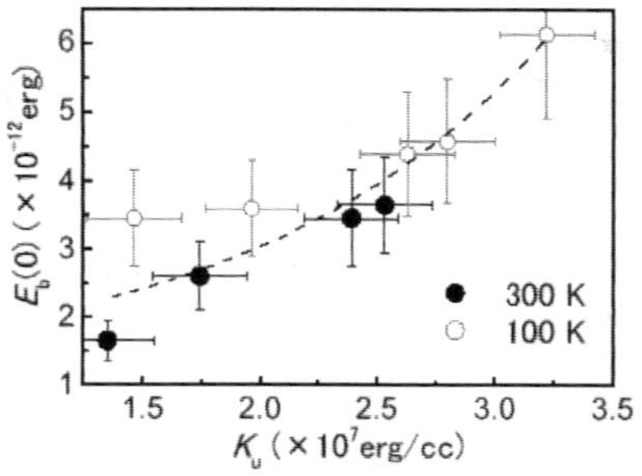

Figure 7 Energy barrier of Co/Pt dots ($d = 200$ nm) as a function of K_u at $T = 100, 300$ K. The broken line is a guide to the eyes.

3.3. Magnetic behavior of Co/Pt nanomagnet in nanosecond regime

By combining the pulse field generator with the AHE device shown in fig.1, the magnetic response of a single Co/Pt nanomagnet to a pulse field is examined. By applying a pulse field with the maximum amplitude of 2.7 kOe (pulse current $I_p = 3.4$ A) to the negatively saturated Co/Pt dot, the magnetization switching solely by the pulse field has been successfully observed. In fig.8, the normalized Hall voltage v_{AHE} after application of positive pulse fields of (a)$\tau_p = 1.8$ ns, (b) 3.8 ns, and (c) 20.8 ns are plotted as functions of I_p. The results for the positive dc field $H_{dc} = 0$ Oe and + 100 Oe are shown for each τ_p. The magnetization of the Co/Pt dot can be switched by the pulse field of $\tau_p = 1.8$ ns even in the absence of the dc bias field. Note that the unsaturated state ($-1 < v_{AHE} < +1$) reflecting the multi-domain structure in the dot is observed for $\tau_p \leq 3.8$ ns in the absence of the dc field ($H_{dc} = 0$ Oe). In contrast, only single domain state is found for longer pulse duration $\tau_p = 20.8$ ns or in the presence of dc field ($H_{dc} = +100$ Oe). The nucleation pulse current I_p^n is almost independent of τ_p, while the saturation pulse current I_p^s increases with reducing τ_p. This facts suggests that the nucleation of a reversed embryo occurs within a very short pulse duration of $\tau_p = 1.8$ ns and is subsequently followed by its gradual growth within several ns. Detailed analyses indicate that its growth rate is nearly proportional to the pulse field amplitude and is on the order of several tens of m/s.[12] This is the first experiment demonstrating the reversal process via the nucleation of a reversed embryo followed by its

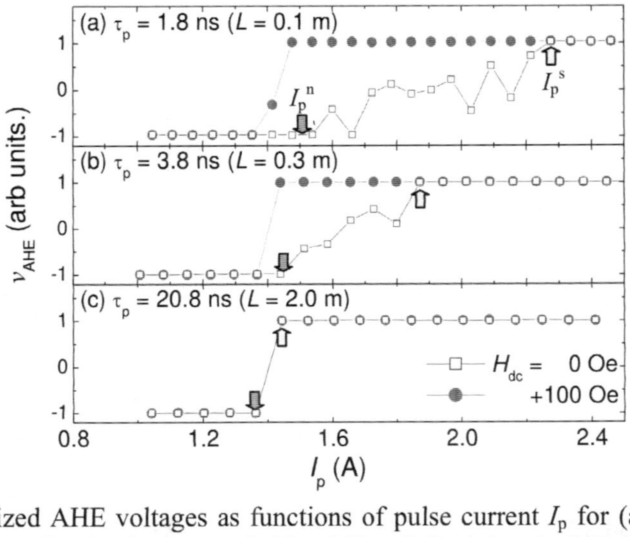

Fig.8 Normalized AHE voltages as functions of pulse current I_p for (a) $\tau_p = 1.8$ns, (b) 3.8ns and (c) 20.8ns under the dc reverse fields of $H_{dc} = 0$ Oe (□) and +100 Oe (•). The pulse field amplitude H_p per current I_p is 0.8 kOe/A. The green and yellow arrows indicate the critical current I_p^n for nucleation and I_p^s for positive saturation, respectively.

gradual expansion in a single nanomagnet. It has also been clarified that the pulse field and the dc bias field can be treated as a simple sum even in the nanosecond regime and the time dependence of magnetic properties can be described very well by the Neel-Arrhenius model.[12]

4. Summary

We have newly developed highly sensitive magnetic detection method utizing the anomalous Hall effect (AHE), which makes it possible to observe the magnetic behavior of a single nanomagnet. By using this method, we have systematically clarified the magnetic properties of $L1_0$ FePt and Co/Pt multilayer in the nannometer size scale. By combining the AHE device and a pulse field generator, we have unambiguouly demonstrated that the magnetization reversal in a submicron Co/Pt dot proceeds via a very fast nucleation of a reversed embryo followed by its gradual expansion. Our final goal is to develop a real time spin dynamics measurement system for a single nanomagnet , and the attempt is in progress.

Acknowledgment
The authors wish to express sincere thanks to Prof. Nakatani of Osaka University for his encouragement and kind help. This work was partly supported by "Research and Development for Next-Generation Information Technology of MEXT," Grant-in-Aid for Scientific Research from MEXT, Special Education and Research Expenses from Post-Silicon Materials and Devices Research Alliance, and the Storage Research Consortium in Japan.

References

[1] Dieny B, Speriosu V S, Parkin S S P, Gurney B A, Wilhoit D R and Mauri D 1991 *Phys. Rev. B* **43** 1297
[2] Gallagher W J and Parkin S S P 2006 *IBM J. Res. & Div.* **50** 5
[3] Kiselev S I, Sankey J C, Krivorotov I N, Emley N C, Schoelkopf R J, Buhrman R A and Ralph D C 2003 *Nature* **425** 380
[4] Datta S and Das B 1990 *Appl. Phys. Lett.* **56** 665
[5] Aharoni A 2001 *Introduction to the theory of magnetism* (Oxford University, London)
[6] Okamoto S, Kitakami O, Kikuchi N, Miyazaki T, Shimada Y and Takahashi Y K 2003 *Phys. Rev. B* **67** 094422
[7] Wernsdorfer W, Hasselbach K, Benoit A, Barbara B, Mailly D, Tuaillon J, Perez J P, Dupuis V, Dupin J P, Guiraud G and Perex A 1995 *J. Appl. Phys.* **78** 7192; Wernsdorfer W and Sessoli R 1999 *Science* **284** 133
[8] Kent A D, von Molnar S, Gieder S and Awschalom D D 1994 *J. Appl. Phys.* **76** 6656
[9] Kikuchi N, Okamoto S, Kitakami O, Shimada Y and Fukamichi K 2003 *Appl. Phys. Lett.* **82** 4313; Okamoto S et al. 2005 Scripta Mater. **53** 395
[10] Kikuchi N, Okamoto S and Kitakami O 2008 *J. Appl. Phys.* **103** 07D511; Okamoto S, Kikuchi N, Kato T, Kitakami O, Mitsuzuka K, Shimatsu T, Muraoka H, Aoi H and Lodder J C 2008 *J. Magn. Magn. Mater.* **320** 2874 (2008)
[11] Kikuchi N, Kato T, Okamoto S, Kitakami O, Tezuka N and Sugimoto S 2008 *J. Appl. Phys.* **103** 07C510; Okamoto S, Kato T, Kikuchi N, Kitakami O, Tezuka N and Sugimoto S 2008 *J. Appl. Phys.* **103** 07C501
[12] Ito A, Kikuchi N, Okamoto S and Kitakami O 2009 *IEEE. Trans. Mang.* (in press); Kikuchi N, Okamoto S and Kitakami O 2009 *J. Appl. Phys.* (in press)
[13] Okamoto S, Nishiyama K, Kitakami O and Shimada Y 2001 *J. Appl. Phys.***90** 4085; Okamoto S, Kikuchi N, Kitakami O, Shimada Y and Fukamichi K 2002 *Phys. Rev. B* **66** 024413
[14] Sakuma A, Tanigawa S and Tokunaga M 1990 *J. Magn. Magn. Mater.* **84** 52
[15] Givord D, Tenaud P and Viadieu T 1988 *IEEE Trans. Magn.* **24** 1921
[16] Miyazaki T, Kitakami O, Okamoto S, Shimada Y, Akase Z, Murakami Y, Shindo D, Takahashi Y K and Hono K 2005 *Phys. Rev. B* **72** 144419

Magnetic logic devices composed of permalloy dots

Ryoichi Nakatani[1], Hikaru Nomura[1] and Yasushi Endo[2]

[1]Division of Materials and Manufacturing Science, Graduate School of Engineering, Osaka University, 2-1 Yamadaoka, Suita, Osaka 565-0871, Japan
[2]Department of Electrical and Communication Engineering, Graduate School of Engineering, Tohoku University, Sendai 980-8579, Japan

nakatani@mat.eng.osaka-u.ac.jp

Abstract. Magnetic logic devices have been investigated by micromagnetics simulation and experiment. The simulation shows that the magnetic logic devices composed of 4 elliptical permalloy dots perform both the NAND and NOR logic operations. The experiments indicate that the distance between adjacent dots must be shorter than 50 nm and 70 nm along the long axis and the short axis of the dots, respectively. The micro-fabricated test devices show the above logic operations.

1. Introduction

Spin-electronics devices such as magnetic random access memories[1-2], microwave oscillators[3] and spin torque diode[4] have been investigated as future magnetic functional devices. The magnetic logic devices[5] have also been investigated. However, the reported logic devices used many magnetic dots, and thus they could not be small. Recently, we have proposed new structures for magnetic logic devices, based on the theoretical investigation using the Landau-Lifshitz-Gilbert equation[6]. The magnetic logic devices consist of 4 elliptical permalloy (Ni-20at%Fe) dots. The size of each dot is 80 nm x 40 nm, and the distance between adjacent dots is 20 nm. Because of the narrow distance, the neighboring magnetic dots are magnetically coupled due to magnetostatic interactions, and the coupling produces both the NAND and NOR logic operations. Moreover, the small sizes and narrow distances allow the high-density arrangement, so that the size of the logic devices can be small. However, there is a difficulty to micro-fabricate the narrow distances between adjacent dots.

In this study, we experimentally investigate the effect of the distance between adjacent dots on the magnetostatic interactions between the dots. If larger distance allows the magnetostatic interactions, the logic devices can be easily micro-fabricated.

2. Experimental procedures

The Ni-20at%Fe (permalloy) dot arrays were micro-fabricated by electron beam lithography, electron beam evaporator and lift-off method on Si(100) substrates with thermally oxidized layers. The magnetic fields of 40 kA/m were applied to the substrates in order to introduce uni-axial anisotropy along the long axis of the elliptical dots. The thickness of the dots was 10 nm. The aspect ratio of the elliptical dots was 1:2. The long axis was changed between 200 and 1600 nm. The distance between the adjacent dots was also changed.

The magnetization processes of the dot arrays were observed using a longitudinal magneto-optical Kerr effect (MOKE) magnetometry at room temperature. The magnetic field was applied along the long axis of the dots, and was between -1 and 1 kOe. The shape of the dot arrays was observed by scanning electron microscopy (SEM) and magnetic force microscopy (MFM) observed the magnetic states of the dot arrays. Magnetic configuration computations were conducted using the Landau-Lifshitz-Gilbert (LLG) equation and the finite element method[7]. The following parameters for Ni-20at%Fe were assumed: saturation magnetization M=1 T, exchange constant A=1.05x10^{-11} J/m, uniaxial anisotropy constant K_u=100 J/m^3.

3. Results and discussion

3.1. Optimum structure obtained using micromagnetics simulation

The optimum structure of magnetic logic device obtained using micromagnetics simulation is shown in Fig.1[6]. The magnetic logic device consists of 4 elliptical permalloy dots. The size of each dot is 80 nm x 40 nm, and the distance between adjacent dots is 20 nm. The three dots "Input A-C" are used to input data. Each dot for input data magnetically couples with the dot (output Z) to output data. The total of three magnetostatic interactions produce the logic operations.

The simulated magnetic configurations of logic device are shown in Fig. 2. The green color means the leftward magnetization, and it means information of zero. On the other hand, the red color means the rightward magnetization, and it means information of one. There are 8 input patterns as shown in Fig. 2.

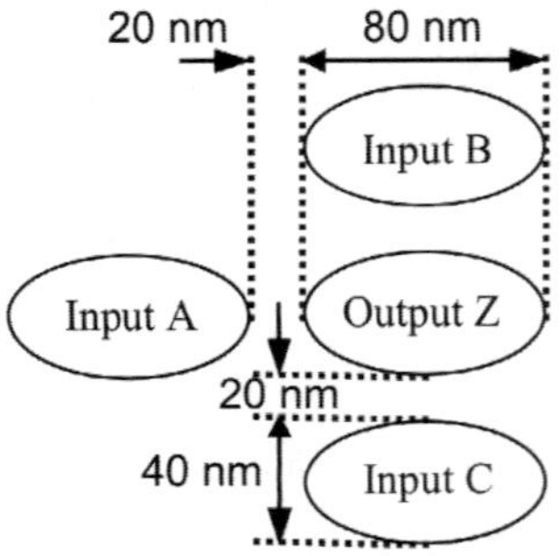

Figure 1. An optimum structure of magnetic logic device obtained using micromagnetics simulation.

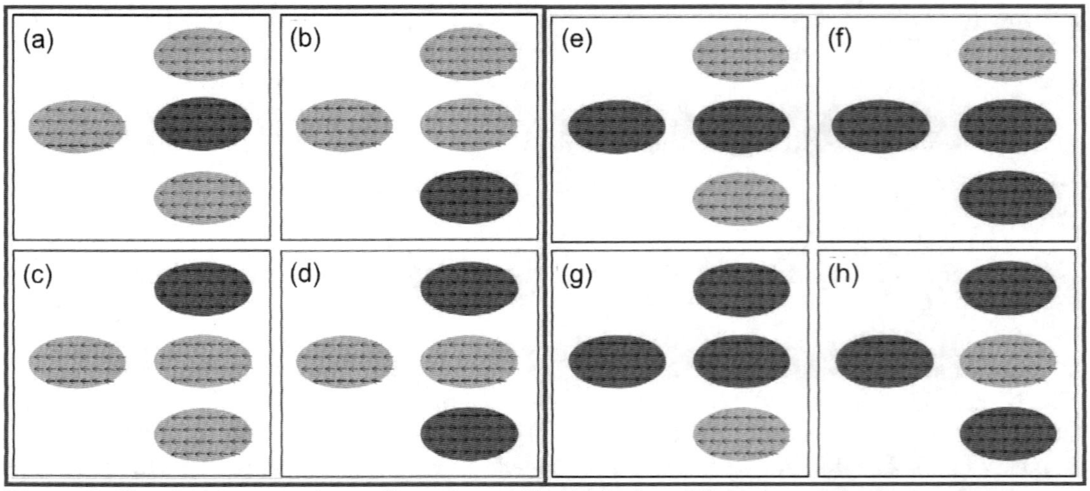

Figure 2. Simulated magnetic configurations of logic device. The green and red mean the leftward and rightward magnetization, and also mean information of zero and one, respectively.

The simulated results are summarized in Table 1. As shown in Table 1, when the input A is "zero", there are 4 patterns for the inputs B and C. The inputs B and C change the output Z. The output Z becomes the result of "NOR" operation for the inputs B and C. On the other hand, when the input A is "one", there are also 4 patterns for the inputs B and C. The output Z becomes the result of "NAND" operation for the inputs B and C. As a result, the input A decides the kind of logic operation, "NOR" or "NAND". It can be good advantage for the magnetic logic devices.

3.2. Magnetostatic interactions between magnetic dots

As mentioned above, the micromagnetics simulation indicates the operation of the magnetic logic device composed of 4 elliptical permalloy dots. The optimized distance between adjacent dots is 20 nm. The distance cannot be easily micro-fabricated. Therefore, we experimentally investigated the limit of the distance between adjacent dots that cause the magnetostatic interactions.

As shown in Fig. 1, there are two sequences in the magnetic logic devices. They are longitudinal and transverse. Figure 3 shows the schematic illustration and magnetization curves for the longitudinal sequence. In the sequence, the magnetostatic interaction between adjacent dots is parallel, and thus the stray field is applied along the anti-parallel direction to the applied magnetic field before the switching of the magnetization, as shown in Fig. 3(a). If the stray field H_{st} is zero, the magnetization of dots switches at the intrinsic switching field $H_{sw,int}$. On the other hand, when the stray field H_{st} is not zero, the stray field decreases the total field applied to the magnetic dots. The decrease of total field increases the switching field H_{sw} which is applied magnetic field to switch the magnetization of the dots. Therefore, the switching field H_{sw} becomes high as shown in Fig. 3.

Table 1. Simulated results of logic device.

A	B	C	Z	
0	0	0	1	NOR
0	0	1	0	
0	1	0	0	
0	1	1	0	
1	0	0	1	NAND
1	0	1	1	
1	1	0	1	
1	1	1	0	

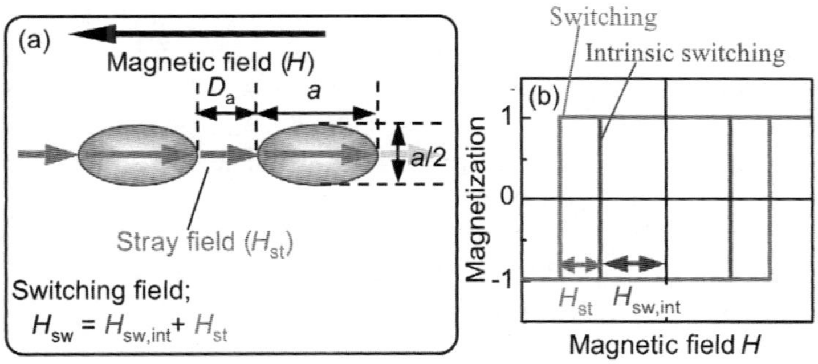

Figure 3. (a) schematic illustration and (b) magnetization curves for the longitudinal sequence.

Figure 4 shows changes in the switching field due to the distance between adjacent dots D_a[8]. The long axis *a* of the dots is changed from 200 nm to 1600 nm. The field is applied along the long axis of the dots. As shown in Fig.4, the switching field increases at 50 nm as the distance decreases. Therefore, it is understood that the magnetoststic interactions is produced when the distance is 50 nm and less than 50 nm. It is longer than the optimized distance by simulation.

Figure 5 shows the schematic illustrations and magnetization curves for the transverse sequence. In the sequence, the magnetostatic interaction between adjacent dots is antiparallel. As shown in Fig. 5 (a), the stray field caused by upper dot is applied to the lower dot, and the direction is parallel to the applied magnetic field before the first switching. Therefore, the stray field helps the first switching, and thus the first switching field decreases as shown in (c). The magnetizations of the half of dots switch at the first switching field, and thus the magnetic configuration of the dots becomes anti-parallel after the first switching. After the first switching, as shown in Fig. 5 (b), the stray field caused by the switched lower dots is applied to the upper dot. The direction of stray field is anti-parallel to the applied magnetic field between the first switching and the second switching. Therefore, the stray field disturbs the second switching, and thus the second switching field increases as shown in (c).

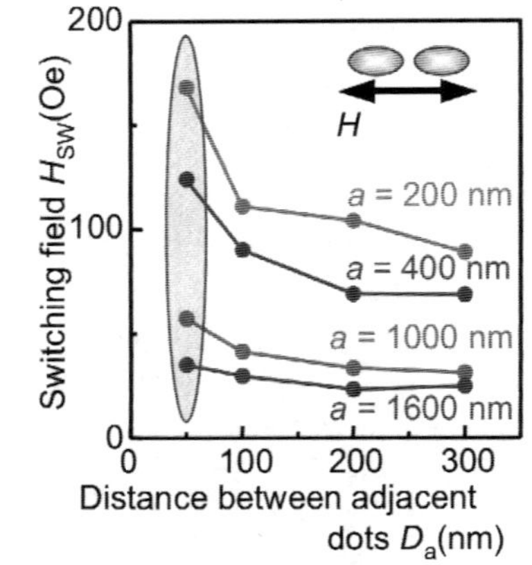

Figure 4. Changes in the switching field due to distance between adjacent dots.

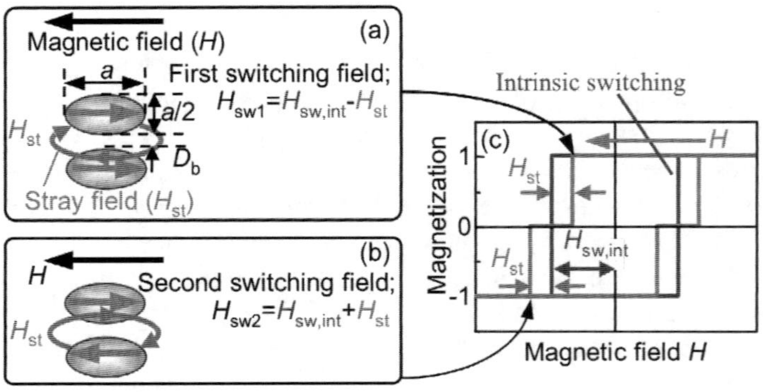

Figure 5. (a), (b) schematic illustrations and (c) magnetization curves for the transverse sequence.

Figure 6 shows changes in the second switching field due to the distance between adjacent dots D_b. The long axis a of the dots is changed from 200 nm to 1600 nm. The field is applied along the long axis of the dots. As shown in Fig. 6, the switching field increases at 70 nm as the distance decreases. Therefore, it is understood that the magnetoststic interaction is produced when the distance is 70 nm and less than 70 nm. It is longer than the optimized distance by simulation.

Figure 7 shows the MFM images and schematic illustrations of micro-fabricated test devices. We have made a lot of test devices that structures are shown in (e), and observed their MFM images as shown in (a)-(d). When we select the MFM images that input A is "one", all images have the pattern shown in (a)-(d), which indicate the "NAND" operations. On the other hand, when we select the MFM images that input A is "zero", all images indicate the "NOR" operations (not shown here).

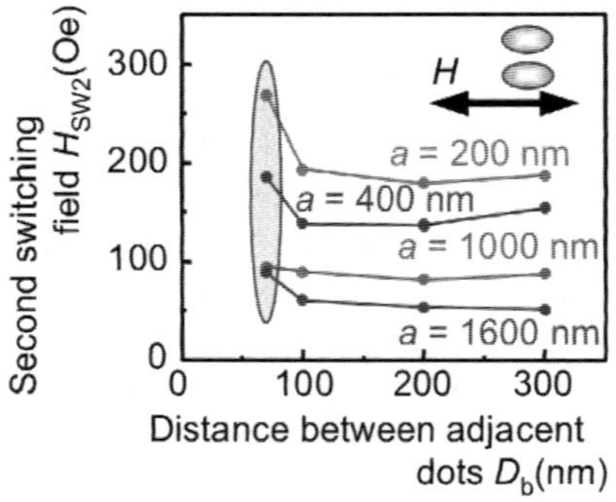

Figure 6. Changes in the second switching field due to distance between adjacent dots.

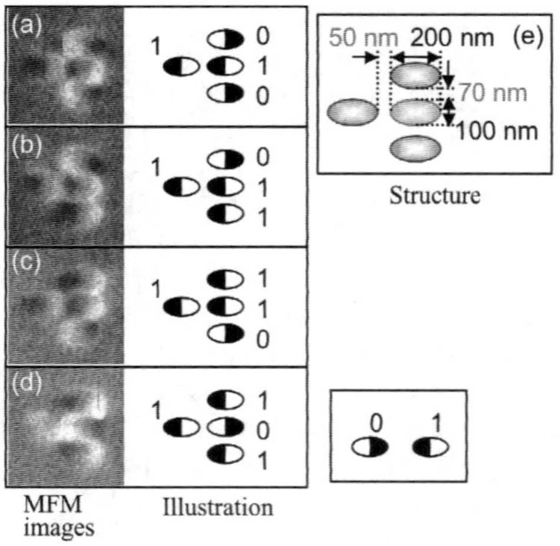

Figure 7. MFM images and schematic illustrations of micro-fabricated test devices.

4. Summary

We have investigated the magnetic logic devices. The micromagnetics simulation shows that the magnetic logic devices composed of 4 elliptical permalloy dots perform both the NAND and NOR logic operations. However, the optimized distances between adjacent dots are 20 nm in the results of simulation. The distances of 20 nm are very difficult to be micro-fabricated. On the other hand, the experimental results show that the distance between adjacent dots must be shorter than 50 nm and 70 nm along the long axis and the short axis of the dots, respectively. The experimental results indicate that it is not so difficult to micro-fabricate the magnetic logic devices.

Acknowledgments

This work was partly supported by Global COE program (Center of Excellence for Advanced Structural and Functional Materials Design) from the Ministry of Education, Culture, Sports, Science and Technology (MEXT), Japan. This work was also partly supported by Strategic Information and Communications R&D Promotion Programme (SCOPE) from the Ministry of Internal Affairs and Communications (MIC).

References

[1] Tehrani S, Slaughter J M, Chen E, Durlam M, Shi J and DeHerrera M 1999 *IEEE Trans. Magn.* **35** 2814
[2] Parkin S S P, Roche K P, Samant M G, Rice P M, Beyers R B, Scheuerlein R E, O'Sullivan E J, Brown S L, Bucchigano J, Abraham D W, Lu Y, Rooks M, Trouilloud P L, Wanner R A and Gallagher W J 1999 *J. Appl. Phys.* **85** 5828
[3] Rippard W H, Pufall M R, Kaka S, Russek S E and Silva T J 2004 *Phys. Rev Lett.* **92** 027201
[4] Tulapurkar A A, Suzuki Y, Fukushima A, Kubota H, Maehara H, Tsunekawa K, Djayaprawira D D, Watanabe N and Yuasa S 2005 *Nature* **438** 339
[5] Cowburn R P and Welland M E 2000 Science **287** 1466
[6] Haque S A, Yamamoto M, Nakatani R and Endo Y 2004 *J. Magn. & Magn. Mater.* **282** 380
[7] Scheinfein M R, and Blue J L 1991 *J. Appl. Phys.* **69** 7740
[8] Endo Y, Fujimoto H, Kawamura Y, Nakatani R and Yamamoto M 2008 *IEEE Trans. Magn.* **44** 2718

Journal of Physics: Conference Series **165** (2009) 012031 doi:10.1088/1742-6596/165/1/012031

Magnetic properties of Mn-Zn ferrite thin films fabricated by pulsed laser deposition

Hideki Etoh[1,2], Junichi Sato[2], Yoshiteru Murakami[2], Akira Takahashi[2] and Ryoichi Nakatani[1]

[1] Division of Materials and Manufacturing Science, Graduate School of Engineering, Osaka University, 2-1 Yamadaoka, Suita, Osaka 565-0871, Japan

[2] Advanced Technology Research Laboratories Corporate Research and Development Group, Sharp Corporation, 2613-1 Ichinomoto-cho, Tenri, Nara 632-8567 Japan

E-mail: etoh-hideki@sharp.co.jp

Abstract. We investigated the basic properties of Mn-Zn ferrite thin films fabricated by pulsed laser deposition (PLD) with the aim of controlling their saturation magnetizations (M_s) and electrical resistivities. The M_s and electrical resistivities varied dependent on the substrate temperature during deposition and post-anneal conditions. X-ray diffraction (XRD) measurement revealed the difference is due to the formation of a Mn-Zn ferrite and an Fe_2O_3 phase, which is attributed to the amount of oxygen in the post anneal atmosphere.

1. Introduction

In the field of magnetic recording, a thermally assisted magnetic recording method is proposed to overcome the thermal agitation limit of recording media to achieve a higher recording density beyond Tb/inch^2 [1]. In this method, it is important to apply the magnetic recording field and the laser beam to the same position on the recording media. Although it is effective to incorporate a magnetic core in the recording head to apply a sufficiently large recording field, the core could possibly interfere with the optical path of the laser beam. Mn-Zn ferrite is a transparent magnetic material with a high M_s and high resistivity [2]. It is a candidate for use as a magnetic core that will not interrupt the laser beam. In this paper, we report basic properties of Mn-Zn ferrite thin films fabricated by PLD. The aim is to adopt this material as the magnetic core of a thermally assisted magnetic recording head.

2. Experiment

Mn-Zn ferrite thin films were fabricated by PLD using a Nd:YAG laser at a wavelength of 266 nm and operated at 10 Hz at a pressure of 1 x 10^{-4} Pa. The target was an oxide ceramic of Mn-8at%, Zn-23at%, Fe-69at%. Films were deposited onto Si substrates with oxidized surfaces. The target and substrates were rotated during depositions. The film thickness was fixed at 50 nm. We tested various procedures of fabricating samples. In one procedure samples are deposited with a substrate temperature of 350 to 500 °C and not annealed after deposition, in another procedure samples are deposited at room temperature and annealed after deposition at a pressure of 1 x 10^{-4} Pa with an annealing temperature of 350 to 500 °C for 20 minutes, and in the other procedure samples are deposited at room temperature and annealed after deposition in an air atmosphere with an annealing temperature of 350 to 500 °C for

5 minutes. Magnetic properties were measured by a vibrating sample magnetometer and crystalline structures were analyzed by X-ray Diffraction (XRD). Electrical resistivities were measured by a two probe method and compositions of metal ions were measured by X-ray Fluorescence (XRF) analysis.

3. Results and discussions

3. 1. X-ray diffraction

Figure 1 shows XRD patterns of samples deposited at various substrate temperatures at a pressure of 1 x 10^{-4} Pa and not annealed after deposition. In the sample deposited at a substrate temperature of room temperature, peaks of Mn-Zn ferrite were not observed but peaks of Fe_2O_3 were observed. Peaks of Fe_2O_3 were not observed at substrate temperatures of 450 °C and 500 °C, while (1 1 l) peaks of Mn-Zn ferrite were observed at substrate temperatures of 400 °C or higher. Figure 2 shows XRD patterns of samples deposited at a substrate temperature of room temperature and annealed after deposition at various annealing temperatures at a pressure of 1 x 10^{-4} Pa. With an increase in annealing temperature the peaks of Fe_2O_3 disappeared, but (3 1 1) peaks of Mn-Zn ferrite were observed at annealing temperatures of 350 °C or higher. Figure 3 shows XRD patterns of samples deposited at a substrate temperature of room temperature and annealed after deposition at various annealing temperatures in an air atmosphere. Peaks of Fe_2O_3 were observed at annealing temperatures between 350 to 500 °C and (1 1 l) peaks of Mn-Zn ferrite were observed at 400 °C or higher.

Mn-Zn ferrite is generally expressed by the formula MFe_2O_4. The amount of oxygen per cation needed for the formation of MFe_2O_4 and Fe_2O_3 is 1.33 and 1.5, respectively. In the samples deposited at various substrate temperatures at a pressure of 1 x 10^{-4} Pa and samples annealed after deposition at various annealing temperature at a pressure of 1 x 10^{-4} Pa, oxygen was not provided during heating samples. With an increase in annealing temperature Fe_2O_3 was less crystallized than MFe_2O_4 because Fe_2O_3 needs more oxygen than MFe_2O_4. On the other hand, in samples annealed after deposition at various annealing temperature in an air atmosphere, oxygen was provided during annealing. So with an increase in annealing temperature MFe_2O_4 and Fe_2O_3 were crystallized.

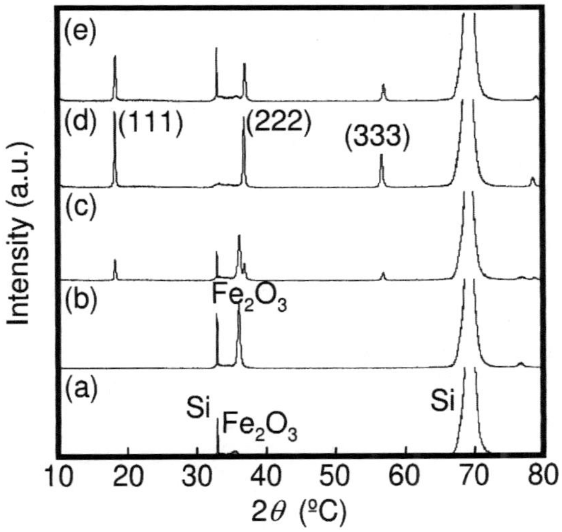

Figure 1. XRD patterns of samples deposited at various substrate temperatures at a pressure of 1 x 10^{-4} Pa (a) room temperature, (b) 350 °C, (c) 400 °C, (d) 450 °C, (e) 500 °C.

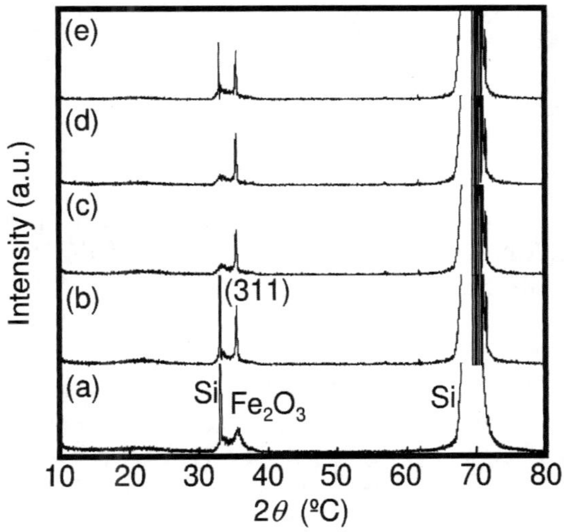

Figure 2. XRD patterns of samples annealed after deposition at various annealing temperature at a pressure of 1 x 10^{-4} Pa (a) room temperature, (b) 350 °C, (c) 400 °C, (d) 450 °C, (e) 500 °C.

3. 2. Magnetic properties

Figure 4 shows the heating temperature dependence of saturation magnetization (M_s) for Mn-Zn ferrite at various heating conditions. For samples deposited at various substrate temperatures at a pressure of 1 x 10^{-4} Pa the sample at the substrate temperature of room temperature and 350 °C showed little M_s. Magnetization appeared at a substrate temperature of 400 °C or higher and the sample deposited at a substrate temperature at 450 °C had the maximum M_s of 0.34 T. For samples annealed after deposition at a pressure of 1 x 10^{-4} Pa, magnetization appeared at the annealing temperature of 350 °C or higher. The sample annealed at 450 °C had the maximum M_s of 0.58 T. For samples annealed after deposition in an air atmosphere, magnetization appeared at the 400°C or higher and the sample annealed at 400 °C had the maximum M_s of 0.24 T.

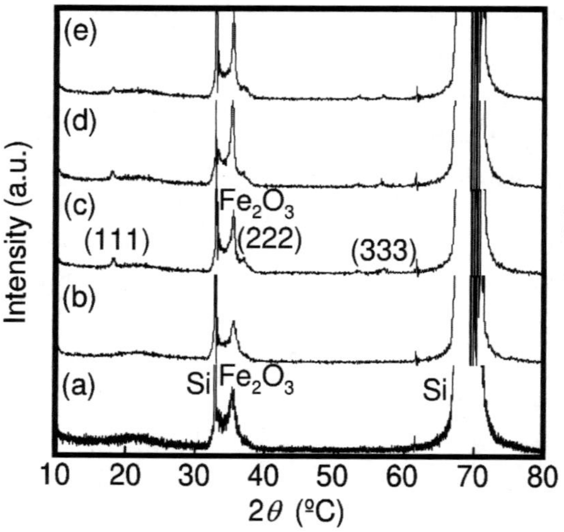

Figure 3. XRD patterns of samples annealed after deposition at various annealing temperature in an air atmosphere (a) room temperature, (b) 350 °C, (c) 400 °C, (d) 450 °C, (e) 500 °C.

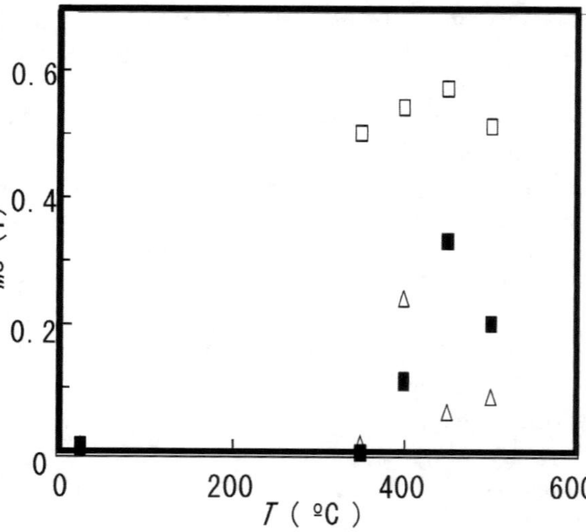

Figure 4. Heating temperature dependence of M_s. The closed squares denote samples deposited at various substrate temperatures at a pressure of 1 x 10^{-4} Pa. The open squares denote samples annealed after deposition at various annealing temperature at a pressure of 1 x 10^{-4} Pa . The open triangles denote samples annealed after deposition at various annealing temperature in an air atmosphere.

From XRD patterns of samples deposited at various substrate temperatures at a pressure of 1 x 10^{-4} Pa, Mn-Zn ferrite crystallization proceeded as the substrate temperature was increased to 450 °C. The Mn-Zn ferrite crystallite degraded with a further increase in the substrate temperature. So M_s of samples deposited at various substrate temperatures at a pressure of 1 x 10^{-4} Pa had a peak value at a substrate temperature of 450 °C. From XRD patterns of samples annealed after deposition at various annealing temperature at a pressure of 1 x 10^{-4} Pa the peak of Fe_2O_3 is not observed, but (3 1 1) peaks of Mn-Zn ferrite were observed at annealing temperatures of 350 °C or higher, while from XRD patterns of samples annealed after deposition in an air atmosphere peaks of Fe_2O_3 were observed at annealing temperatures between 350 to 500 °C. Because Mn-Zn ferrite is ferromagnetic and Fe_2O_3 is nonmagnetic, M_s of samples annealed at a pressure of 1 x 10^{-4} Pa is larger than that of samples annealed after deposition in an air atmosphere. The difference of M_s between samples deposited at various substrate temperatures at a pressure of 1 x 10^{-4} Pa and samples annealed after deposition at various annealing temperature at a pressure of 1 x 10^{-4} Pa is thought to be partly caused by difference of oxygen amount of Mn-Zn ferrites.

3. 3. Electric properties

Figure 5 shows the electrical resistivity of Mn-Zn ferrite at various heating conditions. Resistivities of samples deposited at various substrate temperatures at a pressure of 1 x 10^{-4} Pa and samples annealed after deposition at various annealing temperature at a pressure of 1 x 10^{-4} Pa decreased with an increase in the heating temperature. The resistivity of samples annealed after deposition in an air atmosphere increased with an increase in the annealing temperature. The resistivity of samples annealed after deposition in an air atmosphere at 450 °C and 500 °C could not be measured because of a measurement limit. They were more than 133 Ωcm.

The resistivity of Fe_2O_3 is high at ~10^6 Ωcm [3]. The composition of the sample deposited at a substrate temperature of room temperature and not annealed after deposition was $Mn_{10.3}Zn_{18.6}Fe_{71.1}O_x$ in our composition measurement. Because the amount of Fe was more than two times larger than the sum of Mn and Zn, a part of the Fe was replaced by Mn or Zn. As a result the sample consisted of $MnZnFe_2O_4$ and $FeFe_2O_4$ (Fe_3O_4). The resistivity of Fe_3O_4 is very low at ~10^{-3} Ωcm which is much lower than that of Fe_2O_3 [4]. Because the amount of Fe_2O_3 of the sample annealed after deposition in an air atmosphere increased as described above, the resistivity of the sample annealed after deposition in an air atmosphere increased with an increase in the annealing temperature. On the other hand because the amount of Fe_2O_3 in the sample annealed after deposition at a pressure of 1 x 10^{-4} Pa decreased, the resistivity of the sample decreased as the annealing temperature increased.

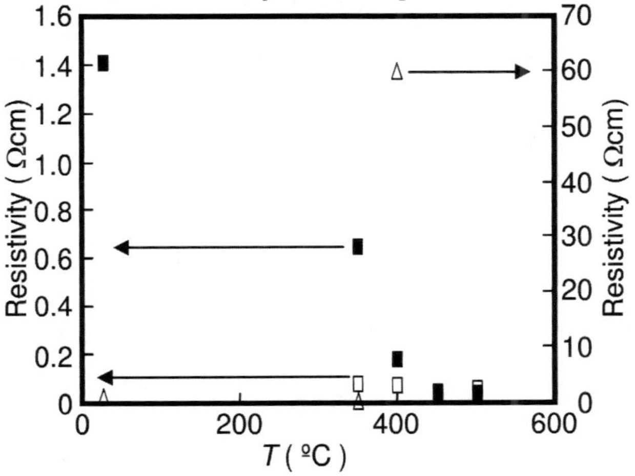

Figure 5. The heating temperature dependence of resistivity. The closed squares denote samples deposited at various substrate temperatures at a pressure of 1 x 10^{-4} Pa. The open squares denote samples annealed at various annealing temperature at a pressure of 1 x 10^{-4} Pa after deposition. The open triangles denote samples annealed at various annealing temperature in an air atmosphere after deposition.

4. Conclusions

Basic properties of Mn-Zn ferrite thin films that were fabricated by pulsed laser deposition were evaluated to investigate their possible application as new magnetic cores in thermally assisted magnetic recording. It was found that M_s and the resistivity of Mn-Zn ferrites depend on the amount of oxygen in the annealing atmosphere. A higher M_s of 0.58 T and a lower resistivity of 0.05 Ωcm were obtained for samples annealed after deposition at a pressure of 1 x 10^{-4} Pa and a relatively lower M_s of 0.24 T and a higher resistivity of 60 Ωcm were obtained for samples annealed after deposition in an air atmosphere. These results indicate that Mn-Zn ferrites with a sufficiently high M_s (0.3 T) and a sufficiently high resistivity (10^2 Ωcm) can be obtained by controlling the partial pressure of oxygen in the annealing atmosphere. They may thus be used as practical magnetic core materials in a thermally assisted magnetic recording system.

References

[1] Miyanishi S, Iketani N, Takayama K, Innami K, Suzuki I, Kitazawa T, Ogimoto Y, Murakami Y, Kojima K and Takahashi A 2005 *IEEE Trans. Magn.* **41** 2817

[2] Williams M C, Chrisey B D, Lubits P, Grabowski S K and Cotell M 1994 *J. Appl. Phys.* **75** 1676

[3] Kumari S, Tripathi C, Singh P A, Chauhan D, Shrivastav R, Dass S and Satsangi R V 2006 *Curr. Sci.* **91** 1062

[4] Yanagihara H, Hasegawa M, Kita E, Wakabayashi Y, Sawa H and Siratori K 2006 *J. Phys. Soc. Jpn.* **75** 054708

Differences in structure and magnetic behavior of Mn-AlN films due to substrate material

Takanobu Sato[1], Yasushi Endo[2], Fumiyoshi Kirino[3] and Ryoichi Nakatani[1]

[1]Division of Materials and Manufacturing Science, Graduate School of Engineering, Osaka University

[2]Depertment of Electrical and Communication Engineering, Graduate School of Engineering, Tohoku University

[3]Conservation of Cultural Property, Graduate School of Fine Arts, Tokyo National University of Fine Arts and Music

takanobu.sato@mat.eng.osaka-u.ac.jp

Abstract. The structure and magnetic behavior of Mn-AlN ($Al_{1-x}Mn_xN$, x = 0.03, 0.04) films deposited on thermally oxidized Si (001) substrates and sapphire (0001) substrates were studied. Mn-AlN films deposited on each substrate had a würtzite-type AlN phase with a preferentially oriented c-axis. Mn-AlN films that were deposited on Si (001) substrate exhibited paramagnetic behavior. In addition to paramagnetic behavior, weak ferromagnetic behavior with curie temperatures higher than room temperature were observed for Mn-AlN films deposited on sapphire (0001) substrates.

1. Introduction

Diluted magnetic semiconductors (DMSs) have attracted much attention as materials for spin-dependent electronic devices. In particular, Mn-AlN films are promising DMSs because of their ferromagnetic behavior at room temperature [1]. Recently, we reported a Mn-AlN film deposited on a thermally oxidized Si (001) substrate that did not show ferromagnetic behavior but did show paramagnetic behavior [2]. This difference might be related to a difference in the substrate material because a ferromagnetic Mn-AlN has previously been deposited on a sapphire (0001) substrate [1]. However, the effect of substrate material on the structure and magnetic behavior of the Mn-AlN film was not investigated in detail.

In this study, we investigate the structure and magnetic behavior of Mn-AlN ($Al_{1-x}Mn_xN$, x = 0.03, 0.04) films deposited on thermally oxidized Si (001) substrates and sapphire (0001) substrates. In particular, we discuss the difference in magnetic behavior which is due to the difference in the substrate materials.

2. Experimental procedure

Mn-AlN ($Al_{1-x}Mn_xN$, x = 0.03, 0.04) films were fabricated by reactive dc magnetron sputtering onto thermally oxidized Si (001) substrates and sapphire (0001) substrates at substrate temperatures (Ts) of

573 K and 773 K. The base pressure was better than 5×10^{-7} Pa. The pressure of the Ar and N_2 mixture was kept at 0.5 Pa and the mixture ratio was 1 : 1. The thickness of the films was 250 nm.

The structure of these films was characterized by X-ray diffraction (XRD). The magnetic behavior of these films was investigated by a superconducting quantum interference devices (SQUID) magnetometer. The valence state of Mn was estimated using X-ray photoelectron spectroscopy (XPS). The Mn concentration was measured by energy-dispersive X-ray analysis (EDX).

3. Results and Discussion

High-angle XRD profiles of $Al_{0.96}Mn_{0.04}N$ films deposited on thermally oxidized Si (001) substrates and $Al_{0.97}Mn_{0.03}N$ films deposited on sapphire (0001) substrates are shown in Figure 1. In addition to diffraction peaks of the substrate, diffraction peaks from a würtzite-type AlN (0002) and (0004) were observed in each case. Although, the diffraction peak derived from the würtzite-type AlN (0006) was not observed for $Al_{0.96}Mn_{0.04}N$ films deposited on thermally oxidized Si (001) substrates at a substrate temperature (Ts) of 773 K. This result shows that $Al_{1-x}Mn_xN$ (x = 0.03, 0.04) films deposited on these substrates have a würtzite-type AlN phase with a preferentially oriented c-axis.

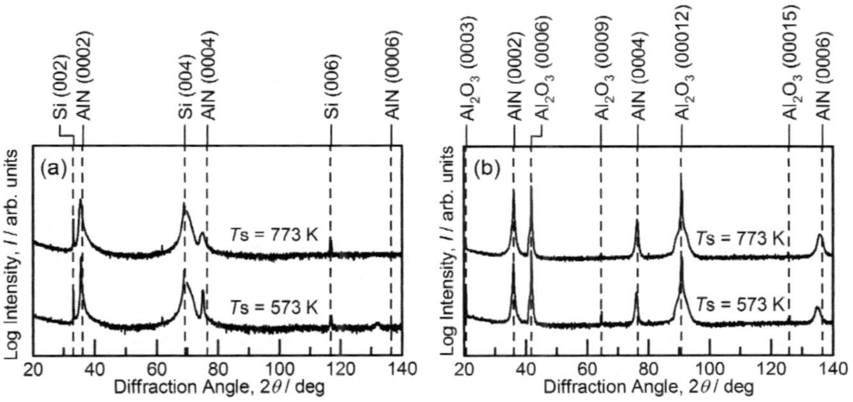

Figure 1. High-angle XRD profiles of (a) $Al_{0.96}Mn_{0.04}N$ films deposited on thermally oxidized Si (001) substrates and (b) $Al_{0.97}Mn_{0.03}N$ films deposited on sapphire (0001) substrates.

To investigate the crystallographic orientation in the film plane, we obtained pole figures of AlN($10\bar{1}1$) for the $Al_{0.96}Mn_{0.04}N$ films that were deposited on thermally oxidized Si (001) substrates and of $Al_{0.97}Mn_{0.03}N$ films deposited on sapphire (0001) substrates as shown in Figure 2 (a) – (d). For $Al_{0.96}Mn_{0.04}N$ films deposited on thermally oxidized Si (001) substrates, the intensity increased slightly at $\chi = 28°$ regardless of ϕ. This result suggests that $Al_{0.96}Mn_{0.04}N$ films deposited on thermally oxidized Si (001) substrates have a random orientation in the film plane. On the other hand, the intensity increased at $\chi = 28°$ and $\phi = 110, 170°$ for $Al_{0.97}Mn_{0.03}N$ films deposited on sapphire (0001) substrates. To determine the epitaxial relationship between $Al_{0.97}Mn_{0.03}N$ films and their sapphire (0001) substrates, we used the pole figure of $Al_2O_3(11\bar{2}3)$ as shown in Figure 2(e). In this pole figure, a high intensity is observed at the same ϕ as AlN($10\bar{1}1$). This indicates that $Al_{0.97}Mn_{0.03}N$ films deposited on sapphire (0001) substrates have an epitaxial relationship of AlN[$10\bar{1}0$] // Al_2O_3[$11\bar{2}0$].

The temperature dependence of magnetization at a magnetic field of 5 kOe for $Al_{0.96}Mn_{0.04}N$ films deposited on thermally oxidized Si (001) substrates and $Al_{0.97}Mn_{0.03}N$ films deposited on sapphire (0001) substrates are shown in Figure 3. In every case, the magnetization decreased markedly with increasing temperature. This result indicates that $Al_{1-x}Mn_xN$ (x = 0.03, 0.04) films deposited on all substrates exhibit paramagnetic behavior. However, relatively high magnetization is still apparent for $Al_{0.97}Mn_{0.03}N$ films deposited on sapphire (0001) substrates at 300 K.

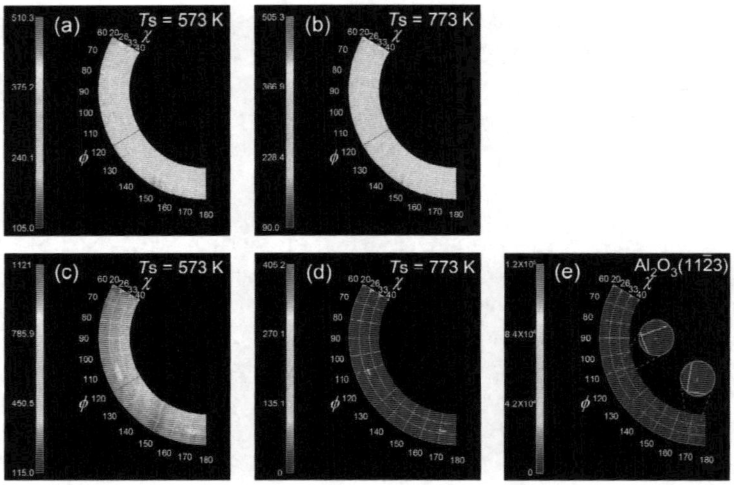

Figure 2. Pole figures of AlN(10$\bar{1}$1) for (a), (b) $Al_{0.96}Mn_{0.04}N$ films deposited on thermally oxidized Si (001) substrates and (c), (d) $Al_{0.97}Mn_{0.03}N$ films deposited on sapphire (0001) substrates. (e) Pole figure of $Al_2O_3(11\bar{2}3)$ for $Al_{0.97}Mn_{0.03}N$ films deposited on sapphire (0001) substrates.

To more precisely investigate the magnetic behavior of $Al_{0.96}Mn_{0.04}N$ films deposited on thermally oxidized Si (001) substrates and $Al_{0.97}Mn_{0.03}N$ films deposited on sapphire (0001) substrates, themagnetization curves at 300 K were measured as shown in Figure 4. For $Al_{0.96}Mn_{0.04}N$ films deposited on thermally oxidized Si (001) substrates, no sharp change in magnetization ((a) and (b)) was observed near the zero magnetic field. This suggests that $Al_{0.96}Mn_{0.04}N$ films deposited on thermally oxidized Si (001) substrates do not exhibit ferromagnetic behavior at 300 K. On the other hand, for $Al_{0.97}Mn_{0.03}N$ films deposited on sapphire (0001) substrates, a sharp change in magnetization ((c) and (d)) was observed in the magnetic field range of ±1 kOe. This suggests that $Al_{0.97}Mn_{0.03}N$ films deposited on sapphire (0001) substrates exhibit ferromagnetic behavior in addition to paramagnetic behavior.

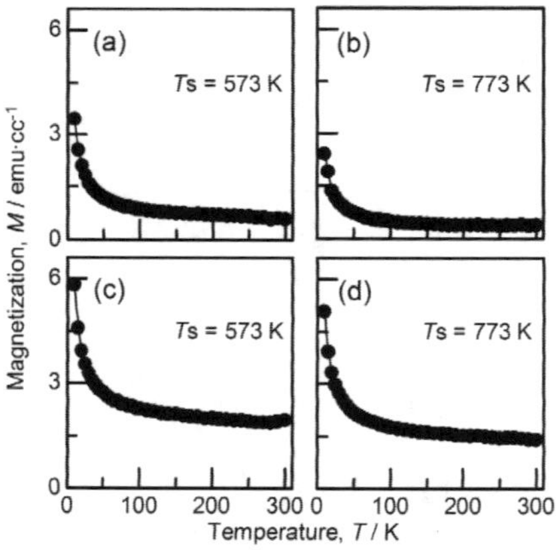

Figure 3. Temperature dependence of magnetization for (a), (b) $Al_{0.96}Mn_{0.04}N$ films deposited on thermally oxidized Si (001) substrates and (c), (d) $Al_{0.97}Mn_{0.03}N$ films deposited on sapphire (0001) substrates.

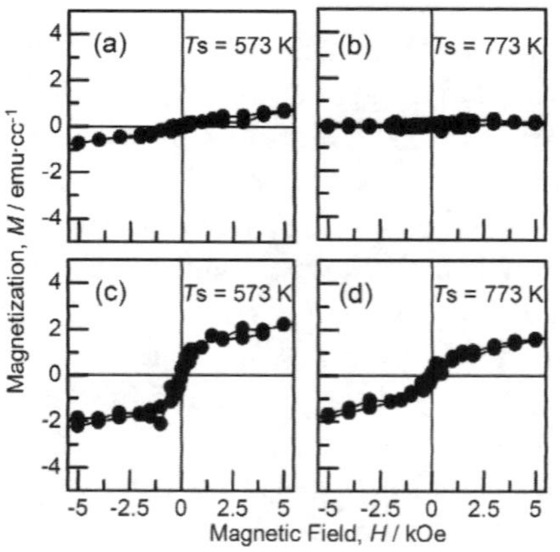

Figure 4. Magnetization curves at 300 K for (a), (b) $Al_{0.96}Mn_{0.04}N$ films deposited on thermally oxidized Si (001) substrates and (c), (d) $Al_{0.97}Mn_{0.03}N$ films deposited on sapphire (0001) substrates.

To discus the origin of the difference in magnetic behavior, the valence state of Mn was confirmed by XPS as shown in Figure 5. The binding energy of Mn $2p_{3/2}$ in $Al_{0.97}Mn_{0.03}N$ films that were deposited on sapphire (0001) substrates is slightly lower than that in $Al_{0.96}Mn_{0.04}N$ films deposited on thermally oxidized Si (001) substrates. This reveals that the valence of Mn decreases when the substrate is changed from a thermally oxidized Si (001) substrate to a sapphire (0001) substrate. The change in ferromagnetic behavior that was caused by the change in valence state of Mn has been reported for Mn-GaN [3]. Our results are similar to their results. For this study, however, the existence of a secondary phase of fine particles below the detection limit of XRD cannot be completely ignored. Therefore, we considered this possibility. Possible secondary phases are antiferromagnetic Mn_6N_5, Mn_3N_2, Mn_2N and ferrimagnetic Mn_4N. Among these materials, ferrimagnetic Mn_4N which has a low valence of Mn is the possible origin of ferromagnetic behavior.

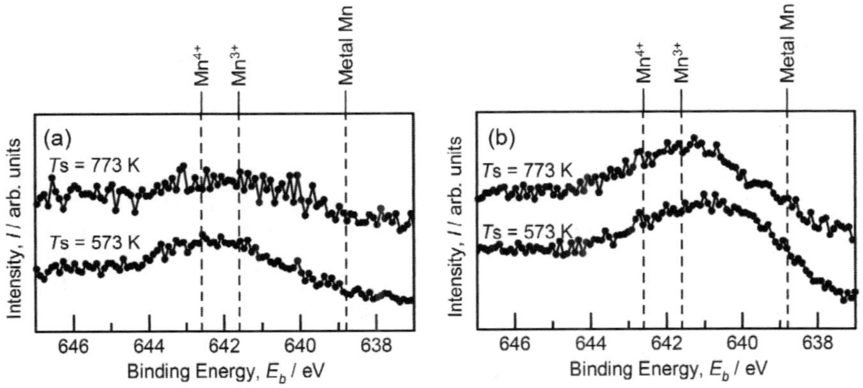

Figure 5. Mn $2p_{3/2}$ core level XPS spectrum for (a) $Al_{0.96}Mn_{0.04}N$ films deposited on thermally oxidized Si (001) substrates and (b) $Al_{0.97}Mn_{0.03}N$ films deposited on sapphire (0001) substrates.

4. Conclusion

We investigated the structure and magnetic behavior of Mn-AlN ($Al_{1-x}Mn_xN$, x = 0.03, 0.04) films deposited on thermally oxidized Si (001) substrates and sapphire (0001) substrates. $Al_{0.96}Mn_{0.04}N$ films deposited on thermally oxidized Si (001) substrates are paramagnetic while $Al_{0.97}Mn_{0.03}N$ films deposited on sapphire (0001) substrates are a mix of ferromagnetic and paramagnetic. This difference might be related to the difference of the valence state of Mn due to the substrate material.

Acknowledgements

This study was supported by Priority Assistance for the Formation of Worldwide Renowned Centers of Research - The Global COE Program (Project: Center of Excellence for Advanced Structural and Functional Materials Design) from the Ministry of Education, Culture, Sports, Science and Technology (MEXT), Japan. The authors would like to thank Professor Shinji Fujimoto of Osaka University for performing the XPS measurement.

References

[1] Frazier R, Thaler G, Overberg M, Gila B, Abernathy C R and Pearton S J 2003 *Appl. Phys.Lett.* **83** 1758

[2] Sato T, Endo Y, Kawamura Y, Kirino F, Nakatani R and Yamamoto M 2008 *IEEE Trans. Magn. in press*

[3] Sonoda S, Tanaka I, Yamamoto T, Yuba Y, Akasaka Y, Yoshida K, Aoki M, Asari M, Araki T, Nanishi Y, Kindo K and Hiro H 2007 *Appl. Phys. Lett.* **90** 012504

Role of underlayer for segregated structure formation of CoCrPt-SiO_2 granular thin film

Y Hirayama[1], I.Tamai[1], I Takekuma[1], and R Nakatani[2]
[1] Central Research Laboratory, Hitachi, Ltd., Odawara, Kanagawa, 256-8510, Japan
[2] Graduate School of Engineering, Osaka University, Suita, Osaka, 565-0871, Japan

E-mail: yoshiyuki.hirayama.ec@hitachi.com

Abstract. We have investigated the role of the Ru underlayer for the segregated structure formation of CoCrPt-SiO_2 granular thin film to improve the recording performance of CoCrPt-SiO_2 perpendicular magnetic recording media. The coercivity of the granular film decreases with decreasing Ru layer thickness, which comes from the change in the segregated structure. Formation of an oxide grain boundary is needed to obtain higher coercivity. According to a morphology analysis, the Ru layer roughness is closely related to the coercivity. The results indicate that the oxide grain boundary formation of the granular layer is enhanced by the roughness of the Ru underlayer. The shapes of the Ru grains probably serve as a template in CoCrPt grain growth and oxide grain boundary formation. As one of the methods for enhancing the roughness of the Ru underlayer, an island shape growth on the low surface energy materials is applicable. The roughness of the Ru layer was increased using a Pd/MgO/Pd seed layer and high coercivity was successfully obtained even when the Ru layer thickness was 3 nm.

1. Introduction

CoCrPt-SiO_2 granular thin film is well known as a recording layer of perpendicular magnetic recording media [1–3]. To obtain a high signal to noise ratio (SNR), magnetically isolated CoCrPt grains are required. Magnetically isolated grains can be attained by forming an oxide grain boundary surrounding the CoCrPt grains, called a "segregated structure", of granular film. To control the segregated structure of the recording layer, the role of the underlayer is important as well as the sputtering condition of the CoCrPt-SiO_2 granular film. Ru film is well known as a suitable underlayer for obtaining a c-axis oriented Co-alloy film [1–7] and a segregated structure of granular film. However, a thick Ru layer is usually used for obtaining better performance. To reduce the magnetic spacing between the head and soft-magnetic layers in the medium, the Ru layer thickness should be reduced. However, the coercivity of the granular medium often decreases with decreasing Ru layer thickness [7]. To improve the recording performance of the granular media, we should overcome the trade-off between Ru layer thickness and recording layer properties. We investigated the role of the Ru underlayer in the formation of a segregated structure of CoCrPt-SiO_2 granular film to overcome this trade-off relationship and to improve the performance of CoCrPt-SiO_2 perpendicular recording media.

2. Experimental procedure

A $CoCrPt-SiO_2$ perpendicular medium was prepared on a glass disk substrate. An amorphous CoTaZr film was deposited as a soft-magnetic underlayer. A Ta seed layer and a Ru underlayer were deposited to control the c-axis orientation of the recording layer. An additional seed layer, Pd/MgO/Pd, were inserted between Ta and Ru layers in experiments for roughness enhancement. The typical sputtering process condition of the Ru layer is a low deposition rate (0.2 nm/sec), which we call process A in this study. We also prepared samples by applying higher deposition rates of 1.0 (process B) and 6.0 nm/sec (process C) to clarify the role of the Ru layer. The $CoCrPt-SiO_2$ recording layer was deposited by co-sputtering of CoCrPt and SiO_2. A carbon nitride protective layer was then deposited on the surface of the recording layer. All the sputtering processes were carried out without heating.

The surface roughness (R_a) of the Ru layer was measured using an atomic force microscope (AFM). These AFM samples were prepared without recording layer deposition. The roughness we measured is the height difference between grain center and boundary. However, the real grain boundary depth is hard to measure due to the limited penetration depth of AFM cantilever. Therefore, the measured roughness is not real but relative value. The magnetic properties were measured using a Kerr effect magnetometer, vibrating sample magnetometer (VSM), and torque magnetometer. Samples without a soft-magnetic underlayer were prepared for VSM and torque measurements. The perpendicular uniaxial magnetic anisotropy energy was calculated by subtracting the shape anisotropy energy from the measured energy. The microstructures were observed using a high-resolution transmission electron microscope (TEM).

3. Results and discussion

3.1. Ru layer thickness impact on magnetic properties of granular film

Figures 1(a) and 1(b) show TEM plan-view images of Ru and $CoCrPt-SiO_2$ layers observed for a typical perpendicular medium with a 17-nm-thick Ru layer. The average grain pitches of Ru and CoCrPt were 8.1 and 7.8 nm, respectively, according to grain size analysis. This result indicates that there is one-to-one grain growth between the Ru and $CoCrPt-SiO_2$ layers. The cross-sectional TEM image, shown in figure 2, supports this relationship between Ru and $CoCrPt-SiO_2$ layers. Therefore, grain size and orientation of the Ru layer greatly impact those of the $CoCrPt-SiO_2$ layer [2]. In this case, CoCrPt grains are well isolated by oxide grain boundaries as shown in figure 1(b).

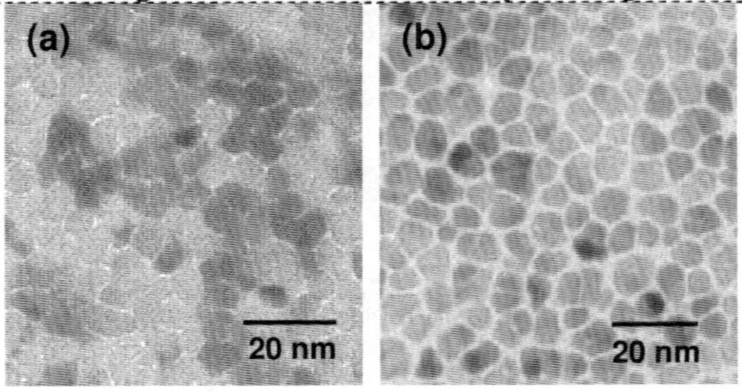

Figure 1. TEM plan-view images observed for (a) Ru and (b) $CoCrPt-SiO_2$ layers of perpendicular medium with 17-nm-thick Ru layer.

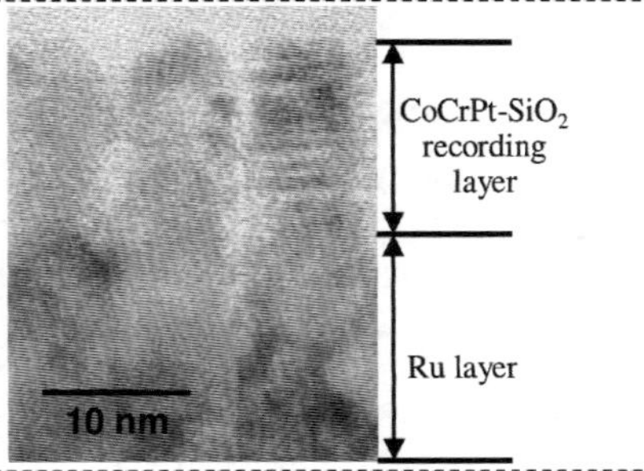

Figure 2. Cross-sectional TEM image observed for $CoCrPt-SiO_2$ and Ru layers.

The magnetic properties were measured for samples with various Ru layer thicknesses using a VSM and torque magnetometer. Figure 3 shows the dependence of coercivity (H_c) on the Ru layer thickness. The coercivity decreased with the Ru layer thickness. This effect is the same as results reported elsewhere [7]. Figure 3 also shows the change of loop slope around coercivity, which we call α. The increase of α with reducing Ru layer thickness comes from the increase in inter-granular exchange coupling. Figure 4 shows the uniaxial magnetic anisotropy energy (K_u) and the saturation magnetization (M_s) measured for the same sample. Both K_u and M_s remained almost the same when the Ru layer thickness was decreased. The results indicate that insufficient phase separation is not the cause of the coercivity reduction and that the difference in the segregated structure, which is probably the width of oxide grain boundary, is a possible cause. The width of oxide grain boundary affects inter-granular exchange coupling and coercivity.

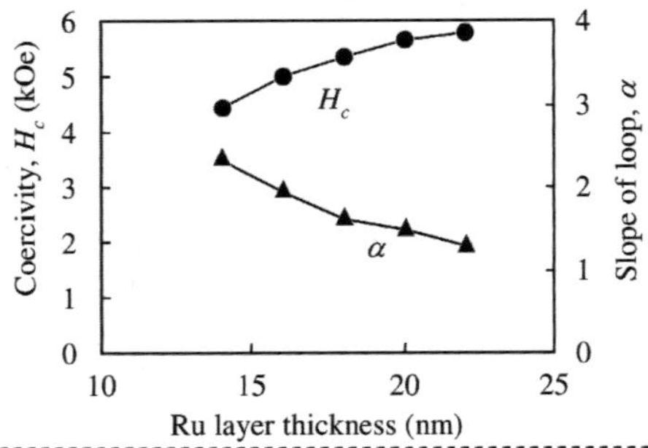

Figure 3. Dependence of H_c and α on Ru layer thickness.

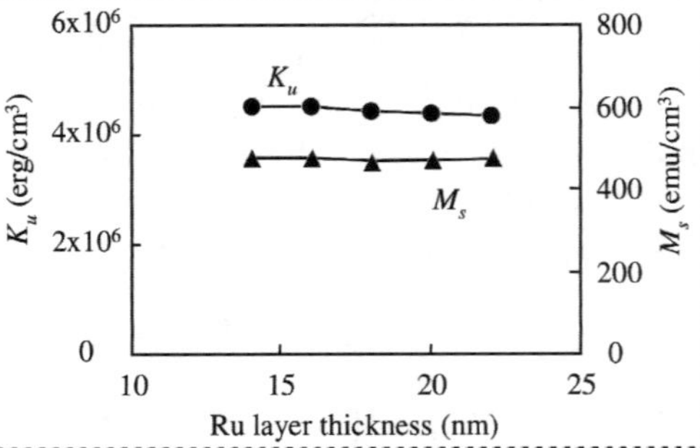

Figure 4. Dependence of K_u and M_s on Ru layer thickness.

3.2. Relationship between Ru layer roughness and magnetic properties

To clarify the mechanism of coercivity reduction observed in the samples with a thin Ru layer, the surface morphologies of the Ru films were observed using AFM. The films were deposited using three process conditions. Figure 5 shows the dependence of the roughness (R_a) on the Ru layer thickness. Here, the roughness indicates the surface morphologies of Ru grains. When the thickness was increased, the roughness increased for each of the deposition process conditions. Figure 6 shows the relationship between Ru layer roughness (R_a) and coercivity (H_c) measured using a Kerr effect magnetometer. The CoCrPt-SiO_2 recording layers were deposited using the same process conditions on the various Ru intermediate layers. These results show that coercivity is closely related to the roughness - the higher the roughness, the higher the coercivity. The two Kerr loops depicted in Figure 6 were measured for samples with Ru thicknesses of 5 and 20 nm. When the roughness was increased, the coercivity increased, and the loops became shared in shape. Such changes in the Kerr loop can be explained from reductions in the inter-granular exchange coupling of the recording layer. Therefore, the oxide grain boundary formation of the granular recording layer is apparently enhanced by the roughness of the Ru layer. The shapes of the Ru grains probably serve as a template in CoCrPt grain growth and oxide grain boundary formation. Therefore, we should increase the roughness of the Ru grains as well as creating proper diameters and a good c-axis orientation to improve the properties of the recording layer.

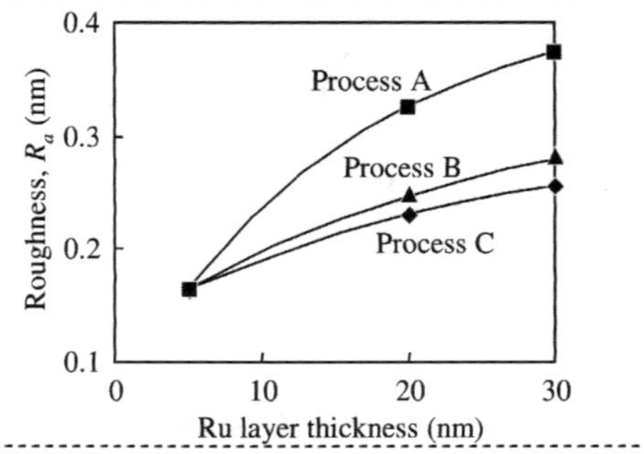

Figure 5. Relation between thickness and roughness of Ru films deposited using three kinds of process conditions.

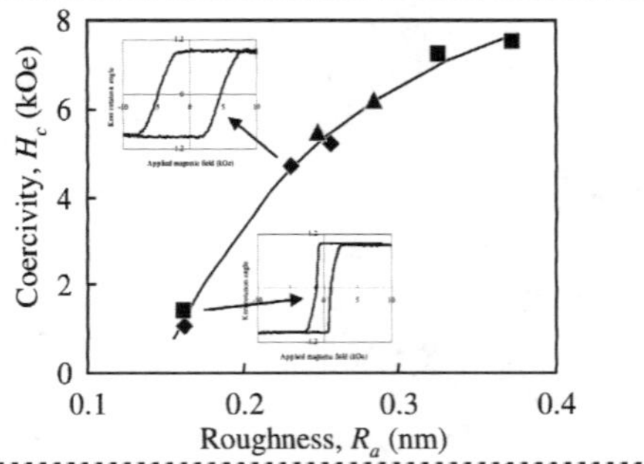

Figure 6. Relation between coercivity and Ru layer roughness.

3.3. Oxide seed layer

When an ultra-thin metal layer is formed on an oxide layer that has a low surface energy, island growth is expected [8-9]. This island shaped layer will help enhance the roughness of Ru layer surface and the segregated structure formation of CoCrPt-SiO_2 layer. We prepared a stacked structure of Pd (1.5 nm)/MgO (1 nm)/Pd (1.5 nm) and inserted it below the Ru layer. Then, we successfully obtained a Ru layer roughness (R_a) of 0.37 nm even when the Ru thickness was only 3 nm. The roughness enhancement by the Pd/MgO/Pd seed layer resulted in high coercivity even when the total thickness of the underlayer was less than 10 nm, as shown in figure 7. The results indicate that the surface roughness of the Ru underlayer contributes to increased coercivity and to an enhanced segregation structure of CoCrPt-SiO_2 granular film. The oxide seed layer is one of the promising candidates for overcoming the trade-off between Ru layer thickness and recording layer properties.

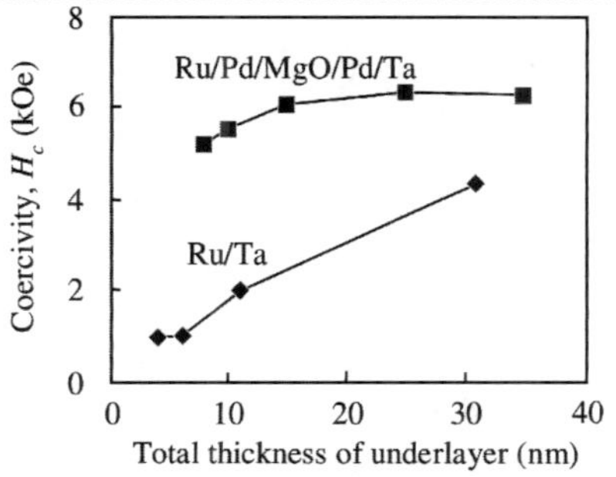

Figure 7. Dependence of coercivity on underlayer total thickness measured for Ru/Ta and Ru/Pd/MgO/Pd/Ta underlayers.

4. Conclusion

We have investigated role of the Ru underlayer for the segregated structure formation of CoCrPt-SiO_2 granular film. The coercivity of the granular film decreases with decreasing Ru layer thickness, which probably comes from the change in the segregated structure. Formation of an oxide grain boundary is needed to obtain higher coercivity. Judging from the AFM analysis results, the shapes of the Ru grains serve as a template in CoCrPt grain growth and oxide grain boundary formation. As one of the methods for enhancing the Ru layer roughness, introduction of a Pd/MgO/Pd seed layer was investigated. The Ru layer roughness was increased and high coercivity was successfully obtained even when the Ru layer thickness was 3 nm.

Acknowledgment

The authors would like to thank R. Araki of the Central Research Laboratory, Hitachi Ltd. for TEM observation.

References

[1] Oikawa S, Takeo A, Hikosaka T and Tanaka Y, 2000 *IEEE Trans. Magn.* **36** 2393
[2] Oikawa T, Nakamura M, Uwazumi H, Shimatsu T, Muraoka H and Nakamura Y, 2002 *IEEE Trans. Magn.* **38** 1976
[3] Ikeda Y, Supper N, Do H, Takano K and Lengsfield B, 2003 *IEEE Trans. Magn.* **39** 2341
[4] Krishnan K, Honda Y, Hirayama Y and Futamoto M, 1994 *Appl. Phys. Lett.* **64** 1499
[5] Zheng M, Choe G, K. Johnson K, Gao L and Liou S, 2002 *IEEE Trans. Magn.* **38** 1979
[6] Saito S, Itagaki N, Hoshi F and Takahashi M, 2003 *J. Appl. Phys.* **93** 8182
[7] Velu E, Malhotra S, Bertero G and Wachenschwanz D, 2003 *IEEE Trans. Magn.* **39** 668
[8] Hirayama Y, Tamai I, Takekuma I and Hosoe Y, 2003 *IEEE Trans. Magn.* **39** 2282
[9] Nakagawa H, Takekuma I, Nemoto H, Takahashi Y, Hirayama Y, Nishida Y and Hosoe Y, 2003 *IEEE Trans. Magn.* **39** 2311

Superparamagnetism of ultrathin Co film on antiferromagnetic Cr_2O_3 layer

Y Shiratsuchi, T Nakatani, R Nakatani and M Yamamoto

Department of Materials Science and Engineering, Graduate School of Engineering, Osaka University, 2-1 Yamadaoka, Suita, Osaka 5650871, Japan.

E-mail: shiratsuchi@mat.eng.osaka-u.ac.jp

Abstract. The effect of antiferromagnetic Cr_2O_3 thin film on the magnetic properties of ultrathin Co film has been investigated. To achieve the investigation, we have also investigated the fabrication of Cr_2O_3 thin film of high quality using MBE. The crystalline quality of Cr oxide film strongly depends on the in-plane epitaxial variants of Cr(110) before the oxidation. We have successfully fabricated Cr_2O_3(0001) film by oxidizing three fold-symmetric Cr(110) film. On Cr_2O_3(0001) thin film, the magnetization of Co is stabilized parallel to the Cr spin direction below the Néel temperature of Cr_2O_3.

1. Introduction

Magnetization of nano-magnets essentially fluctuates due to the thermal agitation [1]. Since the thermal agitation of magnetization makes the magnetization direction difficult to fix to the certain direction, it is recognized the problem to overcome for the device application of nano-magnets. The thermal agitation of magnetization is often discussed using the ratio $\Delta E/k_BT$, where ΔE is the energy barrier for the magnetization reversal, k_B is the Boltzmann constant and T is the absolute temperature. For $\Delta E/k_BT \ll 25$, the net magnetization vanishes, namely the nano-magnet is in the superparamagnetic state. Since ΔE is expressed by the effective magnetic anisotropy and the volume of magnetic entity through the relationship $\Delta E = K_{eff} \cdot V$, in order to suppress the thermal agitation of magnetization in keeping the volume in the nano-size, it is necessary to increase the effective magnetic anisotropy K_{eff}. K_{eff} includes the surface/interface anisotropy as well as the bulk anisotropy like the magneto-crystalline anisotropy. Especially, for the nano-magnets, the surface/interface anisotropy can dominate K_{eff} owing to the large surface/interface contribution. In this viewpoint, we have investigating the control of magnetism of nano-magnet using the interface anisotropy. Among the various surface/interface effect, we utilize the interface effect between the antiferromagnet and the ferromagnet, known as the exchange anisotropy. Since it is well known that the exchange anisotropy depends on the spin direction of antiferromagnetic layer [2, 3], in order to clarify the detailed effect of antiferromagnetic layer, it is suitable to use the antiferromagnet in which the spin direction is controllable. In this study, we chose Cr_2O_3 as an antiferromagnet because Cr_2O_3 has the highly anisotropic crystal structure and thus, the spin direction could be controlled by altering crystallographic orientation. Especially for thin film system, controlling crystallographic orientation is easily possible by using epitaxial growth. In this report, using Co as a ferromagnet we report the effect of antiferromagnetic Cr_2O_3 layer on the magnetic properties of ultrathin Co film.

[1] To whom any correspondence should be addressed.

α-Cr_2O_3 has the corundum structure which is consisted by 12 Cr layers and 6 O layers. Corundum structure is schematically shown in figure 1. In each Cr layer, the spin orients parallel to c axis and is antiferromagnetically aligned in the neighboring Cr layer. This implies that for α-Cr_2O_3 thin film, the spin alignment at the surface/interface is controllable by changing crystallographic orientation. In fact, it has been reported that both the strength and symmetry of exchange anisotropy is changed by altering crystallographic orientation of Cr_2O_3 for the polycrystalline permalloy / Cr_2O_3 bilayer in which the crystallographic orientation of Cr_2O_3 is altered by (0001), $(11\bar{2}0)$ and $(1\bar{1}02)$ [4].

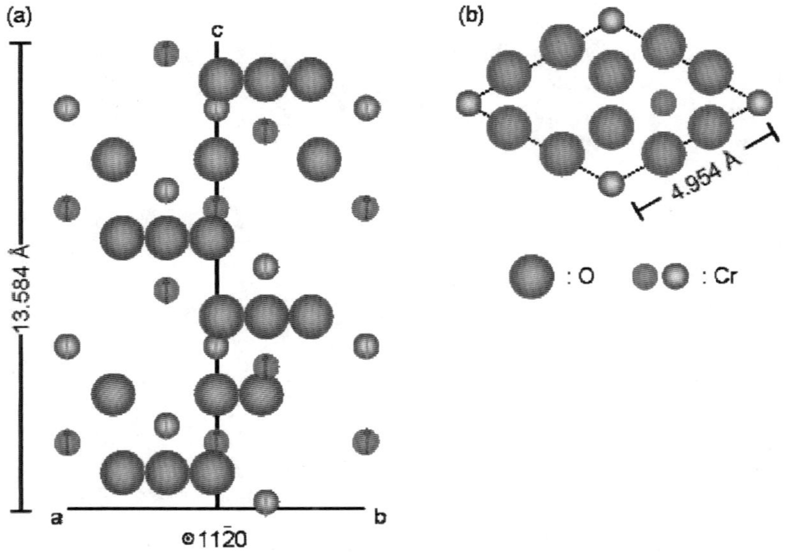

Figure 1. Schematic representation of the crystal structure of α-Cr_2O_3 for (a) $<11\bar{2}0>$ and (b) <0001> projection, respectively. (colour online)

As mentioned above, Cr_2O_3 is an antiferromagnet, and it is known that Cr_2O_3 shows the magneto-electronic (ME) effect, as well [5]. ME effect means that the magnetization is induced by applying electronic field. ME effect of Cr_2O_3 is originated from the combined effects of the low crystallographic symmetry due to the corundum structure and the c-axis oriented Cr spin. Owing to the recent progress in spin-electronics, controlling magnetism using the electronic field has attracted attention, and ME effect of Cr_2O_3 is also being the subject of intense research activity [6-8]. Change of exchange bias by the ME effect of Cr_2O_3 has been experimentally observed for Co/Pt superlattice on Cr_2O_3(111) substrate [6]. However, in this report, since Cr_2O_3 is utilized as 0.7 mm-thick substrate, the required voltage for ME effect is quite large, and it would be necessary to observe ME effect using Cr_2O_3 thin film for the device application.

For the control of antiferromagnetic spin alignment and the utilization of ME effect, it is necessary to fabricate the Cr_2O_3 thin film of high quality. In this report, we have also investigated the fabrication of highly oriented Cr_2O_3 thin film using Molecular Beam Epitaxy (MBE) system.

2. Experimental

2.1. Sample preparation

Cr oxide thin films were prepared by oxidizing Cr thin film in ultra-high-vacuum (UHV) chamber. Cr thin films before the oxidation was prepared by VG-80M molecular beam epitaxy (MBE) system. The base pressure before and during growth was typically below 4×10^{-9} and 5×10^{-8} Pa, respectively. The deposition temperature of Cr was varied 323 to 473 K. The thickness of Cr before oxidation was fixed to 20 nm. Three types of α-Al_2O_3 surfaces having (0001), ($11\bar{2}0$), and ($1\bar{1}02$) were used as the substrate, because α-Al_2O_3 has the corundum structure same to α-Cr_2O_3. The substrate was annealed in air at 1273 K to flatten the surface before introducing the vacuum system. After that, it was cleaned in UHV chamber by annealed for several hours at 1173 K prior to the Cr deposition.

After depositing Cr, the samples were transferred to the oxidation chamber, which is connected to the growth chamber under UHV, without exposing air. At the oxidation chamber, Cr thin film was oxidized by annealing at 973 K in O_2 of 3×10^{-3} Pa. Note that the pressure of oxidation chamber before introducing O_2 was below 9×10^{-9} Pa.

Then, the sample was again transferred to the growth chamber, and Co was deposited on the oxidized Cr film at 473 K by MBE. The thickness of Co is fixed to 1.0 nm. To investigate the magnetic properties mentioned below, the films have to be exposed to air. In order to prevent the surface oxidation, 10 nm-thick Au capping layer was deposited on Co at room temperature.

2.2. Structural and magnetic characterization
The structure of thin films was investigated using reflection high energy electron diffraction (RHHED) and X-ray diffraction (XRD). Investigation using RHEED were performed *in-situ* at growth chamber. Observation of RHEED was carried out at 30 kV with an incident angle 1°. XRD profiles were obtained in polar (φ) scan for the quantitative investigation of crystal structure.

Magnetization measurements of thin films were performed by superconducting quantum interference (SQUID) magnetometry. Our investigation encompassed the temperature dependence of field-cooled (FC), ZFC, and remanent (RM) magnetization in the temperature range of 5 to 300 K under the magnetic fields of 50 Oe. The measurements were performed during the heating process. Especially, two types of RM are measured after FC and ZFC, called as the thermo-remanent magnetization (TRM) and the isothermal remanent magnetization (IRM), respectively [9, 10]. For instance, when TRM was measured, the sample was cooled from the enough high temperature, R.T. to the measurement temperature under the certain field. After reaching measurement temperature, the field was removed and the magnetization was measured. In the case of IRM, the similar procedure was done after ZFC procedure. TRM and IRM give the stable magnetization after removing field and the activated magnetization by applying field, respectively.

3. Results and Discussions

3.1. Structure of Cr oxide thin film
The crystalline quality of Cr oxide film depends on the surface structure of Cr thin film before oxidization. RHEED patterns of Cr thin film before and after oxidization are shown in figure 2. In our preparation method described in Section 2.1, all studied Cr thin films before oxidation possess bcc structure with (110) orientation perpendicular to the film plane. However, the in-plane crystallographic symmetry due to the epitaxial variants changes by the growth condition. In this work, two types of growth are observed. One is shown as RHEED pattern shown in figure 2(a); the formation of three-fold symmetric Cr(110). This growth manner is observed only for Cr grown on α-Al_2O_3(0001) and α-Al_2O_3($11\bar{2}0$) at 473 K. The other growth manner is the formation of Cr(110) variants having more than six variants, shown in figure 2(b). The latter growth occurs for Cr grown on α-Al_2O_3($1\bar{1}02$) and Cr on α-Al_2O_3(0001) and α-Al_2O_3($11\bar{2}0$) at 323 K.

RHEED pattern of Cr oxide thin film are strongly depends on the number of Cr(110) variants before oxidation. As shown in figure 2(c), the clear diffraction spots are observed by oxidizing Cr(110) having three variants. The observed patterns are well matched to the expected reciprocal lattice map of the corundum structure. Since RHEED observation is not suitable for the quantitative analysis, the quantitative information of crystal structure is obtained from the XRD measurement. As a result, adopting Bragg condition for α-Cr_2O_3($1\bar{1}04$), the distinct diffraction peak are observed in φ scan with the six fold-symmetry (not shown). However, α-$Cr_2O_3(\bar{1}104)$ diffraction show the three fold-symmetry for the single crystalline α-Cr_2O_3, and thus our film should form the twinned structure. These results strongly suggest that the fabricated Cr oxidized thin film is the twinned α-$Cr_2O_3(0001)$ thin film.

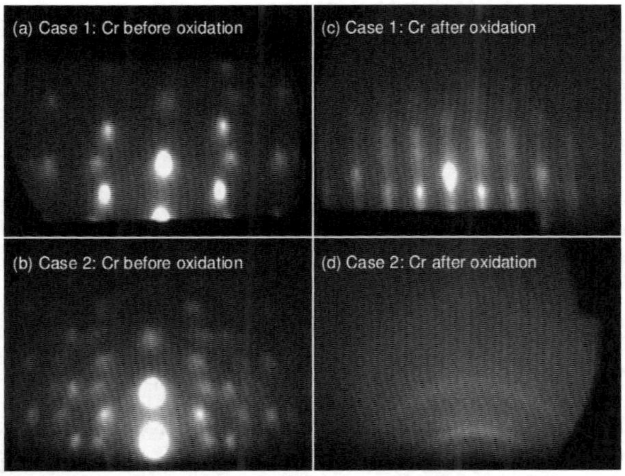

Figure 2. $[11\bar{2}0]_{Al2O3}$-azimuthal RHEED patterns of Cr thin film (a), (b) before and (c), (d) after oxidation. The results are on 20.0 nm-thick Cr grown on α-$Al_2O_3(0001)$ substrate at the different temperature.

On the contrary, as shown in figure 2(d), RHEED patterns are the ring pattern by oxidizing Cr(110) having more than six variants. This means that the Cr oxide films are poly-crystalline. In this case, since the quantitative analysis is difficult as explained above, the crystal structure is not determined.

Here, the reason for the formation of α-$Cr_2O_3(0001)$ by oxidizing three fold-symmetric Cr(110) is discussed. The formation of α-$Cr_2O_3(0001)$ is explained by the epitaxial relationship between α-$Cr_2O_3(0001)$ and Cr(110). The atom arrangements of α-$Cr_2O_3(0001)$ and Cr(110) are schematically represented in figure 3. α-$Cr_2O_3(0001)$ lattice and Cr(110) lattice is matched in following orientation.

$$\alpha\text{-}Cr_2O_3(0001)\langle 1\bar{1}00\rangle \,//\, Cr(110)\langle 001\rangle$$

In this orientation, the lattice misfit is 0.70% assuming lattice constant of bulk value. This orientation relationship is similar to the Nishiyama-Wassermann relationship which is well known as that of fcc(111) and bcc(110). Note that fcc(111) and hcp(0001) have the identical atom arrangement in topmost layer, and thus the appearance of similar relationship is reasonable. This result implies that

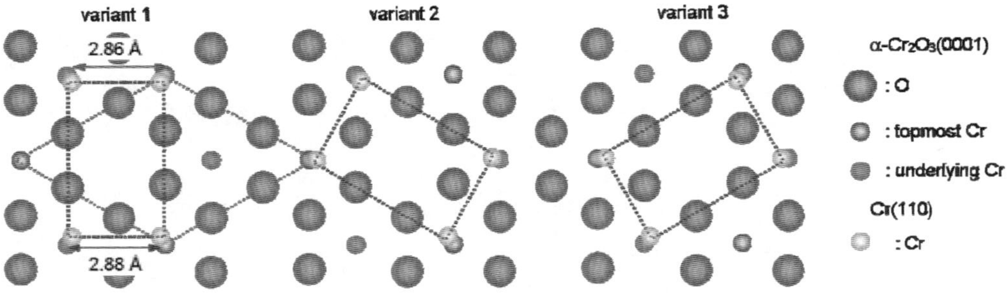

Figure 3. Scheme of orientation relationship between α-$Cr_2O_3(0001)$ and bcc Cr(110). Assuming the bulk value as the lattice constant, the lattice misfit is 0.70 % for the shown orientation relationship.(colour online)

the formation of α-Cr_2O_3 having the specific orientation requires the suitable atomic arrangement of Cr before oxidation. In other words, controlling the crystallographic orientation of α-Cr_2O_3 thin film, namely spin arrangement at interface would be possible by changing the orientation of Cr film before oxidation.

3.2. Magnetic properties of ultrathin Co film on $Cr_2O_3(0001)$ thin film

Before investigating the effect of Cr spin direction on the magnetic properties of ultrathin Co film, the magnetic state of Co at room temperature is shown. Magnetization curve and temperature dependence of magnetization of ultrathin Co film grown on $Cr_2O_3(0001)$ are shown in figure 4. The magnetic field was applied parallel to the film plane. In the magnetization curve at room temperature, neither remanent magnetization nor coercivity is observed, and the magnetization is hard to saturate above 20 kOe (figure 4(a)). Furthermore, in the temperature dependence of magnetization, the clear deviation of FC magnetization and ZFC magnetization is observed below 200 K (figure 4(b)). These results indicate that the magnetization of Co at room temperature is highly fluctuated due to the thermal energy. However, the magnetic properties shown in figure 4, especially the temperature dependence of magnetization contain the distinct difference from the ideal superparamagnet. In figure 4(b), the peak of magnetization is observed in both FC magnetization and ZFC magnetization, while the peak would appear only in ZFC magnetization for the ideal superparamagnet. The observed peaks are explained by considering that the peak is originated from Néel temperature of the fabricated Cr_2O_3 thin film. Assuming that the Cr spin direction in the fabricated Cr_2O_3 thin film is parallel to c-axis, the in-plane magnetic field means the perpendicularly applied field to the Cr spin direction. In the case, FC magnetization and ZFC magnetization below Néel temperature would orient to perpendicular to the film plane, reflecting the under-lying Cr spin direction. Increasing temperature and approaching Néel temperature, the magnetization of Co rotate to the applied field direction owing to the weakened anti-ferromagnetic order of Cr spin. Based on this discussion, it is considered that the peak in FC magnetization and ZFC magnetization is recognized as the Néel temperature of the fabricated Cr_2O_3 thin film.

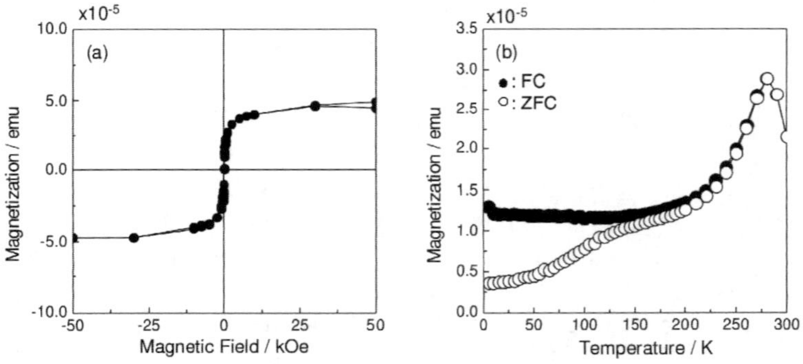

Figure 4. (a) Magnetization curve measured at room temperature and (b) temperature dependence of magnetization for 1.0 nm-thick Co film on $Cr_2O_3(0001)$. The magnetic field is applied parallel to the film plane.

The above discussion would be verified from the temperature dependence of magnetization in which the magnetic field is applied perpendicular the film plane, namely parallel to the expected Cr spin direction. Figure 5 show the temperature dependence of FC magnetization, ZFC magnetization, TRM and IRM. As shown in

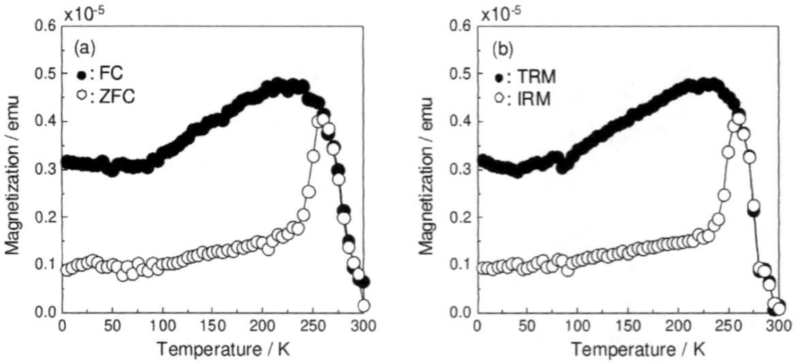

Figure 5. Temperature dependence of (a) FC magnetization and ZFC magnetization, (b) TRM and IRM. for 1.0 nm-thick Co film on $Cr_2O_3(0001)$. The magnetic field is applied perpendicular to the film plane.

figure 5(a), the peak is observed in only the ZFC magnetization, and not in FC magnetization. According to the above discussion, the field parallel to the Cr spin would not cause the magnetization rotation around the Néel temperature for FC magnetization. Hence, the observed feature agree with the above discussion. Furthermore, as show in figure 5(b), TRM and IRM are completely identical to FC magnetization and ZFC magnetization below the Néel temperature, respectively. These results indicate that the magnetization of Co is stable for the Cr spin direction below the Néel temperature even after removing field. In fact, above the Néel temperature when the antiferromagnetic order of Cr spin is destroyed, TRM(IRM) deviate from FC(ZFC) magnetization, and decrease rapidly, respectively.

As mentioned above, while the magnetization of Co on Cr_2O_3(0001) thin film is stabilized parallel to the Cr spin direction below the Néel temperature, Co rapidly becomes the superparamagnetic state above the Néel temperature. Considering that the blocking temperature of Co grown on α-Al_2O_3(0001) substrate is approximately 90 K [11], the ferromagnetic state above 100 K for Co on Cr_2O_3(0001) would be caused by the antiferromagnetic order of Cr_2O_3. However, since the magnetic state of nano-magnets is influenced by the nano-structure such as the particle and the continuous film, the further detailed investigation, especially the correlation to the nano-structure is necessary.

4. Summary

The fabrication of Cr_2O_3 thin film using MBE system and the magnetic properties of ultrathin Co film on Cr_2O_3 have been investigated. The crystalline quality of Cr oxide film is affected by the epitaxial growth of Cr film before oxidation. α-Cr_2O_3(0001) are fabricated by oxidizing Cr(110) film having the three equivalent in-plane variants. The formation of α-Cr_2O_3(0001) is explained by the epitaxial relationship known as Nishiyama-Wasserman relationship. Using fabricated α-Cr_2O_3(0001) film, the effect of Cr spin on the magnetic properties of Co has been examined. In α-Cr_2O_3(0001), Cr spin would align perpendicular to the film plane, and the aligned Cr spin affect the magnetic properties of ultrathin Co film. The magnetization direction of Co is stabilized to the Cr spin direction below the Néel temperature which is approximately 280 K in our case. Above the Néel temperature, decreasing antiferromagnetic order of Cr spin, the magnetization of Co starts to fluctuate against the thermal agitation and Co become superparamagnetic. Comparing results to that of Co on α-Al_2O_3(0001), it is considered that the magnetization of Co is stabilized by the antiferromagnetic spin of Cr.

Acknowledgements

This study was partly supported by Encouragement of Young Scientists (B) and Priority Assistance for the Formation of Worldwide Renowned Centers of Research - The Global COE Program (Project: Center of Excellence for Advanced Structural and Functional Materials Design) from the Ministry of Education, Culture, Sports, Science and Technology (MEXT), Japan.

References

[1] Shiratsuchi Y, Yamamoto M and Bader S D 2007 *Prog. Surf. Sci.* **82** 121
[2] Meiklejon W H and Bean C P 1956 *Phys. Rev.* **102** 1413
[3] Nogués J and Shuller I K, 1999 *J. Magn. Magn. Mater.* **192** 203
[4] Dho J, Blamire M G and Chi E O 2005 *Phys. Rev. B* **72** 224421
[5] Rado G T and Folten V T 1961 *Phys. Rev. Lett.* **7** 310
[6] Borisov P, Hochstrat A, Chen X, Kleemann W and Binek C 2005 *Phys. Rev. Lett.* **94** 117203
[7] Laukin V, Skumryev V. Martí X, Hrabovsky D, Sánchez F, García-Cuenca M V, Ferrater C, Varela M, Lüders U, Bobo J F and Fontcuberta J 2006 *Phys. Rev .Lett.* **97** 227201
[8] Chu Y H, Martin L W, Holcomb M B, Gajek M, Han S H, He Q, Balke N, Yang C H, Lee D, Hu W, Zhan A, Yang P L, Fraile-Rodríguez A, Scholl A, Wang A X and Ramesh R 2008 *Nature Mat.* **7** 478
[9] Dormann J L, Fiorani D and Tronc E 1997 *Adv. Chem. Phys.* **98** 283
[10] Chantrell R W, El-Hilo M and O'Grady K 1991 *IEEE Trans. Magn.* **27** 3570
[11] Shiratsuchi Y, Murakami T, Endo Y and Yamamoto M 2005 *Jpn. J. Appl. Phys.* **44** 8456

Journal of Physics: Conference Series **165** (2009) 012035 doi:10.1088/1742-6596/165/1/012035

On magnetic properties of $BiCrO_3$ and $BiMnO_3$

Alexei A Belik[1] **and Eiji Takayama-Muromachi**
International Center for Materials Nanoarchitectonics (MANA), National Institute for Materials Science (NIMS), 1-1 Namiki, Tsukuba, Ibaraki 305-0044, Japan

E-mail: Alexei.Belik@nims.go.jp

Abstract. We investigated magnetic properties of $BiCrO_3$ having the *C2/c* symmetry and $Bi_{0.9}Y_{0.1}CrO_3$ having the *Pnma* symmetry with $a = 5.5451(1)$ Å, $b = 7.7185(2)$ Å, $c = 5.3882(1)$ Å. $Bi_{0.9}Y_{0.1}CrO_3$ shows magnetic anomalies at T_N = 154 K due to an antiferromagnetic transition with weak ferromagnetism and frequency-dependent anomalies near 50 K with very large temperature shifts on both the real and imaginary parts of the ac susceptibilities (43 K at 0.5 Hz and 56 K at 300 Hz on χ''). Based on the results for $Bi_{0.9}Y_{0.1}CrO_3$, we could understand some peculiarities in magnetic properties of $BiCrO_3$. In particular, an orthorhombic modification of $BiCrO_3$ was found to exist in the monoclinic form. We also discuss magnetic properties of $BiMnO_3$ that shows more complicated behavior.

1. Introduction

Bi-containing perovskites have attracted much attention during the last decade as multiferroic and lead-free ferroelectric materials because of the presence of the stereochemically-active lone electron pair of a Bi^{3+} ion [1]. In multiferroic systems, two or all three of (anti)ferroelectricity, (anti)ferromagnetism, and ferroelasticity are observed in the same phase [1].

Despite of the presence of Bi^{3+} ions in simple $BiMO_3$ compounds (M = transition metals), only $BiFeO_3$ (space group *R3c*) and $BiCoO_3$ (space group *P4mm*) definitely have polar structures at room temperature (RT) and ambient pressure (AP) (see Refs in [2]). $BiNiO_3$ crystallizes in the centrosymmetric space group *P*-1 (see Refs in [2]). $BiCrO_3$ and $BiMnO_3$ have the *C2/c* symmetry [2, 3]. Despite of substantial work on $BiMO_3$, there are still some puzzling and unresolved problems. The structural and magnetic properties of $BiFeO_3$ are especially interesting [4, 5]. In the case of $BiMnO_3$ and $BiCrO_3$, the situation is also not simple. The static magnetic properties (dc magnetization and specific heat) seem to be explained and understood well. While the dynamic magnetic properties show additional anomalies.

In $BiCrO_3$, four anomalies of magnetic origin were found near 40, 75, 109, and 111 K [6]. The long-range antiferromagnetic order with weak ferromagnetism occurs at T_N = 109 K. The ac susceptibilities showed that the transition near T_N is a two-step transition. Additional frequency-independent broad anomalies were observed on the real part of the ac susceptibilities near 75 K. Below 40 K, frequency-dependent anomalies with very large temperature shifts were observed on both the real and imaginary parts of the ac susceptibilities. In $BiMnO_3$, even more anomalies were found on the imaginary part of the ac susceptibilities, at 5, 15, 84, 99, 104, and 112 K [7, 8].

In this work, we try to understand and explain at least some of the magnetic anomalies in $BiCrO_3$ and $BiMnO_3$.

[1] To whom any correspondence should be addressed.

2. Experimental details

The synthesis of $BiCrO_3$ and $BiMnO_3$ is described in [2, 3]. Single-phase $Bi_{0.9}Y_{0.1}CrO_3$ was synthesized from a stoichiometric mixture of Bi_2O_3, Y_2O_3, and Cr_2O_3 at 6 GPa and 1550 K for 90 min in an Au capsule. dc magnetic susceptibilities, $\chi = \mathbf{M/H}$, were measured on a SQUID magnetometer (Quantum Design, MPMS) between 2 and 300 K in an applied field of 100 Oe under both zero-field-cooled (ZFC) and field-cooled (FC; on cooling) conditions. Frequency dependent ac susceptibility measurements at zero static magnetic field were performed with a Quantum Design MPMS instrument from 250 to 2 K at frequencies (*f*) from 0.5 to 300 Hz and an applied oscillating magnetic field (H_{ac}) of 5 Oe.

3. Results and discussion

Figure 1 shows the inverse magnetic susceptibilities (measured at 100 Oe) of two $BiCrO_3$ samples prepared at slightly different conditions. Both samples were single-phase according to X-ray powder diffraction (XRD) data. In addition to the previously reported anomalies [6], we found an anomaly near 165 K on the FC curves. This anomaly can be detected only with a small magnetic field (e.g., 100 Oe), and it is completely suppressed by a large field (e.g., 10 kOe). The thermoremanent magnetization (TRM) curves as a function of temperature for $BiCrO_3$ showed more clearly the anomalies near 165 K (figure 2). The TRM curves were measured at zero magnetic field on heating after cooling the sample from 270 K to 2 K at 1 kOe. The origin of the anomaly at 165 K in $BiCrO_3$ could be understood only after magnetic studies of $Bi_{0.9}Y_{0.1}CrO_3$ having the *Pnma* symmetry with a = 5.5451(1) Å, b = 7.7185(2) Å, c = 5.3882(1) Å. Figure 3 shows the real and imaginary parts of the ac susceptibilities of $Bi_{0.9}Y_{0.1}CrO_3$. This sample shows anomalies at T_N = 154 K due to an antiferromagnetic transition with weak ferromagnetism and frequency-dependent anomalies near 50 K with very large temperature shifts on both the real and imaginary parts of the ac susceptibilities (43 K at 0.5 Hz and 56 K at 300 Hz on χ''). $YCrO_3$ has T_N = 142 K. That is, T_N decreases in orthorhombic solid solutions $Bi_{1-x}Y_xCrO_3$ with increasing x. Therefore, the anomaly at 165 K in $BiCrO_3$ most probably corresponds to the T_N of the orthorhombic (O-) modification of $BiCrO_3$.

The O-modification of $BiCrO_3$ is stable above 420 K at AP [2, 6]. The O-phase is also formed at the synthesis conditions (that is, at high pressure and high temperature). During the transition from the orthorhombic to monoclinic modification in $BiCrO_3$ [9] or isostructural $BiScO_3$ [10] nanodomains (~10 nm) are formed. The O-modification of $BiCrO_3$ can remain, for example, in the domain boundaries and can be an integral part of the system. The O-modification still remained (by magnetization measurements) after cooling $BiCrO_3$ from 550 K to RT in 40 h at AP. That is, very slow cooling did not eliminate the O-phase. All reflections on XRD and neutron powder diffraction patterns of $BiCrO_3$ can be indexed in space group *C*2/*c* (at RT). However, careful Rietveld analysis of XRD data showed that fits of some reflections were not good. Only the introduction of the second orthorhombic *Pnma* modification greatly improved the fits and *R* factors. The weight fraction of the *Pnma* modification (a = 5.5672(4) Å, b = 7.7416(8) Å, c = 5.4762(3) Å) is about 10-15 % in $BiCrO_3$. Therefore, the Rietveld analysis is consistent with the magnetization measurements. The magnetization measurements showed that sample 2 contained a larger amount of the O-modification compared with sample 1 (figures 1 and 2).

The frequency-dependent anomalies in $Bi_{0.9}Y_{0.1}CrO_3$ near 50 K with very large temperature shifts are much more pronounced than those of $BiCrO_3$ [6]. Therefore, their origin in $BiCrO_3$ seems to come from the presence of the *Pnma* modification. These data raise another question how to explain such large 'superparamagnetic-like' [6] temperature shifts in $Bi_{0.9}Y_{0.1}CrO_3$. We note that the superparamagnetic-like anomalies in magnetic properties are sometimes explained by a phase separation. It is a possible scenario for orthorhombic $Bi_{0.9}Y_{0.1}CrO_3$ where domains of the monoclinic phase may emerge and move at low temperatures. In orthorhombic $YCrO_3$, only anomalies at T_N were found on the ac susceptibilities without any anomalies at lower temperatures.

To summarize data on $BiCrO_3$, we believe that magnetic anomalies near 40 K and newly found anomalies near 165 K are originated from the intrinsic presence of the O-modification in the matrix of the monoclinic modification. The anomalies at T_N = 109 K are related to a long-range G-type antiferromagnetic order of the *C*2/*c* phase (with weak ferromagnetism and magnetic moments aligned along the *b* direction [2]). The double-peak structure on the χ'' vs *T* curves near T_N [6] can probably be explained by small compositional inhomogeneity or a specific phase separation during the transition from the O- to monoclinic modification, but more studies are needed. The anomalies near 75 K are originated from the spin rotation away from the *b* axis keeping the G-type antiferromagnetic structure [11]. Therefore, almost all the magnetic anomalies in $BiCrO_3$ can be explained.

Here we give some comments on magnetic properties of $BiMnO_3$ [7, 8]. The very weak anomalies at 104 and 112 K that can be clearly detected only on the χ'' vs *T* curves [8] can be explained by the presence of other perovskite phases [12] (these temperatures are shifted with substitutions: 102 and 107 K in $BiMn_{0.97}Al_{0.03}O_3$ [13] and 114 and 117 K in $BiMn_{0.95}Fe_{0.05}O_3$ [14]). However, the increase of the magnetic transition temperature in $BiMnO_3$ was observed only in the case of Fe substitution [15]. It is possible that there is inhomogeneity in the oxygen content of $BiMnO_3$. However, no bulk single-phase examples are known for $BiMnO_{3\pm\delta}$ where the magnetic transition temperature would increase. The anomalies at 99-100 K in $BiMnO_3$ are well understood and originated from long-range ferromagnetic order (see Refs in [3]).

Two monoclinic modifications of $BiMnO_3$ are known at AP: the so-called phase I with orbital order and phase II stable between 474 and 768 K [3]. Orbital order disappears in phase II [3] or in another interpretation, there is different orbital order in phase II [16, 17]. Phase II can be stabilized by substitutions in $BiMn_{1-x}M_xO_3$ with M = Al, Sc, Cr, Fe, and Ga and $0.05 \le x \le 0.2$ [15]. The T_C drops to about 80 K in phase II of $BiMn_{1-x}M_xO_3$ (except for M = Fe). The magnetic anomalies were observed near 84 K in $BiMnO_3$ [7, 8]. This temperature agrees well with the magnetic transition temperatures of phase II. The additional magnetic anomalies (along with T_C = 109 K) were observed near 98 K in $BiMn_{0.95}Fe_{0.05}O_3$ with the phase I structure [14], and the T_C drops to 98 K in phase II of $BiMn_{0.9}Fe_{0.1}O_3$ [15]. Therefore, it seems that the χ'' vs *T* curves of $BiMnO_3$ [7, 8] and $BiMn_{0.95}Fe_{0.05}O_3$ [14] show the presence of phase II in the main phase I. Several scenarios can explain the appearance of phase II. First, it may appear during the quenching and pressure release after the synthesis similar to the appearance of the O-phase in $BiCrO_3$. Second, it may appear due to compositional (oxygen) inhomogeneity and small non-stoichiometry. Third, the orbital order of phase I may be rather unstable and fluctuate between different states [17]. The effect of the orthorhombic $BiMnO_3$ phase should not be ruled out because T_N = 80 K of orthorhombic weakly ferromagnetic $Bi_{0.8}La_{0.2}MnO_3$ and T_N = 100 K of $Bi_{0.5}La_{0.5}MnO_3$ (but interpolation gives T_N lower than 80 K for a hypothetical orthorhombic $BiMnO_3$). It is obvious that more work will be needed to fully understand the magnetic properties of $BiMnO_3$.

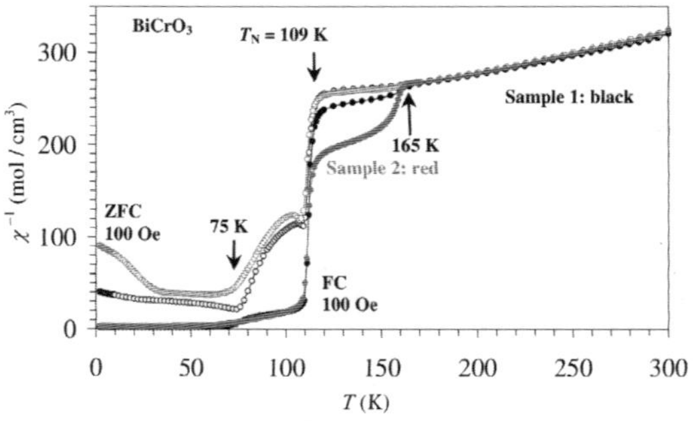

Figure 1. Inverse ZFC (open circles) and FC (filled circles) magnetic susceptibilities for two $BiCrO_3$ samples measured with H = 100 Oe.

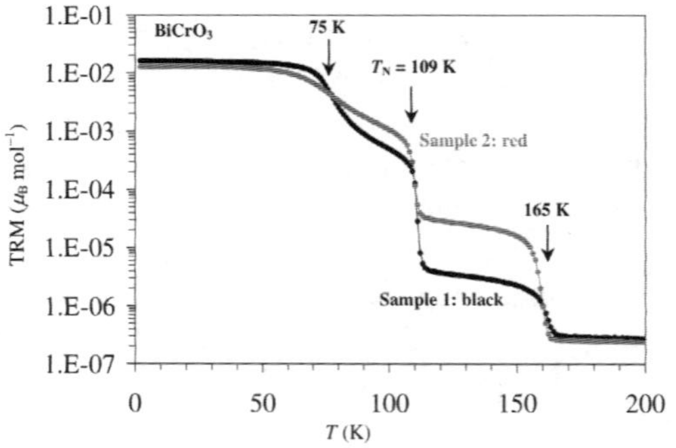

Figure 2. TRM curves for two $BiCrO_3$ samples.

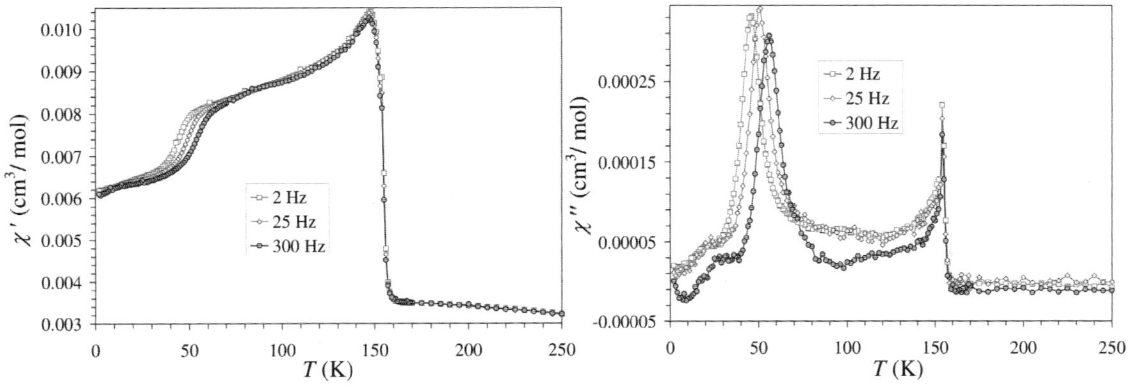

Figure 3. Real (χ') and imaginary (χ'') parts of the ac susceptibilities of orthorhombic $Bi_{0.9}Y_{0.1}CrO_3$ measured with H_{ac} = 5 Oe and at H_{dc} = 0 Oe. The curves at three frequencies (2, 25, and 300 Hz) are shown for the clarity.

References

[1] Ramesh R and Spaldin N A 2007 *Nat. Mater.* **6** 21.
[2] Belik A A, Iikubo S, Kodama K, Igawa N, Shamoto S and Takayama-Muromachi E 2008 *Chem. Mater.* **20** 3765.
[3] Belik A A, Iikubo S, Yokosawa T, Kodama K, Igawa N, Shamoto S, Azuma M, Takano M, Kimoto K, Matsui Y and Takayama-Muromachi E 2007 *J. Am. Chem. Soc.* **129** 971.
[4] Haumont R, Kornev I A, Lisenkov S, Bellaiche L, Kreisel J and Dkhil B 2008 *Phys. Rev. B* **78** 134108.
[5] Palai R, Katiyar R S, Schmid H, Tissot P, Clark S J, Robertson J, Redfern S A, Catalan G and Scott J F 2008 *Phys. Rev. B* **77** 014110.
[6] Belik A A, Tsujii N, Suzuki H and Takayama-Muromachi E 2007 *Inorg. Chem.* **46** 8746.
[7] Belik A A and Takayama-Muromachi E 2006 *Inorg. Chem.* **45** 10224.
[8] Belik A A and Takayama-Muromachi E 2008 *J. Phys.: Condens. Matter* **20** 025211.
[9] Goujon C, Darie C, Bacia M, Klein H, Ortega L and Bordet P 2008 *J. Phys.: Conf. Ser.* **121** 022009.
[10] Trolier-McKinstry S, Biegalski M D, Wang J, Belik A A, Takayama-Muromachi E and Levin I 2008 *J. Appl. Phys.* **104** 044102.
[11] Darie C, Goujon C, Bacia M, Klein H, Toulemonde P, Bordet P and Suard E 2009 *Solid State Sci.* doi:10.1016/j.solidstatesciences.2008.12.004.
[12] Montanari E, Righi L, Calestani G, Migliori A, Gilioli E and Bolzoni F 2005 *Chem. Mater.* **17** 1765.
[13] Belik A A and Takayama-Muromachi E 2008 *Mater. Res. Bull.* **43** 3179.
[14] Belik A A, Hayashi N, Azuma M, Muranaka S, Takano M and Takayama-Muromachi E 2007 *J. Solid State Chem.* **180** 3401.
[15] Belik A A and Takayama-Muromachi E 2007 *Inorg. Chem.* **46** 5585.
[16] Yang C H, Koo J, Song C, Koo T Y, Lee K B and Jeong Y H 2006 *Phys. Rev. B* **73** 224112.
[17] Solovyev I V and Pchelkina Z V 2008 *New J. Phys.* **10** 073021.

International Conference on Advanced Structural and Functional Materials Design 2008 IOP Publishing
Journal of Physics: Conference Series **165** (2009) 012036 doi:10.1088/1742-6596/165/1/012036

Development of a magnetic force microscopy for magnetically soft materials

Hikaru Nomura, Tatsuro Nagayama, Daisuke Yokota, and Ryoichi Nakatani

Division of Materials and Manufacturing Science, Graduate School of Engineering, Osaka University, 2-1 Yamadaoka, Suita, Osaka 565-0871, Japan

nomura@mat.eng.osaka-u.ac.jp

Abstract. We propose a magnetic force imaging technique based on MFS. This method is directly visualize a two dimensional motion of the domain structure of the sample, without a disturbance of the magnetic structure of the sample by MFM tip stray field, and improve the interpretation of the result of MFS. Magnetic force imaging by MFS is useful to investigate magnetic domain structures and magnetization switching mechanism of magnetically soft materials.

1. Introduction

Magnetic force microscopy (MFM) [1] is one of the most powerful techniques to study magnetic phenomena for logic functionality such as magnetic logistic gate [2], magnetic domain wall logic element [3]. In general, magnetically soft materials are used for these elements.

There are two important issues in the imaging of magnetically soft structures by MFM: (1) an effect of the magnetic stray field from the MFM tip and (2) a cross-talk between the topography and the magnetic force image [4]. To reduce these effects, several operation modes are developed, the most common of which are a constant-height mode and a tapping/lift mode.

The constant-height mode maintains the tip position on constant height value in the non-contact region. Thus the MFM tip stray field does not disturb the magnetic structure on the sample surface. However obtained image reflects not only the magnetic structure but also the topography of the sample. The tapping/lift mode obtains the topography and magnetic force image separately. The topography is obtained in the tapping mode, whereas the magnetic force image is obtained by tracing the surface with constant offset height which is determined from the topography. Therefore the tapping/lift mode can reduce the cross-talk between the topography and the magnetic force image. However, during the topographic measurement, the MFM tip contacts to the sample surface and the magnetic stray field from the MFM tip disturbs the magnetic structures of the sample [4].

Recently, Y. Endo, et al proposed a Magnetic field sweeping (MFS) MFM, which employs a MFM tip as a magnetic interaction force detector while sweeping the external magnetic field [5]. By using this method, we can eliminate an effect of the magnetic stray field from the MFM tip and a cross-talk between the topography and the magnetic force image, simultaneously. However, MFS only measure a single point of the sample, and interpretation of the result is still difficult.

In this study, we propose a magnetic force imaging technique based on MFS. This method is directly visualize the two dimensional motion of the domain structure of the sample and improve the interpretation of the result of MFS.

2. Magnetic force imaging by MFS

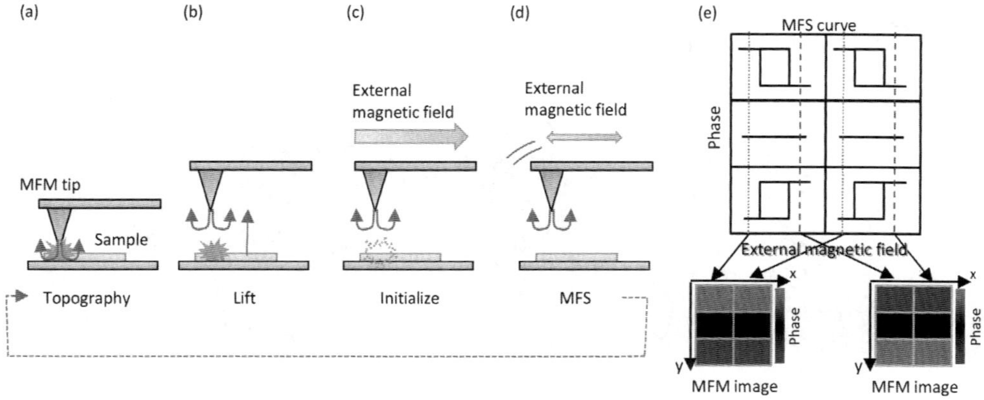

Figure 1. Schematic diagram of magnetic force imaging by MFS.

We implemented the MFS technique in a magnetic force imaging. The procedure for this method include the following step: First, to obtain the topography, the tip scans the surface by tapping mode AFM (Fig. 1(a)). Second, to separate the van del Waals interaction force and magnetic interaction force, the tip is lift-up to a constant lift-height above the previously recorded topography, where the long range magnetic interaction force is dominant and the van del Waals interaction force becomes constant value (Fig. 1(b)). Third, to eliminate the disturbance of the magnetic state of the sample by the MFM tip stray field, the magnetization state of the sample is initialized by applying a large external magnetic field (Fig. 1(c)). Finally, MFS measurement is performed to measure the magnetization process of the sample by sweeping the external magnetic field (Fig. 1(d)). The magnetic force image in each external magnetic field is restructured from MFS curve and x-y location information of each MFS measurements (Fig. 1(e)). To minimize the effect of the thermal drift of the tip-sample distance, this procedure is performed per each scan point.

3. Experimental Procedure

As a sample, Ni-Fe nanowires with constriction was fabricated on thermally oxidized Si(100) substrates by electron-beam lithography, DC magnetron sputtering, and a lift-off technique. All experiments were performed in vacuum condition with a commercial AFM (SII-A300) at room temperature. The probe we used was a CoPtCr-coted cantilever (SI-MF40LM) with spring constant of k=36 N/m, a free resonance frequency of 295.840 kHz, and Q factor of 3462. The topography was measured at free amplitude of 72 nm, amplitude shift of 8.2 nm. In magnetic force image measurement by MFS, the free amplitude of the MFM cantilever was 14 nm and a lift-height was 10 nm. The magnetization state of the sample was initialized by applying an external field of -500 Oe, and then the external magnetic field was increased from -500 Oe to +500 Oe (256 points). The scan area and the number of measuring points are 1000 nm× 2000 nm and 20 points × 40 points, respectively.

4. Results and Discussions

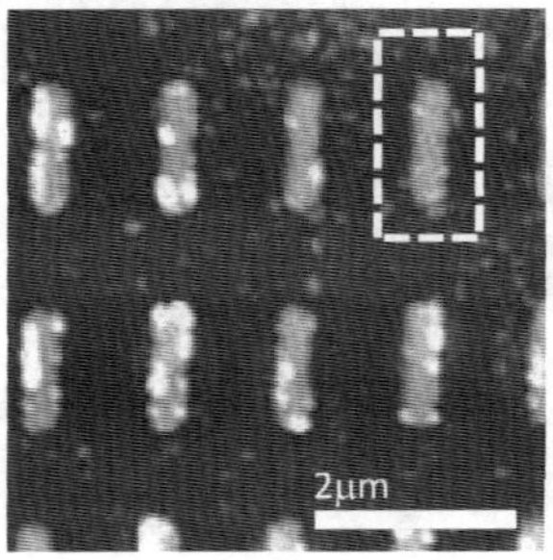

Figure 2. Topography of Ni-Fe nanowire with constriction by tapping mode AFM. (5μm× 5μm)

Figure 2 shows the topography of the Ni-Fe nanowire with constriction by tapping mode AFM (5μm × 5μm). Typical dimension of Ni-Fe nanowires are a length of 2080 nm, a width of 380 nm and constriction width of 325 nm. Magnetic force image by MFS was performed with the nanowire indicated by dots-lines.

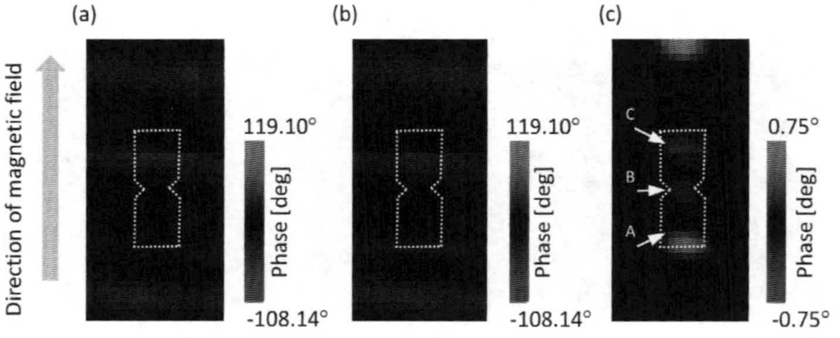

Figure 3. Magnetic force image of Ni-Fe nanowire with constriction by MFS in a zero field. (a) raw image, (b) calculated thermal drift of the phase signal (c) thermal drift compensated image.

Figure 3(a) shows a raw magnetic force image of The Ni-Fe nanowire with constriction by MFS in a zero field. In Fig. 3(a) we can see the corrugation which originates in the thermal drift of the phase signal of the MFM cantilever. To eliminate the effect of the thermal drift in the phase signal, we calculated and subtract the thermal drift of the phase signal by averaging the phase signal in the external magnetic field of -500 Oe to -360 Oe and +360 Oe to 500 Oe at each scan points, where the sample magnetization states were saturated (Fig. 3(b)). As a result, we can clearly see that the magnetic stray field from the sample at (A) the lower-end, (B) the constriction structure and (C) the upper-end of the nanowire, respectively (Fig. 3(c)) without disturbance of the magnetic structure of the sample by MFM tip stray field.

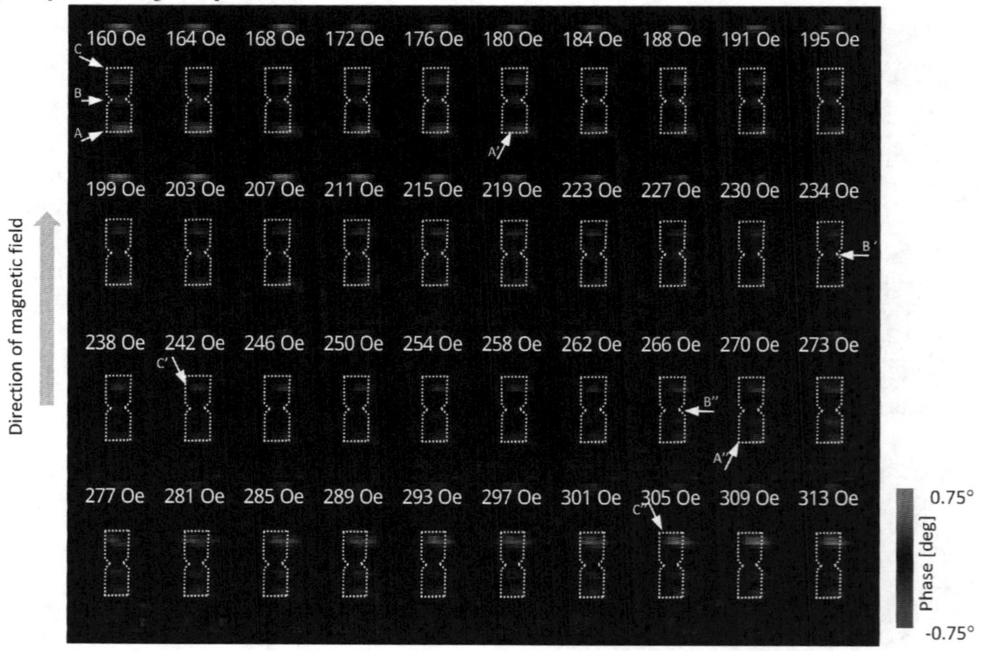

Figure 4. Magnetic force image of Ni-Fe nanowire with constriction by MFS. The external magnetic field increases from +160 Oe to +313 Oe.

Figures 4 show drift compensated magnetic force image of The Ni-Fe nanowire with constriction by MFS. The external magnetic field was increased from -313 Oe to +313 Oe. The magnetization state of (A) the lower-end, (B) the constriction structure and (C) the upper-end of the nanowire are switched in the external magnetic field from (A') +180 Oe to (A'') +270 Oe, from (B') + 234 Oe to (B'') +266 Oe, and from (C') 242 Oe to (C'') + 305 Oe, respectively. The switching process was observed in order of the lower-end, constriction structure and upper-end of the nanowire. These results show that the two dimensional motion of the domain structure in Ni-Fe nanowire can be directly observed using magnetic force image by MFS.

5. Conclusion

We performed a magnetic force imaging technique based on MFS with Ni-Fe nanowire with constriction. This method is directly visualize the two dimensional motion of the domain structure of the sample without the disturbance of the magnetic structure of the sample by MFM tip stray field, and improve the interpretation of the result of MFS. Magnetic force imaging by MFS is useful to investigate magnetic domain structures and magnetization switching mechanism of magnetically soft materials.

6. Acknowledgements

This work was supported by The Global COE Program (Project: Center of Excellence for Advanced Structural and Functional Materials Design) from the Ministry of Education, Culture, Sports, Science and Technology, Japan and also partly supported by Strategic Information and Communications R&D Promotion Programme (SCOPE) from the Ministry of Internal Affairs and Communications (MIC), Japan.

References

[1] Y. Martin and H. K. Wickramasinghe, Appl. Phys. Lett. **50**, 455 (1987).
[2] S. A. Haque, M. Yamamoto, R. Nakatani, Y. Endo, Sci. Technol. Adv. Mater. **5**, 79 (2004).
[3] D. A. Allwood, G. Xiong, C. C. Faulkner, D. Atkinson, D. Petit and R. P. Cowburn, Science **309**, 1688 (2005).
[4] J. G. Ford , M. G. Leonard , J. Swiss and G. C. Gainer, Phys. Rev. B. **66**, 024423 (2002).
[5] Y. Endo, Y. Matsumura, H. Fujimoto, R. Nakatani and M. Yamamoto, Jpn, J. Appl. Phys. **46,** L898 (2007)

TiO_2 nanotube layers with metallic nanoparticles

Hiroaki Tsuchiya[1], Nozomu Tokuoka, Shinji Honda, Yuji Shinkai, Yasutomo Shimizu and Shinji Fujimoto

Division of Materials and Manufacturing Science, Graduate School of Engineering, Osaka University, 2-1 Yamada-oka, Suita, Osaka 565-0871 Japan

E-mail: tsuchiya@mat.eng.osaka-u.ac.jp

Abstract. The present work reports the deposition behavior of Pd or Ag particles on TiO_2 nanotube layers with an average diameter of 100 nm and length of 500 nm. The deposition of Pd or Ag particles is carried out by photo-induced reduction of metal ions. TiO_2 nanotube layers are exposed to UV irradiation in an electrolyte containing Pd ions. Pd deposition behavior strongly depends on the crystal structure of TiO_2 nantoube layers. As-anodized nanotube layers are in the amorphous state and can be converted to anatase without any changes in structural integrity. On as-formed amorphous nanotube layer, no Pd particles are deposited whereas in the case of anatase nanotube layer many Pd particles are formed by UV irradiation in $PdCl_2$ electrolyte. However, the size of Pd particles is larger than the diameter of nanotubes and increases with irradiation time. In order to circumvent this problem, TiO_2 nanotube layer is immersed in an electrolyte, then rinsed with deionized water, followed by exposure to UV irradiation. This results in the deposition of Ag nanoparticles with the size of 10 nm on and inside nanotubes. Photocurrent measurements reveal that the presence of Ag nanoparticles on TiO_2 nanotube layer enhances the absorption efficiency of dye, leading to higher photocurrent.

1. Introduction

Since Masuda et al. reported that the fabrication of aluminum oxide layers with highly ordered parallel-aligned, hexagonally configurated nanopores in 1995 [1], extensive research activities have been achieved due to a wide range of their applications. For example, the highly ordered porous structures have targeted magnetic recording media, optical devices and functional electrodes by filling pores with metals, semiconductors or polymers [2-5]. The highly ordered porous aluminum oxide layers are fabricated by a simple, but optimized electrochemical oxidation, so-called anodization of metallic aluminum in suitable acidic electrolytes such as sulfuric acid, phosphoric acid or oxalic acid. Anodization of other valve metals in these electrolytes results in the formation of flat compact oxide layers or highly random microporous layers depending anodizing voltage. In 1999 Zwilling et al. reported that oxide layers consisting of nanotube arrays can be formed by anodization of titanium in a fluoride containing electrolyte [6]. Since then, the formation of TiO_2 nanotube layers has attracted much attention due to many promising applications of TiO_2 such as photocatalyst and dye-sensitized solar cells. In order to improve these properties of TiO_2, controlling the geometry of TiO_2 nanotube layers has been attempted. As a result, as shown in Fig. 1, it becomes clear that tube length and

[1] To whom any correspondence should be addressed.

diameter can be tuned by tailoring electrochemical conditions. For example, anodization in acidic electrolyte containing a small amount of fluoride ions typically leads to the formation of TiO_2 nanotube layers with a thickness up to several hundred nanometers as shown in Figs. 1(a) and 1(d). On the other hand, TiO_2 nanotube layers with a thickness higher than 2 μm can be formed in a buffered neutral electrolyte containing NaF and more higher aspect ratio nanotubes with smooth tube walls can be fabricated by using viscous organic electrolytes as shown in Figs. 1(b, e) and 1(c, f), respectively. Key to achieve high aspect ratio TiO_2 nanotubes is controlling pH profile and diffusion profile within tubes during anodization [7-11]. Albu et al. also reported that a variety of morphology of TiO_2 layers such as nanolace, nano-bamboo and nano-branching can be fabricated by the precise control of electrochemical conditions [12]. These novel nanostructued TiO_2 layers, due to their highly defined geometry combined with a large surface area, have shown for example to be highly effective for a range of applications including photocatlaysis [13], solar cells [14] and biological interactions with cells [15].

Recently, the presence of metallic particles on TiO_2 layers improves photocatalytic properties of TiO_2 layers [16, 17]. In the present work we examine loading behavior of metallic particles on TiO_2 nanotube layers by different approaches in order to improve functionalities of TiO_2 nanotubular electrodes.

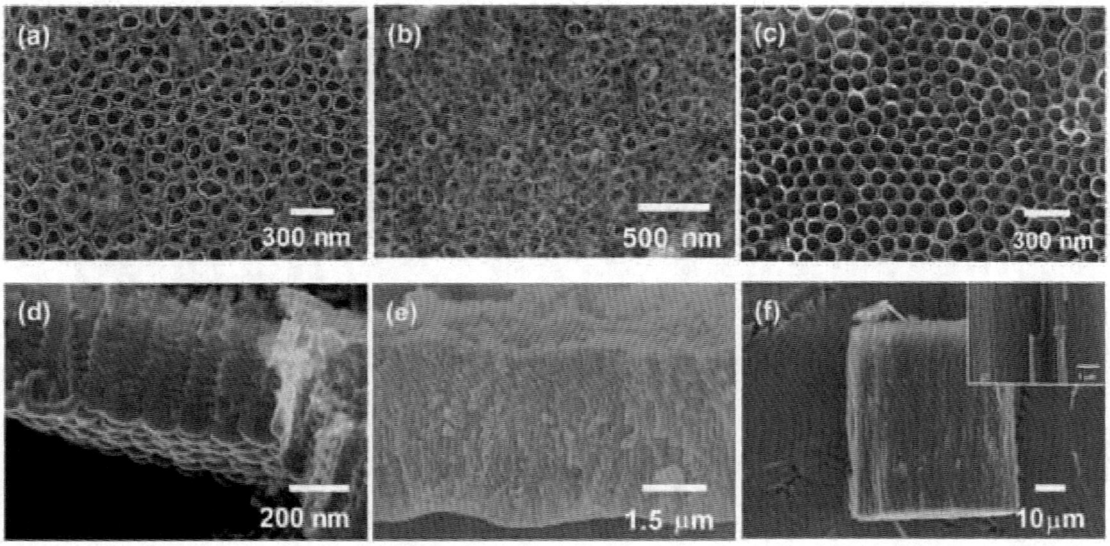

Figure 1. Top-view (a-c) and cross-sectional (d-f) images of anodic oxide layers by anodizaiton of titanium in different electrochemical conditions; (a, d) anodized at 20 V in 1M H_2SO_4 + 0.15wt.% HF, (b, e) anodized at 20 V in 1M Na_2SO_4 + 0.5wt.% NaF, (c, f) anodized at 60 V in ethylene glycol + 0.27M NH_4F.

2. Experimental

In the present work we used TiO_2 nanotube layers with an average diameter of 100 nm and length of 500 nm as shown in Fig. 1(a, d). The detail for preparation of the layers is as follows. Samples were titanium (99.6 % purity) in the form of foils with a thickness of 0.1 mm. Prior to anodizaiton, the samples were degreased by sonication in acetone, isopropanol and methanol, successively, then rinsed with deionized water and dried with a nitrogen stream.

For the anodization, an electrochemical cell with a three-electrode configuration was used. The samples were contacted by a Cu plate and then pressed against an O-ring in the electrochemical cell, exposed to an electrolyte. A platinum plate and a Ag/AgCl electrode are served as a counter electrode and a reference electrode, respectively. The electrolyte used for the tubular layer formation was 1M H_2SO_4 containing 0.15 wt.% HF, prepared from reagent grade chemicals and deionized water. The

anodization consists of a potential sweep from the open circuit potential to 20 V, followed by holding the potential at 20 V for several hours. The anodization was carried out using a high-voltage potentiostate (Hokuto denko, H-3001A) at room temperature under an aerated non-stirred condition. After the anodization, the samples were rinsed with deionized water and dried with a nitrogen stream.
In the present work we tried photo-assisted deposition in order to load metallic nanoparticles on the TiO_2 nanotube layers. The TiO_2 nanotube layers were soaked in electrolytes containing metal ions for some period, and then exposed to UV irradiation to induce photocatalytic reduction of metal ions to metal in the form of nanoparticle. For morphological characterization of the samples, a field-emission scanning electron microscope (FE-SEM, JEOL JSM-6500) and a transmission microscope (TEM, Hitachi-800) were used. Cross-sectional images of nanotube layers were obtained from mechanically scratched samples, at locations where some pieces of the layers on the samples cracked off.
We evaluated phtoresponse of metallic nanoparticle loaded, dye-sensitized TiO_2 nanotube layers in order to evaluate effects of metallic nanoparticles. For dye-sensitization, the TiO_2 nanotube layers with and without metallic nanoparticles were sensitized for several hours in N3 dye dissolved in methanol with a concentration of 300 μM. After the sensitization, photocurrent spectra were recorded at 500 $mV_{Ag/AgCl}$ with a step of 10 nm in the range from 500 nm to 800 nm. The electrolyte for the photocurrent measurements was acetonitrile containing 1.5 mM I_2 and 15 mM KI.

3. Results and discussion

We have examined a variety of metals for metallic nanoparticles deposition on TiO_2 nanotube layers. In the present work we targeted especially on Pd and Ag. Figure 2 shows top-view SEM images after UV irradiation for 30 minutes onto as-formed TiO_2 nanotube layer in 0.2M $PdCl_2$ electrolyte containing a small amount of methanol. It is clear that no nanoparticles can be observed, that is, no reduction of Pd ions to metallic Pd takes place. This can be attributed to the crystal structure of as-formed TiO_2 nanotube layer. Figure 3(a) shows XRD pattern obtained for as-formed TiO_2 nanotube layer. It is apparent that as-formed layer shows only the peaks from a titanium substrate. Therefore, as-formed TiO_2 nanotube layer is in amorphous state and therefore exhibits low photocatalytic reactivity due to fast recombination of electron-hole pairs. However, as-formed amorphous TiO_2 nanotube layer can be converted to crystalline by annealing. Figure 3(b) shows XRD pattern after annealing at 450 °C for 3 hours. The annealed TiO_2 nanotube layer exhibits clearly the crystalline signature of anatase. Note that even after annealing and conversion to anatase, TiO_2 nanotube layer retains its structural integrity [18]. Figure 4 exhibits Pd loading behavior on anatase TiO_2 nanotube layer in the same electrolyte. As apparent in Fig. 4(a, b), Pd particles are successfully deposited on the anatase TiO_2 nanotube layer after 30 min. UV irradiation. The size of the Pd particles is approximately 2 μm and larger than the diameter of nanotubes. Further irradiation does not affect the number of Pd particles deposited on the layer, but increases the size of the particles as shown in Fig. 4(c, d). Cross-sectional TEM

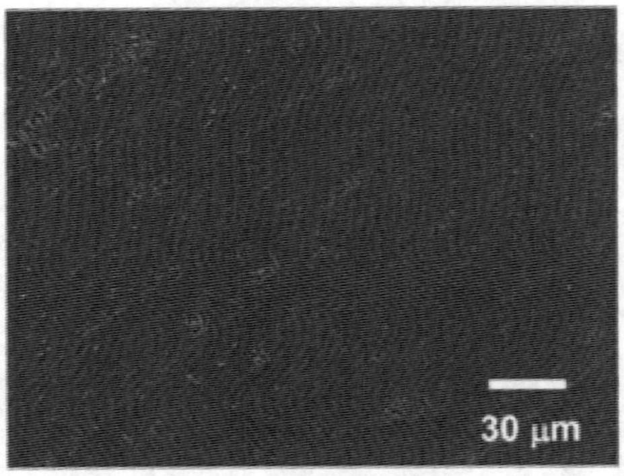

Figure 2. Top-view SEM image of as-formed TiO_2 nanotube layer surface after the photo-induced Pd deposition treatment for 30 min..

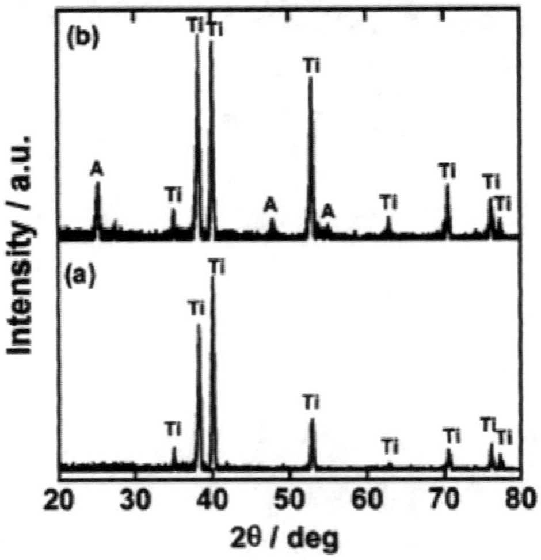

Figure 3. XRD patterns obtained from (a) as-formed and (b) annealed TiO_2 nanotube layers.

observations also revealed that no trace of Pd deposition inside nanotubes was detected. From these results one can deduce that once metallic nanoparticles are initiated on TiO_2 nanotube layer, the nanoparticles act as photo-excited electron collection sites, leading to the situation in which more reduction of Pd ions to metallic Pd can occur at the sites and as a result the particles can increase their sizes without increasing the number of Pd particles.

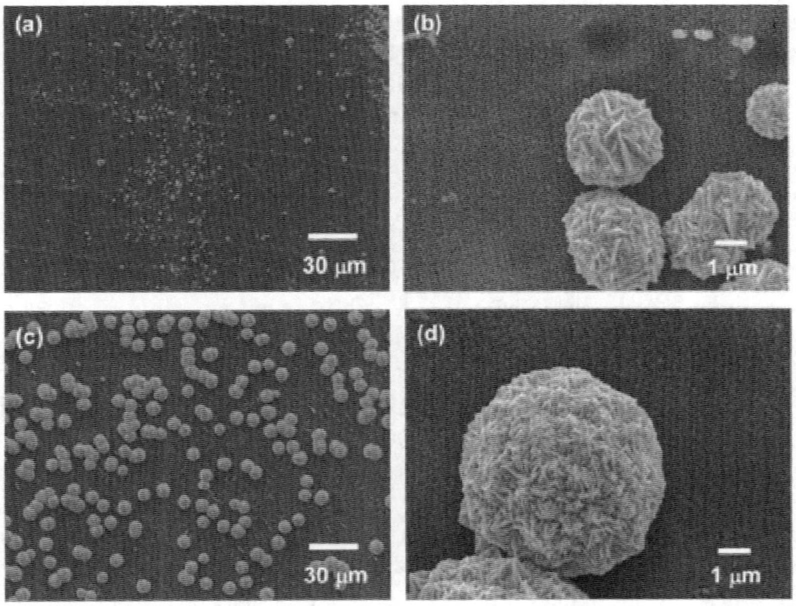

Figure 4. Top-view SEM images of anatase TiO_2 nanotube surfaces after the Pd deposition treatment for (a, b) 30 min. and (c, d) 3 hours.

In order to circumvent this problem, TiO_2 nanotube layers were soaked in electrolytes containing metal ions, then slightly rinsed with deionized water, finally followed by exposure to UV irradiation. This approach resulted in highly dispersed metal nanoparticle deposition on TiO_2 nanotube layers. A variety of metallic nanoparticles such as Pd, Pt and Ag were successfully deposited on TiO_2 nanotube layers. One example obtained for Ag deposition is shown in Fig. 5. Figures 5(a) and 5(b) are top-view SEM and cross-sectional TEM images of anatase TiO_2 nanotube layer after Ag deposition, respectively. It is clear from Fig. 5(a) that Ag nanoparticles with a diameter of approximately 10 nm are deposited on the nanotube layer in highly dispersed manner. Cross-sectional TEM observation reveals that the nanoparticles are deposited not only on the top of the tubes but also inside the tubes. Therefore, it is found that Ag nanoparticles can be successfully deposited on TiO_2 nanotube layer.

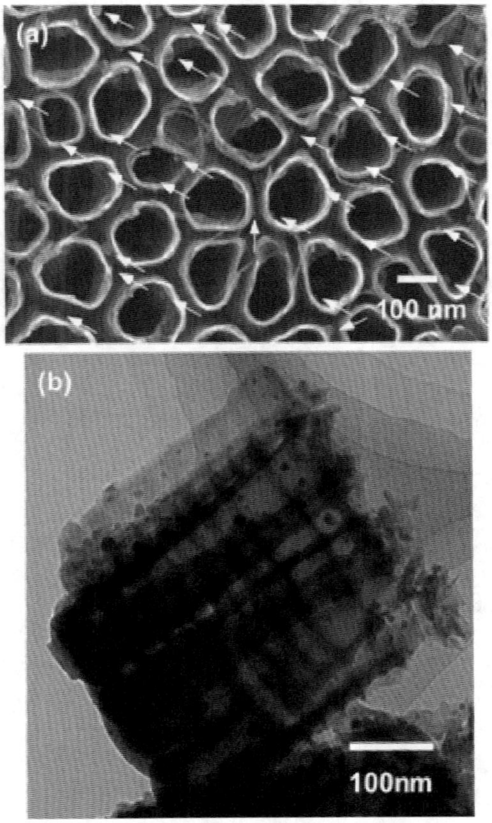

Figure 5. (a) Top-view SEM and (b) cross-sectional TEM images of TiO_2 nanotube layers after the Ag deposition treatment. The arrows in (a) indicate the deposition sites of Ag particles.

Figure 6 shows photocurrent spectrum for Ag nanoparticle loaded TiO_2 nanotube layer. Spectrum obtained for nanotube layer without Ag nanoparticles is also included in the figure as comparison. Both layers were dye-sensitized for 1 hour. It is clear that the photocurrent spectra showed a peak at a wavelength of 570 nm and photocurrent was drastically enhanced by the presence of Ag nanoparticles on TiO_2 nanotube layer. The enhanced photocurrent by Ag nanoparticles may be attributed to the increased absorption efficiency of dye induced by plasmon resonance effect of Ag nanoparticles. Photocurrent spectra increased with increasing sensitization time due to the increased amount of chemisorbed dye. However, the increment of maximum photocurrent significantly decreased at the short sensitization time and then leveled off. Actually, extended sensitization led to the increased amount of chemisorbed dye. However, the increased amount of dye thickens the chemisorbed dye layer even on Ag nanoparticles. The thick chemisorbed layer attenuates the plasmon resonance effect of Ag nanoparticles, resulting in the decrease of photocurrent. In order to improve solar energy conversion efficiency, the surface area of TiO_2 layer should be increased. Therefore, the effect of surface area is considered using longer nanotubes with the same diameter of 100 nm. As mentioned above, the length of nanotubes can be controlled by tailoring electrochemical conditions. In the present work TiO_2 layer consisting of longer nanotubes was grown in 1M Na_2SO_4 containing 0.5 wt.% NaF for 3 hours, leading to 1.5 μm thick TiO_2 nanotube layer. After the growth of the layer, Ag nanoparticles were deposited under the same condition. Photocurrent spectrum obtained for dye-sensitized 1.5 μm thick TiO_2 nanotube layer with Ag nanoparticles is plotted in Fig. 6(b). As comparison, photocurrent spectrum for Ag nanoparticle loaded, 500 nm thick nanotube layer is also included in the figure. It is clear that the longer nanotube layer exhibits larger photocurrent as expected. Therefore, it is found that by a combination of the high surface area of TiO_2 electrode and the presence of Ag nanoparticles higher solar energy conversion efficiency can be obtained. However, optimizations are required in terms of length, diameter of nanotubes and size of Ag nanopartilces in order to obtain higher conversion efficiency.

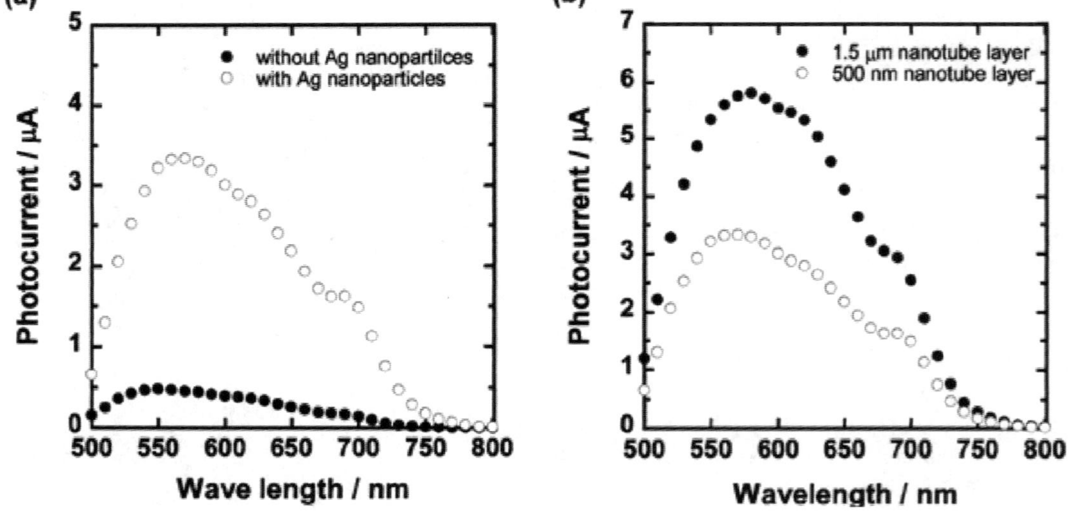

Figure 6. (a) Photocurrent spectra obtained from 500 nm thick TiO_2 nanotube layers with and without Ag nanoparticles. (b) Effect of tube length on photocurrent spectrum obtained from Ag-nanoparticle loaded TiO_2 nanotube layer.

4. Conclusions

In the present work, we examined the deposition behavior of metallic particles on anodic TiO_2 nanotube layers by photo-induced reduction of metal ions. The deposition behavior strongly depended on the crystal structure of nanotube layers – on as-formed amorphous layer, no deposition took place whereas on annealed anatase nanotube layer metallic particles were formed on the entire surface. However, when the deposition was carried out in an electrolyte, the size of deposited metallic particles was larger than the diameter of nanotubes. On the other hand, in the case that TiO_2 nanotube layers were rinsed with deionized water after removing from an electrolyte, then exposed to UV irradiation, metallic particles with the size of 10 nm were deposited on and inside nanotubes. The presence of Ag nanoparticles strongly enhanced photucrrent spectrum of dye-sensitized TiO_2 nanotube layers. The increment of photocurrent was varied depending on the sensitization time and the length of nanotubes.

Acknowledgements
The present work is supported by "Priority of Assistance of the Formation of Worldwide Renowned Centers of Research – Global COE Program (Project: Center of Excellence for Advanced Structural and Functional Materials Design)" for the Ministry of Education, Sports, Culture, Science and Technology of Japan. The financial support by the Grant-in-Aid (Project No. 18686901, No. 20613008) from the Ministry of Education, Culture, Sports, Science and Technology of Japan is appreciated. A part of the present work was carried out using a facility in the Research Center fur Ultra-High Voltage Electron Microscopy, Osaka University.

References

[1] Masuda H and Fukuda K 1995 *Science* **268** 1466
[2] Preston C K and Moskovitz M 1993 *J. Phys. Chem.* **97** 8495
[3] Masuda H, Yada K and Osaka A 1998 *Jpn. J. Appl. Phys.* **37** L1340
[4] Li A P, Müller F, Birner A, Nielshc K and Gösele U 1999 *Adv. Mater.* **11** 483
[5] Masuda H, Hasegawa K and Ono S 1997 *J. Electrochem.* Soc. **144** L127
[6] Zwilling V, Darque-Ceretti E, Boutry-Forveille A, David D, Perrin M Y and Aucouturier M 1999 *Surf. Interface Anal.* **27** 629
[7] Beranek R, Hildebrand H and Schmuki P 2003 *Electrochem. Solid-State Lett.* **6** B12
[8] Macak J M, Tsuchiya H and Schmuki P 2005 *Angew. Chem. Int. Ed.* **44** 2100
[9] Ghicov A, Tsuchiya H, Macak J M and Schmuki P 2005 *Electrochem. Commun.* **7** 505
[10] Macak J M, Tsuchiya H, Taveira L, Aldabergerova and S, Schmuki P 2005 *Angew. Chem. Int. Ed.* **44** 7463
[11] Albu S P, Ghicov A, Macak J M and Schmuki P 2007 *phys. stat. sol. (RRL)* **1** R65
[12] Albu S P, Kim D and Schmuki P 2008 *Angew. Chem. Int. Ed.* **47** 1916
[13] Zlamal M, Macak J M, Schmuki P and Josef Krýsa 2007 *Electrochem. Commun.* **9** 2822
[14] Macak J M, Tsuchiya H, Ghicov A and Schmuki P 2005 *Electrochem. Commun.* **7** 1133
[15] Park J, Bauer S, von de Mark K and Schmuki P 2007 *Nano Lett.* **7** 1686
[16] Jin S and Shiraishi F 2004 *Chem. Eng. J.* **97** 203
[17] Szabó-Bárdos E, Pétervári E, El-Zein V and Horváth A 2006 *J. Photochem. Photobiol. A Chem.* **184** 221.
[18] Beranek R, Tsuchiya H, Sugishima T, Macak J M, Taveira L, Fujimoto S, Kisch H and Schmuki P 2005 *Appl. Phys. Lett.* **87** 243114

Hydrothermal synthesis of visible light sensitive vanadium doped TiO_2 photocatalysts and their applications for the degradation of isobutanol diluted in water

Takashi Kamegawa*, Kazuaki Sugimura, Keiichi Maki, Kohsuke Mori, Hiromi Yamashita

Division of Materials and Manufacturing Science, Graduate School of Engineering, Osaka University, 2-1 Yamadaoka, Suita, Osaka 565-0871, Japan.

E-mail: kamegawa@mat.eng.osaka-u.ac.jp

Abstract. Vanadium doped TiO_2 photocatalysts (V/TiO_2) were synthesized by sol-gel method and hydrothermal treatment using tetraisopropyl orthotitanate (TPOT) and vanadium acetylacetonate in the presence and absence of NH_4F. In comparison to the vanadium non-doped TiO_2 (pure TiO_2), V/TiO_2 exhibited the long-tailed absorption in the visible light region above 380 nm. Degradation of isobutanol diluted in water proceeded efficiently on V/TiO_2 even under visible light (λ > 430 nm) irradiation, although pure TiO_2 did not show any photocatalytic activities. It was also found that the hydrothermally synthesized V/TiO_2 (HT-V/TiO_2) in the presence of NH_4F possessed high crystallinity of anatase phase and hydrophobic surface, showing the higher photocatalytic activities as compared to those of sol-gel V/TiO_2 as well as HT-V/TiO_2 prepared in the absence of NH_4F.

1. Introduction

In a few decades, various kinds of photocatalysts have been widely investigated for applications in air and water purification as well as the production of clean energy resources [1]. Especially, the photocatalytic degradation of organic pollutants diluted in water is of vital interest to practical utilizations. Among them, titanium dioxide (TiO_2) has attracted much attention due to its strong oxidation ability, nontoxicity and low cost, however, TiO_2 only show the fascinating photocatalytic properties under UV light irradiation, which limited the use of natural solar light below 3 %. Therefore, the development of TiO_2, which can work not only under UV but also visible light, has become one of the most interesting topics for design of efficient and environmentally friendly photocatalytic reaction systems. Along these lines, a number of research groups have attempted to create a bathochromic shift of band gap energy by doping of transition metals into TiO_2 [2]. Moreover, Asahi et al. demonstrated that the doping of nitrogen into TiO_2 attained the formation of visible light sensitive TiO_2 due to the narrowing the band gap [3].

On the other hand, our research has already shown that the surface hydrophobic modification of TiO_2 as well as catalysis supports such as zeolite and mesoporous materials by using fluorine led to the enhancement of the adsorption of organic compounds from aqueous solutions and gas phase [4,5].

In the present study, we designed vanadium doped TiO_2 photocatalyst (V/TiO_2) in the presence and absence of NH_4F by a sol-gel method as well as a hydrothermal treatment, aiming at the development of multifunctional materials showing high photocatalytic activity even under visible light irradiation.

The photocatalytic activities were estimated by the degradation of isobutanol diluted in water. In particular, the influences of the preparation methods, the amounts of doping vanadium species as well as the addition of NH_4F on the photocatalytic activities have been examined in detail.

2. Experimental

Tetraisopropyl orthotitanate (TPOT: $Ti(OC_3H_7)_4$), vanadium acetylacetonate ($V(CH_3COCHCOCH_3)_3$), ethanol and NH_4F were used as received. The starting solutions were prepared by the addition of NH_4F in H_2O to the mixed TPOT-$V(CH_3COCHCOCH_3)_3$-ethanol solutions. The molar ratio of starting solutions were as follows; Ti : H_2O : C_2H_5OH : NH_4F : V = 1 : 5 : 5 : 0.05 : X (X = 0, 0.002, 0.01, 0.02). The hydrothermal syntheses of V/TiO_2 (denoted as HT-V/TiO_2(X)) were performed using conventional autoclave. The starting mixture was transferred into an autoclave and kept at 433 K for 48 h. In the case of sol-gel V/TiO_2, above starting mixture was stirred at 298 K for 24 h and then washing in water at 353 K for 24 h. After centrifugation, washing and drying, both types of samples were calcined at 823 K for 5 h.

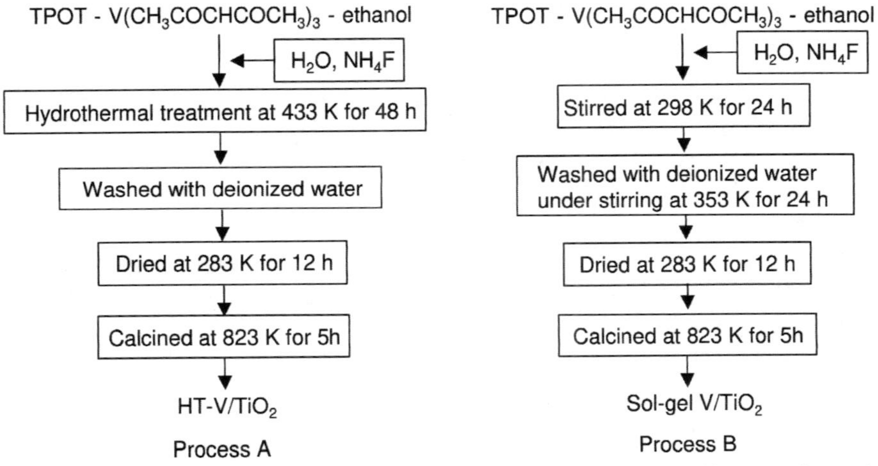

Scheme 1. Preparation processes of vanadium doped TiO_2 photocatalysts by the hydrothermal treatment (Process A) and the conventional sol-gel method (Process B).

X-ray diffraction of all samples was measured by Rigaku RINT2500 diffractometer with Cu Kα radiation (λ = 1.5406 Å). The UV-vis absorption spectra were recorded with Shimadzu UV-2200A spectrophotometer.

The photocatalytic activity of samples was evaluated by the degradation of isobutanol diluted in water. The fixed amount of catalysts (50 mg) was dispersed in 25 ml of isobutanol solution (2.61 mmol/L). After stirring under dark conditions for 30 min, the solution was bubbled by oxygen for another 30 min. Then visible light irradiation (λ > 430 nm) was carried out using 500 W Xe-lamp and colored filters (HOYA; L-43). The progress of the reactions was monitored by gas chromatography analysis (GC-14B, Shimadzu). Water adsorption isotherms of the catalysts were measured at 293 K using a conventional vacuum system.

3. Results and discussions

Figure 1 shows the XRD patterns of HT-V/TiO_2 with different contents of vanadium and sol-gel TiO_2 after calcinations at 823 K. All samples only showed the diffraction peaks due to the presence of TiO_2 anatase phase, suggesting that vanadium oxide species were highly dispersed within TiO_2 without the formation of large vanadium oxide clusters under the hydrothermal treatment. The formation of vanadium oxide clusters was also hardly observed in the case of sol-gel V/TiO_2 (data not shown). The intense diffraction peak assigned to the (101) reflection of TiO_2 anatase phase was decreased with increasing of vanadium contents (Fig. 1(b-e)), however, the peak intensity of HT-

V/TiO_2 was high as compared to that of sol-gel TiO_2 (Fig. 1(a)), indicating that the highly crystalline V/TiO_2 was successfully synthesized by the hydrothermal treatment. Moreover, the BET surface areas of HT-V/TiO_2 were almost same in the series of samples (ca. 50 m^2/g), which was larger than that of sol-gel TiO_2 (36 m^2/g). By applying the Scherrer's equation for (101) reflection of TiO_2 anatase phase, the crystalline size of HT-V/TiO_2 could be determined to be 19 nm (HT-V/TiO_2(0)), 21 nm (HT-V/TiO_2(0.002)), 18 nm (HT-V/TiO_2(0.01)) and 19 nm (HT-V/TiO_2(0.02)), respectively. The crystalline sizes were hardly affected by the contents of vanadium oxide species.

Figure 2 shows the UV-vis spectra of HT-V/TiO_2 and the commercially available TiO_2 (ST-01, anatase). The absorption edge of vanadium non-doped TiO_2 (HT-V/TiO_2(0)) and ST-01 (Fig. 2(a,e)) was located in the bad gap position of anatase TiO_2 at around 380 nm (3.2 eV), which was good coincidence with the results of XRD investigations. On the other hand, as shown in Fig. 2(b-d), vanadium doped TiO_2 (HT-V/TiO_2(0.002, 0.01, 0.02)) exhibited the new long-tailed absorption in the visible light region above 380 nm, which increases with increasing the amounts of vanadium oxide species. This visible light absorption can be ascribed to the formation of the impurity energy level within the band gap of TiO_2. The same phenomena were also observed in the case of sol-gel V/TiO_2 (data not shown).

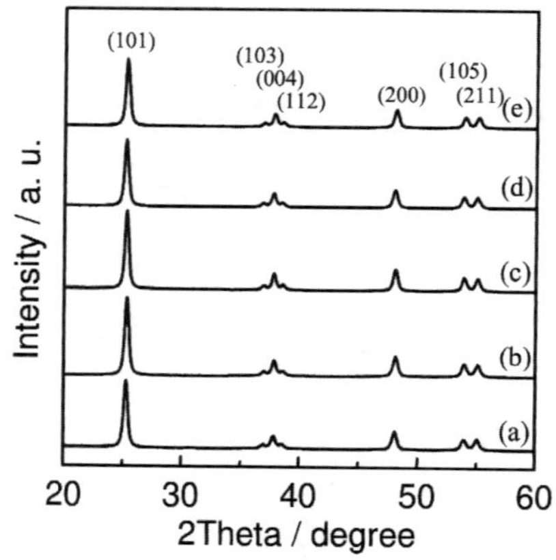

Figure 1. XRD patterns of (a) sol-gel TiO_2 and (b-e) HT-V/TiO_2(X) (X = (b) 0, (c) 0.002, (d) 0.01, (e) 0.02).

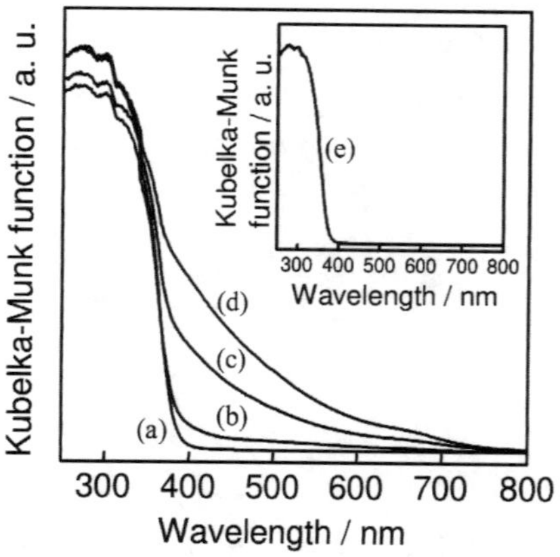

Figure 2. UV-vis spectra of (a-d) HT-V/TiO_2(X) (X = (a) 0, (b) 0.002, (c) 0.01, (d) 0.02) and (e) TiO_2 (ST-01).

According to the above characterizations, highly crystalline visible light sensitive HT-V/TiO_2 can be expected to show high photocatalytic activities even under visible light irradiation as compared to those of sol-gel V/TiO_2. Therefore, the reaction rates for the photocatalytic degradation of isobutanol diluted in water were investigated under visible light ($\lambda > 430$ nm), aiming at the applications for environmentally friendly photocatalytic reaction systems. As shown in Fig. 3, HT-V/TiO_2(0) did not show any photocatalytic activities, however, the degradation of isobutanol was proceeded on HT-V/TiO_2(0.002, 0.01, 0.02) even under visible light irradiation. The reaction rate increased upon an increase in the contents of vanadium species while passing through a maximum (HT-V/TiO_2(0.01)) and then drastically decreasing in the region of the higher vanadium contents. These results suggested that the reaction rate was effectively improved by the small amounts of vanadium doping into TiO_2. It was also found that the reaction rate on HT-V/TiO_2(0.01) was about 2.4 times larger than that on sol-gel V/TiO_2(0.01), indicating that the large surface area as well as the high crystallinity of the anatase phase were important factors for the preparation of highly functional TiO_2 photocatalyst.

Furthermore, the effect of the addition of NH_4F into starting solutions for the hydrothermal treatment on the photocatalytic activities was evaluated by the degradation of aqueous isobutanol. The addition of NH_4F in the starting solutions led to the formation of highly crystalline TiO_2. Moreover, as shown in Fig. 4, the comparative study using HT-V/TiO_2(0.01) prepared in the absence and presence of NH_4F (F = 0.0, 0.05) revealed that HT-V/TiO_2(0.01)(F = 0.05), which possessed high crystallnity of TiO_2 phase, showed the higher photocatalytic activity than that on HT-V/TiO_2(0.01)(F = 0.0) under visible light. On the other hand, the amount of adsorbed water on HT-V/TiO_2(0.01)(F = 0.05) was much smaller than that on HT-V/TiO_2(0.01)(F = 0.0). Thus, the less hydrophilic surface will enhance the transfer of organics from water to catalysis surface, which also led to the higher photocatalytic activities of HT-V/TiO_2(0.01)(F = 0.05). A more detailed study of the mechanisms behind these photocatalytic reactions is now underway.

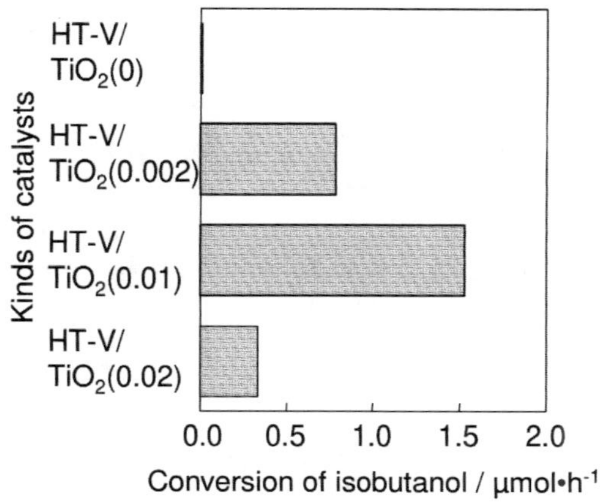

Figure 3. Photocatalytic degradation of isobutanol diluted in water on HT-V/TiO_2 with different contents of vanadium under visible light (λ > 430 nm) irradiation.

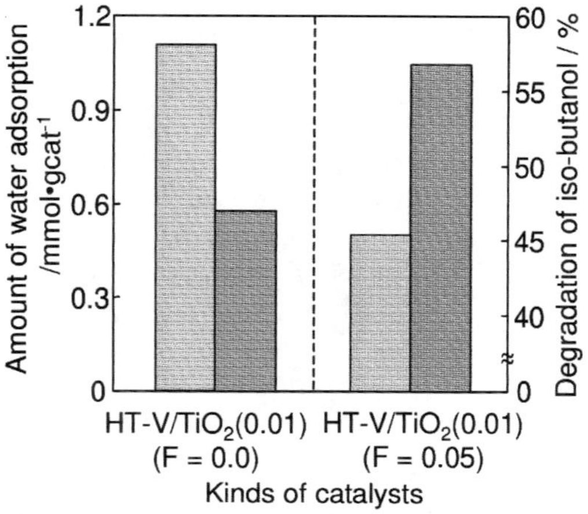

Figure 4. The amount of water adsorption (left bar) as well as the photocatalytic activities (right bar) for the degradation of isobutanol in water under visible light (λ > 430 nm) irradiation.

4. Conclusions

In conclusion, vanadium doped TiO_2, which exhibited the photocatalytic activities even under visible light irradiation, were successfully synthesized by the hydrothermal treatment with high crystallinity and large surface area. The reaction rate for the degradation of isobutanol on HT-V/TiO_2(0.01) exhibited about 2.4 times larger that on sol-gel V/TiO_2(0.01), showing that the high crystallinity of TiO_2 phase as well as the large surface area plays an important role in this reaction. Moreover, it was found that the addition of NH_4F in the starting solutions led to the formation of hydrophobically modified V/TiO_2(HT-V/TiO_2(0.01)(F = 0.05)), which exhibited the higher photocatalytic activity as compared to that on HT-V/TiO_2(0.01) prepared in the absence of NH_4F.

Acknowledgements

This study was supported by Priority Assistance for the Formation of Worldwide Renowned Centers of Research - The Global COE Program (Project: Center of Excellence for Advanced Structural and Functional Materials Design) from the Ministry of Education, Culture, Sports, Science and Technology (MEXT), Japan.

References

[1] Fujishima, A., Rao, T.N., Tryk, D.A., *J. Photochem. Photobio. C: Photochem. Rev.*, **1**, 1 (2000).
[2] Yamashita, H., Anpo, M., *Curr. Opin. Solid State Mater. Sci.*, **7** 471 (2004).
[3] Asahi, R., Morikawa, T., Ohwaki, T., Aoki, K., Taga, Y., *Science*, **293**, 269 (2001).
[4] Yuan, S., Kawasaki, S., Mori, K., Yamashita, H., *Res. Chem. Intermed.*, **34**, 331 (2008).
[5] Kuwahara, Y., Kamegawa, T., Mori, K., Yamashita, H., *Chem. Commun.*, 4783 (2008).

Application of microwave to the synthesis of nanosized metal and alloy catalysts on titanium dioxide supports

Sayoko Shironita, Tomoya Takasaki, Takashi Kamegawa, Kohsuke Mori, and Hiromi Yamashita*

Division of Materials and Manufacturing Science, Graduate School of Engineering, Osaka University, 2-1 Yamada-oka, Suita, Osaka 565-0871, Japan

E-mail: yamashita@mat.eng.osaka-u.ac.jp

Abstract. Nanosized Pd and $PdAu_x$ alloy particles were deposited on various titanium dioxides (TiO_2; anatase + rutile, anatase, and rutile) using a simple and unique microwave-assisted method. The structure of the obtained samples was characterized by X-ray diffraction (XRD), N_2 adsorption-desorption and X-ray absorption fine structure (XAFS) methods. Results of XAFS indicated that $PdAu_x$ alloy particles had formed on TiO_2 supports. Catalytic activity was estimated by a direct synthesis reaction of hydrogen peroxide (H_2O_2) from hydrogen and oxygen gases at room temperature. A high reaction rate could be obtained using these $PdAu_x$ alloy catalysts compared with conventional Pd catalysts.

1. Introduction

Microwave dielectric heating has attracted a great deal of attention as a new and promising method in organic and inorganic syntheses [1]. It is widely accepted that microwave irradiation enables rapid, uniform and energy efficient heating. This heating is also applicable to the simple preparation of metallic nanostructured materials [2]. For example, Ru, Ag, Pt, PtNi, Cu nanoparticles, ZrO_2, Ag nanowires and zeolites have been prepared using microwave heating [3-9]. As a supported system, carbon-supported and polymer stabilized colloids have been synthesized [10-13]. Recently, the microwave synthesis of Au and Pd nanoparticle catalysts on CeO_2 supports and their high catalytic activities for the CO oxidation reaction were reported [14]. Although a large number of studies dealing with the preparation of metal colloids and clusters have been reported, no reports exist concerning the microwave-assisted synthesis of metal particles supported on titanium dioxides (TiO_2).

Many reports describe a direct synthesis of hydrogen peroxide (H_2O_2) from hydrogen and oxygen gases using Pd and PdAu nanoparticles on supports [15-19]. H_2O_2 is a very important industrial oxidant and is currently produced by the sequential hydrogenation and oxidation of an alkyl anthraquinone as an intermediate. There are problems associated with the anthraquinone route and these include the cost of the quinone solvent system and the requirement for periodic replacement of anthraquinone due to hydrogenation.

In this work, nanosized Pd metal and $PdAu_x$ alloy particles were prepared using microwave irradiation on three types of TiO_2 with different crystal phases (anatase + rutile, anatase and rutile) as support materials. The obtained catalysts were characterized by XRD, BET and XAFS measurements. Additionally, the direct synthesis of H_2O_2 from hydrogen and oxygen gases at room temperature was performed and their catalytic activity is discussed.

2. Experimental

2.1. Materials

Three types of TiO_2 were used as supports, JRC-TIO-4 (equivalent to Degussa P-25; anatase : rutile = 7 : 3), JRC-TIO-6 (rutile) and JRC-TIO-8 (anatase) were kindly supplied from the Japan Catalysis Society of Japan. $PdCl_2$ and $HAuCl_4$ were purchased from Wako Pure Chemical Ind., Ltd.

2.2. Preparation and characterization of catalysts

The Pd metal precursor and various titanium oxide materials were irradiated by microwave (500 W, 2450 ± 30 MHz, MWO-1000S, Tokyo Rikakikai Co. Ltd.) for 15 min with stirring. After centrifugation the sample was dried at 383 K for 12 h, calcined at 723 K for 5 h and reduced by H_2 (20 mL/min) at 473 K for 1 h. The specific surface area measurement was performed using a BEL-SORP max (Bel Japan, Inc.) at 77 K. The sample was degassed under vacuum at 473 K before data collection. X-ray fluorescence measurements were performed using an EDX-720 (Shimadzu). Au L_{III}-edge XAFS spectra were recorded using a fluorescence-yield collection technique at the beam line BL01B1 station with an attached Si (311) monochromator at SPring-8, JASRI, Harima, Japan (prop. No. 2008A1366, 2008A1457). EXAFS data were normalized by fitting the background absorption coefficient around the energy region higher than the edge of about 35–50 eV with the smoothed absorption of an isolated atom. EXAFS data were examined using the Rigaku EXAFS analysis program. Fourier transformation (FT) of k^3-weighted normalized EXAFS data were performed over the $3\ Å < k/Å^{-1} < 12\ Å$ range to obtain the radial structure function. To compare catalytic activities of Pd and $PdAu_x$, alloy catalysts (0.1 g) were prepared by microwave irradiation. The direct synthesis of hydrogen peroxide (H_2O_2) from hydrogen and oxygen (80 mL/min, $H_2 : O_2 = 1 : 1$) was carried out under atmospheric pressure at room temperature.

3. Results and discussion

From the elemental analysis of samples, it was found that the precursor of Pd metal can be successfully deposited on all three types of TiO_2 supports under microwave irradiation of the TiO_2 powder slurry in an aqueous $PdCl_2$ solution without using any reducing reagents or surfactants. The surface area of support and the loading amount of metals are summarized in Table 1. The amount of Pd loading was not proportional to the surface area of supports. It was found that the amount of Pd loading depended on the crystal phase and that the rutile phase allowed easier loading than the anatase phase. In addition, alloy particles were successfully prepared on three types of TiO_2 by the same procedure using two different metal salt precursors ($PdCl_2$ and $HAuCl_4$). In this binary system, a similar trend was observed for the deposition of Pd metal particles on each type of TiO_2 support but the amount of Au metal particles increased with increasing surface area of TiO_2 supports.

The X-ray absorption fine structure (XAFS) was performed to investigate formation of the $PdAu_x$ alloy particle on TiO_2 supports by the microwave irradiation method. Figure 1 shows Au L_{III}-edge Fourier Transform of Extended X-ray Absorption Fine Structure (FT-EXAFS) spectra of samples. Au foil and $PdAu_3$ were used as reference materials. The spectrum of each sample was similar to that of $PdAu_3$ but different to that of Au foil. $PdAu_x$ alloy particles were thus formed on TiO_2 supports by the

Table 1. The surface area of supports and the amount of metal loading.

Support (Phase of TiO_2)	Surface area of supports (m^2/g)	Pd (wt%)	Pd / Au (wt%)
TIO-4 (anatase + rutile)	50	0.42	0.34 / 0.40
TIO-6 (rutile)	100	0.78	0.44 / 0.82
TIO-8 (anatase)	338	0.51	0.39 / 1.12

microwave irradiation method. As shown in Figure 2, the bulk Au foil sample showed a characteristic white line [20] whereas there was no intrinsic white line for $PdAu_x$ samples. These results suggest that in this microwave irradiation method a nucleus of metal alloy nanoparticles is created. Concerning the structure of the $PdAu_x$ alloy particles, it is reasonable to consider that most of the Au atoms are preferentially deposited in the core region while the Pd atoms are preferentially deposited in the shell region. Because the reduction rate of Au ions is faster than that of Pd ions they are reduced first and constitute the core [21]. Alloy particles were thus formed by the reduction of Pd ions on the surface of the Au core.

To compare the catalytic activity of Pd and the $PdAu_x$ alloy that was deposited on TiO_2, synthesis of H_2O_2 from hydrogen and oxygen gases was carried out under atmospheric pressure and at room temperature, as shown in Figure 3. The $PdAu_x$ alloy deposited on TiO_2 (anatase + rutile) catalyst showed the highest activity among the examined catalysts. We found that the anatase phase was better than the rutile phase. For $PdAu_x$ alloy catalysts, catalytic activities of all supports were increased by Au addition. Although the amount of loaded Au metal was different for each TiO_2 support, a similar catalytic enhancement was observed for all $PdAu_x$ catalysts. These results suggest that the amount of loaded Au does not significantly affect the catalytic activity but Au addition was effective for the synthesis of H_2O_2.

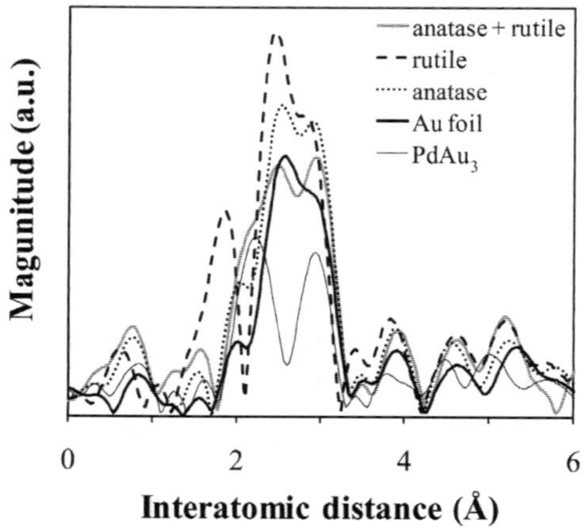

Figure 1. Au L_{III}-edge FT-EXAFS spectra of TiO_2 catalysts and reference samples.

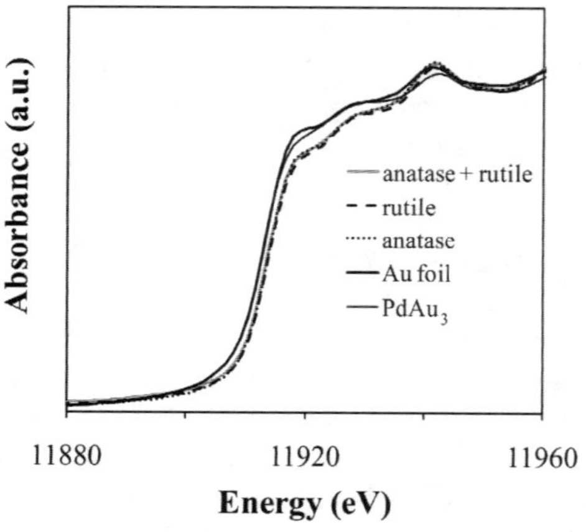

Figure 2. Au L-$_{III}$ edge XANES spectra of TiO_2 catalysts and reference samples.

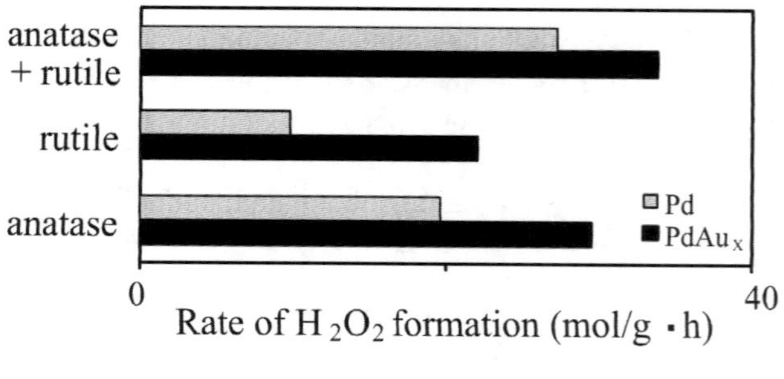

Figure 3. Rate of H_2O_2 formation using TiO_2 catalysts.

4. Conclusions

Nanosized Pd and $PdAu_x$ particles were successfully deposited on various titanium dioxides using a simple and unique microwave irradiation method. Results of XAFS spectra indicated that the obtained $PdAu_x$ particles formed an alloy structure. These catalysts were used for the direct synthesis of H_2O_2 from hydrogen and oxygen gases. The $PdAu_x$ that was deposited on titanium dioxide comprised a mixed phase of anatase and rutile and exhibited the highest activity for this catalytic reaction.

Acknowledgment
We are grateful for financial support from Priority Assistance for the Formation of Worldwide Renowned Centres of Research – The Global COE Program (Project: Centre of Excellence for Advanced Structural and Functional Materials Design) from the Ministry of Education, Culture, Sports, Science and Technology (MEXT), Japan. X-ray adsorption experiments were performed at the SPring-8 BL01B1 station (2008A1366, 2008A1457)).

[1] Kappe C O 2004 *Angew. Chem. Int. Ed.* **43** 6250
[2] Tsuji M, Hashimoto M, Nishizawa Y, Kubokawa M and Tsuji T 2005 *Chem. Eur. J.* **11** 440
[3] Yan X, Liu H and Liew K Y 2001 *J. Mater. Chem.* **11** 3387
[4] Zhu J -F and Zhu Y -J 2006 *J. Phys. Chem. B* **110** 8593
[5] Deivaraj T C, Chen W and Lee J Y 2003 *J. Mater. Chem.* **13** 2555
[6] Yamamoto T, Wada Y, Sakata T, Mori H, Goto M, Hibino S and Yanagida S 2004 *Chem. Lett.* **33** 158
[7] Liang J, Deng Z, Jiang X, Li F and Li Y 2002 *Inorg. Chem.* **41** 3602
[8] Gou L, Chipara M and Zaleski J M 2007 *Chem. Mater.* **19** 1755
[9] Hwang Y K, Chang J -S, Park S -E, Kim D S, Kwon Y -U, Jhung S H, Hwang J -S and Park M S 2005 *Angew. Chem. Int. Ed.* **44** 556
[10] Chen W X, Lee J Y and Liu Z 2002 *Chem. Commun.* 2588
[11] Chen W, Lee J Y and Liu Z 2004 *Chem. Lett.* **33** 474
[12] Zhang X, Jiang W, Song D, Liu J and Li F 2008 *Mater. Lett.* **62** 2343
[13] Yu W, Tu W and Liu H 1999 *Langmuir* **15** 6
[14] Glaspell G, Fuoco L and El-Shall M S 2005 *J. Phys. Chem. B* **109** 17350
[15] Landon P, Paul J. Collier P J,Carley A F, Chadwick D, Papworth A J, Burrows A, Kielyd C J and Hutchings G J 2003 *Phys. Chem. Chem. Phys.* **5** 1917
[16] Lunsford J H 2003 *J. Catal.* **216** 455
[17] Choudhary V R, Samanta C and Choudhary T V 2006 *Appl. Catal. A: General* **308** 128
[18] Yamashita H, Mori K 2007 *Chem. Lett.* **36** 348
[19] Yamashita H, Mori K, Shironita S, Horiuchi Y 2008 *Catal. Surv. Asia* **12** 88
[20] Liu F, Wechsler D and Zhang P 2008 *Chem. Phys. Lett.* **461** 254
[21] Edwards J K, Solsona B E, Landon P, Carley S F, Herzing A, Kiely C J and Hutchings G J 2005 *J. Catal.* **236** 69

Fabrication of core-shell Ti-containing silica encapsulating FePt particles as a spherical nanocatalyst

Kohsuke Mori, Kohei Sugihara Yuichi Kondo and Hiromi Yamashita

Division of Materials and Manufacturing Science, Graduate School of Engineering, Osaka University, 2-1 Yamada-oka, Suita, Osaka 565-0871, Japan

E-mail: yamashita@mat.eng.osaka-u.ac.jp

Abstract. The encapsulation of fcc FePt nanoparticles (NPs) having a mean diameter of ca. 6 nm with the Ti-SiO_2 thick shell gives a new nanocomposite (FePt@Ti-SiO_2), which serves as an efficient nanocatalyst for the liquid-phase selective epoxidation. Recovery from the reaction mixture was readily attained by applying an external permanent magnet, and the spent catalyst could be recycled without any appreciable loss in activity.

1. Introduction

The development of promising catalysts with well-defined active structures that enable selective chemical transformations and can be completely separated is a paramount challenge [1]. It is widely accepted that the activity of a solid catalyst suspended in a liquid phase can benefit greatly from the use of smaller catalyst particles (< 1 μm) to avoid mass-transfer limitations [2]. However, the difficulties in recovering small particles from the reaction mixture severely circumvent their industrial applications, which can lead to blocking of filters and valves in chemical plants. Although nanoparticle-based catalysts possessing extremely large surface areas are also key components of catalytic activity [3-5], the separation step becomes a more troublesome issue as the size of the particles decrease to the nanometer scale.

With this in mind, we have developed Ti-containing silica (Ti-SiO_2) encapsulating superparamagnetic fcc FePt NPs (FePt@Ti-SiO_2). The inner magnetic FePt cores were shielded from the external environment by the impermeable coating, which prevents sintering and agglomeration of metal NPs under catalytic conditions. The catalytically active Ti-oxide moieties are located on the external surface of the coating and considered to be highly dispersed at the atomic level. This report describes the synthesis and characterization of the novel FePt@Ti-SiO_2 nanocomposite and its evaluation as a nanocatalyst for the oxidation of organic compounds using H_2O_2.

2. Experimental

2.1 The synthesis of FePt NPs: FePt NPs were prepared using standard airless techniques on a Schlenk line according to a procedure modified from that reported by Chen et al [6]. Platinum acetylacetonate (1 mmol, $Pt(acac)_2$, Wako Pure Chemical Ind., Ltd.) was mixed with 20 mL of dibenzyl ether (Tokyo Kasei Kogyo Co., Ltd) in a 200 mL three-neck flask. The mixture was degassed for 1 h and then heated to 120°C under Ar. Iron pentacarbonyl (2 mmol, $Fe(CO)_5$, Aldrich), 10 mmol of oleic acid (Wako Pure Chemical Ind., Ltd.), and 10 mmol of oleylamine (Aldrich) were then injected with mechanical stirring. The reaction mixture was heated at a rate of ~5°C/min to 310°C and maintained for 1 h. The black colored solution was allowed to cool to room temperature and NPs were precipitated with excess ethanol and collected by centrifugation. After washing several times with ethanol, the precipitate was uniformly dispersed in hexane. Elemental analysis confirmed that the average composition of the NPs was $Fe_{48}Pt_{52}$.

2.2 The ligand exchange procedure of the FePt NPs with (3-mercaptpropyl)-trimethoxysilane: (3-Mercaptpropyl)-trimethoxysilane (2 mL, 3-MPTS, Aldrich) was added to the 100 mL of hexane solution of as-synthesized FePt NPs (0.13 g). The mixture was stirred for 15 min, and precipitated with ethanol and collected by centrifugation. After washing several times with ethanol, the precipitate was stored in ethanol.

2.3 The synthesis of FePt@Ti-SiO_2: A 500 mL ethanol solution of the modified FePt NPs with 3-MPTS (0.05 g) was mixed with a 200 mL of ethanol solution containing tetraethyl orthosilicate (20 μL, TEOS, Wako Pure Chemical Ind., Ltd.) and tetrapropylorthotitanate (1 μL, TPOT, Wako Pure Chemical Ind., Ltd.). After stirring for 12 h at room temperature, the encapsulated particles were repeatedly centrifuged and washed using excess ethanol, giving a FePt@Ti-SiO_2.

3. Results and discussion

The synthetic procedure for the FePt@Ti-SiO_2 is illustrated in **Figure 1**. Uniformly distributed FePt NPs were synthesized by decomposition of $Fe(CO)_5$ and reduction of $Pt(acac)_2$ in the presence of oleic acid and oleylamine. Next, 3-MPTS was added to the hexane solution of as-synthesized FePt NPs and stirred. Ligand exchange with 3-MPTS made the FePt NPs dispersible in the alcoholic solution. Finally, precursors of Ti-SiO_2 (TEOS and TPOT) were added to the FePt NPs dispersed in ethanol and stirred. The use of 3-MPTS was motivated by the following reasons; 3-MPTS is soluble in alcoholic solvent and would readily bind to the surface of the metal particles via thiolate linkage. Furthermore, trimethoxysilane moieties facilitate formation of the Ti-SiO_2 phase by subsequent sol-gel polymerization.

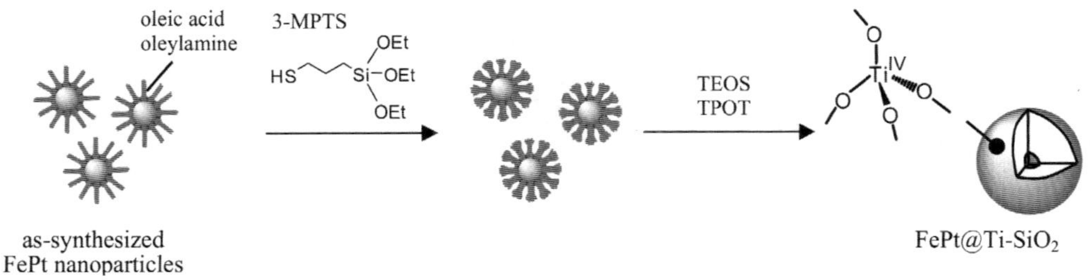

Fig. 1 Schematics of the procedure for the synthesis of FePt@Ti-SiO_2.

The intensities of the diffraction peaks in the XRD patterns (Cu Kα radiation) of the FePt@Ti-SiO_2 are low, presumably due to the smaller size of the particles. However, they exhibited clear peaks due to the fcc FePt around 40.0°, 46.7°, and 68.2°, corresponding to the (111), (200), and (220) reflections, respectively. Assuming that the FePt NPs are spherical in shape, the average crystalline size after encapsulation was calculated to be 5.9 nm by applying Scherrer's equation for the (111) reflection.

Figure 2 shows a typical TEM image of the core-shell FePt@Ti-SiO_2. These spherical particles have 45~50 nm thick Ti-SiO_2 shell that were formed by TEOS and TPOT with FePt NPs. Each Ti-SiO_2 sphere encapsulates one FePt NPs, although a small fraction of the particles possess either two or zero FePt NPs. The average diameter (d_{ave}) of FePt

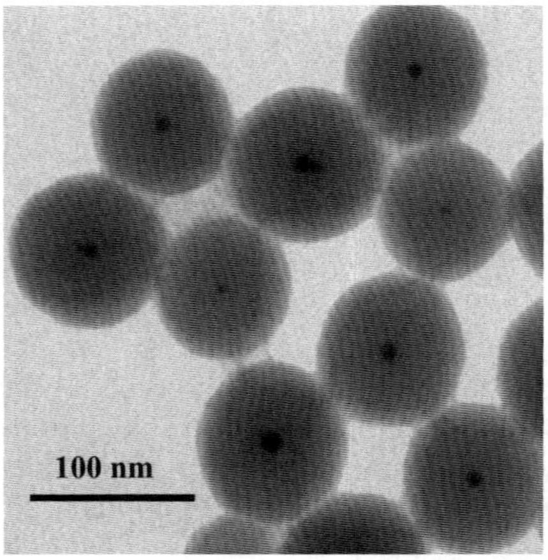

Fig. 2 TEM imageof FePt@Ti-SiO_2.

NPs were determined to be ca. 6 nm. This observation is consistent with the results of XRD analysis, as described earlier, verifying the single-crystalline nature of the particles.

The magnetic properties were investigated using a SQUID magnetometer. As shown in **Figure 3**, the isothermal magnetization curve of the FePt@Ti-SiO_2 at 300 K displayed a rapid increase with increasing applied magnetic field, due to superparamagnetic relaxation [7]. This indicates that the thermal energy can overcome the anisotropy energy barrier of the individual particles, and the net magnetization of these nanoparticle assemblies is zero in the absence of an external field. Hysteresis was absent with zero remanence (M_r) and coercivity (H_c), and the saturation magnetization (M_s) reached up to 0.58 emu/g for FePt@Ti-SiO_2. With decreasing temperature, the magnetization of both samples increased and exhibited a symmetric hysteresis loop at 5 K (M_r = 0.44 emu/g, H_c = 750 Oe), indicating a transition from superparamagnetic to ferrimagnetic behavior.

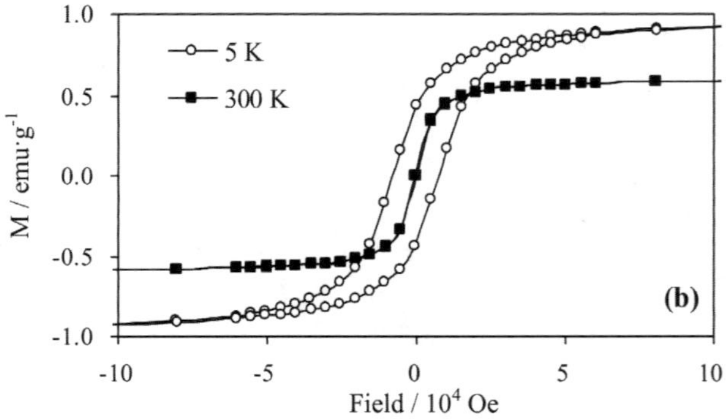

Fig. 3 Field-dependent magnetization curve for FePt@Ti-SiO_2 measured at 300 K (■) and 5 K (○).

Figure 4A shows the Ti K-edge XANES spectra of the FePt@Ti-SiO_2, Ti(O^iPr)$_4$ and TiO_2 powder (P25: anatase 80%, rutile 20%). The spectrum for TiO_2 powder shows several well-defined pre-edge peaks attributable to the anatase TiO_2 in a symmetric octahedral environment [8]. On the other hand, the FePt@Ti-SiO_2 and Ti(O^iPr)$_4$ were found to exhibit an intense single pre-edge peak due to the lack of an inversion center in the regular tetrahedron structure [8]. Observation of this intense single pre-edge peak indicates that the Ti-oxide moieties in the FePt@Ti-SiO_2 have tetrahedral coordination [9]. The FT-EXAFS data (**Figure 4B**) shows a peak suggestive of a contiguous Ti–O–Ti bond for the TiO_2 powder. On the other hand, the FePt@Ti-SiO_2 showed only a strong peak at around 1.6 Å, which could be assigned to the neighboring oxygen atoms due to a Ti–O bond.

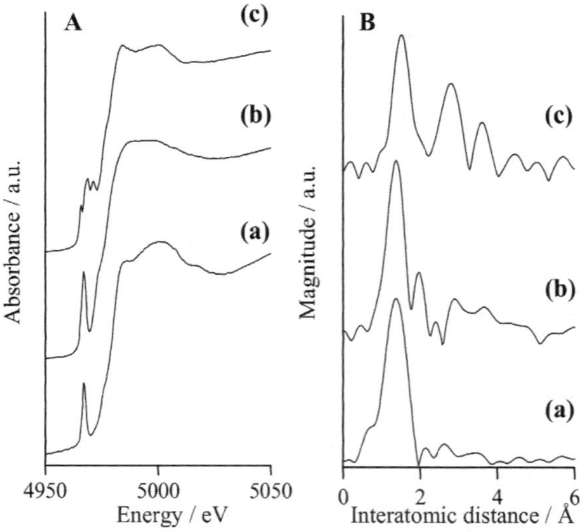

Fig. 4 (A) Ti K-edge XANES spectra and **(B)** FT-EXAFS spectra for (a) FePt@Ti-SiO_2, (b) Ti(O^iPr)$_4$, and (c) TiO_2 powder (P25).

We explored the catalytic ability of the FePt@Ti-SiO_2 in the epoxidation of cyclooctene using 30% hydrogen peroxide as a simple test reaction. Results of this reaction indicated that the corresponding cyclooctene oxide was obtained in >99% selectivity and the turn over number (TON) based on Ti approached 48 after 24 h (**Table 1**). More important advantage of the FePt@Ti-SiO_2 is the facile recovery from the reaction mixture and the high reusability. Upon completion of the oxidation reaction, the magnetic properties of the FePt@Ti-SiO_2 can afford a straightforward means to isolate the catalyst from the colloidal solution. By applying an external permanent magnet, the catalyst was attracted. The recovered catalyst could then be recycled in the epoxidation of cyclooctene at least three times while maintaining identical inherent activity to the initial run.

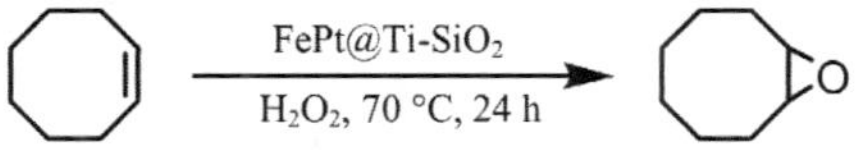

Table 1. Oxidation of cyclooctene using FePt@Ti-SiO_2 and recycling experiment

Run	1st	2nd	3rd	4th
TON	48	48	48	48
Select. / %	>99	>99	>99	>99

4. Concluding remarks

A novel spherical nanocatalyst, FePt@Ti-SiO_2, was synthesized by adopting a two-step coating method via a ligand exchange and a sol-gel process. Our designed architecture enabled the powerful combination of useful functionality, superparamagnetism that was attracted to a magnetic field, as well as a catalytically active site that exhibited prominent performance for specific catalytic reactions. The present strategy provides great flexibility in the selection of the magnetic core, as well as catalytically active center. Introduction of these features into nanoparticulate systems enables achievement desired superparamagnetic response as well as particular catalytic reactions.

5. Acknowledgements

This work was supported by Priority Assistance for the Formation of Worldwide Renowned Centers of Research – The Global COE Program (Project: Center of Excellence for Advanced Structural and Functional Materials Design) from the Ministry of Education, Culture, Sports, Science and Technology (MEXT), Japan. The authors appreciate Dr. Eiji Taguchi and Prof. Hirotaro Mori at the Research Center for Ultra-High Voltage Electron Microscopy, Osaka University for assistance with TEM measurements. The X-ray adsorption experiments were performed at the KEK-PF (2007G031, 2007G077).

References

[1] Wijngaarden R J and Kronberg A 1998 Westerterp, K. R. in *Industrial Catalysis: Optimizing Catalysts and Processes* (Wiley-VCH: Weinheim)
[2] Teunissen W, Bol A A and Geus J W 1999 *Catal. Today* **48** 329.
[3] El-Sayed M A 2001 *Acc. Chem. Res.* **34** 257.
[4] Mori K, Hara T, Mizugaki T, Ebitani K and Kaneda K 2004 *J. Am. Chem. Soc.* **126** 10657.
[5] Astruc D, Lu F and Aranzaes J R, 2005 *Angew. Chem. Int. Ed.* **44** 7852.
[6] Chen M, Liu J P and Sun S H, 2004 *J. Am. Chem. Soc.* **126** 8394.
[7] Coey J M, Khalafalla D, 1972 *Phys. Status. Solidi. A* **11** 229.
[8] Yamashita H, Ichihashi Y, Anpo M, Louis C and Che M 1996 *J. Phys. Chem.* **100** 16041.
[9] Mori K, Kondo Y, Morimoto S and Yamashita H 2008 *J. Phys. Chem. C* **112** 2593.

Journal of Physics: Conference Series **165** (2009) 012041 doi:10.1088/1742-6596/165/1/012041

Synthesis of Y_2O_3-doped CeO_2 nanocrystals and their surface modification

Yuji Goto, Kazuyuki Takahashi, Takahisa Omata and Shinya Otsuka-Yao-Matsuo

Division of Material and Manufacturing Science, Graduate School of Engineering, Osaka University, 2-1 Yamada-oka, Suita, Osaka 565-0871, Japan

E-mail: omata@mat.eng.osaka-u.ac.jp

Abstract. Colloidal undoped and Y_2O_3-doped CeO_2 nanocrystals (NCs) with diameters of 2.1~2.6 nm were synthesized by the thermolysis of cerium and yttrium acetylacetonate dissolved in oleylamine. A surface modification of the NCs, *i.e.*, removing the oleylamine adsorbed on the NCs surfaces was attempted by heating to 200 °C under an oxygen atmosphere and by treatment at 200 °C in pure water and H_2O_2 aqueous solution in an autoclave. Heating under an oxygen atmosphere resulted in a complete removal of oleylamine. However, the particle size increased to above 5 nm after heating. For the autoclave treatment, especially in the H_2O_2 aqueous solution, the oleylamine was almost removed without distinct particle growth.

1. Introduction

Since the discovery of enhanced Li^+-conductivity in LiI and porous Al_2O_3 nanocomposites [1] the enhanced ionic conductivity at interfaces and/or near-interfaces has been drawing attention as a nanosize effect that appears in solid electrolytes. Ionic conduction of materials in which significant interfaces were introduced by nanostructuring has been extensively investigated. One of the most important uses of ionic conductors is in fuel cells. When the oxide-ionic conductivity of the solid electrolyte in solid oxide fuel cells (SOFC) is enhanced by the nanosize effect the operating temperature of the SOFC can be decreased. The SOFC will thus be far more cost efficient and thus an improved solution for energy and environmental issues. An improvement of one or two orders of magnitude in conductivity compared with a microcrystalline sinter has previously been reported for yttria-stabilized zirconia (YSZ) and aliovalently doped ceria (doped CeO_2) thin films that consisted of nanocrystals (NCs) with a 15 ~ 30 nm grain size [2,3]. The conductivity was found to increase with a decrease in grain size. These observations apparently indicated that conductivities of YSZ and doped CeO_2 were enhanced by an increase in grain boundaries because of the decrease in the grain size. A detailed mechanism of this enhanced conductivity has not been clarified yet although several mechanisms such as the space-charge layer effect near grain boundaries, mechanical stress effect and protonic conduction at grain boundaries have been proposed. To study the size dependent electrical conductivity NCs with a sharp particle size distribution, *i.e.*, monodispersed NCs, are needed. This is because the size dependency of NCs may be masked if a broad particle size distribution is used. Previously, we reported the synthesis of colloidal YSZ NCs by a nonhydrolytic sol-gel method and the

NCs obtained were highly monodipersed and suitable for the investigation of size dependent properties [4]. In this study, we synthesized colloidal Y_2O_3-doped CeO_2 (YDC) NCs by thermolysis of cerium and yttrium acetylacetonate in oleylamine. The oleylamine adsorbed on NCs surfaces affects the measurement of precise size dependent conductivity. We thus also attempted a surface modification of NCs, *i.e.*, removing the oleylamine adsorbed on NCs surfaces.

2. Experimental

Initial solutions were prepared by the dissolution of cerium and yttrium acetylacetonate [$Ce(acac)_3 \cdot nH_2O$ and $Y(acac)_3 \cdot nH_2O$] into oleylamine at 50 °C for 30 min under an Ar atmosphere. Solutions were heated at 200 or 250 °C for 10~45 min. After the reaction, NCs were extracted from the colloidal solutions by addition of methanol and ethanol and successive centrifugation. Obtained phases were characterized by powder XRD and Raman spectroscopy. Chemical compositions were determined by ICP atomic emission spectroscopy (ICP-AES). The particle size distribution was evaluated by small angle X-ray scattering (SAXS). Removal of the oleylamine molecules, adsorbed on NCs surfaces, was attempted using two methods. One was an O_2 treatment at 200 °C. The other was autoclave treatment in pure water and H_2O_2 aqueous solution at 200 °C. The NCs powder was loaded into a Teflon-lined autoclave and heated under autogenous pressure. The remaining oleylamine content after these treatments was evaluated by TG-DTA under an oxygen atmosphere.

3. Result and discussion

Figure 1 shows XRD profiles of the as-synthesized NCs synthesized at 250 °C. The chemical formulae in the figure indicate the compositions determined by the ICP-AES. XRD profiles were broad because of the small crystal size. The average crystal size evaluated from the diffractions ranged from 2.1 to 2.6 nm. All diffractions were indexed as those of the cubic fluorite form. However, we could not conclude from XRD results whether the Y_2O_3 was successfully doped into the CeO_2 lattice or not because the precise lattice parameter could not determined because of its extremely broad profile. Figure 2 shows the Raman spectra of as-synthesized NCs. For undoped CeO_2 NCs a broad Raman band at ~615 cm^{-1} was observed in addition to a band at 465 cm^{-1} which is attributed to the fluorite structure. According to a previous study, the band results from intrinsic oxygen vacancies [5], *i.e.*, the oxygen vacancy generated by the partial reduction of Ce^{4+} to Ce^{3+}. For YDC samples a broad Raman

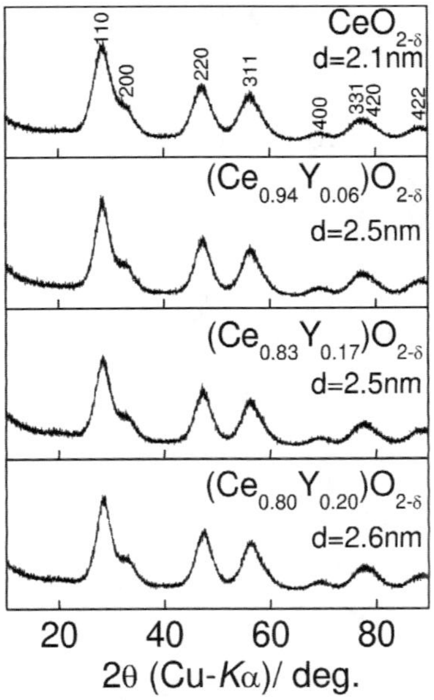

Figure 1. XRD profiles of as-synthesized CeO_2 and YDC NCs. Compositions indicated are values determined by ICP-AES.

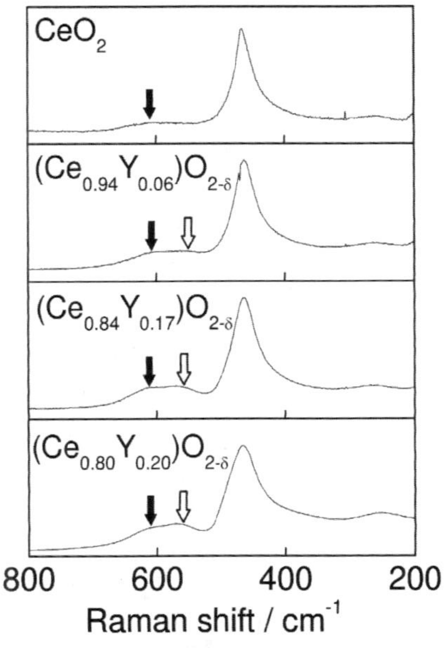

Figure 2. Raman spectra of as-synthesized CeO_2 and YDC NCs. The excitation wavelength was 532 nm.

band appeared at ~565 cm^{-1} and its intensity increased with an increase in the Y_2O_3-concentration. The band is attributed to an extrinsic oxygen vacancy [5], *i.e.*, an oxygen vacancy introduced by the substitution of tetravalent Ce^{4+} with trivalent Y^{3+}. This means that Y_2O_3 was successfully doped into the fluorite-type CeO_2 lattice. Thus we concluded that Y_2O_3 did not separately precipitate and the dissolution of Y_2O_3 into the CeO_2 lattice progressed during the NCs formation reaction.

Figure 3 (a) shows the volume distribution of the particle size for CeO_2 NCs synthesized at 200 °C for 10 min. The distribution was evaluated from SAXS data. The maximum distribution, d_{max}, was observed at 2.4 nm and the average diameter, d_{ave}, was 2.5 nm. The obtained NCs had a very sharp distribution in the range from 1.5 to 3.5 nm, *i.e.*, highly monodispersed NCs. The NCs are thus suitable for the investigation of size dependent properties. The high resolution TEM image in Figure 3 (b) showed that particles were spherical and well crystallized. Their diameter was approximately 2.5 nm, which is consistent with the size determined from SAXS data.

Figure 4 shows TG-DTA results of the as synthesized $(Ce_{0.8}Y_{0.2})O_{2-\delta}$ NCs. A clear exothermic peak and a steep mass decrease at ~200 °C because of the combustion of oleylamine were observed. We thus attempted to remove the oleylamine by heating the NCs in O_2 atmosphere at 200 °C. The oleylamine rapidly burned out because of the catalytic activity of the YDC. The temperature of the NCs might have risen to higher than 300 °C. The particle size increased to above 5 nm after heating. Consequently, heating the NCs is not a viable method for the elimination of oleylamine adsorbed on NCs surfaces. Figure 5 shows TG-DTA results of NCs after autoclave treatment. Compared with results from the as-synthesized powder (Figure 4) the exothermic peak of the DTA curve and the mass decrease in the TG curve were drastically reduced after autoclave treatment in pure water (Fig. 5 (a)). After treatment in an aqueous H_2O_2 solution the exothermic peak in DTA and a mass decrease in TG were hardly observed. The particle size after the treatment was 3.3 nm; it was slightly larger than that of the sample before the autoclave treatment. Although the particle size distribution

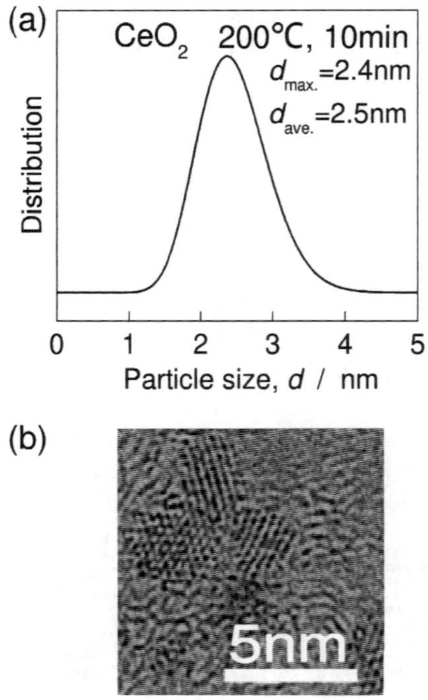

Figure 3. (a) Distribution of the particle size determined from SAXS for CeO_2 NCs synthesized at 200 °C for 10 min. d_{ave} and d_{max} indicate average and maximum distribution diameters, respectively. (b) high resolution TEM image of CeO_2 NCs.

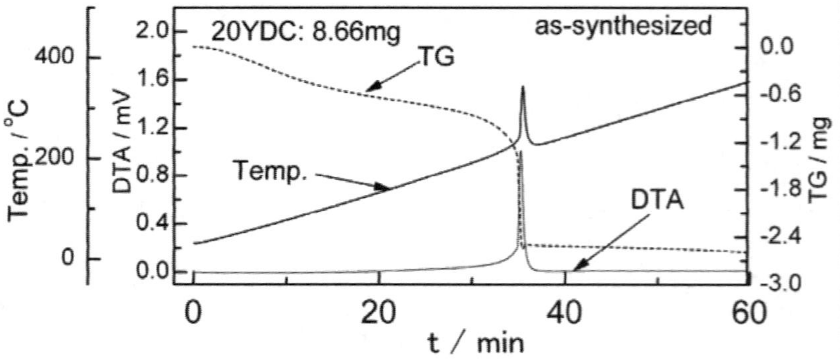

Figure 4. TG-DTA under an oxygen atmosphere of as-synthesized $(Ce_{0.8}Y_{0.2})O_{2-\delta}$ NCs.

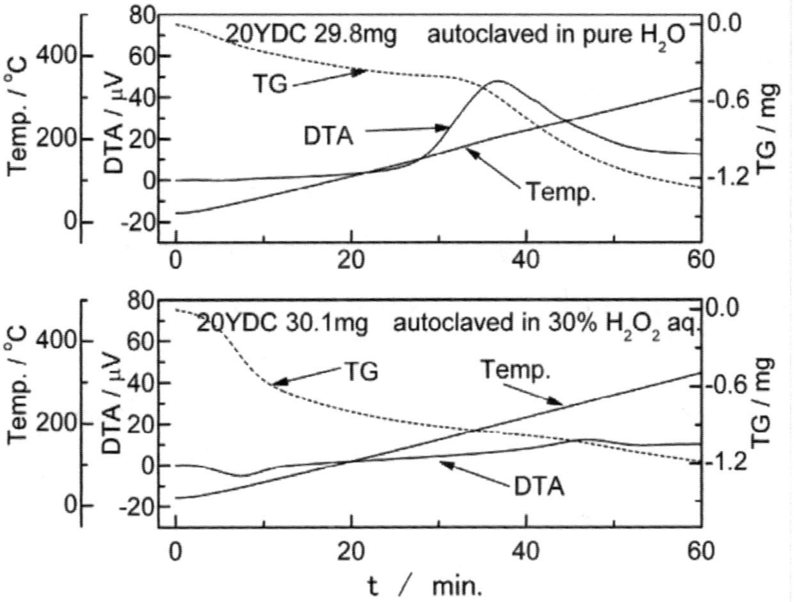

Figure 5. TG-DTA results under an oxygen atmosphere of $(Ce_{0.8}Y_{0.2})O_{2-\delta}$ NCs after autoclave treatment in pure H_2O and H_2O_2 aq.

was not evaluated, it was suggested that broadening of the size distribution was suppressed in this treatment based on the slight increase in the average particle size. Thus we conclude that the autoclave treatment in H_2O_2 aqueous solution is a suitable surface modification method to suppress grain growth.

4. Conclusion

Colloidal Y_2O_3-doped CeO_2 NCs with ~2.5 nm diameter were successfully synthesized by the thermolysis of cerium and yttrium acetylacetonate dissolved in oleylamine. NCs obtained were spherical and well crystallized and the particle size distribution was sufficiently sharp, *i.e.*, highly monodispersed. When NCs powder was heated to 200 °C under an oxygen atmosphere the oleylamine adsorbed on the NCs surfaces were completely removed. However, the particle size increased to above 5 nm after heating. Autoclave treatment in aqueous H_2O_2 solution resulted in a better removal of the oleylamine without distinctly increasing the particle size. We conclude that the thermolysis of cerium and yttrium acetylacetonate is a suitable technique to synthesize doped CeO_2 NCs with single nanometer grains and an autoclave treatment in aqueous H_2O_2 solution results in suitable NCs for the investigation of size dependent electrical conductivity.

Acknowledgement

This work was supported by a Grant-in-Aid for Scientific Research on Priority Area, "Nanoionics (439)" and Priority Assistance for the Formation of Worldwide Renowned Centers of Research - The Global COE Program "Center of Excellence for Advanced Structural and Functional Materials Design" from the Ministry of Education, Culture, Sports, Science and Technology (MEXT), Japan.

References

[1] Liang C C, 1973 *J. Electrochem. Soc.* **120** 1289.
[2] Kosacki I, Suzuki T, Petrovsky V, Anderson H U, 2000 *Solid State Ionics* **136-137** 1225.
[3] Kosacki I, Rouleau C M, Becher P F, Bentley J, Lowndes D H, 2005 *Solid State Ionics* **176** 1319.
[4] Goto Y, Omata T, Otsuka-Yao-Matsuo S, 2009 *J. Electrochem. Soc.* **156** K4.
[5] Dohčević-Mitrović Z D, Grujić-Brojčin M, Šćepanović M, Popović Z V, Bošković S, Matović B, Zinkevich M, Aldinger F, 2006 *J. Phys.: Condens. Matter* **18** S2061.

Journal of Physics: Conference Series **165** (2009) 012042 doi:10.1088/1742-6596/165/1/012042

Large-area and low-damage processes for hybrid flexible device fabrications with reactive high-density plasmas driven by multiple low-inductance antenna modules

Yuichi Setsuhara [1,2], **Kosuke Takenaka** [1,2], **Ken Cho** [3] **and Jeon G. Han** [4]

[1] Joining and Welding Research Institute, Osaka University, 11-1 Mihogaoka, Ibaraki, Osaka 567-0047, Japan

[2] JST, CREST

[3] Division of Materials and Manufacturing Science, Graduate School of Engineering, Osaka University, 2-1 Yamadaoka, Suita, Osaka 565-0871, Japan

[4] Center for Advanced Plasma Surface Technology, Sungkyunkwan University, Suwon 440-746, Korea

E-mail: setsuhara@jwri.osaka-u.ac.jp

Abstract. Plasma generation and control technologies for meters-scale ultra-large-area plasma sources have been developed with multiple low-inductance antenna (LIA) modules, as a promising candidate of ultra-large-area and high-density plasma sources for next-generation processing of hybrid flexible devices. Properties of argon-oxygen mixture plasmas sustained with multiple LIA units have been investigated and surface modifications of polymer substrates using the plasmas have been performed. Ion energy distribution at the sheath edge of the argon-oxygen mixture plasmas showed considerable suppression of ion energies as small as or less than 10 eV. We have examined effect of plasma exposure on surface modification and/or degradation of polymer. Surface analysis of polytetrafluoroethylene (PTFE) exhibited nano-surface modification of the polymer surface without suffering degradation of molecular structures beneath the nano- surface layer.

1. Introduction

Low-damage and nano-surface processes are of key importance for development of next-generation devices including hybrid flexible devices or electronics on polymers. For successful fabrication of devices on polymers, it is significant to develop low-damage plasma sources for fine control of interface between the polymer substrate and inorganic functional films without suffering degradations due to ion bombardment. Furthermore, for enhancement of production efficiency and/or cost reduction in fabrication of these devices, it is significant to develop meters-scale/ultra-large area uniform plasma reactor. In enlargement of source size exceeding a meter, however, plasma distributions hence processing profiles become inherently non-uniform primarily due to non-uniform power deposition profile caused by standing-wave effects.

In order to overcome these constraints, we have developed low-damage plasma technologies with low-inductance antenna (LIA) modules to drive inductively coupled plasmas (ICPs), which can solve the problems in a straightforward manner via low-voltage operation of ICPs to generate high-density and extremely low-damage plasmas for meters-scale ultra-large-area processing [1,2]. Our unique plasma technology is based on principles of multiple operation and integrated control of LIA modules.

In this article, we present issues involved in meters-scale large-area plasma generation and low-damage plasma production with multiple LIA modules. The issues involved in designing of ultra-large area plasma sources are discussed in terms of simulation aided methods based on experimental justifications to demonstrate feasibility of plasma sources exceeding 2 meters for next generation FPD processes. Properties of argon and argon-oxygen mixture plasmas are reported especially for evaluation of ion energy distribution and effects of plasma exposure onto polymer surface will be discussed based on surface analysis of chemical-bonding states via hard x-ray photoelectron spectroscopy (HXPES) at SPring8 (national SOR facility in Japan).

2. Plasma technologies employing multiple low-inductance antenna modules

Major specifications required for the plasma sources to be employed in large-area processes include a) controllability of plasma profile, b) high-density plasma production for raising throughput, c) low plasma potential for high quality processing and d) low-pressure operation for enhanced efficiency of materials-gas consumption and suppression of dust formation. Among them, the specifications b) - d) are mainly dependent on the discharge schemes and/or the plasma excitation frequency, which are to be selected according to requirements in the materials processing. Therefore, many efforts have been focused on the issues to scale up the conventional sources, which meet the process requirements.

However, considering design issues for plasma production with radio-frequency (RF) power sources to scale up the source size exceeding a meter, power deposition profiles and hence process profiles become inherently non-uniform especially due to standing wave effects and/or edge effects, which can not be avoided with increasing source size when the source employs power coupling elements with a scale length equivalent to or as long as the 1/4 wavelength of the RF-power transmission [3-5].

In order to overcome the constraints via active control of power-deposition profile and hence the plasma uniformity, we have developed plasma control technology as illustrated in Fig. 1 [1,2]. The LIA consists of a U-shaped antenna conductor, which is fully covered with dielectric tubing for complete isolation from the plasma [1,2]. With this configuration and integrated control of each module, the RF power supplied to each of the LIA modules can be controlled actively. Our proposal of the unique source configuration and control method can provide breakthroughs to solve the problems with the standing-wave effects.

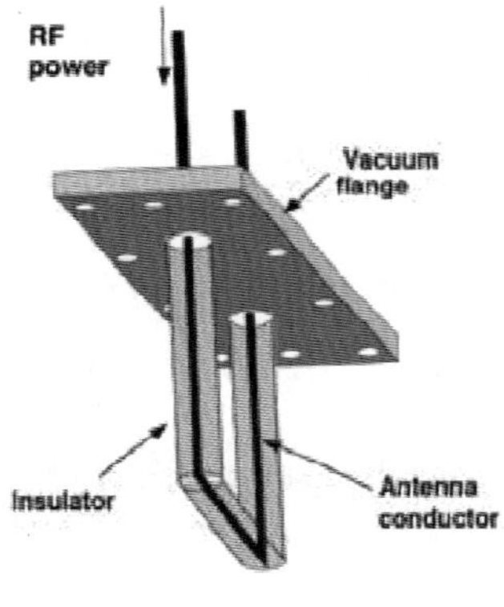

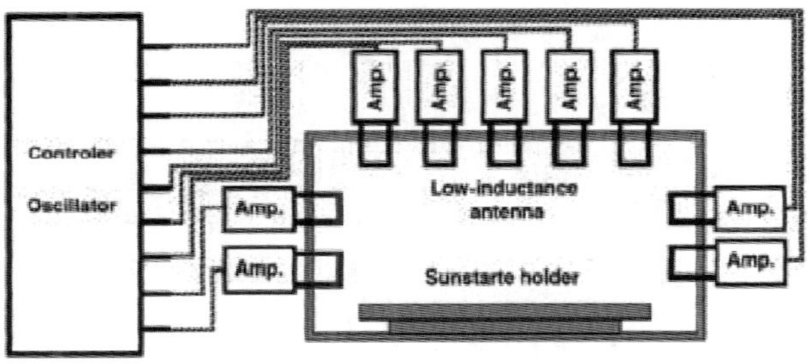

Figure 1. Schematic illustrations of LIA unit and plasma control system with integrated and independent power control.

Journal of Physics: Conference Series **165** (2009) 012042 doi:10.1088/1742-6596/165/1/012042

3. Simulation-aided designing of meters-scale large-area plasma source

Designs of ultra-large area plasma sources were examined with simulation codes consisting of electromagnetic module, plasma fluid module and electron Monte Carlo module, where argon (Ar) plasmas sustained with 13.56 MHz HF power were simulated. Distributions of plasma density, electron temperature and neutral density were obtained from plasma fluid module, which solved continuity equation and electron energy conservation equation under drift-diffusion approximation.

In the present investigation, the plasma fluid simulations were carried out for a rectangular process chamber with an internal square area of 300 cm and an internal height of 40 cm, as shown in Fig. 2. Ar plasmas at a pressure of 20 mTorr were simulated by installing LIA modules (40-cm width, 16-cm height) on the top plate located at z = 40 cm and plasma profiles were examined at a substrate holder located at z = 0 cm.

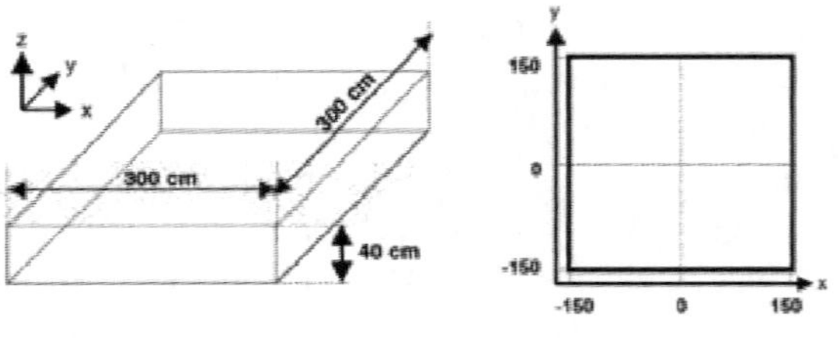

Figure 2. Schematic illustration of process chamber for simulation.

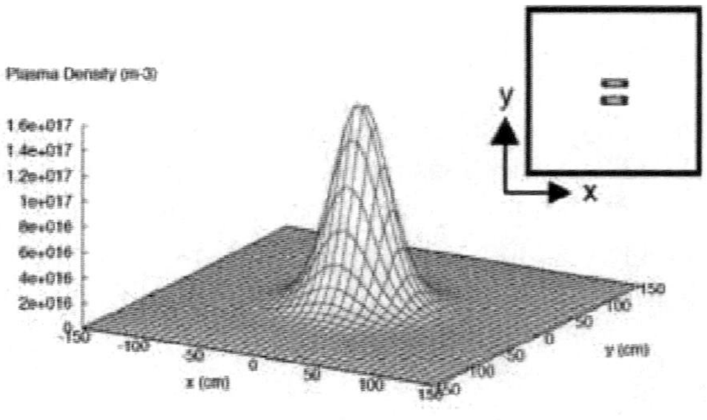

Figure 3. Plasma density profile simulated for two LIA modules located at the center of the top plate.

Characteristics of plasma density profile sustained with the LIA module were examined by installing two LIA modules, as shown in Fig. 3, in order to evaluate appropriate spacing between LIA modules for obtaining uniform plasmas in multiple antenna configurations. The numerical result plotted in Fig. 3 indicates that proper spacing between LIA modules should be about ~ 50 cm, which is equivalent to FWHM of the profile, for attaining large-area uniformity in the multiple LIA configurations.

Based on the principle mentioned above, a variety of configurations for multiple LIA regime were examined to attain uniform plasma-density profile. One example of designing ultra-large-area source is shown in Fig. 4, in which RF-power inputs to each of LIA modules were set in the range of 500-1200 W. In this example, uniform plasma profile over 200-cm rectangular area was obtained with a non-uniformity of 5.5%.

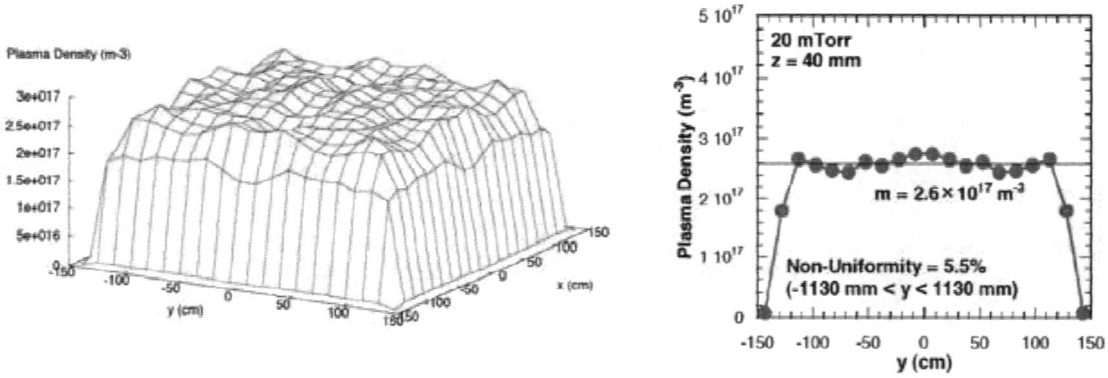

Figure 4. Plasma density profile simulated for 300cm x 300cm plasma source.

4. Properties of Ar-O_2 mixture plasmas and low-damage surface modification of polymers

Plasma modifications of polymers have attracted great interests in a variety of fields to utilize polymer materials for advanced engineering applications [6,7] including device fabrications on polymer substrates.

In the plasma processing of polymers, interactions of ions and radicals impinging onto the polymer surface play important roles. For successful treatment of polymers, it is significant to develop low-damage plasma sources in order to control interface between the polymer substrate and functional films without suffering degradations or chemical-structure change of polymers via ion bombardment due to elemental loss and/or bond breaking of polymers. The ion-induced degradation of polymers may cause unwanted change in desired characteristics of polymers as flexible materials; e.g. carburization of polymers due to loss of H, O and N atoms from their chemical chains enhances hardness and brittleness, which in turn lower flexibility in the modified layers and/or adhesion strength of the films formed on them. Thus special attentions should be paid in controlling and/or suppressing the ion damage. Generally, the extent of ion damage in RF plasmas can be evaluated by the ion energies, which are roughly equivalent to potential difference in the sheath between plasma potential and the potential at the substrate surface. As one of the effective plasma sources with suppressed plasma damage, we have developed ICP sources sustained with LIA modules, which allowed a low-voltage operation of ICPs to generate large-area and high-density plasmas [1,2].

In this section, we report on properties of argon-oxygen mixture plasmas sustained with multiple LIA modules for low-damage processing of polymers. For evaluation of surface-modification characteristics and/or plasma damage during processing of polymers with reactive argon-oxygen mixture plasmas, nondestructive depth-profile analysis of chemical-bonding states has been carried out via hard x-ray photoelectron spectroscopy (HXPES).

4.1. Experimental

Schematic diagram of the chamber installed with a set of 8 LIA units is shown in Fig. 5. The LIA unit consists of a U-shaped antenna conductor, which was fully covered with dielectric tubing for complete isolation from the plasma [1,2]. Eight LIA units with a 70 mm width and a 160 mm height were mounted on the top flange of the discharge chamber and were coupled to a 1000-W RF power generator at 13.56 MHz via a matching network, to which all the LIA units were connected in parallel. The plasma chamber had a 500 mm inner diameter and a 200 mm height, which was connected to a diffusion chamber with a 500 mm inner diameter and a 400 mm height. A water-cooled substrate holder was placed at a distance of 297 mm from the top flange. The base pressure of the chamber evacuated with a turbomolecular pump was 3 x 10^{-4} Pa and argon-oxygen plasmas were produced at a total pressure of 2.6 Pa. Here, a partial pressure ratio R in the argon-oxygen mixture gas is defined as $R = P_O / (P_{Ar} + P_O)$, in which P_O and P_{Ar} denote partial pressures of oxygen and argon, respectively.

Kinetic energy distributions of ions impinging onto ground potential from argon-oxygen mixture plasmas were measured using a mass-separated ion-energy analyzer (EQP500, Hiden) mounted on sidewall as shown in Fig. 5. Plasma potential and floating potential were measured with a cylindrical Langmuir probe, which was inserted radially at an axial position of 280 mm below the top flange.

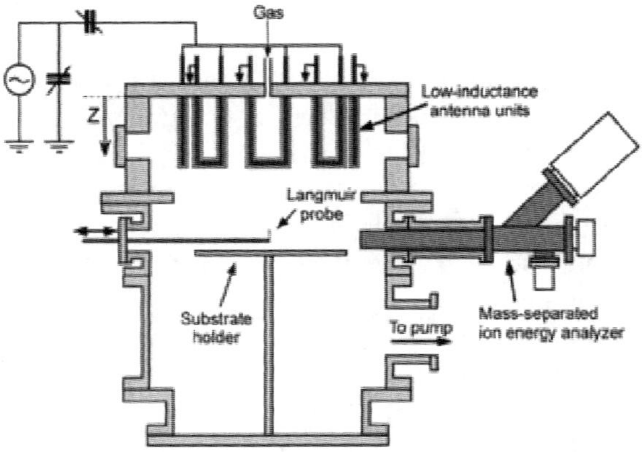

Figure 5. Schematic diagram of plasma chamber with multiple low-inductance antenna units.

Chemical bonding states in the nano-surface layer of polymers exposed to plasmas were analyzed using HXPES at photon

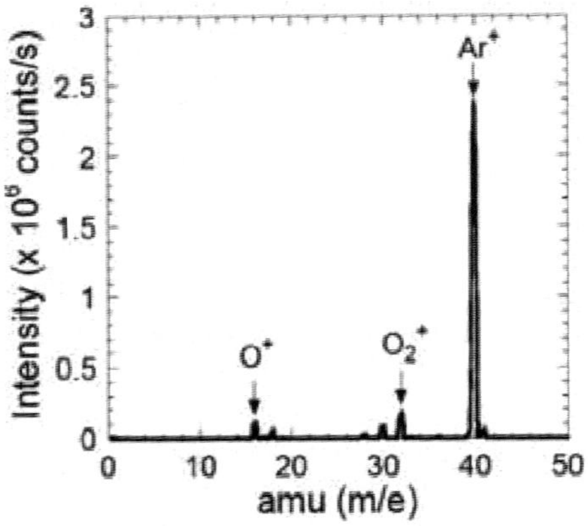

Figure 6. Ion mass spectrum for argon-oxygen plasma sustained at $R = 3\%$.

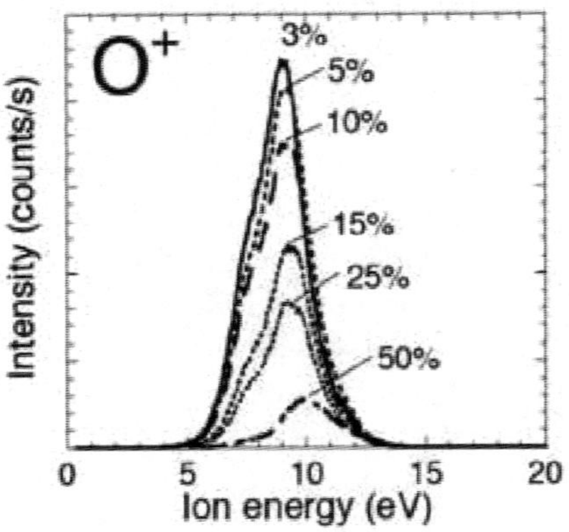

Figure 7. Kinetic energy distribution of O^+ ions with oxygen partial pressure as parameter.

energy of 7.94 keV. The HXPES measurements were carried out at the national synchrotron radiation facility SPring-8 of the Japan Synchrotron Radiation Research Institute (JASRI). In the chemical analysis of polymer surface with x-ray photoelectron spectroscopy, it is of great significance to employ x-rays in the hard x-ray region because the photoelectrons with hard x-rays have sufficiently longer mean free path (about 20 nm) than that with the conventional characteristic x-ray photons (a few nm). This feature with the HXPES makes it possible to analyze the depth profile of the chemical bonding states via variation of glancing angles for photoelectrons detection from ~ 90 degrees (surface normal) for analysis of deep regions (~ 20 nm) down to about 10 degrees for analysis of shallow regions (a few nm) without degradation of the bonding states due to ion milling as in the conventional XPS analysis.

4.2. Results and discussion

First, a typical mass spectrum of the ionic species measured with the mass-separated ion-energy analyzer is shown in Fig. 6 for the argon-oxygen mixture plasma sustained at a partial pressure ratio of 3% ($R = 3\%$). The mass spectrum shows that the major ionic species in the argon-oxygen mixture plasma are Ar^+, O_2^+ and O^+ ions, indicating that reactivity of the plasma is enhanced via electron-impact ionization, excitation and dissociation of oxygen molecules in the discharge.

Figure 7 shows variations of kinetic energy distribution of atomic oxygen ions (O^+) measured in the argon-oxygen mixture plasmas sustained at an RF power of 1000 W as a parameter of oxygen partial pressure ratio R. The ion energy distributions observed in the present experiments exhibit considerably suppressed ion energies as small as or less than 10 eV, which is considered to be mainly due to low-voltage operation of ICPs via low inductance antenna units. With decreasing R, the peak ion energy of the energy spectrum is found to decrease slightly from 9.8 eV at $R = 100\ \%$ to 8.1 eV at $R = 3\%$. These results suggest that the ICPs proposed in the present investigation are promising for a variety of polymer processes, which require considerable low-damage processing.

Furthermore, potential drop from the plasma potential to the floating potential was evaluated using Langmuir probe measurements. The potential drops measured by subtracting the floating potential from the plasma potential were as low as 6 V, which is slightly smaller than the ion kinetic energy distribution shown in Fig. 7 since the floating potential was measured to be a few volts. Considering that the potential drop is almost equivalent to the energy of ion bombardment during the film growth with plasma CVD on non-conductive substrate such as glass and polymers, the results of ion-energy distribution and potential-drop measurements demonstrate that ICPs sustained with LIA units are attractive as plasma sources for polymer processes, which require low-damage processing with markedly reduced plasma damage by ion bombardment.

As a demonstration for plasma modification of polymers, the argon-oxygen mixture plasma sustained at an oxygen partial pressure ratio R = 3% was exposed to the surface of polytetrafluoroethylene (PTFE) substrate for 10 min and the surface chemical bonding states were analyzed via HXPES. Figure 8 shows C1s HXPES spectra measured for the PTFE sample after exposure to the argon-oxygen mixture plasma with photoelectron detection angle as a parameter. The HXPES spectra indicate that the CF_2 bond (F-C-F) or main chemical bond of the PTFE is maintained from deeper region of about 20 nm to nano-surface layer, while the CO bond increases with decreasing detection angle of photo electrons. The results indicate that nano-surface modification (CO bond formation) of PTFE surface is performed without suffering degradation of major molecular structures of PTFE beneath the surface nano-layer possibly due to low-damage plasma processing with the present plasma source, in which the energy of ion bombardment onto the PTFE surface was measured to be as small as 6 eV.

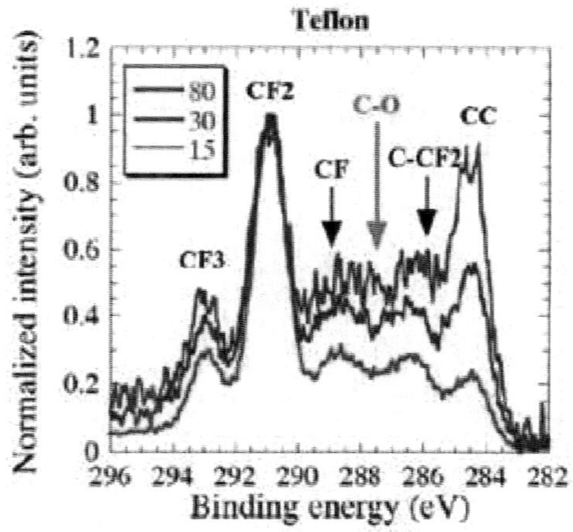

Figure 8. HXPES C1s spectra for PTFE sample exposed to argon-oxygen plasma.

5. Summary

Plasma generation and control technologies for fabrication of hybrid flexible devices have been developed with multiple low-inductance antenna (LIA) modules. Feasibility of ultra-large-area and uniform plasma sources was demonstrated via numerical prediction with plasma-fluid simulation; indicating that uniform plasma profile over 200-cm rectangular area could be designed with a non-uniformity of 5.5% in the plasma source with 3m x 3m scales to meet requirements in the next-generation processes. Ion energy distributions of atomic oxygen ions in argon-oxygen mixture plasmas showed considerable suppression of ion energies as small as or less than 10 eV. The HXPES measurements of PTFE surface after exposure to the plasma exhibited nano-surface modification of polymer surface without suffering degradation of molecular structures beneath the nano-surface layer.

Acknowledgments

This work was supported partly by The Global COE Program "Center of Excellence for Advanced Structural and Functional Materials Design" from the Ministry of Education, Culture, Sports, Science and Technology (MEXT), Japan. The HXPES measurements were carried out at the BL47XU in the SPring8 with the approval of the Japan Synchrotron Radiation Research Institute (JASRI) (Proposal No. 2007B0005 and 2008A0005).

References

[1] Setsuhara Y, Shoji T, Ebe A, Baba S, Yamamoto N, Takahashi K, Ono K. and Miyake S 2003 Surf. Coatings. Tehcnol. **174-175** 33
[2] Setsuhara Y, Takenaka K, Ebe A and Nishisaka K 2007 Plasma Process. Polym. **4** S628
[3] Perrin J, Schmitt J, Hollenstein C, Howling A and Sansonnens L 2000 Plasma Phys. Control. Fusion **42** B353
[4] Wu Y and Lieberman M A 2000 Plasma Sources Sci. Technol. **9** 210
[5] Khater M H and Overzet L J 2000 Plasma Sources Sci. Technol. **9** 545
[6] Naddaf M, Balasubramanian C, Alegaonkar P S, Bhoraskar V N, Mandle A B, Ganeshan V, Bhoraskar S V 2004 Nucl. Instr. and Meth. in Phys. Res. B **222** 135.
[7] Kim S H, Na S W, Lee N E, Nam Y W, Kim Y H 2005 Surf. Coat. Technol. **200** 2072.

Journal of Physics: Conference Series **165** (2009) 012043 doi:10.1088/1742-6596/165/1/012043

Effect of hydrogen on stability of vacancy in $LaNi_5$

Masataka Mizuno [1, 2], Hideki Araki [1, 2] and Yasuharu Shirai [3]

[1] Center for Atomic and Molecular Technologies, Graduate School of Engineering, Osaka University, 2-1 Yamadaoka, Suita, Osaka 565-0871, Japan
[2] Division of Materials and Manufacturing Science, Graduate School of Engineering, Osaka University, 2-1 Yamadaoka, Suita, Osaka 565-0871, Japan
[3] Department of Materials Science and Engineering, Graduate School of Engineering, Kyoto University, Sakyo, Kyoto 606-8501, Japan

mizuno@mat.eng.osaka-u.ac.jp

Abstract. First-principles electronic structure calculations have been performed in order to investigate the effect of hydrogen on the stability of Ni vacancy after hydrogen desorption in $LaNi_5$. Our calculated binding energy of hydrogen to the Ni vacancy indicates that up to two H atoms are trapped at the Ni vacancy. The formation energy of the Ni vacancy trapping H atoms was found to decrease with the number of H atoms and to turn negative when two H atoms are trapped. These results suggest that the Ni vacancy is energetically stabilized by trapping two H atoms in hydrogen desorption process.

1. Introduction

The intermetallic compound $LaNi_5$ possesses exceptional properties that are suitable for hydrogen storage materials. The hydrogen absorption process accompanies large volume expansions, which causes the formation of lattice defects. Our positron lifetime measurement revealed that not only dislocations but also vacancies and vacancy clusters are formed by the initial hydrogen absorption process in $LaNi_5$ and excess vacancies are still contained in the specimen after hydrogen desorption [1]. Furthermore, we investigated the recovery process of the excess vacancies to find that vacancy migration and annihilation are observed around 423 – 673 K [2]. This temperature range corresponds to the release temperature of residual hydrogen in the specimen after hydrogen desorption in $LaNi_5$ [2]. Residual hydrogen in $LaNi_5$ is, therefore, most likely trapped by vacancies induced by hydrogen absorption-desorption process. Calculated positron lifetimes of $LaNi_5$ indicate that vacancies formed by hydrogen absorption-desorption process are composed of Ni vacancies [3,4]. However, calculated formation energy of the Ni vacancy in $LaNi_5$ is about 1.0 eV [5], which indicates that the concentration of vacancies is 10^{-17} at ambient temperature. These results suggest that H atoms trapped at the Ni vacancy energetically stabilize the Ni vacancy.

In the present work, we examine the effect of H atoms on the stability of the Ni vacancy after hydrogen desorption in $LaNi_5$ using first-principles electronic structure calculations. The heats of solution of hydrogen both to interstitial sites and to the Ni vacancy are calculated in order to evaluate binding energies of hydrogen to the Ni vacancy. We also calculate formation energies of the Ni vacancy taking into account the H atoms trapped at the Ni vacancy.

2. Computational method

We employed the Vienna *ab initio* simulation package (VASP) [6,7] with generalized gradient approximation proposed by Perdew and Wang [8]. Potential based upon the all-electron projector augmented wave (PAW) method were used [9,10]. The intermetallic compound $LaNi_5$ crystallizes in the hexagonal $CaCu_5$ type structure, which belongs to the space group *P*6/mmm, with lattice parameters of a = 5.0228 Å and c = 3.9826 Å [11] as shown in figure 1. First, we calculated the equilibrium lattice parameters of $LaNi_5$ using the kinetic energy cutoff of 350 eV and a 15×15×19 *k*-mesh in the Monkhorst-Pack scheme with the unit cell including six atoms. We obtained the equilibrium lattice parameter of a = 5.0107 Å and c = 3.9690 Å, which well reproduces the experimental values. The structure of $LaNi_5$ consists of stacking of two types of layers: one is composed of La atoms at the 1*a* sites and Ni atoms at the 2*c* sites, the other is composed of Ni atoms on the 3*g* sites.

Our results using positron lifetime measurements and theoretical calculations ruled out the possibility of the formation of La vacancies during hydrogen absorption [3,4]. In addition, in our previous work, it was found that the formation energy of the Ni vacancy at the 2*c* site is about 0.8 eV lower that that at the 3*g* site [5]. The Ni vacancy at the 2*c* is, therefore, considered in the present work.

In the case of intermetallic compounds, vacancy formation process accompanies formation of other types of defects or precipitation of other phases in order to maintain the chemical composition. The formation of Ni vacancies in $LaNi_5$ proceeds via the reaction [5]:

$$0 \rightarrow V_{\mathrm{Ni}} + \frac{1}{7}(\mathrm{Ni}_{dumbbell} + \mathrm{LaNi}_5)\,. \tag{1}$$

Two Ni atoms removed as vacancies occupy one La site as a Ni-Ni dumbbell along the *c*-axis. One La atom removed for the formation of a Ni-Ni dumbbell requires extra La + 5Ni sites. One Ni-Ni dumbbell, therefore, accommodates seven Ni atoms. On the basis of this process, the formation energy of Ni vacancy can be calculated as

$$\begin{aligned}\tilde{E}^{\mathrm{Ni}}_{vac} &= \left[E^{\mathrm{LaNi}_5}_{tot}(N_{\mathrm{La}}, N_{\mathrm{Ni}}-1) - E^{\mathrm{LaNi}_5}_{tot}(N_{\mathrm{La}}, N_{\mathrm{Ni}})\right] \\ &+ \frac{1}{7}\left[E^{\mathrm{LaNi}_5}_{tot}(N_{\mathrm{La}}-1, N_{\mathrm{Ni}}+2) - E^{\mathrm{LaNi}_5}_{tot}(N_{\mathrm{La}}, N_{\mathrm{Ni}})\right] + \frac{1}{7}E^{\mathrm{LaNi}_5}_{tot}(1,5)\end{aligned} \tag{2}$$

The first term is the energy difference between the supercell including a Ni vacancy and the perfect supercell. The second term is the energy difference between the supercell including a Ni dumbbell and the perfect supercell. The last term is the energy of additional lattice site. In the hydrogen desorption process, Ni vacancies traps hydrogen atoms. The formation of Ni vacancy, therefore, proceeds by the reaction:

$$\frac{n}{2}\mathrm{H}_2 \rightarrow (V_{\mathrm{Ni}} + n\mathrm{H}) + \frac{1}{7}(\mathrm{Ni}_{dumbbell} + \mathrm{LaNi}_5)\,. \tag{3}$$

The formation energy of Ni vacancy trapping hydrogen atom can be calculated as

$$\begin{aligned}\tilde{E}^{\mathrm{Ni}}_{vac} &= \left[E^{\mathrm{LaNi}_5}_{tot}(N_{\mathrm{La}}, N_{\mathrm{Ni}}-1, N_{\mathrm{H}}) - E^{\mathrm{LaNi}_5}_{tot}(N_{\mathrm{La}}, N_{\mathrm{Ni}})\right] \\ &+ \frac{1}{7}\left[E^{\mathrm{LaNi}_5}_{tot}(N_{\mathrm{La}}-1, N_{\mathrm{Ni}}+2) - E^{\mathrm{LaNi}_5}_{tot}(N_{\mathrm{La}}, N_{\mathrm{Ni}})\right] + \frac{1}{7}E^{\mathrm{LaNi}_5}_{tot}(1,5) - \frac{N_{\mathrm{H}}}{2}E^{\mathrm{H}_2}_{tot}\,.\end{aligned} \tag{4}$$

Here $E(N_{\text{La}}, N_{\text{Ni}} - 1, N_{\text{H}})$ denotes the energy of a supercell with $N_{\text{La}} + N_{\text{Ni}} - 1$ atoms including one Ni vacancy trapping N_{H} H atoms. The last term is the total energy of hydrogen molecule.

The heat of solution of H atom to a Ni vacancy or to an interstitial site is calculated as the energy difference from the energy of H atom in hydrogen molecule:

$$\text{Vacancy:}\quad E_{Sol}^{\text{H}} = \frac{1}{N_{\text{H}}}\left\{E_{tot}^{\text{LaNi}_5}(N_{\text{La}}, N_{\text{Ni}} - 1, N_{\text{H}}) - \left[E_{tot}^{\text{LaNi}_5}(N_{\text{La}}, N_{\text{Ni}} - 1) + \frac{N_{\text{H}}}{2}E_{tot}^{\text{H}_2}\right]\right\}, \qquad (5)$$

$$\text{Interstitial:}\quad E_{Sol}^{\text{H}} = E_{tot}^{\text{LaNi}_5}(N_{\text{La}}, N_{\text{Ni}}, \text{H}_{int}) - \left[E_{tot}^{\text{LaNi}_5}(N_{\text{La}}, N_{\text{Ni}}) + \frac{1}{2}E_{tot}^{\text{H}_2}\right]. \qquad (6)$$

Here $E(N_{\text{La}}, N_{\text{Ni}}, \text{H}_{int})$ denotes the energy of a supercell with $N_{\text{La}} + N_{\text{Ni}}$ atoms including one H atom at an interstitial site.

The binding energy of N_{H} H atoms to a Ni vacancy is defined as the energy gain when the Ni vacancy trapping N_{H} *-1* H atoms traps one H atom from an interstitial site:

$$\begin{aligned} E_b^{\text{H}} &= \left[E_{tot}^{\text{LaNi}_5}(N_{\text{La}}, N_{\text{Ni}} - 1, N_{\text{H}}) + E_{tot}^{\text{LaNi}_5}(N_{\text{La}}, N_{\text{Ni}})\right] \\ &\quad - \left[E_{tot}^{\text{LaNi}_5}(N_{\text{La}}, N_{\text{Ni}} - 1, N_{\text{H}} - 1) + E_{tot}^{\text{LaNi}_5}(N_{\text{La}}, N_{\text{Ni}}, \text{H}_{int})\right] \end{aligned} \qquad (7)$$

In the present work, we employed the 2x2x2 supercells composed of 8 La and 40 Ni lattice sites for calculating each term of the above equations. Whereas relaxations of the atomic positions and lattice parameters were allowed for the supercells including a Ni vacancy and a Ni dumbbell, the lattice parameters were fixed for the supercells including a Ni vacancy trapping H atoms.

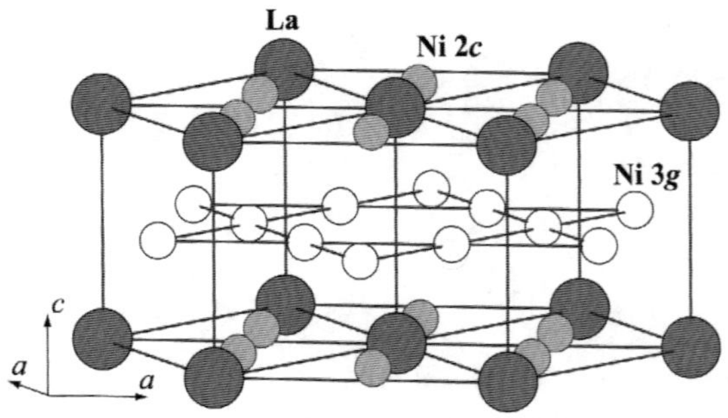

Figure 1. Crystal structure of $LaNi_5$.

3. Results and discussions

3.1. Stability of hydrogen at interstitial sites

Figure 2 shows local structures around interstitial sites in $LaNi_5$. The $3f$ and $12n$ sites are O-sites surrounded by four Ni atoms and two La atoms. The $4h$, $12o$ and $6m$ sites are T-sites surrounded by four Ni atoms, three Ni atoms and one La atom, and two Ni atoms and two La atoms, respectively. The calculated heats of solution of H atom to these interstitial sites are listed in table 1. The most stable site is the $12n$ site that is the off-center O-site surrounded by four Ni atoms and two La atoms. This result agrees with the previous work [12]. We employ the heat of solution of hydrogen at the $12n$ site as the energy of a H atom at the interstitial site in $LaNi_5$ for the calculation of binding energy of hydrogen to a Ni vacancy.

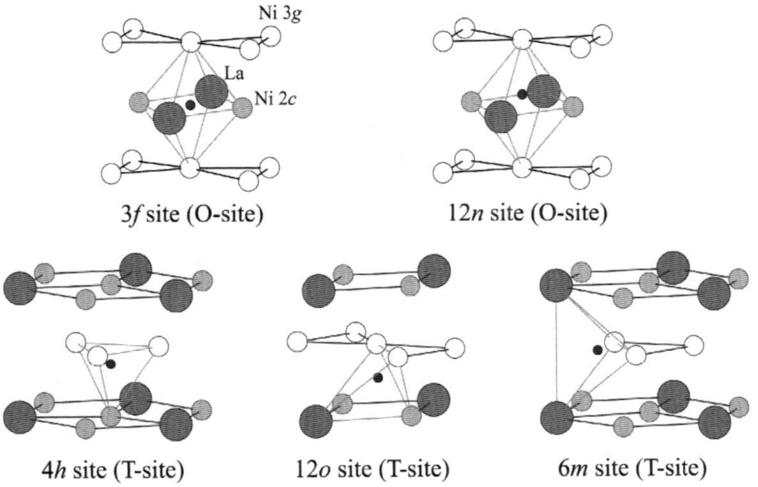

Figure 2. Interstitial sites in $LaNi_5$.

Table 1. Heats of solution of interstitial sites in $LaNi_5$.

Site	Type	Heats of solution (eV)
3*f*	O-site	−0.156
12*n*	O-site	−0.162
4*h*	T-site	0.003
12*o*	T-site	−0.004
6*m*	T-site	0.089

3.2. Stability of hydrogen at Ni vacancy

There are no experimental data reported for the position of H atoms trapped at a Ni vacancy in $LaNi_5$. We, therefore, need to determine the initial position of H atoms by considering local structures around a Ni vacancy at the 2*c* site. The Ni vacancy at the 2*c* site are surrounded by three Ni atoms at the 3*g* sites in the upper layer, three Ni atoms at the 2*c* sites and three La atoms in the same layer, and three Ni atoms at the 3*g* site in the lower layer. The H atoms in $LaNi_5$ hydride are mainly bonded to Ni atoms. Considering the arrangement of the Ni atoms around the Ni vacancy and the position of H atoms in $LaNi_5$ hydride, there are two possible positions for one H atom at the Ni vacancy: one is bonded to three Ni atoms at the 3*g* sites and the other is a bridging site between Ni 3*g* and Ni 2*c* atoms. We calculated the heat of solution of one H atom for both configurations and obtained the heats of solution of -0.62 eV and -0.24 eV for the H atom bonded to three Ni 3*g* and the H atom on the bridging site, respectively. This result suggests that one or two H atoms are trapped near the three Ni 3*g* atoms and further H atoms occupy the bridging sites as shown in figure 3. According to this rule we configured the initial positions of up to five H atoms at the Ni 2*c* vacancy and calculated the heats of solution and binding energies.

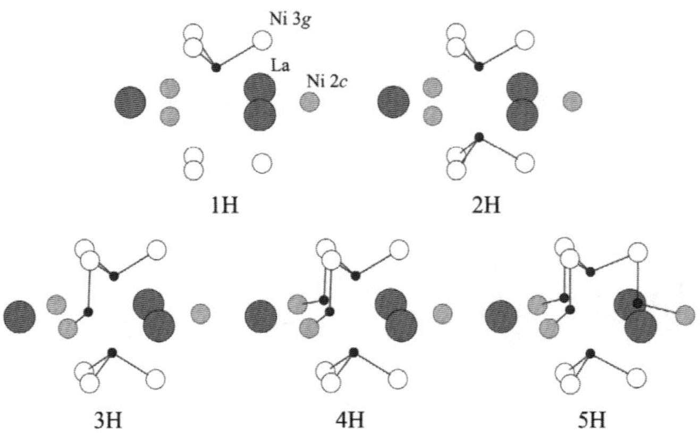

Figure 3. Positions of hydrogen atoms trapped at a Ni vacancy on the 2*c* site.

Figure 4 (a) shows the calculated heats of solution to a Ni vacancy at the 2*c* site as a function of the number of trapped H atoms. The heats of solution of *V*-H and *V*-H_2 are about -0.6 eV that is about 0.5 eV lower than that of H at the interstitial site. For *V*-H_3, *V*-H_4 and *V*-H_5, the heat of solution increases with the number of H atoms, but still lower than that of H at the interstitial site. This is because the high-energy sites, which are the bridging sites between Ni 2*c* and Ni 3*g* atoms, are occupied *V*-$H_{n>2}$. The calculated binding energies of H atoms to a Ni vacancy at the 2*c* site are plotted in figure 4 (b). Negative (positive) values indicate energy gain (penalty). Whereas *V*-H and *V*-H_2 complexes gain energy by trapping H atoms, *V*-$H_{n>2}$ complexes lose energy because of the increase in the heat of solution shown in figure 4 (a). This result clearly indicates that the number of H atoms trapped at the Ni vacancy is up to two atoms.

3.3. Effect of hydrogen on the stability of Ni vacancy.

Figure 5 shows the calculated formation energies of the Ni vacancy at the 2*c* site as a function of the number of trapped H atoms. There is a steep drop for *V*-H and *V*-H_2 complexes and the formation energy of the Ni vacancy turns negative at *V*-H_2 that is the most stable vacancy-hydrogen complex as shown in figure 4. The Ni vacancy at the 2*c* site is, therefore, energetically stabilized by trapping two H atoms in the hydrogen desorption process.

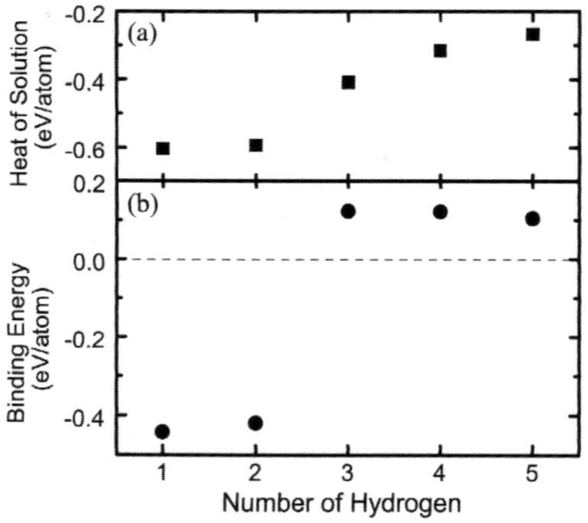

Figure 4. Calculated solution energies (a) and binding energies (b) to a Ni vacancy at the 2*c* site as a function of the number of trapped H atoms.

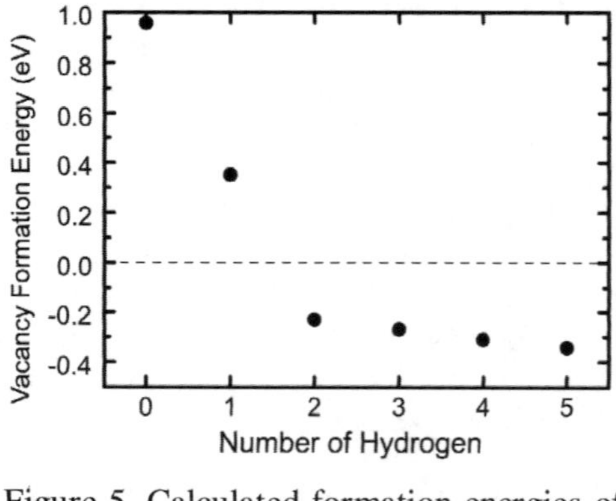

Figure 5. Calculated formation energies of a Ni vacancy trapping H atoms at the 2*c* site as a function of the number of trapped H atoms.

Our recent study using positron lifetime measurement indicates that the formation of vacancies during hydrogen absorption-desorption cycling is reduced by the addition of Sn [13]. The Sn atom seems to weaken the stabilizing effect of hydrogen on the Ni vacancy. We will address this problem in the near future.

4. Conclusions

We have investigated the effect of H atoms on the stability of the Ni vacancy in $LaNi_5$ using first-principles calculations. Up to two H atoms occupy the low-energy sites and further H atoms occupy the high-energy sites at the Ni vacancy. The heat of solution of hydrogen to the Ni vacancy begins to increase when more than two H atoms are trapped. Therefore, the binding energy of H atoms to the Ni vacancy is negative when one or two H atoms are trapped at the Ni vacancy. On the other hand, the formation energy of the Ni vacancy trapping H atoms decreases with the number of trapped H atoms and turns to negative by trapping two H atoms. These results suggest that the Ni vacancy is energetically stabilized by trapping two H atoms after hydrogen desorption in $LaNi_5$.

Acknowledgements

We would like to thank Dr. Koji Sakaki (AIST) for many interesting discussions. This work supported by 'Priority Assistance of the Formation of Worldwide Renowned Centers of Research – The Global COE Program (Project: Center of Excellence for Advanced Structural and Functional Materials Design)' from the Ministry of Education, Culture, Sports, Science and Technology of Japan, and in part by a Grant-in-Aid for Scientific Research from Ministry of Education, Culture, Sports, Science and Technology of Japan.

References

[1] Shirai Y, Araki H, Mori T, Nakamura W and Sakaki K 2002 *J. Alloys Compd.* **330-332** 125
[2] Sakaki K, Araki H and Shirai Y 2002 *Mater. Trans.* **43** 1494
[3] Mizuno M, Sakaki K, Araki H and Shirai Y 2003 *J. Alloys Compd.* **356-357** 186
[4] Mizuno M, Tonomori R, Sakaki K, Araki H and Shirai Y 2004 *Solid State Ionics* **172** 149
[5] Mizuno M, Araki H and Shirai Y 2008 *J. Phys.: Condens. Matter* **20** 275232
[6] Kresse G, and Furthmuller J 1996 *Comput. Mater. Sci.* **6** 15
[7] Kresse G and Furthmuller J 1996 *Phys. Rev. B* **54** 11169
[8] Perdew J P and Wang Y 1992 *Phys. Rev. B* **45** 13244
[9] Kresse G and Joubert D 1999 *Phys. Rev. B* **59** 1758
[10] Blöchl, P E 1994 *Phys. Rev. B* **50** 17953
[11] Thompson P, Reilly J J and Hastings J M *1987 J. Less-Common Met.* **129** 105
[12] Tatsumi K, Tanaka I, Inui H, Tanaka K, Yamaguchi M and Adachi H 2001 *Phys. Rev.* B **64** 184105
[13] Araki H, Date R, Sakaki K, Mizuno M and Shirai Y 2007 *Phys. Stat. Sol.(c)* **4** 3510

Frozen-in vacancies in PVD-Cu films with improved high-pressure reflowability studied using a slow positron beam

A Yabuuchi[1], D Kubo[1], M Mizuno[1,2], H Araki[1,2], T Onishi[3] and Y Shirai[4]

[1]Division of Materials and Manufacturing Science, Graduate School of Engineering, Osaka University, 2-1 Yamadaoka, Suita, Osaka 565-0871, Japan

[2]Center for Atomic and Molecular Technologies, Graduate School of Engineering, Osaka University, 2-1 Yamadaoka, Suita, Osaka 565-0871, Japan

[3]Materials Research Laboratory, Kobe Steel Ltd., 5-5 Takatsukadai 1-chome, Nishi-ku, Kobe, Hyogo 651-2271, Japan

[4]Department of Materials Science and Engineering, Graduate School of Engineering, Kyoto University, Yoshida-Honmachi, Sakyo-ku, Kyoto 606-8501, Japan

atsushi.yabuuchi@mat.eng.osaka-u.ac.jp

Abstract. Recently, a new process has been proposed for fabricating a LSI interconnection; filling trenches and via holes with Cu using high-pressure annealing treatment. It is already known that a Cu film produced by physical vapor deposition (PVD) has a lower reflowability compared to a Cu film produced by electrochemical deposition (ECD). Additionally, it has also been recognized that the addition of Sb to the PVD-Cu film improves the reflowability. However, the factors responsible for the reflowability of Cu films have not yet been studied. In this work, we evaluated a PVD pure-Cu film and a PVD Cu-0.5at%Sb film by using a slow positron beam. Addition of Sb led to the introduction of lattice defects in the as-deposited film. These defects that were observed in the PVD-CuSb dilute alloy film were identified as frozen-in vacancies that were produced during deposition.

1. Introduction

Copper is currently used extensively as large-scale integration (LSI) interconnections because of its lower electrical resistivity and high resistance to electromigration damage. Cu interconnects are fabricated by embedding Cu into trenches and via holes by electrochemical deposition (ECD) and this is known as the Dual-Damascene process. However, the accelerated reduction of the feature size in LSI devices is making it difficult to completely fill high aspect-ratio structures with Cu. The international technology roadmap for semiconductors (ITRS) 2007 requires that a Metal 1 interconnect layer with a 64 nm wiring pitch will be in manufacturable by 2013. Therefore, a post-deposition high-pressure annealing process is proposed for the fabrication of Cu interconnections. It has been recognized that a physical vapor deposition (PVD) Cu film has low reflow performance compared to an ECD Cu film [1] despite Cu reflowability being an important characteristic for this process. It has been also found that the addition of Sb to the PVD-Cu film improves its reflowability.

A positron beam may be used as a unique probe to study lattice defects, in particular vacancy-type defects. When positrons are implanted into solids, they are annihilated by electrons and emit two 511 keV γ-rays. In perfect crystals, positrons delocalize in interstitial sites and annihilate with electrons there since positrons are strongly repelled by ion cores because of a Coulomb repulsion. In vacancies, vacancy clusters or dislocations, which have a lower density of ion cores than their surroundings, positrons are trapped in the defects and annihilate with electrons in the defects. These defects are detected by a decrease in the Doppler broadening of the annihilation radiation because of the absence of core electrons which have a higher momentum distribution than valence electrons. Generally, the resultant changes in Doppler broadening spectra are characterized by the S parameter which reflects the ratio change of annihilation with low momentum valence electrons. The value of S for the annihilation of positrons trapped by vacancies is larger than that for positrons annihilated from the interstitial sites. Recent positron annihilation spectroscopy studies have suggested that ECD-Cu films have a larger amount of vacancy clusters than PVD-Cu films [2-4]. In this paper, we investigate lattice defects in a PVD-CuSb dilute alloy film compared to a PVD-pure-Cu film by measuring the Doppler broadening of the annihilation radiation using a slow positron beam.

2. Experimental

Samples used in our experiments consisted of Cu (1 μm)/TaN (50 nm)/Si or Cu-0.5at%Sb (1 μm)/TaN (50 nm)/Si. TaN films were deposited on Si substrates by the DC magnetron sputtering technique. After deposition a 1-μm pure-Cu film or a 1-μm Cu-0.5at%Sb dilute alloy film was sputtered onto TaN/Si at room temperature. Sputtering conditions were as follows: the distance between the Cu target and the substrate was 55 mm, the Ar pressure was 0.25 Pa and the discharge power was 260 W. For comparison, a bulk-Cu (purity: 4*N*) sample was also characterized. This sample was annealed at 1073 K for 2 h under an Ar atmosphere.

These as-deposited PVD-Cu films and the annealed bulk-Cu sample were irradiated with a slow positron beam to measure the Doppler broadening spectra of the annihilation radiation as a function of the incident positron energy E. For each value of E, a spectrum of 5×10^5 total counts was measured. The S parameter was defined as the number of annihilation events over an energy range of 511 ± 0.76 keV, which reflects the low momentum part of the spectrum. Positrons were implanted into samples at energies in the range of 0.025 – 30 keV. The positron lifetime in the bulk-Cu sample was also measured at room temperature using a conventional positron annihilation lifetime spectrometer with a time resolution of 182 ps (FWHM). Approximately 1×10^6 total counts were accumulated for each lifetime spectrum.

Furthermore, to observe the annealing behavior of lattice defects formed during deposition, isochronal annealing in steps of 15 min/50 K was performed. A monoenergetic positron beam with an energy of 15 keV was then used to irradiate each sample and Doppler broadening spectra of the annihilation radiation with 2×10^6 total counts were acquired for each annealing temperature. All of the measurements were carried out under ultra-high vacuum (UHV) conditions at room temperature.

3. Results and discussion

3.1. The S parameter as a function of the incident positron energy E

S parameters of as-deposited pure-Cu and Cu-Sb dilute alloy films are shown in Figure 1 as a function of incident positron energy E. The result for the annealed bulk-Cu sample is also shown. At low E (≈ 0 keV), higher S values were observed for all samples. This is caused by positronium formation at the surface of Cu [5-6]. Pure-Cu and Cu-Sb dilute alloy films had S values at $E > 20$ keV that tended to increase with increasing E. This was due to partial implantation of positrons into Si substrates in this energy range. For the annealed bulk-Cu sample, the S value saturated in the range of $E = 25 – 30$ keV. The positron lifetime for this sample was 111 ps. This value agreed with the reported lifetime of positrons that are delocalized and annihilated in interstitial sites [3]. This indicates that the S value of the

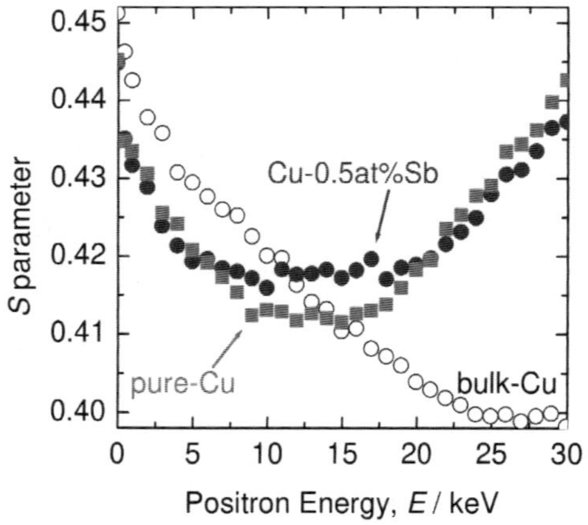

Figure 1. The S parameter as a function of incident positron energy E for Cu(1 μm) or Cu-Sb(1 μm)/TaN(50 nm)/Si samples and annealed bulk Cu(4N). Error bars are roughly the same size as the symbols.

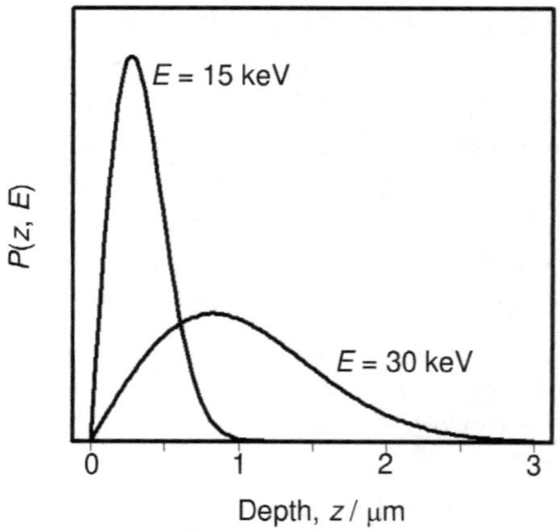

Figure 2. Positron implantation profiles $P(z, E)$ in copper calculated for two incident positron energies, E = 15 or 30 keV.

annealed bulk-Cu sample, observed at E = 25 – 30 keV, was the characteristic value of S for a Cu perfect crystal.

For the pure-Cu film, values of S were almost constant in the range of E = 9 – 17 keV. This means that the S value in this energy range corresponds to the annihilation of positrons in the 1 μm pure-Cu film. A previous study on conventional positron annihilation lifetime spectroscopy indicated that almost all positrons which were implanted into a PVD-pure-Cu film were trapped at grain boundaries and annihilated there because of submicron grain sizes [3]. This was the reason for the higher value of S for a pure-Cu film compared to the value of a Cu perfect crystal. Moreover, the value of S for a Cu-Sb dilute alloy film was higher than the value for a pure-Cu film in the same range of E = 9 – 17 keV. This indicates that the Cu-Sb alloy film contains large amounts of defects compared to a pure-Cu film.

3.2. Change in the S value during isochronal annealing

Figure 2 shows positron implantation profiles $P(z, E)$ in copper calculated for two incident positron energies, E = 15 or 30 keV. The calculation formula is described elsewhere [7-8]. This calculation result showed that almost all positrons were annihilated in the Cu film when positrons were implanted with energy of 15 keV. To put it another way, we do not need to consider the effects of interfaces or substrates. For this reason, we measured the S value at E = 15 keV for each annealing temperature.

Figure 3 shows the change in S value for pure-Cu and Cu-Sb dilute alloy films before and during isochronal annealing. The most prominent feature of these results is

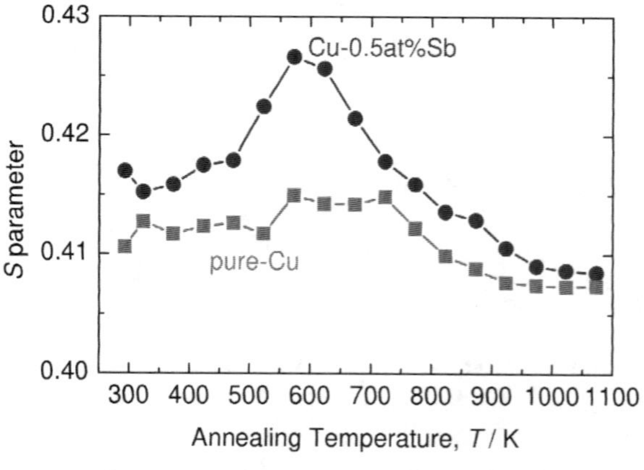

Figure 3. The change in S value for pure-Cu and Cu-Sb dilute alloy films before and during isochronal annealing. Error bars are roughly the same size as the symbols.

that the S value of the Cu-Sb alloy film increases significantly at around 573 K and then decreases. In general, lattice defects monotonically decrease as annealing temperature increases. Nevertheless, a significant increase of the *S* value for the Cu-Sb alloy film was observed. This can be explained by the migration and clustering of excess vacancies. The Cu-Sb film thus contains a large amount of vacancies compared to a pure-Cu film. It is well known that the binding energy between a Sb atom and a Cu mono-vacancy has a relatively high value (0.32 eV experimental value [9], 0.39 eV calculated value [3]). Therefore, we can interpret this drastic change in the *S* value for the Cu-Sb dilute alloy film in the following way: vacancies were introduced during deposition by the high binding energy between a Sb and Cu mono-vacancy. In isochronal annealing experiments, the introduced excess vacancies started to dissociate from Sb atoms at around 573 K and aggregated into vacancy clusters which led to an increase the *S* value. Further annealing at 623 K or at higher temperature resulted in a decrease of vacancy clusters.

4. Summary

Doppler broadening spectra of annihilation radiation have been measured for PVD-pure-Cu and PVD-Cu-0.5at%Sb dilute alloy films to determine the behavior of lattice defects. Defects observed in the Cu-Sb film were identified as frozen-in vacancies that were formed during deposition and which were introduced by the high binding energy between a Sb atom and a Cu mono-vacancy. We conclude that these excess frozen-in vacancies in the PVD-CuSb dilute alloy film causes good reflowability in the post-deposition high-pressure annealing process.

Acknowledgement

This study was supported by Priority Assistance for the Formation of Worldwide Renowned Centers of Research – The Global COE Program (Project: Center of Excellence for Advanced Structural and Functional Materials Design) from the Ministry of Education, Culture, Sports, Science and Technology (MEXT), Japan.

References

[1] Onishi T, Fujii H, Yoshikawa T, Munemasa J, Inoue T and Miyagaki A 2003 Thin Solid Films **425** 265
[2] Uedono A, Suzuki T and Nakamura T 2004 J. Appl. Phys. **95** 913
[3] Mizuno M, Kihara T, Araki H, Shirai Y and Onishi T 2007 phys. stat. sol. (c) **4** 3350
[4] Uedono A, Mori K, Ito K, Imamizu K, Hachiya T, Kamijo H, Hasunuma M, Kaneko H, Toyoda H, Ohdaira T and Suzuki R 2007 Jpn. J. Appl. Phys. **46** L483
[5] Mills A P Jr. 1978 Phys. Rev. Lett. **41** 1828
[6] Schultz P J and Lynn K G 1988 Rev. Mod. Phys. **60** 701
[7] Vehanen A, Saarinen K, Hautojärvi P and Huomo H 1987 Phys. Rev. B **35** 4606
[8] Asoka-Kumar P, Lynn K G and Welch D O 1994 J. Appl. Phys. **76** 4935
[9] Bosse H, Sager A, Lühr-Tanck W and Hehenkamp T 1986 J. Phys. F: Met. Phys. **16** 1337

International Conference on Advanced Structural and Functional Materials Design 2008 IOP Publishing
Journal of Physics: Conference Series **165** (2009) 012045 doi:10.1088/1742-6596/165/1/012045

Structure analysis of complex metallic alloys in the Eu-Ag-In system by X-ray diffraction

Cesar Pay Gómez*[1], Yoshiki Morita[2], Akiji Yamamoto[3] and An Pang Tsai[2,4]

[1]International Center for Young Scientists, International Center for Materials NanoArchitectonics (ICYS-MANA), National Institute for Materials Science (NIMS) 1-1 Namiki, Tsukuba, Ibaraki, 305-0044, JAPAN
[2]Institute of Multidisciplinary Research for Advanced Materials (IMRAM), 2-1-1 Katahira, Tohoku University, Aoba-ku, Sendai 980-8577, JAPAN
[3]National Institute for Materials Science (NIMS) 1-1 Namiki, Tsukuba, Ibaraki, 305-0044, JAPAN
[4]National Institute for Materials Science (NIMS) 1-2-1 Sengen, Tsukuba, Ibaraki, 305-0047, JAPAN

E-mail: GOMEZ.Cesar@nims.go.jp

Abstract. In this work we investigate the complex atomic structures of two CMAs in the Eu-Ag-In system, these are quasicrystal approximants and their structures are similar to those of the 1/1 and 2/1 approximants found in the Yb-Cd system. The similarities between the Yb-Cd and Eu-Ag-In systems indicate that the RE-Ag-In systems in several ways can be treated as pseudo-binary. The investigated phases however also show some interesting differences; when Cd is replaced by the elements Ag and In partial chemical disorder is introduced into the system. As a response, the positional disorder which is usually observed in certain atomic positions for these types of approximants tends to disappear. The complete structures refined from single crystal data are described in detail and the differences with other related phases are highlighted. The 2/1 approximant in the Eu-Ag-In system is the first to exhibit such a degree of structural perfection in terms of positional order, a feature which makes this phase very suitable for further studies of chemical order and also can give insight into the local structures of related disordered phases. Electron diffraction studies on the 1/1 approximant phase have previously shown the presence of weak superlattice reflections, indicating a superstructure similar to that of the related phase Eu_4Cd_{25}. However, no superlattice reflections could be observed by single crystal diffraction and it is thus concluded that the phase has a disordered average structure with possible short to medium range superlattice ordering.

1. Introduction

Complex Metallic Alloys or CMAs for short are characterized by very large unit cells and the presence of well-defined atomic clusters. Quasicrystals are an intriguing subgroup of CMAs that possess long-range atomic order but no periodicity in 3-dimensional space. Conventional methods of structure determination based on X-ray diffraction techniques rely on the prerequisite of lattice periodicity, thus quasicrystals have represented a challenging problem for crystallographers from the very beginning.

Another group of CMAs which are directly related to quasicrystals are their so-called approximants. These are periodic materials in the traditional sense and are related to the corresponding quasicrystals in several ways; mainly they contain the same kind of atomic clusters but in a periodic arrangement. Approximant structures furthermore have strict mathematical relations to the parent quasicrystal, these will restrict the designation (1/1, 2/1 etc.), cell parameters and the symmetry of a specific approximant.[1] Approximants play an important role for understanding the atomic structures of related quasicrystals, since their structures can be determined by conventional X-ray diffraction techniques.

The phases investigated as part of this work are two different approximants in the Eu-Ag-In system. These approximants are related to the family of quasicrystals and approximants in the Yb-Cd system, where the first stable binary icosahedral quasicrystal was found.[2] The typical cluster structure of Yb-Cd type approximants and quasicrystals is displayed in **figure 1 a-e**. There are two known Yb-Cd type approximant structures; 1/1 $YbCd_6$ and 2/1 $Yb_{13}Cd_{76}$, they contain essentially identical atomic clusters but in two different periodic arrangements, as seen in **figure 2 a** and **b**.[3-5] Similar approximant structures are known to form in all the binary RE-Cd systems and also ternary systems by replacing Yb by other Rare Earth (RE) elements and Cd by pairs of other elements such as Ag-In, Au-Sn etc.[6, 7] A complete structure model that could be refined against experimental X-ray data was recently obtained for the binary Yb-Cd quasicrystal which serves as a prototype for all the related ternary phases.[8] However, the related ternary phases often suffer from chemical disorder as a consequence of their ternary nature. Although the title compounds naturally show many similarities with the parent phases in the Yb-Cd system there are also several important differences that will be discussed in detail. The initial strategy for synthesising RE-Ag-In approximants involved the replacement of Cd in the parent RE-Cd approximants by half amounts of Ag and In. In the RE-Cd systems 1/1approximants generally form with the stoichiometry $RECd_6$, thus the corresponding phases in the RE-Ag-In system were prepared with the stoichiometries $REAg_3In_3$. It has been noted experimentally however, that several ternary approximants have deviating compositions. For example, the ideal composition for the 1/1 approximant in the Yb-Ag-In system is $Yb_{14}Ag_{47}In_{39}$. This already implies that the phase does not form by substituting Cd by Ag and In atoms statistically distributed over the Cd sites in the structure. It would appear that certain atomic sites have a preference for either Ag or In.[9] This chemical order can however not be analysed using standard X-ray diffraction techniques due to the negligible difference in scattering power between Ag and In at conventional wavelengths.

In the RE-Cd systems there are two 1/1 approximants that have RE:Cd ratios deviating from the ideal 1:6 stoichiometry. These are the phases Ce_6Cd_{37} and Eu_4Cd_{25}.[10, 11] These two phases exhibit superlattice ordering due to the incorporation of interstitial Cd atoms into cube-shaped cavities in the structure. The insertion of these additional atoms causes the RE:Cd stoichiometry to deviate from the ideal 1:6 ratio and further causes the formation of superstructures by steric effects on the Cd_4 tetrahedron which is the innermost component of the atomic clusters in these phases (**figure 1 a**). The purpose of the present study was thus to investigate the effect that the replacement of Cd by Ag and In would have on the superstructure of the 1/1 Eu_4Cd_{25} approximant and to study the atomic structure of the 2/1 approximant in the Eu-Ag-In system. Other authors have observed superlattice ordering by TEM in the 1/1 Eu-Ag-In approximant that arises as a $2a\times2a\times2a$ doubling of the basic cubic cell, and has an F-centred lattice.[12] This is completely in accordance with the refined superstructure of the 1/1 Eu_4Cd_{25} approximant.

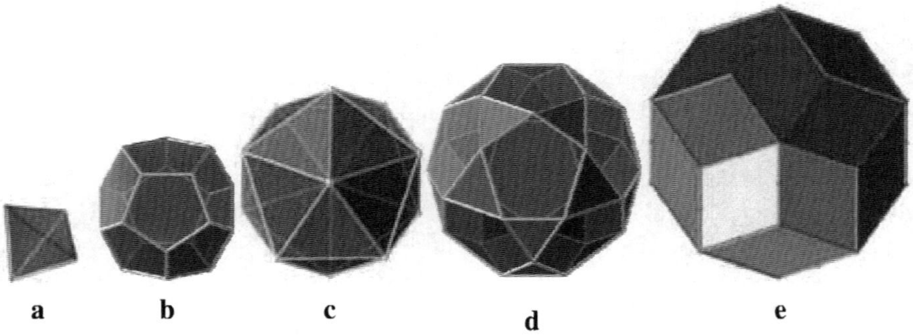

Figure 1. Atomic shell structure of the basic cluster in a typical Yb-Cd type 1/1 approximant. These atomic shells reside one inside the other. From left to right we see the inner disordered Cd tetrahedron (4 atoms, occurring in different orientations, **a**), a Cd dodecahedron (**b**), a Yb icosahedron (**c**), a Cd icosidodecahedron (**d**), and the outer Cd triacontahedron with atoms at all vertices and edge centers (**e**).

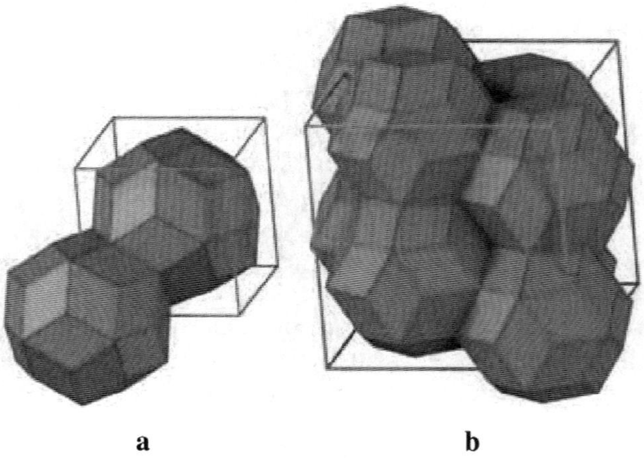

Figure 2. Cluster arrangements in typical Yb-Cd type 1/1 approximant **a**, and 2/1 approximant **b**.

2. Experimental

Reaction mixtures of high purity metals with the relative compositions $Eu_{16}Ag_{42}In_{42}$ for the 1/1 approximant phase and $Eu_{15}Ag_{42.5}In_{42.5}$ for the 2/1 approximant were mixed in stainless steel ampoules and sealed by arc-welding in argon atmosphere. The ampoules were then inserted into an electric muffle furnace and then kept at 1273K for 1h followed by annealing at 823K for 72 h and subsequent slow cooling to ambient temperature inside the furnace. Although the samples contained substantial amounts of the target phases, powder patterns could confirm the presence of unknown impurities in the samples. Other investigations show that the ideal Ag:In composition ratios generally deviate from 1:1 in several RE-Ag-In approximant phases, which tend to be richer in Ag.[6] This most likely accounts for the observed impurities in the reaction products. All reaction products were initially identified by powder X-ray diffraction (M03XHF, Mac Science). Single crystal data were collected on a Rigaku R-axis Rapid II single crystal diffractometer operated at 50 kV, 40 mA with MoKα radiation and graphite monochromator. Indexing and empirical absorption correction was performed using the included Rigaku software package. All Ag/In atoms were refined as Cd, since the difference in scattering power between these elements is negligible at the wavelength used for the data collection. The program JANA2006 was used for the structure refinements.[13] The structures were visualized using the programs VESTA 2.0.1 and Caligari Truespace 7.51.[14, 15].

3. Results and Discussion

Crystallographic data for the 1/1 and 2/1 Eu-Ag-In approximant phases are shown in **table 1** and refined atomic parameters for the 1/1 and 2/1 approximants respectively in **table 2** and **table 3**.

Refined comp.	$Eu(Ag/In)_{6.268}$	$Eu_{13}(Ag/In)_{79}$
Coll. Temperature [K]	298	298
Space group	Im-3	Pa-3
Lattice const. a [Å]	15.6222	25.2823
Cell volume [$Å^3$]	3812.6	16160.3
Z	24	8
Calc. Density [gcm^{-3}]	8.9505	8.921
Abs. Coeff. μ [mm^{-1}]	30.133	30.169
Indep. Refl.	817	8236
Indep. Refl. Obs.[$I > 3\sigma(I)$]	754	4413
R_{int} (obs/all)	12.73 / 12.81	9.54 / 13.12
Parameters	56	338
$R1/wR2$ (obs)	2.51 / 7.47	4.73 / 8.53
$R1/wR2$ (all)	2.88 / 7.65	12.29 / 9.82
GOF on F^2 (obs/all)	1.79 / 1.76	1.95 / 1.62
$\Delta\rho_{min}$, $\Delta\rho_{max}$ [e/$Å^3$]	-1.45 / 3.51	-4.59 / 6.22

Table 1. Crystallographic data for the Eu-(Ag/In) 1/1 (left column), and 2/1 (right column) approximants. Mixed Ag/In positions were refined as Cd.

atomic position	occupancy	x	y	z	U_{eq} ($Å^2$)
Eu1	1	0	0.29782(3)	0.18656(3)	0.01368(18)
(Ag/In)1	0.395(7)	0.06676(13)	0.06676(13)	0.06676(13)	0.0522(10)
(Ag/In)2	1	0	0.09055(5)	0.24475(6)	0.0272(3)
(Ag/In)3	0.400(5)	0.1216(2)	0.1216(2)	0.1216(2)	0.0429(10)
(Ag/In)4	1	0.34769(4)	0.10822(4)	0.19736(4)	0.0220(2)
(Ag/In)5	1	0	0	0.40971(7)	0.0298(4)
(Ag/In)6	1	0	0.34788(5)	0.40673(5)	0.0164(2)
(Ag/In)7	1	0	0.5	0.30907(7)	0.0259(4)
(Ag/In)8	0.369(3)	0.27215(15)	0.27215(15)	0.27215(15)	0.0436(7)
(Ag/In)9	0.600(5)	0.33475(11)	0.33475(11)	0.33475(11)	0.0287(5)
(Ag/In)10	0.092(6)	0.0616(12)	0	0.0932(12)	0.090(9)

Table 2. Refined atomic parameters for the average substructure of the Eu-(Ag/In) 1/1 approximant.

atomic position	occupancy	x	y	z	U_{eq}($Å^2$)
Eu1	1	0.04326(3)	0.15486(3)	0.34228(3)	0.0133(2)
Eu2	1	0.03945(3)	0.34629(3)	0.47138(3)	0.0133(2)
Eu3	1	0.15447(3)	0.33519(3)	0.26648(3)	0.0135(2)
Eu4	1	0.27193(3)	0.34799(3)	0.46623(3)	0.0150(2)
Eu5	1	0.46111(3)	0.46111(3)	0.46111(3)	0.01569(16)
(Ag/In)1	1	0.00186(4)	0.15652(5)	0.09795(4)	0.0185(3)
(Ag/In)2	1	0.05982(5)	0.05670(5)	0.25003(5)	0.0230(4)
(Ag/In)3	1	0.05676(5)	0.24853(4)	0.24784(5)	0.0189(3)
(Ag/In)4	1	0.09191(4)	0.30762(4)	0.15551(5)	0.0189(3)
(Ag/In)5	1	0.15146(10)	0.19083(10)	0.25565(12)	0.0322(4)
(Ag/In)6	1	0.15588(5)	0.28645(5)	0.49891(5)	0.0223(4)
(Ag/In)7	1	0.24750(4)	0.24750(4)	0.24750(4)	0.0196(2)
(Ag/In)8	1	0.02980(5)	0.36850(5)	0.08479(6)	0.0318(4)
(Ag/In)9	1	0.02562(5)	0.35798(5)	0.23770(5)	0.0226(4)
(Ag/In)10	1	0.06138(4)	0.28773(4)	0.35359(4)	0.0146(3)
(Ag/In)11	1	0.09121(4)	0.46770(4)	0.43803(4)	0.0152(3)
(Ag/In)12	1	0.15580(4)	0.35826(4)	0.39696(4)	0.0169(3)

(Ag/In)13	1	0.21413(4)	0.45064(4)	0.24270(4)	0.0162(3)
(Ag/In)14	1	0.22653(5)	0.47005(5)	0.44504(5)	0.0250(4)
(Ag/In)15	1	0.24904(4)	0.28905(4)	0.35139(4)	0.0153(3)
(Ag/In)16	1	0.06358(4)	0.09315(5)	0.46801(5)	0.0198(3)
(Ag/In)17	1	0.05847(4)	0.40308(4)	0.34458(5)	0.0169(3)
(Ag/In)18	1	0.07723(5)	0.21752(5)	0.44959(5)	0.0259(4)
(Ag/In)19	1	0.09106(5)	0.46336(4)	0.25118(4)	0.0181(3)
(Ag/In)20	1	0.15570(5)	0.13432(5)	0.41227(4)	0.0198(3)
(Ag/In)21	1	0.15496(5)	0.46166(4)	0.34353(5)	0.0200(3)
(Ag/In)22	1	0.25101(4)	0.40597(4)	0.34534(4)	0.0172(3)
(Ag/In)23	1	0.34916(4)	0.34916(4)	0.34916(4)	0.0139(2)
(Ag/In)24	1	0.34312(5)	0.44433(5)	0.40566(4)	0.0210(3)
(Ag/In)25	1	0.02443(17)	0.48011(16)	0.16458(14)	0.0611(17)
(Ag/In)26	1	0.09669(5)	0.09669(5)	0.09669(5)	0.0189(2)
(Ag/In)27	1	0.09490(5)	0.19250(5)	0.15535(5)	0.0301(4)
(Ag/In)28	1	0.15614(13)	0.23250(15)	0.3599(3)	0.091(2)
(Ag/In)29	1	0.03349(6)	0.03349(6)	0.03349(6)	0.0311(3)

Table 3. Refined atomic parameters for the structure of the Eu-(Ag/In) 2/1 approximant.

In contrast to previous TEM studies, the single crystal X-ray diffraction experiments on the corresponding 1/1 Eu-Ag-In phase performed as part of this work do not indicate the presence of any superstructure and only the average structure could be refined. The average structure reveals substantial occupancy of the interstitial atomic positions within the cubic cavities which are responsible for the formation of the superstructure in the related Eu_4Cd_{25} phase, these interstitial atomic positions are denoted Ag-In8 in **table 2**. Powder samples containing the 1/1 Eu-Ag-In phase were measured at the SPring-8 synchrotron facility and although weak additional peaks could be seen in the powder patterns, these could not be indexed with the expected supercell. The rather strong intrinsic intensity variations among these additional peaks when compared with other powder samples having somewhat different sample history led to the conclusion that these peaks arise due to unidentified impurities in the samples. The fact that superlattice ordering could be detected by electron diffraction but not by single crystal or synchrotron powder XRD indicates that the 1/1 Eu-Ag-In phase exhibits short to medium range superlattice ordering and that the size of the superlattice domains is too small in relation to the coherence length of the used X-rays, but can obviously be detected by selected area electron diffraction (SAED). The refinement result on the average structure confirm that indeed the phase is a 1/1 approximant of the Yb-Cd family and the observed disorder along the 3-fold axes of the cubic unit cell can be seen as a weighted average atomic decoration between the Eu_4Cd_{25} superstructure and a standard Yb-Cd type 1/1 approximant. The refined average basic structure is seen in **figure 3 a**, where the disorder and partially occupied positions along the 3-fold axes is clearly seen.

Single crystals of the 2/1 Eu-Ag-In approximant could be isolated from a multiphase sample and analysed by single crystal X-ray diffraction. The results show that this 2/1 approximant also belongs to the Yb-Cd family of 2/1 approximants, but with a unique degree of structural perfection. Other known 2/1 approximants of this family will display severe positional disorder (in particular the binary phases $Yb_{13}Cd_{76}$ and $Ca_{13}Cd_{76}$) or partial occupancies in the atomic positions defining the central tetrahedron of the atomic cluster.[3, 5] It would seem that the introduction of Ag and In instead of Cd allows for the formation of the 2/1 approximant in this system (no 2/1 approximant has been found in the Eu-Cd system) and furthermore, the partial chemical disorder that is introduced in the ternary 2/1 Eu-Ag-In approximant apparently replaces the positional disorder that is usually found in these phases. As a consequence of the perfect order of the tetrahedron in the centre of the cluster, the next atomic shell (the dodecahedron, **figure 1 b**) is severely distorted as seen in **figure 3 b**.

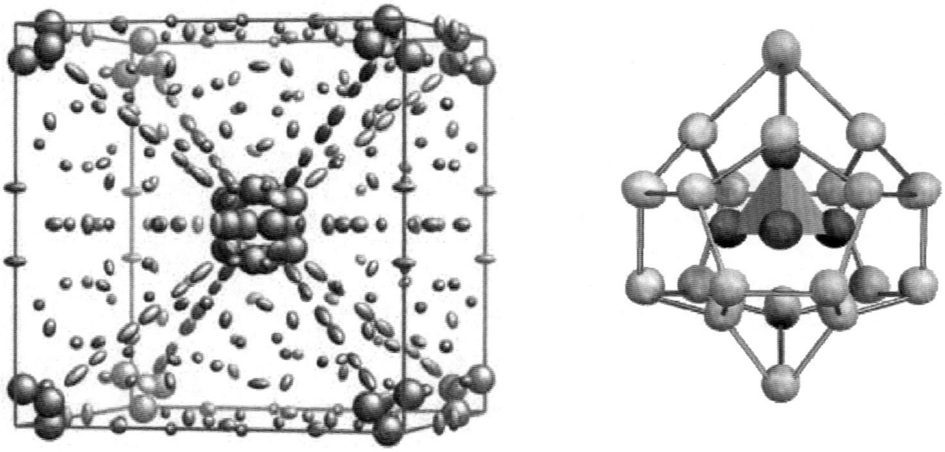

a b

Figure 3. In **a**, the cell-content of the Eu-Ag-In 1/1 approximant is displayed with thermal ellipsoids at the 70% probability level. Eu atoms are seen in red, Ag/In in grey. The disorder along 3-fold axes and in the central tetrahedron is clearly seen. In **b**, we see the central tetrahedron and surrounding dodecahedron in the cluster of the Eu-Ag-In 2/1 approximant. Atoms are displayed as electron density iso-surfaces at the 8 $eÅ^{-3}$ level, tetrahedron atoms are seen in red. In this phase the tetrahedron has no disorder but the dodecahedral cage is severely distorted.

Chemical and positional disorder are two of the major difficulties encountered when analysing the structures of quasicrystals and approximants. Most quasicrystals are ternary and usually suffer from both chemical and positional disorder. The most efficient way to counter these difficulties is by combined structural studies of both quasicrystals and approximants. Although the same problems are encountered in approximant phases, they can be treated by well established classical methods of structure analysis. In this context the structure of the 2/1 Eu-Ag-In approximant determined as part of this work plays a crucial role: although the partial chemical order between Ag and In has not yet been analysed, this phase is certainly the best choice among ternary 2/1 approximants in the RE-Ag-In systems for such analyses since the structure is essentially devoid of positional disorder. Thus the troublesome correlations between occupational parameters and atomic displacement parameters that are otherwise encountered in the structure refinement can be completely avoided. In order to study the chemical order between Ag and In however, it is necessary to use special techniques such as anomalous X-ray diffraction. Furthermore the well ordered structure of this phase can be used to model and understand the local structures of other positionally disordered 2/1 approximants.

4. Acknowledgements

The authors would like to thank Prof. S. Lidin for valuable discussions and Prof. H. Yamane for his assistance during the single crystal measurements. This work was partially funded by the Japan Science and Technology Agency, Solution Oriented Research for Science and Technology (JST-SORST), and by World Premier International Research Center Initiative (WPI Initiative) on Materials Nanoarchitectonics, MEXT, Japan.

References

[1] Janot C, 1994 *Quasicrystals A Primer*, second ed., Oxford University Press Inc., Grenoble
[2] Tsai A P, Guo J Q, Abe E, Takakura H, Sato T J, 2000 *Nature* **408** 537

[3] Gómez C P, 2003 *Order and disorder in the RE-Cd and related systems*, Stockholm University, Stockholm,54
[4] Gómez C P, Lidin S, 2003 *Phys. Rev.*B **68** 024203(1)
[5] Gómez C P, Lidin S, 2001 *Angew. Chem., Int. Ed. Engl.* **40** 4037
[6] Morita Y, Tsai A P, 2008 *Jap. J. Appl. Phys.* **47** 7975
[7] Ruan J F, Kuo K H, Guo J Q, Tsai A P, 2004 *J. Alloys Compd.* **370** L23
[8] Takakura H, Gómez C P, Yamamoto A, de Boissieu M, Tsai A P, 2007 *Nat. Mater.* **6** 58
[9] Gómez C P, Ohhashi S, Cui C, Yamamoto A, Tsai A P, 2008 *Unpublished work*
[10] Armbrüster M, Lidin S, 2000 *J. Alloys Compd.* **307** 141
[11] Gómez C P, Lidin S, 2004 *Chem. Eur. J.* **10** 3279
[12] Nishimoto K, Muraki M, Tamura R, 2008 *Z. Kristallogr. Conf. Proc. ICQ10.*
[13] Petricek V, Dusek M, Palatinus L, 2006 *JANA 2006*, Praha
[14] Momma K, Izumi F, 2006 *Commission on Crystallogr. Comput., IUCr Newslett.* **7** 106.
[15] Ormandy R, 2007 *Truespace* 7.51, Caligari Corporation, Mountain View

Journal of Physics: Conference Series **165** (2009) 012046 doi:10.1088/1742-6596/165/1/012046

Fabrication of GDL microporous layer using PVDF for PEMFCs

Sung Bum Park, Sungjin Kim, Yong-il Park and Myung-Hoon Oh

Department of Information & Nano Materials Engineering, Kumoh National Institute of Technology, 1 Yangho-dong, Gumi, Gyeongbuk 730-701, Korea.
E-mail: ohmyung@kumoh.ac.kr

Abstract. The Gas Diffusion Layer(GDL) of fuel cell, are required to provide both delivery of reactant gases to the catalyst layer and removal of water in either vapor or liquid form in typical PEMFCs. In this study, the fabrication of GDL containing Micro Porous Layer(MPL) made of the slurry of PVDF mixed with carbon black is investigated in detail. Physical properties of GDL containing MPL, such as electrical resistance, gas permeability and microstructure were examined, and the performance of the cell using developed GDL with MPL was evaluated. The results show that MPL with PVDF binder demonstrated uniformly distributed microstructure without large cracks and pores, which resulted in better electrical conductivity. The fuel cell performance test demonstrates that the developed GDL with MPL has a great potential due to enhanced mass transport property due to its porous structure and small pore size.

1. Introduction

In proton-exchange membrane fuel cells(PEMFC), gas diffusion layer(GDL) serves as current collectors that allow ready access of fuel and oxidant to the anode and the cathode catalyst surface.

Usually, the catalyst layer is reinforced by a thick porous electrode support layer. In PEM fuel cell, this electrode support layer is called a gas diffusion layer(GDL). The GDL protects the delicate catalyst structure, provides mechanical strength, allows easy gas access to the catalyst, and enhances electrical conductivity[1,2]. The gas Diffusion Layers (GDLs) of fuel cell are required to provide both delivery of reactant gases to the catalyst layer and removal of water in either vapor or liquid from in typical PEMFCs. A Microporous layer (MPL) reduces the contact resistance between catalyst layer and macroporous substrate [3,4]. Primary purpose of MPL is water management, as they provide effective wicking of liquid water from the cathode catalyst layer into the diffusion media. It also may have utility in reducing electrical contact resistance with the adjacent catalyst layer. The properties of microporous layers can be adjusted by changing the carbon, specifically the particle and agglomerate structure, and the hydrophobicity. Owing to its superior thermal ability and resistance to chemical degradation, PTFE is used in majority of gas diffusion electrodes as a hydrophobic binder. Though, PTFE is insoluble in any known solvent; it can be prepared at high temperature sintering processes from powder or suspension, which results in increase in the cost of PEMFC. With the aim of producing a low-cost, easy to prepare gas diffusion electrodes with favorable chemical and electrical properties, a method for obtaining gas diffusion electrodes based on poly(vinylidene fluoride) (PVDF)-carbon blends via phase inversion was firstly detailed by Cabasso et al.[5] PVDF matrix demonstrates lower physical and chemical properties than PTFE; especially PVDF has lower melting temperature (160~180℃) than PTFE(327℃). Also, it was proved that PVDF coated on carbon cloth shows good impact and compressive strength, as well as improved wear resistance and higher thermal conductivity. Besides, PVDF has more advantages than PTFE, including low cost, easy processing property and etc.

In this study, the fabrication of GDL containing Micro Porous Layer(MPL) made from the slurry of PVDF(Polyvinyliden Fluoride) mixed with carbon black is investigated in detail. The microstructure of the developed GDL containing MPL was analyzed by SEM/EDX by observing the surface and cross section of the specimen. The properties, such as electric conductivity, permeability and porosity were measured. The amount of carbon loading to get optimum thickness was defined, and the performance of the fuel cell using developed GDL with MPL was evaluated.

2. Experimental procedure

Commercial carbon cloth was cut out into pieces and washed in acetone to remove dust or undesirable matters. Later, the washed carbon cloths were dried at 50℃ for 2hours. 30wt% of PTFE emulsion (Du Pont) was diluted to make 3wt% aqueous solutions. The dried carbon cloths were dipped entirely in the different PTFE solutions for 30 seconds and dried again at room temperature for 5hours. After then, the samples were heat-treated in 350℃ at air.

In order to prepare carbon slurry for MPL, carbon powder (Vulcan XC-72R) was mixed with PVDF in the ultrasonic bath for 2hours. The obtained carbon slurry was coated onto one side of woven carbon cloth gas diffusion-backing layer(E-TEK Division) by doctor blade method, and successively dried at 80℃ for 30min. The MPL coated GDL sample was heat treated at 180℃ for 1hours to distribute PVDF homogeneously through the MPL. The microstructure of the developed GDL containing MPL was analyzed in SEM/EDX by observing the surface and cross section of the specimen. The properties such as electric conductivity and pore size were measured. The bubble point method using a capillary flow porometer CFP-1500AEL (PMI, USA) was selected for the measurement of the mean flow pore diameter. The amount of carbon loading to get optimum thickness was defined, and the performance of the fuel cell using developed GDL with MPL was evaluated. The performance of the developed single cell was evaluated at 80℃ in a fuel cell test station, using H_2 and O_2 gases as fuel gas with 500 sccm and 300 sccm flow rates, respectively. Nafion® 117 with 5 x 5cm^2 was used as a membrane and the active area of electrode was 5cm^2.

3. Results and Discussion

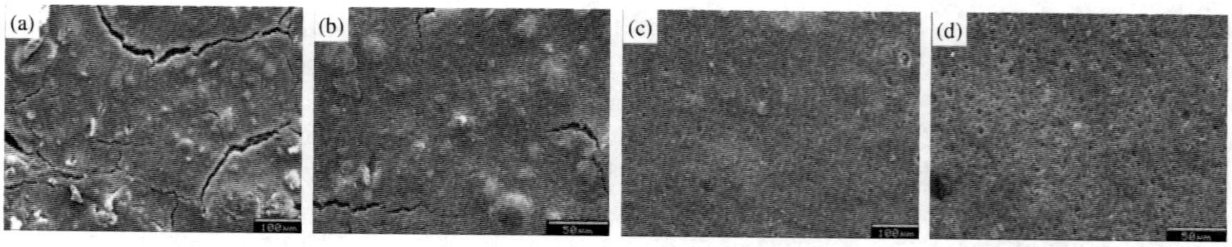

Figure 1. Scanning Electrode Microscopy(SEM) images of the commercial PTFE MPL-coated GDL(a-b) and PVDF-based MPL-coated GDL. (c-d))

SEM photographs of the surface morphology of a microporous layer with PTFE and PVDF binders are shown in Figure 1. The microstructure of the microporous layer with PTFE binder was characterized by a large number of coarse cracks and pores on the surfaces (Figure 1(a-b)).

The presence of large cracks and open pores on the surface of diffusion layer isn't desirable because they lead to poor electrical conductivity. Besides, the Pt catalyst particles might fall into the open cracks and pores which results in reduced support for the catalyst.

In the case of the microporous layer with PVDF binder, the microstructure of the diffusion layer was uniformly distributed on the layer (Figure 1(c-d)). The microstructure with uniform distribution and without large crack and pore was desirable to get better electrical conductivity, as well as good support of the Pt catalyst.

Figure 2 shows the MPL resistance and gas permeability as a function of PVDF concentration. As it can be seen that the conductivity and gas permeability were increased with decreasing the PVDF contents. It was observed that the MPL on carbon cloth change the physical properties of GDL and changed the cell performance. The analysis of results proved that the lower ratio of PVDF resulted in lower resistance. However, a steady increase in the gas permeability values was observed with decreased ratio of PVDF. This phenomena could be explained by low porosity due to small amount of PVDF binder. Thus, it can be assumed that the further decrease of PVDF ratio would result on lower resistance, but in higher gas permeability values.

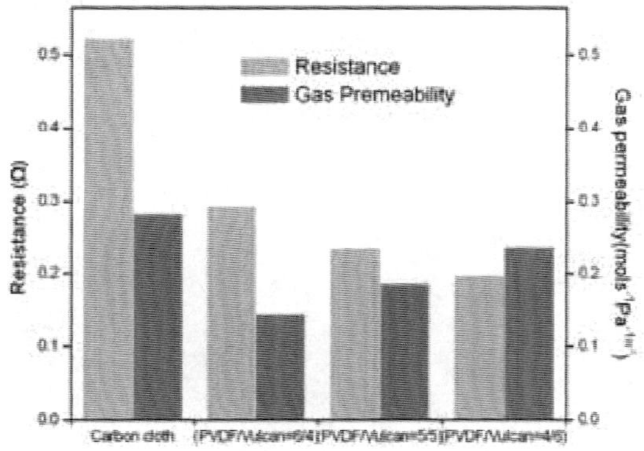

Figure 2. Effect of PVDF concentration on MPL surface resistance and gas permeability.

The pore size distribution in PTFE-based MPL and PVDF-based MPL is illustrated in Figure 3(a) and summarized in table shown in Figure 3(b). The pore size of commercial non-treated carbon cloth was 81.40 ㎛. The analysis showed that using traditional PTFE binder resulted in considerable decrease of pore size distribution down to 48.97 ㎛. However, there are some drawbacks of using PTFE binder, such as relatively higher surface resistance and lower gas permeability values. These drawbacks can be eliminated by using PVDF binder instead. The results indicate that the concentration ratio of PVDF/Vulcan was key parameter for controlling the pore size distribution. When using high concentration of PVDF, the pore size could be successfully decreased down to 67.30 ㎛. Though, as shown in Figure 2, high concentration of PVDF resulted in relatively higher surface resistance and lower gas permeability, which were undesirable for fuel cell performance. Decreasing the PVDF concentration to 40% allowed to decrease the pore size as low as 53.73 ㎛, which was close to one of traditional PTFE-based MPL.

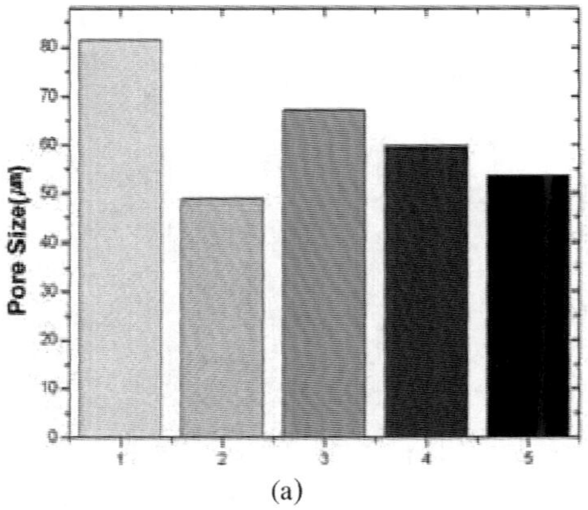

(a)

	Sample	Average Pore size(㎛)
1	Non treated carbon cloth	81.4042
2	PTFE-based Microporous layer (3wt.% PTFE treated carbon cloth)	48.9753
3	PVDF-based Microporous layer (PVDF/Vulcan=6/4) (3wt.% PTFE treated carbon cloth)	67.3055
4	PVDF-based Microporous layer (PVDF/Vulcan=5/5) (3wt.% PTFE treated carbon cloth)	59.7666
5	PVDF-based Microporous layer (PVDF/Vulcan=4/6) (3wt.% PTFE treated carbon cloth)	53.735

(b)

Figure 3. The diagram (a) and table (b) of pore size distribution measured in PTFE-based MPL and PVDF-based MPL with different PVDF concentration

During the cell operation, gas mixture is delivered by convention from cathode towards catalyst through GDL. Therefore, good mass transportation is very important to enhance the cell performance. In order to examine the effect of binder type on fuel cell performance, different types of gas diffusion layers were prepared. Figure 4 shows cell potential-current density plots and power density plots of non-treated carbon cloth GDL, PTFE MPL-coated GDL and PVDF-based MPL coated GDL. From the plots it can clearly be observed that the cell performance varied with different microporous layer. Fuel cell performance analysis showed that PVDF-based MPL demonstrated considerably higher power density values than those of other samples. This phenomenon could be explained by appropriate in-plane and through-plane microstructure of PVDF-MPL coated GDL. Due to symmetric porous structure and small pore size, the in-plane and through-plane mass transport in PVDF-MPL coated GDL could be enhanced.

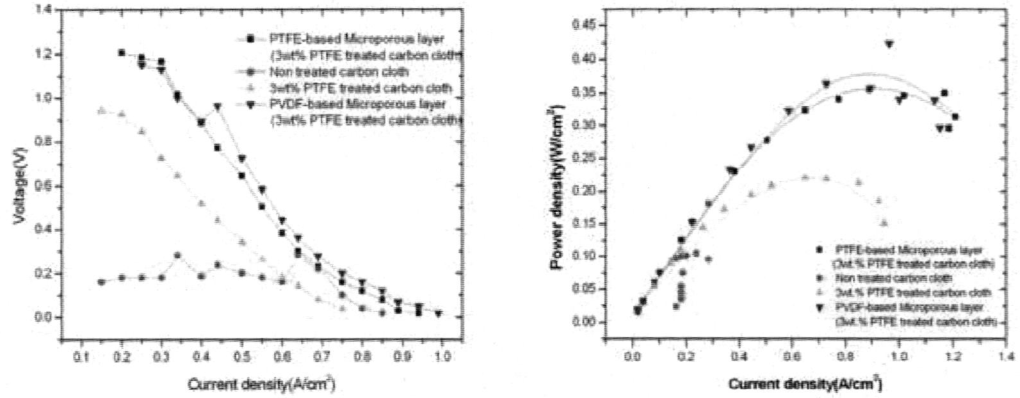

Figure 4. Potential-current density plots and power density plots of non-treated carbon cloth GDL, PTFE MPL-coated GDL and PVDF-based MPL coated GDL.

4. Conclusions

Different types of GDL containing Microporous layer have been prepared from slurries. The effect of binder type and PVDF concentration ratio on resistance, gas permeability and cell performance has been investigated. The PEM fuel cell performance test demonstrated that the presence of PVDF-based MPL potentially reduced mass transport losses within the GDL at current densities, thus, resulted in enhancement of the cell performance.

Acknowledgment
This paper was supported by Research Fund, Kumoh National Institute of Technology.

References
[1] O'Hayre R, Cha S w, Colella W, Prinz F B, Fuel Cell Fundamentals
[2] Yoon G H, Park S B, Kim E H, Oh M-H, Cho K-S, Jeong S W, Kim S, Park Y-I 2007 J Electroceram DOI 10.1007/s10832-007-9321-1
[3] Hung T F, Huang J, Chuang H J, Bai S H, Lai Y J, Chen-Yang Y W 2008 J. Power Sources **184** 165-171
[4] Ong A L, Bottino A, Capannelli G, Comite A 2008 J. of Power Sources, **183**, 62-68
[5] Cabasso I., Y. Yuan and X. Xu 1998 'US 5,783,325: Gas Diffusion Electrodes Based on Poly(Vinylidene Fluoride) Carbon Blends', Patent issued

Movement of solder fillers because of the unevenness of interfacial tension in self-organization assembly process

Koushi Ohta, Kiyokazu Yasuda, Michiya Matsushima and Kozo Fujimoto

Graduate School of Engineering, Osaka University, 2-1 Yamadaoka, Suita, Osaka 565-0871, Japan

kou-ohta@mapse.eng.osaka-u.ac.jp

Abstract. Soldering or anisotropic conductive adhesives (ACAs) are used for assembling of devices with high-density area-array terminals. However, solder bump processes are high-cost and interconnects of ACAs are less-reliable. The purpose of this study is to develop a novel method. Self-organization assembly method that uses active resin containing solder fillers may allow reliable interconnects at low-cost. Fundamental process of Self-organization assembly are movement, coalescence, and wetting of molten fillers in resin. The focus of this paper is on the movement of fillers. In-situ observations of coalescence behavior for 40μmϕ molten fillers in the resin revealed irregular movement of the fillers at a velocity of several μm/s. Numerical analysis, using improved volume fraction method, indicated that a 10% degradation of interfacial energy on one side of a 40μmϕ filler could move the filler at a velocity of several mm/s. This degradation of the interfacial energy was resulted the remaining oxide film.

1. Introduction

Most of highly-integrated electronic components with narrow-pitch area-array terminals are assembled by solder bumps formed by printing. However, the process designs need tries and errors [1]. A high accuracy for voids is also required [2]. On the other hand, the interconnects using ACAs are not reliable because the conductivity depends on only mechanical contacts [3].

Figure 1 shows schematic diagrams of Self-organization assembly method [4]. As shown in (a), resin containing solder fillers was supplied between components and substrates allover the terminal patterns. The melting point (m.p.) of the solder was lower than curing temperature (T_c) of the resin. The temperature was controlled between the m.p. and the T_c, as shown in (b). Solder fillers then move, coalesce with each other and are transported to the wettable terminals. The resultant conductive pathways are shown in (c) and they are similar to solder bumps in morphology. Reheating at higher than the T_c made the resin underfill. Interconnects formed in this low cost process, as for ACAs, may achieve high reliability and high degree of flexibility against design change of devices or substrates.

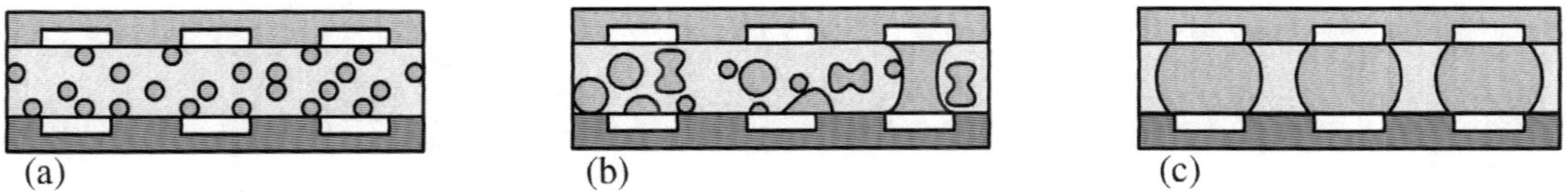

Figure 1. Schematic views of Self-organization assembly process. (a) supplied resin containing solder particles, (b) movement, coalescence, and wetting of particles, (c) formed conductive paths.

Fundamental processes of Self-organization assembly at temperatures higher than the m.p. of the solder are movement, coalescence, and wetting of molten fillers. In-situ observation of filler coalescence were carried out. These observations clarified the irregular movement of fillers in the active resin that led to coalescence and wetting. The focus of this paper is on this movement of the fillers. The unevenness of interfacial tension that was involved in the elimination of the oxide films were evaluated by a numerical study as the possible mechanism of the movement.

2. Materials

Eutectic Sn-Bi spherical particles were used as the material to produce a conductive path. The m.p. of Sn-Bi solder is 138 °C. Average particle sizes of the solder fillers used in this study was 10 or 40 μmϕ. A newly developed thermosetting epoxy resin was used as a container for the fillers. Temperature characteristics of the resin viscosity were measured. To simulate the actual temperature change from assembly experiments the temperature was held at 150 °C. The resin begins to cure in a few minutes after the solder melts. A carboxylic acid contained in the resin as the curing agent acts as an activator to eliminate the oxide film around the fillers.

3. In-situ observations of fillers movement

3.1. Experimental procedure

Figure 2 shows the instruments used for in-situ observations on filler movement. Resin containing 10 vol%, 40 μmϕ fillers was used. To eliminate filler wetting effects, cover glasses of low wettability were used. 100 μmϕ stainless wires were cramped between two cover glasses together with the resin as spacers. Transmitted light through a hole in a metal plate enabled clear observations.

Figure 3 shows a typical heating profile. Before heating to 145 °C to melt the fillers, a pre-heating treatment to eliminate oxides around fillers at 80 °C was carried out for 4 min.

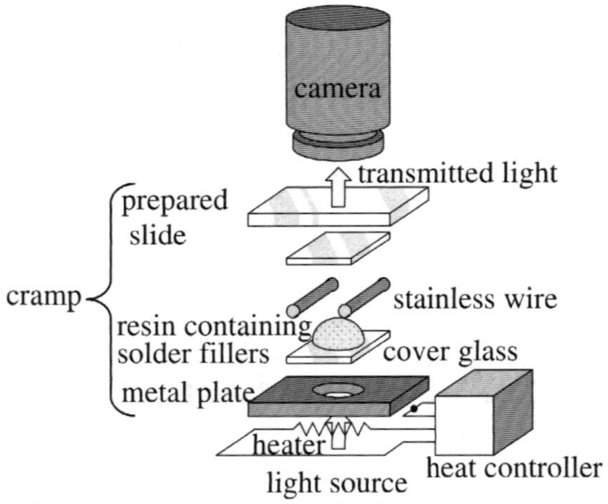

Figure 2. Instruments for in-situ observations on coalescence of many particles.

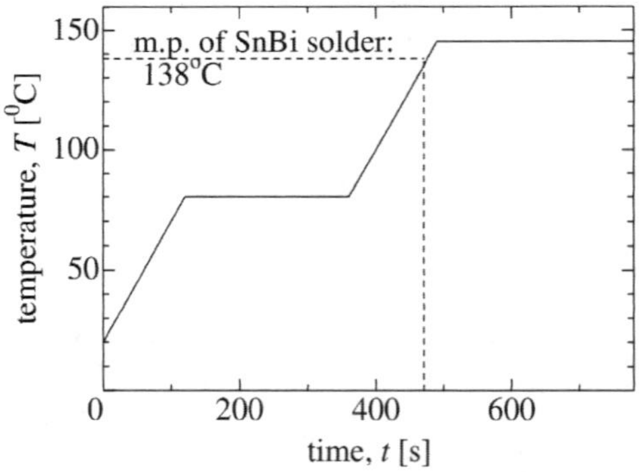

Figure 3. Heating profile for in-situ observations.

3.2. Results and discussions

Figure 4 shows the trajectory of a typical filler. Figure 5 shows the temporal change of the coalescence frequency with a blue line. This frequency was measured by the shrinkage in the filler number in sight per unit time. Figure 5 shows also the velocity of the filler focused in figure 4, by a red line. Heating was started at t=0 s and the measurement interval was Δt=1 s as shown in both figures.

From t=0 s to t=460 s, at which time the m.p. of the solder was reached, movement of the fillers was hardly observed.At the m.p. of the solder neighboring fillers began to coalesce. This coalescence induced flow in the resin and induced further coalescences. However, by t=466 s, this movement ceased. The reason for this is that as the distance between fillers increased because of coalescence further coalescence became difficult. From approximately t=500 s, the fillers showed irregular movement at a velocity of several μm/s. The fillers showed repeat of aggregation and separation from the clusters. Figure 4 shows active filler movement through some congestions. As the coalescence frequency was low untill t=525 s, this active movement could not have been caused by coalescence. After t=525 s, the frequency of the coalescence increased again. This increase is thought to be due to fillers movement which was not caused by coalescence when the oxides around the fillers eliminated enough.

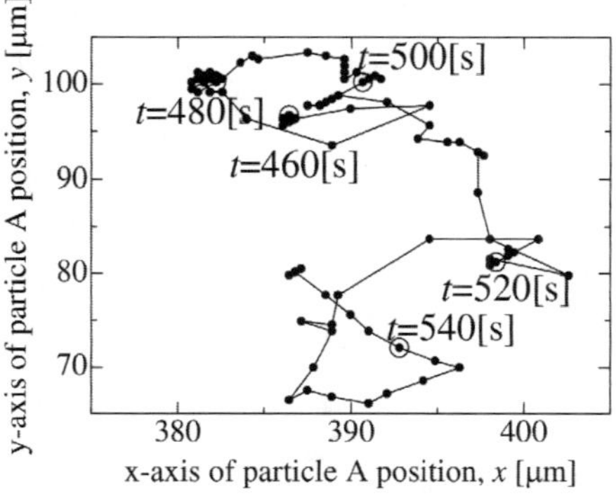

Figure 4. Trajectory of a solder filler.

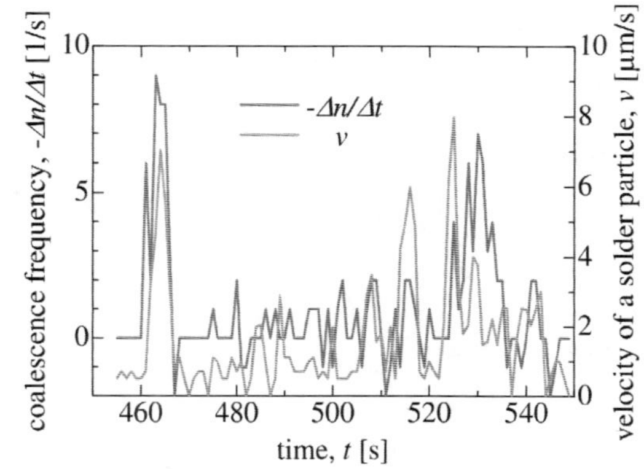

Figure 5. Coalescence frequency of solder fillers and velocity of a filler.

4. Clarification of fillers movement mechanism

The concentration gradient, Brownian movement, and heat convection were discussed as possible mechanisms of filler movement. These are popular mechanism of particle movement in fluids. These mechanisms do not depend on whether the particles are solids or liquids. Since the fillers hardly moved before melting these mechanisms are not plausible as the mechanism of filler movement in this study.

A self-running mechanism for oil droplets in water with added surfactant [5] has been reported. This phenomenon was thought to be caused by an unevenness of interfacial energy around a droplet due to chemical Marangoni effect. It is well-known that the fillers have deep scars on the surface resulted from sieve classification. Unevenness of the interfacial energy around molten filler on this system was thought to be caused by the thickness distribution of oxide films.

Quantification of the unevenness of interfacial energy was attempted using a numerical study based on fluid dynamics. The mass conservation law and the momentum conservation law were analyzed through an improved Volume of Fluid (VOF) method [6]. To model the unevenness using the Continuum Surface Force (CSF) model [7], the interfacial energy of the filler in the resin was defined as interfacial tension in the following equation:

$$\gamma = \begin{cases} \gamma_0 & (\text{at } x < x_g) \\ \gamma_0 - \Delta\gamma & (\text{at } x > x_g) \end{cases} \tag{1}$$

Where γ_0 is interfacial tension between the bare solder and the resin, $\Delta\gamma$ is the degradation of the interfacial tension because of oxides on the filler and x_g is an x-coordinate of the center of gravity for the filler. Under the assumption that the density of the solder is constant, x_g can be obtained by the following equation:

$$x_g = \left(\int_{x=0}^{X} x \times \rho \, dV \right) \Big/ \left(\int_{x=0}^{X} \rho \, dV \right) = \left(\sum_{n=1}^{N} x(n) \times VOF_1(n) \right) \Big/ \left(\sum_{n=1}^{N} VOF_1(n) \right) \tag{2}$$

Where n is element number and VOF_1(n) is the VOF ratio of the solder at nth element.

A 40 μmϕ droplet of molten solder was positioned in the resin. This model, as shown in Figure 6 (a), simulates the filler when the oxides on the left half of the filler are eliminated. The unevenness of the interfacial tension makes the static fluid pressure in the left half of the filler higher than that in the right half. The density and the viscosity of the resin are $1.2\text{x}10^3$ kg/m^3 and $1.0\text{x}10^{-1}$ Pa s, respectively.

The density and the viscosity of the solder are 8.5x10^3 kg/m^3 and 1.8x10^{-3} Pa s, respectively. The interfacial tension between the bare molten solder and the resin, γ_0=5.7x10^{-1}[N/m].

Figure 6 shows the behavior of the filler movement at $\Delta\gamma$=0.5γ_0. The filler moved to the right without much deformation. Figure 7 quantitatively shows effects of the unevenness of the interfacial tension. x_1 represents the rightmost side of the filler and x_2 represents the leftmost side. At higher $\Delta\gamma$, when simulating the fillers with a thicker oxide film on the right half, the filler moved faster. Even the filler of $\Delta\gamma$=0.1γ_0 moved at a velocity of several mm/s which is much faster than the movement of the observed fillers. The unevenness of interfacial tension is thought to be a plausible mechanism for filler movement in the resin.

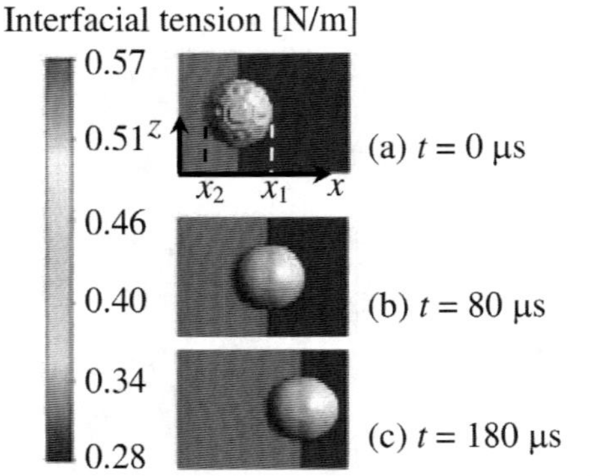

Figure 6. Analysis results on effects of interfacial tension distribution, $\Delta\gamma$= 0.5γ_0.

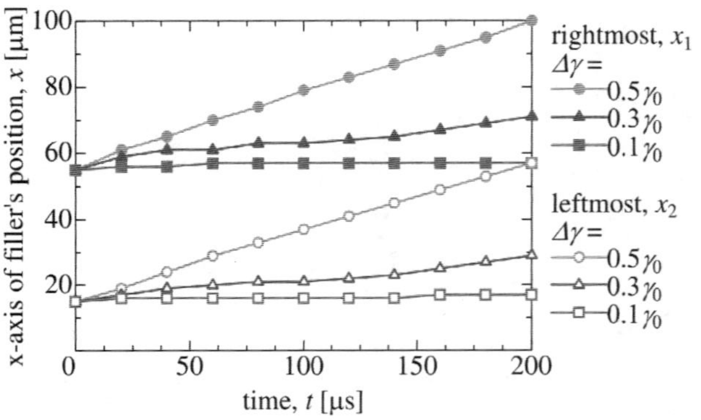

Figure 7. Quantitative evaluation on effects of interfacial energy unevenness.

5. Conclusion

The focus of this study was to determine the fundamental processes and mechanism of the Self-organization assembly process for the movement of fillers in resin. Based on the in-situ observation of the movement of many molten fillers in an active resin, the unevenness of the interfacial energy around a filler and other phenomena are quantitatively discussed as a plausible. The following results were obtained from this study.

Through in-situ observations of the coalescence behavior of the fillers in a resin, irregular movements of the fillers at a velocity of several μm/s after melting were observed. Before melting, the fillers hardly moved.

The numerical studies using an improved VOF method, revealed that a 10% degradation of the interfacial tension at the right half of the filler can move the filler at a velocity of several mm/s.

Acknowledgements

This study was supported by Priority Assistance for the Formation of Worldwide Renowned Centers of Research - The Global COE Program (Project: Center of Excellence for Advanced Structural and Functional Materials Design) from the Ministry of Education, Culture, Sports, Science and Technology (MEXT), Japan.

References

[1] Pan J, Tonkay G L, Storer R H, Sallade R M and Leandri D J 2004 *IEEE Trans. Electron. Packag. Manufact.* **27** 125-32
[2] Yunus M, Srihari K, Pitarresi J M Primavera A 2003 *Microelectron. Rel.* **43** 2077-86
[3] Sun M 2001 *Microelectron. J.* **32** 197-203
[4] Ohta K, Yasuda K, Matsushima M and Fujimoto K 2007 *Solid State Phenomena*, **124-126** 543-6
[5] Nagai K, Sumino Y, Kitahata H and Yoshikawa K 2005 *Phys. Rev.* E **71** 065301(R)-1-4
[6] Hirt C W and Nichols B D 1981 *J. Comput. Phys.* **39** 201-25
[7] Blackbill J U, Kothe D B and Zemach C 1992 *J. Comput. Phys.* **100** 335-54

Silica-based microstructures on nonplanar substrates by femtosecond laser-induced nonlinear lithography

M Mizoshiri[1], H Nishiyama[1], J Nishii[2] and Y Hirata[1]

[1]Division of Materials and Manufacturing Science, Graduate School of Engineering, Osaka University, 2-1 Yamadaoka, Suita, Osaka 565-0871, Japan

[2]National Institute of Advanced Industrial and Science Technology, 1-8-31 Midorigaoka, Ikeda, Osaka 563-8577, Japan

mizoshiri@mapse.eng.osaka-u.ac.jp

Abstract. We developed a technique for the formation of nonplanar surfaces of inorganic optical materials by a combined process of nonlinear lithography and plasma etching. This technique can be used to fabricate structures even on non-flat substrates, which is difficult using current semiconductor technology. Three-dimensional patterns were written directly inside a positive-tone photoresist using femtosecond laser-induced nonlinear optical absorption. The patterns were then transferred to underlying nonplanar substrates by the ion beam etching technique. For the lithographic process, we obtained a minimum feature size of 900 nm, which is below the diffraction limit. We demonstrated the fabrication of silica-based hybrid diffractive-refractive lenses. Fresnel zone plates with smooth surfaces were obtained on convex microlenses. When a 633-nm-wavelength He-Ne laser was coupled normally to the hybrid lens, the primary focal length was measured as 630 μm. This hybridization shifted the focal length by 200 μm, which agreed with the theoretical value. Our process is useful for the precise fabrication of nonplanar structures based on inorganic materials.

1. Introduction

Micro-optical elements are becoming increasing important for compact and functional components in optical imaging, optical pickup systems and so on. Hybrid elements have received particular attention because they are useful for the creation of new photonic functions. For example, multifocal and achromatic lenses can be obtained by combining refractive and diffractive lenses because of the opposite wavelength dispersion between both elements. To fabricate hybrid structures, microfabrication on nonplanar substrates is required. However, it is difficult to obtain such structures using current semiconductor technology because of the difficulty in achieving a uniform resist coating. The fabrication of polymeric diffractive elements on bulk convex lenses was demonstrated by maskless ultraviolet laser lithography [3]. However, specially modified stage systems are needed. A spray-coating technique was also reported to obtain a uniform resist thickness, even on nonplanar structures. The spatial resolution is several tens of micrometers [4]. In these techniques, complex exposure systems are required because single-photon processes are employed.

Nonlinear optical processes such as multiphoton absorption and tunneling ionization occur only around the focal volume when femtosecond laser pulses are tightly focused into transparent materials.

Consequently, polymerization, densification and an increase in refractive index can be induced directly inside materials. By utilizing such nonlinear optical processes, three-dimensional (3-D) microsculptures of resin including photonic crystals and so on, have been formed using photosensitive resins such as SCR-500 and SU-8 [5-7]. Direct laser writing of 3-D waveguides inside glass materials has also been reported [8].

In this paper, we report femtosecond laser lithography-assisted micromachining (FLAM), which is a combined process of nonlinear lithography and subsequent plasma etching for fabricating structures onto nonplanar substrates. Silica-based diffractive-refractive hybrid microlenses were obtained using FLAM.

2. Experimental

We used a femtosecond fiber laser for this process. The wavelength, pulse duration and repetition rate were 780 nm, 68 fs and 50 MHz, respectively. The laser beam was focused by an objective lens with a numerical aperture (NA) of 0.5. The focal spot diameter and focal depth, as measured by the knife-edge method, were 2.6 and 14 μm, respectively. Silica glass plates of 1 mm thickness and silica microlens arrays were used. The diameter, height, curvature radius and focal length of each individual lens were 240, 18.9, 380, and 830 μm, respectively. We used a chemically amplified positive-tone photoresist PMER P-CA1000PM (Tokyo Ohka Kogyo Co., LTD.). Silica was etched by electron cyclotron resonance (ECR) plasma using CHF_3 gas.

3. Results and discussion

3.1. Femtosecond laser lithography-assisted micromachining

Figure 1 schematically shows the FLAM process for the fabrication on nonplanar structures. We begin with spin-coating of the positive tone resist PMER onto nonplanar substrates. Then, the patterns are directly written inside the resist, using femtosecond laser-induced nonlinear optical absorption. After post-exposure baking and development, the patterns are transferred to the underlying nonplanar substrates by plasma etching. Subsequently, we obtain nonplanar microstructures of inorganic materials after removal of the residual resist. Unlike previously reported femtosecond laser microsculptures, a positive-tone resist is desirable for use in FLAM because unwanted plasma damage can be avoided to the surroundings.

In the standard semiconductor process, ultraviolet light is absorbed from the resist surfaces. Therefore, a uniform resist coating is required to obtain well-defined patterns. When the resist is coated upon nonplanar substrates, the resist thickness varies from area to area mainly because of surface tension. As a result, it is difficult to form precise structures even with highly accurate laser control. In contrast, we can directly expose the internal region of the resist using the nonlinear process in FLAM. Unlike standard lithography, the pattern width is mainly determined by the diameters of the region where the nonlinear process occurs, and does not depend on the exposed position in the resist. Therefore, after plasma etching, we obtain precise inorganic structures even on nonplanar substrates.

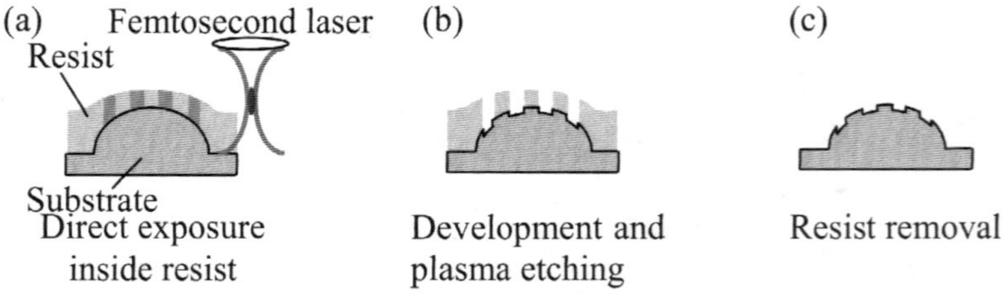

Figure 1. A schematic illustration of FLAM on nonplanar substrates.

3.2. Nonlinear lithographic processes

Cross sectional patterns of resists on silica glass plates are shown in Figure 2(a). The focal position for writing each pattern is indicated in the figure. The substrate surface is defined as the origin of the z-axis, which is the optical axis. These patterns were formed without laser scanning the laser spot along the z-axis. The average laser power and writing speed were 18 mW and 150 μm/s, respectively. It is apparent that the patterns were formed directly inside the resist under the condition that the focal position was at $z =$ 3.5 μm. PMER is transparent at wavelengths $\lambda > 480$ nm. This suggests that multiphoton absorption is responsible for the absorption at a laser wavelength of 780 nm.

The dependence of z-positions of the laser focal spot on pattern heights is shown in Figure 2(b). The dashed line indicates the resist thickness on the substrate. As shown in the inset, the pattern height and width were determined as the groove depth and surface width, respectively. We scanned the laser spot inside a resist of 13 μm thickness at a wring speed of 100 μm/s. The pattern widths were approximately 2–3 μm. The pattern heights decreased linearly with the focal position. The disagreement of the trend near the resist surface is most likely due to the shift in the focal spot, which originated from the difference in the refractive index between air and the resist. For our exposure setup, a minimum width of 900 nm was obtained at a power of 21 mW and writing speed of 1500 μm/s.

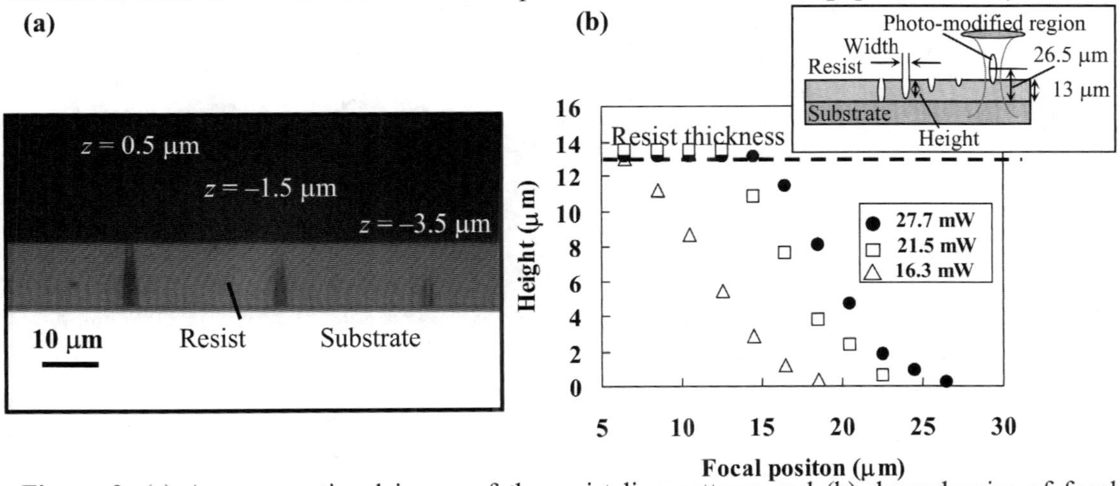

Figure 2. (a) A cross sectional image of the resist line patterns and (b) dependencies of focal positions on pattern heights.

3.3. Hybrid lenses

To obtain silica-based diffractive-refractive hybrid microlenses, binary Fresnel lenses were fabricated on convex microlenses. The focal length of the hybrid lens can be designed using eq. (1) [9]:

$$\frac{1}{f_h} = \frac{1}{f_m} + \frac{1}{f_F} \tag{1}$$

where f_h, f_F, and f_m are the primary focal lengths of a hybrid lens, a Fresnel lens, and a convex lens, respectively. From eq. (1) the focal length of a hybrid lens should be 618 μm when a Fresnel lens with a focal length of 2420 μm is formed on the convex lens. The radius r_m of the m-th zone of the binary Fresnel lens can be expressed by $r_m = \sqrt{m f_F \lambda_0}$ for an incident wavelength of 633 nm. Figure 3 schematically shows the laser writing procedure of the hybrid lenses. The average laser power and writing speed were 20 mW and 300 μm/s, respectively. Fresnel lenses consist of a series of concentric rings of different radii. The separation between adjacent rings (dr) was 2.0 μm, which was comparable in size to the diameter of the spot. To penetrate a thick resist (> 20 μm), multiple patterns of the Fresnel lens, centered on the same axis (z-axis), were written at different heights in the resist. The layer spacing was 3.0 μm, which was smaller than the height of the photo-modified region. Figure 4 shows SEM images of a silica-based diffractive-refractive hybrid microlens after pattern transfer. We obtained well-defined silica structures with smooth surfaces on curved substrates. The ECR power, bias voltage, and gas pressure were 100 W, 700 V and 2.0×10^{-2} Pa, respectively. The etching depth was approximately 1 μm. The primary focal length was 630 μm when a 633 nm wavelength He-Ne laser light was coupled normally to the hybrid lens. This hybridization shifted the focal length by 200 μm, which is consistent with the theoretical value of 212 μm. In the lithographic process, the patterns were written with dr = 2.0 μm. This relatively large spacing made it different to an ideal Fresnel lens structure resulting in a diffraction angle shift. An improvement in the FLAM process can result in the fabrication of more functional devices with 3-D surfaces on inorganic materials.

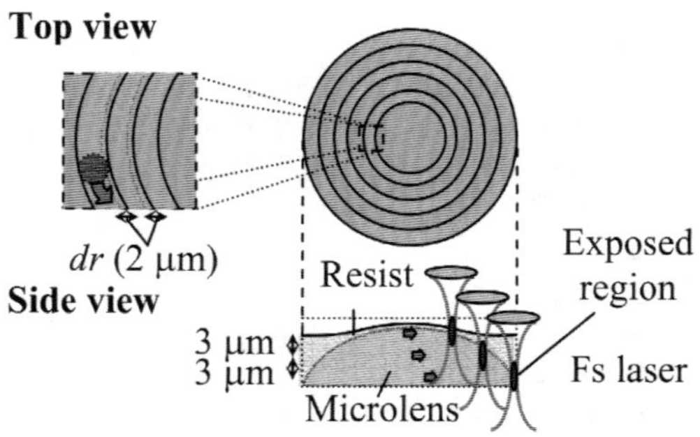

Figure 3. A schematic diagram for the laser writing of a micro-Fresnel lens onto a convex

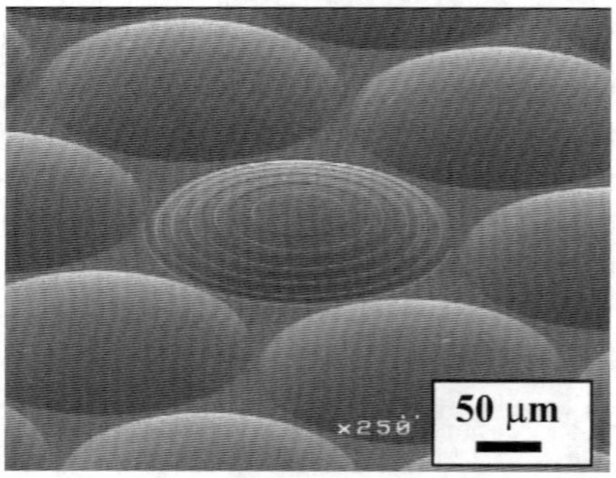

Figure 4. SEM images of a silica-based hybrid lens.

4. Conclusion

We report on FLAM, which is a combined process of nonlinear lithography and plasma etching, for the fabrication of nonplanar structures on inorganic materials. The fabrication of silica-based diffractive-refractive hybrid microlenses was demonstrated. The silica Fresnel lenses were created on 18.9 μm-thick convex lenses. This hybridization shifted the focal spot by 200 μm, which agreed with the theoretical value. Improvement of the FLAM process can lead to the fabrication of more functional devices with 3-D surfaces on inorganic materials.

Acknowledgements

This work was partly supported by Grants-in-Aid Nos. 19760034 and 18360354 and Priority Assistance for the formation of Worldwide Renowned Centers of Research—The Global COE Program (Project: Center of Excellence for Advanced Structural and Functional Materials Design) from the Ministry of Education, Culture, Sports, Science and Technology, Japan.

References

[1] Stone T and George N 1988 *Appl. Opt.* **27** 2960
[2] Sweeney D W and Sommargren G E 1995 *Appl. Opt.* **34** 2469
[3] Radtke D and Zeitner U D 2007 *Opt. Express* **15** 1167
[4] Singh V K, Sasaki M and Hane K 2007 *Jpn. J. Appl. Phys.* **46** 6449
[5] Maruo S, Nakamura O and Kawata S 1997 *Opt. Lett.* **22** 132
[6] Kawata S, Sun H –B, Tanaka T and Takada K 2001 *Nature* **412** 697
[7] Mizeikis V, Seet K K, Joudkazis S and Misawa H 2006 *Opt Lett.* **29** 2061
[8] Davis K M, Miura K, Sugimoto N and Hirao K 1996 *Opt. Lett.* **21** 1729
[9] Rossi M, Kunz R E and Herzig H P 1995 *Appl. Opt.* **34** 5996

Periodic structures in GeO_2-B_2O_3-SiO_2 glass films fabricated using ultraviolet laser pulses

Hiroaki Nishiyama[1], Junji Nishii[2] and [1]Yoshinori Hirata[1]

[1]Division of Materials and Manufacturing Science, Graduate School of Engineering, Osaka University, 2-1 Yamada-oka, Suita, Osaka, Japan.

[2]National Institute of Advanced Industrial Science and Technology, 1-8-31 Midorigaoka, Ikeda, Osaka, Japan

E-mail: hiroaki@mapse.eng.osaka-u.ac.jp

Abstract. We firstly observed the surfaces of periodic structures consisting of Ge nanoparticles in channel waveguide cores of GeO_2-B_2O_3-SiO_2 glasses. Such periodic structures were formed only by direct laser writing and subsequent annealing. Refractive index patterns were written in the glass films by irradiation with KrF excimer laser pulses. Then, the patterns were inverted by thermo-induced precipitation of Ge nanoparticles at unirradiated areas. Brown channel cores were obtained due to predominant precipitation of the nanoparticles in the cores. The periodic structures of 530 nm pitch and microchannels were directly observed by using wet-etched samples. The periodic structures were confirmed only inside the cores.

1. Introduction

Photonic band-gap (PBG) devices have been recognized as one of key components for integrated photonic systems. The PBG devices, which have artificial structures whose refractive index is periodically modulated, can control optical properties of propagation light. Due to their potential applications such as optical band-pass filters, high-Q resonators, integrated photonic sensors, display systems, and so on, PBG devices have been intensively studied in the last decade [1-3]. However, the fabrication of fine periodic structures which can work in optical communication wavelength is still a challenging topic.

Most PGB devices are produced by the semiconductor technology. To obtain submicron structures, we must use highly expensive electron-beam lithography system. In this system, electron beam exposure inside the vacuum chamber and complicated resist process are required.

On the other hand, direct laser writing techniques of photonic structures have attracted much attention. By these techniques, photonic devices such as optical filters and diffractive lenses can be directly written in glasses and polymeric materials [4-6]. Although this technique is highly flexible and does not require a vacuum chamber, laser-induced changes of refractive index are reported to be small and thermally less stable [7]. For example, the refractive index changes induced by ultraviolet laser exposure in optical fiber cores are as small as 10^{-4} even though photosensitizing process is used. Furthermore, the index changes start to decay by annealing at 150°C.

Recently, we have reported that crystalline Ge nanoparticles could be patterned in GeO_2-B_2O_3-SiO_2 (GBS) glass, which is often used as optical fiber and waveguide cores, by a combined process of ultraviolet laser exposure and thermal annealing [5,8,9]. We also demonstrated the formation of periodic structures consisting of the nanoparticles in the cores of waveguide channels, which were fabricated by the semiconductor technology. The refractive index changes of such nanoparticle structures are approximately ten times larger than those of conventional direct laser writing technique. Furthermore, there are no changes of the refractive index changes even after annealing at 500°C [8].

In the present paper, we firstly report the observation of the surfaces of PBG elements consisting of Ge nanoparticles inside channel cores, which are formed only by direct laser writing and subsequent annealing.

2. Experimental

GBS thin glass films with SiO_2 upper cladding layers were deposited on 1-mm thick SiO_2 substrates at 400°C by the plasma-enhanced chemical vapor deposition method. Respective thickness of GBSs and upper layers were approximately 4 and 1 μm. Liquid sources of $Si(OC_2H_5)_4$, $Ge(OCH_3)_4$ and $B(OC_2H_5)_3$ were used as raw materials for SiO_2, GeO_2 and B_2O_3, respectively. Photoinduced periodic refractive index changes were written by irradiation with KrF excimer laser of 248 nm wavelength through the phase mask of 1060 nm pitch at room temperature. Refractive indices of the films were measured by the prism coupling method using 632.8-nm light source. Thermal annealing was carried out in a tube furnace with nitrogen atmosphere. The sample surfaces were observed by FE-SEM.

3. Results and discussion

Figure 1 shows changes of refractive indices of the GBS films by annealing at 600°C. Here, we used two films; an as-deposited GBS film and a homogeneously irradiated GBS. The laser fluence and pulse number were 180 mJ/cm^2 and 27000, respectively. The refractive index increased by 3.1×10^{-3} after this irradiation. The index difference between both films was remarkably decayed by annealing only for 10 min by thermal relaxation. However, refractive indices increased by annealing longer than 10 min. The amount of increase for the unirradiated film was larger than that of the irradiated one, resulting in inversion of the refractive index between both films. The inverted refractive index difference was highly stable and no changes were observed even after further annealing at 500°C. Irradiation of higher fluence enhanced the difference of refractive index induced by the inversion.

We previously reported that periodic refractive index changes formed in GBSs by the direct laser writing technique were drastically enhanced by annealing at 600°C [8]. In the films, crystalline Ge nanoparticles of 20—40 nm diameters are precipitated by annealing at 600°C. This precipitation can be suppressed by laser exposure prior to the annealing. Due to this suppression, Ge nanoparticles, which have high refractive indices, were periodically precipitated, resulting in large refractive index modulation [8]. The inversion of refractive index in Fig. 1 is most likely due to this suppression effect of the nanoparticles by laser exposure.

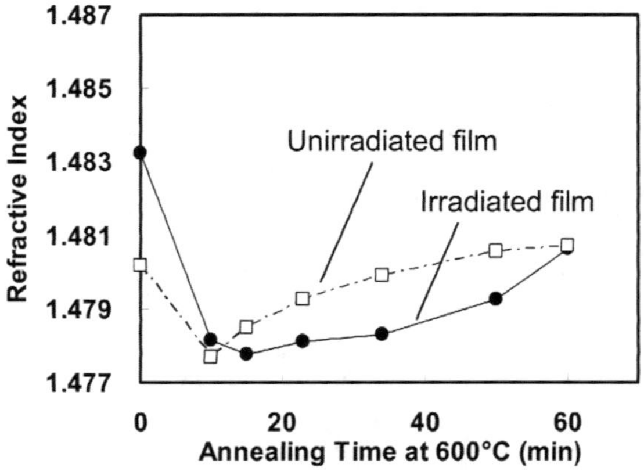

Figure 1. Changes of refractive indices of unirradiated films and homogeneously irradiated films as a function of annealing time at 600°C. The refractive indices were measured using 632.8 nm light source.

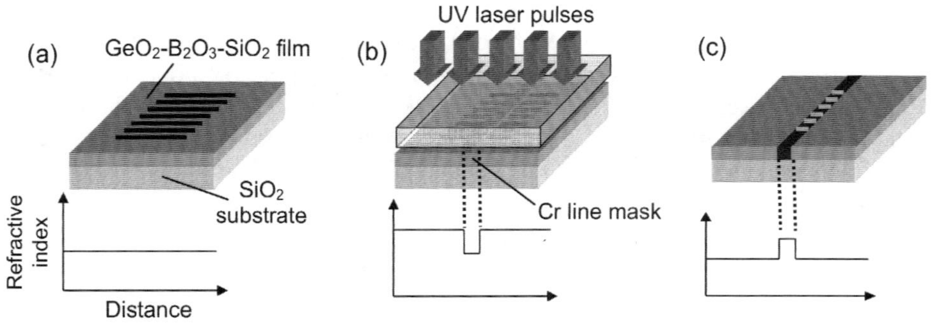

Figure 2. Fabrication processes for periodic structures consisting of Ge nanoparticles inside channels. (a) Formation of photoinduced periodic patterns, (b) homogeneous exposure through a Cr mask, and (c) thermal annealing at 600°C for 20 min.

Figure 2 presents the fabrication processes for periodic structures in channel waveguides. These photonic structures consist of Ge nanoparticles and work as PGB devices. We also indicate the expected cross-sectional distributions of refractive index at each step. First, 5-mm-long periodic structures are written by laser exposure through a phase mask. The period of the periodic structures is 530 nm. Then, homogeneous exposure is performed through a Cr mask with a straight line pattern of 7 μm width on a SiO_2 substrate. The laser fluence and pulse number are 180 mJ/cm^2 and 27000, respectively. As a final step, annealing at 600°C is carried out for 20 min. We can expect that the refractive index patterns are inverted by this annealing, resulting in the formation of periodic structures in 4-μm-high and 6-μm-wide channel cores.

An optical microscope image of the waveguide structures is shown in Fig. 3(a). We can see that a brown straight line was induced in the film. The line width was approximately 7 μm, which was consistent with the width of the Cr line mask. In our films, intense absorption bands are induced in ultraviolet-visible wavelength ranges by annealing at 600°C. According to previous studies, these

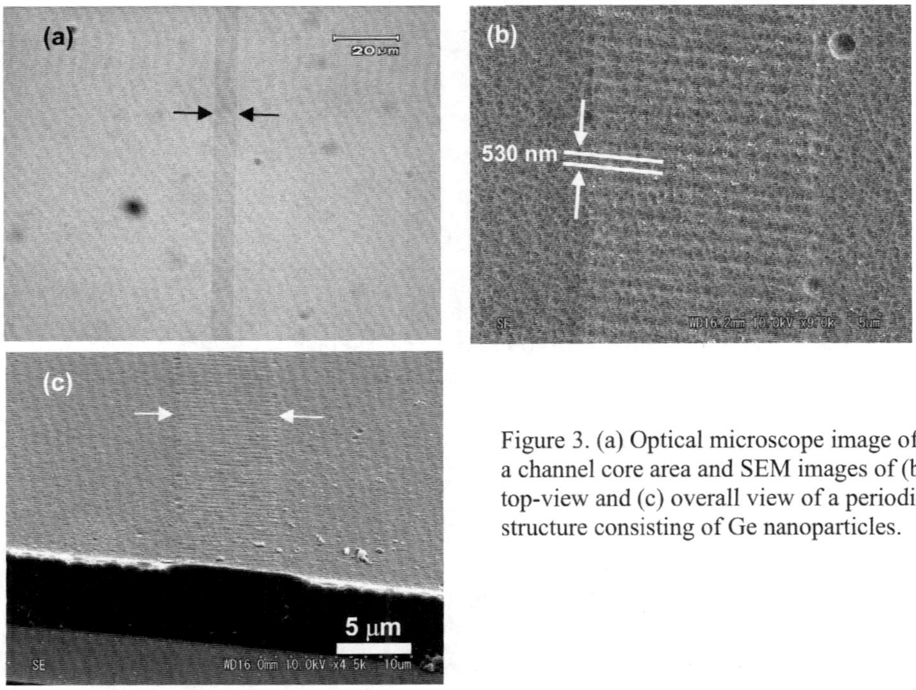

Figure 3. (a) Optical microscope image of a channel core area and SEM images of (b) top-view and (c) overall view of a periodic structure consisting of Ge nanoparticles.

absorption changes are originated from Ge nanoparticles [10]. As the diameter of the nanoparticles decreases, the absorption edge remarkably shifts to shorter wavelength due to the size effect. The edge of our GBSs is located at about 300 nm wavelength, and absorption shoulder spreads to visible wavelength. The laser exposure suppresses the precipitation of the nanoparticles. Therefore, we can consider that the concentration of the nanoparticle at the straight line area became higher than that of the surroundings, and brown-colored straight line appeared due to the absorption shoulder of the Ge nanoparticles. Figures 3 (b) and (c) present SEM images of top view and overall view of the samples after HF wet etching treatment. Although it was difficult to recognize individual nanoparticle by this SEM observation because the annealing time was as short as 20 min, the channel core and periodic structures in the films were visible. We could not confirm as-fabricated structures because their surfaces were smooth. The width of channel core and period of the structures were 7 μm and 530 nm, respectively, which were identical with those of the photoimprited patterns in Fig. 2. Bright lines in the periodic structures are likely to be aligned aggregated nanoparticles. Note that the periodic structures at clad areas (outside areas of the cores in the film) disappeared. This erase is most likely due to exposure with higher flunece in Fig. 2(c) than that in Fig. 2(a). By this exposure, photoinduced refractive index contrast in Fig. 2(a) decayed because larger refractive index changes can be induced by higher fluence. This confinement of the periodic structures in the core areas can avoid coupling of propagation light to clad and leaky modes, which is important for practical applications. The PBG properties of the periodic structures were reported in our previous works [5]. These structures exhibited high diffraction efficiencies and no changes of their optical properties were observed even after further annealing up to 500ºC. Our laser-assisted patterning technique of the nanoparticles is useful for the fabrication of more complicated functional photonic structures.

4. Conclusions

We firstly observed the periodic structures consisting of Ge nanoparticles in microchannels formed only by direct laser writing and subsequent thermal annealing. These photonic structures were induced by the inversion of photoimprinted patterns during the annealing. Brown-colored channel cores of 7 μm widths were observed, indicating that Ge nanoparticles were mainly precipitated in the core areas. By using HF etching of sample surfaces, we could see the periodic patterns of 530 nm pitch and microchannel directly. The periodic structures were induced only inside the cores.

Acknowledgements

This work was partly supported by Priority Assistance for the Formation of Worldwide Renowned Centers of Research – The Global COE Program (Project: Center of Excellence for Advanced Structural and Functional Materials Design) from the Ministry of Education, Culture, Sports, Science, and Technology, Japan

References

[1] Yablonvitch E 1993 *J. Opt. Soc. Am. B.* **10**, 283.
[2] Noda S, Tomoda K, Yamamoto N and Chtinan A 2000 *Science* **289**, 604.
[3] Hill K O and Meltz G 1997 *J Lightwave Technol.* **15**, 1263.
[4] Davis K M, Miura K, Sugimoto N and Hirao K 1996 *Opt. Lett.* **21**, 1729.
[5] Nishiyama H, Miyamoto I, Matsumoto S, Saito M, Kintaka K and Nishii J 2004 *Opt. Exp.* **12**, 4589.
[6] Nishiyama H, Mizoshiri M, Kawahara T, Nishii J and Hirata Y 2008 *Opt. Exp.* **16**, 17288
[7] Douay M, Xie W X, Taunay T, Bernage P, Niay P, Cordier P, Poumellec B, Dong L, Bayon J F, Poignant H and Delevaque R 1997 *J Lightwave Technol.* **15,** 1329.
[8] Nishiyama H, Miyamoto I, Matsumoto S, Saito M, Fukumi K, Kintaka K and Nishii J 2004 *Appl. Phys. Lett.* **85** 3734.
[9] Nishiyama H, Miyamoto I and Hirata Y 2007 *Opt. Exp.* **15**, 2047.
[10] Yang H, Yao X, Xie S, Wang X, Liu S, Fang Y, Gu X and Wang F 2005 *Opt. Mater.* **27,** 725.

International Conference on Advanced Structural and Functional Materials Design 2008 IOP Publishing
Journal of Physics: Conference Series **165** (2009) 012050 doi:10.1088/1742-6596/165/1/012050

Magnetocaloric effect in ferromagnetic Heusler shape-memory alloys

Antoni Planes, Lluís Mañosa

Departament d'Estructura i Constituents de la Matèria. Facultat de Física. Universitat de Barcelona. Diagonal 647, 08028 Barcelona, Catalonia

E-mail: toni@ecm.ub.es

Mehmet Acet

Experimentalphysik, Universität Duisburg-Essen, D-47048 Duisburg, Germany

Abstract. In this paper the magnetocaloric behaviour of Ni-Mn-based Heusler alloys is discussed in relation to their shape-memory and superelastic properties. We show that the magnetocaloric effect in these materials originates from two different contributions. The first, associated with the mechanism which makes feasible the magnetically induced rearrangement of martensite variants, controls the magnetocaloric effect at low applied fields, while the latter is dominant at higher fields and is essentially related to the possibility of magnetically inducing the martensitic transition. The occurrence of the inverse magnetocaloric effect associated with these two contributions is also considered.

1. Introduction

Ferromagnetic Heusler alloys undergoing a martensitic transition show giant magnetostriction, which is related to either a field-induced reorientation of martensitic twin-related variants (the magnetic shape-memory effect) or to magnetic superelasticity, which originates from the possibility of inducing the transition by means of a magnetic field. Prototypical examples of systems showing magnetic the shape-memory effect and magnetic superelasticity are Ni_2MnGa [1] and Mn-rich $Ni_2Mn_{1.4}In_{0.6}$ [2] respectively. Associated with the martensitic transition, these materials also display interesting magnetocaloric effects which are related to the thermal response to magnetization changes. These changes are commonly induced by the application or removal of an externally applied magnetic field and the magnetocaloric effect is measured as a temperature change when the field is applied adiabatically or as an entropy change when it is applied isothermally [3].

In the present paper we will discuss the magnetocaloric properties exhibited by Heusler alloys of the Ni-Mn-based family. These materials display either conventional or inverse magnetocaloric effect depending on specific composition and, in some cases, on the strength of the applied field. We will show that this behaviour is a direct consequence of the coupling between magnetic and structural degrees of freedom.

2. General features and results

In what follows the magnetocaloric effect will be quantified by means of the isothermal entropy change which results from the application of a magnetic field H. It can be expressed as,

$$\Delta S(T,H) = \int_0^H \left(\frac{\partial M}{\partial T}\right)_H dH\,, \tag{1}$$

where the Maxwell relation $(\partial S/\partial H)_T = (\partial M/\partial T)_H$ has been used. Therefore, $\Delta S(T,H)$ can be obtained from magnetization measurements performed in a conveniently selected range of temperatures. Fig. 1 shows $M(H)$ isothermal curves for an almost stoichiometric Ni_2MnGa single crystal in a temperature range about the martensitic transition ($T_M \simeq 180K$).

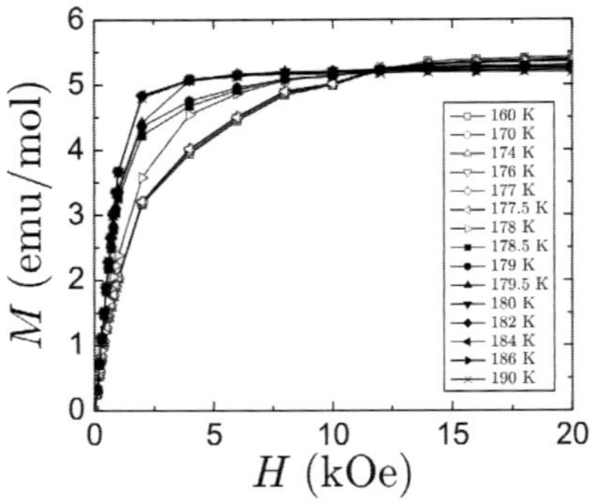

Figure 1. $M(H)$ curves at temperatures in the range of the martensitic transition (from 160 K to 190 K). H is applied along the [001] direction.

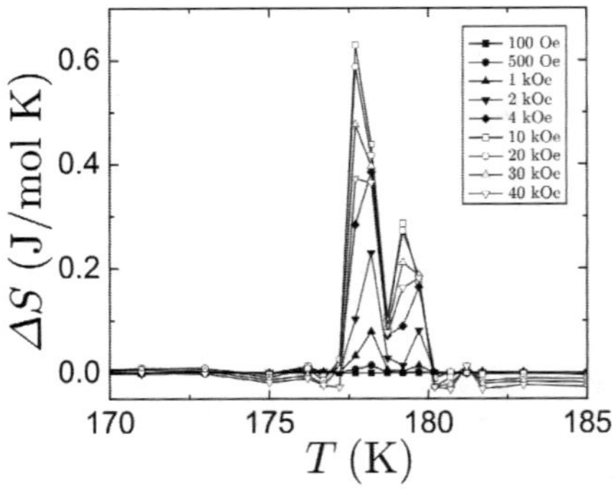

Figure 2. Field-induced entropy change as a function of temperature at selected values of the maximum applied field along the [001] direction.

From the magnetization data in Fig. 1 the field-induced entropy change is obtained as a function of temperature and selected values of H by using eq. (1). The derivative and integral involved in the calculations were performed numerically. Results are shown in Fig. 2. Interestingly, the $\Delta S(H,T)$ curves show peaks that arise from discontinuities in the transformation path. These discontinuities are characteristic of martensitic transitions in shape-memory alloys [4, 5]. In order to compare the magnetocaloric effect in different Heusler systems, it is convenient to calculate an average field-induced entropy change over the transition range ΔT, for each H. This expression is obtained as,

$$\langle \Delta S(H) \rangle = \frac{1}{\Delta T} \int_{\Delta T} \Delta S(T,H) dT. \tag{2}$$

In Fig. 3 we show this average field-induced entropy change for five alloys of compositions close to the $Ni_{2+x}Mn_{1-x}Ga$ line. Along this line, with increasing x, T_M increases and approaches the Curie temperature T_c (for $x \simeq 0.2$, $T_M \simeq T_c$). Except for alloy with $T_M \simeq T_c$ (alloy 5), the entropy change first increases with H, reaches a positive maximum, and decreases linearly for high fields. In particular, the initial increase of $\langle \Delta S \rangle$ is rather large for alloys 1 and 2 with composition closer to the 2-1-1 stoichiometry . It has been shown that this increase is related to the decrease of $\Delta M(H)$ observed at low applied fields, which is a consequence of the magnetostructural interplay driven by the strong uniaxial magnetocrystalline anisotropy of the martensitic phase [6, 7]. At higher fields, $\langle \Delta S(H) \rangle$ decreases linearly with increasing field. In alloy 5 an almost linear decrease of $\langle \Delta S(H) \rangle$ is observed in the full range of the field. Actually,

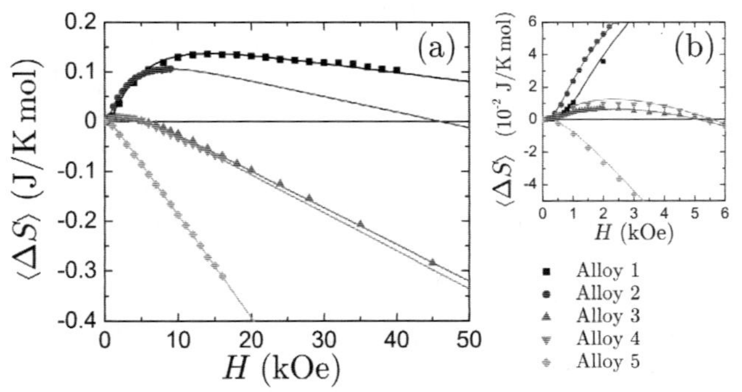

Figure 3. (a) Averaged field-induced entropy change as a function of the applied field H for alloys of compositions close to the $Ni_{2+x}Mn_{1-x}Ga$ line. x increases from alloy 1 ($x \simeq 0$) to alloy 5. For alloy 5, $T_M \simeq T_c$. (b) Expanded view of the low-field region.

the linear decrease is due to the fact that for this alloy family the magnetic moment of the martensite is larger than the magnetic moment of the high temperature cubic phase.

Besides Ni-Mn-Ga, martensitic transformations have been reported in other ferromagnetic Heusler alloys. Within the Ni-Mn-Z family, Ni-Mn-Sn and Ni-Mn-In display very interesting properties. These alloys undergo a martensitic transition in the composition region $Ni_2Mn_{2-4x}X_{4x}$, with $x < 0.25$ (Mn-rich region) far from stoichiometry. The interesting composition range is a narrow interval close to $e/a \simeq 8$ where, on cooling, these systems first become ferromagnetic and on further cooling undergo a martensitic transition [8]. For $Ni_2Mn_{1.4}Sn_{0.6}$ and $Ni_2Mn_{1.44}In_{0.56}$ an average field induced entropy change has been obtained following the same procedure as in the Ni-Mn-Ga family. Results are shown in Fig. 4. In these systems $\langle \Delta S(H) \rangle$ linearly increases within the full range of the applied field. This inverse magnetocaloric effect is a consequence of the decrease of the magnetic moment which takes place at the magnetostructural transition. Interestingly, this mechanism enables the reverse martensitic transition to be induced by means of an applied field (the change of transition temperature with the applied magnetic field is particularly strong in the case of the Ni-Mn-In alloy). The drop in the magnetic moment has been shown to originate from the tendency of the excess Mn atoms (with respect to the 2-1-1 stoichiometry) to introduce antiferromagnetic exchange coupling caused by the change in the Mn-Mn distance as the martensitic phase gains stability. It is worth noting that a large drop of magnetic moment in Ni-Mn-In is basic feature that makes possible the magnetic superelastic behaviour displayed by this alloy [2]. In contrast, as the magnetic anisotropy is low in Ni-Mn-In and Ni-Mn-Sn, field-induced reorientation of martensitic variants is not operative in these systems [9].

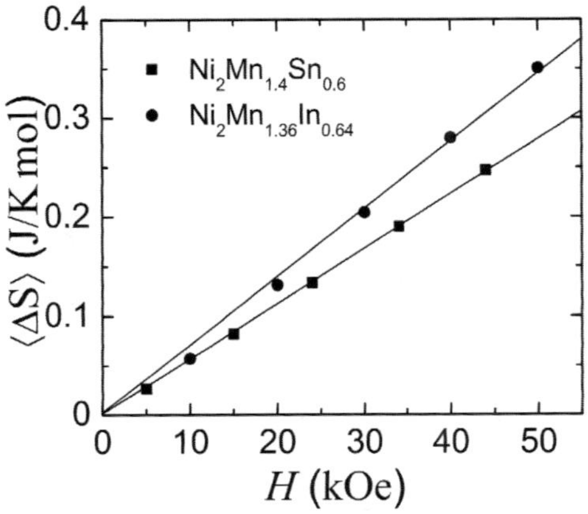

Figure 4. Average field-induced entropy change as a function of the field H for $Ni_2Mn_{1.4}Sn_{0.6}$ (squares) and $Ni_2Mn_{1.36}In_{0.64}$ (circles).

3. Discussion and summary

Results presented in this paper show that the magnetocaloric effect associated with the martensitic transition in Ni-Mn based Heusler alloys is determined by two contributions. The former is an extrinsic contribution which is a consequence of the magnetostructural interplay at the length scale of martensitic variants driven by the magnetic anisotropy of the martensitic phase. The latter is intrinsic and is controlled by the change of magnetization taking place at the transition. In nearly stoichiometric Ni_2MnGa the first contribution is dominant, while it plays a minor role in Ni-Mn-Ga alloys undergoing the martensitic transition close to or above the Curie point. The extrinsic contribution is also irrelevant in Ni-Mn based alloys such as Ni-Mn-Sn and Ni-Mn-In. In these alloys, however, due to transition-induced antiferromagnetism, the magnetocaloric effect is inverse. Therefore, in these materials cooling occurs by adiabatic magnetization.

Acknowledgments

This work has received financial support from CICyT (Spain), project MAT2007-61200, DURSI (Catalonia), Project No. 2005SGR00969 and Deutsche Forschungsgemeinschaft (No. SPP1239).

References

[1] Ullako K et al. 1996 *Appl. Phys. Lett.* **69** 1966
[2] Krenke T et al. 2007 *Phys. Rev. B* **75** 104414.
[3] Brück E 2008 *Handbook of magnetic materials* vol. 17, ed. by K.H.J. Buschow, Elsevier Science, Amsterdam, 1999, pp. 235–291.
[4] Pérez-Reche F J et al. 2003 *Phys. Rev. B* **69** 064101.
[5] Sullivan M R et al. 2004 *Phys. Rev. B* **70** 094428.
[6] Marcos J et al. 2002 *Phys. Rev. B* **66** 224413.
[7] Marcos J et al. 2003 *Phys. Rev. B* **68** 094401.
[8] Moya X et al. 2006 *Mater. Sci. Engng. A* **438-440** 911.
[9] Aksoy S et al. 2008 to be published.

Effects of magnetic field on martensitic transformations

Tomoyuki Kakeshita and Takashi Fukuda

Department of Materials Science and Engineering, Graduate School of Engineering, Osaka University, 2-1 Yamada-oka, Suita, Osaka 565-0871, Japan

kakeshita@mat.eng.osaka-u.ac.jp

Abstract. Magnetic field is one of intensive variables which influence phase transformations in solids. In the present study we will show how the characteristics of martensitic transformations are influenced by magnetic field, such as transformation temperature, morphology and kinetics.

1. Introduction

Martensitic transformation is one of typical first-order structural phase transitions without atom diffusion and has been widely studied to determine its characteristics from physical, metallographical and crystallographic view points[1]. According to studies by Patel and Cohen[2] and Otsuka et al.[3], some aspects of martensitic transformation such as transformation temperatures, crystallography and morphology of the product martensites are strongly influenced by external field, such as temperature, hydrostatic pressure and uniaxial stress. Thus, studies on the effects of these external fields are important to understand the essential problems of the martensitic transformation such as thermodynamics, kinetics and the origin of the transformation. Magnetic field is one of such external fields because some difference in magnetic moment between the parent and martensite states exists[4]. We have especially investigated the effect of magnetic field on martensitic transformation in many alloys. Typical results are presented in the paper.

2. Effect of magnetic field on martensitic transformation temperature

We schematically show how the transformation temperature is influenced by magnetic field in Figure 1. Gibbs chemical free energy of the parent phase G^{P} is lower than that of the martensite phase G^{m} above equilibrium temperature T_0. Martensitic transformation starts at M_s which is below T_0 because the interfacial energy and strain energy are needed to initiate the transformation. Under a magnetic field, the G^{m} decreases mainly due to the magnetostatic energy (change in G^{P} is neglected here for simplicity), resulting in increase in T_0. Also, the M_s under the magnetic field increases to M_s' if we assume that the martensitic transformation under the magnetic field occurs at the temperature where G^{m}-G^{p} under magnetic field is the same as that under no magnetic field at M_s.

In the following, we will show the results for Invar Fe-Ni alloys, disordered and ordered Fe-Pt Invar alloys and non-Invar Fe-Ni-C alloys[5-7]. Their structural changes are basically those from fcc to bcc. High-field magnetization measurements were performed at Research Center for Materials Science at Extreme Conditions, Osaka University [8]. The solid marks in figure 2 show the change in martensitic transformation temperature ΔMs (= M_s'–M_s) for Fe-Ni, Fe-Pt and Fe-Ni-C alloys. It is known form the figure that ΔM_s increases with increasing magnetic field for all the alloys examined, and the rate of the increase against the magnetic field depends on the alloy system.

We have proposed [9] the following equation to estimate the relation between the critical magnetic field and the transformation temperature:

$$\Delta G(M_s) - \Delta G(M_s') = -\Delta M(M_s')H_c - \frac{1}{2}\chi_h{}^p H_c{}^2 + \varepsilon_0\left(\frac{\partial\omega}{\partial H}\right)H_c B\,, \qquad (1)$$

where $\Delta G(M_s)$ and $\Delta G(M_s')$ is the difference in Gibbs chemical free energy between the parent and martensite phases at M_s and M_s' temperatures, respectively, $\Delta M(M_s')$ is $M^m(M_s')$–$M^p(M_s')$ (where M^m and M^p are the spontaneous magnetization of the martensite and parent phases, respectively), $\chi_h{}^p$ is the high magnetic field susceptibility in the parent phase, ε_0 is the volume change associated with martensitic transformation, ω is the forced volume magnetostriction and B is the parent bulk modulus. The first term on the right-hand side of Eq.(1) represents the energy due to the magnetostatic effect, which was already proposed by Krivoglaz and Sadovsky[4]. The second and the third terms are introduced by our group, which represent the energies due to the high-field susceptibility and forced volume magnetostriction effects, respectively. The second term in Eq.(1) is related to the induced magnetic moment in the parent phase by applying magnetic field and is originally the same as magnetostatic energy like the first term. The third term in Eq.(1) is quite effective for the Invar alloys because ω in Invar alloys are larger by one order than those of non-Invar alloys.

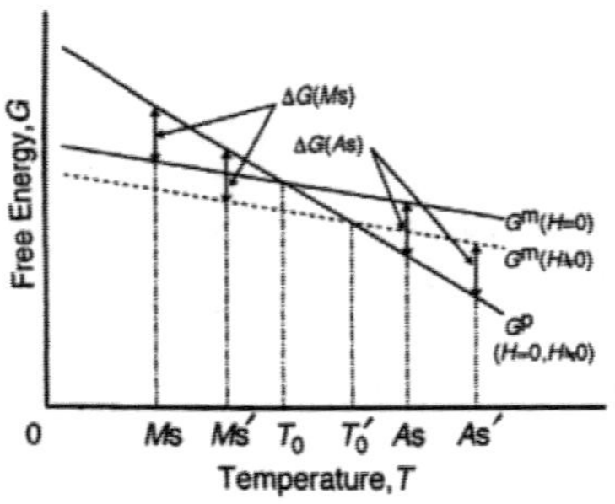

Figure 1. Schematic Gibbs chemical free energy as a function of temperature under a magnetic field.

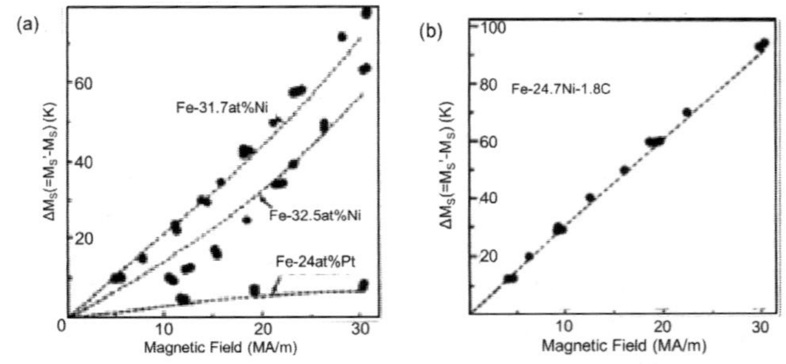

Figure 2 ΔM_s vs H_c relation for Invar Fe-Ni and Fe-Pt alloys (a), non-Invar Fe-Ni-C (b). Solid marks are measured values and dotted lines are calculated relations.

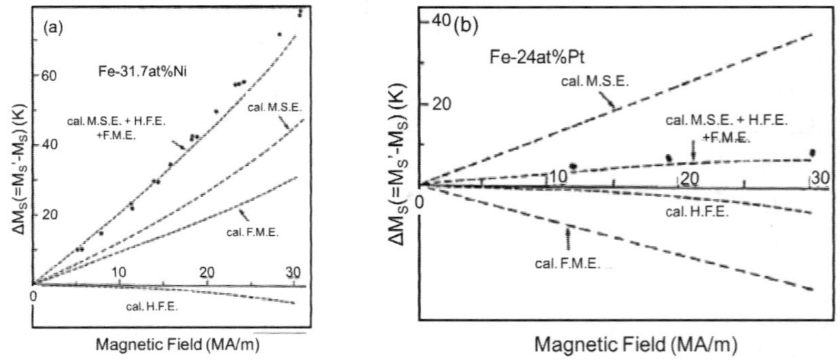

Figure 3. Shift of M_s as a function of magnetic field for Invar Fe-31.7Ni (a), and Invar ordered Fe-24Pt alloys (b).

The calculated relation between ΔM_s and magnetic field based on Eq.(1) are shown in figure 2 with dotted lines, which are in good agreement with the experimental ones for all the alloys. We also show the calculated magnetic field dependence of transformation temperature for each term in Eq.(1) in order to know which term is effective in changing the transformation temperature by magnetic field. Typical calculated results for the Invar Fe-31.7at.%Ni alloy and the Fe-24.0at.%Pt alloy with $S = 0.8$ are shown in figure 3, where the dotted lines denoted by M.S.E., H.F.E., F.M.E. and (M.S.E.+H.F.E.+F.M.E.) are the H_c vs M_s' relations calculated for the magnetostatic, high field susceptibility, forced volume magnetostriction and their total effects, respectively. As known form the figure, the calculated relations (M.S.E.+H.F.E.+F.M.E.) are in good agreement with the experimental ones for both alloys. It should be noted that the shift of M_s due to the forced magnetostriction effect is nearly of the same order as that due to the magnetostatic effect for both alloys, and the shift of M_s due to this effect is a decrease in the ordered Fe-Pt alloy, but an increase in the Fe-Ni alloy. This difference is due to the fact that the volume change associated with the martensitic transformation in the ordered Fe-Pt alloy is negative, but positive in the Fe-Ni alloy. Incidentally, in Eq.(1) we neglect the energies due to the magnetic anisotropy because these effects are small in many iron-based alloys. Recently, however, the magnetocrystalline anisotropy was found to play an important role in magnetic field-induced martensitic transformation in Co-Fe alloy[10] and in Ni-Mn-Ga alloy[11].

3. Magnetoelastic Martensitic Transformation

In alloys exhibiting a thermoelastic martensitic transformation, each martensite crystal elastically responds to temperature to maintain the balance between thermal and elastic energies. If a uniaxial stress is applied to such an alloy at a temperature above A_f and released, the forward and reverse transformation occur by a loading cycle. That is, the martensite is induced only while the stress is loaded. In analogy to this behavior, it is expected that if a magnetic field is applied to an alloy above A_f and removed, martensite can be induced only while the magnetic field is applied. We found this type of martensitic transformation in an ausaged Fe-31.9Ni-9.8Co-4.1Ti (at.%) shape memory alloy and termed it as a magnetoelastic martensitic transformation[12] in analogy to the thermoelastic transformation.

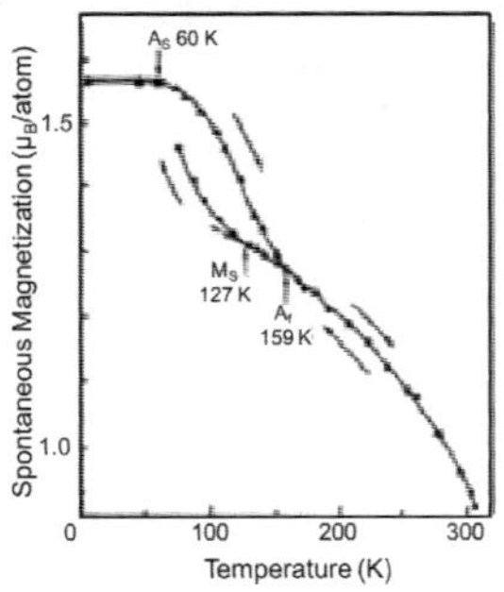

Figure 4. Spontaneous magnetization as a function of temperature in an ausaged Fe-Ni-Co-Ti alloy.

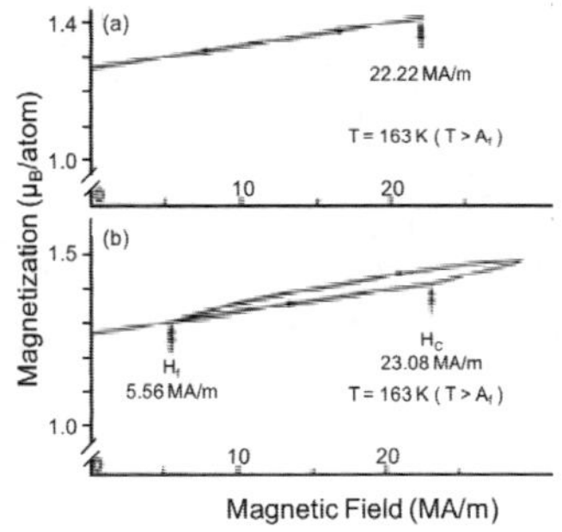

Figure 5. M(t)-H(t) curves for an ausaged Fe-Ni-Co-Ti alloy at 163K above A_f.

Figure 4 shows the spontaneous magnetization of the parent and the martensite phases as a function of temperature, which was obtained by magnetization measurements under low magnetic field. The difference in spontaneous magnetization between the two phases is about $0.3\mu_B$/atom at M_s, which is of the same order as that in a previous Fe-32.5at.%Ni alloy. M_s and A_f shown with arrows are determined to be about 127 and 159K, respectively and A_s is 60K. A pulsed high magnetic field was applied to a sample at a temperature above A_f, 163K (ΔT=(T-M_s)=36K) and typical $M(t)$-$H(t)$ curves obtained are shown in figure 5(a) and (b). It is noted in figure 5(a) that there is no hysteresis in magnetization when a pulsed magnetic field of 22.22MA/m is applied and removed. This means that 22.22MA/m is lower than the critical field, H_c, for inducing the martensitic transformation, and therefore no martensitic transformation occurs under the magnetic field of 22.22MA/m. Then, a higher magnetic field was applied, and the obtained $M(t)$-$H(t)$ curve is shown in (b), which reveals a hysteresis of magnetization. That is, when a magnetic field is applied, the rate of increase of magnetization against magnetic field changes at H_c=23.08MA/m, as indicated with an arrow, and when the magnetic field is removed, the increased magnetization returns to the initial value at about H_f=5MA/m indicated with another arrow. This means that martensitic transformation is induced at H_c and its reverse transformation is completed at H_f. These observations show that the magnetoelastic martensitic transformation is realized in the ausaged Fe-Ni-Co-Ti alloy.

4. Effect of Magnetic Field on Morphology of Martensites

Figure 6 shows optical micrographs of thermally-induced martensites formed by cooling a little below the M_s temperature in Fe-Ni-C alloys ((a), (d) and (g)) and those of magnetic field-induced ones ((b), (c), (e), (f), (h) and (i)). The formation temperature T, ΔT (=T-M_s) and H are shown below each photograph. Despite the different formation temperatures, the Fe-28.7Ni-1.8C mass% alloy exhibits thin plate morphology and other two alloys do lenticular one. It is noted that the martensite morphology is different even if the martensites are formed nearly the same temperature, as seen from the comparison of (c) and (d). This result is contradictory to the proposition that the martensite morphology in Fe-Ni-C alloys is decided only by the formation temperature[13]. The reason for this difference is not known yet.

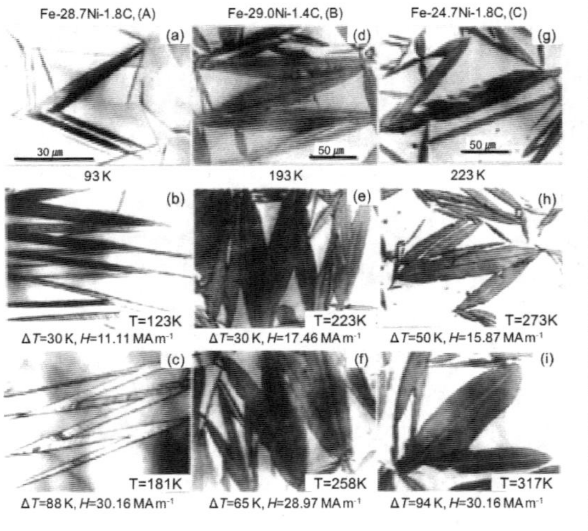

Figure 6. Optical micrographs of thermally-induced martensites ((a), (d), (g)) and magnetic field induced-ones ((b), (c), (e), (f), (h), (i)) in Fe-Ni-C alloys.

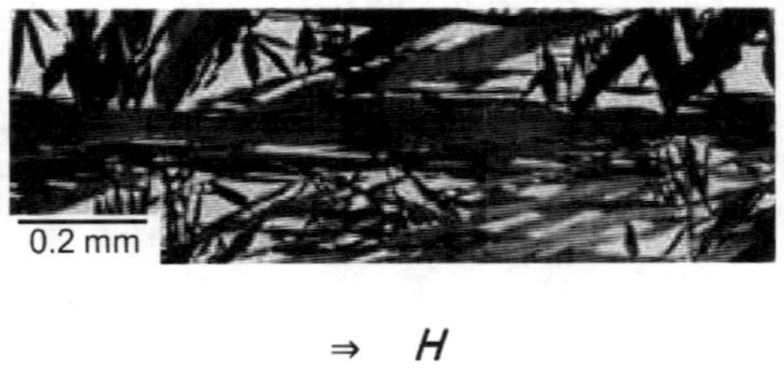

Figure 7. Morphology of field-induced martensite in a single crystal of Fe-31.6Ni (at.%) alloy.

Figure 7 is an optical micrograph showing the morphology of magnetic field-induced martensites in Fe-31.6Ni (at.%) alloy single crystal. It is known from the figure that several martensite plates grow nearly parallel to the direction of magnetic field. Such a directional growth of martensite plates does not appear in thermally-induced martensite. Therefore, the directional growth appears to be characteristic of magnetic field-induced martensites. We also notice that other plates terminate at the directionally grown plates. This means that the directionally grown plates were first formed and the other plates followed. The reason for such lengthwise growth of martensite plates under magnetic field is not clear, but shape magnetic anisotropy effect plays an important role.

5. Kinetics of Martensitic Transformation

Martensitic transformations are frequently classified into two groups with respect to kinetics: athermal and isothermal transformations. The former type has a well defined transformation temperature M_s and occurs spontaneously at M_s, while the latter does not have M_s but occurs after some finite incubation time during isothermal holding.

Fe-Ni-Mn is an alloy system in which both the athermal and isothermal transformations occurs depending on composition. For example, Fe-31.4Ni-0.5Mn alloy has a clear M_s temperature at 195 K, while Fe-29.4Ni-3.9Mn (mass%) alloy does not. The latter alloy exhibits a C-curve in the *TTT* diagram as shown in figure 8. We applied a pulsed magnetic field on these alloys. Figure 9 (a) is a *M-H* curve of the former alloy obtained 23 K above its M_s, and (b) is a *M-H* curve of the latter alloy at 4.2 K. Note that the same increase in magnetization is observed in both alloys. The increase in magnetization is due to martensitic transformation. The time required for the formation of all the martensite is about 20μs. This result means that isothermal transformation in the Fe-29.4Ni-3.9Mn alloy changes to athermal one. On the basis of this finding, we have constructed a phenomenological theory which gives a unified explanation for the two transformations[14]. The dotted curve in figure 8 is calculated C-curve by using this model, which agree well with experimental result.

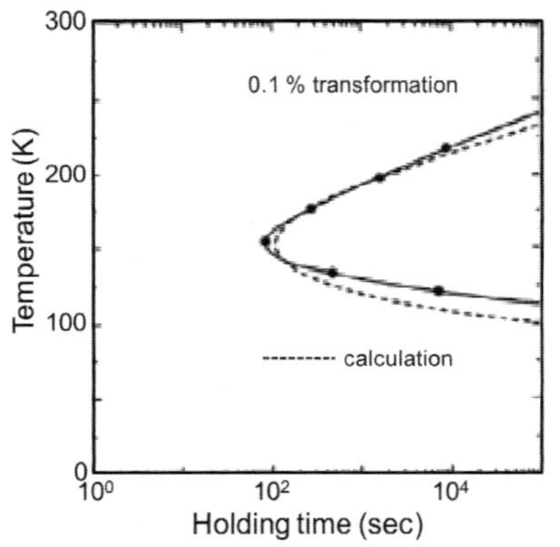

Figure 8. *TTT* diagram of the isothermal martensitic transformation in Fe-29.4Ni-3.9Mn (mass%) alloy.

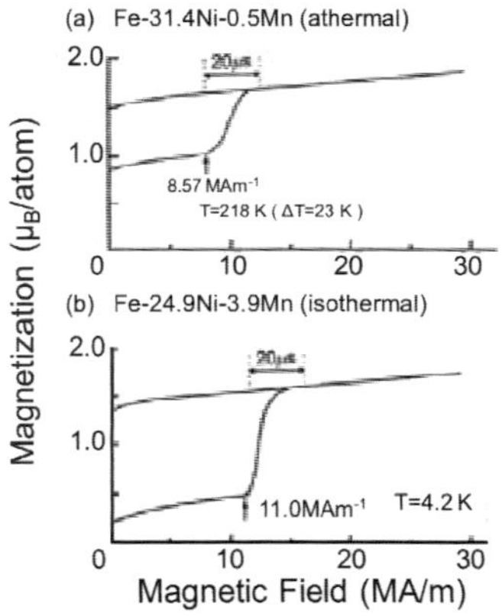

Figure 9. Magnetization vs magnetic field curves showing field-induced transformation.

6. Giant Magnetic Field-Induced Strain in Ferromagnetic Shape Memory Alloys

Shape memory alloys exhibit a large recoverable strain of several percent in association with rearrangement of martensite variants. Recently, it was found by Ullakko et al. that the large strain can be induced by applying a magnetic field in ferromagnetic shape memory alloy of Ni_2MnGa. After that, similar results were found in other shape memory alloys, such as Fe-Pd alloys[16, 17] and in Fe_3Pt[18]. In this section, we will show our results of magnetic field-induced strain (MFIS) in these alloys, and explain the condition for realizing the large magnetic field-induced strain.

Figure 10 shows the MFIS obtained in single crystals of Fe-31.2Pd, Fe_3Pt and Ni_2MnGa. In all the measurements, the specimen was cooled down to 4.2 K or 77 K without applying magnetic field. Then the magnetic field was applied in the $[001]_P$ direction. This direction corresponds to easy axis or hard axis depending on variants. The strain starts to appear at a field of about 0.25MA/m and then saturates. The strain is due to rearrangement of martensite variants, which has been confirmed by optical microscopy or by x-ray diffraction.

The reason for the rearrangement of martensite variants by magnetic field is essentially explained by magnetocrystalline anisotropy constant K_u and the twinning stress. The magnetic field will exert a magnetic shear stress across the twinning plane, and the maximum of it should be $|K_u|/s$, where s is twinning shear. In order to realize the rearrangement of martensite variants, $|K_u|/s$ should be larger than the stress required for twinning plane movement. We have confirmed that this condition is actually realized when the MFIS occurs in the three alloys shown in figure 10 regardless of temperature.

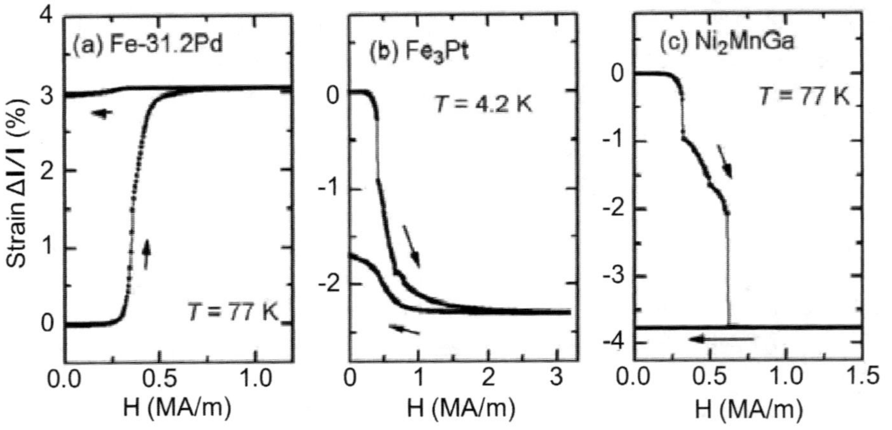

Figure 10. The longitudinal magnetic field-induced strain along the $[001]_P$ direction of Fe-31.2Pd at 77 K(a), Fe_3Pt at 4.2 K (b), and Ni_2MnGa at 77 K(c).

Acknowledgements
The present work was partly supported by Global COE program (Center of Excellence for Advanced Structural and Functional Materials Design), Osaka University.

References

[1] Nishiyama Z 1978 *Martensitic transformations* (Academic Press, New York)
[2] Patel J R and Cohen M 1953 Acta Metall. **1** 531
[3] Otsuka K, Sakamoto H and Shimizu K 1976 Acta Metall. **24** 585
[4] Krivoglaz M A, Sadovsky V D 1964 Fizika Metal Metall. **18** 502
[5] Kakeshita T, Shimizu K, Funada S and Date M 1985 Acta Metall. **33** 1381
[6] Kakeshita T, Shimizu K, Funada S and Date M 1984 Trans. Jpn. Inst. Met. **25**, 837
[7] Kakeshita T, Shimizu K, Kijima S, Yu T and Date M 1985 Trans. Jpn. Inst. Met. **26**, 630
[8] Date M, 1986 *Physics in High Magnetic Fields* (Springer, Berlin) pp. 44.
[9] Kakeshita T and Shimizu K 1986 *Proc. ICOMAT-86*, (JIM) pp. 230.
[10] Tanaka T, Takahashi M and Kadowaki S, 2000 Mater. Trans. JIM **41** 950
[11] Kim J-H, Inaba F, Fukuda T and Kakeshita T 2006 Acta Mater. **44** 439
[12] Kakeshita T, Shimizu K, Maki T, Tamura I, Kijima S and Date M 1985 Scripta Metall. **19** 973
[13] Maki T, Shimooka M, Umemoto M and Tamura I 1971 J. Japan Inst. Metals. **35**, 1073
[14] Kakeshita T, Kuroiwa K, Shimizu K, Ikeda K, Yamagishi A, Date M 1993 Mater. Trans. JIM **34** 423.
[15] Ullakko K, Huang J K, Kantner C, O'Handley R C and Kokorin V V 1996 Appl. Phys. Lett. **69** 1966
[16] James R D and Wuttig M 1988 Philos. Mag. **A77** 1273
[17] Sakamoto T, Fukuda T, Kakeshita T, Takeuchi T and Kisho K 2003 J. Appl. Phys. **93**, 8647
[18] Kakeshita T, Takeuchi T, Fukuda T, Tsujiguchi M, Saburi T, Oshima R and Muto S, 2000 Appl. Phys. Lett. **77** 1502

Rearrangement of crystallographic domains driven by magnetic field in ferromagnetic Ni_2MnGa and antiferromagnetic CoO

Tomoyuki Terai, Motoyoshi Yasui, Masataka Yamamoto and Tomoyuki Kakeshita
Department of Materials Science and Engineering, Graduate School of Engineering, Osaka University, 2-1 Yamadaoka, Suita, Osaka 565-0871, Japan

E-mail: terai@mat.eng.osaka-u.ac.jp

Abstract. We have investigated the rearrangement of crystallographic domains (martensite variants) in Ni_2MnGa ferromagnetic shape memory alloy and CoO antiferromagnetic oxide by applying magnetic field up to 8.0 MA/m. From the result of optical microscope observation of Ni_2MnGa single crystal, when a magnetic field is applied along $[001]_p$ (p represents a parent phase), the rearrangement of crystallographic domains occurs and the single domain state is obtained below T_{Ms} = 202 K. The same rearrangement occurs but partially when a magnetic field is applied along $[110]_p$. On the other hand, when a magnetic field is applied along $[111]_p$, the rearrangement does not occur. In case of the CoO single crystal, when a magnetic field is applied along $[001]_p$ below T_{Ms} = 293 K, the rearrangement occurs at 170 K $\leq T \leq$ 293 K, but does not occur at $T <$ 170 K. When a magnetic field is applied along $[110]_p$ and $[111]_p$, the rearrangement does not occur below T_{Ms}. In order to explain the rearrangement in the alloy and the oxide, we have evaluated the magnetic shear stress, τ_{mag}, which is derived from the difference in magnetic energy among crystallographic domains and have compared it with the shear stress required for the twinning plane movement, τ_{req}. As a result, we have found that the rearrangement occurs when the value of τ_{mag} is larger than or equal to the value of τ_{req} for the present alloy and oxide.

1. Introduction

Recently, it has been found that a large strain of several percent appears in some ferromagnetic shape memory alloys by applying magnetic field[1-6]. This large magnetic field-induced strain in ferromagnetic shape memory alloys is not due to conventional magnetostriction, but due to the rearrangement of crystallographic domain (martensite variant) of the martensite phase under a magnetic field. This phenomenon is of quite interest because magnetic field influences not only its intensive variable of magnetization but also another variable of strain. Thus, in a sense, a ferromagnetic shape memory alloy is a kind of multi-ferroic materials. Concerning the condition for the occurrence of such rearrangement, we have discussed about the energy evaluation and derived the condition for the occurrence of the rearrangement of crystallographic domains driven by magnetic field, as described below. That is, we introduce a shear stress acting across a twinning plane driven by magnetic field (We call it a magnetic shear stress), τ_{mag}, considering the fact that the twinning plane movement certainly occurs by magnetic field[3,4]. Using the value of τ_{mag}, the condition is that the value of τ_{mag} is equal to or larger than the shear stress required for the twinning plane movement, τ_{req}. The value of τ_{mag} is expressed as

$$\tau_{mag} = \frac{\Delta U_{mag}}{s} \geq \tau_{req}, \tag{1}$$

where ΔU_{mag} is the magnetic energy difference per unit volume between the two crystallographic domains separated by the twinning plane, and s is the corresponding twinning shear. The condition described above, $\tau_{\mathrm{mag}} \geq \tau_{\mathrm{req}}$, has been confirmed to be appropriate for some ferromagnetic shape memory alloys[1-6].

By the way, considering the condition of the equation (1), we speculate that the rearrangement of crystallographic domains is realized even in antiferromagnetic martensite phases when τ_{mag} is equal to or larger than τ_{req}. In order to the validity of this speculation, we select a NaCl-type 3*d* transition metal monoxide CoO because it is reported that CoO shows a simultaneous magnetic and structural transition from a paramagnetic cubic phase to an antiferromagnetic pseudo tetragonal phase[7,8]. Then, we investigate the rearrangement of crystallographic domain driven by magnetic field in antiferromagnetic CoO and show the validity of this condition in CoO. For comparison, we also investigate the rearrangement in a typical ferromagnetic shape memory alloy of Ni_2MnGa.

2. Experimental Procedure

Single crystals of Ni_2MnGa and CoO were prepared by a floating zone method. They were cut into a parallelepiped for optical microscope observations after determining the orientation by a Laue camera. The specimen was mounted on a cooling/heating stage and inserted into the center of a magnetic field. In order to determine the shear stress required for the twinning plane movement, τ_{req}, the tensile tests along the $[001]_p$ (p represents a parent phase) direction were performed by using an Instron-type machine.

3. Results

We have observed the rearrangement of crystallographic domains (martensite variants) driven by magnetic field in Ni_2MnGa ferromagnetic shape memory alloy which transforms to a low temperature phase (martensite phase) at 202K. Optical microscope observations under a magnetic field up to 1.2 MA/m at 77 K ≤ T ≤ 300 K have been made and a typical result at 77 K is shown in figure1. As known from the figure, when a magnetic field is applied along $[001]_p$, the rearrangement of crystallographic domains occurs and the single domain state is obtained (figure 1(a) and (b)). The same

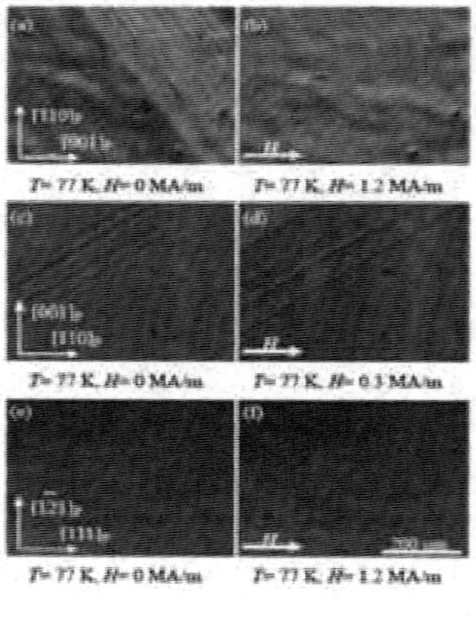

Figure 1. A series of optical micrographs of Ni_2MnGa under a magnetic field at 77 K. The field direction is along $[001]_P$ (a), (b); $[101]_P$ (c), (d); and $[111]_P$ (e), (f).

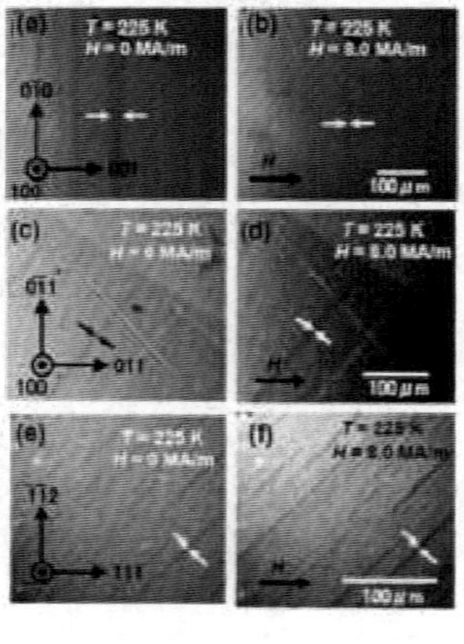

Figure 2. A series of optical micrographs of CoO under a magnetic field at 225 K. The field direction is along $[001]_P$ (a), (b); $[101]_P$ (c), (d); and $[111]_P$ (e), (f).

rearrangement occurs but partially when a magnetic field is applied along $[110]_p$ (figure 1(c) and (d)). On the other hand, when a magnetic field is applied along $[111]_p$, the rearrangement of crystallographic domains does not occur (figure 1(e) and (f)). In case of the CoO single crystal, CoO transforms from the parent phase to the martensite phase at 293K. Optical microscope observations under a magnetic field up to 8.0 MA/m have been made and a typical result at 225 K is shown in figure.2. As known from the figure, when a magnetic field is applied along $[001]_p$, the rearrangement of crystallographic domains occurs (figure 1(a) and (b)). This rearrangement occurs at temperatures in the range of 170 K $\leq T \leq$ 293 K, but does not occur at $T <$ 170 K. When a magnetic field is applied along $[110]_p$ and $[111]_p$, the rearrangement of crystallographic domains does not occur (figure 1(c) ~ (f)) below T_{Ms}. These results show that the rearrangement of crystallographic domains is certainly realized in CoO as same as in Ni_2MnGa and has magnetic field direction and temperature dependences.

4. Discussion

We evaluate the magnetic shear stress, $\tau_{mag} = \Delta U_{mag}/s$, in Ni_2MnGa as in the previous study[3]. The value of ΔU_{mag} is obtained from the area enclosed by the two magnetization curves: along hard magnetization axis and along easy magnetization axis. Typical magnetization curves of Ni_2MnGa are shown in figure 3. The maximum of ΔU_{mag} is equal to the uniaxial magnetocrystalline anisotropy constant $|K_u|$ in the case that a magnetic field is applied to $[001]_p$ direction. The value of s is given by the lattice parameters as $s = \{1-(c/a)^2\}/(c/a)$. Using the values of ΔU_{mag} and s, the values of magnetic shear stress, τ_{mag}, are evaluated and shown in figure 4. The value of τ_{req} obtained by tensile tests is also shown in the figure. As known from the figure, when τ_{mag} is nearly equal to or larger than τ_{req}, the rearrangement occurs.

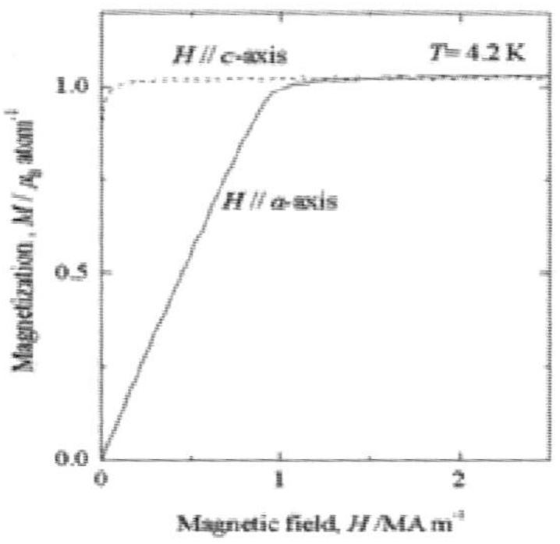

Figure 3. Typical magnetization curves of Ni_2MnGa along easy and hard magnetization axes at 4.2 K.

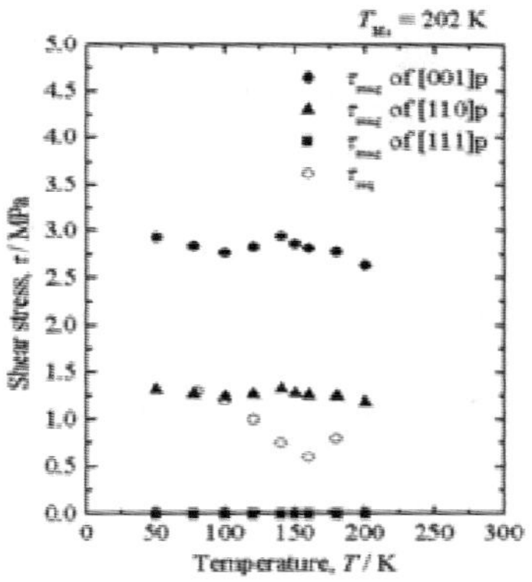

Figure 4. Temperature dependence of τ_{mag} and τ_{req} in Ni_2MnGa.

We also evaluate the magnetic shear stress in CoO. Considering that the magnetic susceptibility, χ, of CoO depends on the magnetic field direction, the value of ΔU_{mag} is evaluated as $\{(\chi_a-\chi_c)H^2\}/2$, where χ_a and χ_c are the magnetic susceptibility along a- and c-axes of tetragonal martensite phase in case that magnetic field is applied along $[001]_p$ direction. Typical magnetization curves along a- and c-axes are shown in figure 5. The value of s is given by the lattice parameters as $s = \{1-(c/a)^2\}/(c/a)$. Using the values of ΔU_{mag} and s, the values of τ_{mag} are calculated when the magnetic field of 8.0 MA/m is applied along $[001]_p$, $[110]_p$ and $[111]_p$ directions. The result is shown in figure 6. The value of τ_{req} obtained by tensile tests is also shown in the figure. As known from the figure, when τ_{mag} is equal to or larger than τ_{req}, the rearrangement occurs, as in ferromagnetic Ni_2MnGa. As a result, we have confirmed the validity of the condition, $\tau_{mag} \geq \tau_{req}$ even in antiferromagnetic CoO.

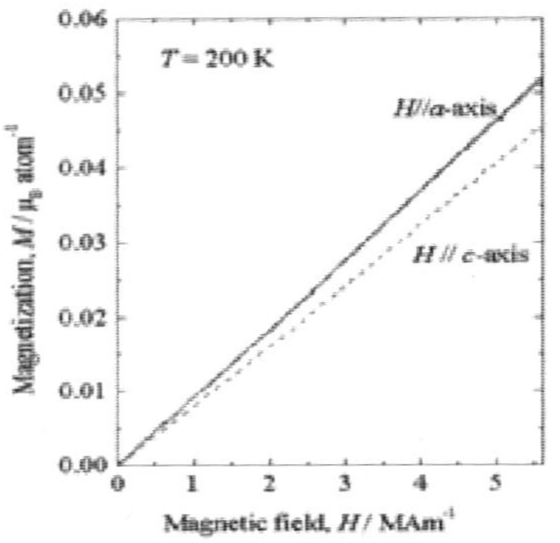

Figure 5. Typical magnetization curves of CoO along *a*- and *c*-axes of tetragonal phase at 200 K.

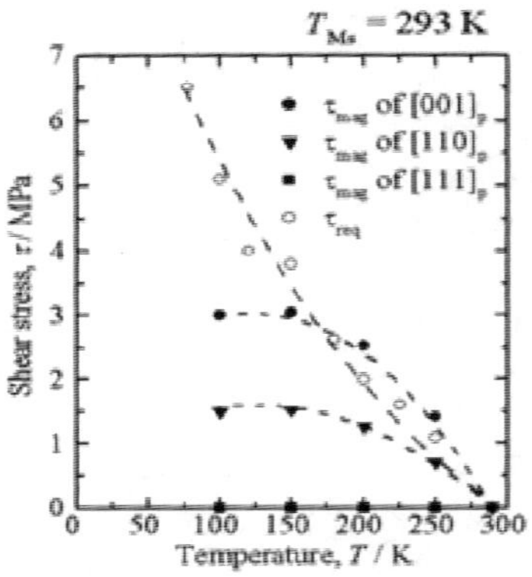

Figure 6. Temperature dependence of τ_{mag} and τ_{req} in CoO.

5. Conclusions

We have investigated the rearrangement of crystallographic domains (martensite variants) driven by magnetic field in ferromagnetic Ni_2MnGa and antiferromagnetic CoO. As a result, we have confirmed that when τ_{mag} is larger than or equal to τ_{req}, the rearrangement occurs in antiferromagnetic CoO, as in Ni_2MnGa. This shows that the rearrangement of crystallographic domain is able to be explained quantitatively by τ_{mag} and τ_{req} irrespective of the magnetism of materials.

Acknowledgments

This study was supported by Priority Assistance for the Formation of Worldwide Renowned Centers of Research - The Global COE Program (Project: Center of Excellence for Advanced Structural and Functional Materials Design) from the Ministry of Education, Culture, Sports, Science and Technology (MEXT), Japan.

References

[1] Ullakko K, Huang J K, Kantner C, O'Handley R C, Kokorin V V, 1996 *Appl. Phys. Lett.* **69** 1966
[2] Murray S J, Marioni M A, Kukla A M, Robinson J, O'Handley R C, Allen S M, 2000 *J. Appl. Phys.* **87** 5774
[3] Fukuda T, Sakamoto T, Terai T, Kakeshita T and Kishio K 2003 *Mat. Res. Soc. Symposium - Proceedings* **785** 421
[4] Kakeshita T, Takeuchi T, Fukuda T, Tsujiguchi M, Saburi T, Oshima R, Muto S 2000 *Appl. Phys. Lett.* **77** 1502
[5] James R D and. Wuttig M, 1998 *Phil. Mag.* A **77** 1273
[6] Sugiyama M, Oshima R and Fujita F E 1986 *Trans. Jpn. Inst. Metals* **27** 719
[7] Tombs N C and Rooksby H P 1950 *Nature* **165** 442
[8] Greenwald S and Smart J S 1950 *Nature* **166** 523

Journal of Physics: Conference Series **165** (2009) 012053 doi:10.1088/1742-6596/165/1/012053

Unusual pseudoelastic behaviour of Fe_3Ga shape memory alloys

H Y Yasuda[1], M Aoki[1], Y Oda[1], K Fukushima[1] and Y Umakoshi[2]

[1]Division of Materials and Manufacturing Science, Graduate School of Engineering, Osaka University, 2-1 Yamadaoka, Suita, Osaka 565-0871, Japan
[2]National Institute for Materials Science, 1-2-1 Sengen, Tsukuba, Ibaraki 305-0047, Japan

hyyasuda@mat.eng.osaka-u.ac.jp

Abstract. Pseudoelastic behaviour of Fe-23.8at%Ga single crystals compressed with different loading axes at room temperature was examined focusing on the activated deformation mode. In the crystals, {101}<111> slip and {211}<111> twin were mainly activated depending on the loading axis. Pseudoelastic behaviour of the crystals depended strongly on the deformation mode. If {101}<111> slip was operative, sole and paired 1/4<111> superpartial dislocations moved dragging antiphase boundaries (APB). During unloading, the APB pulled back the superpartials due to its tension resulting in pseudoelasticity. In contrast, twinning and untwinning of {211}<111> pseudo-twins also led to large strain recovery accompanying a serrated flow during loading and unloading. It is suggested the energy of the pseudo-twin interface was the driving force for the twinning pseudoelasticity.

1. Introduction

Shape memory alloys generally exhibit the following two phenomena: shape memory effect and pseudoelasticity [1]. By pseudoelasticity, an applied strain during loading recovers during unloading. So, pseudoelasticity is applied to potable phone antenna, eyeglass frame and so on. In general, thermoelastic martensitic transformation is responsible for pseudoelasticity; stress-induced martensitic transformation during loading and the reverse transformation during unloading result in strain recovery. However, Fe_3Al single crystals with the $D0_3$ structure showed giant pseudoelasticity which is not related to martensitic transformation but peculiar dislocation motion [2-4]. In the crystals, 1/4<111> superpartial dislocations moved individually dragging antiphase boundaries (APB) during loading. During unloading, APB pulled back the superpartials, resulting in strain recovery. Thus, this phenomenon is called “APB pseudoelasticity”.

Recently, we found that Fe_3Ga single crystals with the $D0_3$ structure also demonstrated giant pseudoelasticity at room temperature [5, 6]. More recently, three kinds of pseudoelasticity were found to appear in Fe-23.8at%Ga single crystals tensile-deformed at room temperature, depending on the degree of $D0_3$ order [6]. So, we call the phenomena “multimode pseudoelasticity” which has not been observed before. Figure 1 shows a schematic illustration of the mechanism of the pseudoelasticities. In

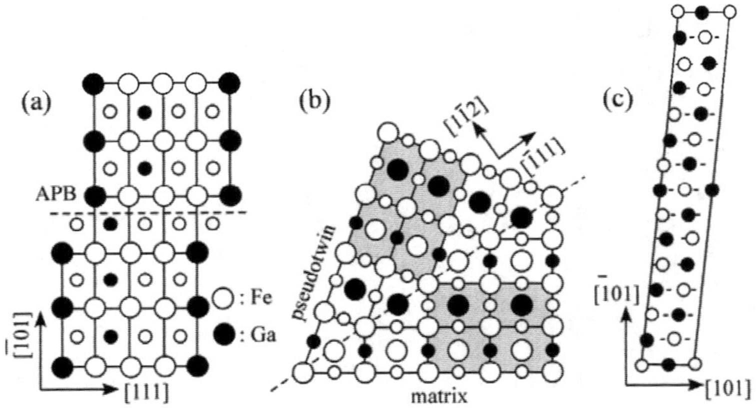

Figure 1. Schematic illustrations of deformation mode in Fe_3Ga single crystals with the $D0_3$ structure. (a) {101}<111> slip, (b) {211}<111> pseudo-twin (2.2^T-type) and (c) 14M martensite.

Fe_3Ga single crystals, APB pseudoelasticity took place associated with {101}<111> slip, which is similar to Fe_3Al single crystals (figure 1 (a)). The APB tension on {101} plane was large enough to pull back the 1/4<111> superpartial dislocations during unloading. {211}<111>-type twins were also formed in the course of tensile deformation in the crystals fully ordered to the $D0_3$ phase as shown in figure 1 (b). The $D0_3$ matrix and the twins did not have mirror symmetry at the twin plane. In general, an atomic shuffling occurs to maintain the mirror symmetry. However, the atomic shuffling never took place in Fe_3Ga crystals. Christian et al. [7] defined the {211} twins as 2.2^T-type pseudo-twins. The interface energy of the pseudo-twins was higher than that of perfect twins. Thus, the energy of the twin boundaries was the driving force for untwinning during unloading. Furthermore, if the degree of $D0_3$ order in Fe_3Ga was low, stress-induced transformation of 14M martensites with 14-layered periodic structure occurred in the crystals (figure 1 (c)). The reverse transformation during unloading led to the shape recovery of Fe_3Ga single crystals. The free energy difference between the parent and martensite phases resulted in the pseudoelasticity.

It is well known that twinning deformation depends strongly on loading axis and stress sense such as tension and compression. In the present study, the pseudoelastic behaviour of Fe_3Ga single crystals compressed with different loading axes at room temperature were investigated focusing on the orientation dependence and tension-compression asymmetry of the pseudo-twins.

2. Experimental procedure

Fe-23.8at%Ga single crystals were grown by a floating zone method at a growth rate of 5 mm h^{-1}. After homogenisation at 1373 K for 48 h, the crystals were annealed at 873 K for 10 h for $D0_3$ ordering. By the ordering treatment, possible activated deformation modes in compression were limited to {101}<111> slip and {211}<111> pseudo-twins while martensitic transformation never occurred [6]. The compression specimens with gauge dimension of $2.5 \times 2.5 \times 5.5$ mm^3 were prepared by spark machining. The loading axes of the compression specimens were selected to be $[\bar{1}49]$ (A), $[\bar{3}58]$ (B) and $[\bar{1}44]$ (C) as shown in Figure 2. Schmid factors (SF) for $(\bar{1}01)[111]$ slip and $(\bar{2}11)[111]$ pseudo-twin are listed in table 1. It should be noted that twinning deformation generally shows tension-compression asymmetry, $(\bar{2}11)[111]$ pseudo-twin is activated in compression. From the previous study, the critical resolved stresses (τ_{CRSS}) for $(\bar{1}01)[111]$ slip and $(\bar{2}11)[111]$ pseudo-twin were determined to be 224 and 225 MPa, respectively. If the Schmid law hold true, the yield stress (σ_Y) corresponding to each deformation mode can be predicted by the following equation,

$$\sigma_Y = \frac{\tau_{CRSS}}{SF}, \quad (1)$$

Table 1. Schmid factor (SF) for $(\bar{1}01)[111]$ slip and $(\bar{2}11)[111]$ twin at different loading axes.

Orientation	SF for $(\bar{1}01)[111]$ slip	SF for $(\bar{2}11)[111]$ twin
A	0.500	0.433
B	0.458	0.457
C	0.433	0.500

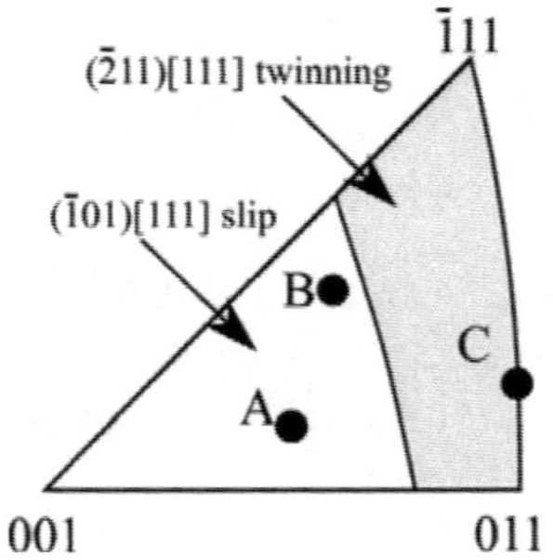

Figure 2. Compression axes for Fe-23.8at%Ga single crystals. Predicted deformation mode is also shown in the figure.

Using equation (1), one can predict loading axis favorable for the activation of $(\bar{1}01)[111]$ slip and $(\bar{2}11)[111]$ pseudo-twin as shown in figure 2. Near [011] orientation, $(\bar{2}11)[111]$ pseudo-twin is activated before $(\bar{1}01)[111]$ slip. Compression tests were done in air at room temperature at a cross-head speed of 0.05 mm/min corresponding to an initial strain rate of 1.7×10^{-4} s^{-1}. After loading to a plastic strain (ε_p) of 5 %, the load was subsequently released at the same cross-head speed. After the compression tests, surface relief due to the slip or pseudo-twin was observed by an optical microscope. Deformation substructure was also observed using a transmission electron microscope (TEM) at an accelerating voltage of 300 kV.

3. Results

Figure 3 shows stress-strain (S-S) curves of Fe-23.8at%Ga single crystals compressed to a plastic strain of 5 % at room temperature. The pseudoelastic behaviour of these crystals depends strongly on loading axis. At A orientation where $(\bar{1}01)[111]$ slip is active, perfect pseudoelasticity appears showing smoothly curved S-S curve. This means no twinning occurred at A orientation which is consistent with theoretical calculation shown in figure 2. Moreover, there is also no evidence that proves occurrence of pseudo-twinning at B orientation. It is also noted that strain recovery at B

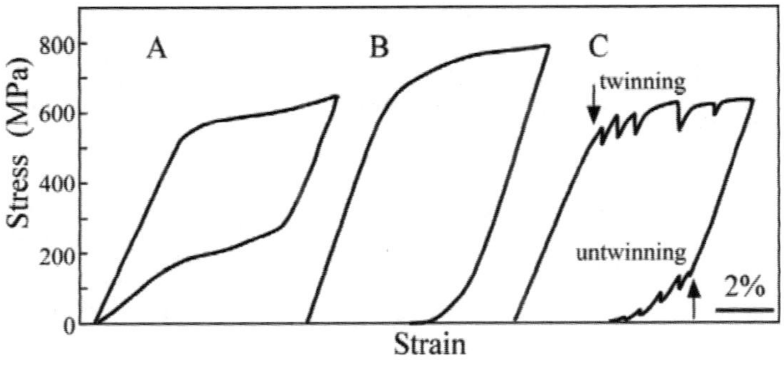

Figure 3. Stress-strain curves of Fe-23.8at%Ga single crystals compressed to ε_p = 5 % at A, B and C orientations at room temperature.

Figure 4. Surface relief due to deformation in Fe-23.8at%Ga single crystals compressed to ε_p = 5 % at A (a), B (b) and C (c) orientations.

orientation is quite small compared with A orientation. On the other hand, a serrated flow is observed in the S-S curve for C orientation. It is also interesting to note that the serration also appears in the unloading curve. This suggests twinning and untwinning during loading and unloading resulted in pseudoelasticity at C orientation though the strain recovery was not perfect.

Optical micrographs of specimen surface at A, B and C orientations are shown in figure 4. At A orientation, fine slip traces parallel to primary $(\bar{1}01)$ slip plane are observed as shown in figure 4 (a). The contrast of the slip traces is very low since the slip traces considerably disappeared during unloading. On the other hand, $(\bar{1}01)$ coarse slip traces remain after compression at B orientation (figure 4 (b)). In contrast, there exists coarse bands associated with $(\bar{2}11)$ pseudo-twins in addition to $(\bar{2}11)$ and its cross slips (figure 4 (c)). Thus, $(\bar{1}01)[111]$ slip and $(\bar{2}11)[111]$ pseudo-twin were activated depending on loading axis, which is in good agreement with figure 2. Thus, APB pseudoelasticity appeared at A and B orientations while the crystals compressed at C orientation revealed twinning pseudoelasticity. Strictly saying, the Schmid law (i.e. equation (1)) did not fully hold for $(\bar{1}01)[111]$ slip due to the complicated core structure of <111> screw dislocations. However, one could roughly estimate the activated deformation mode from figure 2.

4. Discussion

In the present study, APB and twinning pseudoelasticities were observed to appear in F-23.8at%Ga single crystals compressed at room temperature. The mechanism of pseudoelasticities was discussed in detail.

At A and B orientations, only APB pseudoelasticity appeared. Figure 5 shows dislocation configuration and atomic arrangement of APB in the $D0_3$ structure. From TEM observation, two types of dislocation configuration were observed. Uncoupled and paired 1/4[111] superpartial dislocations moved dragging the nearest-neighbour and next-nearest neighbour APB (NNAPB and NNNAPB), respectively as shown in figure 5 (a) and (b). The atomic arrangements on {101} plane for perfect crystals, near the NNAPB and NNNAPB are shown in figure 5 (c), (d) and (e), respectively. The energies of NNAPB (γ_{NN}) and NNNAPB (γ_{NNN}) depend strongly on the number of wrong bonds and are given by, [8]

$$\gamma_{NN} = \frac{2\sqrt{2}}{a_0^2}\left\{4V_1S_1^2 + V_2\left(S_2^2 - 4S_1^2\right)\right\}, \qquad (2)$$

$$\gamma_{NNN} = \frac{2\sqrt{2}}{a_0^2}\left\{2V_2S_2^2\right\}, \qquad (3)$$

where a_0 is the lattice constant, V_1 and V_2 are the first- and second-nearest ordering energies and S_1 and S_2 are the first- and second-nearest degrees of order, respectively. It should be noted that S_1 and S_2 are

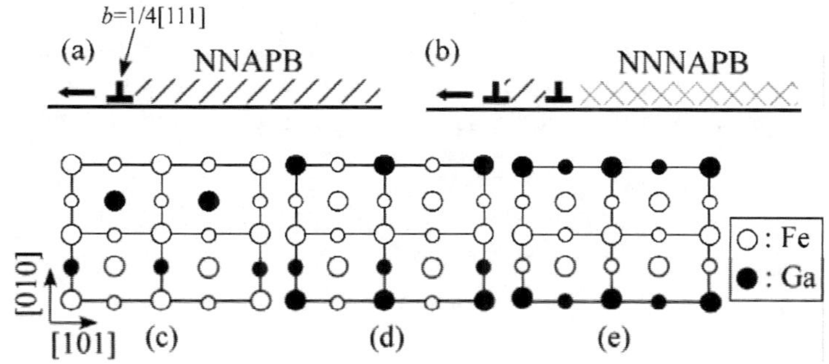

Figure 5. Schematic illustrations of dislocation configuration in Fe_3Ga crystal; (a) 1/4<111> superpartial dragging the NNAPB, (b) paired 1/4<111> superpartials dragging the NNNAPB. The atomic arrangement on $(\bar{1}01)$ plane for (c) $D0_3$ perfect crystal, (d) NNAPB and (e) NNNAPB.

0.5 and 1 for the perfect $D0_3$ lattice, respectively. In addition, tensions of NNAPB (τ_{NN}) and NNNAPB (τ_{NNN}) are written as [4]

$$\tau_{NN} = \frac{\gamma_{NN}}{b}, \tag{4}$$

$$\tau_{NNN} = \frac{\gamma_{NNN}}{2b}, \tag{5}$$

where b is the magnitude of 1/4[111] superpartials. Substituting each parameter for equations (2)~(5), τ_{NN} and τ_{NNN} were calculated to be 268 and 204 MPa, respectively. Since the APB tension was much higher than frictional stress of 1/4[111] superpartials, APB could pull back the superpartials during unloading, which resulted in APB pseudoelasticity. In fact, to-and-fro motion of 1/4<111> superpartials during loading and unloading was confirmed to occur by in-situ TEM observation [6].

At A orientation, perfect pseudoelasticity appeared while strain recovery at B orientation was small, though $(\bar{1}01)[111]$ slip was a major deformation mode at both orientations. Frequency of cross-slip of 1/4[111] screw dislocation is higher at B orientation than at A orientation. The cross-slip segments suppressed the reversible motion of the dislocations resulting in the decrease in strain recovery.

Twinning pseudoelasticity associated with twinning and untwinning of $(\bar{2}11)[111]$ pseudo-twins appeared at C orientation. The $(\bar{2}11)[111]$ pseudo-twin showed tension-compression asymmetry; the twins were formed only in compression. It is also noted that $(1\bar{1}2)[\bar{1}11]$ pseudo-twins were observed to appear in the crystals tensile-deformed at A orientation. Such tension-compression asymmetry is the common feature of {211}<111>-type twins in bcc metals. The interface energy for the pseudo-twins (γ_T) and the tension (τ_T) are given by, [8]

$$\gamma_T = \frac{2\sqrt{2}}{\sqrt{3}a_0^2}\left\{V_1\left(4S_1^2 - S_2^2\right) + V_2\left(2S_2^2 - 5S_1^2 - 0.25 + S_1\right)\right\}, \tag{6}$$

$$\tau_T = \frac{\gamma_T}{t}, \tag{7}$$

where t is the twin shear for $(\bar{2}11)[111]$ pseudo-twin. τ_T was calculated to be 275 MPa, comparable with τ_{NN} and τ_{NNN}. Therefore, it is strongly suggested that the high interface energy of the pseudo-twins resulted in the twinning pseudoelasticity.

5. Conclusions

APB and twinning pseudoelasticities in $D0_3$-ordered Fe-23.8at%Ga single crystals compressed at room temperature were investigated. The operative pseudoelastic mode could be controlled by changing the compressive axis. The tensions of APB and twin interfaces were the driving force for APB and twinning pseudoelasticities in Fe-23.8at%Ga single crystals, respectively.

Acknowledgments
This work was supported by a Grant-in Aid for Scientific Research from the Ministry of Education, Culture, Sports, Science and Technology (MEXT) of Japan. This study was supported by Priority Assistance for the Formation of Worldwide Renowned Centers of Research - The Global COE Program (Project: Centre of Excellence for Advanced Structural and Functional Materials Design) from the MEXT, Japan.

References

[1] Otsuka K and Wayman CM 1998 *Shape Memory Materials* (Cambridge: Cambridge University Press)
[2] Kubin LP, Fourdeux A, Guedou JY and Rieu J 1982 *Phil. Mag.* A **46** 357
[3] Yasuda HY, Nakano K, Nakajima T and Umakoshi Y 2003 *Acta Mater.* **51** 5101
[4] Yasuda HY, Nakajima T, Nakano K, Yamaoka K, Ueda M and Umakoshi Y 2005 *Acta Mater.* **53** 5343
[5] Yasuda HY, Aoki M and Umakoshi Y 2007 *Acta Mater.* **55** 2407
[6] Yasuda HY, Oda Y, Aoki M, Fukushima K and Umakoshi Y 2008 *Intermetallics* in press
[7] Christian JW and Laughlin DE 1988 *Acta Metall.* **36** 1617
[8] Leamy HJ and Kayser FX 1969 *Phys. Stat. Sol.* **34** 765

Incommensurate structures of intermediate phase and martensite phase in Ni_2MnGa

H Kushida[1], K Fukuda[1], T Terai[1], T Fukuda[1], T Kakeshita[1], T Ohba[2], T Osakabe[3], K Kakurai[3], K Kato[4]

[1] Department of Materials Science and Engineering, Graduate School of Engineering, Osaka University, 2-1 Yamada-oka, Suita, Osaka 565-0871, Japan

[2] Department of Material Science, Shimane University, 1060 Nishikawatsu-cho, Matsue, Shimane 690-8504, Japan

[3] Quantum Beam Science Directorate, Japan Atomic Energy Agency (JAEA), 2-4 Sirane Sirakata Tokai-mura, Naka-gun, Ibaraki 319-1195, Japan

[4] Japan Synchrotron Radiation Research Institute (Spring-8), Sayo-gun, Hyogo 679-5198, Japan

hiroaki.kushida@mat.eng.osaka-u.ac.jp

Abstract. Neutron diffraction measurements using single crystal and powder synchrotron X-ray diffraction measurements revealed that an intermediate (I-) phase and a martensite (M-) phase in Ni_2MnGa have incommensurate modulated structures. The modulation vectors of the I- and M-phases are nearly equal to $\boldsymbol{q} = 0.341 \langle 2\ 2\ 0\rangle^*$ at 210 K and $\boldsymbol{q} = 0.427 \langle 2\ 2\ 0\rangle^*$ at 100 K, respectively. Moreover, displacements of Ni, Mn and Ga atoms in both the I- and M-phase are expressed by sinusoidal waves with the same phase and almost the same amplitudes.

1. Introduction

A Heusler-type intermetallic compound of Ni_2MnGa is known to transform from a $L2_1$-type parent phase (P-phase) to an intermediate phase (I-phase) at about 250 K [1, 2] and then to a martensite phase (M-phase) at about 200 K [3]. The first report on the crystal structure of the M-phase was made by Webster *et al.* [3]. They suggested a pseudo-tetragonal structure with $c/a < 1$ for the M-phase. After that the frequently mentioned 5-layered or 10-layered modulated structure (5M or 10M) for the M-phase was reported by Martynov *et al.* [4] and by Pons *et al.* [5]. On the other hand, a 7-layered or 14-layered modulated structure (7M or 14M) for the M-phase was reported by Brown *et al.* [6]. Moreover, an incommensurate modulated structure for the M-phase was suggested by Zheludev *et al.* using neutron diffraction [7] and by Righi *et al.* using powder X-ray diffraction [8]. Concerning the structure of the I-phase, a 3-layered or 6-layered modulated structure (3M or 6M) was reported by Zheludev *et al.* [7] and by Brown *et al.* [6]. An incommensurate structure for the I-phase was suggested by Tsuchiya *et al.* using an electron diffraction pattern [9].

Despite the many investigations described above, the crystal structures of the I-phase and the M-phase have not been clarified yet. In this study, therefore, we investigated the crystal structure of the I-phase and M-phase by combining neutron and synchrotron X-ray diffraction measurements.

2. Experimental procedure

For neutron diffraction experiments, a single crystal of Ni_2MnGa was grown by a floating zone method and was heat-treated at 1173 K for 24 h for homogenization and then at 923 K for 24 h to obtain a highly ordered $L2_1$-type structure. A specimen with a dimension of $5.0 \times 1.9 \times 1.9$ mm^3 and all edges parallel to $<0\ 0\ 1>_P$ was cut from the single crystal. Neutron diffraction measurements were made with a triangular spectrometer (TAS-1) using a non-polarized neutron beam at the 2G beam-line of JRR-3 at Japan Atomic Energy Agency (JAEA). A wavelength of 0.236 nm was selected for the neutron experiments. Horizontal collimation used was of open - 40′ - sample - 40′ - 80′. A pyrolytic graphite filter was used to attenuate higher-order harmonic contamination. Measurements were made at 245 K, 210 K, 200 K and 100 K by sweeping the scattering vector $\boldsymbol{q} = [h\ 2{-}h\ 0]^*_P$ in the range of $-0.1 \le h \le 2.1$.

For synchrotron diffraction measurements, the ingot of homogenized Ni_2MnGa was crushed into a powder specimen and it was subjected to ordering and annealing heat-treatment at 923 K for 3 h to eliminate the residual strain. Synchrotron X-ray diffraction measurements of powder specimens were made with a Debye-Scherrer-type imaging plate camera using a wavelength of 0.0500 nm at the BL02B2 beam-line at Japan Synchrotron Radiation Research Institute (SPring-8). Measurements were made during a successive cooling process at 280 K, 220 K, 190 K, 160 K, 130 K and 90 K.

3. Results

Figure 1(a) and (b) shows neutron diffraction profiles of the I-phase in Ni_2MnGa, which were obtained at 245 K and 210 K, respectively. In Figure 1(a) two satellite reflections are seen at the incommensurate positions of $h = 0.341$ and 1.659 between the fundamental reflections of $h = 0$ and 2. When the specimen was cooled to 210 K the satellite reflection positions remained almost constant but their intensities increased as shown in Figure 1(b). Figure 1(c) and (d) show profiles of the M-phase and were obtained at 200 K and 100 K, respectively. In Figure 1(c), four satellite reflections are seen at incommensurate positions of $h = 0.425$, 0.854, 1.146 and 1.575 between the fundamental reflections. When the specimen was cooled to 100 K (in Figure 1(d)) the satellite reflections moved slightly towards $h = 0.427$, 0.861, 1.139 and 1.573 while their intensities increased slightly. These results suggest that the I- and M-phases have incommensurate structures and the satellite intensities of the I- and M-phase are dependent on temperature, and also suggest that satellite position of the M-phase

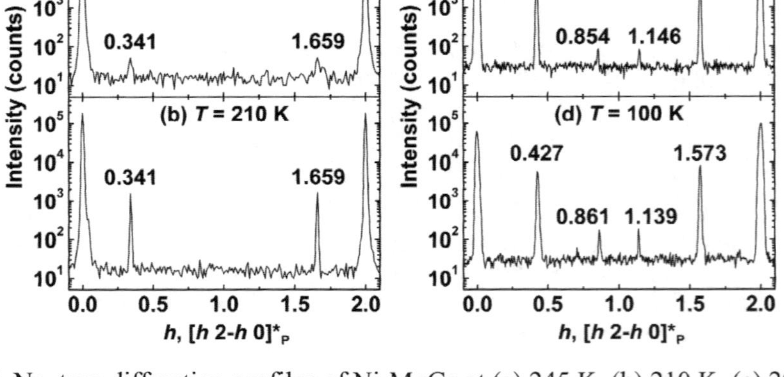

Figure 1 Neutron diffraction profiles of Ni_2MnGa at (a) 245 K, (b) 210 K, (c) 200 K and (d) 100 K.

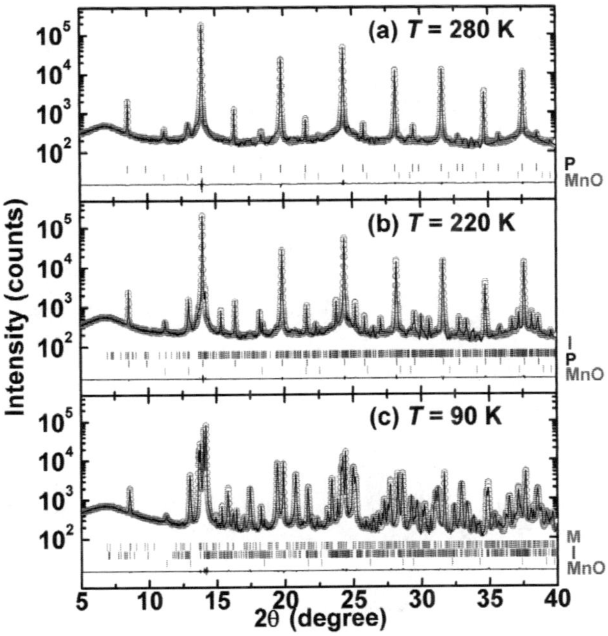

Figure 2 Powder synchrotron X-ray diffraction profiles of Ni_2MnGa at (a) 280 K, (b) 220 K and (c) 90 K. Observed and calculated intensities are indicated with open red circles and the solid black line, respectively. Short lines and the solid line below the diffraction profiles indicate peak positions of each phase (black, blue, red and green indicate P, I, M and MnO) and the difference between observed and calculated intensities.

depend on temperature but that of the I-phase is independent of temperature.

To obtain atom positions for the two phases, powder synchrotron measurements were carried out. Results are shown in Figure 2. At 280 K (Figure 2(a)) the profile is indexed by typical $L2_1$-type peaks of the P-phase and a small amount MnO impurity. The appearance of MnO is due to oxidation of the powder specimen during the heat-treatment. When the specimen is cooled to 220 K (Figure 2(b)), new peaks of the I-phase appear with those of the P-phase and of MnO. At temperatures of 190 K or lower peaks of the P-phase completely disappear and new peaks of the M-phase appear with those of the I-phase and of MnO. A typical example at 90 K is shown in Figure 2(c). The coexistence of the M- and I-phases at 90 K is due to the so-called size effect of the martensitic transformation temperature.

4. Discussion

In order to investigate atom positions, we carry out the Rietveld analysis using the results of powder synchrotron X-ray diffraction measurements. For the Rietveld analysis, we introduce hypothetical commensurate structures for the I- and M-phases although they have incommensurate structures as described above. That is, in neutron diffraction experiments satellite positions of the I-phase are close to $h = 13/38$ and 63/38. Therefore, we introduce a 76-layered structure with a space group of *P2/m* for the I-phase. Similarly, the satellite positions of the M-phase are close to 3/7, 6/7, 8/7 and 11/7. Therefore, we introduce a 14-layered structure with a space group of *P2/m* for the M-phase. In addition, we assume that all atoms move only within the basal plane corresponding to $(1\ 1\ 0)_P$ along the $[1\ \bar{1}\ 0]_P$ direction in association with the martensitic transformations. This behaviour is also seen in the movement of atoms in copper based shape memory alloys [10, 11]. We fixed the monoclinic angle β of the I- and M-phase to 90.0° and 90.2° considering the results of TEM observations [5].

Using these hypothetical structures and assumptions we performed Rietveld analyses using RIETAN-2000 [12] and results are shown in Figure 2 with the diffraction profiles described above. Diffraction profiles were completely fitted by the $L2_1$-type P-phase for Figure 2(a) by assuming the coexistence of the I- and P-phases for (b) and by assuming the coexistence of the M- and I-phases for (c). These results mean that the hypothetical structures of the I- and M-phases which were introduced from the results of neutron diffraction using a bulk single crystal are consistent with actual crystal structures of the I- and M-phases in the powder specimen. Incidentally, the amount of MnO impurity was less than 0.5 mass%.

Considering the Ni, Mn and Ga atom positions obtained by Rietveld analyses we determined the atom displacements Δx from the position of the parent phase. The displacement Δx of Ni, Mn, Ga atoms can be fitted by the following sinusoidal curve

$$\Delta x(n) = A_{\mathrm{I}} \sin(2\pi \times 13n/76) \qquad (1)$$

for the I-phase. Here n stands for the layer number of the basal plane. Also they can be fitted by a sinusoidal curve of

$$\Delta x(n) = A_{\mathrm{M}} \sin(2\pi \times 3n/14) \qquad (2)$$

for the M-phase. Modulation amplitudes of Ni, Mn and Ga atoms in the I-phase, A_{I}, are almost 0.03 at 220 K and those in the M-phase, A_{M}, are almost 0.06 at 90 K. The modulation amplitude of each atom for the I-phase, A_{I}, increased from 0.03 to 0.04 as the temperature was decreased from 220 K to 90 K. It should be noted that the modulation amplitude of the I-phase is smaller than that of the M-phase even at the same temperature. This result corresponds to the difference in satellite intensity of neutron diffraction between the I- and M-phase.

5. Conclusions

Both the intermediate and the martensite phases of Ni_2MnGa have incommensurate modulated structures. The modulation vectors of the I- and M-phases are nearly equal to $\boldsymbol{q} = 13/76 \langle 1\ 1\ 0\rangle^{*}$ and $\boldsymbol{q} = 3/14 \langle 1\ 1\ 0\rangle^{*}$, respectively. Moreover, displacements of Ni, Mn and Ga atoms in both the I- and M-phase are expressed by sinusoidal waves with the same phase and almost the same amplitudes. The results obtained by neutron diffraction and synchrotron x-ray diffraction are consistent.

6. Acknowledgements

This study was supported by Priority Assistance for the Formation of Worldwide Renowned Centres of Research - The Global COE Program (Project: Centre of Excellence for Advanced Structural and Functional Materials Design) from the Ministry of Education, Culture, Sports, Science and Technology (MEXT), Japan.

7. References

[1] Kokorin V V, Chernenko V A, Cesari E, Pons J and Segui C 1996 *J. Phys.: Condens. Matter* **8** 6457.
[2] Planes A, Odradó E, Comas A G and Manõsa L 1997 *Phys. Rev. Lett.* **79** 3926
[3] Webster P J, Ziebeck K R A, Town S L and Peak M S 1984 *Phil. Mag.* B **49** 295.
[4] Martynov V V and Kokorin V V 1992 *J. Phys. France* III **2** 739.
[5] Pons J, Chernenko V A, Santamarta R and Cesari E 2000 *Acta Mater.* **48** 3027.
[6] Brown P J, Crangle J, Kanomata T, Matsumoto M, Neumann K U, Ouladdiaf B and Ziebeck K R A 2002 *J. Phys.: Condens. Matter* **14** 10159.
[7] Zheludev A, Shapiro S M, Wochner P and Tanner L E 1996 *Phys. Rev.* B **54** 15045.
[8] Righi L, Albertini F, Calestani G, Pareti L, Paoluzi A, Ritter C, Algarabel P A, Morellon L and Ibarra M R 2006 *J. Sol. Stat. Chem.* **176** 3525.
[9] Tsuchiya K, Yamamoto K, Hirayama T, Nakayama H, Todaka Y and Umemoto M 2003 *Electron Microscopy: Its role in Materials Science (The Mike Meshii symposium)* ed Weertman J R, Fine M, Fabor K, King W and Liaw P (TMS: The Minerals, Metals & Materials Society) pp 305-312.
[10] Tendeloo G V, Chandrasekaran M and Lovey F C 1986 *Metall. Mater. Trans.* A **17** 2153.
[11] Otsuka K, Ohba T, Tokonami M and Wayman C M 1993 *Scr. Metall. Mater.* **29** 1359.
[12] Izumi F and Ikeda T 2000 *Mater. Sci. Forum* **321-324** 198-203.

Journal of Physics: Conference Series **165** (2009) 012055 doi:10.1088/1742-6596/165/1/012055

Influence of a magnetic field on microstructure formation in $L1_0$-type ferromagnetic intermetallics

Sahar Farjami, Takashi Fukuda and Tomoyuki Kakeshita

Department of Materials Science and Engineering, Graduate School of Engineering, Osaka University, 2-1, Yamadaoka, Suita, Osaka 565-0871, Japan

Email: farjami@mat.eng.osaka-u.ac.jp

Abstract. The influence of a magnetic field on microstructure formation through a disorder-order transformation has been investigated in Co-Pt and Fe-Pd alloys. Single crystals of disordered Co-50Pt(at%) and Fe-55Pd(at%) were subjected to an ordering heat-treatment under a magnetic field. When the ordering heat-treatment is performed without applying a magnetic field, three equivalent variants are formed. On the other hand, when the ordering heat-treatment is performed under a magnetic field of 0.5 T (in CoPt) - 4 T (in Fe-55Pd) and higher as applied along the [001] direction of the disordered phase, a single variant with an easy axis along the field direction is obtained. The induced anisotropy energy of the ordered phase under a magnetic field of 1 T was 4.1 $kJ \cdot m^{-3}$ at 773 K for CoPt and 45.3 $kJ \cdot m^{-3}$ at 673 K for Fe-55Pd.

1. Introduction

Control of microstructure through solid-solid phase transformation under a magnetic field is a promising method to improve the properties and design of new materials. In fact, many studies on the effect of a magnetic field have been done, especially on diffusionless transformation (martensitic transformation) [1-3]. Diffusional transformation, however, has not been studied much.

Disorder-order transformation (A1-type → $L1_0$-type) in ferromagnetic intermetallics such as Co-Pt and Fe-Pd alloys is a representative diffusional transformation. Microstructure formation is expected to be influenced by a magnetic field because the ordered phase ($L1_0$-type, tetragonal) consists of three equivalent variants with an "easy" *c*-axis [4-6], and this is shown in figure 1. Under the influence of a magnetic field a difference exists in the magnetocrystalline anisotropy energy among these three variants. The consequence of this is that only a preferred variant for energy can grow under the field's influence.

In this study, therefore, the effect of a magnetic field on microstructure formation during the disorder-order transformation was investigated using single crystals of Co-50at%Pt and Fe-55at%Pd alloys. To determine the influence of a magnetic field on microstructure formation we carried out a two-step ordering heat-treatment. The first step of ordering was done under a magnetic field of up to 10 T and applied to the $[001]_{A1}$ direction to initiate the ordering under the effect of the magnetic field. This was followed by a second step of ordering without using a magnetic field that would result in a disorder-order transformation. We found that a single variant of both alloys was obtained when a magnetic field of 0.5 T or higher was used for CoPt and when 4 T or higher was used for Fe-55Pd if applied at an early stage of the ordering.

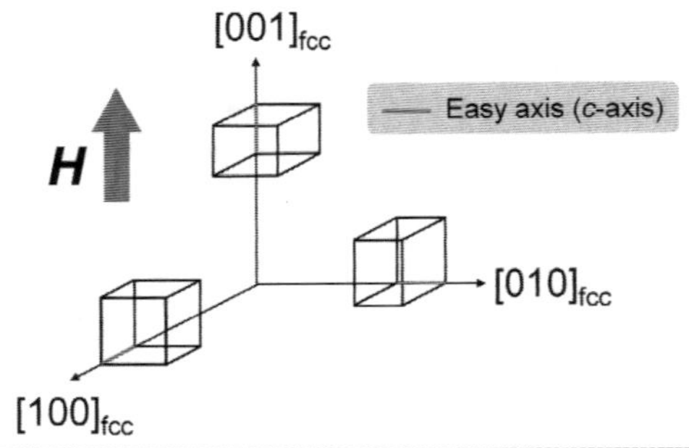

Figure 1. The three variants of the $L1_0$ phase with an easy *c*-axis.

2. Experimental procedure

Ingots of Co-50Pt(at%) and Fe-55Pd(at%) were prepared by arc melting. We selected Fe-55at%Pd alloy instead of Fe-50at%Pd alloy to obtain the single $L1_0$-phase considering the phase diagram [7]. A single crystal of each alloy was grown by a floating zone method using purified argon gas flow. Homogenization was carried out at 1273 K for 168 h followed by ice-water quenching. Cubic and rectangular specimens with each edge parallel to the $<001>_{A1}$ direction were cut. After cutting, the specimens were heat-treated at 1273 K (CoPt) and 1173 K (Fe-55Pd) for 1 h followed by ice-water quenching to obtain the disordered phase.

Ordering heat-treatments under a magnetic field were made by using a furnace inserted into a superconducting magnet. The specimen was located at the centre of the hot zone which is also the centre of the magnetic field. A two-step ordering heat-treatment was done: the first step of ordering was under a magnetic field of up to 10 T applied to the $[001]_{A1}$ direction of the disordered cubic specimen. The second step was an ordering at an elevated temperature without a magnetic field.

3. Results

Characteristic temperatures of CoPt and Fe-55Pd such as the Curie temperature and the disorder-order transformation temperature have been determined by resistivity measurements in disordered and ordered specimens. These characteristic temperatures are summarized in table 1.

Table 1. Characteristic temperatures of CoPt and Fe-55Pd. $T_c^{(d)}$ and $T_c^{(o)}$ are the Curie temperature of the disordered and ordered phases, $T^{(d\text{-}o)}$ and $T^{(o\text{-}d)}$ are the disorder-order and the order-disorder transformation temperatures, respectively.

	$T_c^{(d)}$	$T_c^{(o)}$	$T^{(d\text{-}o)}$	$T^{(o\text{-}d)}$
CoPt	850 K	730 K	1045 K	1095 K
Fe-55Pd	740 K	670 K	955 K	1025 K

Considering these characteristics, the temperature conditions of the two-step ordering heat-treatment were: the first step of ordering was at 773 K for 0.5 h (CoPt) and 673 K for 1 h (Fe-55Pd) under a magnetic field of up to 10 T. The second step of ordering was at 1023 K for 3 h (CoPt) and 773 K for 24 h (Fe-55Pd). Magnetization curves were measured at 300 K along the X, Y and Z-directions after the first and the second step of ordering for CoPt as well as Fe-55Pd and are shown in figures 2 and 3, respectively. It is noted from figures 2(a) and 3(a) (measured after the first step under a magnetic field of 10 T) that the Z-direction is slightly easier to magnetize. This implies that the fraction of the variant whose easy axis lies in the Z-direction (bold curve) is slightly larger than that of the X- and Y-variants (dotted and dashed curves). Figures 2(b) and 3(b) clearly show that a single variant of both alloys is certainly obtained after the second step of ordering without the influence of a magnetic field. These results suggest that the magnetic field is especially effective during the early stage of ordering for the selective formation of a preferred variant.

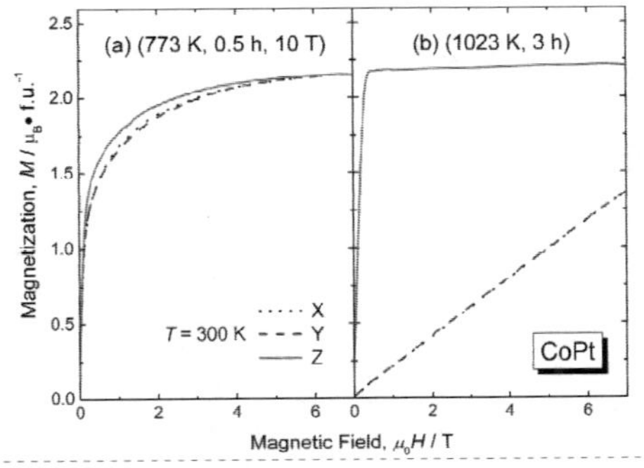

Figure 2. Magnetization curves of CoPt after (a) ordering at 773 K for 0.5 h under a magnetic field of 10 T and (b) successive ordering at 1023 K for 3 h.

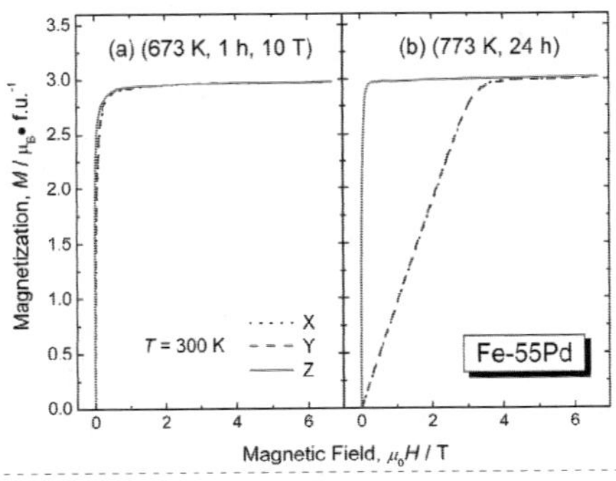

Figure 3. Magnetization curves of Fe-55Pd after (a) ordering at 673 K for 1 h under a magnetic field of 10 T and (b) successive ordering at 773 K for 24 h.

We also investigated the effect of magnetic field-strength on the selective formation of a preferred variant. That is, the first step of ordering was carried out under different magnetic field-strengths of up to 10 T. Figure 4 shows the fraction of the Z-variant (f_z) after the second step of ordering as a function of magnetic field as applied in the first step for CoPt (open circles) and Fe-55Pd (open squares). The figure clearly shows that f_z increases with increasing magnetic field and reaches 100 % at about 0.5 T for CoPt and 4 T for Fe-55Pd. We notice from the figure that there is an increase in the f_z at about 0.3 T for CoPt and 3 T for Fe-55Pd which indicates the existence of a critical magnetic field to obtain a single variant for both alloys.

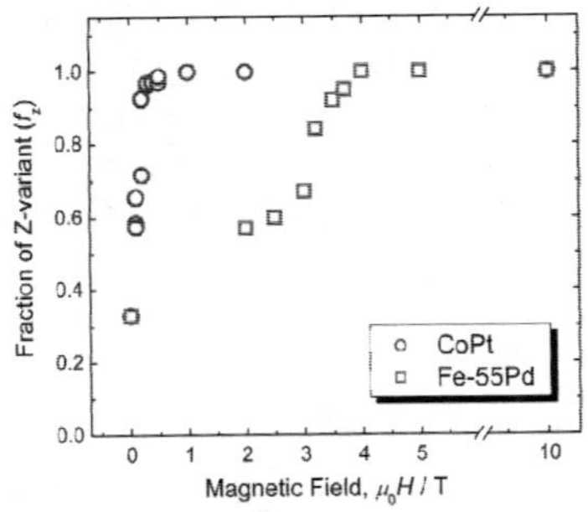

Figure 4. The fraction of Z-variant (f_z) after the second step of ordering as a function of magnetic field for CoPt and Fe-55Pd.

Results from resistivity measurements, as summarized in table 1, indicate that the ordering temperature in the first step of 773 K for CoPt and 673 K for Fe-55Pd is higher than the Curie temperature of the ordered phase and less than the Curie temperature of the disordered phase. This means that the ordered phase is paramagnetic and the disordered phase is ferromagnetic at these temperatures. To obtain the induced magnetocrystalline anisotropy energy at the ordering temperature of the first step the magnetization curves were measured along the easy and hard axes of the single variant up to a magnetic field of 1.8 T. Figures 4 and 5 show magnetization curves of CoPt and Fe-55Pd as measured along the easy axis, *c*-axis (bold curve) and hard axis as well as the *a*-axis (dashed curve). From the enclosed area between the easy and hard axes up to a magnetic field of 1 T the induced magnetocrystalline anisotropy energies were 4.1 and 45.3 kJ·m^{-3} for CoPt and for Fe-55Pd, respectively. The induced anisotropy energies calculated here are a thousand times smaller than the uniaxial magnetocrystalline anisotropy constant of CoPt at 298 K, $K_u = 4.5\times10^7$ erg·cm^{-3} [8] and Fe-55Pd at 4.2 K, $K_u \sim 1.7\times10^7$ erg·cm^{-3} [9].

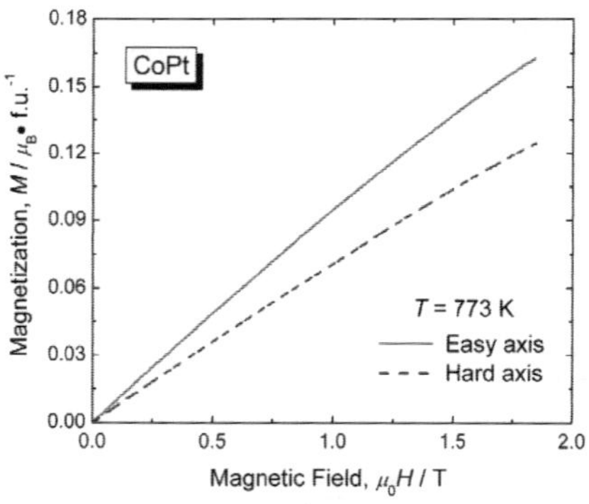

Figure 5. Magnetization curves of the single variant CoPt measured at 773 K along the easy and hard axes.

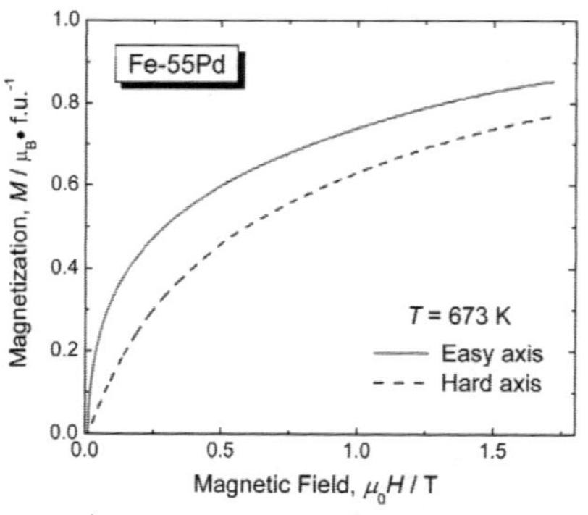

Figure 6. Magnetization curves of the single variant Fe-55Pd measured at 673 K along the easy and hard axes.

4. Conclusions

A single variant of Co-50at%Pt and Fe-55at%Pd alloys was obtained by a two-step ordering heat-treatment. A magnetic field was applied in the first step of ordering while the second step was done without a magnetic field. The results suggest that the magnetic field is especially effective during the early stage of ordering for the selective formation of a preferred variant. The magnetic field-strength dependence of single variant formation indicates that there seems to be a critical magnetic field value required to obtain a single variant for both alloys. The induced anisotropy energy of the ordered phase under a magnetic field of 1 T is 4.1 kJ·m^{-3} at 773 K for CoPt and 45.3 kJ·m^{-3} at 673 K for Fe-55Pd.

Acknowledgements

The authors would like to thank Prof. S. Nishijima of Osaka University for providing the superconducting magnet for this work. This study was supported by Priority Assistance for the Formation of Worldwide Renowned Centres of Research-The Global COE Program (Project: Centre of Excellence for Advanced Structural and Functional Materials Design) from the Ministry of Education, Culture, Sports, Science and Technology (MEXT), Japan.

References

[1] Ullakko K, Huang J K, Kantner C, O'Handley R C and Kokorin V V 1996 *Appl. Phys. Lett.* **69** 1966
[2] Heczko O, Lanska N, Soderberg O and Ullakko K 2002 *J. Magn. Magn. Mater.* **242–245** 1446
[3] Kakeshita T, Kuroiwa K, Shimizu K, Ikeda T, Yamagishi A and Date M 1993 *Mater. Trans. JIM* **34** 423
[4] Newkirk J B, Geisler A H, Martin D L and Smoluchowski R 1950 *Trans. AIME J. Metals* **188** 1249
[5] Vlasova N I, Kandaurova G S and Shchegoleva N N 2000 *Fiz. Met. Metalloved* **90** 31
[6] Tanaka K, Ichitsubo T, Amano M, Koiwa M and Watanabe K 2000 *Mater. Trans. JIM* **41** 917
[7] Massalski B, Murray J L, Bennett L H and Baker H 1986 *Binary alloy phase diagrams*, Metals Park, Ohio, American Society for Metals, Vol. 1, p.1093.
[8] Shima H, Oikawa K, Fujita A, Fukamichi K and Ishida K 2005 *Appl. Phys. Lett.* **86** 112515
[9] Shima H, Oikawa K, Fujita A, Fukamichi K, Ishida K and Sakuma A 2004 *Phys. Rev.* B **70** 224408

Anomalies in physical properties and diffuse scatterings for iron-doped Ti-Pd shape memory alloys

M Todai[1], T Fukuda[1] and T Kakeshita[1]
Department of Materials Science and Engineering, Graduated School of Engineering, Osaka University, 2-1 Yamadaoka, Suita, Osaka 565-0871, Japan

E-mail: misuharu.todai@mat.osaka-u.ac.jp

Abstract. Transformation behaviour in Ti-(50-x)Pd-xFe shape memory alloys was investigated by electrical resistivity and specific heat measurements. The alloys with x = 14, 16 and 18 exhibit an obvious first order martensitic transformation but the alloys with $x \geq 19$ do not exhibit a first order transformation down to 4.2 K. However, the alloys with $x \geq 19$ show a negative temperature coefficient in electrical resistivity below a specific temperature T_{min}. In addition, specific heat measurements at cryogenic temperatures revealed that the Debye temperature of the first order transformation-suppressed state is significantly lower than that of the martensite state. These anomalies are suggested to be caused by an instability of the electronic structure of the B2-phase.

Introduction

Intermetallic compounds of TiPd are shape memory alloys that exhibit a thermoelastic martensitic transformation from the B2-type structure to the B19-type (2H) structure at about 800K [1]. Because of the high transformation temperature, TiPd is expected to be a shape memory alloy that actuates at high temperatures [2]. When 3d-transition elements are added in TiPd the M_s temperature changes remarkably [3-5]. Enami *et al.* found that martensitic transformation temperatures are controlled over a wide temperature range from 300 K to 800 K by substituting Fe for Pd [6,7]. Through the study, they reported that Ti-(50-x)Pd-xFe alloys have several martensite phases such as 2H, 9R, 4H and incommensurate martensite. Moreover, they found that electrical resistivity exhibits a negative temperature coefficient (NTC) above the martensitic transformation start temperature. Furthermore, they reported that diffuse satellites appear at incommensurate positions in the electron diffraction pattern. They called this state a premartensite phase. These diffuse satellites in the premartensite phase were then confirmed by Li *et al.* [8] and Murakami *et al.* [9] using Ti-36Pd-14Fe and Ti-34Pd-16Fe alloys. They also observed a tweed microstructure, rod-like streak and a nanoscale domain-like microstructure in this phase. However, the origin of these precursor phenomena is not clear yet.

This study was undertaken develop an understand of the origin of NTC in electrical resistivity and the diffuse satellite in Ti-(50-x)Pd-xFe alloys. It was thus desirable to examine physical properties in a wide temperature range for the alloys exhibiting NTC and the diffuse satellite. These experiments require the alloys, in which the martensitic transformation is suppressed, to be near absolute zero temperature. Therefore, we investigated the temperature dependence of their physical properties, which are closely related to their electronic structure. The studied properties were the electrical resistivity and the specific heat and the compounds of interest were Ti-(50-x)Pd-xFe alloys with an iron content of x = 19, 20 and 22 at.%, and whose the martensitic transformations are expected to be suppressed. For comparison, we examined Ti-(50-x)Pd-xFe alloys with an iron content of x = 14, 16 and 18 that do have a martensitic transformation. Compositions of the first two alloys are the same as those reported by Enami *et al.*

Experimental procedure

Ingots of Ti-(50-x)Pd-xFe (14, 16, 18, 19, 20, 22) alloys (in at.%) were prepared by arc melting and were homogenized at 1273 K for 24 h in quartz tubes. Specimens for all measurements described below were cut from these ingots, heat treated at 1273 K for 1 h and then quenched in iced water. The oxidized surface layer was removed by electropolishing. Electrical resistivity (ρ) was measured by a standard four probe method at a cooling and heating rate of about 1 K/min and a constant current density of about 0.1 MA/m^2. Specific heat was measured using a relaxation method in a PPMS (Quantum Design). For each measurement of specific heat the specimen was heated by two per cent of its absolute temperature and the heat capacity was calculated from the relaxation time of cooling by thermal conduction. We denote each alloy by its iron content: the Ti-36Pd-14Fe alloy is denoted 14Fe alloy, for example.

Results and discussion

Figure 1 shows results of some resistivity measurements. Figure 1 (a) shows the electrical resistivity of the 14Fe alloy as measured by the cooling and heating processes where the resistivity was normalized by the value at 600 K. In the cooling process of the alloy, the resistivity exhibits a local minimum at T_{min} = 426 K as indicated by an arrow. It then exhibits a sharp increase at M_s because of the martensitic transformation as indicated by a double arrow. The martensitic phase of the 14Fe alloy is incommensurate according to a previous study [3]. After the increase, the resistivity decreases linearly down to 10 K. In the heating process, the resistivity shows a sharp decrease because of a reverse transformation with an obvious hysteresis between the cooling and heating processes. The result for the 14Fe alloy is in good agreement with that reported by Enami *et al.* [3]. Similar resistivity curves were obtained for the 16Fe and 18Fe alloys. Figure 1 (b) shows the electrical resistivity for the 19Fe alloy measured in the cooling and heating processes where the resistivity was normalized by the value at 350 K. This alloy does not exhibit a sharp increase or a hysteresis between the cooling and heating processes. This alloy, however, exhibits a local minimum of resistivity T_{min} of 310 K. Below T_{min}, it exhibits NTC in electrical resistivity like the Ti-(50-y)Ni-yFe alloys with $y \geq 6$ [10]. Similar resistivity curves were obtained for the 20Fe and 22Fe alloys. Figure 1 (c) is a magnification of the electrical resistivity curve of the 14Fe alloy near the martensitic transformation temperature. In addition to the sharp increase at M_s, we notice a slight bend of the resistivity curve at M_s' in the cooling process. This suggests that there is an additional transformation. To study this behaviour further, we measured the electrical resistivity only in the vicinity of M_s' and the result is shown in Fig. 1 (d). There is an obvious hysteresis between the cooling and heating processes suggesting that a weak first order transformation occurs at M_s'. Similar results were obtained for the 16Fe and 18Fe alloys. It is thus likely that a new phase, designated the X-phase exists between M_s' and M_s.

Results of characteristic temperatures obtained in this study are summarized in Fig. 2. In this figure, the temperatures obtained by Enami *et al.* [3] are shown by open marks. In addition, the crystal structures identified by Enami *et al.*, Nishida *et al.* and Murakami *et al.* are also shown. We recognize in the figure that T_{min} decreases almost linearly with an increase in iron content. We also observe that M_s and M_s' decrease nearly linearly with an increase in iron content up to x = 18 and then suddenly disappear between x = 18 and x = 19.

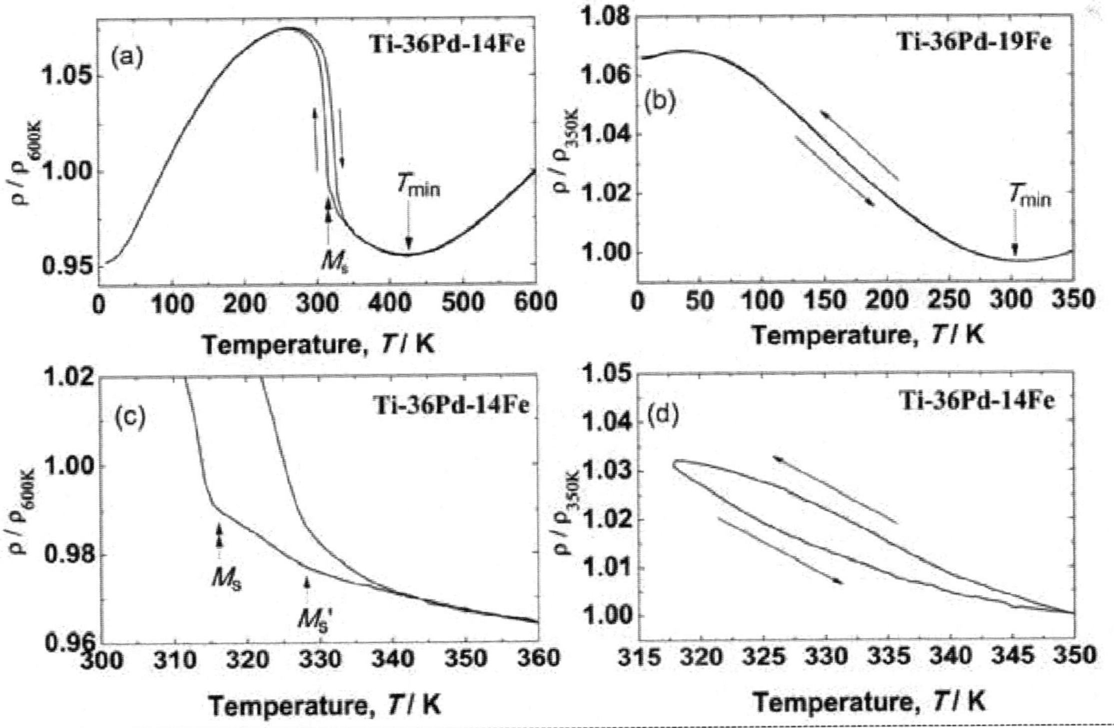

Figure 1. Temperature dependence of the resistivity of (a) Ti-36Pd-14Fe, (b) Ti-31Pd-19Fe, (c) near M_s in Ti-36Pd-14Fe and (d) a thermal hysteresis associated with a first order transformation at M_s' in Ti-36Pd-14Fe.

To understand the nature of NTC in electrical resistivity further, we measured the specific heat C_p at cryogenic temperatures using the 19Fe alloy and the result is plotted as C_p / T vs T^2 in Fig. 3. In this figure, the same plot for the 14Fe alloy, which is martensite at low temperatures, is also shown for comparison. From the slope in Fig. 3, the Debye temperature Θ_D is calculated to be 221 K for the 19Fe alloy and 262 K for the 14Fe alloy. The low value of Θ_D in the 19Fe alloy suggests that its lattice has low elastic constant at low temperature. In addition, the electronic specific heat coefficient γ is calculated to be 5.4 mJ/mol K^2 for the 19Fe alloy and 4.5 mJ/mol K^2 for the 14Fe alloy. The high value of γ in the 19Fe alloy suggests that its electronic state is unstable. The high γ in the 19Fe alloy suggests that the electron-phonon interaction is large in this alloy. Similar results of low Θ_D and high γ have been reported in Ti-(50-*y*)Ni-*y*Fe alloys by the authors [10]. Thus, the origin of NTC and diffuse scattering in Ti-(50-*x*)Pd-*x*Fe alloys can be considered to be due to a nesting effect and a large electron-phonon interaction as in Ti-(50-*y*)Ni-*y*Fe alloys [11,12]. An electronic structure calculation of Ti-(50-*x*)Pd-*x*Fe alloys is required for further discussion and is now in progress.

Conclusion

We examined the transformation behaviour of Ti-(50-*x*)Pd-*x*Fe alloys ($14 \leq x \leq 22$) by electrical resistivity and specific heat measurements. The 14Fe, 16Fe and 18Fe alloys exhibit a first order martensitic transformation preceded by a weak first order transformation. Transformation temperatures are found to decrease with increasing iron content. On the other hand, these first order transformations are suppressed in the 19Fe, 20Fe and 22Fe alloys. They, however, exhibit a negative temperature coefficient in electrical resistivity. Debye temperature of the transformation-suppressed state is lower than that of the martensite state. This behaviour resembles that observed in Ti-(50-*y*)Ni-*y*Fe alloys.

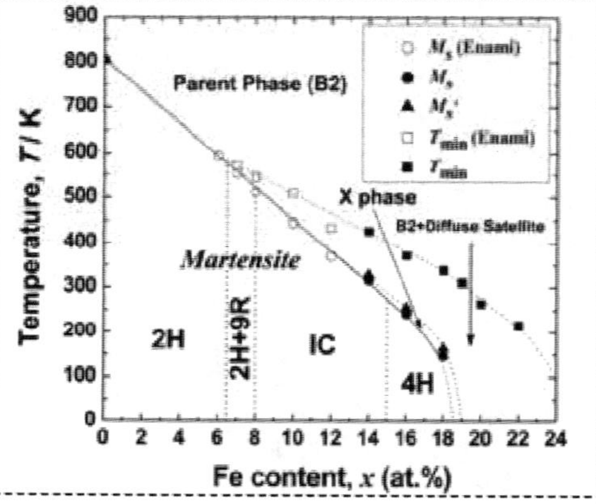

Figure 2. Iron content dependence of the transformation temperature for Ti-(50-x)Pd-xFe alloys. Solid marks represent this study. Open marks are those reported previously [3].

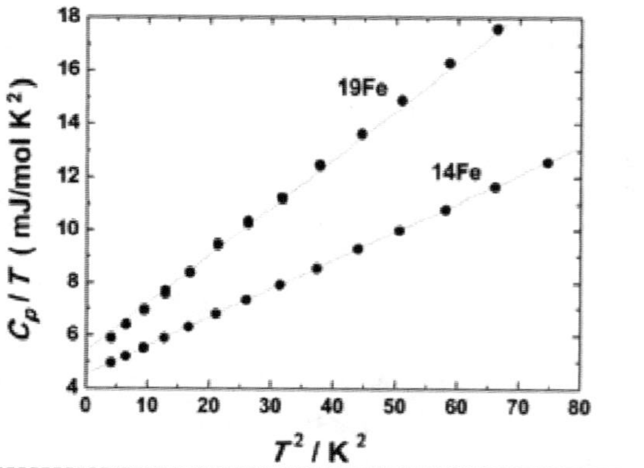

Figure 3. The C_p / T vs T^2 relation in the low temperature region of the Ti-36Pd-14Fe alloy and the Ti-31Pd-19Fe alloy.

Acknowledgements
This work is partly supported by Grant-in-Aid for Scientific Research (Kiban (B), No.18360331) and Priority Assistance for the Formation of Worldwide Renowned Centres of Research–The Global Century COE Program (Project: Centre of Excellence for Advanced Structural and Functional Materials Design) from MEXT of Japan.

References

[1] Donkersloot H C and Van Vucht J H N 1970 *J.Less Common Met.* **20** 83
[2] Otuka K, Oda K, Ueno Y, Piao M, Ueki T and Horikawa H 1993 *Scripta Metall. Et Mater,* **29** 1355
[3] Enami K, Yoshida T and Nenno S 1986 *Proc.ICOMAT-86* 103
[4] Enami K, Horii K, and Takahashi J 1989 *ISIJ International.* **29** 430
[5] Hosoda H, Enami K, Kamio A and Inoue K 1996 *J. of Intelligent Materials Systems and structures.* **7** 312
[6] Enami K and Hoshiya T 1993 *Proc. 1st Europ. East-West Symp. on Materials and processes, Int. J. Material and product Technology.* **8** 361
[7] Enami K, Seki H and Nenno S 1986 *Tetsu-to-Hagane (J. of Iron and Steel Inst. Jpn.)* **72** 103
[8] li S, Nishida M, Murakami Y and Shindo D 2003 *J. Phys. IV France* **112** 1035
[9] Murakami Y Kidu S and Shindo D 2003 *J. Phys. IV France* **112** 1031
[10] Choi M.-S. Fukuda T and Kakeshita T 2005 *Scripta Mater* **53** 869
[11] Bihlmayer G, Eibler R and Neckel A 1992 *Phys.Chem,* **96** 1626
[12] Yamamoto T, Fukuda T and Kakeshita T 2006 *Mater. Trans,* **47** 594

Grain size effect on martensitic transformation behavior in Fe-Ni invar alloys

Yoshikazu Aikawa[1], Tomoyuki Terai[2], and Tomoyuki Kakeshita[2]

[1]Metal Powder Manufacturing & Sales Division, Sanyo Special Steel Co.,Ltd, 3007 Nakashima, Shikama-ku, Himeji 672-8677, Japan

[2]Department of Material Science and Engineering, Graduate School of Engineering, Osaka University, 2-1 Yamadaoka, Suita, Osaka 565-0871, Japan

Abstract. We investigated the grain size effect on martensitic transformation behavior in Fe-30at.%Ni powder and ribbon specimens. The powder specimen with a particle size of 5 μm does not show an athermal martensitic transformation but does show an isothermal martensitic transformation after an incubation time of about 10^4 s at 205 K. On the other hand, the powder specimen with a particle size of 20 μm shows an athermal martensitic transformation at 150 K. The value of M_s is much lower than that of the single crystal and of bulk specimens. However, the M_s temperature of a ribbon specimen with an average grain size of 15 μm is found to be almost identical to that of the single crystal and of bulk specimens. Considering these results, the athermal martensitic transformation is suppressed by the decrease in particle size if grains do not have grain boundaries.

1. Introduction

It is well known that grain size influences various physical properties. For example, nanosized particles of Au and Ag are found to have lower melting points than their respective bulk metals because of changes in surface energy [1,2]. Grain size is also known to have an influence on the martensitic transformation of ZrO_2 powder [3] and Fe-Ni powder [4]. However, detailed transformation behavior and the transformation mechanism are not clear yet.

For this paper, we investigated the effect of grain size on the martensitic transformation temperature and its kinetics using an Fe-30at.%Ni powder and ribbon specimens. As a result, we found that the behavior of the martensitic transformation with fine powder specimens is different from that of a ribbon and of coarse powder specimens.

2. Experimental Procedure

Powder and rapidly quenched ribbon specimens of Fe-30at%Ni were prepared. The powder was produced by an argon gas atomization process and classified to average particle sizes of 5 μm and 20 μm. Specimens of rapidly quenched ribbon were prepared by melt-spun methods with a copper roll at a rotation speed of 15.7m/s.

The microstructure of each specimen was observed by optical microscope. To identify the phase of each specimen, an X-ray diffraction measurement was made. To investigate magnetic properties and martensitic transformation temperatures, magnetic susceptibility measurements were made using a SQUID magnetometer at temperatures from 300 K to 5 K.

3. Results and Discussion

3.1. Microstructure and crystal structure

Figure 1 shows a cross section of the microstructure for each specimen using an optical microscope after etching. The powder specimen with an average particle size of 5 μm had no grain boundary and the other powder specimen with an average particle size of 20 μm generally had no grain boundary either although a few powder specimens had boundaries. The ribbon specimen is a polycrystal with a grain size of about 15 μm.

Crystal structures of these specimens were confirmed by an X-ray diffraction measurement. Results are shown in Fig.2. As shown in the figure, only the fcc phase is apparent and the bcc phase is not observed in all the specimens.

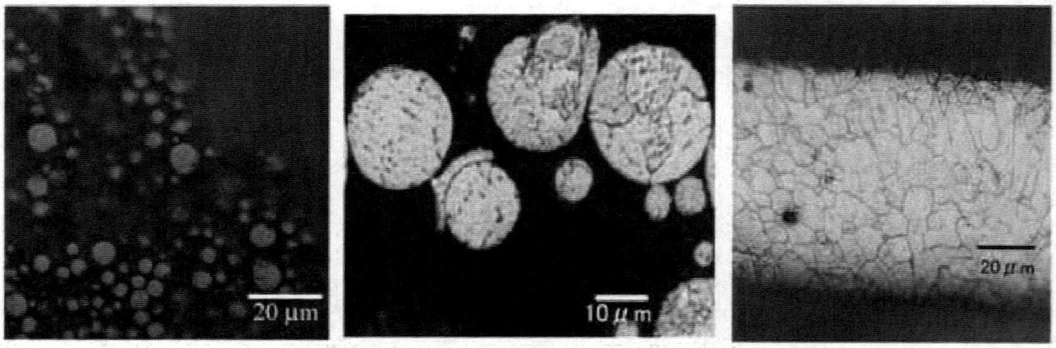

Fig.1 A series of optical microscope photograph of Fe-30at.%Ni.

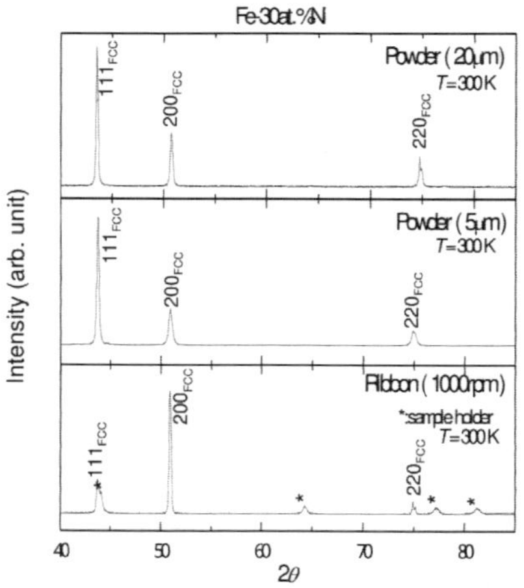

Fig. 2. Powder X-ray diffraction patterns of Fe-30at.%Ni.

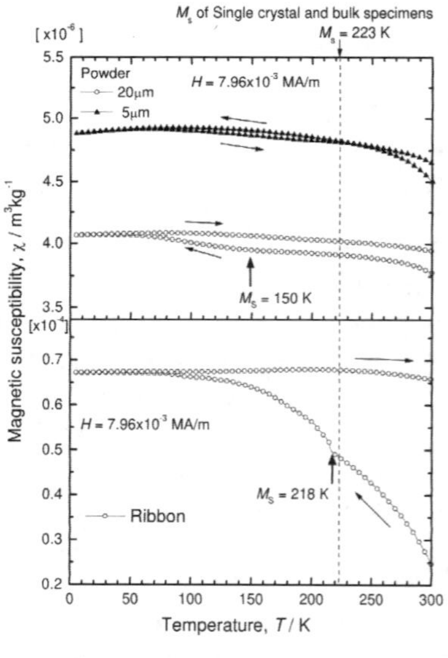

Fig. 3 Temperature dependence of magnetic susceptibility.

3.2. Magnetic susceptibility

To investigate magnetic properties and the martensitic transformation temperature, we measured the temperature dependence of the magnetic susceptibility for each specimen. Results are shown in Fig.3. It should be noted that the magnetic susceptibility of the powder specimen with a particle size of 5 μm gradually increases as the temperature decreases. This indicates that the specimen does not show an athermal martensitic transformation. On the other hand, the powder specimen with a particle size of 20 μm shows a sudden increase at 150 K because of an athermal martensitic transformation. This value of M_s is much lower than those of the single crystal and bulk specimens as obtained in a previous study [5]. The susceptibility of the ribbon specimen, with grain boundaries, gradually increased as the temperature decreased and the sudden increase at 218 K is due to an athermal martensitic transformation. This M_s is almost identical to that in a previous study [5]. On the other hand, we investigated in previous work that the effect of quench speed, internal stress and deficiency on magnetic susceptibility with ribbon specimens by various rotation speed and found that these factors hardly influenced on the behaviour of magnetic susceptibility[6]. These results clearly suggest that the particle size of the specimen influences the transformation behavior but the presence of a boundary does not.

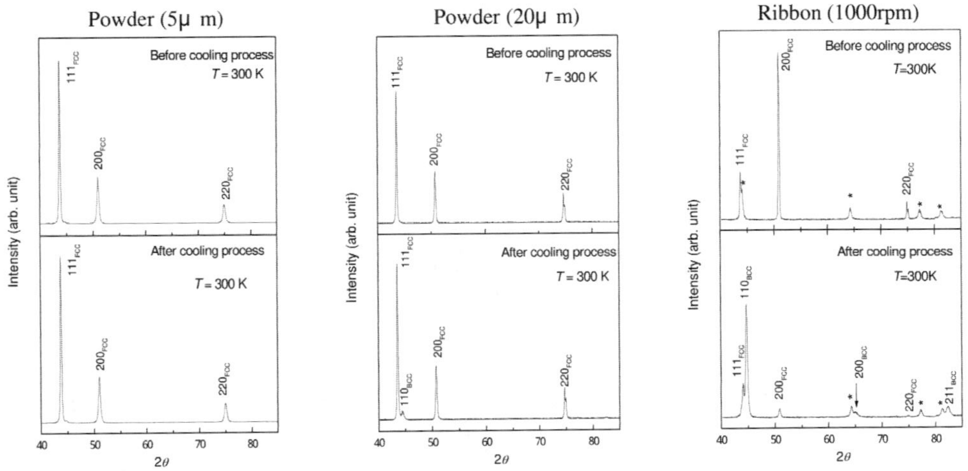

Fig. 4. X-ray diffraction patterns for each specimen before and after cooling to 5K.

Figure 4 shows X-ray diffraction patterns of each specimen before and after cooling to 5K. The powder specimen with a particle size of 5 μm does not show a bcc phase even after cooling. The powder specimen with a particle size of 20 μm shows a small amount of bcc phase after cooling while the ribbon specimen shows a large amount of bcc phase after cooling. These results agree well with the magnetic susceptibility measurements shown in Fig. 3.

For the powder specimen with a particle size of 5 μm the martensitic transformation behavior is not athermal but is isothermal because a hysteresis is present in the susceptibility curves for the cooling and the heating process, as shown in Fig. 3. We thus measured the time dependent nature of magnetic susceptibility for the powder specimen with a particle size of 5 μm at 205 K. The result is shown in Fig. 5. As shown in this figure, the magnetic susceptibility increases slightly after an incubation time of about 10^4 s. This result indicates that the 5 μm powder certainly undergoes an isothermal martensitic transformation.

Considering the results described above, we can say that the athermal martensitic transformation is suppressed by a decrease in particle size if no grain boundary is present. Further experiments are required to study this phenomenon.

3.3. Spontaneous magnetization of the Fe-30at%Ni fcc phase

In this study, the athermal martensitic transformation was suppressed in the 5 μm grain size powder. We thus measured the spontaneous magnetization of the Fe-30at%Ni fcc phase at a very low temperature using this fine grain size powder. A typical result at 5 K is shown in Fig. 6. As a result, the spontaneous magnetization of the fcc phase was found to be 114 emu/g (1.16 μ_B/atom) by drawing a line from the curve at the higher magnetic field as shown with the dotted line in Fig. 6. The value of the spontaneous magnetization is almost identical to that obtained in a previous study (1.1 μ_B/atom) [5].

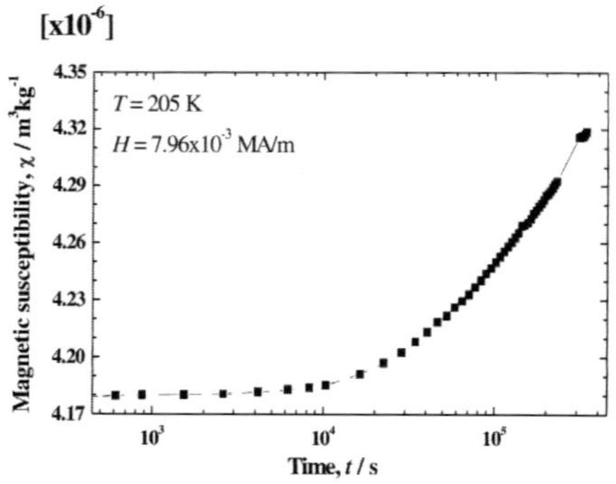

Fig. 5 Time dependence of magnetic susceptibility for the 5μm powder.

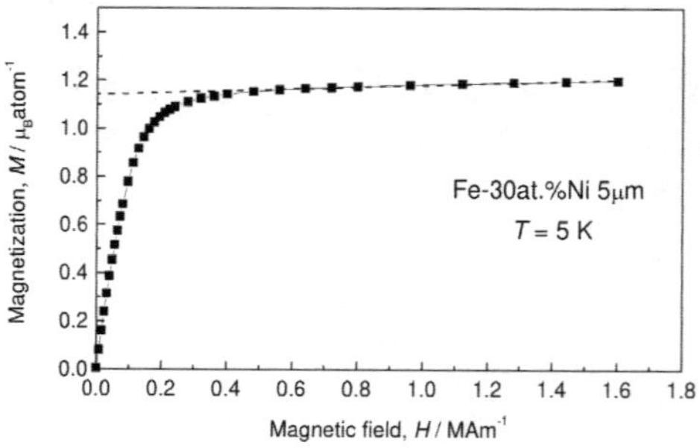

Fig. 6. Spontaneous magnetization of the fcc phase at 5 K.

4. Conclusions

We investigated the effect of grain size on the martensitic transformation temperature and its kinetics using a Fe-30at.%Ni powder and ribbon specimens. As a result, a fine grain size powder of 5 μm with a single crystal shows an isothermal martensitic transformation. On the other hand, the sample with a grain size of about 20μm with a single crystal shows an athermal martensitic transformation but its M_s is lower than that of the bulk specimen. The same result is obtained for the ribbon specimen with a grain size of about 15 μm, but its M_s is almost at the same temperature as that of the bulk specimen. These results suggest that the particle size of the specimen influences the transformation behavior and the presence of a boundary does not.

The spontaneous magnetization of the fcc phase is found to be 114 emu/g(1.16 μ_B/atom). The value of spontaneous magnetization is found to be almost identical to that found in a previous study [5].

Acknowledgements

This study was supported by Priority Assistance for the Formation of Worldwide Renowned Centers of Research - The Global COE Program (Project: Center of Excellence for Advanced Structural and Functional Materials Design) from the Ministry of Education, Culture, Sports, Science and Technology (MEXT), Japan.

References

[1] Ph.Buffat and J-P.Borel, Phys. Rev.A 13, p2287, 1976.
[2] T.Castro et.al., Phys. Rev.B 42, p8548, 1990.
[3] Hayakawa et.al J.Phys. 112, p1103, 2003.
[4] Kati et.al J.Phys.Soc. Japan 27, p536, 1962.
[5] T. Kakeshita et.al. Maeterials Transactions JIM 34, p423, 1993.
[6] Unpublished data.

Martensitic transformation behaviour in sensitized SUS304 austenitic stainless steel during isothermal holding at low temperature

Jae-hwa Lee, Takashi Fukuda and Tomoyuki Kakeshita

Department of Materials Science and Engineering, Graduate School of Engineering, Osaka University, 2-1 Yamada-oka, Suita, Osaka 565-0871, Japan

ljhcom@mat.eng.osaka-u.ac.jp

Abstract. We investigated martensitic transformation behaviour in sensitized SUS304 austenitic stainless steel to determine the stability of the austenitic phase at low temperatures. We found that a specimen that was sensitized at 973 K for 100 h exhibits an isothermal martensitic transformation when the specimen is held in the temperature range between 60 and 260 K. We constructed a time-temperature-transformation (*TTT*) diagram corresponding to the formation of 0.5 vol. % α'-martensite. A magnetization measurement was used to evaluate the volume fraction of α'-martensite. The *TTT* diagram shows a double-C curve with two noses located at about 100 and 200 K. *In-situ* optical microscope observations reveal that the double C-curve is due to two different transformation sequences. That is, the upper part of the C-curve is due to a direct $\gamma \rightarrow \alpha'$ martensitic transformation and the lower part of the C-curve is due to a successive $\gamma \rightarrow \varepsilon' \rightarrow \alpha'$ martensitic transformation. The direct $\gamma \rightarrow \alpha'$ transformation occurs in the vicinity of grain boundaries while the successive $\gamma \rightarrow \varepsilon' \rightarrow \alpha'$ transformation occurs near the centre of grains. A scanning electron microscope observation reveals that carbide particles of $M_{23}C_6$ are formed in the grain boundaries. The concentration difference between the centre of the grains and regions near grain boundaries is the reason for the difference in the isothermal transformation sequence for the sensitized SUS304 stainless steel.

1. Introduction

Austenitic stainless steels are characterized by good corrosion resistance, excellent mechanical properties, superior weldability and nonmagnetic characteristics [1-2]. SUS304 stainless steel is extensively used in equipment for cryogenic applications such as tanks, piping systems and other equipment for handling condensed gases. The welding process is an essential factor for the use of this steel in these applications as thermal effects can result in sensitized zones where carbide precipitation ($M_{23}C_6$) at grain boundaries and chromium depletion in the vicinity of grain boundaries may occur [3-4]. Recently, we found that sensitized SUS304 stainless steel exhibits an isothermal martensitic transformation at cryogenic temperatures [5-6]. However, its time-temperature-transformation (*TTT*) diagram which contains important information relating to its isothermal transformations has not been constructed yet.

For this study we constructed a *TTT* diagram of the isothermal martensitic transformation in sensitized SUS304 stainless steel.

2. Experimental Procedure

The material used in the study was SUS304 austenitic stainless steel with a composition of 0.06C-0.67Si-1.01Mn-0.029P-0.009S-8.5Ni-18.10Cr-Bal.Fe (mass %). Specimens of 3 × 3 × 1 mm were cut from the sheet and solution-treated at 1323 K for 0.5 h in vacuum followed by quenching in ice water. Most specimens were sensitized by heat treatment at 973 K for 100 h. The oxidized surface layer was then eliminated by electropolishing in an electrolyte composed of 85 % C_2H_5OH and 15 % $HClO_4$ by volume. Isothermal holding was carried out without a magnetic field in the temperature range of 60 to 260 K. The volume fraction of the α′-martensite that was formed by isothermal holding was obtained by a magnetization measurement as in a previous study [7]. The microstructure of the specimen was investigated by a scanning electron microscopy (SEM) and the change in morphology during isothermal holding was observed using an *in-situ* optical microscopy.

3. Results and Discussion

The microstructure near grain boundaries is shown in Figure 1 for as-sensitized SUS304 stainless steel. Some particles are present along the grain boundaries. These particles are mainly $M_{23}C_6$ type, as confirmed by electron diffraction and EDS analysis. From this observation, the concentration of chromium between the centre of grains and near grain boundaries has been partially altered by the formation of $M_{23}C_6$ carbide.

To investigate the magnetic properties and the martensitic transformation we carried out a magnetic susceptibility measurement in the temperature range between 4.2 and 300 K. Figure 2 shows the temperature dependence of the magnetic susceptibility for sensitized SUS304 stainless steel. The value of χ starts to increase at about 260 K during the cooling process as indicated by "A". Such an increase in the value of χ means that the ferromagnetic α′-martensite is formed during the cooling process. During the heating process the value of χ starts to increase at about 60 K indicating that α′-martensite is also formed during the heating process. This result suggests that the martensitic transformation of the sensitized SUS304 stainless steel proceeds isothermally in the temperature range between 60 and 260 K. Furthermore, we notice that the value of χ increased by about 125 K during the cooling process and is indicated by "B". This result implies that two types of martensitic transformations exist for the sensitized SUS304 stainless steel. The value of χ has a peak at about 40 K because of a paramagnetic to anti-ferromagnetic transition of the γ-phase.

To investigate the isothermal nature further we conducted isothermal holding experiments at several temperatures between 60 and 260 K to construct a *TTT* diagram. We then carried out magnetization measurements at room temperature to evaluate the amount of α′-martensite from the spontaneous magnetization, as described elsewhere [7]. We obtained the time required for the formation of 0.5 vol. % α′-martensite for each temperature. Using these times we constructed the *TTT* diagram for α′-martensite as shown in Figure 3. It should be noted that the *TTT* diagram shows a double C-curve with two noses located at about 100 and 200 K as indicated by the arrows. This result is completely

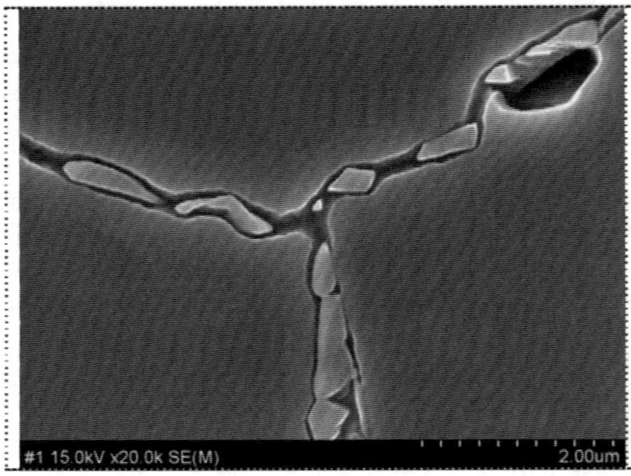

Figure 1. SEM observation result of as-sensitized SUS304 stainless steel.

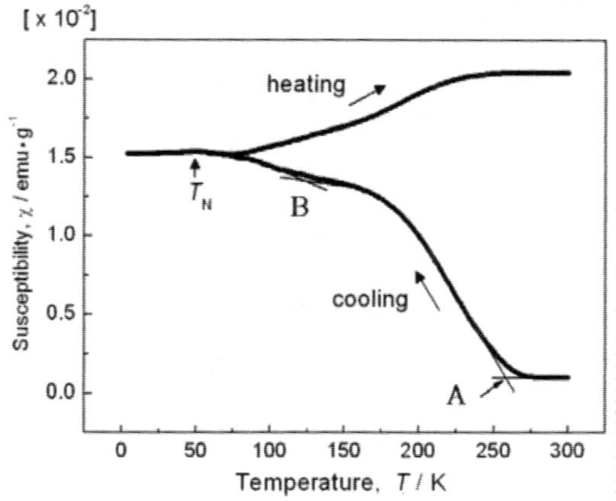

Figure 2. The magnetic susceptibility curve of sensitized SUS304 stainless steel. The measurement was made during the cooling process and also during the heating process.

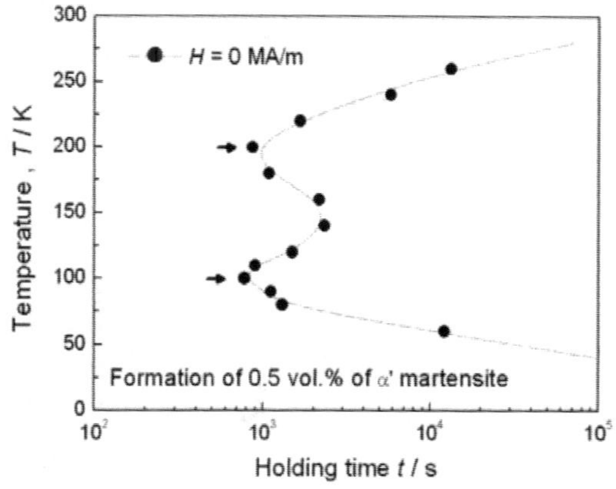

Figure 3. The *TTT* diagram of the isothermal martensitic transformation for sensitized SUS304 stainless steel. The dashed line is a guide.

different from the *TTT* diagram of solution-treated SUS304L stainless steel in which only one nose is present [7].

To determine why two noses appear in the *TTT* diagram we conducted *in-situ* optical microscopy during isothermal holding at the nose temperatures of 100 and 200 K. Figure 4 shows a series of optical micrographs taken during the isothermal holding at 200 K. After isothermal holding for 30 min wedge-shaped plates were formed directly from the γ-phase in the vicinity of the grain boundaries. These plates are α′-martensite because the wedge-shaped morphology is characteristic of α′-martensite [8]. The amount of α′-martensite gradually increased near the grain boundaries as the isothermal holding time increased. The above result implies that the upper part of the double C-curve is related to the direct $\gamma \rightarrow \alpha'$ martensitic transformation that was induced isothermally in the vicinity of a grain boundary during the isothermal holding.

On the other hand, Figure 5 shows a series of optical micrographs taken during isothermal holding at 100 K and the nose temperature of the lower part of the double C-curve. After isothermal holding for 10 min, a banded plate (characteristic of ε′-martensite) gradually appears near the centre of the grains as indicated by the dashed rectangle in Figure 5 (b). This result suggests that the $\gamma \rightarrow \varepsilon'$ martensitic transformation proceeds isothermally. Figure 5 (c) shows a micrograph of the

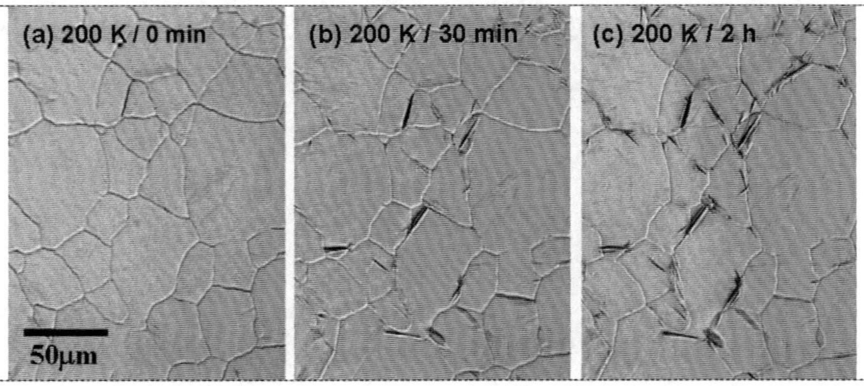

Figure 4. A series of *in-situ* optical micrographs of sensitized SUS304 stainless steel showing a direct $\gamma \rightarrow \alpha'$ transformation during isothermal holding at 200 K.

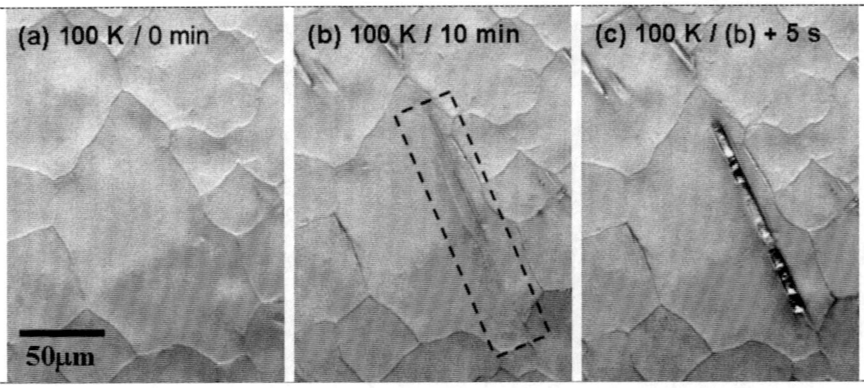

Figure 5. A series of *in-situ* optical micrographs of sensitized SUS304 stainless steel showing a successive $\gamma \rightarrow \varepsilon' \rightarrow \alpha'$ transformation during isothermal holding at 100 K.

microstructure taken 5 sec after capturing Figure 5 (b). We notice that α'-martensites instantaneously forms within the banded ε'-martensite. The $\varepsilon' \rightarrow \alpha'$ martensitic transformation thus proceeds athermally. This result confirms that the lower part of the double C-curve is related to the successive $\gamma \rightarrow \varepsilon' \rightarrow \alpha'$ martensitic transformation that was induced near the centre of the grains during the isothermal holding experiment.

The reason for appearing two types of isothermal transformation sequence in the sensitized SUS304 stainless steel will be due to the difference in concentration by sensitization heat-treatment.

4. Conclusions

We investigated martensitic transformation behaviour in sensitized SUS304 austenitic stainless steel at low temperatures and the following results were obtained.

(i) Sensitized SUS304 stainless steel exhibits an isothermal martensitic transformation when a specimen is held in a temperature range between 60 and 260 K.

(ii) The *TTT* diagram of the martensitic transformation shows a double C-curve with two noses located at about 100 and 200 K because of two different transformation sequences: the upper and lower parts of the double C-curve are assigned to a direct $\gamma \rightarrow \alpha'$ martensitic transformation in the vicinity of the grain boundaries and the successive $\gamma \rightarrow \varepsilon' \rightarrow \alpha'$ martensitic transformation near the centre of grains, respectively.

Acknowledgements

This study was supported by Priority Assistance for the Formation of Worldwide Renowned Centres of Research - The Global COE Program (Project: Centre of Excellence for Advanced Structural and Functional Materials Design) from the Ministry of Education, Culture, Sports, Science and Technology (MEXT), Japan.

References

[1] D. C. Larbalestier and H. W. King 1973 *Cryogenics* **13** 160
[2] D. T. Read and R. P. Reed 1981 *Cryogenics* **21** 415
[3] S. M. Bruemmer and L. A. Charlot 1986 *Scr. Metall.* **20** 1019
[4] E. A. Trillo and L. E. Murr 1999 *Acta Mater.* **47** 235
[5] T. Fukuda, T. Kakeshita and K. Kindo 2006 *Mater. Sci. Eng.* **A438-440** 212
[6] J. H. Lee, T. Fukuda, T. Kakeshita and K. Kindo2007 *Mater. Trans. JIM* **48** 2833
[7] J. H. Lee, T. Fukuda and T. Kakeshita 2008 *Mater. Trans. JIM* **49** 1937
[8] A. K. De, D. C. Murdock, M. C. Mataya, J. G. Speer and D. K. Matlock 2004 *Scr. Mater.* **50** 1445

Journal of Physics: Conference Series **165** (2009) 012059 doi:10.1088/1742-6596/165/1/012059

Investigation of a quadrupolar interaction in DyCu by elastic constants measurement

Motoyoshi Yasui[1], Tomoyuki Terai[1], Tomoyuki Kakeshita[1] and Masayuki Hagiwara[2]

[1]Division of Materials and Manufacturing Science, Osaka University, 2-1 Yamadaoka, Suita, Osaka 565-0871, Japan

[2]KYOKUGEN, Osaka University, 1-3 Machikaneyama Toyonaka, Osaka 560-8531, Japan

E-mail: motoyoshi.yasui@mat.eng.osaka-u.ac.jp

Abstract. We have investigated a quadrupolar interaction by measuring the temperature dependence for elastic constants of $C_B = (C_{11} + 2C_{12})/3$ (Γ_1-symmetry), $C' = (C_{11} - C_{12})/2$ (Γ_3-symmetry) and C_{44} (Γ_5-symmetry) in DyCu. We have evaluated the quadrupolar interaction coefficient $K(\mathbf{0})$ by analyzing critical fields of metamagnetic transitions. The following results are obtained; (i) C_{44} in the paramagnetic phase exhibits softening at temperatures near T_N while C_B and C' do not. This indicates that the dominant components of the quadrupolar moment are O_{yz}, O_{zx} and O_{xy} with Γ_5-symmetry. (ii) From an analysis of critical fields of metamagnetic transitions based on a mean-field approximation we have determined the quadrupolar interaction coefficient to be $K(\mathbf{0})|Q|^2/k_B = -16.8$ K. The negative value of $K(\mathbf{0})$ indicates that the quadrupolar interaction has an antiferro-type of Γ_5-symmetry.

1. Introduction

The rare earth intermetallic compound DyCu that has a CsCl-type cubic crystal structure shows a magnetic transition from an antiferromagnetic (AFM) phase to a paramagnetic (PM) phase at $T_N = 63.0$ K. The AFM phase has a triple-$\boldsymbol{q}$ structure (propagation vector: $\boldsymbol{q} = \langle 1/2\ 1/2\ 0\rangle 2\pi/a$, where a is the crystal lattice parameter) with magnetic moments of Dy^{3+} being parallel to the four equivalent <1 1 1> directions [1,2]. When a magnetic field is applied to the AFM phase, multi-step metamagnetic transitions occur [2-5]. These transitions have a field-direction dependence, that is, magnetization curves in the [0 0 1], [1 1 0] and [1 1 1] directions exhibit two-step, three-step and four-step transitions, respectively. The origin of the complicated magnetic structure and metamagnetic transitions are reported to be due to the coexistence of exchange and quadrupolar interactions [1-6]. However, there have been a few studies on the quadrupolar interaction type (Γ_3 or Γ_5-symmetry) and on the quantitative analysis for quadrupolar interaction coefficients.

In this study, we have measured the temperature dependence of elastic constants for DyCu to investigate the quadrupolar interaction [5,7]. We have also evaluated the exchange interaction coefficient and the quadrupolar interaction coefficient by an analysis of critical fields of metamagnetic transitions.

E-mail: motoyoshi.yasui@mat.eng.osaka-u.ac.jp

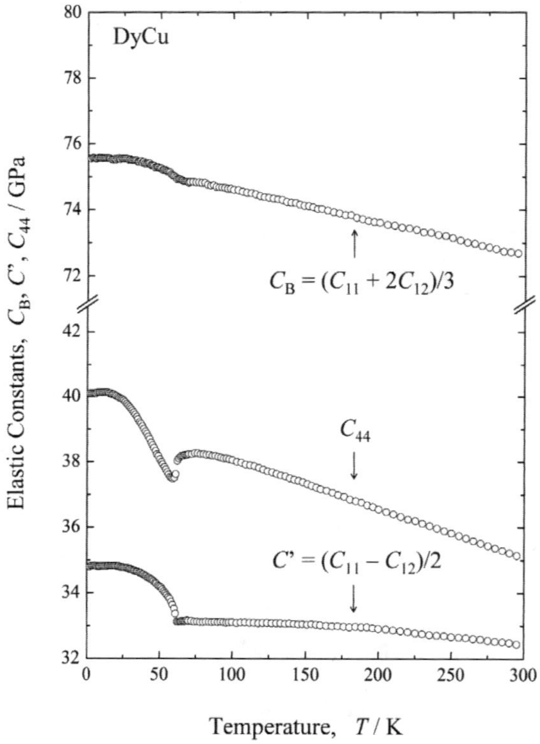

Figure 1. Temperature dependence of elastic constants $C_B = (C_{11} + 2C_{12})/3$, $C' = (C_{11} - C_{12})/2$ and C_{44} of DyCu.

2. Experimental procedure

An ingot of DyCu was prepared by arc-melting using Dy (99.9 %) and Cu (99.99 %) as starting materials. A single crystal rod of DyCu was grown by a floating zone method at a growth rate of 5.0 mm h^{-1}. The specimen used for elastic constants measurement was cut into $3.17 \times 3.29 \times 3.45$ mm^3 and it weighted 318.53 mg. Its surfaces were parallel to the {1 0 0} plane. The temperature dependence of the elastic constants were measured by using a rectangular parallelepiped resonance method, which enables the determination of all elastic constants simultaneously [7,8]. This was done in the temperature range of 2 - 300 K. Resonance spectra were obtained in the range of 200 - 800 kHz with a step of 0.1 kHz and the number of resonance peaks used for the analysis was about 50 - 55 at all temperatures.

3. Results

Figure 1 shows the temperature dependence of elastic constants of $C_B = (C_{11} + 2C_{12})/3$ (Γ_1-symmetry), $C' = (C_{11} - C_{12})/2$ (Γ_3-symmetry) and C_{44} (Γ_5-symmetry). It should be noted that C_{44} in the PM phase shows a small softening at temperatures near $T_N = 62.0$ K, while C_B and C' do not show any softening. Considering that C_{44} in the present system has Γ_5-symmetry, we can say that the quadrupolar interaction is Γ_5-symmetry type and the dominant components of the quadrupolar moment are O_{yz}, O_{zx} and O_{xy}. Using this interaction symmetry type, we propose quadrupolar structures for DyCu and discuss the quadrupolar interaction coefficients in the next section.

4. Discussion

First of all, we propose quadrupolar structures for the AFM phase and the first field-induced metamagnetic phase in the [0 0 1] and [1 1 1] directions (FM-I_{001} and FM-I_{111}) considering the

Table 1. Magnetic structures and quadrupolar structures of AFM, FM-I_{001} and FM-I_{111} phases.

Phase	Magnetic structure, $S_{hkl} = [S_x, S_y, S_z]$		Quadrupolar structure, $Q_{hkl} = [O_{yz}, O_{zx}, O_{xy}]$	
AFM	$\boldsymbol{S}_{000} = (1/\sqrt{3})\,[1\ 1\ 1]$	$\mathrm{S}_{011} = (1/\sqrt{3})\,[1\ \bar{1}\ \bar{1}]$	$\boldsymbol{Q}_{000} = (1/\sqrt{3})\,[1\ 1\ 1]$	$\boldsymbol{Q}_{011} = (1/\sqrt{3})\,[1\ \bar{1}\ \bar{1}]$
	$\boldsymbol{S}_{100} = (1/\sqrt{3})\,[1\ \bar{1}\ \bar{1}]$	$\boldsymbol{S}_{101} = (1/\sqrt{3})\,[\bar{1}\ 1\ \bar{1}]$	$\boldsymbol{Q}_{100} = (1/\sqrt{3})\,[1\ \bar{1}\ \bar{1}]$	$\boldsymbol{Q}_{101} = (1/\sqrt{3})\,[\bar{1}\ 1\ \bar{1}]$
	$\boldsymbol{S}_{010} = (1/\sqrt{3})\,[\bar{1}\ 1\ \bar{1}]$	$\boldsymbol{S}_{110} = (1/\sqrt{3})\,[\bar{1}\ \bar{1}\ 1]$	$\boldsymbol{Q}_{010} = (1/\sqrt{3})\,[\bar{1}\ 1\ \bar{1}]$	$\boldsymbol{Q}_{110} = (1/\sqrt{3})\,[\bar{1}\ \bar{1}\ 1]$
	$\boldsymbol{S}_{001} = (1/\sqrt{3})\,[\bar{1}\ \bar{1}\ 1]$	$\boldsymbol{S}_{111} = (1/\sqrt{3})\,[1\ 1\ 1]$	$\boldsymbol{Q}_{001} = (1/\sqrt{3})\,[\bar{1}\ \bar{1}\ 1]$	$\boldsymbol{Q}_{111} = (1/\sqrt{3})\,[1\ 1\ 1]$
FM-I_{001}	$\boldsymbol{S}_{000} = (1/\sqrt{3})\,[1\ 1\ 1]$	$\mathrm{S}_{011} = (1/\sqrt{3})\,[\bar{1}\ 1\ 1]$	$\boldsymbol{Q}_{000} = (1/\sqrt{3})\,[1\ 1\ 1]$	$\boldsymbol{Q}_{011} = (1/\sqrt{3})\,[1\ \bar{1}\ \bar{1}]$
	$\boldsymbol{S}_{100} = (1/\sqrt{3})\,[\bar{1}\ 1\ 1]$	$\boldsymbol{S}_{101} = (1/\sqrt{3})\,[1\ \bar{1}\ 1]$	$\boldsymbol{Q}_{100} = (1/\sqrt{3})\,[1\ \bar{1}\ \bar{1}]$	$\boldsymbol{Q}_{101} = (1/\sqrt{3})\,[\bar{1}\ 1\ \bar{1}]$
	$\boldsymbol{S}_{010} = (1/\sqrt{3})\,[1\ \bar{1}\ 1]$	$\boldsymbol{S}_{110} = (1/\sqrt{3})\,[\bar{1}\ \bar{1}\ \bar{1}]$	$\boldsymbol{Q}_{010} = (1/\sqrt{3})\,[\bar{1}\ 1\ \bar{1}]$	$\boldsymbol{Q}_{110} = (1/\sqrt{3})\,[1\ 1\ 1]$
	$\boldsymbol{S}_{001} = (1/\sqrt{3})\,[\bar{1}\ \bar{1}\ \bar{1}]$	$\boldsymbol{S}_{111} = (1/\sqrt{3})\,[1\ 1\ 1]$	$\boldsymbol{Q}_{001} = (1/\sqrt{3})\,[1\ 1\ 1]$	$\boldsymbol{Q}_{111} = (1/\sqrt{3})\,[1\ 1\ 1]$
FM-I_{111}	$\boldsymbol{S}_{000} = (1/\sqrt{3})\,[1\ 1\ 1]$	$\mathrm{S}_{011} = (1/\sqrt{3})\,[\bar{1}\ 1\ 1]$	$\boldsymbol{Q}_{000} = (1/\sqrt{3})\,[1\ 1\ 1]$	$\boldsymbol{Q}_{011} = (1/\sqrt{3})\,[1\ \bar{1}\ \bar{1}]$
	$\boldsymbol{S}_{100} = (1/\sqrt{3})\,[\bar{1}\ 1\ 1]$	$\boldsymbol{S}_{101} = (1/\sqrt{3})\,[1\ \bar{1}\ 1]$	$\boldsymbol{Q}_{100} = (1/\sqrt{3})\,[1\ \bar{1}\ \bar{1}]$	$\boldsymbol{Q}_{101} = (1/\sqrt{3})\,[\bar{1}\ 1\ \bar{1}]$
	$\boldsymbol{S}_{010} = (1/\sqrt{3})\,[1\ \bar{1}\ 1]$	$\boldsymbol{S}_{110} = (1/\sqrt{3})\,[1\ 1\ \bar{1}]$	$\boldsymbol{Q}_{010} = (1/\sqrt{3})\,[\bar{1}\ 1\ \bar{1}]$	$\boldsymbol{Q}_{110} = (1/\sqrt{3})\,[\bar{1}\ \bar{1}\ 1]$
	$\boldsymbol{S}_{001} = (1/\sqrt{3})\,[1\ 1\ \bar{1}]$	$\boldsymbol{S}_{111} = (1/\sqrt{3})\,[1\ 1\ 1]$	$\boldsymbol{Q}_{001} = (1/\sqrt{3})\,[\bar{1}\ \bar{1}\ 1]$	$\boldsymbol{Q}_{111} = (1/\sqrt{3})\,[1\ 1\ 1]$

following two assumptions.

- Only the quadrupolar components of O_{yz}, O_{zx} and O_{xy} are considered.
- The most symmetrical structure with respect to the magnetic field direction is considered.

Proposed quadrupolar structures are shown in table 1 and the magnetic structures as determined by neutron diffractions are also shown [1-3,10].

By considering the magnetic and quadrupolar structures in table 1, we quantitatively evaluate the exchange and quadrupolar interaction coefficients using measured critical fields based on fitting the following total Hamiltonian $\hat{H}_{\mathrm{tot}}$, consisting of the exchange interaction term $\hat{H}_{\mathrm{exch}}$, the quadrupolar interaction term $\hat{H}_{\mathrm{quad}}$ and the Zeeman term $\hat{H}_{\mathrm{Zeeman}}$:

$$\hat{H}_{\mathrm{tot}} = \hat{H}_{\mathrm{exch}} + \hat{H}_{\mathrm{quad}} + \hat{H}_{\mathrm{Zeemn}}. \tag{1}$$

The exchange interaction term $\hat{H}_{\mathrm{exch}}$ is given by:

$$\begin{aligned} \hat{H}_{\mathrm{exch}} &= \sum_{\mathrm{n.n.}} -2J_1(S_{x,i}S_{x,j} + S_{y,i}S_{y,j} + S_{z,i}S_{z,j}) + \sum_{\mathrm{n.n.n.}} -2J_2(S_{x,i}S_{x,j} + S_{y,i}S_{y,j} + S_{z,i}S_{z,j}) \\ &= -2J(\mathbf{0})(S_{x,i}S_{x,j} + S_{y,i}S_{y,j} + S_{z,i}S_{z,j}), \end{aligned} \tag{2}$$

where J_1 and J_2 are the exchange interaction coefficients of the first and second nearest nearest-neighbours, respectively, $S_{x,i}$, $S_{y,i}$ and $S_{z,i}$ are the components of magnetic moment for the i-site and $J(\mathbf{0})$ is the Fourier transform of J_1 and J_2. The quadrupolar interaction term $\hat{H}_{\mathrm{quad}}$ is written as:

$$\begin{aligned} \hat{H}_{\mathrm{quad}} &= \sum_{\mathrm{n.n.}} -K_1(O_{yz,i}O_{yz,j} + O_{zx,i}O_{zx,j} + O_{xy,i}O_{xy,j}) + \sum_{\mathrm{n.n.n.}} -K_2(O_{yz,i}O_{yz,j} + O_{zx,i}O_{zx,j} + O_{xy,i}O_{xy,j}) \\ &= -K(\mathbf{0})(O_{yz,i}O_{yz,j} + O_{zx,i}O_{zx,j} + O_{xy,i}O_{xy,j}), \end{aligned} \tag{3}$$

where K_1 and K_2 are the quadrupolar interaction coefficients of the first and second nearest-neighbours, respectively, $O_{yz,i}$, $O_{zx,i}$ and $O_{xy,i}$ are the components of the quadrupolar moment for the i-site and $K(\mathbf{0})$ is the Fourier transform of K_1 and K_2. The Zeeman term is given by:

$$\hat{H}_{\mathrm{Zeemn}} = g_{\mathrm{J}}\mu_{\mathrm{B}}\mu_0(S_x H_x + S_y H_y + S_z H_z), \quad (4)$$

where g_{J} is the Landé g-factor, μ_{B} is the Bohr magneton, μ_0 is the magnetic permeability in vacuum and H_x, H_y and H_z are components of the applied magnetic field. We then determine the exchange and quadrupolar interaction coefficients $J(\mathbf{0})$ and $K(\mathbf{0})$ using the critical fields of metamagnetic transitions from the AFM phase to the FM-I_{001} phase ($\mu_0 H_{\mathrm{C1,001}} = 6.6$ T at $T = 4.2$ K) and from the AFM phase to the FM-I_{111} phase ($\mu_0 H_{\mathrm{C1,111}} = 8.0$ T at $T = 4.2$ K). This was based on $\hat{H}_{\mathrm{tot}}$ between the AFM and FM-I_{001} phases and between the AFM and FM-I_{111} phase which are equal at the critical fields when a mean-field approximation is used. As a result, we obtain the following values:

$$J(\mathbf{0})\,|\,S\,|^2 / k_{\mathrm{B}} = -53.7\ \mathrm{K} \quad \text{and} \quad K(\mathbf{0})\,|\,Q\,|^2 / k_{\mathrm{B}} = -16.8\ \mathrm{K}, \quad (5)$$

where S and Q are the absolute values of magnetic moment and quadrupolar moment and k_{B} is the Boltzmann constant. The negative value of $K(\mathbf{0})$ indicates that the quadrupolar interaction has an antiferro-type of Γ_5-symmetry. This result is agrees with the experimental result reported by Aléonard *et al.* where the magneto-striction was isotropic in the transition from the PM phase to the AFM phase [2].

5. Conclusions

We have investigated the temperature dependence of elastic constants for DyCu and determined that the quadrupolar interaction has a Γ_5-symmetry type. Considering the quadrupolar interaction symmetry we have proposed quadrupolar structures and evaluated the quadrupolar interaction coefficient, $K(\mathbf{0})|Q|^2/k_{\mathrm{B}} = -16.8$ K, by an analysis of critical fields of metamagnetic transitions. The negative value of $K(\mathbf{0})$ indicates that the quadrupolar interaction has an antiferro-type of Γ_5-symmetry.

Acknowledgements

This study was supported by Priority Assistance for the Formation of Worldwide Renowned Centers of Research – The Global COE Program (Project: Center of Excellence for Advanced Structural and Functional Materials Design) from the Ministry of Education, Culture, Sports, Science and Technology (MEXT), Japan.

References

[1] Wintemberger M, Belakhovsky M and Pierre J 1971 *Phys. Status Solidi B* **48** 705
[2] Aléonard R, Morin P and Rouchy J 1984 *J. Magn. Magn. Mater.* **46** 233
[3] Amara M, Morin P and Bourdarot F 1997 *J. Phys. Condens. Matter.* **9** 7441
[4] Amara M and Morin P 1998 *J. Alloys Compd.* **549** 275
[5] Morin P and Schmitt D *Ferromagnetic Materials* vol 5 (Elsevier, Amsterdam, 1990)
[6] Kakeya I, Kakeshita T, Kindo K, Yamamoto Y and Saburi T 1999 *J. Phys. Soc. Jpn.* **68** 1025
[7] Levy P M 1973 *J. Phys. C* **6** 3545
[8] Demarest Jr H H 1971 *J. Acous. Soc. Am.* **49** 768
[9] Ohno I 1976 *J. Phys. Earth* **24** 355
[10] Yasui M, Terai T, Kakeshita T, Matsuda M, Metoki N and Nojiri H 2008 *J. Appl. Phys.* **103** 07B710

Electrochemical formation of Dy alloy films in a molten LiCl-KCl-$DyCl_3$ system

H Konishi[1], T Usui[1], T Nohira[2] and Y Ito[3]
[1]Division of Materials and Manufacturing Science, Graduate School of Engineering, Osaka University, 2-1 Yamadaoka, Suita, Osaka 565-0871, Japan
[2]Department of Fundamental Energy Science, Graduate School of Energy Science, Kyoto University, Sakyo-ku, Kyoto 606-8501, Japan
[3]Department of Environmental Systems Science, Graduate School of Engineering, Doshisha University, Kyotanabe, Kyoto 610-0321, Japan

E-mail: konishi@mat.eng.osaka-u.ac.jp

Abstract. As to the electrochemical formation of Dy-Ni alloy films in a molten LiCl-KCl-$DyCl_3$ system at 700 K, the growth of $DyNi_2$ film and behavior of anodic dissolution of Dy from the formed $DyNi_2$ film were investigated. The $DyNi_2$ films were formed by potentiostatic electrolysis at 0.55, 0.62 and 0.70 V with Ni electrodes. The growth rates of $DyNi_2$ films are higher at less noble potential, i.e., 0.47 μm min^{-1} at 0.55 V, 0.32 μm min^{-1} at 0.62 V and 0.14 μm min^{-1} at 0.70 V. From RBS analysis, it was suggested that the Dy-Ni alloy film was formed for 10 or 30 s during electrodepositing Dy at 0.30 V with a Ni electrode. Moreover, the growth rate of Dy-Ni alloy film was faster than that of Dy-Fe alloy film. Anodic electrolysis of the formed $DyNi_2$ film with thickness of 15 μm was conducted at 0.90 V, 1.30 V and 1.90 V, respectively. The formed $DyNi_2$ were transformed to other phases, i.e., $DyNi_3$, $DyNi_5$ and Ni, by selective anodic dissolution of Dy. The transformed Ni film was about 10 μm in thickness and had a porous structure with a pore diameter of 1~2 μm.

1. Introduction

Rare earth (RE) - transition metal (TM) alloy films attract great interest of many researchers for functional properties, e.g., excellent magnetic, hydrogen absorbing / permeable and others. Generally, these RE-TM alloy films are manufactured by PVD, CVD, ion implantation and ion plating. However, in near future, novel methods to produce more functionalized RE-TM alloy films will be required. Among novel methods, the authors focused on the molten salt electrochemical process. The molten salt electrochemical process has been studied by the authors [1, 2, 3]. And the authors reported the noticeable phenomenon in the RE-TM alloy films [1]. In the case of $DyNi_2$ formation, the alloy film of 60 μm thickness was formed merely in 2 hours at 700 K. The growth rate of the $DyNi_2$ film, 60 μm in 2 hours at 700 K, was extremely high, when considering conventional solid phase diffusion at the temperature lower than the half of melting point T($DyNi_2$) = 1446 K [4]. The growth rate was found to be extremely fast and strongly affected by electrochemical parameters such as potential and current density. The phenomenon was termed "electrochemical implantation", since the phenomenon was difficult to be explained by ordinary concepts of electrodeposition followed by solid phase diffusion. Furthermore, they reported that the anodic polarization of the formed $DyNi_2$ resulted in the rapid dissolution of Dy to form other Dy-Ni phases. The obtained phase and morphology were dependent on applied potential. The phenomenon is regarded as "electrochemical displantation".

From these backgrounds, in this study, regarding the electrochemical formation of Dy-Ni alloy films in a molten LiCl-KCl-$DyCl_3$ system at 700 K, the growth of $DyNi_2$ film and behavior of anodic dissolution of Dy from the formed $DyNi_2$ film were investigated.

2. Experimental

All chemicals were anhydrous reagent grade. The LiCl-KCl eutectic (LiCl : KCl = 58.5 : 41.5 mol%; Wako Pure Chemical Co., Ltd.) was introduced in a high purity alumina crucible (99.5 wt% Al_2O_3; SSA-S grade, NIKKATO Co., Ltd.), and kept under vacuum for more than 72 hours at 473 K to remove water. All experiments were performed in LiCl-KCl eutectic melts under dry argon atmosphere at 700 K. $DyCl_3$ (99.9 %, Kojundo Chemical Laboratory Co., Ltd.) of 0.50 mol% was added directly to the melts. To remove residual water and metal impurities, pre-electrolysis was conducted at an electrolysis voltage of 2.80 V with an Al plate cathode and glassy carbon anode. A chromel-alumel thermocouple was used for temperature measurement. The working electrodes were Mo (5 mm × ϕ 1 mm, 99.95 %, Nilaco Co., Ltd.) and Ni (5 mm × ϕ 1 mm, 99.99 %, Nilaco Co., Ltd.) wires for the investigation of electrochemical behavior in this system. For the formation of Dy-Ni alloy samples, rectangular shaped plates of Ni (10 mm × 5 mm × 0.2 mm, 99.7 %, Nilaco Co., Ltd.) were used as the working electrodes. The reference electrode was a silver wire immersed in LiCl-KCl containing 1 mol% of AgCl, placed in a Pyrex glass tube with thin bottom to maintain electrical contact with the melt. The potential of this reference electrode was calibrated with reference to that of a Li^+ / Li electrode, which was prepared by electrodepositing Li metal on a Mo wire. All the potentials given hereafter are referred to Li^+ / Li electrode potential on a Mo wire. The counter electrode was a glassy carbon rod (50 mm × ϕ 5 mm, Tokai Carbon Co., Ltd.) or an Al rod (50 mm × ϕ 5 mm, 99.9 %, Nilaco Co, Ltd.). A potentio / galvanostat (Solartron: SI1287) was used for cyclic voltammetry and open-circuit potentiometry. The Dy-Ni alloy samples were prepared by potentiostatic electrolysis using the same apparatus. After the electrolysis, the samples were rinsed with distilled water, and analyzed by XRD (Rigaku Denki Co., Ltd.: ultraX 18HB) with CuKα line at 50 kV and 300 mA. Cross-sections of these samples were also observed by SEM (Hitachi Co., Ltd.: S-2300). Moreover, these samples were analyzed by RBS (Rutherford backscattering spectrometry).

3. The growth of $DyNi_2$ film

It has been found that the equilibrium potential values of 0.762, 0.913, 1.181 and 1.344 V obtained from several independent open-circuit potential measurements at 700 K correspond to coexisting phase states of ($DyNi_2$+$DyNi_3$), ($DyNi_3$+Dy_2Ni_7), (Dy_2Ni_7+$DyNi_5$) and ($DyNi_5$+Ni), respectively [3]. Combining this result with the Dy(III)/Dy potential value, 0.47 V (vs. Li^+/Li) [1], the $DyNi_2$ was found to be stable at least in the potential region of 0.55-0.75 V. The following experiment was conducted in this potential region. In order to investigate the dependence of applied potential on the $DyNi_2$ film growth, potentiostatic electrolysis were conducted at 0.55, 0.62 and 0.70 V with Ni electrodes at 700 K. The electrolysis times were changed from 0.5 to 5.0 h at all potentials. Since all potential values are within the stable potential region of $DyNi_2$, Dy metal does not deposit at these potentials. Figure 1 shows the obtained transient curve of measured cathodic current density at 0.55 V as an example. Although large cathodic current of 15 mA cm^{-2} flowed at the beginning, the currents decreased to the constant value of about 8 mA cm^{-2} in 1.0 h. Similar transient curves were observed in the cases, at 0.62 and 0.70 V, except the constant current value became smaller as the potential value increased. All samples obtained after electrolysis were identified as $DyNi_2$ by XRD analysis. The relationship between thickness of the formed $DyNi_2$ films and the electrolysis time at 0.55, 0.62 and 0.70 V is shown in figure 2. Almost linear relation is seen for all potentials. Moreover, the slopes, which mean growth rates, are higher at less noble potential, i.e., 0.47 μm min^{-1} at 0.55 V, 0.32 μm min^{-1} at 0.62 V and 0.14 μm min^{-1} at 0.70 V.

Concerning the observed linear growth of $DyNi_2$ film, the following explanation can be proposed. According to the obtained transient curve in figure 1, as the measured current directly shows the film growth rate, the film growth at the initial period (within 15 min) seems to follow the parabolic rate law, but afterwards the growth changes to follow the linear rate law.

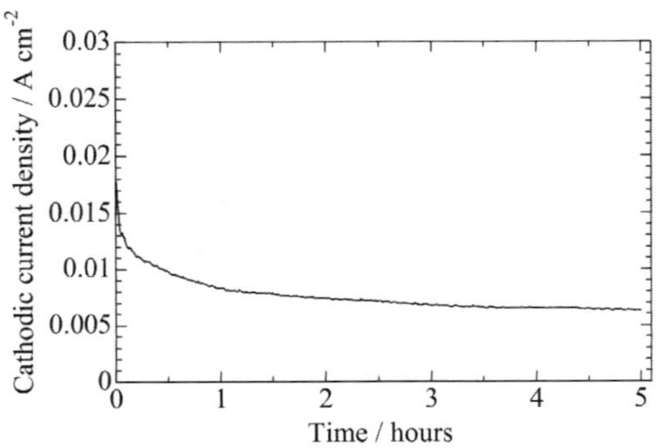

Figure 1. Transient curve of measured cathodic current density for a Ni electrode by potentiostatic electrolysis at 0.55 V for 5 h at 700 K.

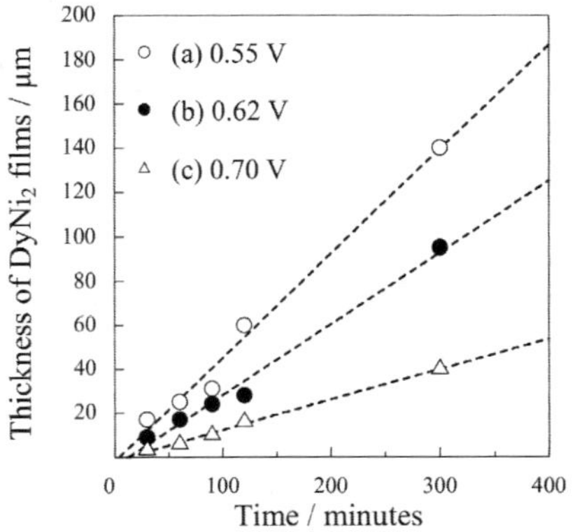

Figure 2. The relationship between the thickness of $DyNi_2$ films and the electrolysis time at 700 K: (a) 0.55 V, (b) 0.62 V and (c) 0.70 V.

4. RBS analysis of Dy alloy films

The Dy alloy films obtained by potentiostatic electrolysis were analyzed by RBS. RBS is the powerful method in analyzing the composition and thickness of thin films. RBS with beam energy of 2.0 MeV incident He^{2+} was performed on a tandem type accelerator. Figure 3(a) shows the spectra for the samples prepared by electrodepositing Dy at 0.30 V with Ni electrodes. The counts for Dy at about 1.8 MeV increases with the electrolysis time, which means that Dy is electrodeposited. The rising shoulder at about 1.5 MeV indicates the existence of Ni. The increasing of counts from 1.5-1.8 MeV corresponds to the growth of Dy-Ni alloy film. From this result, it was suggested that the Dy-Ni alloy film was formed for 10 or 30 s. On the other hand, the spectra for the samples obtained by electrodepositing Dy at 0.30 V with Fe electrodes as the comparative samples are shown in figure 3(b). The counts for Dy also increase with the electrolysis time. However, the counts from 1.5-1.8 MeV with Fe electrodes are lower than those with Ni electrodes. This result indicated that the growth rate of Dy-Ni alloy film was faster than that of Dy-Fe alloy film.

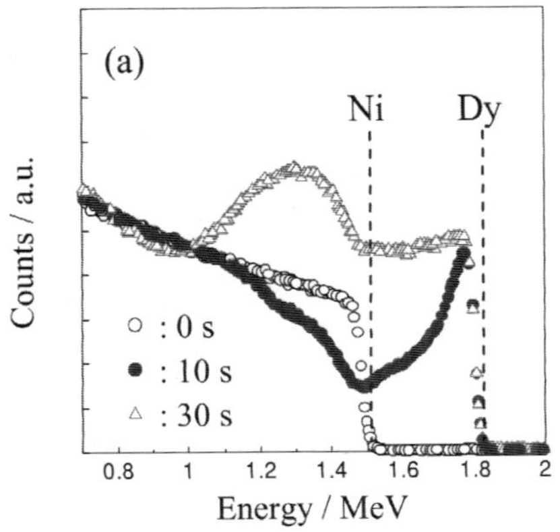

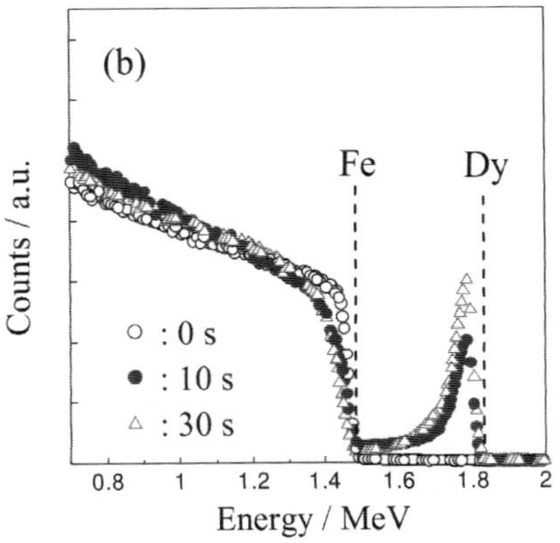

Figure 3. RBS spectra of the samples prepared by potentiostatic electrolysis at 0.30 V for 0, 10 and 30 s with (a) Ni and (b) Fe electrodes.

5. Selective anodic dissolution of Dy from the $DyNi_2$ alloy films

Selective anodic dissolution of Dy from the formed $DyNi_2$ film was performed at 0.90, 1.30 and 1.90 V, respectively. The $DyNi_2$ film was prepared by potentiostatic electrolysis at 0.55 V for 20 min at 700

K. Thickness of the $DyNi_2$ film was found to be about 15 μm by cross-sectional SEM measurement (figure 4(a)). These potential values, 0.90, 1.30 and 1.90 V, were chosen as intermediate values between the adjoining plateaus. The original $DyNi_2$ was transformed to the Dy-poor phase ($DyNi_3$) at 0.90 V for 1 h. In figure 4(b) the thickness of the $DyNi_3$ film was about 10 μm. The alloy film had some cracks and pores, that were thought to be caused by the volume change during the selective anodic dissolution of Dy. The identified phase of sample obtained at 1.30 V for 3 h was $DyNi_5$. The thickness of the obtained film was found to be about 10 μm in figure 4(c). The obtained film contained more micropores than the $DyNi_3$ film. It was found that the alloy phase of the sample obtained at 1.90 V for 3 h was transformed to Ni. In figure 4(d), the transformed film was about 10 μm and had a porous structure with a pore diameter of 1~2 μm.

When quite noble potential step is applied to the starting $DyNi_2$ film, dissolution of Dy into the melts rapidly occurs and large anodic currents are observed. Therefore, the author considered that such rapid transformation process should be related to the feature of morphological change during anodic dissolution. Since Dy is considered to mainly diffuse through microscopic cracks and/or grain boundaries in the $DyNi_2$, its diffusion behavior is assumed to be strongly affected by the number of microscopic cracks and/or grain boundaries.

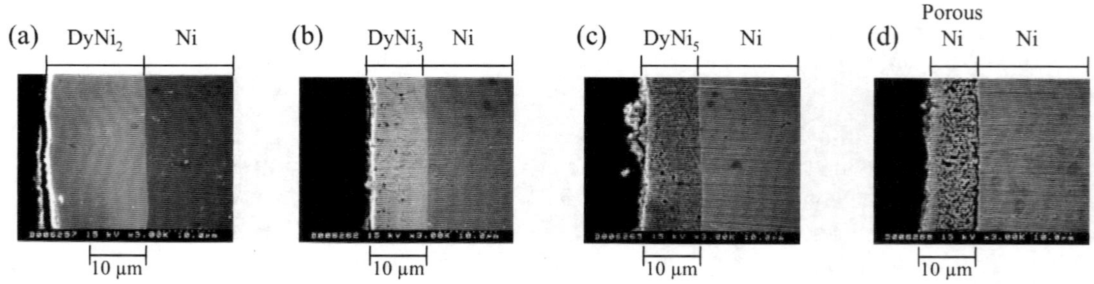

Figure 4. Cross-sectional SEM images of the samples prepared by potentiostatic electrolysis with a Ni electrode at (a) 0.55 V for 20 min and by anodic dissolution of Dy with the formed $DyNi_2$ electrode at (b) 0.90 V, (c) 1.30 V and (d) 1.90 V.

6. Conclusions

Concerning the electrochemical formation of Dy-Ni alloy films in a molten LiCl-KCl-$DyCl_3$ system at 700 K, the growth of $DyNi_2$ film and behavior of anodic dissolution of Dy from the formed $DyNi_2$ film were investigated. The $DyNi_2$ films were formed by potentiostatic electrolysis at Ni electrodes at 0.55, 0.62 and 0.70 V. The growth rates were higher at less noble potential, i.e., 0.47 μm min^{-1} at 0.55 V, 0.32 μm min^{-1} at 0.62 V and 0.14 μm min^{-1} at 0.70 V. It was suggested that the Dy-Ni alloy film was formed for 10 or 30 s during electrodepositing Dy at 0.30 V with the Ni electrode from RBS analysis. The formed $DyNi_2$ were transformed to other phases, i.e., $DyNi_3$, $DyNi_5$ and Ni, by selective anodic dissolution of Dy. The transformed Ni film was about 10 μm and had a porous structure with a pore diameter of 1~2 μm.

Acknowledgements

This study was supported by Priority Assistance for the Formation of Worldwide Renowned Centers of Research - The Global COE Program (Project: Center of Excellence for Advanced Structural and Functional Materials Design) from the Ministry of Education, Culture, Sports, Science and Technology (MEXT), Japan.

References

[1] Konishi H, Nohira T and Ito Y 2001 *J. Electrochem. Soc.*, **148**(7) C506.
[2] Konishi H, Nohira T and Ito Y 2002 *Electrochim. Acta*, **47**(21) 3533.
[3] Konishi H, Nohira T and Ito Y 2003 *Electrochim. Acta*, **48**(10) 1403.
[4] T. B. Massalski Editor 1990 ASM International *Binary Alloy Phase Diagrams*, Vol. 2 1540.

New multifunctional lightweight materials based on cellular metals – manufacturing, properties and applications

Günter Stephani, Peter Quadbeck, Olaf Andersen

Department of Cellular Metals

Fraunhofer-Institute for Manufacturing and Advanced Materials Research

Winterbergstrasse 28, D-01277 Dresden, Germany

E-mail: guenter.stephani@ifam-dd.fraunhofer.de

Abstract. Cellular metallic materials are a new class of materials which have been the focus of numerous scientific studies over the past few years. The increasing interest in cellular metals is due to the fact that the introduction of pores into the materials significantly lowers the density. These highly porous materials also possess combinations of properties which are not possible to achieve with other materials. Besides the drastic weight and material savings that arise from the cell structure, there are also other application-specific benefits such as noise and energy absorption, heat insulation, mechanical damping, filtration effects and also catalytic properties. Cellular metallic materials are hence multi-functional lightweight materials.

1. Introduction

Cellular metal materials feature properties and combination of properties far away from that one of their basic materials. Besides the drastic weight and material savings arising from the cellular structure, features such as noise and energy absorption, low heat conductivity and excellent mechanical damping properties, good filtration capability and catalytic properties demonstrate their striking application-specific benefits. Hence, in the past various technologies of manufacturing and applications of cellular metals have been explored [1,2].

Commonly, manufacturing technologies involve either the use of liquid metals in combination with blowing agents or bubbles, or the use of solid state methods. The term "solid state method" refers to the processing in the solid state. However, in some cases limited amounts of liquid phase appear in the process, i.e. during liquid phase sintering or brazing of single cells. Typically, solid state methods require a sintering step before the final cellular metallic structure is obtained. Thus, powder metallurgical methods play a dominant role in this area. The formation of the cellular structure can either be accomplished by building it up from single cells, which do not necessarily have to be in the final metallic state. Alternatively, the structure can be built up from bulk. The formation of the cells themselves can either be carried out with the help of a lost core, or in a coreless manner [3].

At Fraunhofer IFAM, different manufacturing technologies basing on solid state methods have been developed [4]. The present paper demonstrates the main fabrication routes and applications.

2. Technologies

The physical properties of cellular materials basically depend on their structure, their cell size and their porosity. Since these materials cover various types of structures and a wide range of pore sizes and porosities, miscellaneous applications were tested. In detail, metal fibre structures and network-like structures with open cell porosity as well as metal hollow sphere structures with more or less closed cell porosity were developed: produced cell sizes in the range of $10^{-2} - 10^{1}$ mm and porosities in the range of 50 – 96 pct. Figure 1 shows the classification concerning the cell size of these cellular materials.

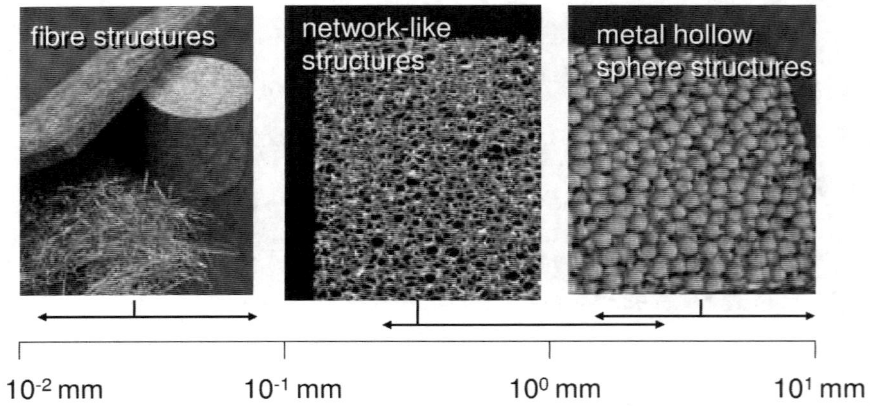

Figure 1. Types of structures and range of cell sizes for different cellular metals

2.1. Open porous metal fibre structures

In order to obtain open-porous fiber structures, crucible melt extraction is used to yield short metal fibers with equivalent diameters as low as 50 μm from almost arbitrary metals and alloys. Highly porous components can be made from such fibers by suitable deposition and sintering methods. This method is applicable to a wide variety of different metals and alloys. A thorough review of the manufacturing route can be found elsewhere [5].

2.2. Open cell network-like structures

Open cell metal foams have been produced by the replication method using polyurethane sponge as a template. The replication method essentially consists of three production steps: First a reticulated polyurethane sponge is coated by slurry infiltration. In the next step, the template is thermally removed and the debindered brown structure is sintered. The method allows the manufacture of cell sizes in the range of 0.4 – 5.0 mm and porosities of 75 – 95 pct [6].

2.3. Hollow Sphere Structures

The manufacturing of metal hollow sphere structures is characterized by a three step including:

- powder coating of an styrofoam template
- shaping of the structure using the coated styrofoam spheres
- debindering and sintering

A feature of the MHS (metal hollow sphere) technology is its suitability for a wide range of materials. A broad cell size spectrum from 0.5 - 10 mm is possible whereas the cell size distribution is very narrow. The cell wall thickness can be varied from approx 20 μm to 1,000 μm. The walls can be sintered fully dense as well as porous.

3. Characteristic properties

Cellular metal materials exhibit key benefits for applications where stringent operating conditions require extraordinary combinations of properties. Due to their porous structure, they are extremely lightweight (up to 95 pct porosity), combined with high specific strength and stiffness. At the same time, they feature low heat conductivities (approx. 1-5 pct of the matrix material) and sound absorption coefficients comparable to the values of polymeric damping material. Open cell metal foams are permeable, combined with high specific inner surface. This set of properties is the basis for a variety of applications.

4. Applications

4.1. Stirling Engine

Stirling engines for the combined production of heat and electricity are currently being developed by several companies. One of them is Enerlyt Technik GmbH of Potsdam, Germany. The Stirling cycle relies on the alternating heating up and cooling down of a fixed amount of gas. This is accomplished with the help of the so-called regenerator which acts as a temporary heat accumulator. To date, commercial wire mesh is cut, stapled and sintered together in order to build regenerators that are able to withstand the oscillating gas flow at temperatures up to 800 °C.

The overall efficiency of the Stirling engine is determined by the qualitiy of the heat transfer and the pressure drop in the regenerator, therefore a large degree of freedom in the design of the regenerator is desirable. Enerlyt Technik GmbH has thus carried out stationary simulations of the heat regeneration process with the help of the SolidWorks / FlowWorks software package. Fig. 2 illustrates the optimization approach. First, the heat transfer due to the flow around one fiber is modelled. Then, a volume element containing a number of fibers is constructed, this way determining the porosity of the structure. This allows to take into account a) the interaction between the gas flow around a single fiber and the flow around the neighbouring fibers, and b) the influence of the fiber orientation at a given porosity. The volume elements are then taken to model the regenerator in the Stirling engine.

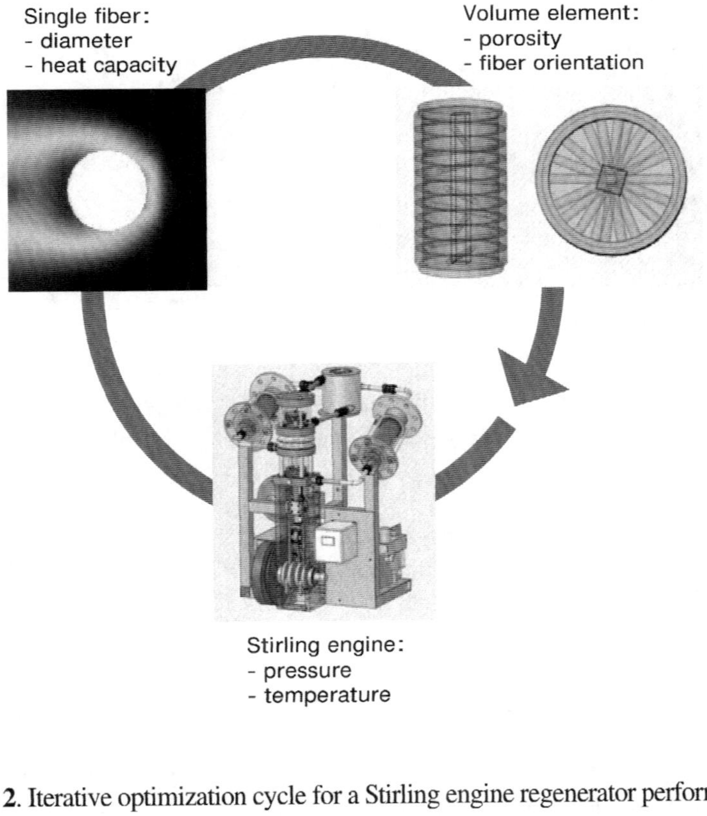

Figure 2. Iterative optimization cycle for a Stirling engine regenerator performance

A first performance optimum was detected at high porosities of approx. 90 pct, using small fibers with a diameter of 50 µm. For a regenerator of dia. 80 mm x length 80 mm, a heat transfer power of 19.6 kW corresponds to a power loss of only 28 W due to the pressure drop. However, recent simulations revealed that even lower losses can be achieved by switching to much larger fiber diameters and a lower porosity of the fiber structure. Further potential lies in the orientation of the fibers along the direction of flow.

Past tests of sintered short fiber structures in different Stirling engines have already shown that the overall efficiency reaches the order of the theoretically expected performance. Currently, field testing is underway and will be finished 2009/2010.

4.2 Porous Surface Burner

Porous surface burners for different kinds of fuels have very advantageous properties: compact design, very high power modulation and extremely low emissions. These properties are achieved by stabilising the combustion next to a macroporous surface structure (Fig. 3a). To date, these structures are made mainly from special ceramics. For non-stationary applications such as the auxiliary power unit (APU), the resistance to mechanical vibrations is not always sufficient. A metallic solution is therefore desireable.

The required longevity calls for an advanced high temperature resistant material. The properties of the well-known ferritic iron-chromium-aluminum alloys, containing up to 7 % aluminum, can be further enhanced by raising the aluminium content up to 15 %. This is only possible with melt extracted fibers Fig. 3b shows results of oxidation testing of sintered fiber structures in air at 1000 °C for 1000 hrs. It can be seen that both compositions show the same typical parabolic mass gain due to the aluminum oxide scale formation. However, the aluminum reservoir of the 5 % Al composition is consumed after roughly 600 hours, whereas the aluminum reservoir of the 15 % Al composition will last up to 2500 hours.

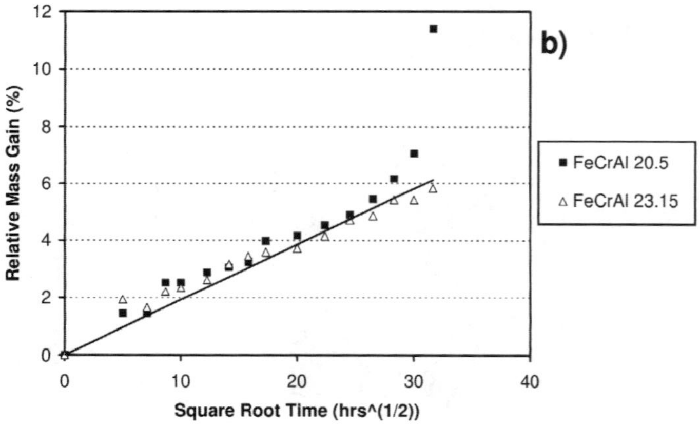

Figure 3. Burner test rig with IFAM fiber burner plate (Fig. a) and relative mass gain over time due to oxidation in air at 1000 °C for sintered fiber structures with a mean porosity of 75 % and different aluminum contents (Fig. b)

The preliminary experiments carried out with IFAM fiber mats towards an application as flame holders are very promising [7]. Future work concentrates on the fine tuning of the mat properties such as: low pressure drop, even gas distribution, high maximum temperature resistance, and good heat transport properties.

4.3. Biomaterials: bone implant material

Bone replacement of large defects (eg. due to osteoporotic fractures) is an urgent clinical problem. Recently, such defects are filled by autogenic bone or massive metal implants. Former needs an additional surgical intervention, which is fraught with risk. Latter suffer from a stiffness mismatch between bone and implant (*stress shielding*), leading to implant loosening. Network-like metal foams exhibit a natural bone-like structure (Figure 4), which enables in grow of bone cells and blood vessels.

Thus, open cell metal foams out of steel and titanium have been developed. Since the stiffness of the cellular metal is comparable to that one of the bone material, stress shielding is inhibited as shown in Fig.5. Recently, the new material is tested as vertebra replacement in a sheep model.

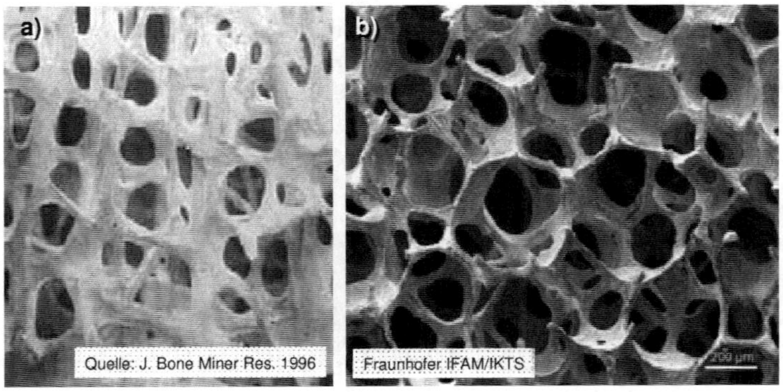

Figure 4. Spongeous bone (Fig. a) and open cell metal foam (Fig. b) show highly similar structures.

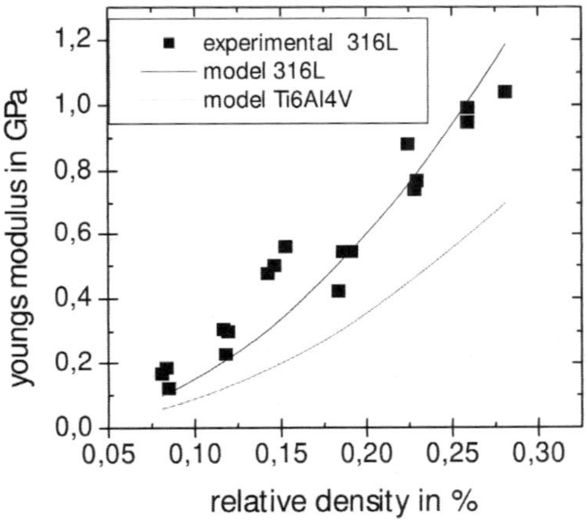

bone structure		youngs modulus [GPa]	compression strength [MPa]
cortical bone	Femur	12	180
	Radius	12	200
	Tibia	14	130
trabecular bone	Femur (Kondyle)	2.9	68
	Vertebra	1.5	4.1

Figure 5. Mechanical properties of open cell metal foam (316L) in comparison to different types of bone

4.4 Thermal insulation

Because of its high temperature strength molybdenum is frequently used in kiln engineering. e.g., hot plates, heating element support and shielding plates are manufactured out of molybdenum. Since the price of molybdenum rose by a factor 7 within the last 5 years, material costs became critical for manufacturer of industrial furnaces.

Consequently, the Fraunhofer Institutes IFAM-DD and IKTS in cooperation with H.C.Starck Hermsdorf GmbH developed a lightweight molybdenum foam for heat insulation applications (Fig. 6), using the low heat conductivities of about 1-5 % of the basic material. Network-like structures with porosities up to 95 pct and cell sizes of 0.8-1.2 mm are synthesized.

Figure 6. Open cell molybdenum foam as insulation material in industrial furnaces, using the low heat conductivity of the material.

In first tests, the heat insulation capability of the new material was tested in an industrial vacuum furnace. The molybdenum foam was laminated by a thin molybdenum foil. In comparison, to the conventional shielding plates, the temperatures at the cold zones only show differences of about 2 pct, when the molybdenum foam heat insulation was used. At the same time, the mass of the heat insulation package was reduced by factor 4.

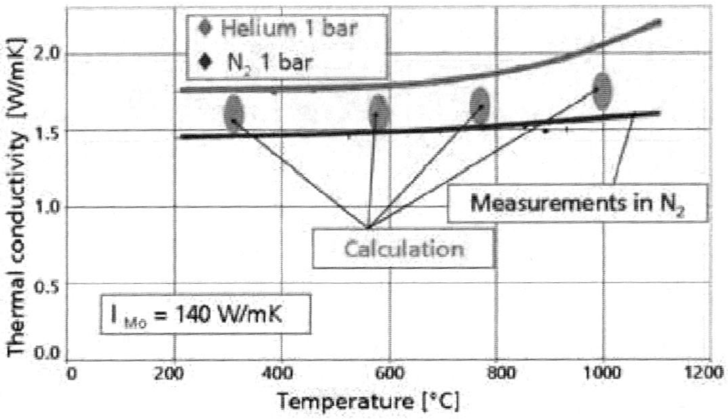

Figure 7. Heat conductivity of metal hollow spheres. The curves show good correlation between model calculations and experimental findings.

In the same way, metal hollow spheres (MHS) can be used in thermal insulation applications especially in high temperature environment. The heat conductivity values of the material ranges between 0.5 and 2 $W \cdot m^{-1} \cdot K^{-1}$. Furthermore the thermal conductivity of Molybdenum hollow sphere sandwich structures was calculated by FEM and verified by experiments afterwards (see Fig. 7). The results were in good correlation with the experimental findings.

5. Conclusion

Numerous technologies have been developed in order to manufacture cellular metal materials. Because of the unique properties of the material (low density, excellent sound absorption, thermal insulation, energy absorption, high specific surface area, mechanical damping), a variety of new applications for the vehicle industry (diesel soot filtration, energy absorption, sound dissipation, heat insulation), for machine constructions (acoustic and mechanical damping, weight reduction), and for process engineering (weight reduction, sound absorption, catalytic reaction, thermal insulation, filtration, infrared burners) have been tested.

References

[1] Davies, G.J., Zhen, S.; J. Mat. Sci 13, 1983, p 1899
[2] Banhart, J., Progress in Materials Science 46, 2001, p 559
[3] Andersen, O., Stephani, G, in: Handbook of Cellular Metals, Ed. Degischer, P., Kriszt, B., (Viley VCH, Weinheim, Germany)
[4] Stephani, G, Andersen, O., Göhler, H., Kostmann, C., Kümmel, K., Quadbeck, P., Reinfried, M., Studnitzky, U., Waag, U., Adv. Eng. Mat. Vol. 8, 2006, 847
[5] Andersen, O., Kostmann, C., Stephani, G, Conference Proceedings European Congress and Exhibition on Powder Metallurgy, Valencia, Spain, Octobre 20-22, 2003, Vol. 2, p 361
[6] Quadbeck, P., Stephani, G,_Kümmel, K., Adler, J., Standke, G, Materials Science Forum 534/536, 2007, p 1005
[[7] Mößbauer, S., Wawrzinek, K., Andersen, O.: Conference Proceedings CELLMET 2005, Dresden, Germany,May 18-20, 2005 (Fraunhofer IRB Verlag) p. 12

Design and synthesis of thin palladium membranes on porous metal substrate for hydrogen extraction

Z Shi[1], J A Szpunar[1] and S Wu[2]

[1]Department of Materials Engineering, McGill University, M. H. Wong Building, 3610 University Street, Montreal, QC, Canada H3A 2B2
Email: jerzy.szpunar@mcgill.ca, zhongliang.shi@mail.mcgill.ca
[2]AH2T Inc., 225 President Kennedy Ave., PK6230, Montreal, QC, Canada H2X 3Y8

Abstract. Membrane separation is regarded nowadays as a preferred method for production of purified hydrogen. Palladium (Pd) is an attractive membrane material due to its ability to dissociate molecular hydrogen into atoms. It is usually deposited on the porous substrate that can provide good mechanical support and reduce the thickness of the membrane for maximizing hydrogen permeability. Pd membrane used for hydrogen separation must be thin enough to increase hydrogen flux and reduce cost while remaining thick enough to retain adhesion, attrition resistance and mechanical integrity during high temperature cycles. In this paper, the progress of electroless deposition of Pd around the pore area at surface of porous stainless steel was recorded and a bridge structure that was formed during the membrane deposition around the pore area of the substrate was illustrated. After that, the porous substrate was modified using micro-or nano-size metal or metal oxide particles in order to reduce pore size in the substrate surface. The experimental results obtained from hydrogen permeation through the Pd membranes having the thickness from 400 nm to 18 μm built on both modified and original porous stainless steel substrates demonstrate that these thin membranes are solid and they can be used at the temperature of 550°C and hydrogen pressure difference of 3.447×10^5 Pa. The proposed processing will allow optimizing the design and fabrication of thin Pd membranes on different porous substrates for hydrogen separation.

1. Introduction

Growing environmental concerns and global interest in the development of "hydrogen economy" are leading to an increased demand for hydrogen and a revival of interest in methods for separation and purification of hydrogen from gas mixtures [1]. The interaction between Pd and hydrogen has been studied extensively and applied widely in hydrogenation catalysts and chemical engineering because Pd absorbs hydrogen easily and Pd membranes have hydrogen permeability, chemical compatibility and excellent hydrogen selectivity. [2-6] So far, Pd membranes were always fabricated in the processes of electroless, electrochemical and sputtering depositions. Most efforts have been focused on how to reduce the membrane thickness for maximizing hydrogen flux and how to improve the membrane service time through alloying. [7-10] No matter what kind of process is employed, the fabricated Pd membrane should be dense enough for hydrogen extraction from a mixture of gases. For this reason, the preparation of an ultra-thin and pin-hole free Pd membrane may attract novel engineering applications. As we know, the mechanism of hydrogen permeation through a Pd membrane involves dissociative adsorption and recombinative desorption at the surface, as well as diffusion through the membrane. The flux of hydrogen permeating through the membrane is mainly dependent on the membrane thickness. So far, there was no report showing how a membrane was built on porous area of a substrate and how the membrane could be controlled through the deposition process. These questions will be answered in this paper. The purposes of this paper will demonstrate how a Pd membrane forms on a porous stainless steel substrate, and answer the question why the membrane is

[1] To whom any correspondence should be addressed: zhongliang.shi@mail.mcgill.ca

dense and how an ultra-thin Pd membrane can be designed and fabricated using an electroless deposition process.

2. Experimental

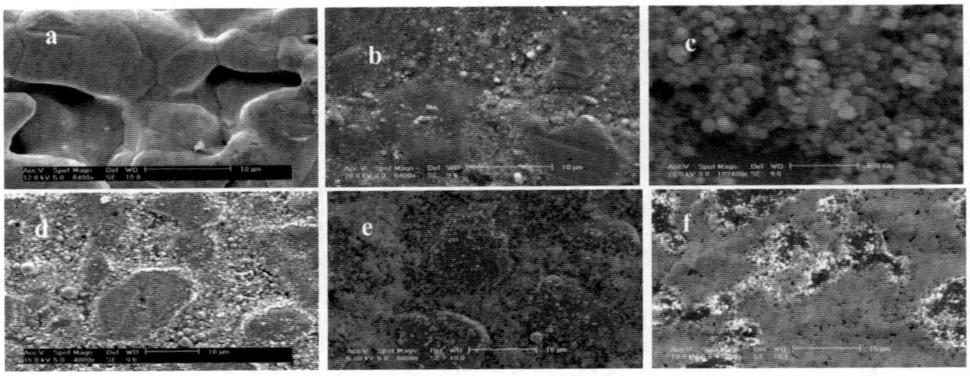

Figure 1. SEM micrographs showing the surface microstructures of 0.2 μm grade porous stainless steel before and after modification using nano Ni powders or Nano Ni and α-Al_2O_3 ((a) Original surface, (b-d) Modified surface structures using nano Ni powders before and after sintering at 600°C for 2 hrs, (e, f) Modified surface structures using nano Ni and α-Al_2O_3 powders before and after sintering at 600°C for 2 hrs)

A 0.2 μm grade porous 316L stainless steel plate, purchased from Mott Metallurgical Corporation, was used as a substrate to support Pd membrane. The samples used have the sizes of 20 mm x 20 mm or ϕ 45 mm. These small size substrate samples were used to investigate the microstructure of Pd deposits in the initial stages of deposition and at different concentration of $PdCl_2$ in the plating baths. The bigger size substrate samples were used to deposit Pd membranes for hydrogen permeation tests. The substrates were cleaned in an ultrasonic bath with acetone for 15 min. One-side surfaces in some bigger porous substrate samples were soaked using micro- or nanosize metal or metal oxide powders (such as Ni, Al_2O_3 or V_2O_5 nano powders) and modified to be smooth in order to be built ultra-thin Pd membranes on them, as shown in Figure 1. Figure 1 (a) shows the original surface of the porous stainless steel substrate. The pore size measured on the substrate surface was in the range of 0.2~10 μm. Figures 1 (b-f) illustrate the microstructures of the substrate surface after soaking in nano size powder fluid and sintering at elevated temperature. The micrographs presented in Figures 1 (b-d) illustrate modified surface of the substrate with nanosize Ni powders (60 nm) before and after sintering at the temperature of 600°C for 2 hrs. The similar micrographs are shown in Figures 1(e, f), these substrates were treated using nanosize Ni powders (60 nm) and α-Al_2O_3 (50 nm) (the volume ratio 1:1) and sintered at the temperature of 600°C for 2 hrs. It can be seen that the pores at the substrate surfaces are fully filled and the surfaces become smooth. These surfaces were activated by an acid and rinsed with deionized water. After such treatments, the substrate samples were transferred immediately into the electroless plating baths for the deposition of Pd. The deposition time was from 30 s to 6 hrs at a constant temperature of 60°C and the baths contained different concentration of $PdCl_2$. The composition of the Pd plating baths using hydrazine (N_2H_4) as a reducing agent and $Na_2EDTA.2H_2O$ as a chelating agent was described in Table 1. Microstructural characteristics of deposited Pd samples were observed and analyzed using a PHILIPS XL30 FE SEM (Field Emission Scanning Electron Microscopy) equipped with an EDS (Energy Dispersive Spectroscopy, GENESIS 2000 X-ray Microanalysis System). The gas-tightness of the membranes was examined and no leakage was detected with nitrogen gas at the pressure of 3.447×10^5 Pa (50 psi) from room temperature to 550°C. The hydrogen permeation tests for the deposited membranes were carried out in a permeation cell at the temperature from 25°C to 550°C and hydrogen pressure difference from 1.379×10^5 Pa to 3.447×10^5 Pa. Hydrogen flux was measured at the temperature range from 300°C to 550°C using a soap bubble meter. The experimental apparatus used for gas permeation tests was similar to the one described elsewhere. [11]

Table 1. The composition of electroless plating solution and some experimental parameters

Component or parameters	Concentration and temperature
$PdCl_2$	1.8~2 g/l, 2.4 g/l, 3.0 g/l, 4.2 g/l
$Na_2EDTA.2H_2O$	40.1 g/l
$NH_3.H_2O$ (28%)	198 ml/l
N_2H_4 (1M)	5.6 ml/l
PH	10 ~ 10.4
Temperature	60 °C

3. Results and discussion

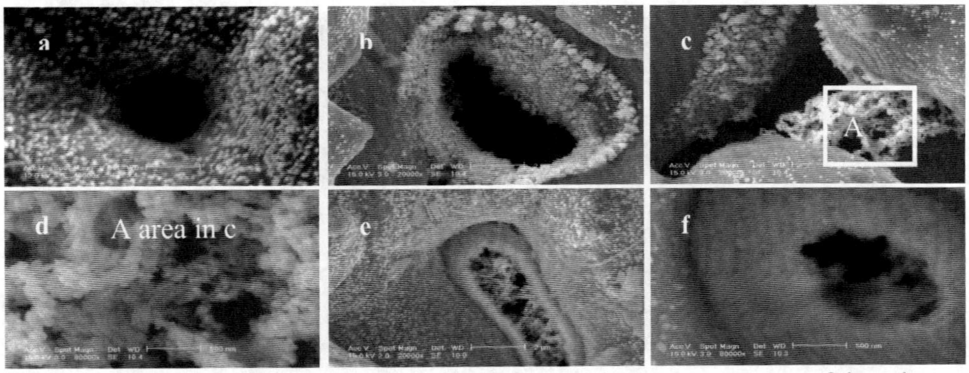

Figure 2. SEM micrographs showing the palladium deposits on the pore area of the substrate at different time (The deposition time: (a) 45 s, (b) 90 s, (c-e) 150 s and (f) 300 s)

SEM images in Figure 2 show the Pd deposits on the pore areas at different deposition time from 45 s to 300 s. A great number of Pd nanoparticles (diameter about 100 nm) were deposited around the pore at time of 45 s as shown in Figure 2(a). With the extension of deposition time to 90 s, a Pd layer was built on the wall of the pore, as shown in Figure 2(b). Figures 2(c-e) show that network structures have already been created at some areas on the pores after the deposition time of 150 s. With increasing the deposition time to 300 s, Pd membrane almost covered completely the pore area, as illustrated in Figure 2(f). The surfaces and cross-sections of deposited membrane after the deposition time of 60 min are shown in Figure 3 (a-d). Micrographs in Figure 3(a, b) illustrate surface microstructure of the membrane after the deposition time of 60 min. The cross sections in Figure 3(c, d) clearly show how the membrane was built on the porous steel substrate. We can see that the Pd membrane is dense and relatively uniform and that the cross section is stretched across the pores of the substrate like a bridge. The thickness of the Pd membrane was about 3.5 μm.

As shown in Figures 1(b-d), the surface of porous stainless steel substrate was modified to become more flat using nanosize nickel powders (60 nm) before and after sintering at 600°C for 2 hrs. After this treatment, Pd was deposited in the plating bath with $PdCl_2$ concentration of 1.8~2 g/l for about 20 min. The cross section images illustrate that the membrane deposited on the modified substrate was continuous and solid, as shown in Figure 3(e, f).

SEM micrographs in Figure 4 showed the Pd nanoparticles deposited on the stainless steel substrate for the same time of 150 s but for different concentration of $PdCl_2$ in the plating baths. One can see that the Pd nanoparticles were relatively well distributed on the solid surface area (without pore area), and the particle size was varying with the concentration of $PdCl_2$ in the plating bath. The higher the concentration of $PdCl_2$ in the plating bath was, the smaller the Pd nanoparticles were. However, the size of Pd nanoparticles did not change much with the extension of deposition time at the same concentration of $PdCl_2$ in the plating bath. However, the size of deposited Pd nanoparticles for different concentration of $PdCl_2$ at the same deposition time of 150 s, as shown in Figures 4(a) and (b), has changed significantly. One can see that the size of deposited Pd nanoparticles changed from 30~50 nm to 50~70 nm when the concentration of $PdCl_2$ in the plating bath was reduced from 4.2 g/l to 3 g/l. Similar observations for the $PdCl_2$ concentrations of 2.4 g/l and 1.8~2.0 g/l demonstrate that the size of deposited Pd nanoparticles respectively increased to 70~100 nm and 100~130 nm, as shown in Figures 4(c) and (d). This experimental result is different from presented in some other reports that claimed the size of deposited nanoparticles is increasing with the deposition time. [12-15]

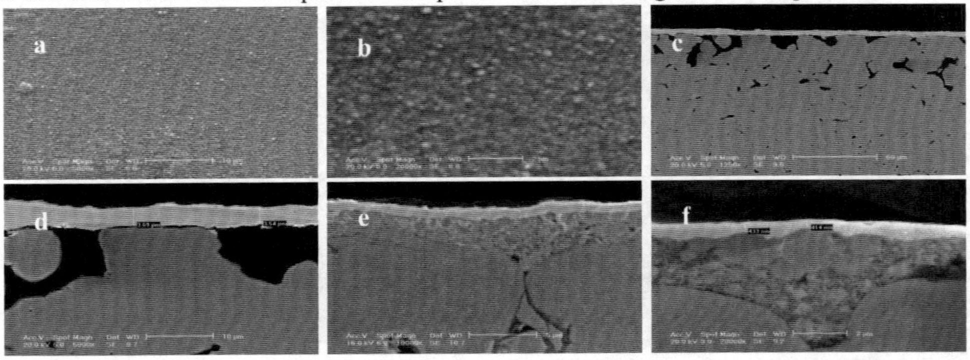

Figure 3. SEM micrographs showing the surface and cross-section structures of the Pd membrane deposited on the porous stainless steel substrates ((a, b) Surface microstructures of Pd membrane after the deposition of 60 min, (c, d) Cross-section of the membrane, and (e, f) Cross-section of the membrane built on modified porous steel substrate after the deposition of 20 min)

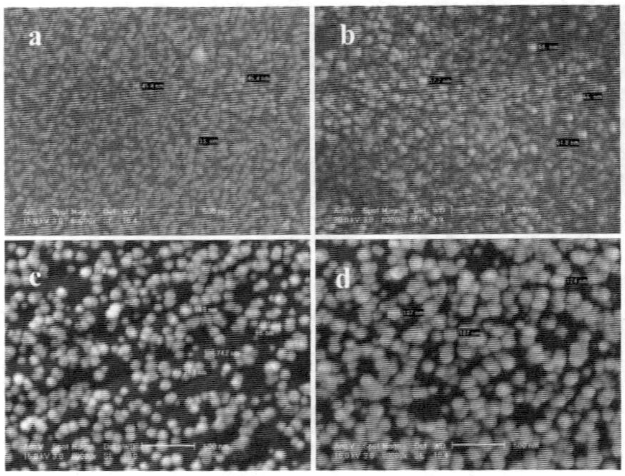

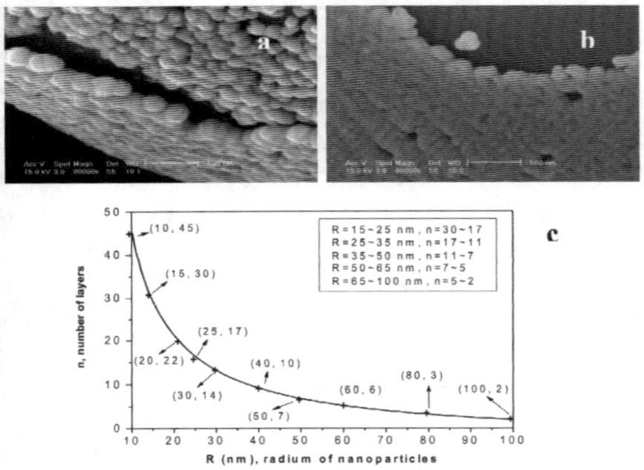

Figure 4. SEM micrographs showing Pd nanoparticles on the surface of stainless steel substrate for different concentration of $PdCl_2$ in the plating bath under the same time of 150 s (Concentration of $PdCl_2$ in the plating bath: (a) 4.2 g/l, (b) 3.0 g/l, (c) 2.4 g/l, and (d) 1.8~2 g/l)

Figure 5. SEM micrographs showing nanoparticle-monolayer membranes deposited on the substrate surface at the deposition time of 150 s (a, b) and the relationship between radius of nanoparticles and number of nanoparticle layers (c) ($PdCl_2$ concentration: (a) 2 g/l, (b) 2.4 g/l)

As demonstrated in the literature [15], in order to obtain nanosize metallic particles, a high supersaturation concentration is often required, which are best achieved when starting with highly reactive metallic species and very strong reducing agents. If the fraction of atoms consumed in the nucleation step is high, the increase in particle size due to the diffusional growth that immediately follows the nucleation burst is drastically limited and only slightly larger particles than the nuclei can be obtained. When the concentration of $PdCl_2$ in the plating baths increases, this chemical reaction is carried out from left to right and the reduction of Pd is fast. Pd deposition and nucleation on the substrate is also fast. In the case of instantaneous nucleation on the stainless steel substrate, all the nuclei form instantaneously and subsequent growth of Pd deposits is limited. The deposited particle size is obviously small with increasing the concentration of $PdCl_2$ in the plating baths. Therefore, the size of Pd nanoparticles deposited on the substrate in the initial stages can be controllable through the concentration of $PdCl_2$ in plating bath. The formation mechanism of Pd nanoparticles deposited on the porous steel substrate and their size control are being further investigated.

$$2Pd\,[NH_3]_4^{2+} + N_2H_4 + 4OH^{-1} \rightarrow 2Pd^{o} + N_2 + 4H_2O + 8\,NH_3 \qquad (1)$$

SEM micrographs in Figures 5(a, b) showed that nanoparticle-monolayer Pd membranes made of nanoparticles were deposited on the wall of a pore area and surface of the substrate. The thickness of Pd membrane in this case in Figure 5(a) was about 100 ~ 130 nm which was the same as the diameter of deposited Pd nanoparticles. When the concentration of $PdCl_2$ in the plating bath increased to 2.4 g/l, the thickness of Pd layer decreased to 70~100 nm, as shown in Figure 5(b). One can see that the Pd nanoparticles are packed and well-organized one by one. The nanoparticle-monolayer are dense except that some small areas were not fully filled. Such monolayer membranes made of nanoparticles cannot guarantee that they are dense enough and ready for hydrogen purification. However, the membrane assembled by a few layers should be dense enough for extracting hydrogen from a mixture of gases. As illustrated in Figure 4, the size of Pd nanoparticles can be effectively controlled by the concentrations of $PdCl_2$ in the plating bath. Therefore, the thickness of an ultra-thin membrane built on a relatively smooth substrate can be well controlled by the electroless deposition process.

Since the Pd membrane is essentially built from the nanoparticles, one can assume that the same size Pd nanoparticles are packed in the simple geometrical arrays. The thickness of a membrane is T (nm), where n is the number of nanoparticles layer, and R is the radius of palladium nanoparticles. The relationship between T, n and R can be expressed in the equations (2) and (3).

$$T=2R+(n-2)*0.414R \qquad (2)$$

$$n=(T/R-1.172)/0.414 \qquad (3)$$

If the thickness of an ultra-thin Pd membrane is 200 nm, the layer number n, and the radius of particle is R, the relation is shown in Figure 5(c). For example, the membrane having thickness of 200 nm is approximately assembled by 45 nanoparticle-layers if the radius of nanoparticle is 10 nm. As the particle size in Figures 3(e, f) and 4(d) was about 100~130 nm and the thickness of the membrane built on modified substrate ranged from 300 nm to 400 nm, according to the equation (3), this membrane should have 9 ~ 17 nanoparticle layers. This type of membrane having the diameter of ϕ= 45 mm was used for hydrogen permeation tests. The tested area for hydrogen permeation at this membrane was 42 mm in diameter. The membrane was solid and dense because no leakage was detected when the gas tightness of the membrane was examined using nitrogen gas at gas pressure difference of $3.447\text{x}10^5$ Pa. This test was carried from room temperature to 550°C. The membrane thickness was about 300~400 nm as shown in Figure 3(f). Hydrogen flux through this membrane is a function of hydrogen pressure and test temperature, as shown in Figures 6(a, b). One can see that hydrogen flux transported through the membrane linearly rises with increasing the temperature and hydrogen pressure. The experimental results demonstrate that these membranes can be used for hydrogen extraction up to the temperature of 550°C and hydrogen pressure difference of $3.447\text{x}10^5$ Pa.

The thickness of such continuous and dense membrane is controlled by the diameter of the particles. Assuming that the ultra-thin membrane is packed from 3 one-particle layers, the dense membrane thickness T can be calculated as equation (4).

$$T=2.414R \qquad (4)$$

So, if the diameter of Pd nanoparticles is controlled effectively, the thickness of a dense membrane deposited on a relatively smooth substrate surface will be controlled.

Based on the observation of Pd deposition progress on the porous steel substrate, the minimum thickness of a dense Pd membrane is around 1.5~2 μm. Over 100 Pd membrane samples having thickness from 1.5 to 18 μm were prepared for hydrogen permeation tests. All samples were dense and no leakage was detected as they were examined by nitrogen gas at the absolute pressure difference of

3.447×10^5 Pa and the temperature from 300 °C to 550°C. These experimental results show that the Pd membranes were dense. After the leakage tests, hydrogen fluxes through the membranes were measured under the absolute pressure difference of hydrogen gas at 3.447×10^5 Pa and for the temperatures from 300°C to 550°C. The average hydrogen fluxes are shown in Figure 7. The experimental results reveal that hydrogen flux is dependent on the membrane thickness. Thinner membrane has higher hydrogen flux. If the pores at the substrate surface are modified and the surface becomes smooth, an ultra-thin Pd membrane will be easily built.

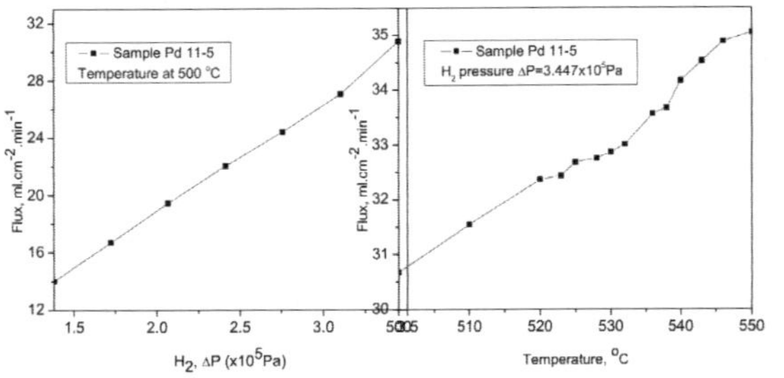

(a) (b)

Figure 6. Hydrogen flux vs. hydrogen pressure difference (a) and temperature (b)

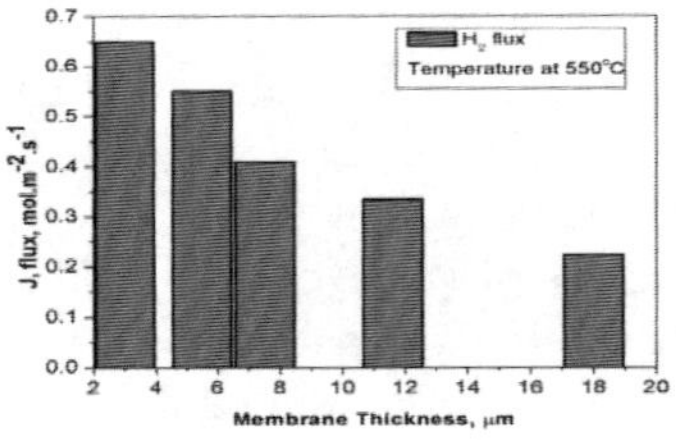

Figure. 7 Hydrogen flux is as a function of the thickness of Pd membranes deposited on the 0.2 μm grade porous steel substrate

4. Conclusions

Based on the observation of Pd membrane formation on porous steel substrate, the thickness of the membrane can be effectively controlled by the electroless process. The size of nanoparticles from which the membrane is built depends on the concentration of $PdCl_2$ in the plating bath. The smaller the size of Pd nanoparticles, the thinner the dense Pd membrane can be built. Hydrogen permeability test, supported by the observation of microstructure and theoretical analysis, will allow optimizing the design of ultra-thin Pd membranes by modifying the surface microstructure and porosity.

REFERENCES

[1] van Rijn C J M. 2004 Nano and Micro engineered membrane technology, Membrane Science and Technology Series, 10, ELSEVIER, p267
[2] Mitsui T, Rose M K, Fomin E, Ogletree D F and Salmeron M 2003 *Nature* **422** 705
[3] Lewis F A, Magennis J P, McKee S G and Ssebuwufu P J M 1983 *Nature* **306** 673
[4] Sykes E C H, Fernandez-Torres L C, Nanayakkara S U, Mantooth B A, Nevin R M and Weiss P S 2005 *Proc. Natl. Acad. Sci.* **102** 17907
[5] Sachs C, Pundt A, Kirchheim R, Winter M, Reetz M T and Fritsch D 2001 *Phys. Rev.* B. **64** 075408
[6] Ward T L and Dao T 1999 *J. Membr. Sci.* **153** 211
[7] Rothenberger K S, Cugini A V, Howard B H, Killmeyer R P, Ciocco M V, Morreale B D, Enick R M, Bustamante F, Mardilovich I P and Ma Y H 2004 *J. Membr. Sci.* **244** 55
[8] Gielens F C, Tong H D, van Rijn C J M, Vorstman M A G and Keurentjes J T F 2004 *J. Membr. Sci.* **243** 203
[9] Howard B H, Killmeyer R P, Rothenberger K S, Cugini A V, Morreale B D, Enick R M and Bustamante F 2004 *J. Membr. Sci.* **241** 207
[10] Hou K and Hughes R 2003 *J. Membr. Sci.* **214** 43
[11] Lee D W, Lee Y G, Nam S E, Ihm S K and Lee K H 2003 *J. Membr. Sci.* **220** 137
[12] Fransaer J L and Penner R M 1999 *J. Phys. Chem.* B **103** 7643
[13] Hirsch T, Zharnikov M, Shaporenko A, Stahl J, Weiss D, Wolfbeis O S and Mirsky V M 2005 *Angew. Chem. Int. Ed.* **44** 6775
[14] Fu Y, Zhang L and Zheng J 2005 *J. Nanosci. Nanotech.* **5** 558
[15] Goia D V 2004 *J. Mater. Chem.* **14** 451

Journal of Physics: Conference Series **165** (2009) 012063 doi:10.1088/1742-6596/165/1/012063

Fabrication of porous aluminium with directional pores through thermal decomposition method

H Nakajima, S Y Kim and J S Park
The Institute of Scientific and Industrial Research, Osaka University, Ibaraki, Osaka 567-0047, Japan

E-mail: nakajima@sanken.osaka-u.ac.jp

Abstract. Lotus-type porous metals were fabricated by unidirectional solidification in pressurized gas atmosphere. The elongated pres are evolved by insoluble gas resulted from the solubility gap between liquid and solid when the melt is solidified. Recently we developed a novel fabrication technique, in which gas compounds are used as a source of dissolving gas instead of the high pressure. In the present work this gas compound method was applied to fabrication of lotus aluminium. Hydrogen decomposed from calcium hydroxide, sodium bicarbonate and titanium hydride evolves cylindrical pores in aluminium. The porosity is about 20%. The pore size decreases and the pore number density increases with increasing amount of calcium hydroxide, which is explained by increase in pore nucleation sites.

1. Introduction

The porous and foamed metals exhibit various characteristics such as an inherent low density and large surface area, which are different from bulk metals. These metals are expected to be utilized as light-weight materials, catalysts, electrodes, vibration and acoustic energy damping materials, impact energy absorption materials etc.[1] Recently lotus-type porous metals, a new type of porous metals have attracted much attention because of the long cylindrical pores aligned in one direction, which are fabricated by unidirectional solidification process in pressurized gas atmosphere such as hydrogen[2,3]. The pores are evolved from insoluble gas when the molten metal dissolving the gas is solidified. The lotus-type porous metals have not only the properties of conventional porous metals but also the unique properties originating from their directional pores. In particular, these metals exhibit superior mechanical properties to the conventional porous metals whose pore shape is isotropic and nearly spherical[4]. Thus, the metals are attracting considerable attention for various industrial applications[3]. Shapovalov[2] and the present authors' group[3] fabricated lotus-type porous metals with homogeneously distributed pores using the mold casting technique (Gasar method) in high pressure gas. Although this technique is a simple process, it is difficult to control the solidification velocity and as a result the pore structure with uniform pore size and porosity cannot be produced for the metals with lower thermal conductivity. In order to improve this shortcoming, the continuous zone melting technique [5] and continuous casting technique [6] were developed by the present authors. Since these techniques can control the solidification velocity, various kinds of lotus metals with homogeneously distributed long pores are fabricated.

Although such fabrication techniques were quite advanced, we still have a big technical barrier that high-pressure hydrogen gas must be used. The use of high-pressure hydrogen gas may be risky to lead to inflammable and explosive accidents if oxygen is mixed. Therefore, it is desirable not to use such high-pressure hydrogen when fabricating lotus metals. In order to solve this difficulty, we proposed an alternative but very simple method to fabricate such lotus-type porous metals by utilizing a thermal decomposition method (TDM) of compounds containing gas elements in a non-hydrogen atmosphere under nearly atmospheric pressure [7]. The present paper reports to fabricate lotus-type porous aluminium to use the TDM.

1.1 Melting and solidification processes

Shapovalov and Nakajima et al. fabricated lotus metals containing homogeneously distributed pores using a mold casting technique (Gasar method) under a high-pressure hydrogen as shown Fig.1(a). Although this technique is a simple process, it is difficult to control the solidification velocity, which affects the pore morphology. Although the heat from the melt is easily dissipated to the water-cooled plate during the solidification process, the cooling becomes slow in the upper part of the solidified ingot, and thus the pores become coarse. Therefore, large-sized lotus metals with a uniform pore size and porosity, particularly metals and alloys with low thermal conductivity such as stainless steel, cannot be fabricated by this technique. In order to overcome this shortcoming, we invented a continuous zone melting technique [5].

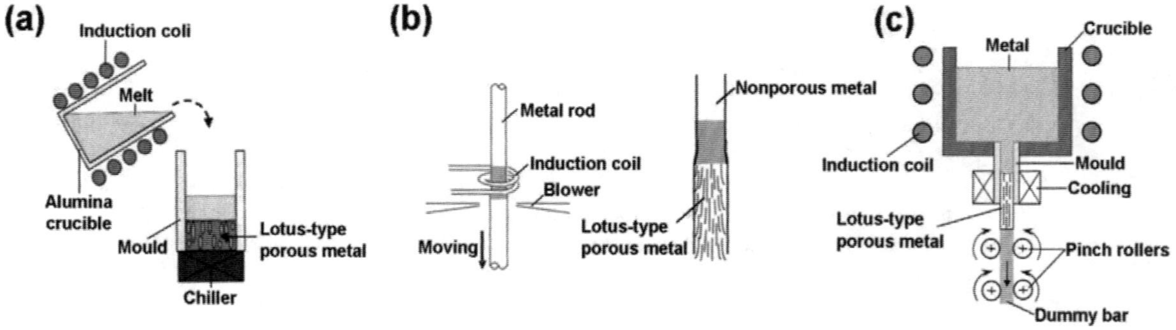

Figure 1. Schematic drawings of fabrication apparatus for high-pressure gas method. (a) mold casting technique, (b) continuous zone melting technique and (c) continuous casting technique.

Figure 1(b) shows the schematic drawing of the continuous zone melting technique. By using this technique, we were able to fabricate lotus metals and alloys with a low thermal conductivity, which possess long cylindrical pores distributed homogeneously. A part of the metal rod is melted by an induction heating coil. Gas is dissolved into the melt from the surrounding atmosphere. When the melt is moved downward from the coil to be solidified, insoluble gas pores are evolved in the solidification direction. This technique has a benefit to control the solidification velocity by changing the transference velocity. However, this technique is not suitable for the mass production of the specimen rod because the size of the ingot that can be fabricated is restricted. Therefore, we considered the continuous casting technique [6] as a new fabrication method for the mass production of the lotus-type porous metals.

The continuous casting technique is extensively used as a mass-production method for ferrous and nonferrous metals and alloys. In this conventional continuous casting process, the solidified ingot can pass through the mold smoothly due to the solidification shrinkage from the melt. However, in the present process, a large volume expansion is arisen when the solidified ingot is passed through the mold. We initially thought that such an expansion would bring us a difficulty to stack the ingot to the mold and therefore this technique could not be applied to the fabrication of the lotus-type porous metals. However, we later realized that such a large expansion inherent from the pore evolution was released to push the volume toward the copper part of the molten metal; the melt can accommodate large strain of the solidified ingot. That is why the continuous casting technique is applicable for the fabrication of lotus-type porous metals.

Through this technique, the solidification velocity can be controlled by the transference velocity in the hydrogen gas atmosphere. Since the pore morphology of the lotus-type porous metal is related to the solidification velocity, it is suggested that the pore morphology of the lotus-type porous metals can be easily controlled by the technique. We successfully fabricated the lotus-type porous copper and reported the method for controlling its pore size and porosity [3].

1.2 Gas dissolution processes

(1) Gas dissolution from high pressure gas

As mentioned above lotus metals can be fabricated by unidirectional solidification in high pressure gas atmosphere. The pores evolve by insoluble gas resulted from the solubility gap between liquid and solid when the melt is solidified. According to the Sieverts' law, the concentration C_x of gas (hydrogen, nitrogen or oxygen) soluble in the liquid and solid is expressed as

$$C_x = K\sqrt{P_{X2}} \quad (1)$$

where K and P_{X2} are constant and the pressure of gas X. In order to obtain high porosity, the solubility gap

$$\Delta C_x = C_x(liquid) - C_x(solid) \quad (2)$$

should be larger. Therefore, high pressure of the gas atmosphere is usually used.

(2) Gas decomposition from gas compounds

As mentioned, the use of high-pressure gas is indispensable for fabricating the lotus metals. Hydrogen gas is explosive when a small amount of oxygen is mixed so that its use is not convenient for industrial mass production. Thus, safety procedure must be necessary to fabricate the lotus metals. In order to avoid such difficulty, it is desirable to use an alternative technique.

It is noticed that some compounds containing gas elements are very useful to supply a gas source into molten metals. For example, hydrogen can be dissolved into molten metals through thermal decomposition reaction

$$TiH_2 \leftrightarrow Ti + 2H\ .$$

If hydrogen can be dissolved in the level as much as the saturated solubility in the melt, insoluble hydrogen evolves gas pores as a result of rejection of hydrogen solution in the solid/liquid interface during unidirectional solidification. Figure 2 shows a schematic drawing of the principle for fabricating lotus metals by the mold casting technique through thermal decomposition of gas compounds.

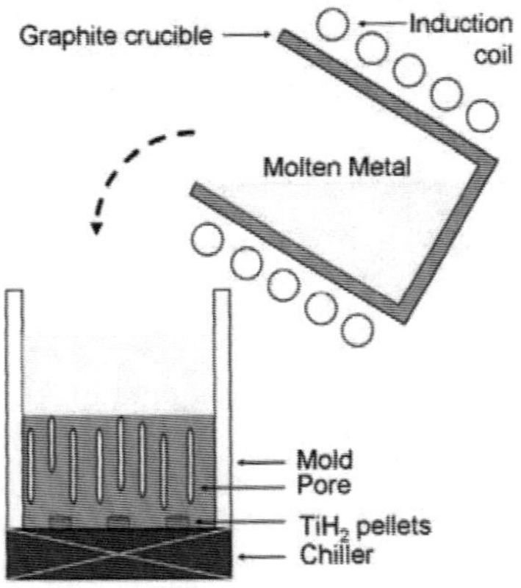

Figure 2. Schematic drawing of the principle to fabricate lotus metals by thermal decomposition of gas compounds

2. Experimental procedure

Pure aluminum (99.99%) about 100 g was melted in a graphite crucible by an induction heating coil in vacuum. The bottom of the mold was copper plate cooled down by circulated water chiller, while the side was made of stainless steel sheet. 0.2 g of the compounds as calcium hydroxide, sodium bicarbonate and titanium hydride were usually wrapped with aluminum foils and were set on the bottom plate of the mold.

The temperature of the liquid in the graphite crucible was monitored by an infrared pyrometer (model IR-AP, Chino Co.). After pouring the molten aluminum at 1023 K in the crucible into the mold, hydrogen decomposed from the compounds dissolves in the melt, which is then solidified so that insoluble hydrogen precipitates to evolve the hydrogen pores. In order to investigate dependence of the pore size and porosity on the mass of calcium hydroxide, the amount of the compounds was changed from 0.1 g to 1.0 g.

When the solidification was taken place unidirectionally, the lotus aluminum with directional cylindrical pores was fabricated. The size of the obtained ingot was about 25 mm in diameter and about 60 mm in height. The specimens were cut using a spark-erosion wire cutting machine (Model LN1W, Sodick Co.) in the both directions parallel and perpendicular to the solidification direction. Each cross-section was polished with a series of emery papers and was observed using an optical microscope. The pore diameter was measured from the cross-section perpendicular to the solidification direction. The porosity (p) was evaluated from the following equation:

$$p(\%) = (1 - \frac{\text{Apparent density of porous aluminum}}{\text{Desity of nonporous aluminum}}) \times 100 \quad (3)$$

The apparent density of the individual specimen was calculated by measuring both the weight and the apparent volume of each specimen. The shape and size of pores were observed with an optical microscope and evaluated with image analyzer (Win ROOF, Mitani Co.).

3. Results and discussion

Figure 3 shows the microstructure of lotus aluminum with aligned pore parallel and perpendicular to the solidification direction. The pore morphologies perpendicular to the solidification direction are shown in the upper photos, while those parallel to the solidification direction are shown in lower photos. In all specimens the aligned pores parallel to solidification direction were observed. It is considered that the cylindrical gas pores are evolved by TDM from the compounds during solidification of aluminum containing hydrogen gas.

Table 1 compiles the decomposition reactions of three compounds in the aluminum melt. Calcium hydroxide and sodium bicarbonate decompose into vapor and other compounds and then, the vapor decomposes to hydrogen and metallic oxide. On the other hand, titanium hydride decomposes into hydrogen and titanium. These decomposed gas element can dissolve in aluminum melt. When the melt is solidified, lotus aluminum can be produced.

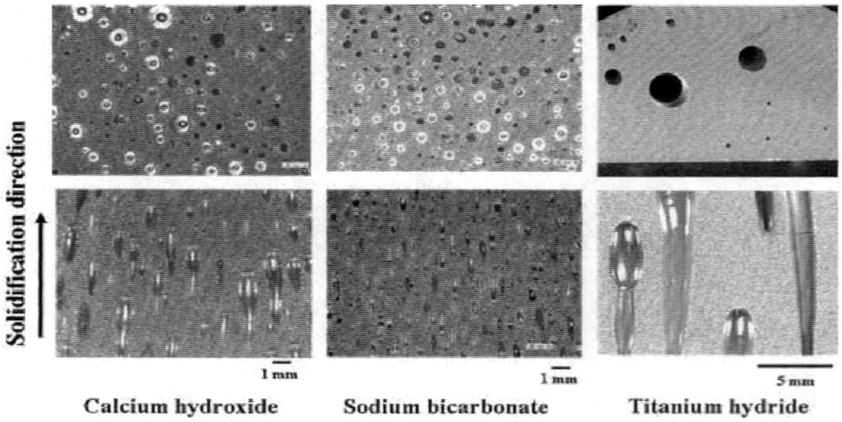

Figure 3. Optical micrographs of lotus aluminium fabricated by mold casting technique with different compounds in vacuum at 1023K.

Figure 4 shows the pore size and porosity of lotus aluminum with different compounds using mold casting technique in vacuum. The porosity of lotus aluminum was similar, being as much as 20% regardless of kinds of the compounds. The pore size was small as much as 300-400 μm and its distribution was homogeneous in porous aluminum through TDM of calcium hydroxide and sodium bicarbonate, while the pore size was large as much as 1000 μm and its distribution is not uniform through TDM of titanium hydride.

Calcium hydroxide or sodium bicarbonate decomposes into compounds (CaO, Na_2CO_3+CO_2) and vapor, the latter of which furthermore decomposes into metallic oxide and hydrogen. Small pore size with homogeneous distribution was observed through these reactions. On the other hand, titanium hydride decomposes directly into titanium and hydrogen. Large pores with inhomogeneous distribution were observed. It is not clear at present, but evolution of small pores with homogeneous distribution may be attributed to existence of the oxide particles which may serve as the nucleation sites of the pores.

Figure 5 shows the pore size and porosity of lotus aluminum with different amounts of calcium hydroxide at 1023K. Although the porosity is almost constant, the average pore size decreases with increasing amount of calcium hydroxide. Figure 6 also shows that the number density of pores increases with increasing amount of calcium hydroxide. It is suggested from these results that high density of the nucleation sites for pores is resulted from the increase in the amounts of the compound; the number density of metallic oxide CaO increases by further addition of calcium hydroxide.

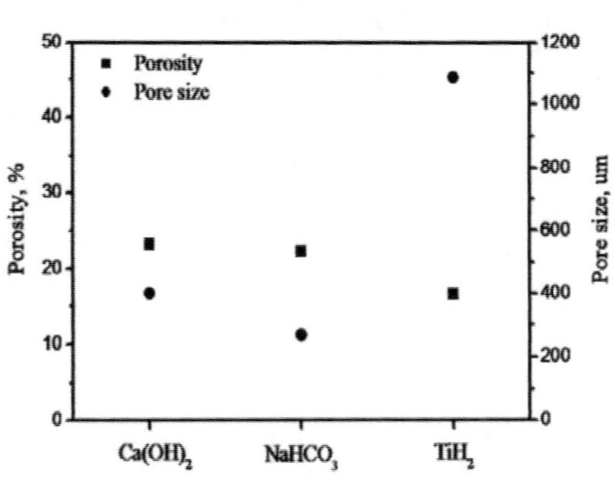

Figure 4. Pore size and porosity of lotus aluminium fabricated by mold casting technique using compounds

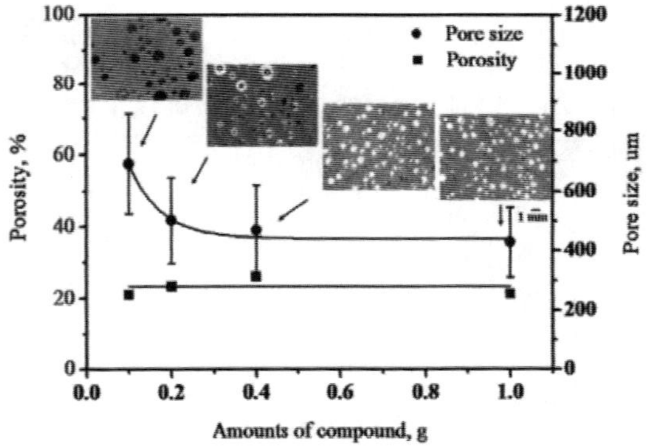

Figure 5. Change in porosity and pore size of lotus aluminium as a function of amount of calcium hydroxide in vacuum at 1023K.

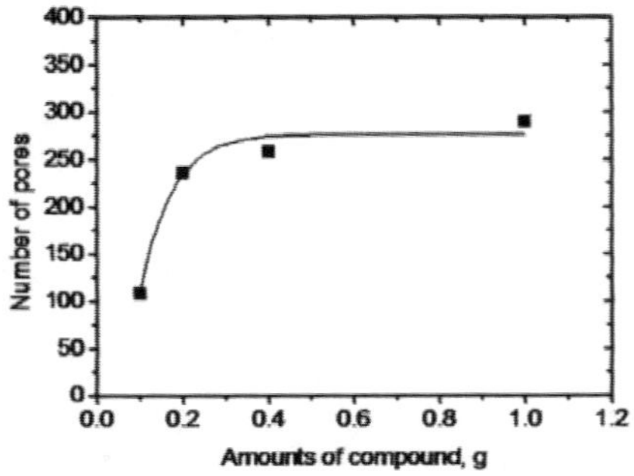

Figure 6. Change in number of pores of lotus aluminium as a function of calcium hydroxide.

4. Conclusions

Lotus aluminum was fabricated through thermal decomposition method of compounds containing gas elements by unidirectional solidification in vacuum. The results are summarized as follows.

The pores with homogenous distribution were evolved by decomposition of the compound, calcium hydroxide and sodium bicarbonate.

The porosity of lotus aluminum is similar to about 20 % regardless of different compounds in vacuum. The pore size decreases and the number of pores increases with increasing amount of calcium hydroxide. For homogenous pore structure of lotus aluminium, the oxide particles are important for the pore nucleation sites.

Acknowledgement

This present work was supported by Global COE Program (Project: Center of Excellence for Advanced Structural and Functional Materials Design) and the Grant-in-Aid for Scientific Research (S) from the Ministry of Education, Sports, Culture, Science and Technology of Japan.

Reference

[1] Banhart J 2001 *Prog. Mater. Sci.* **46** 559
[2] Shapovalov V 1998 *Proc. MRS Symp.* **52** 281
[3] Nakajima H 2007 *Prog. Mater. Sci.* **52** 1091
[4] Hyun S K, Murakami K and Nakajima H 2001 *Mater. Sci. Eng. A* **299A** 241
[5] Ikeda T, Aoki T and Nakajima H 2005 *Metall.Mater Trans.A* **36A** 77
[6] Park J S, Hyun S K, Suzuki S and Nakajima H 2007 *Acta Mater.* **55** 5646
[7] Nakajima H and Ide T 2008 *Metall.Mater Trans.A* **39A** 390

Fabrication of lotus-type porous copper through thermal decomposition of titanium hydride

T Ide and H Nakajima

The Institute of Scientific and Industrial Research, Osaka University, Ibaraki, Osaka 567-0047, Japan

E-mail:Ide23@sanken.osaka-u.ac.jp

Abstract. Lotus-type porous copper was fabricated by unidirectional solidification through thermal decomposition of titanium hydride. Effects of additive method and additive amount of titanium hydride on pore formation were investigated. The porosity of lotus copper depends on additive method and additive amount of titanium hydride. The pore formation effectively occurs in the method that titanium hydride decomposes in molten copper. For all the additive methods of titanium hydride, the porosity increases and pore diameter does not change with increasing additive amount of titanium hydride. While, for adding large amount of titanium hydride, the porosity became constant. This is because hydrogen solubility in liquid phase does not change owing to bubbling of hydrogen gas.

1. Introduction

A unique type of porous metal called lotus-type porous metal (lotus metal) possessing long cylindrical pores aligned in one direction is focus of great interest in materials science and application[1]. The unique pore structure of lotus metals provides not only various functions but also high strength compared with other porous metals with spherical or irregular shape pores[2]. Therefore, lotus metals have been expected as candidates for new structural and functional materials. However, a large technical barrier existed; in order to fabricate lotus metal, use of pressurized hydrogen gas which is explosive was necessary. To overcome this problem, a new fabrication method of lotus metal was developed through thermal decomposition of compound including gas element (thermal decomposition method)[3]. It is reported that gas solubility in liquid phase is controlled by amount of titanium hydride (TiH_2) in the melt. In order to supply the gas element in the melt, an additive method of compound is also important factor. However, an effect of additive method of compound on pore formation has not been investigated.

In the present study, we fabricate lotus copper by thermal decomposition method through different three additive methods of compound, and investigated effects of additive method and additive amount of compound on pore formation.

2. Experimental procedure

A mass of 200 g of copper (99.99 %) was melted by induction heating in a graphite crucible under 0.1 MPa argon pressure. To supply the gas element in molten metal, a small amount of TiH_2 was added into a mold. The molten copper was poured into the mold coated with a mixture of alumina powder, water glass (54.5% H_2O, 31.2% SiO_2 and 14.3% Na_2O) and water in the weight ratio of 3:3:2 (about

0.5 g). Then, molten copper was cooled and unidirectionally solidified from the bottom plate cooled by water. In the present study, different three additive methods of TiH_2 were used. Firstly, 20 mass% of TiH_2 was added in the mixture. The mixture including TiH_2 was coated on the lateral side of the mold (method A). Amount of the mixture was 2.00, 3.00, 4.00, 5.00 and 6.00 g. In another methods, a few pressed pellets of TiH_2 were used, which were set on the bottom plate of the mold (method B). Amount of TiH_2 was 0.13, 0.25, 0.50, 0.75 and 1.00 g. The TiH_2 pellets were fixed on the lateral side of the mold (method C). The amount was 0.08, 0.10, 0.13 and 0.25 g.

The fabricated porous samples were cut with a spark-erosion wire cutting machine in the directions parallel and perpendicular to the solidification direction. The cross-sections were observed with an optical microscope. The porosity p was determined by the following equation:

$$p = (1-\rho/\rho_0)\times 100,$$

where, ρ and ρ_0 are the relative density of the porous sample and the density of the nonporous copper measured by the Archimedes method, respectively. The pore diameter of the cross-sections perpendicular to the solidification direction was measured with an image analyzer (WinROOF, Mitani Corp., Fukui, Japan).

3. Experimental results and discussion

Figure 1 shows the typical porous structures of cross-sections perpendicular (left) and parallel (right) to the solidification direction of porous copper. For all the additive methods and additive amounts of TiH_2, lotus copper is fabricated. Figure 2 shows the additive amount of TiH_2 dependence of the average pore diameter and porosity for lotus copper fabricated through different additive methods. TiH_2 was added into the melt through (A) the applying the mixture including TiH_2 on the lateral side of the mold, (B) the setting TiH_2 pellets on the bottom plate of the mold and (C) the fixing TiH_2 pellets on lateral side of the mold. For all the additive methods of TiH_2, the same tendencies about the dependence of the average pore diameter and porosity on amount of TiH_2 are found. For adding small amount of TiH_2, the porosity increases with increasing additive amount of TiH_2. However, the porosity became constant when large amount of TiH_2 is added. Since the pores are formed by hydrogen gas released owing to solubility gap of hydrogen between liquid and solid phase, it is thought that the change in porosity with the amount of TiH_2 is caused by the change in

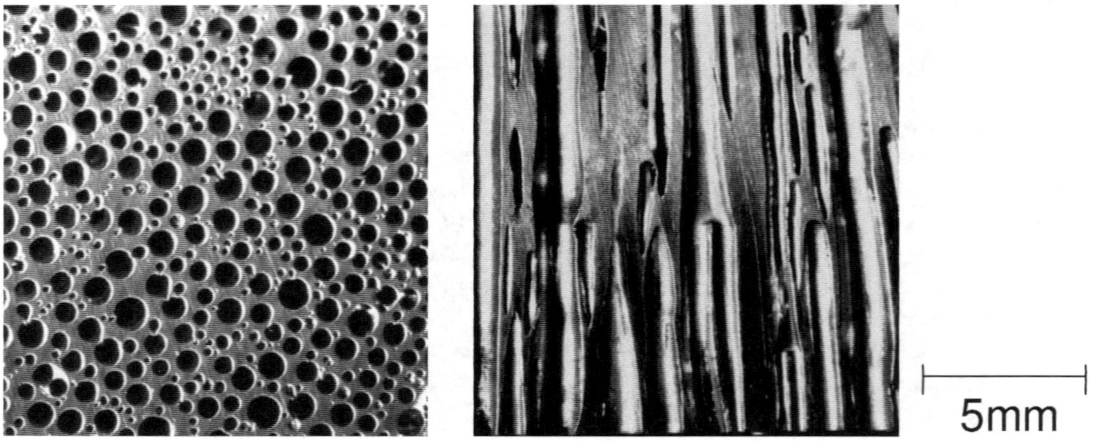

Figure 1. Porous structures of cross-sections perpendicular (left) and parallel (right) to the solidification direction of lotus copper fabricated through thermal decomposition of TiH_2 during unidirectional solidification under argon pressure of 0.1 MPa. (0.13 g of TiH_2 pellet was fixed on the lateral side of the mold, and molten copper was poured into the mold).

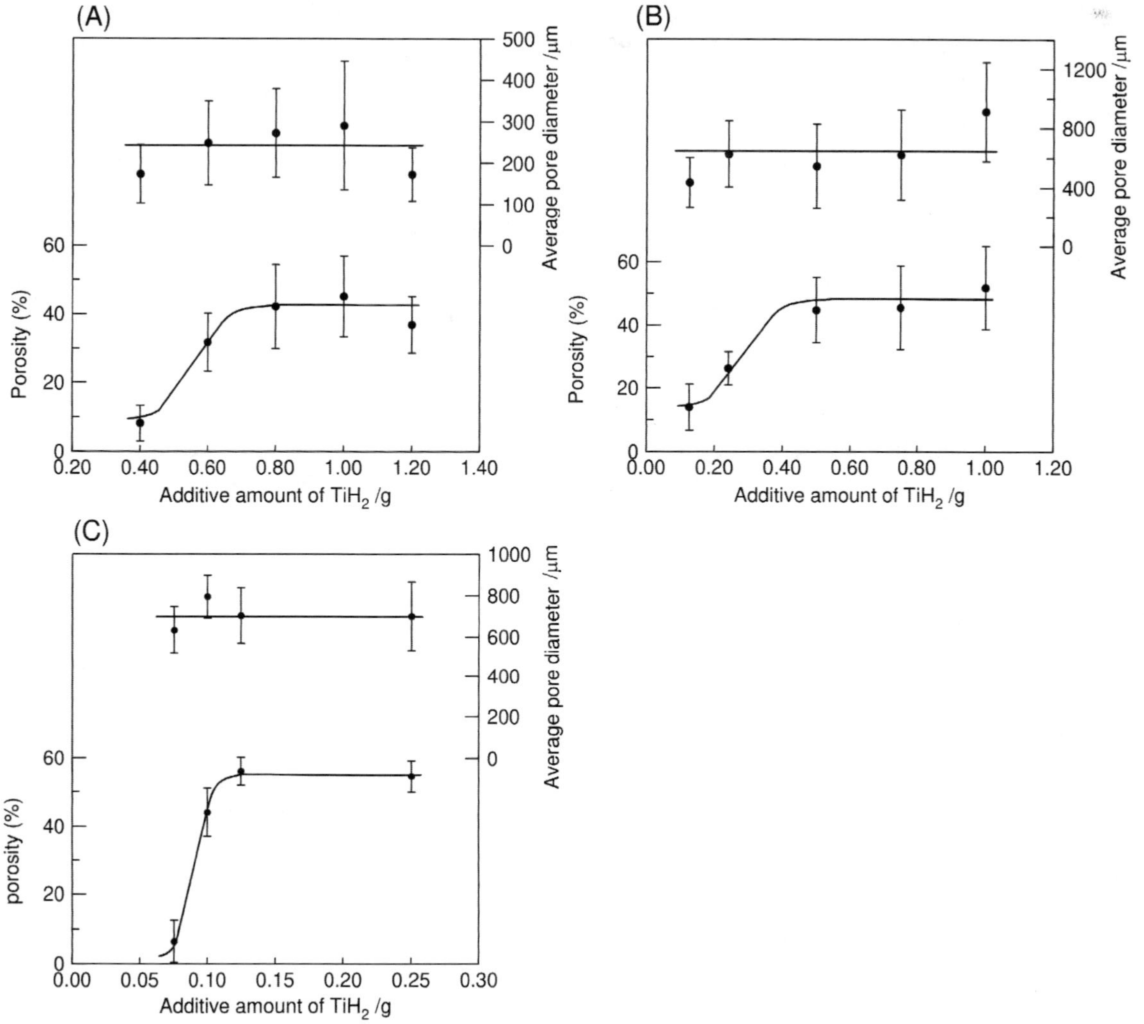

Figure 2. Additive amount of TiH_2 dependence of the average pore diameter and porosity for lotus copper fabricated through different additive methods. TiH_2 was added into the melt through (A) the applying the mixture including TiH_2 on the inside of the mold, (B) the setting TiH_2 pellets on the bottom plate of the mold and (C) the fixing TiH_2 pellets on the lateral side of the mold.

hydrogen solubility in liquid or solid phase. For the thermal decomposition method, the hydrogen solubility in liquid phase is determined by the amount of TiH_2. Therefore, when the small amount of TiH_2 is added in the molten copper, the TiH_2 contributes to the increase of hydrogen solubility in liquid phase dominantly. On the other hand, if the amount of TiH_2 more than the solubility is added in the molten copper, the hydrogen solubility in liquid phase does not change. In the case of method A, most of adding TiH_2 is coated with a mixture that consists of SiO_2 and Al_2O_3. Since the coating of SiO_2[4] or Al_2O_3[5] delays gas release of TiH_2 by several tens of seconds, the decomposition is retarded and the insufficient decomposition takes place. For method B, TiH_2 pellets may float in molten copper owing to lower density than that of molten copper, so that some pellets are not decomposed well. For method C, pore formation most effectively occurs in molten copper. Thus, higher porosity in the case of method C is obtained even by smaller amount of TiH_2.

Journal of Physics: Conference Series **165** (2009) 012064 doi:10.1088/1742-6596/165/1/012064

On the other hand, the average pore diameter does not depend on additive amount of TiH_2, which is the same tendency of the result that the average pore diameter does not depend on additive amount of TiH_2, and depends on argon pressure[3].

4. Conclusions

Lotus-type porous copper is fabricated through hydrogen caused by thermal decomposition of TiH_2. The effects of additive method and additive amount of TiH_2 on pore formation are investigated.

Pore formation effectively occurs in the method that TiH_2 decomposes in molten copper. For all the additive methods of TiH_2, the porosity increases and pore diameter does not change with increasing additive amount of TiH_2. While, for adding large amount of titanium hydride, the porosity became constant. This is because hydrogen solubility in liquid phase does not change.

Acknowledgment

This study was supported by priority Assistance for the Formation of Worldwide Renowned Centers of Research – The Global COE Program (Project: Center of Excellence for Advanced Structural and Functional Materials Design) from the Ministry of Education, Culture, Sports, Science and Technology, Japan.

References

[1] Nakajima H 2007 *Prog.Mater. Sci.* **52,** 1091
[2] Ide T, Tane M, Ikeda T, Hyun S K, and Nakajima H 2006 *J. Mater. Res.* **21,** 185
[3] Nakajima H and Ide T 2008 *Metall. Mater. Trans.* A **39,** 390
[4] Yang Z, Fang J, Ding B 2005 *Colloid Interface Sci.* **290,** 305
[5] Fang J, Ding B, Yang Z, Zhao K, Gu C 2005 *Colloid Interface Sci.* **283,** 1

Fabrication of lotus-type porous magnesium through thermal decomposition of magnesium hydride

M Tane and H Nakajima

The Institute of Scientific and Industrial Research, Osaka University, 8-1 Mihogaoka, Ibaraki, Osaka 567-0047, Japan

mtane@sanken.osaka-u.ac.jp

Abstract. Lotus-type porous magnesium was fabricated through thermal decomposition of MgH_2 powders as a source of hydrogen. Liquid magnesium was cast into a mold in which MgH_2 powders were placed and solidified unidirectionally in the mold, which achieved the growth of unidirectional elongated pores in magnesium matrix. The fabrication method is safer than a conventional method using pressurized hydrogen atmosphere, because the risk of explosion can be avoidable. The effect of the amount of MgH_2 powders and the distance from the bottom of the ingot on the porous structure was investigated, which clarified that the two factors have the large influence on the pore growth.

1. Introduction

Lotus-type porous metals with cylindrical pores oriented in one direction [1,2], gather much interest because of their unique anisotropic pore structure. Since stress concentration hardly occurs for loadings along the longitudinal direction of the pores, the mechanical strength in the longitudinal direction is superior to that of conventional porous metals with isotropic pores[3,4]. Furthermore, the unique structure provides various functions such as energy absorption and sound absorption[1]. Therefore, lotus metals are expected to be used as light-weight structural materials with various functions.

A lotus metal can be fabricated through unidirectional solidification of a metal in pressurized hydrogen atmosphere, when hydrogen solubility gap between the liquid and solid exists in equilibrium state [1,2]. When the liquid metal is solidified unidirectionally, a part of the hydrogen dissolved in the liquid is rejected at the solid-liquid interface owing to the hydrogen solubility gap. Then, the rejected hydrogen forms pores, and the pores grow unidirectionally along the solidification direction. In this method, pressurized hydrogen gas is indispensable for hydrogen to be dissolved in liquid metal. Thus, the method is called as "pressurized gas method".

Recently, Nakajima and Ide developed a new method for fabricating lotus metals without using pressurized hydrogen gas[5]. The fabrication method, so-called "thermal decomposition method", uses hydrogen thermally decomposed from compounds in place of pressurized hydrogen gas for hydrogen to be dissolved in liquid metal. Since pressurized inflammable hydrogen gas is not used in the thermal decomposition method, the fabrication process is safe and the production cost is much lower than that of the conventional method. Therefore, the thermal decomposition method is expected to promote practical applications of lotus metals.

In the present study, lotus magnesium was fabricated with the thermal decomposition method using MgH_2 powders as the hydrogen source, and the effects of the amount of the MgH_2 powders and the distance from the bottom of the ingot on the porous structure were investigated.

2. Experimental procedure

The apparatus for fabricating lotus magnesium consists of a crucible surrounded by a high-frequency induction coil and a cylindrical mold, which are installed in a chamber. The side wall of the mold is made of a stainless steel sheet of 0.10 mm in thickness, and the bottom of the mold is made of a water-cooled copper. The MgH_2 powders (98 mass%) of 0.2, 0.4, or 0.6 g were placed on the bottom of the mold; the diameter of the powders is about 10 - 100 μm. The MgH_2 powders whose onset temperature for decomposition is lower than that of TiH_2 powders were used, because pores were not formed owing to the insufficient decomposition when TiH_2 powders were used. The chamber was filled with argon gas of 0.10 MPa. Pure magnesium (99.9 mass%) of 50 g was melted by induction heating inside the crucible. Then, the melt was poured into the mold. The melt was solidified unidirectionally because only the bottom of the mold was cooled and the lateral side of the mold was made of thin stainless steel sheet which has a small cooling capacity.

The solidified ingots were cut out by a spark erosion cutting machine, and the cross sections parallel and perpendicular to the solidification direction were observed. The porosity, average pore diameter and number density of pores on cross sections perpendicular to the solidification direction was measured by an image analyzer (WinRoof, Mitani Corp.)

3. Results and discussion

Figure 1 shows the pore structures of cross sections of prepared ingots for $M = 0.6$ g, where M is the mass of used MgH_2 powders. The cross sections are (a) parallel and (b) perpendicular to the solidification direction. Cylindrical pores oriented along the solidification direction were formed in the ingots, which was similar among the ingots prepared for $M = 0.2$, 0.4, and 0.6 g. It is confirmed from this result that lotus-type porous magnesium can be fabricated with the thermal decomposition method using MgH_2 powders.

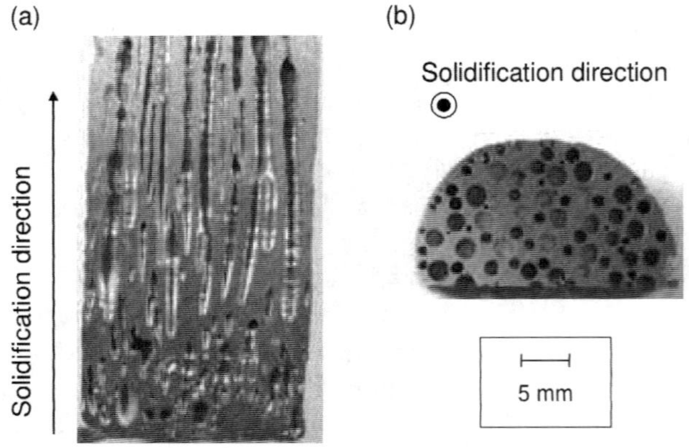

Figure 1. Porous structures of lotus magnesium prepared by thermal decomposition method using MgH_2 powders of 0.6 g. (a) Cross section parallel to the solidification direction. (b) Cross section perpendicular to the solidification direction, where the distance between the cross section and the bottom of the ingot is 25 mm.

In the present thermal decomposition method, a way for supplying melt with hydrogen is different from that in a conventional pressurized gas method. In the present thermal decomposition method, MgH_2 powders are heated by melt poured into the mold. Then, hydrogen is thermally decomposed from the MgH_2 powders. The decomposed hydrogen is dissolved in the melt, while pressurized hydrogen gas in a chamber is dissolved in the melt in the pressurized gas method. The formation of pores from the melt containing hydrogen is similar between the two methods; in the thermal decomposition method, pores elongated along the solidification directions forms during unidirectional solidification owing to the hydrogen solubility gap between the liquid and solid.

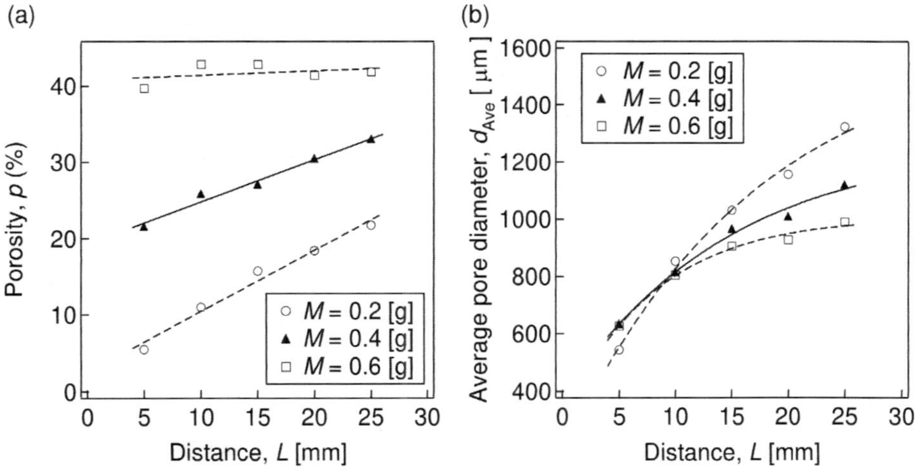

Figure 2. (a)Porosity p and (b) average pore diameter d_{Ave} of lotus magnesium as a function of a distance from the bottom of the ingot, L.

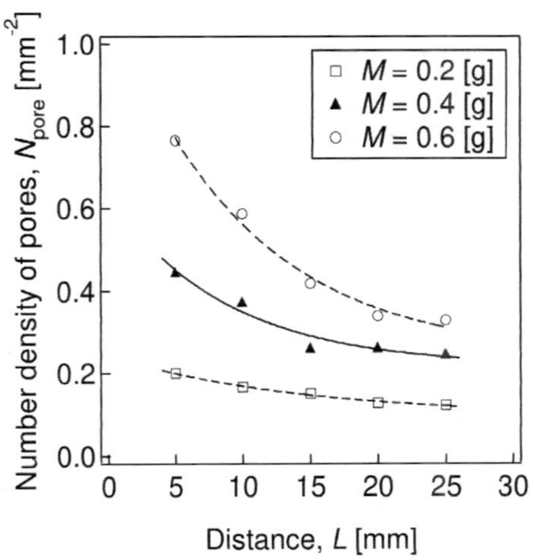

Figure 3. Number density of pores N_{pore} of lotus magnesium as a function of a distance from the bottom of the ingot, L.

Figure 2 (a) shows the porosity p of lotus magnesium as a function of a distance from the bottom of the ingot, L. The porosity increases with increasing L for $M = 0.2$ and 0.4 g, while the porosity for $M = 0.6$ g hardly increases with increasing L. This is because the hydrogen dissolved in liquid, which determines the porosity, is almost saturated at the beginning of the solidification for $M = 0.6$ g. Figure 2 (b) shows the average pore diameter d_{Ave} of lotus magnesium as a function of a distance from the bottom of the ingot, L. Average pore diameter d_{Ave} increases with increasing L. The rate of the increase in d_{Ave} with regard to L decreases with increasing M. Therefore, d_{Ave} for M = 0.6 g is the smallest when L = 15 - 25 mm, while d_{Ave} is almost independent of M when L = 5 and 10 mm.

The number density of pores N_{pore} increases with increasing M as shown in Fig. 3. Since the hydrogen rejected at the solid-liquid interface increases with increasing M, the nucleation site of pores (number density of pores) increases with increasing M. The number density of pores N_{pore} decreases with increasing L for all M values. This is probably caused by the coalescence of small pores formed near the bottom of the ingots into a large pore with advancing pore growth, which occurs to decrease the interface energy between pores and liquid metal. When M is small (M = 0.2 g), it is probable that pores can not easily coalesce, i.e., the single pore growth is dominant. In this case, an increase in porosity causes an increase in average pore diameter. Therefore, an increase in porosity with increasing L provides an steep increase in d_{Ave} with increasing L. On the other hand, when M is large (M = 0.6 g), it is probable that the coalescence of pores is dominant when the distance L is small (near the bottom of the ingot). When the distance L is large, a steady state condition for pore growth is probably achieved. Therefore, the rate of the increase in d_{Ave} with increasing L is almost saturated for large L; the porosity and number density of pores are also saturated for large L.

4. Conclusions

Lotus-type porous magnesium with cylindrical pores oriented in one direction was fabricated with a thermal decomposition method using MgH_2 powders as a compound which provides hydrogen. Liquid magnesium was cast into a mold in which MgH_2 powders were placed and unidirectionally solidified, which achieved the growth of pores elongated along the solidification direction.

The effect of the mass of MgH_2 and distance from the bottom of the ingot on the porous structure was investigated, which clarified that the two factors have the large influence on the porosity, average pore diameter, and number density of pores. The coalescence of pores during pore growth probably affects the change in the porous structure during unidirectional solidification. .

Acknowledgement

This study was supported by Priority Assistance for the Formation of Worldwide Renowned Centers of Research - The Global COE Program (Project: Center of Excellence for Advanced Structural and Functional Materials Design) from the Ministry of Education, Culture, Sports, Science and Technology (MEXT), Japan.

References

[1] Nakajima H 2007 *Prog. Mater. Sci.* **52** 1091.
[2] Nakajima H, Hyun S K, Ohashi K, Ota K and Murakami K 2001 *Colloid Surf. A* **179** 209.
[3] Tane M, Ichitsubo T Nakajima H, Hyun S K and Hirao M 2004 *Acta Mater.* **52** 5195.
[4] Tane M, Ichitsubo T, Hyun S K anb Nakajima H 2005 *J. Mater. Res.* **20** 135.
[5] Nakajima H and Ide T 2008 *Metall. Mater. Trans. A* **39A** 390.

Fabrication of lotus-type porous aluminum using thermal decomposition of magnesium hydroxide

J S Park[1], H Nakajima[1]

The Institute of Scientific and Industrial research, Osaka University, 8-1 Mihogaoka, Ibaraki, Osaka 567-0047, Japan

E-mail: park23@sanken.osaka-u.ac.jp

Abstract. Lotus-type porous aluminium was fabricated using a thermal decomposition of magnesium hydroxide in a vacuum. The effect of addition of magnesium hydroxide on the porosity and the average pore diameter were investigated. The hydrogen dissolved in molten aluminum through thermal decomposition of magnesium hydroxide when the melt solidified evolves pores. Furthermore, magnesia formed through thermal decomposition of magnesium hydroxide in the melt serves as nucleation sites for the pores. The porosity and the number of the pores increase with increasing amount of magnesium hydroxide, and the average pore diameter is not much varied with amount of magnesium hydroxide.

1. Introduction

Porous and foamed metals have been the focus of great interest due to their special features that include low density, high surface area, sound absorption and so on. Thus, these metals are expected to be utilized as light-weight materials, catalysts, electrodes, vibration and acoustic energy damping materials, impact energy absorption materials and so on[1,2]. Especially, porous Al alloys are expected to be used for lightweight structural applications because of low specific weight, high stiffness and high energy absorbing capacity.

One class of the porous metals, lotus-type porous metals possess long cylindrical gas pores aligned in one direction. These have superior mechanical properties originating from unidirectional pores compared with other porous metals with isotropic and spherical pores[3], so that they are attracting attention as new light-weight structural materials.

In many fields of industry such as aircrafts and automobiles, aluminum and its alloys have been developed to be used as structural materials which have good advantages such as light weight and corrosion resistance. If the aluminum and its alloy with the lotus-type structure can be fabricated, light weight structural applications become possible. However, it is difficult to fabricate lotus-type porous aluminum with high porosity utilizing hydrogen gas solubility gap between solid and liquid, since the solubility of the aluminum is small[4]. In order to fabricate the lotus-type porous aluminum with a higher porosity, the new fabrication technique is necessary.

Recently, Nakajima and Ide[5] developed a new fabrication technique using a thermal decomposition method (TDM) of compounds containing gas elements in a non-hydrogen atmosphere under nearly atmospheric pressure. They reported that the metallic oxide particle formed by the thermal decomposition of the compound may serve as the nucleation sites for the hydrogen pores in the solid/liquid interface during unidirectional solidification. It is expected that the pore size and porosity become homogeneous by uniformly distributed nucleation sites. Thus, the TDM is expected to the new fabrication of the lotus-type porous aluminum.

In this study, we fabricated the lotus-type porous aluminum using the thermal decomposition of magnesium hydroxide in order to investigate an effect of the magnesium hydroxide on the porosity and the pore morphology.

2. Experimental procedure

The mold casting apparatus for lotus-type porous aluminum used in this study is shown in figure 1, which consists of melting in the upper part and casting in the lower part. Pure aluminum (99.99%) of about 80 g was melted in a graphite crucible by an induction heating coil in vacuum. The bottom of the mold was copper plate cooled down by circulated water chiller, while the side wall was made of a stainless steel sheet of 0.1 mm in thickness to avoid solidification from the side surface. Magnesium hydroxide was usually wrapped with aluminum foils and was set on the bottom plate of the mold.

The temperature of the liquid in the graphite crucible was monitored by an infrared pyrometer (model IR-AP, Chino Co.). After pouring the molten aluminum at 1173 K in the crucible into the mold, hydrogen decomposed from magnesium hydroxide dissolves in the melt, which is then solidified so that insoluble hydrogen precipitates to evolve the hydrogen pores. Amount of magnesium hydroxide was added from 0.1 g to 0.4 g. The experiments were carried out 3 times for each mass condition.

The fabricated lotus-type porous aluminum was cut using a spark-erosion wire cutting machine (Model LN1W, Sodick Co.) in directions parallel and perpendicular to the solidification direction. Each cross-section was polished with a series of emery papers and was observed using an optical microscope. The pore diameter was measured in the cross-section perpendicular to the solidification by image analyzer (Win ROOF, Mitani Co.). The porosity (p) was evaluated from the following expression:

$$p\ (\%) = \left(1 - \frac{\text{Apparent density of porous aluminum}}{\text{Density of nonporous aluminum}}\right) \times 100 \qquad (1)$$

The apparent density of the individual specimen was calculated by measuring both the weight and the apparent volume of each sample.

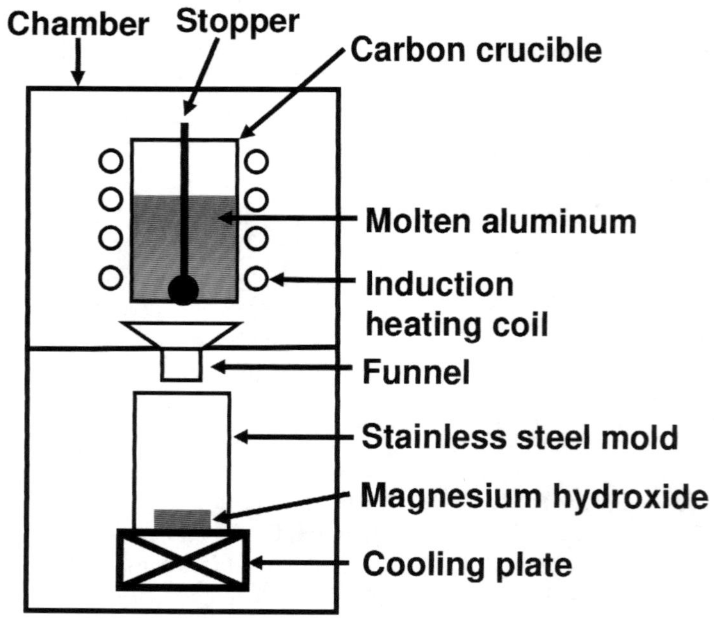

Figure 1. Mold casting technique for fabrication of lotus-type porous aluminum.

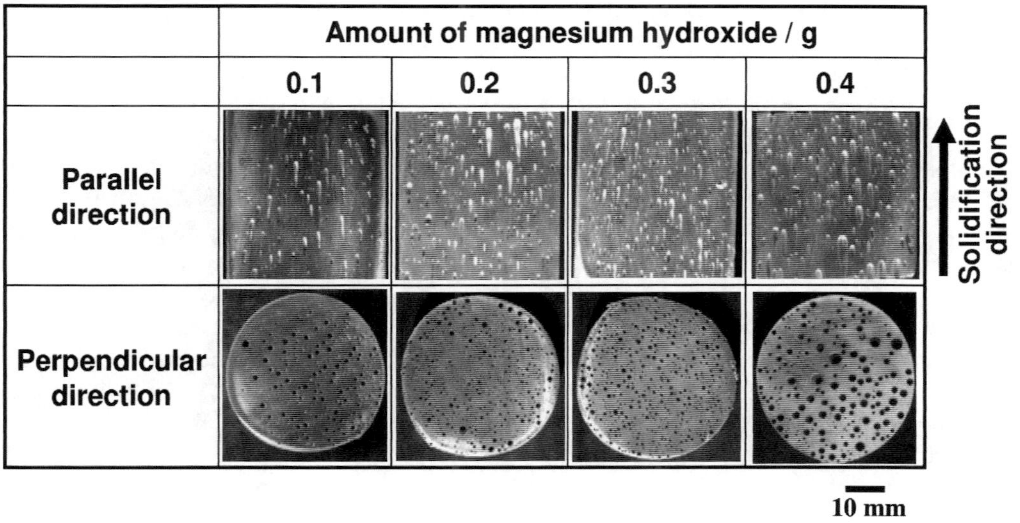

Figure 2. Cross sections perpendicular and parallel to the solidification direction of lotus-type porous aluminium fabricated by decomposition of magnesium hydroxide in vacuum at 1173 K.

3. Results and discussions

Figure 2 shows the cross-sectional views of the lotus-type porous aluminum fabricated by the mold casting technique in the mass range of the magnesium hydroxide from 0.1 to 0.4 g in vacuum. The upper and lower views are parallel and perpendicular to the solidification direction, respectively. The pores are aligned in one direction parallel to the solidification direction, and the number of the pores seems to increase generally with an increase in the amount of magnesium hydroxide.

The porosity measured in each samples are shown in figure 3. The porosity increases with increasing amount of magnesium hydroxide. On the other hand, the variation of the porosity decreases with an increase in the amount of magnesium hydroxide.

Figure 4 shows the average pore diameter of the lotus-type porous aluminum fabricated in the mass range of the magnesium hydroxide from 0.1 to 0.4 g. Although the average pore diameter has large

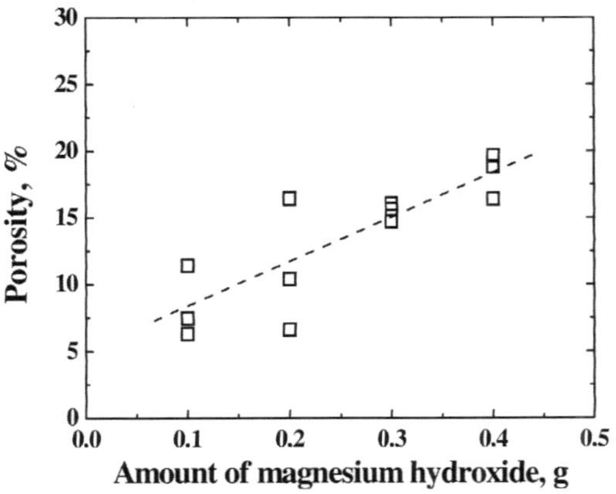

Figure 3. Porosity of lotus-type porous aluminium varied with amount of magnesium hydroxide in vacuum at 1173 K.

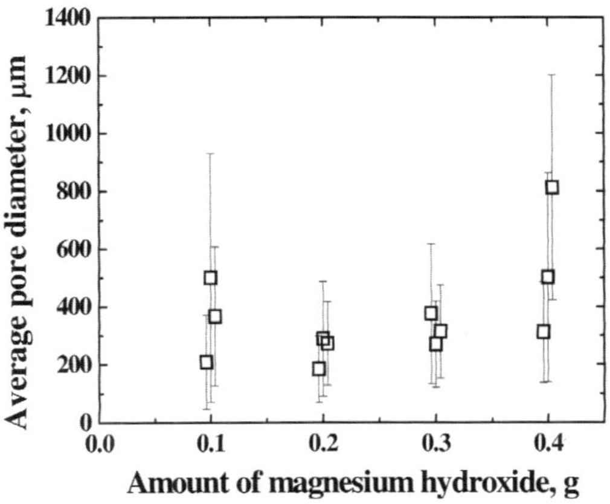

Figure 4. Average pore diameter of lotus-type porous aluminium varied with amount of magnesium hydroxide in vacuum at 1173 K. (Error bars are standard deviations)

standard deviations, the pore size is not varied so much. The results on both the porosity and the average pore diameter in the lotus-type porous aluminum fabricated through the thermal decomposition of magnesium hydroxide are in a good agreement with that in the lotus-type porous copper fabricated through the thermal decomposition of titanium hydride reported by Nakajima and Ide[5].

The porosity in the lotus-type porous metals mainly depends on the amount of the insoluble hydrogen rejected from the solid during unidirectional solidification[3,6]. Since the hydrogen dissolved in the liquid aluminum through the thermal decomposition of magnesium hydroxide increases with an increase in the amount of magnesium hydroxide, the porosity also increases because the amount of the insoluble hydrogen rejected from the solid during the solidification process increases. On the other hand, it is reported that the pore size in the lotus-type porous metals is affected by the undercooling during the solidification process, which can be controlled by the solidification velocity[3,5,6]. Because the solidification velocity is almost same in present study, the average pore diameter is not varied so much and independent of the amount of magnesium hydroxide. In addition, the increased insoluble hydrogen caused by the increase in added magnesium hydroxide is easy to form the pores at magnesia, which may serves as nucleation sites for insoluble hydrogen. Hence, it is considered that the number of the pores increases with increasing amount of magnesium hydroxide.

4. Conclusions

The lotus-type porous aluminum with the longitudinal cylindrical pores aligned in one direction was fabricated by the unidirectional solidification through thermal decomposition of magnesium hydroxide in a vacuum.

The hydrogen dissolved in the aluminum melt through thermal decomposition of magnesium hydroxide evolves the pores at the magnesia formed by the thermal decomposition of magnesium hydroxide in the melt, which may serves as nucleation sites for insoluble hydrogen. Therefore, when the amount of magnesium hydroxide increases, the porosity and the number of the pores increase. The average pore diameter is not varied so much and almost independent of the amount of magnesium hydroxide

Acknowledgement

The present work was supported by Grant-in-Aid for the Creation of Innovations through Business-Academic-Public Sector Cooperation of the Ministry of Education, Culture, Sports, Science and Technology of Japan. This work was also supported by the Global COE Program (Project: Center of Excellence for Advanced Structural and Functional Materials Design) from the Ministry of Education, Sports, Culture, Science and Technology of Japan.

References

[1] Evans A G, Hutchinson J W and Ashby M F 1999 *Prog. in Mater. Sci.* **43** 171
[2] Banhart J 2001 *Prog. in Mater. Sci.* **46** 559
[3] Nakajima H 2007 *Prog. Mater. Sci.* **52** 1091
[4] Opie W R and Grant N J 1950 *Trans. AIME* **188** 1237
[5] Nakajima H and Ide T 2008 *Metal. Mater. Trans.* A **39** 390
[6] Park J S, Hyun S K, Suzuki S and Nakajima H 2007 *Acta Mater.* **55** 5646

Fabrication of a lotus-type porous Al-Si alloy by continuous casting with a thermal decomposition method

T B Kim, S Suzuki, H Nakajima
The Institute of Scientific and Industrial Research, Osaka University,
8-1 Mihogaoka, Ibaraki, Osaka 567-0047, Japan

E-mail: kim23@sanken.osaka-u.ac.jp

Abstract. Lotus-type porous Al-14 wt.%Si alloy was fabricated by continuous casting at a transference velocity of 10 mm·min^{-1} in vacuum and by adding $Ca(OH)_2$ pellets to the crucible. The porosity and the pore diameter increased by varying the amount of $Ca(OH)_2$ from 0.2 g to 0.6 g. In the case of 0.6 g $Ca(OH)_2$, the average porosity was about 30 % and the average pore diameter was about 3.8 mm. XRD patterns of the pellets after continuous casting showed the $Ca(OH)_2$ pellet did not decompose completely during continuous casting. The TG-DTA analysis showed that the $Ca(OH)_2$ pellet decomposes more slowly than $Ca(OH)_2$ powder. These results suggest that the $Ca(OH)_2$ pellets gradually decomposed in the crucible during the continuous casting, which is suitable for the supply of hydrogen over extended periods.

1. Introduction

Porous metals are expected to be used as light-weight materials, catalysts, electrodes, vibration and acoustic energy damping materials, impact energy absorption materials etc. [1]. Porous aluminium alloys have especially attracted attention for use in various applications such as light-weight structural materials. While many kinds of metal foams have spherical pores [1] the lotus-type porous metals have pores aligned in one direction [2-5]. Lotus metals have superior mechanical properties that originate from unidirectional pores which are different from conventional porous metals that consist of almost isotropic and spherical pores [2, 3]. Therefore, fabrication of lotus-type porous aluminium alloys is desired in many industrial fields.

Park et al. reported a lotus-type porous Al-Si alloy that was fabricated using the continuous casting technique in a pressure hydrogen atmosphere [4]. However, the use of pressurized hydrogen gas is expensive because of the high-pressure chamber and special handling is required for the explosive H_2 gas. To address these problems a thermal decomposition method was developed to fabricate lotus metals without using pressurized hydrogen gas [6]. In this thermal decomposition method compounds containing potential gaseous elements are thermally decomposed in liquid metal and the decomposed gas atoms form pores at the solid – liquid interface during solidification because of the gas solubility gap between the solid and the liquid. Until now investigations on the thermal decomposition method have been done using the mould casting technique. However, the mould casting technique is limited by the length of samples and the control of pore distribution.

In this study the thermal decomposition method was applied to the continuous casting technique to control pore morphologies of lotus aluminium alloys. A long and continuous supply of gas to the melt was, however, required for this continuous casting method. Therefore a new additional technique was required for the continuous casting. Since solidification during mould casting is a short process the compound used should decompose quickly into a supply gas. $Ca(OH)_2$ was chosen for this study as the compound for the thermal decomposition reaction because gaseous elements are supplied to the melt. $Ca(OH)_2$ decomposes at a temperature below the melting temperature of aluminium alloys. For continuous casting the compound powder was compacted into pellet form to reduce the surface area. This process was expected to reduce the reaction rate of the compound and thus lengthen the reaction time. This paper reports the fabrication of a lotus Al-Si alloy by the continuous casting technique and by the addition of $Ca(OH)_2$ pellets.

2. Experimental procedure

$Ca(OH)_2$ pellets of 15 mm diameter and 1.25 mm thickness were prepared by compression moulding of a mixture of aluminium powder and $Ca(OH)_2$ powder of equal weight. The compression load was 20 kN and the compression time was 60 s.

Ingots of Al-14 wt% Si (330 g) and pellets that included $Ca(OH)_2$ of 0.2, 0.4, 0.6 g were heated in the graphite crucible by an induction coil in vacuum. After the temperature reached 1023 K the melt was maintained for 300 s to dissolve and diffuse gas into the melt uniformly. The melt was pulled down by the dummy bar of non-porous pure aluminium through a cooled mould (10 mm in thickness and 30 mm in width) at a transference velocity of 10 $mm \cdot min^{-1}$.

Fabricated slabs were cut by a spark-erosion wire cutting machine. Cross-sections were observed with an optical microscope. The porosity and pore diameter were measured on the cross-section perpendicular to the transference direction by the image analyzer (Win ROOF, Mitani Co.).

The decomposition temperature of the $Ca(OH)_2$ pellets was measured under an Ar flow of 150 ml/min and a heating rate of 10 K/min using a TG-DTA (TG-DTA2000SA, Bruker Co.). $Ca(OH)_2$ pellets before and after the continuous casting, aluminium powder and $Ca(OH)_2$ powder were analyzed using XRD (X-ray diffraction). Diffraction patterns were recorded between scattering angles of 10° and 70° at 10°/min (Cu Kα, RINT2000).

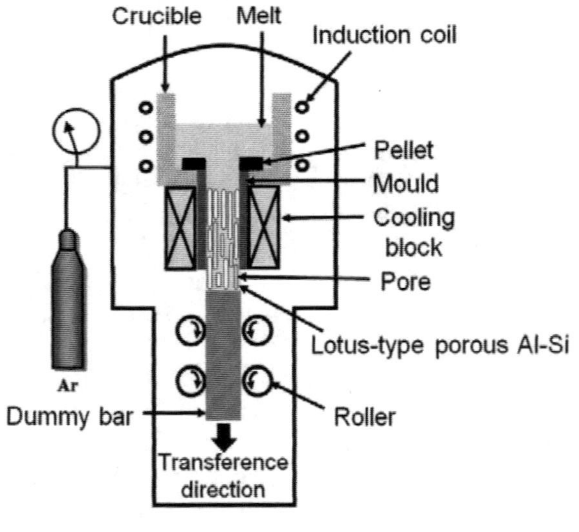

Figure 1. A schematic drawing of the fabrication apparatus for lotus-type porous Al–Si alloys by the continuous casting technique.

3. Results and discussion

Figure 2 (a) shows cross-sections in the perpendicular to the transference direction of the samples cut from the starting point of the transference to 100 mm at an interval of 10 mm. Figure 2(b) shows the cross-section parallel to the transference direction. Figure 2(a) and (b) show that the fabricated samples have cylindrical pores elongated in the transference direction. The sample fabricated with 0.6 g $Ca(OH)_2$ has an average porosity of approximately 30 % and an average pore diameter of about 3.8 mm.

Figures 3(a) and (b) show the effect of added $Ca(OH)_2$ pellets on the average pore porosity and the pore diameter, respectively. Both the porosity and the pore diameter increased as the amount of added

$Ca(OH)_2$ increased. At 0.2 g $Ca(OH)_2$ pores were not generated in the sample. The concentration of dissolved hydrogen does not obtain the equilibrium solubility of hydrogen in the liquid aluminium. Therefore, the amount of hydrogen excluded from the liquid to the solid – liquid interface is too small to generate pores. With an increasing amount of pellets the amount of excluded hydrogen increases and both the porosity and the pore diameter increases. Figures 3(a) and (b) indicate that there is no saturation until 0.6 g $Ca(OH)_2$ was used. Even for 0.6 g $Ca(OH)_2$ the concentration of hydrogen in the melt does not obtain the maximum solubility.

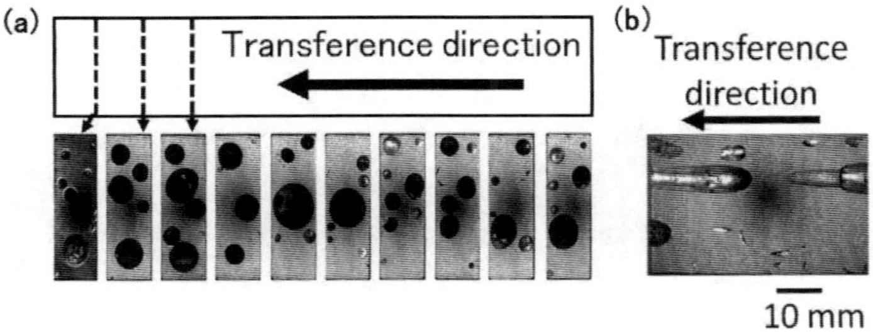

Figure 2. (a) Cross-sectional views parallel and (b) perpendicular to the transference direction of lotus-type porous Al-Si alloy fabricated with 0.6 g $Ca(OH)_2$.

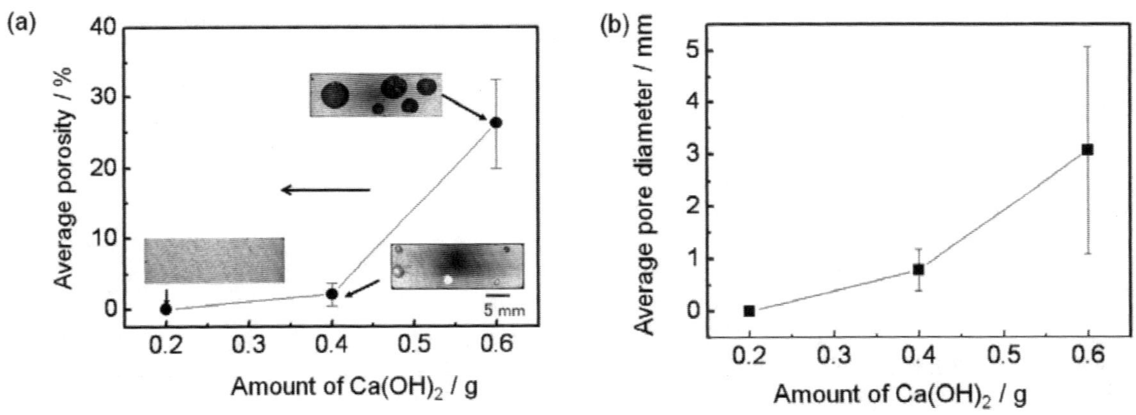

Figure 3. Dependence of the average porosity (a) and the average pore diameter (b) on the amount of added $Ca(OH)_2$.

After the continuous casting, pellets with their original shape were found in the crucible. Figure 4 shows XRD patterns of aluminium powder, $Ca(OH)_2$ powder and $Ca(OH)_2$ pellets before and after the continuous casting. XRD peaks of the pellets before the continuous casting indicate that aluminium and $Ca(OH)_2$ did not change during compression moulding. The XRD pattern of the pellet after the continuous casting shows CaO peaks. CaO probably forms by the following chemical reaction:

$$Ca(OH)_2 \rightarrow CaO + H_2O(g), \qquad H_2O(g) \rightarrow 2H + O \qquad (1)$$

Furthermore, the pellet's XRD shows that $Ca(OH)_2$ was still present after the continuous casting. The $Ca(OH)_2$ pellets, therefore, do not decompose completely during the heating up process and the continuous casting for about 100 min.

Figure 5 shows results of TG-DTA measurements for $Ca(OH)_2$ powder and a $Ca(OH)_2$ pellet before continuous casting. From DTA results the first reaction peak at 723 K is assigned to the decomposition reaction of $Ca(OH)_2$ [7] and the second reaction peak is the melting point of aluminium. From TG results, the amount of decomposed $Ca(OH)_2$ pellet is less than that of $Ca(OH)_2$ powder. The smaller reaction surface area of the compacted pellet reduces the decomposition rate and thus gas may be supplied for a longer time than would be the case for the powder.

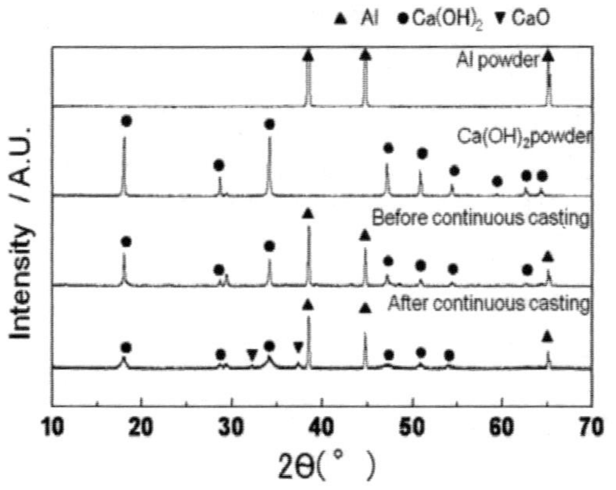

Figure 4. XRD patterns of aluminium powder, $Ca(OH)_2$ powder and a $Ca(OH)_2$ pellet before and after the continuous casting.

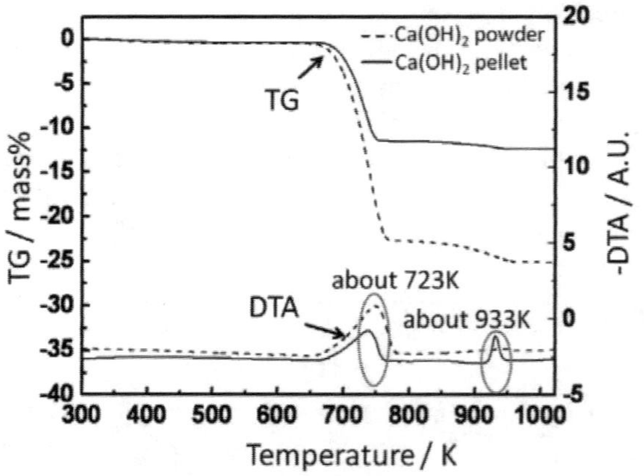

Figure 5. Results of TG-DTA with $Ca(OH)_2$ powder and a pellet.

4. Conclusion

Lotus-type porous Al-Si alloys were fabricated by continuous casting using the thermal decomposition method. The following properties were observed:

1. The average porosity was approximately 30 % and the average pore diameter was about 3.8 mm for the sample that was fabricated by addition of 0.6 g $Ca(OH)_2$.
2. By varying the amount of $Ca(OH)_2$ from 0.2 g to 0.6 g, the porosity and pore diameter increased.
3. As $Ca(OH)_2$ pellets have a lower decomposition reaction rate than $Ca(OH)_2$ powder the pellets may be used as a suitable gas supply source for the continuous casting.

Acknowledgements

This work was supported by the Grants-in-Aid for Young Scientists and the Global COE program (Centre of Excellent for Advanced Structural and Functional Materials Design) from the Ministry of Education, Culture, Sports, Science and Technology. The pure aluminium ingots were supplied by the Light Metal Educational Foundation, Inc.

Reference

[1] Banhart J 2001 *Prog. in Mater. Sci.* **46** 559
[2] Hyun S K, Murakami K and Nakajima H 2001 *Mater. Sci. Eng. A* **299** 241
[3] Nakajima H 2007 *Prog. Mater. Sci.* **52** 1091
[4] Park J S, Hyun S K, Suzuki S and Nakajima H 2008 *Metal. Mater. Trans. A* accepted. Available from: <http://dx. doi.org /10.1007/s11661-008-9710-3>
[5] Shapovalov V I 1996 *Mater. Sci. Forum* **485** 215
[6] Nakajima H, T Ide 2008 *Metall. Mater. Trans. A* **39**A 390
[7] Sawada Y, Ito Y 1994 *Thermochimica Acta* **232** 47

Fabrication of Al-Cu alloy with elongated pores by continuous casting technique

S Suzuki, T-B Kim, H Nakajima

The Institute of Scientific and Industrial Research, Osaka University
Mihogaoka 8-1, Ibaraki, Osaka567-0047, Japan

E-mail:suzuki@sanken.osaka-u.ac.jp

Abstract. Al-4.5wt%Cu was unidirectionally solidified by continuous casting technique under hydrogen pressure of 0.1MPa at the transference velocities ranging from 1 to 50 mm·min^{-1}. The fabricated slabs have microstructures of columnar α-dendrite and eutectic, which are typical for hypo-eutectic Al-Cu alloys. Elongated pores are observed in the eutectic region surrounded by several columnar α-dendrites. The shapes of the pores are affected by that of the surrounding α-dendrites. The average pores diameter is several ten μm smaller than the average dendrite arm spacing, which decreases with increasing solidification rate. Therefore, the pore diameter varies from about 200 μm to 30 μm with increasing transference velocity from 1 to 50 mm·min^{-1}. The porosity of the fabricated samples is in the range between 2 ~ 6 %. There is not significant dependence of the porosity on the transference velocity in the range of the present study. The porosity is similar to the reported value of Al-Si, where the area fraction of eutectic region is smaller than that of α-dendrite.

1. Introduction

Lotus-type porous aluminum alloys (lotus aluminum alloys), which possess small aligned pores in one direction, are expected to be used as lightweight structural materials for transport machinery. In recent years the fabrication methods for lotus aluminum alloys have been developed. Park *et al.* reported that fabrication of lotus Al-Si alloy became possible by continuous casting technique [1]. The lotus aluminum alloys are produced by unidirectional solidifications in pressurized hydrogen atmosphere. The pores are evolved by insoluble hydrogen when the melt dissolving hydrogen is solidified [2,3]. Now it is desired to apply this fabrication technique to various kinds of aluminum alloys. However, the porosity and the pore morphology are affected complicatedly by the solidification conditions and the alloying elements, as known in previous investigations of casting defects [4]. Therefore it is still difficult to predict the pore formation in each aluminum alloy under each solidification condition. Systematic investigations of pore formation are necessary for fabrication of lotus aluminum alloys.

The present study was undertaken to investigate the pore formation mechanism in binary aluminum alloys during unidirectional solidification using continuous casting technique in a pressurized hydrogen gas. As the aluminum alloy, Al-Cu was selected because it is widely used as structural materials and its solidification behavior has been investigated by many researchers [5,6]. This paper reports the relationship between the microstructure and the pore morphology in the Al-Cu slabs fabricated by the continuous casting technique in order to elucidate the pore formation mechanism.

2. Experimental

Figure 1 shows the schematic drawings of the continuous casting technique in this study. Pure aluminum ingots and Al-40wt.%Cu ingots were set in the crucible so as to be Al-4.5wt.%Cu of 500 g in weight. The ingots were melted by induction heating under hydrogen pressure of 0.1 MPa.

The crucible had a rectangular hole (10×30 mm^2) at the bottom, to which the mold was connected. A dummy bar of graphite was placed in the mold before the transference. The temperature of the molten Al-Cu in the crucible was measured by a thermocouple, and was maintained for longer than 600 sec after the temperature reached 1173K in order to dissolve hydrogen sufficiently into the melt.

Then the melt was pulled down by the dummy bar through the mold by the pinch rollers. The transference velocity of the dummy bar was controlled by the rotation of the pinch rollers. The transference velocity was set at 1, 2, 5, 20 and 50 $mm \cdot min^{-1}$. Thus, a long Al-Cu slab with pores was fabricated by the continuous casting technique.

The fabricated Al-Cu slabs were cut using a spark-erosion wire cutting machine (Model LN1W, Sodick Co.) in directions parallel and perpendicular to the transference direction. The cross-sections were polished, etched and observed by an optical microscope. The weight and the volume of a prismatic sample piece cut out from the lotus carbon slab were measured and the density of the sample piece was calculated.

3. Results and discussion

Figure 2 shows the density of the Al-Cu slabs fabricated at the various transference velocities. There seems the density does not depend on the transference velocity in the range between 1 and 50 $mm \cdot min^{-1}$. The average density with the standard deviation is 2.73(± 0.03) x10^3 $kg \cdot m^{-3}$. The porosity p of the fabricated slabs was estimated from the following equation.

$$p\ (\%) = (1 - (\text{density with pores / density without pores})) \times 100\ (\%) \quad (1)$$

The estimated porosity varies between 2 and 6 % when the density of Al-Cu alloy without pore is assumed to be 2.8 x 10^3 $kg \cdot m^{-3}$ [7]. The difference in density between the samples with and without pores is so small that the variation of the estimated porosity is sensitive to the deviation of the density. Therefore, significant dependence of the porosity on the transference velocity is not found in the present study.

The cross-sections of the Al-Cu slabs fabricated at various transference velocities are shown in figure 3 in the direction perpendicular and parallel to the transference direction. The cross-sections show pores elongated in the transference direction with a diameter smaller than several hundred μm.

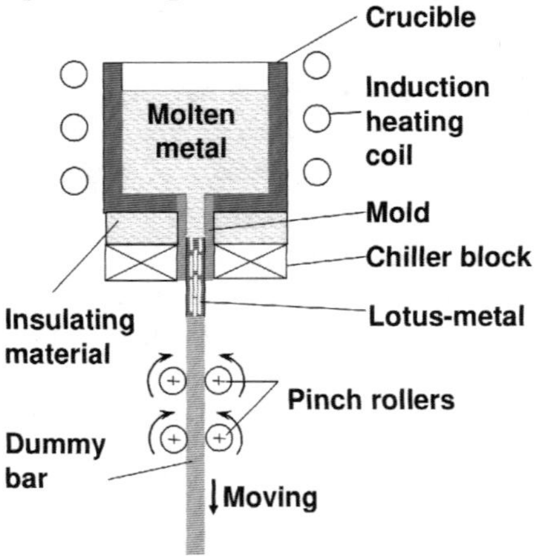

Figure 1. Schematic drawings of continuous casting technique in a pressurized gas chamber for fabrication of lotus metals.

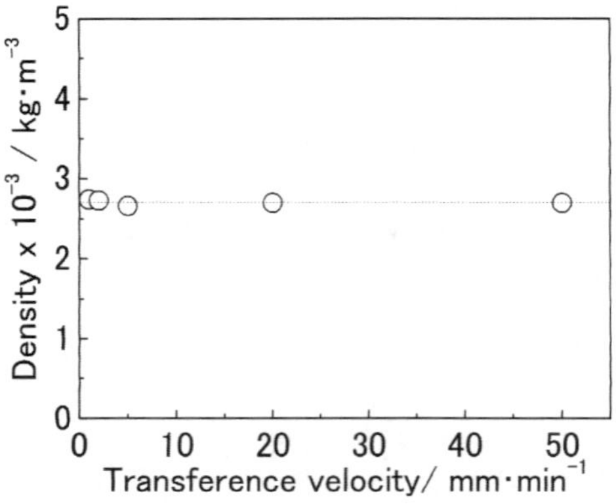

Figure 2. Density of fabricated Al-Cu slab.

The cross-sections show the tendency that pores become smaller with increasing transference velocity.

Figures 4 and 5 show the microstructure and the pores on the cross-sections perpendicular and parallel to the transference direction, respectively. The white and dark phases are α-dendrite and eutectic phase, respectively. The microstructure is typical for hypo-eutectic Al-Cu alloys [4]. The pores are observed in the eutectic region among several dendrites (figure 4) and elongate along the surrounding columnar α-dendrites in the transference direction (figure 5). The shape of the pores is affected by the surrounding α-dendrites.

The primary dendrite arm spacing (DAS), which is the distance between dendrites as shown in figure 4, was measured for more than 50 pairs of dendrites at each transference velocity and the average values and the standard deviations are shown in figure 6. The dendrite arm spacing decreases with increase in the transference velocity due to the increase of the solidification rate.

The statistics of the size of the pores observed on the cross sections are shown in figure 6, as well. These results show that the average pores diameter is several ten μm smaller than the average dendrite arm spacing. These pores are considered to form and grow in the eutectic region surrounded by several primary α-dendrites. It has been reported that the pore forms in the eutectic region in the case of unidirectional solidification of Al-Si alloys in hydrogen atmosphere [1], and Al-Cu in air and in vacuum [8]. The rejected hydrogen atoms during the solidification of α-dendrites are considered to escapes to the mushy zone or the liquid phase so that large pores are not formed. On the other hand, the eutectic region solidifies later than the α-dendrites, the rejected hydrogen atoms during the solidification of the eutectic phase are considered to form at the interface between the liquid and the eutectic phase [1].

The pores become smaller and irregular because the pore growth is disturbed by the surrounding dendrites with decrease in area fraction of the eutectic region in the case of lotus Al-Si [1]. As the Al-4.5wt.% Cu in the present study has a smaller area fraction of eutectic region than that of α-dendrites. Therefore the shape and the size of the pores seem to be limited by the surrounding α-dendrites.

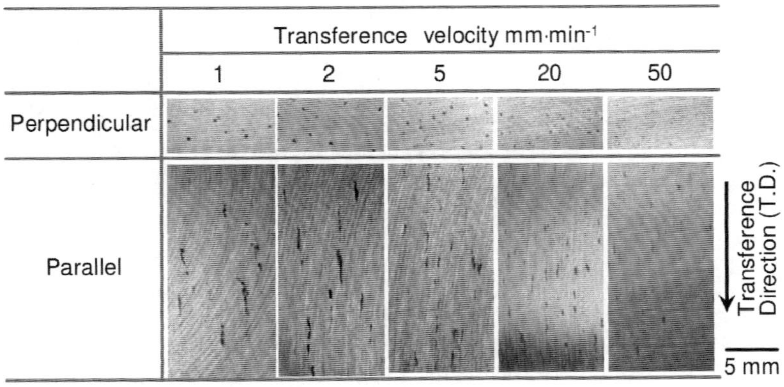

Figure 3. Cross-sections perpendicular and parallel to the transference direction (T.D.) of the fabricated Al-Cu slabs.

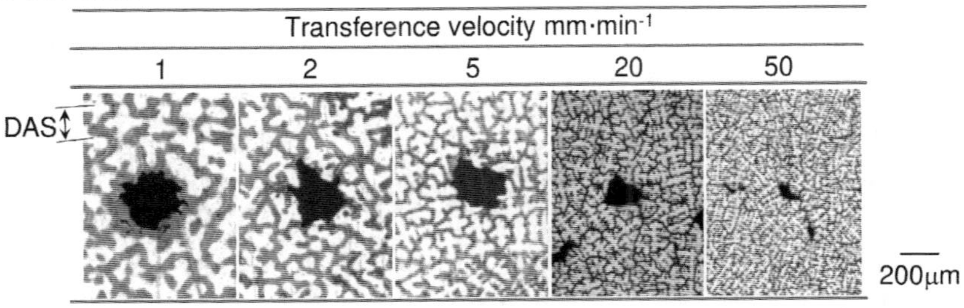

Figure 4. Pores and microstructures on the cross-sections perpendicular to the transference direction.

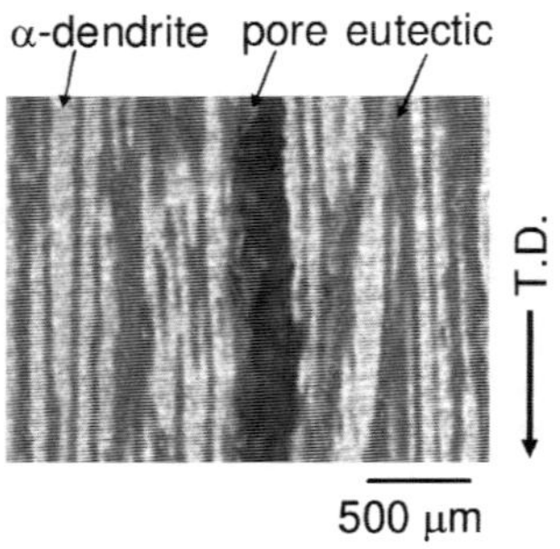

Figure 5. Pores and microstructures on the cross-section parallel to the transference direction (2 mm·min^{-1}).

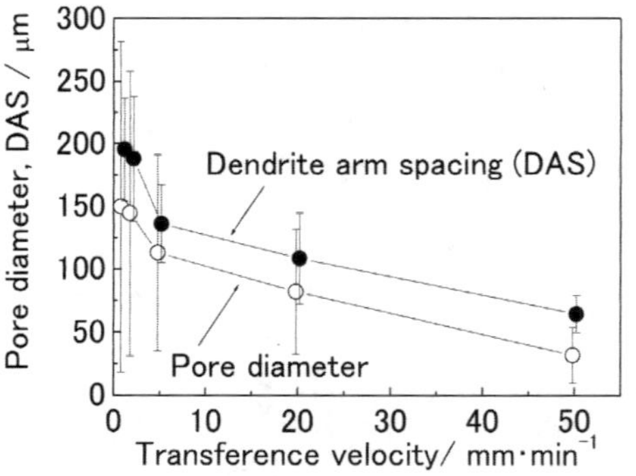

Figure 6. Pore diameter and dendrite arm spacing of the fabricated Al-Cu slab against the transference velocity.

The investigation on the pore formation in hypo-eutectic Al-Si alloys indicated that the porosity increases with increase in Si content, because of the increase in the area fraction of the eutectic region where pores form [1]. The Al-Cu alloy in the present study has a smaller area fraction than that of the α-dendrite and has porosity, which is similar to that of Al-Si alloys with a small area fraction of the eutectic region, such as Al-4wt.% Si.

4. Summary

Al-4.5wt.%Cu slabs with pores were fabricated by unidirectional solidification with using the continuous casting technique in a pressurized hydrogen atmosphere.
(1) The porosity of the fabricated Al-Cu slabs is in the range between 2 and 6 % and has no significant dependence on the transference velocity. The porosity in the present study is similar to the reported value of Al-Si, where the area fraction of eutectic region is smaller than that of α-dendrite.
(2) Pores elongated in the transference direction are observed in the eutectic region among several α-dendrites. The shape of the pores is determined by that of the surrounding α-dendrites. The pore diameter is similar to the primary dendrite arm spacing, which decreases with increasing transference velocity.

Acknowledgement

The present study was supported by the Grants-in-Aid for Young Scientists (B), Iketani Science and Technology Foundation, and Global COE Program (Project: Center of Excellence for Advanced Structural and Functional Materials Design) from the Ministry of Education, Culture, Sports, Science and Technology, Japan. The authors express their appreciation to the Light Metal Educational Foundation, Inc. for supply of the pure Al ingots.

References

[1] Park J S, Hyun S K, Suzuki S and Nakajima H 2008 *Metal. Mater. Trans. A* accepted. Available from: <http://dx.doi.org /10.1007/s11661-008-9710-3>
[2] Nakajima H 2007 *Prog. Mater. Sci.* **52** 1091
[3] Shapovalov V 1998 *Porc. Mater. Res. Soc. Symp* .**521** 281
[4] Samuel A M and Samuel F H 1992 *J. Mater. Sci.* **27** 653
[5] Sharp R M and Flemings M C 1973 *Metall. Trans.* **4** 997
[6] Lee P D and Hunt J D 1997 *Scripta Mater.* **4** 399
[7] Smithells C J 1992 *Metals Reference Book* (Oxford: Butterworth-Heinemann Ltd.) pp 14-14
[8] Nishi S and Kurobuchi T 1974 *J. Jpn. Inst. Light Metals* **24** 245

Fabrication of porous magnesium spinel ($MgAl_2O_4$) with cylindrical pores by unidirectional solidification

Shunkichi Ueno[1], Tatashi Akatsu[2] and Hideo Nakajima[1]
[1]The Institute of Scientific and Industrial Research, Osaka University, 8-1 Mihogaoka, Ibaraki, Osaka 567-0047, Japan
[2]Materials and Structure Laboratory, Tokyo Institute of Technology, 4259, Nagatsuta, Midori-ku, Yokohama 226-8503, Japan

E-mail: ueno23@sanken.osaka-u.ac.jp

Abstract. A porous magnesium spinel ($MgAl_2O_4$) with cylindrical pores was fabricated by unidirectional solidification in pressurized 1%H_2-99%Ar mixed gas. A small amount of Al_2O_3 phase was formed in porous $MgAl_2O_4$ bulk. Two different kinds of pores with large cylindrical and small facet shape were formed in the solidified samples. The former pores were dominant in the porous structure. The cylindrical pores were formed at solid-liquid interface due to hydrogen solubility gap. On the other hand, the small facet shape pores were formed by vaporization of MgO component on the cooling step in the solidification process. The pore length of the cylindrical pores decreased with increasing total pressure due to increasing the chance to nucleate pores during the solidification.

1. Introduction

Porous ceramics with excellent chemical stability at high temperatures is a promising for a filter in corrosive gas flow and in demand for integrated coal gasification combined cycle (IGCC) power generation system [1-3]. $MgAl_2O_4$ ceramic with high thermal and chemical resistances at high temperatures is a candidate for the gas filter in the system [3,4].

The pore size between 10 micron and several hundred microns is suitable for the gas filter application [1]. Several research groups have been proposed the fabrication process of the porous ceramics with open pores. Zhang et al. fabricated porous alumina with unidirectionally aligned continuous pores by sintering of cotton fiber with alumina coat [5]. Isobe et al. prepared porous alumina with oriented cylindrical pores by an extrusion method [6]. However, in these methods, it is difficult to control the porosity and pore size of the porous alumina. The authors have proposed a fabrication process of porous alumina with cylindrical pores by unidirectional solidification in pressurized hydrogen atmosphere [7]. In this method, the porosity and pore size can be controlled by hydrogen partial pressure and total pressure [7].

The present paper reports the fabrication of porous magnesium spinel with cylindrical pores by unidirectional solidification in pressurized 1%H_2-99%Ar mixed gas in order toelucidate the formation mechanism of the pores.

2. Experimental Procedure

For the preparation of $MgAl_2O_4$ feed rod, high purity alumina (99.99% purity, Sumitomo Chemical Co. Ltd.) and magnesia (99.9% purity, Konoshima Chemical Co. Ltd.) powders were used as starting materials. The powders with the stoichiometric ratio were mixed with a binder in water and the sully was poured into a mold. After drying in air, calcination was performed at 1200°C for 7.2 ks in air and then, the feed rod with 8 mm in diameter and 150 mm in length was prepared.

The solidification was carried out using optical floating zone apparatus in 1%H_2-99%Ar mixed gases in the range from 0.1 MPa to 0.7 MPa. Xenon lamp was used as a heating source. The xenon lamp and melting zone put on the focus in the elliptical mirror. The feed rod was hooked on upper shaft and another feed rod was fixed with lower shaft. This melting system was set up in a quartz tube. The environmental gas was introduced into the quartz tube. The transference velocity of the floating zone was fixed to 200 mm/h.

The cross-sections of the samples perpendicular and parallel to the solidification direction were observed by SEM (JEOL JSM-6360T). The porosity of the samples was calculated from the transversal cross section images.

3. Result and Discussion

Figure 1 shows the transversal and longitudinal cross-sections of the samples fabricated in 1%H_2-99%Ar mixed gas of 0.1, 0.2, 0.4, 0.6 and 0.7 MPa. Cylindrical pores are aligned in the solidification direction. In addition to the cylindrical pores, a small amount of small facet shape pores are also formed in the bulk as indicated by arrows.

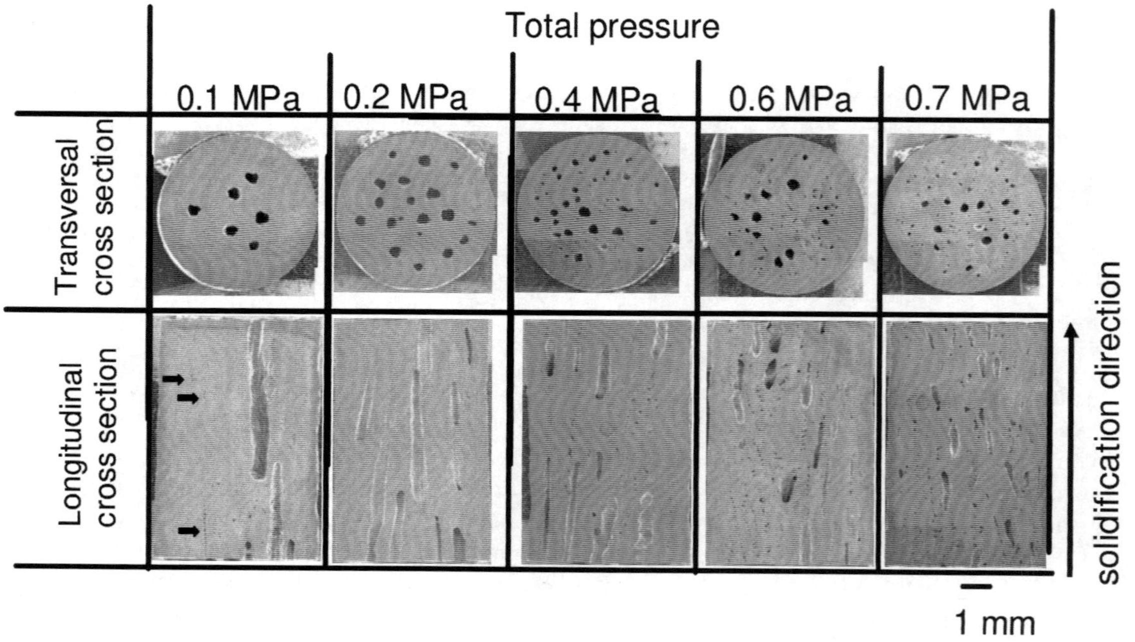

Fig.1 Transversal and longitudinal cross-sections of samples fabricated in 1%H_2-99%Ar mixed gas.

Figure 2 shows the powder X-ray diffraction pattern of the sample fabricated under total pressure of 0.2 MPa. A small amount of Al_2O_3 phase can be detected as secondary phase. This fact indicates that a small amount of MgO component vaporized during the solidification process.

Figure 3 shows the porosity change as a function of the total pressure. The porosity did not change with the total pressure. If the pores are formed at solid-liquid interface due to hydrogen solubility gap between solid and liquid phases, the pressure of inner pore depends on the total pressure according to the following equation (1) and the volume of pores decreases with increasing total pressure according to Boyle's law. Then, the porosity of the samples decreases with increasing total pressure.

$$P = \frac{2\gamma}{R} + P_0 \tag{1}$$

Where γ, R and P_0 denote the surface tension of the liquid $MgAl_2O_4$ phase, the radius of the pore and the total pressure, respectively. However, the porosity dose not changes with total pressure. In the case of formation for porous alumina with cylindrical pores by unidirectional solidification, the porosity decreased with increasing total pressure and increased with increasing hydrogen partial pressure [7]. The pores are formed at solid-liquid interface due to hydrogen solubility gap between solid and liquid phases.

Figure 4 (a) and (b) show the porosity and average pore diameter changes as a function of the total pressure; the porosity and pore size are evaluated for large cylindrical pores excluding the spherical small pores. The porosity and pore size decrease with increasing total pressure and increase with increasing hydrogen partial pressure as shown in Fig.4. Hence, it is confirmed that the cylindrical pores in the samples are formed at solid-liquid interface due to hydrogen solubility gap between solid and liquid phases during the unidirectional solidification as in the case of porous alumina in reference [7].

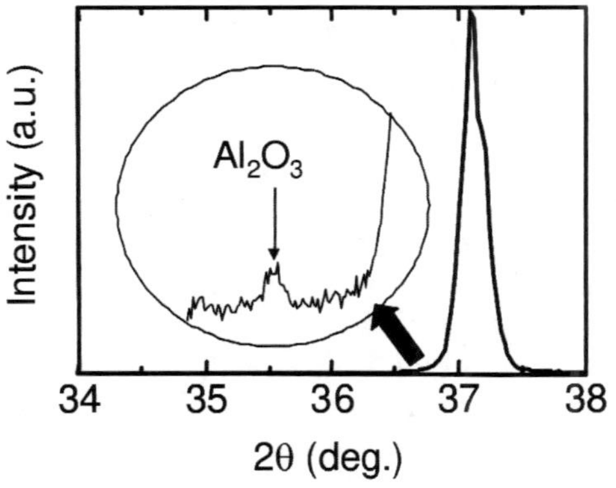

Fig.2 The powder X-ray diffraction patterns of samples fabricated at 0.1 MPa.

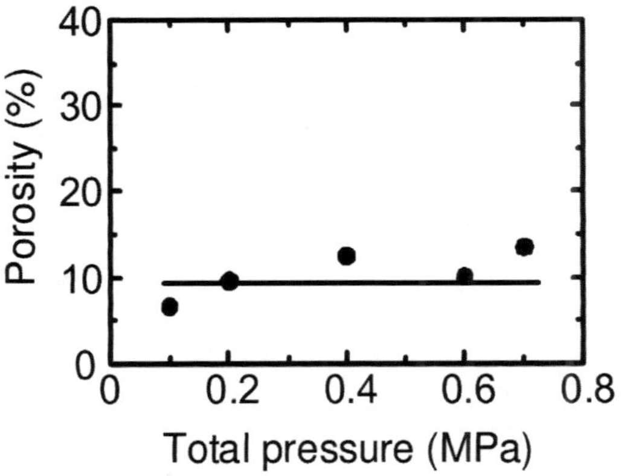

Fig.3 The porosity change as a function of the total pressure.

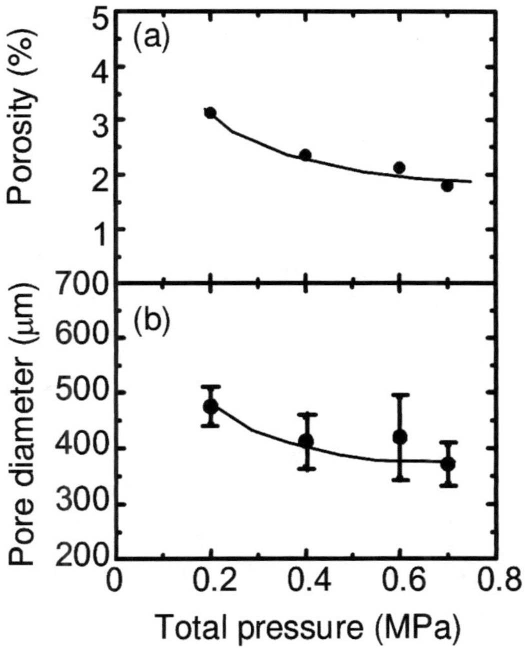

Fig. 4 The porosity (a) and average pore diameter (b) of large cylindrical pores as a function of the total pressure

Obviously, the pore length of the cylindrical pores decreases with increasing the total pressure as shown in Fig.1. When hydrogen solubility in the liquid $MgAl_2O_4$ phase obeys Sieverts' law, hydrogen solubility increases with increasing the total pressure and then, the chance to make the pores increases with the total pressure during the solidification.

Since non-porous $MgAl_2O_4$ was obtained by the solidification under 100%Ar atmosphere, the formation of the small pores is closely related to the hydrogen solubility in the solid phase. It is expected that the hydrogen solubility in solid $MgAl_2O_4$ phase also increases with increasing hydrogen partial pressure according to Sievert' law. Hence, it is considered that the evaporation rate of MgO component from the solid phase below melting point is accelerated by the dissolved hydrogen in the solid phase and then, many small pores are formed in the bulk. The evaporation of MgO component from $MgAl_2O_4$ was reported by Altman [8] and Sakamoto et al. [9]. The magnesium partial pressure according to equation (2) is hundred times larger than that of magnesium oxide due to equation (3) at high temperatures [8].

$$MgO(s) = Mg(g) + \frac{1}{2}O_2(g) \quad (2)$$

$$MgO(s) = MgO(g) \quad (3)$$

It can be understood the reduction reaction of equation (2) is accelerated by the dissolved hydrogen in the solid $MgAl_2O_4$ and the porosity increases with increasing the amount of dissolved hydrogen atom, namely, hydrogen partial pressure.

4. Conclusion

A porous magnesium spinel with cylindrical pores and small facet shape pores can be fabricated by unidirectional solidification in pressurized Ar-H_2 mixed gas. The cylindrical pores are formed at solid-liquid interface due to hydrogen solubility gap between solid and liquid phase. On the other hand, small pores are formed by vaporization of MgO component from the solid $MgAl_2O_4$ on the cooling step after the solidification that is accelerated by dissolved hydrogen in solid $MgAl_2O_4$ phase. The pore length of the cylindrical pores decreases with increasing total pressure due to increasing the hydrogen solubility in the liquid phase.

Acknowledgment

The present work was supported by Materials Science and Technology Research Center for Industrial Creation in Institute of Science and Industrial Research of Osaka University from the Ministry of Education, Culture, Sports, Science and Technology and the Grant-in-Aid for Scientific Research (C) (No. 18560655) by the Japan Society for the Promotion of Science. The compressive test was performed at Tokyo Institute of Technology on Collaborative Research Project of Materials and Structure Laboratory.

References

[1] Rice RW, Porosity of Ceramics, Marcel Decker, New York, 1998
[2] Alvin MA 1998 *Fuel Process. Technol.* **56** 143
[3] Suzuki Y, Kondo N and Ohji T 2003 *J. Am. Ceram. Soc.* **86** 1128
[4] Suzuki Y, Kondo N and Ohji T 2004 *Int. J. Appl. Ceram. Technol.* **1** 76
[5] Zhang GJ, Yang JF and Ohji T 2001 *J. Am. Ceram. Soc.* **84** 1395
[6] Isobe T, Tomita T, Kameshima Y, Nakajima A and Okada K 2006 *J. Europ. Ceram. Soc.* **26** 957
[7] Ueno S, Lin LM and Nakajima H 2008 *J. Am. Ceram. Soc.* **91** 223
[8] Altman RL 1963 *J. Phys. Chem.* **67** 366
[9] Sasamoto T, Hara H and Sata T 1981 *Bull. Chem. Soc. Jpn.* **54** 3327

Structural change and improvement of the mechanical properties of a lotus-type porous copper by wire-brushing

J Lobos[1], S Suzuki[1], H Nakajima[1], Y S Ji[2], H Fujii[2], D Terada[3], N Tsuji[3]

[1]The Institute of Scientific and Industrial Research, Osaka University, Ibaraki, Osaka 567-0047, Japan.
[2]Joining and Welding Research Institute, Osaka University, Ibaraki, Osaka 567-0047, Japan.
[3]Dept. of Adaptive Machine Systems, Graduate School of Engineering, Osaka University, Suita, Osaka 565-0871, Japan.

E-mail: juanlm23@sanken.osaka-u.ac.jp

Abstract. The surface of lotus-type porous copper plates that had cylindrical open pores in the thickness direction (porosity 50.4%, average pore diameter 144.4 μm) were processed by wire-brushing. Open pores on the surface of the lotus copper are closed by a newly formed nonporous thin layer. Electron backscatter diffraction patterns of the processed plate cross section show that the deformed surface consists of ultra-fine grains and that a nonporous layer was formed on the deformation of the surface layer. The Vickers hardness of the wire-brushed lotus copper is higher than that of the wire-brushed non-porous copper. The Vickers hardness increases with the increase in the rate of revolution of the wire–brush due to grain refinement. The increment of the ultimate tensile strength of lotus copper by wire-brushing is larger than of non-porous copper. The increment of ultimate tensile strength of the lotus copper reaches maximum when the newly deformed layer closes all the pores on the surface. These results show that wire-brushing is an effective process for the improvement of mechanical properties for lotus metals.

1. Introduction

Porous metals have been actively studied in recent years because of their promising properties such as light–weight, vibration and acoustic energy damping, and impact energy absorption [1]. However, for many kinds of porous metals the specific strength decreases with increasing porosity because spherical pores cause stress concentrations. On the other hand, lotus-type porous metals with long cylindrical pores aligned in one direction have superior mechanical properties compared with porous metals with spherical pores. The specific strength does not decrease with increasing porosity in the pore growth direction [2]. Therefore, lotus metals are expected to be used as light-weight structural materials.

However, as lotus metals are fabricated by solidification processes, their as-cast microstructure have lower strength than that of commonly used structural materials, which is usually produced by plastic forming. It is, therefore, desirable to deform the metallic matrix of lotus metals without causing a severe change in the porous structure during plastic deformation.

It was reported that the Vickers hardness of lotus copper increases by ECAE (equal-channel angular extrusion) with a remnant porous structure [3], which showed the effect of severe plastic deformation on the improvement of lotus metal strength. Recently, the wire-brushing process is also considered to be an effective method for use in the severe plastic deformation of bulk metals. It was reported that wire-brushed nonporous metals have fine-grained surfaces [4,5]. This process is expected to provide severe plastic deformation to the metallic matrix of lotus metals without destroying the porous structures.

For these reasons, lotus copper plates were processed by wire-brushing. The effects of angular velocity of the wire-brush on mechanical properties such as the Vickers hardness and the ultimate tensile strength were investigated. An improvement of mechanical properties is discussed considering the structural change in the lotus copper plates.

2. Experimental

2.1. Fabrication of lotus copper
A lotus copper ingot was fabricated by the mould casting technique [2]. Copper ingots were melted in a pressurized hydrogen atmosphere of 0.5 MPa. Inside the chamber, the melt was poured into a mould and was solidified unidirectionally. Because of the solubility gap of hydrogen between solid and liquid phases the pores evolved in the solidification direction. The fabrication process of the lotus copper ingot resulted in pore alignment in the direction of the cylinder's axis. The lotus copper had a porosity of 50% and an average pore diameter of 144 μm.

2.2. Samples for the tensile test
The cylindrical lotus copper ingot was sliced into disks of 2.1 mm in thickness and 100 mm in diameter. The pores were orientated in the direction of the thickness and open pores were visible on the surface. Tensile test samples were cut and had a length of 80 mm (25 mm for the test part) and a width of 10 mm (6.5 mm for the test part). Some samples were not processed and were used as reference materials for tests of the mechanical properties.

2.3. Wire–brushing
A schematic illustration of the wire-brushing process used in this study is shown in Fig. 1. A lotus copper plate in the shape of the tensile specimen was fixed on a table. On the surface of the plate, a rotating stainless steel wire cup brush (grade CN-16, length of the wires was 18 mm and the diameter 0.3 mm, the cup diameter was 70 mm) was transferred in the direction of the sample's width. The rate of revolution of the wire–brush was set at 2500, 3750 or 5000 rpm. The transference velocity was set so that the revolutions per unit length, were constant at 100 mm^{-1}. The wire-brush was rotated clockwise as observed from the upper side and the axis of the wire-brush was tilted 7 degrees from the vertical during movement. Only half of the circumferential part of the wires hit the sample.

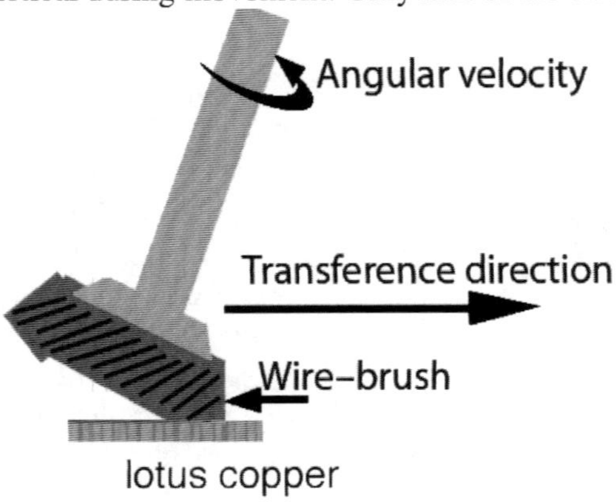

Figure 1. Schematic illustration of the wire-brushing process of a lotus copper plate.

2.4. Evaluation of the processed samples
The Vickers hardness was measured on the processed surface of the lotus copper plates by a hardness tester (AKASHI MVK-III). The testing force was 5~10 N and the testing time was 20 s. Ten points on

the same sample were tested and statistical processing was made.

The tensile tests were carried out using a universal testing machine (INSTRON 5582). The crosshead speed was set at 2 mm min^{-1}. From the obtained stress-strain curve, the ultimate tensile strength was measured. The tensile tests were performed four times for each processing condition and statistical processing was done.

Mechanical properties of nonporous copper plates, before and after the wire-brushing, were also measured as reference data.

The surface of the processed lotus copper was observed with a digital microscope (KEYENCE VHX-200). The microstructure of processed lotus copper plate cross sections was observed with a FE–SEM JEOL 9800 and EBSD analyzer Tex SEM Labo. Inc. MSC-2200.

3. Results and discussion

Figure 2 shows the Vickers hardness of the lotus copper surface and the nonporous copper surface versus the angular velocity of the wire-brush. For both cases, the Vickers hardness increased after the wire-brushing process. The Vickers hardness of lotus copper after wire-brushing at a rate of revolution of 5000 rpm was higher than that of the nonporous copper. The Vickers hardness values of the lotus copper and the nonporous copper before wire-brushing (0 rpm) were similar. These results suggest that the deformation behaviour of the lotus and the nonporous copper are different as explained below.

In the case of the lotus copper, the Vickers hardness increases with an increase in the rate of revolution. For severe plastic deformation, it is known that an increase in strain causes an increase of Vickers hardness. Higher strain is obtained by a higher angular velocity of the wire-brush because of its higher kinetic energy.

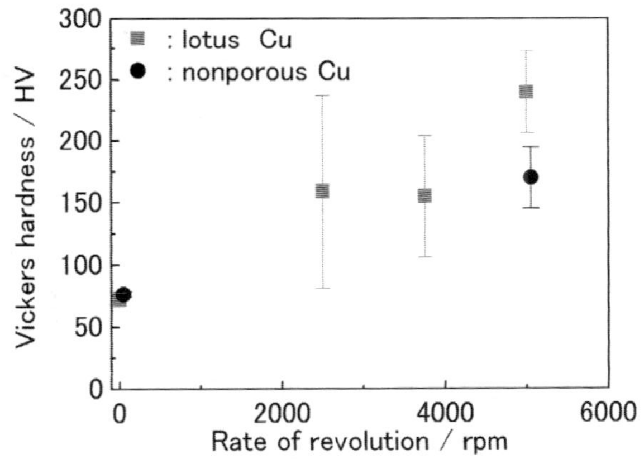

Figure 2. The Vickers hardness on the surface of the lotus copper and the nonporous copper at different rates of revolution.

$\Delta\sigma_{UTS}/\sigma_{UTS}$ = (σ_{UTS} of the wire-brushed sample - σ_{UTS} of the as-cast sample) / (σ_{UTS} of the as-cast sample) (1)

For the nonporous sample, there is no significant change in the strain-stress curves between wire–brushed and as-cast samples, as can be seen in previous work [4,5]. The reason is that the grain refinement in a layer of only several tens of μm in thickness does not affect the mechanical properties of a bulk sample of several mm in thickness. On the other hand, there is an increase in all the samples of the lotus copper. Furthermore, the change in tensile strength is dependent on the angular velocity. The strength increases up to 3750 rpm and then decreases. This trend is different to that of the Vickers hardness.

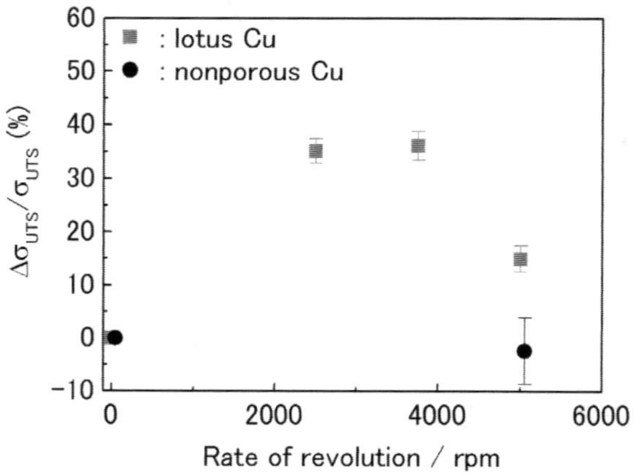

Figure 3. Changes in the ultimate tensile strength ($\Delta\sigma_{UTS}$) of the wire–brushed samples.

To investigate the increase of the tensile strength by wire-brushing and the angular velocity dependence of the tensile strength the surface and the cross sections of the processed lotus copper plate were observed and the grain size was analyzed.

Surfaces of the processed samples were changed by a change in the rate of revolution. At 2500 rpm, many pores were closed and large pores remained after wire-brushing. At 3750 rpm almost all the pores were closed. However, large pores were opened again at 5000 rpm.

The cross section of the processed sample showed that the thickness of the layer, which covers the top of the pores is 80~100 μm. EBSD mapping showed that the grain size is smaller than 80 nm in the top part of this layer, several hundred nm in the middle and a few μm in the bottom. The deformed layer on the metallic part near the pores was around 50 μm in thickness, while that on the metallic part, apart from the pore, was 20μm in thickness. This is similar to the results of wire-brushed nonporous metals [4,5].

These results suggest that metallic parts near the pores can be deformed easily toward the pore side. In the case of lotus copper, a layer of fine grains was generated because of the large strain of the metal near the pores. The strain increases with increasing angular velocity of the wire-brush and then some part of surface was removed at a higher velocity which decreases strength.

4. Summary

A severe plastic deformation by a wire-brushing process increased the hardness and the strength of lotus copper by the formation of a nonporous skin layer with fine-grained crystals that closed the top of the pores. This process forms a sandwich structure consisting of a lotus metal plate with two fine-grained nonporous layers. The effect of angular velocity of the wire-brush on the hardness of the surface and the tensile strength of the processed lotus copper was investigated by considering the structural change.

Acknowledgement

This work was supported by the Global COE Program (Project: Centre of Excellence for Advanced Structural and Functional Materials Design) from the Ministry of Education, Culture, Sports, Science and Technology, Japan and by the Spanish Agency for International Cooperation (AECI).

References

[1] J Banhart 2001 *Prog. in Mater. Sci.* **46** 559
[2] H Nakajima 2007 *Prog. Mater. Sci.* **52** 1091
[3] S Suzuki, H Utsunomiya, H Nakajima 2008 *Mater. Sci. Eng. A* **490** 465
[4] M Sato, N Tsuji, Y Minamino and Y Koizumi 2003 *Mater. Sci. Forum* **246–432** 2753
[5] M Sato, N Tsuji, Y Minamino and Y Koizumi 2003 *Sci. Tech. Adv. Mater.* **5** 14

Mechanical property of lotus-type porous carbon steel fabricated by continuous casting method

Y Kawamura[1], S Suzuki[1], S-Y Kim[1], H Nakajima[1], M Kashihara[2] and H Yonetani[2]

[1] The Institute of Scientific and Industrial Research, Osaka University, Ibaraki 567-0047, Japan
[2] Materials Laboratory, Mori Seiki Co. Ltd, Iga 519-1414, Japan

E-mail:kawamu23@sanken.osaka-u.ac.jp

Abstract. Lotus-type porous carbon steel (AISI1018) with aligned cylindrical pores was fabricated by continuous casting method in a mixture gas atmosphere (P_{N_2}=0.8MPa, P_{Ar}=1.7MPa). Compressive yield strength of the nonporous and the lotus carbon steels was measured in the direction parallel to the transference direction. The compressive stress of the lotus carbon steel is lower than that of the nonporous carbon steel because of the existence of pores. The specific compressive yield strength ($\sigma_{0.2}/\rho$) of as-cast lotus carbon steel is higher than that of as-cast nonporous carbon steel, which is due to the solid solution hardening of nitrogen. Although the microstructure of carbon steel changes from Widmannstätten to homogeneous ferrite and pearlite by normalizing, the yield strength does not change significantly by normalizing. The microstructure of lotus carbon steel changes from Widmannstätten to martensite by quenching so that the yield strength increases significantly.

1. Introduction

Lotus-type porous metals (lotus metals) with cylindrical pores aligned in one direction are attracted much attention because their specific strength does not decrease in the pore growth direction even if the porosity increases [1]. Recently, it became possible to fabricate lotus carbon steel homogeneously with using the continuous casting method in a nitrogen atmosphere [2]. Nitrogen is an important alloying element, which is widely used to improve corrosion resistance and mechanical properties of steels. Therefore, it is expected that the strength of lotus carbon steel increases by solid solution hardening of nitrogen [3]. In addition, it was reported that the mechanical properties of lotus carbon steel, which is fabricated in a hydrogen atmosphere by continuous zone melting method, can be changed by heat treatment [4]. Therefore, the mechanical properties of lotus carbon steel, fabricated in a nitrogen atmosphere by continuous casting method can also be improved.

In this study, lotus carbon steel was fabricated by the continuous casting method in the pressurized mixture gas of nitrogen and argon. The effects of heat treatments on the compressive strength of the lotus carbon steel were investigated.

2. Experimental procedure

Figure 1 shows the schematic illustration of the continuous casting apparatus of a pressurized gas chamber. Low carbon steel (AISI1018, Fe-0.16wt.%C) ingots were melted in the crucible by radio-frequency induction heating in a mixture gas atmosphere (P_{N_2}=0.8MPa, P_{Ar}=1.7MPa). The molten steel was maintained for longer than 600s after the temperature reached about 1873K, and then pulled downwards at the transference velocity of 5mm/min through the mold (10 × 30mm^2) cooled by circulated water. The lotus carbon steel plate was cut using a spark-erosion wire cutting machine (Model LNIW, Sodick Co.) in directions parallel and perpendicular to the transference direction (figure 2). The specific gravities ρ and the porosities of the specimens were evaluated from their weight and volume. The diameter of the pores observed on a cross section perpendicular to the transference direction was measured by the image analyzer (Win Roof, Mitani Co.). The nitrogen concentration of the fabricated lotus carbon steel was analyzed using an oxgen-nitrogen analyzer (Model TC300, LECO Co.). As reference, a nonporous carbon steel was also fabricated in an argon atmosphere (P_{Ar}=2.5MPa).

Specimens for compression tests (7×7×7mm^3) were cut from the plates. The size of a specimen was determined so that the length of a side is larger than 10 times of the average pore diameter (0.45mm). Some of the specimens were compressed without heat treatment. The specimens for heat treatment were heated at 1200K for 3.6ks in an argon atmosphere, and then were cooled either by air-cooling (normalizing) or by water-quenching. Compression tests of the specimens before and after the heat treatments were carried out using a universal test machine (Model 4482, Instron Co. Ltd.). The cross-head speed was set up to be 0.2mm/min and no lubrication was used between the specimen and the spacers. The compression direction was parallel to the transference direction and the engineering stress and the engineering strain were measured. The stress at the plastic strain of 0.2% was defined as compressive yield strength $\sigma_{0.2}$ in the present work.

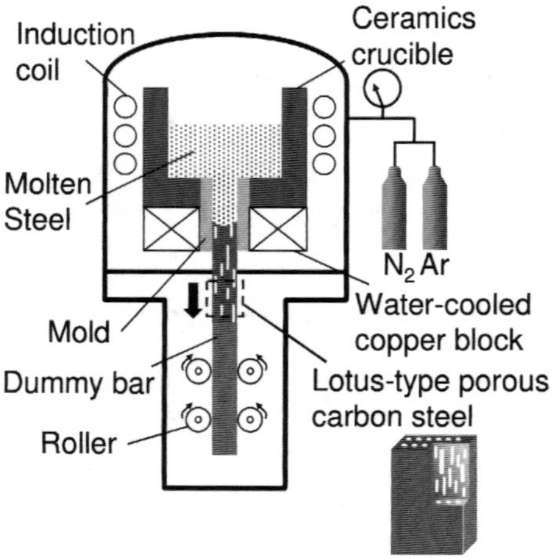

Figure 1. Schematic illustration of the continuous casting apparatus for lotus carbon steel.

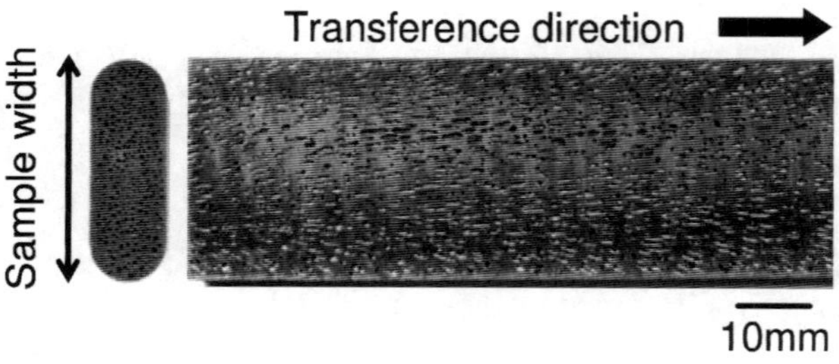

Figure 2. Cross-sections perpendicular (left) and parallel (right) to the transference direction of the lotus carbon steel fabricated at the transference velocity of 5mm/min.

3. Results and discussion

3.1. *Fabricated lotus carbon steel*
Figure 2 shows the cross-sections perpendicular and parallel to the transference direction of the lotus carbon steel fabricated at 5mm/min. It is obvious that long cylindrical pores are aligned in the transference direction. The porosity and the average pore diameter of the lotus carbon steel are about 30% and 453 μ m, respectively.

3.2. *Compressive properties of nonporous and lotus carbon steel*
The compressive stress-strain curves of as-cast nonporous and as-cast lotus carbon steel are shown in

figure 3. The compressive stress of the lotus carbon steel is lower than that of the nonporous carbon steel because of the existence of pores. However, the specific yield strength ($\sigma_{0.2}/\rho$) of the lotus carbon steel, 42.1×10^{-3} MPa·m^3·kg^{-1}, is higher than that of nonporous carbon steel (35.1×10^{-3} MPa·m^3·kg^{-1}). As there is not significant difference in microstructures between the nonporous and lotus carbon steel (Widmannstätten structure [5] shown in Figure 5 (a)), the morphologies of the microstructure does not effect the increase in the strength of lotus carbon steel. The nitrogen concentration in the matrix of the lotus carbon steel is 0.104 wt. %, which is almost equal to the nitrogen concentration which was reported to cause solid solution hardening of the same grade of steel [5]. Therefore, the increase in the stress of the lotus carbon steel may be attributed to the solid solution hardening of nitrogen.

3.3. Influence of heat treatment on the compressive properties

The compressive stress-strain curves of the lotus carbon steel before and after the heat treatments are shown in Figure 4. There is not significant difference in the compressive stress between the normalized lotus carbon steel and the as-cast lotus carbon steel. On the other hand, the stress of the quenched lotus carbon steel is higher than that of the as-cast lotus carbon steel. Figure 6 shows the compressive yield strength of the as-cast nonporous carbon steel and the lotus carbon steel before and after the heat treatments. The yield strength of the normalized lotus carbon steel is almost equal to the as-cast lotus carbon steel although the normalized lotus carbon steel has homogeneous ferrite and pearlite structure (Figure 5 (b)). On the other hand, the strength of the quenched lotus carbon steel is higher than that of the as-cast lotus carbon steel because the quenched lotus carbon steel has martensite structure (Figure 5 (c)). The strength of the quenched lotus carbon steel is higher than that of the as-cast nonporous carbon steel, as shown in figure 6. The specific strength of the quenched lotus carbon steel is 62% higher than that of the as-cast lotus carbon steel. This increase in the specific yield strength is considered to be attributed to the solid solution hardening of nitrogen and the martensitic transformation.

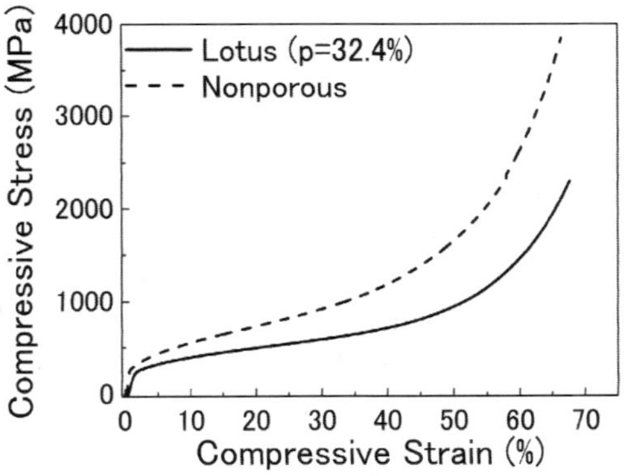

Figure 3. Compressive stress-strain curves of as-cast nonporous and as-cast lotus carbon steel.

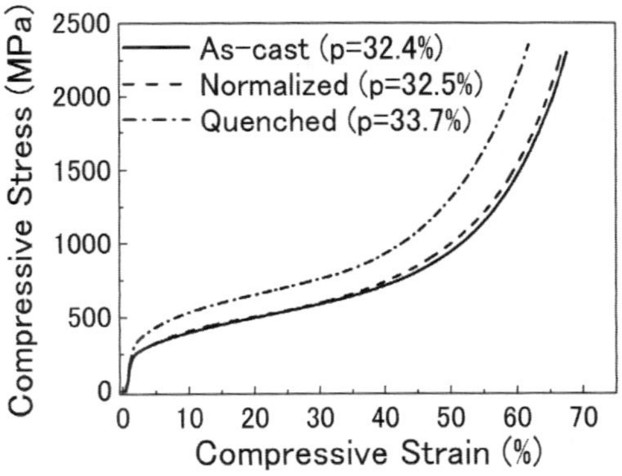

Figure 4. Compressive stress-strain curves of lotus carbon steel before and after the heat treatments.

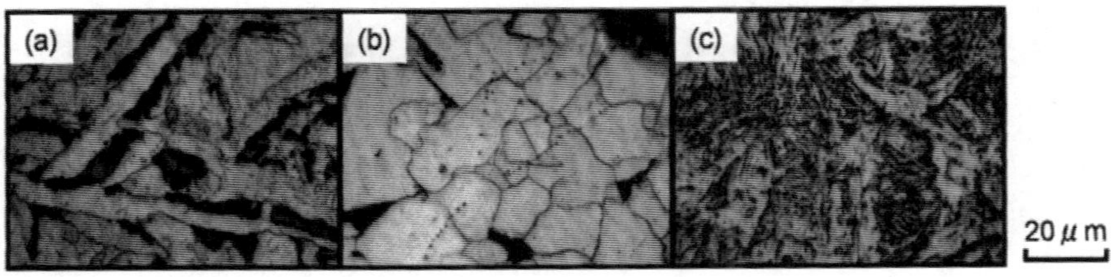

Figure 5. Microstructures of lotus carbon steel: (a) as-cast (Widmannstätten structure), (b) normalized (ferrite and pearlite structure) and (c) quenched (martensite structure).

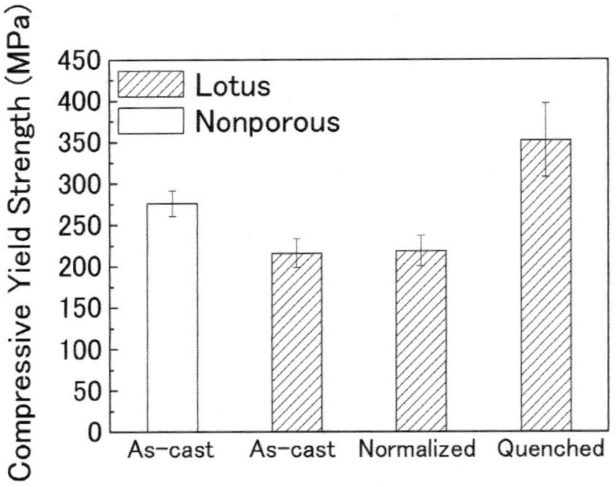

Figure 6. Compressive yield strength of lotus carbon steel before and after the heat treatments and as-cast nonporous carbon steel.

4. Conclusion

- Lotus-type porous carbon steel, whose porosity and average pore diameter are about 30% and 453 μ m , was fabricated in a mixture gas atmosphere (P_{N_2}=0.8MPa, P_{Ar}=1.7MPa).
- The compressive stress of the lotus carbon steel is lower than that of the nonporous carbon steel because of the existence of pores. However, the specific yield strength ($\sigma_{0.2}/\rho$) of the lotus carbon steel is higher than that of nonporous carbon steel by solid solution hardening of nitrogen.
- Although the microstructure of lotus carbon steel changes from Widmannstätten to homogeneous ferrite and pearlite by normalizing, the yield strength does not change significantly. The microstructure of lotus carbon steel changes from Widmannstätten to martensite by quenching, and the yield strength becomes much higher than that of the as-cast lotus carbon steel.

Acknowledgement

The present work was carried out as a part of the program "Advanced Machining System Development Project" supported by New Energy and Industrial Technology Development Organization (NEDO). The present work was also supported by Global COE Program (Project: Centre of Excellence for Advanced Structural and Functional Materials Design) from the Ministry of Education, Culture, Sports, Science and Technology, Japan.

References

[1] Nakajima H 2007 *Prog. Mater. Sci.* **52** 1091
[2] Kashihara M, Yonetani H, Suzuki S, Hyun S K, Kim S Y, Kawamura Y and Nakajima H 2008 *Porous Metals and Metallic Foams (Montreal)* (Pennsylvania:U.S.A/ DEStech) p 201
[3] Kashihara M, Yonetani H, Kobi T, Hyun S K, Suzuki S and Nakajima H 2008 *Tetsu-To-Hagané* **13** 30
[4] Kujime T, Hyun S K and Nakajima H 2006 *Metall. Mater. Trans.* A **37** 393
[5] Rawers J 2000 *Mater. Sci. Eng.* A **276** 91

Structural stability of hollow oxide nanoparticles at high temperatures

R Nakamura and H Nakajima
The Institute of Scientific and Industrial Research, Osaka University, 8-1 Mihogaoka, Ibaraki, Osaka 567-0047, Japan

E-mail: rnakamur@sanken.osaka-u.ac.jp

Abstract. The structural stability of hollow Cu_2O and NiO nanoparticles, which were obtained via oxidation of Cu and Ni nanoparticles in air, was studied by transmission electron microscopy (TEM). Hollow Cu_2O and NiO were observed to have shrunk at 473 and 623 K in annealing under 5.0×10^{-5} Pa, respectively, where the reduction reactions from oxides to metals started. As a result of shrinking associated with reduction, hollow oxides turned into solid metal nanoparticles after annealing at higher temperatures for a long time. On the other hand, hollow oxides shrunk and collapsed through annealing in air at high temperatures. It was found that shrinking of hollow oxides during annealing in air occurs at temperature where the diffusion coefficients of slower diffusing species reach around 10^{-22} m^2s^{-1}. We can conclude that hollow oxide nanoparticles tend to shrink and collapse at high temperatures because hollow structures with extra surface energy are energetically unstable.

1. Introduction

The synthesis and fabrication of nano-scale controlled materials are rapidly developing fields of materials science. Particularly, hollow nanostructures have attracted much attention because their unique shape makes them applicable as delivery vehicles, fillers as well as for catalysis, and could bring about changes in physical and chemical properties [1]. Chemical reaction processes are the most common of the methods to synthesize hollow nanoparticles. However, a different principle to fabricate hollow nanoparticles based on the Kirkendall effect was demonstrated by Yin *et al.* [2]: the initially solid nanocrystals of cobalt turn into hollow spheres of sulfides and oxides when isolated nanocrystals of cobalt are exposed to sulfur and oxygen at relatively low temperatures near 400 K. Thereafter, it was reported by present authors that hollow Zn-[3], Al-, Cu-[4] and Ni-[5] oxides are formed using metal oxidation reactions. Up to now, it has been commonly recognized that the rapid outward diffusion of metal ions from the metal core to the outer oxide shell during oxidation or sulfidation is accompanied by the generation of vacancies and finally a nano-hole is formed in the oxide particles.
According to several theoretical calculations [6,7], hollow nanospheres are thermodynamically unstable; they tend to shrink into solid nanospheres because a solid nanosphere has a lower surface energy than a hollow nanosphere and is therefore more energetically favorable. However, no experimental studies on structural stability of hollow nanoparticles associated with heat treatment have been performed so far. The purpose of the present study is to characterize the structural stability of hollow oxide nanoparticles during reduction and oxidation reactions at high temperatures by transmission electron microscopy (TEM).

2. Experimental procedure

Cu and Ni nanoparticles were evaporated onto an amorphous carbon mounted on a Mo grid or an amorphous Al_2O_3 film mounted on a Pt grid in an electron-beam deposition chamber (the details in [4,5]). Cu and Ni nanoparticles on the substrate were subjected to 423 K and 673 K, respectively, for 3.6 ks in air in an electric furnace and then hollow Cu_2O and NiO nanoparticles were obtained.

The hollow Cu_2O and NiO nanoparticles were annealed at higher temperatures both in vacuum and in air. The isochronal annealing of the hollow Cu_2O and NiO from 423 to 873 K at intervals of 50 K and 1.8 or 3.6 ks in a vacuum was performed in TEM (Hitachi H-800) under 5×10^{-5} Pa and the changes in morphology and structure of them were observed in-situ. On the other hand, the hollow oxide nanoparticles mounted on the substrate were annealed from 523 to 773 K for Cu_2O and from 873 to 923K for NiO for a given time in an electric furnace controlled within ±0.5 K in air. Changes in morphology and crystal structure after oxidation were observed by the TEM.

3. Results and discussion

3.1. Annealing in vacuum

The change in morphology of hollow NiO during isochronal annealing from 573 to 873 K under vacuum in TEM is shown in figure 1. Hollow structures of NiO nanoparticles remain after annealing at 573 K for 1.8 ks, but it can be seen that hollow NiO nanoparticles start to shrink at 623 K for 1.8 ks. The shrinking of nano-holes in hollow NiO proceeds with increasing temperature and all the hollow nanoparticles turn into solid spheres after annealing at 873 K. We observed that the electron diffraction patterns after annealing at 573 K is the same as that before annealing, but the rings corresponding to fcc Ni appeared at 623 K where shrinking of hollow NiO started. In the case of annealing of hollow Cu_2O nanoparticles in vacuum, the behavior of the shrinkage was similar to that of hollow NiO; the shrinking of hollow Cu_2O occurred at 473 K, where the reduction reaction from Cu_2O to Cu started. The starting temperature for shrinking accords with that of the reduction reaction, indicating that shrinking of hollow oxide nanoparticles is closely related to the mass transport during the reduction reaction.

Figures 2 (a) and (b) show TEM images of hollow Cu_2O and NiO nanoparticles after annealing under 5×10^{-5} Pa for 1.8 ks at 573 K and 773 K, respectively. As shown in figure 2 (a), a spherical particle was observed to have formed along the inner surface of hollow particles, as indicated by arrows. A similar behavior can be seen for NiO nanoparticles after annealing at 773 K for 1.8 ks in figure 2 (b). The behavior suggests strongly that the reduced metal atoms are generated at the inner surface of

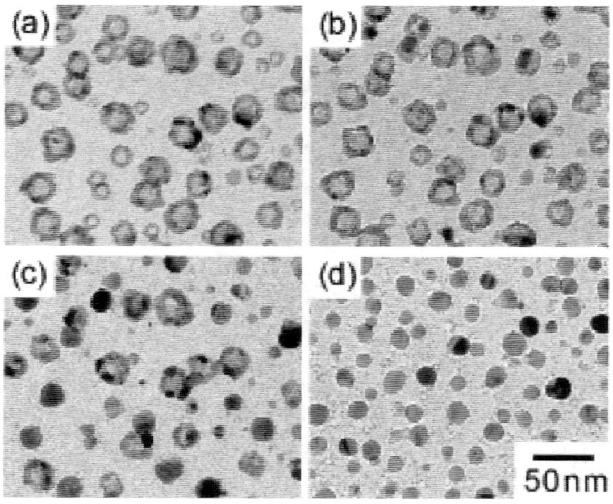

Figure 1. TEM images of hollow NiO nanoparticles during isochronal annealing for 1.8 ks in TEM: (a) before annealing, (b) 573 K, (c) 623 K and (d) 873 K.

hollow oxide and they aggregate via surface diffusion. Formation of metal nanoparticles along the inner surface of hollow oxides during reduction of oxides to metals provides an important clue to consider the mechanism of mass transport in shrinking.

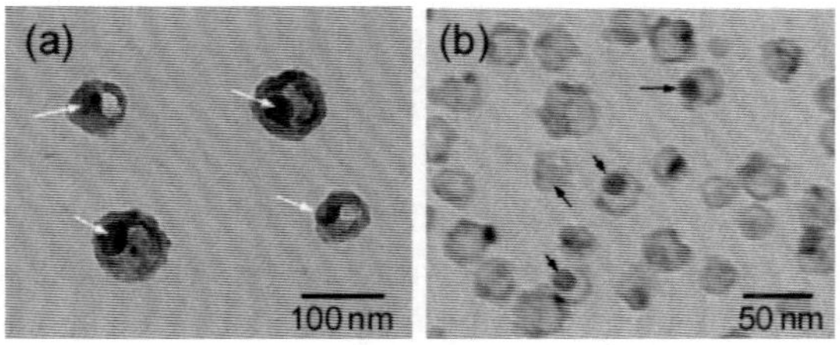

Figure 2. A typical example of changes in morphology of hollow oxide nanoparticles during annealing under 5×10^{-5} Pa: (a) Cu_2O at 573 K for 1.8 ks and (b) NiO at 773 K for 1.8 ks.

According to the kinetic models of reduction of oxides [8], oxygen ions are removed from the lattice of the surface leaving behind an anion vacancy and then reduced zone are formed on the surface at the initial stage of reduction. In general, a dominant process of reduction reaction is considered to be the outward diffusion of oxygen ions via the anion vacancies generated at the surface [9]. Such mass transport results in the inward movement of the metal/reduced oxide interface. This is true of the reduction of hollow oxides; oxygen ions are removed from the outer surface to vacuum. As mentioned earlier, however, the formation of metal nanoparticles along the inner surface of hollow oxides during the reduction of Cu_2O to Cu (figure 2(a)) and NiO to Ni (figure 2(b)), respectively, demonstrates that the reduced Cu and Ni atoms are generated at the inner surface of hollow Cu_2O and NiO. The formation of metal nanoparticles at the inner surface indicates that oxygen ions diffuse from the inner surface migrate to the outer surface and that the reduced zone (i.e. Cu and Ni layers) is formed at the inner surface. After the formation of metal layers, metal atoms aggregate and make a particle in order to fill in the nano-hole. This behavior can be interpreted as the self-assembly of metal atoms which causes the extra energy of the inner surface of a hollow sphere to lower.

3.2. Annealing in air

Figure 3 show BFIs of hollow NiO nanoparticles after annealing in air at (a) 873 K for 36 ks and (b) 923 K for 36 ks. Although the hollow structures were maintained after oxidation at 873 K for 36 ks, almost all hollow particles collapsed into solid spheres at 923 K for 36 ks, as can be seen in figure 3(b). In the case of annealing hollow Cu_2O at high temperatures in air, hollow Cu_2O transforms into CuO at 573 K with the hollow structure being maintained. However, the shrinkage of hollow CuO was observed to begin at 673 K and completely solid CuO was formed after oxidation at 773 K.

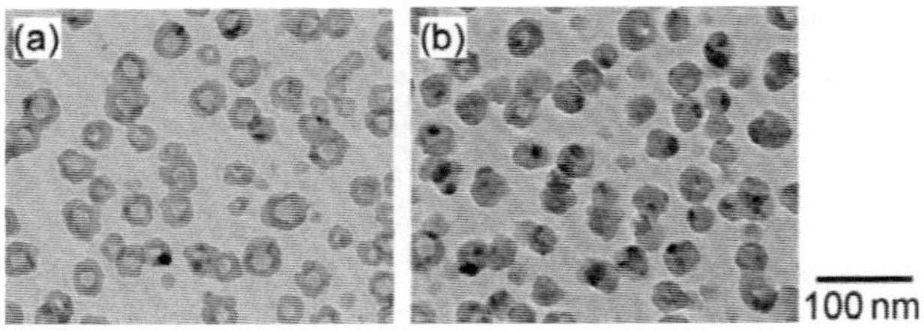

Figure 3. BFIs of hollow NiO nanoparticles after annealing in air (a) at 873 K for 36 ks and (b) at 923 K for 36 ks.

Atkinson and Taylor reported that the activation energy and pre-exponential factor for the self-diffusion of Ni in NiO are 245 $kJmol^{-1}$ and 2.2×10^{-6} m^2s^{-1}, respectively [10]. O'Keeffe and Moore determined those of O in NiO to be 226 $kJmol^{-1}$ and 1×10^{-9} m^2s^{-1}, respectively [11]. At 923 K, where hollow NiO nanoparticles completely collapse, the self-diffusion coefficient of Ni and O in NiO are estimated to be 3×10^{-20} m^2s^{-1} and 1.6×10^{-22} m^2s^{-1}, respectively, based on the Arrhenius equation. On the other hand, the self-diffusion coefficients of Cu [12] and O [13] in CuO at 673 K, where hollow CuO nanoparticles shrink, are 1.3×10^{-22} m^2s^{-1} and 3×10^{-16} m^2s^{-1}, respectively. It is found that the diffusion coefficients of the slower diffusing species, O in NiO and Cu in CuO, at the temperature where shrinking of hollow oxides takes place, are around 10^{-22} m^2s^{-1}. When D is 10^{-22} m^2s^{-1} and t is 3.6 ks, the average diffusion distance, $(Dt)^{1/2}$, is estimated to be of the order of 0.1 to 1 nm, which is smaller than 10 to 20 nm corresponding to the radius of hollow oxide nanoparticles. It seems that, however, the calculated value can be close to several to ten nanometers, though it must be remembered that it was calculated by using the Arrhenius parameters obtained from the diffusion coefficients at high temperatures. Therefore, it is expected that atomic diffusion to fill in the nano-hole with several to ten nanometers in diameter can occur at temperatures where the diffusion coefficient is around 10^{-22} m^2s^{-1}. This result suggests that the idea by Gusak *et al.* [7] that the rate of shrinking is controlled by the slower diffusing species is valid. It should be noted that shrinking of hollow oxides by high-temperature oxidation starts at a temperature where the diffusion coefficient of the slower diffusion species reaches about 10^{-22} m^2s^{-1}.

4. Summary

The shrinking of hollow oxide nanoparticles occurs at a temperature, where the reduction reactions from oxides to metals start in annealing them under 5.0×10^{-5} Pa. Hollow NiO and Cu_2O turn into solid Cu and Ni nanoparticles as a result of the shrinking associated with reduction. It seems that the shrinking of hollow CuO and NiO nanoparticles during annealing in air occurs at temperatures where the diffusion coefficients of slower diffusing ions in the oxides are of the order of 10^{-22} m^2s^{-1}. At high temperatures, where atomic diffusion is active, hollow nanoparticles tend to shrink and collapse because of their unstable structures.

5. Acknowledgement

We would like to thank Prof. H. Mori of Research Center for Ultra-High Voltage Electron Microscopy, Osaka University for our TEM experiment. This work was supported by Priority Assistance for the Formation of World Wide Renowned Centers of Research- The Global COE Program (Project: Center of Excellence for Advanced Structural and Functional Materials Design) and also by Grant-in-Aid for Scientific Research (No. 17106009) from the Ministry of Education, Culture, Sports, Science and Technology, Japan.

References

[1] Xia Y and Halas N J 2005 *MRS Bulletin* **30** 338
[2] Yin Y, Rioux R M, Erdonmez C K, Hughes S, Somorjai G A and Alivisatos A P 2004 *Science* **304** 711
[3] Nakamura R, Lee J G, Tokozakura D, Mori H and Nakajima H 2007 *Mater. Lett.* **61** 1060
[4] Nakamura R, Lee J G, Tokozakura D, Mori H and Nakajima H 2007 *J. Appl. Phys.* **101** 074303
[5] Nakamura R, Lee J G, Mori H and Nakajima H 2008 *Phil. Mag.* **88** 257.
[6] Tu K N and Gösele U 2005 *Appl. Phys. Lett.* **86** 093111
[7] Gusak A M, Zaporozhets T V, Tu K N and Gösele U 2005 *Phil. Mag.* **85** 4445
[8] Kung H H 1989 *Surface Chemistry and Catalysis* (New York: Elsevier)
[9] Sholtz J J and Langell M A 1985 *Surf. Sci.* **164** 543
[10] Atkinson A and Taylor R I 1979 *Phil. Mag.* A **39** 581
[11] O'Keeffe M and Moore W J 1965 *J. Phys. Chem.* **65** 1438
[12] Tu K N, Yeh N C, Park S I and Tsuei C C 1989 *Phys. Rev.* B **39** 304
[13] Rebane J A, Yakovlev N V, Chicherin D S, Tretyakov Y D, Leonyuk L I, Yakunin V G 1997 *J. Mater. Chem.* **7** 2085

Reactive wetting in liquid Cu alloy - carbon and silicon carbide systems

K Nogi, Y Hirata, T Matsumoto and H Fujii
Joining and Welding Research Institute, Osaka University
11-1, Mihogaoka, Ibaraki, Osaka, Japan

E-mail: nogi@jwri. osaka-u. ac.jp

Abstract. Wettability of carbon and SiC by liquid Cu - Si alloys was measured by the sessile drop method and was mainly investigated from a thermo-dynamical view point. Penetration phenomena of liquid Cu – Si alloy into porous carbon was also investigated. It was found that it is possible to expect whether wetting system or non-wetting system from thermo-dynamic calculation.

1. Introduction

It is very important to understand wetting phenomena of liquid metals-ceramics systems in metallurgical, welding and joining fields to obtain high quality and high performance materials. For example, to understand phenomena such as a reaction between liquid alloys and refractory, deoxidation of liquid steel, metal-ceramics joining and a fabrication of metal matrix composites, information on wettability of solid oxide, nitride, carbide and carbon by liquid metals is indispensable. Accordingly, many research articles concerning the wettability of solid ceramics by liquid metals have been published. However, there is a considerable difference in the contact angle between liquid metal on solid ceramics, which is one of an index of wettability.

It is well known that the wettability is affected by many factors such as temperature, partial oxygen pressure in an atmosphere, surface condition of ceramics (surface roughness and surface crystal structure), thermo-dynamical stability of ceramics and metal and alloying element. In Al/Al_2O_3 system, there is a very large difference in contact angle among reported values and Hausner found the difference in contact angle can be attributed to the difference in oxygen partial pressure in an atmosphere [1]. As an example of the effect of surface crystal orientation on wettability, one of the present authors found a characteristic wetting in liquid pure Cu/SiC system [2]. Beta SiC has a hexagonal crystal structure and liquid Cu spreads over on SiC maintaining hexagonal shape. It is well known that thermo-dyanamically active metals such as Ti, Zr and Al also affect the wettability [3]. Small amount of active metals in liquid metals reduce contact angles of liquid metals on ceramics. Furthermore, an increase in temperature usually causes a linear decrease in contact angle. However, when a surface condition such as adsorption and desorption of gas on solid changes with temperature, a change in contact angle reflects the surface condition change. In the liquid metal/diamond system, it was reported that contact angle shows sudden change at the temperature which corresponds to the desorption temperature of hydrogen from diamond surface [4,5].

In the present study, the wettability of carbon and carbide by liquid Cu-Si alloy will be mainly investigated from a view point of thermo-dynamic calculation.

2. Experimental principle

Sessile drop method was employed to measure a contact angle between liquid Cu alloy and carbon and SiC. A schematic illustration of a furnace used is shown in Fig.1. The furnace has some strong points to obtain an accurate contact angle. As graphite heater was used, oxygen potential of an atmosphere can be lowered by the reaction between carbon and O_2 gas in the atmosphere, which yields CO gas. By equipping a sample dropping assembly to the furnace, it is possible to avoid a reaction between liquid Cu alloy and a substrate before starting contact angle measurement. Laser beam irradiation behind liquid Cu alloy drop enables us to obtain a clear profile of the drop. By taking photographs of the drop from 2 directions which is set a right angle, it is possible to confirm a symmetrical shape of liquid drop.

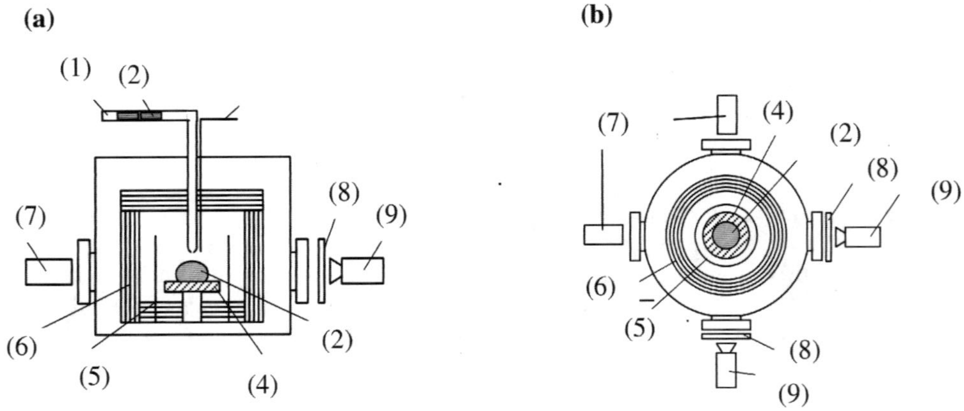

Figure 1 Schematic illustration of a furnace (a) side view, (b) top view
(1) Sample dropping assembly, (2) Cu alloy sample, (3) Thermocouple,
(4) Carbon or SiC substrate, (5) Graphite heater, (6) Mo reflector,
(7) He-Ne laser, (8) Band-pass filter, (9) Digital camera

Two types of glassy carbon substrates whose porosity were 18% and 0% respectively were employed to observe wetting and penetration phenomena. Here after, the Glassy carbon whose porosity is 18% will be called porous carbon and the glassy carbon whose porosity is 0% will be called the dense carbon. SiC substrate was obtained by hot pressing of SiC powder with B_4C as a sintering aid. Alloy samples were made by arc melting of 5N of Cu rod, 6N of Si pieces.

Measurements were conducted at the scheduled temperatures under around 1 X 10^{-3}Pa.

3. Results and Discussion

Figure 2 shows the time dependence of contact angle of liquid Cu - Si alloy on a dense carbon. It can be seen from Fig. 2 that in the case of Cu - 15at.% Si, the contact angle dos not change with time and in the case of Cu – 25at.% Si and Cu – 56at.% Si, the contact angle decreases with a lapse of time and a final contact angle is 40 to 50 degree. The difference of wetting behaviour between liquid Cu - 15 at.% alloy and liquid Cu - 25 at.% and liquid Cu - 56 at.% alloys can be attributed to the difference in a reactivity of liquid Cu - Si alloy with carbon.

It can be expected that SiC is produced at the interface between liquid Cu - Si alloy and carbon by a reaction expressed by equation 1. The standard Gibbs energy of the reaction (1) can be described by equation (2) [6].

Si (in liquid Cu) + C = SiC (1)

$\Delta G^0 = 27100 - 2.73T \log T + 18.1T$ (cal/mol) (2)

where ΔG^0 is the standard Gibbs energy of the reaction and T is the absolute temperature.

It can be calculated from equation (2) that the Si concentration which is required to produce SiC at the interface between liquid Cu - Si alloy and carbon is 18.3 at.% at 1300℃. As Cu - 15at.%Si alloy is smaller than a calculated value of 18.3 at.%, no reaction occurs at the interface and contact angle does not change with time. On the other hand, as liquid Cu - 25 at.% and liquid Cu - 56 at.% alloys is larger than the calculated value of 18.3 at.%, the reaction occurs and SiC is formed at the interface.

Formation of SiC was investigated by XRD at a cross section of Cu - Si alloy and carbon interface. In the case of Cu - 15at.% Si, no SiC was detected and in the Cu – 25at.% Si and Cu – 56at.% Si, on the other hand, SiC was detected as shown in Figure 3.

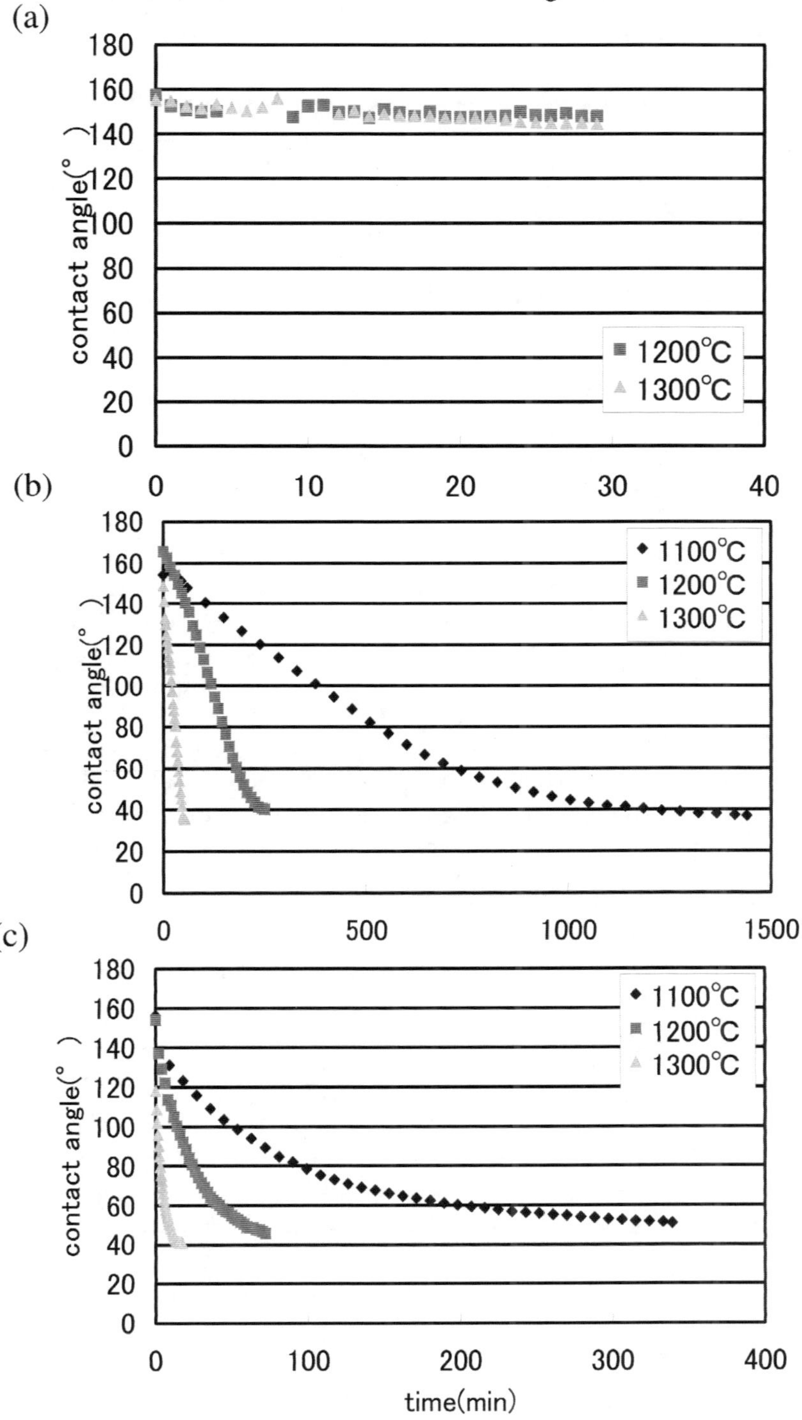

Fig. 2 Change in contact angle of liquid Cu-Si alloy on carbon with time

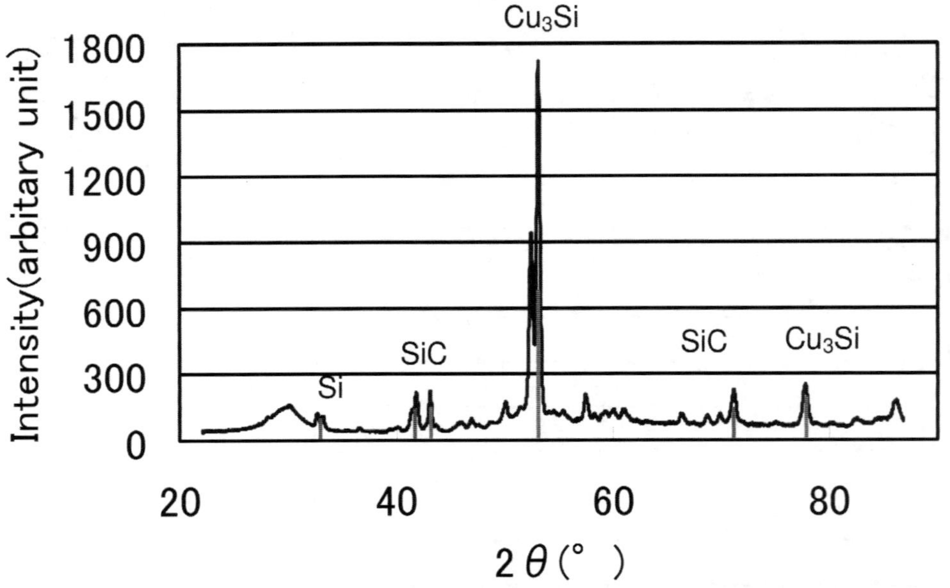

Fig. 3 XRD pattern at the interface between solidified Cu-25at%Si and dense carbon

Figure 4 shows contact angles of liquid Cu - 25 at.% and liquid Cu - 56 at.% alloys on SiC. It is interesting to compare Figure 4 with Figure 3. It can be clarified by comparing Figure 4 with Figure 3 that final contact angles on SiC are almost the same as that on carbon and a time dependence of decreasing in contact angle on SiC is larger than that on C. These results might suggest that reaction between Si and carbon is not a momentary one.

Figure 5 shows change in a shape of liquid Cu - 56at.% alloy drop on porous carbon substrate. It can be seen from Figure 5 that apparent contact angle on porous carbon decreases gradually with time and liquid Cu alloy absolutely penetrates in the porous carbon after 22 minutes. When a viscosity of liquid Cu - Si alloy is low enough to penetrate at high temperature, the results suggest that a low rate of penetration is caused by a low rate of reaction between Si and carbon.

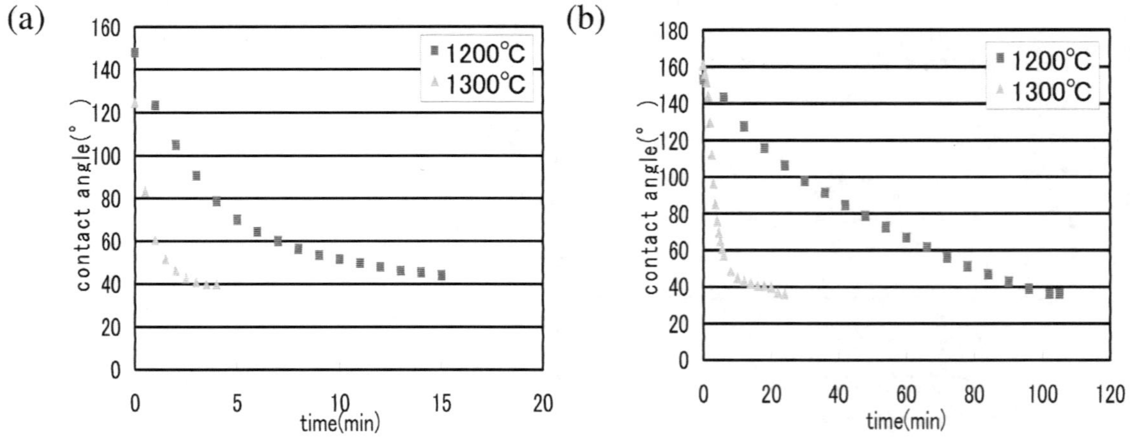

Fig. 4 Change in contact angle of liquid Cu-Si alloy on SiC with time

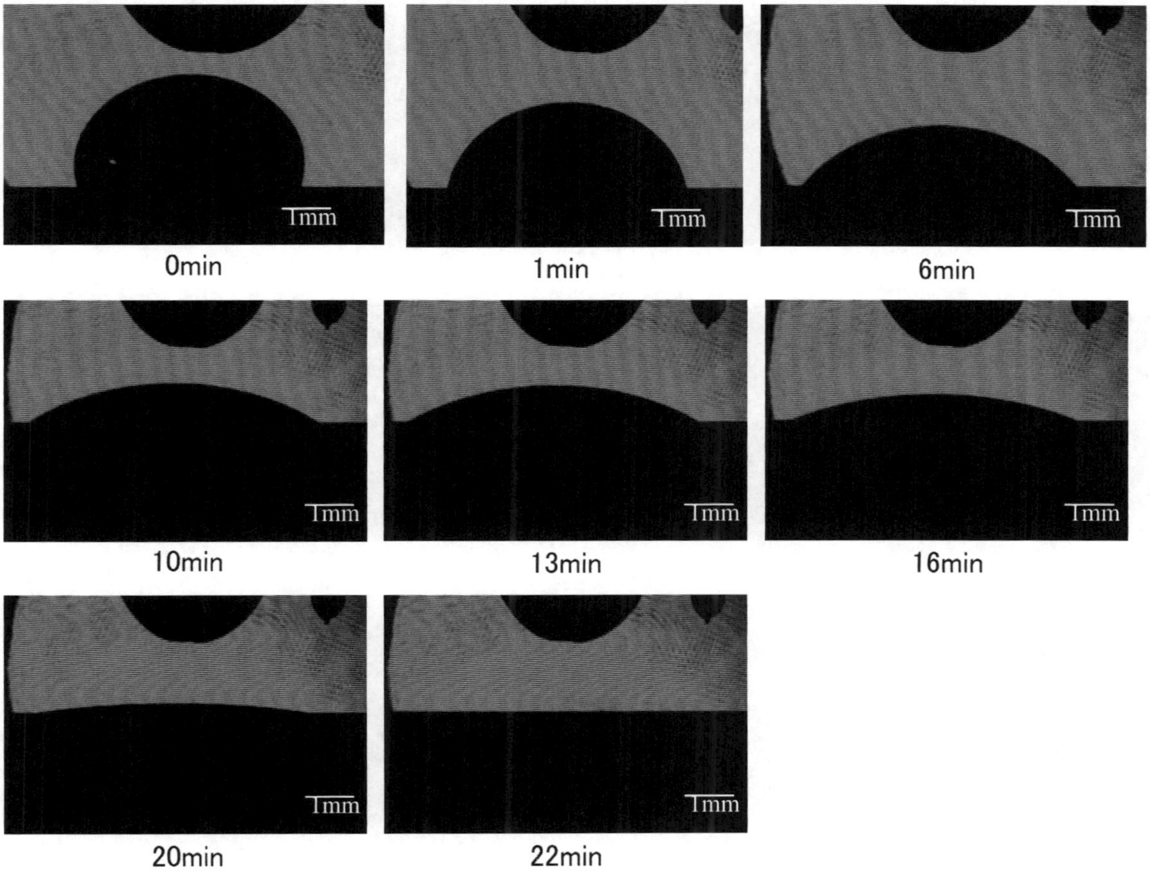

Fig. 5 Change in a shape of liquid Cu-56at.%Si alloy on a porous carbon substrate

4. Conclusion

Wettability of dense and porous carbons by liquid Cu - Si alloy and penetration phenomena of liquid Cu - Si alloy were investigated and obtained the following results.

(1) When Si content exceeds 18.3 at.% in Cu, liquid Cu - Si alloy wets carbon substrate and SiC is formed at the interface between liquid Cu and carbon in the experimental condition.

(2) Penetration of liquid Cu - Si alloy into porous carbon can be observed when liquid Cu - Si alloy wets carbon substrate

(3) It can be concluded from time dependence of contact angle and observation of penetration behavior that reaction rate of Si and carbon is not fast.

Acknowledgement

The research was supported by a Grant-in-Aid for Scientific Research on Priority Area (A) and Priority Assistance for the Formation of Worldwide Renowned Centres of Research – The Gloval COE Program (Project: Centre of Excellence for Advanced Structural and Functional Materials Design) from the Ministry of Education, Culture, Science and Technology (MEXT), Japan.

References

[1] Hausner H, J. Mat. sci. Let. **5** 549 (1986)
[2] Nogi K, Ikeda K, Shimada S, Ogino K, Trans. JIM **52** 663 (1988)

[3] For example, Nogi K, Ikeda K, Ogino K, Trans. JIM **54** 1401 (1990)
[4] Nogi K, Okada Y, Ogino K, Trans. JIM **57** 63 (1993)
[5] Nogi K, Nishikawa H, Fujii H, Hara S, Acta Materiaria **46** 2305 (1998)
[6] Metallurigical Thermo-chemistry, ed. by Kubaschewski O, Alcock C B Pergamon press Oxford New York Toronto Sydney Paris Frankfurt 383 (1979)

A novel metal-to-metal bonding process through in-situ formation of Ag nanoparticles using Ag_2O microparticles

Akio Hirose[1], Hiroaki Tatsumi[1], Naoya Takeda[1], Yusuke Akada[1], Tomo Ogura[1], Eiichi Ide[2] and Toshiaki Morita[2]

[1] Division of Materials and Manufacturing Science, Graduate School of Engineering, Osaka University, 2-1 Yamadaoka, Suita, Osaka 565-0871, Japan
[2]Materials Research Laboratory, Hitachi, Ltd., Hitachi, Ibaraki 319-1292, Japan

hirose@mapse.eng.osaka-u.ac.jp

Abstract. The metal-to-metal bonding has been successfully achieved via the bonding process using Ag metallo-organic nanoparticles at a bonding temperature of around 300°C, which can be alternative to the current microsoldering in electronics assembly using high-temperature solders. However, further reduction of bonding temperature and/or bonding pressure is needed. In the present research, a novel bonding process through in-situ formation of Ag nanoparticles instead of the filler material of the Ag metallo-organic nanoparticles has been developed. The Ag nanoparticles can form by the reduction of Ag_2O particles. In this study, the Ag_2O particles were mixed with triethylene glycol as a reducing agent to form a paste for bonding. The Au coated cylindrical specimens were bonded using the paste. The Ag nanoparticles formed at around 130°C to 160°C through the reduction process of Ag_2O particles with triethylene glycol. The Ag nanoparticles were immediately sintered each other due to a great surface energy per volume. A transmission electron microscope observation revealed that the sintered Ag metallurgically bonded to the Au substrate at around 160°C, and a dense Ag layer formed after further heating. The tensile strength of the joint bonded at 250°C under a bonding pressure of 5MPa was around 60MPa

1. Introduction

A recent rise of working temperature of silicon devices for the use of automotive is driving the substitution of a new bonding process for conventional Pb-containing solders. Additionally, conventional solders are inapplicable to wide bandgap semiconductor devices such as silicon carbide because of a higher operation temperature than silicon semiconductors. There is a strong drive to develop a novel bonding technology for these high temperature electronic applications.

The authors have proposed a novel bonding process using Ag nanoparticles as a bonding material through the viewpoint of the rapid sintering property of nanopartilces at a relatively lower temperature [1, 2]. The Ag nanoparticle is composed of a core Ag nanoparticle and an organic shell protecting from self-consolidating. Rapid sintering of Ag nanoparticles occurs after the release of the organic elements from a bonding layer. The metal-to-metal bonding has been successfully achieved via the bonding process using Ag metallo-organic nanoparticles at a bonding temperature of around 300°C under an applied pressure of 2.5 to 5MPa. However, further reduction of bonding temperature and/or bonding pressure is needed. In the bonding process using Ag metallo-organic nanoparticles the decomposition and removal of the organic matter adhering to the nanoparticles are necessary to bring out the low-temperature sintering property of the nanoparticles. This is a bottleneck of the bonding process. In practical applications, a productive cost of Ag nanoparticles is problem as well.

For above problems, we focused on the Ag nanoparticles generated through the reduction reaction of low-cost silver oxide (Ag_2O) microparticles. In this paper, we discussed the in-situ formation of Ag nanoparticles through the reduction reaction of Ag_2O microparticles and the interfacial bonding mechanism between Ag layer with sintered Ag nanoparticles and Au substrate.

The Ag_2O microparticles were processed to a paste with triethylene glycol (TEG). We analyzed the thermal characteristics of Ag_2O paste and observed the particles heated up to various temperatures using a field emission scanning electron microscope (FE-SEM) and a transmission electron microscope (TEM). We evaluated the tensile strengths of joints bonded with Ag_2O paste and observed an interfacial structure of the joints.

2. Experimental procedure

Figure 1 shows the SEM image of Ag_2O microparticles. The Ag_2O microparticles were processed to a paste with TEG as a reduction agent. We carried out a differential scanning calorimeter (DSC) and a thermogravimetry (TG) in air at the heating rate of 10 ℃/min to investigate the thermal characteristics of the reduction reaction of Ag_2O microparticles with TEG. We analyzed and observed the decomposition process of Ag_2O microparticles and the generation process of Ag nanoparticles using X-ray diffraction (XRD), FE-SEM and TEM. The observed samples were prepared by heating Ag_2O paste to each temperature and quenched with DSC.

Figure 2 shows the shape of a Au/Cu specimen for a bonding test. The thickness of Au layer is about 0.7μm with Ni buffer layer (approximately 2μm in thickness) on Cu specimen. We performed a tensile test of Cu/Au-to-Cu/Au joints using Ag_2O paste bonded at temperature ranging from 140℃ to 400℃ with a bonding pressure of 5MPa and quenched. We observed the interfacial structures of the Cu/Au-to-Cu/Au joints bonded at 160℃ and 400℃ using HR-TEM.

Fig. 1 SEM images of Ag_2O powder.

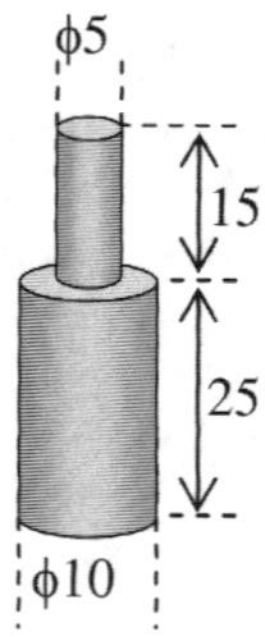

Fig. 2 Shape of joint.

3. Results and discussion

3.1. In-situ formation of Ag nanoparticles from Ag_2O paste

Figure 3 shows the DSC and TG traces upon heating of the Ag_2O paste. An exothermic peak was detected at around 150℃ in the DSC trace, and simultaneously a great weight reduction occurred in the TG trace. These results suggest that a reduction reaction of Ag_2O take place at this temperature range. The XRD analysis was, therefore, applied to the Ag_2O paste heated to various temperatures ranging from 110℃ to 160℃ as shown in Fig. 4. The strength of the peaks from Ag_2O crystal decreased with increasing temperature and disappeared at 160°C. Reversely, the peaks from Ag crystal appeared at 130°C, and became stronger with temperature rise. Therefore, it can be concluded that the reduction of Ag_2O into Ag by TEG starts at 130°Cand finishes at 160°C in the present case.

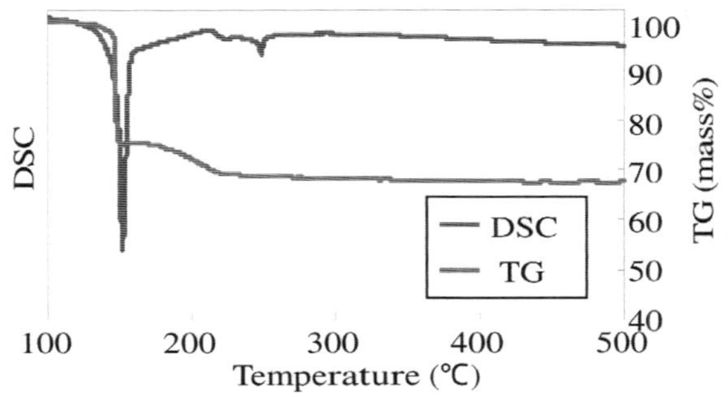

Fig. 3 DSC and TG traces of Ag_2O paste with a heating rate of 10°Cmim.

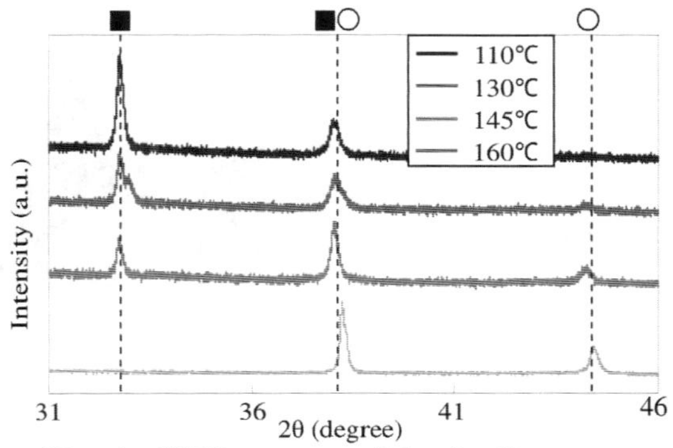

Fig. 4 XRD patterns of the Ag_2O paste heated up to around peak temperatures of the DSC curve with a heating rate of 10°C/min.

Figure 5 shows TEM images of the Ag_2O paste heated to 130°C and quenched. It was confirmed from the selected are diffraction patterns (SADP) that Ag nanoparticles formed in the outer region of the Ag_2O particles. Such formed Ag nanoparticles were immediately sintered each other and became larger in size as shown in Fig. 6 presenting TEM images of the Ag_2O paste heated to 145°C. After heated to 160°C, the Ag_2O particles were completely reduced to Ag particles, which were several hundred nm in size as shown in Fig. 7. Thus, in-situ formation of Ag nanoparticles, which were sintered rapidly, through the reduction reaction of Ag_2O paste at a temperature ranging from 130°C to 160°C was confirmed. It can be utilized to a low-temperature bonding process using nanoparticles.

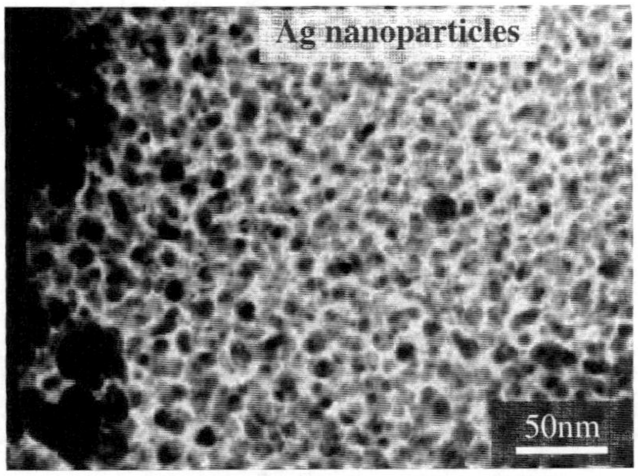

Fig. 5 TEM image of a cross-section of the Ag_2O paste heated up to 130°C using DSC with a heating rate of 10°C/min.

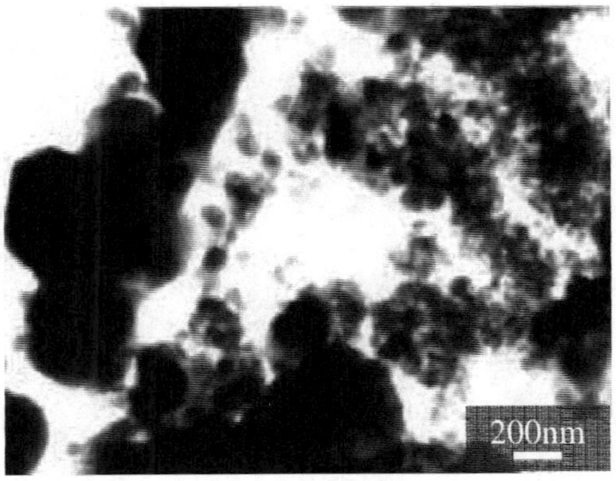

Fig. 6 TEM image of a cross-section of the Ag_2O paste heated up to 145°C using DSC with a heating rate of 10°C/min.

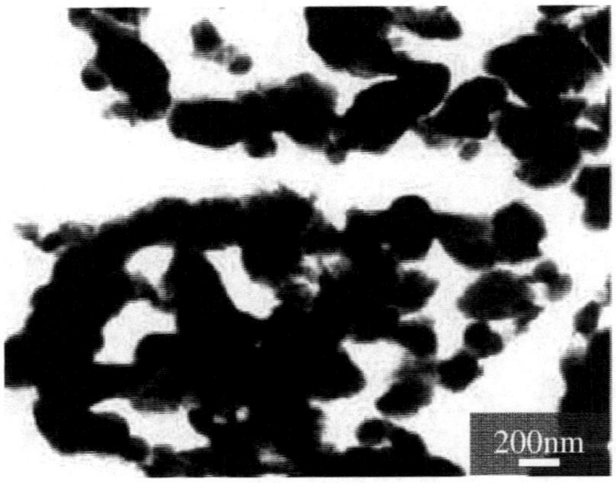

Fig. 7 TEM image of a cross-section of the Ag_2O paste heated up to 160°C using DSC with a heating rate of 10°C/min.

3.2. Relationship between joint strength and bonding temperature

The tensile strength of the Cu/Au-to-Cu/Au joints using Ag_2O paste was plotted against bonding temperature in Figure. 8(a). The joint strength monotonously increased with bonding temperature. A certain joint strength was attained at a bonding temperature above 160°C. The detailed interfacial microstructure of the joint bonded at 160°C was observed using TEM as shown in Fig. 9(a). At this bonding temperature, Ag_2O still remained in the bonding layer as indicated in SADP in Fig. 9(b) and (c). This is caused by a higher heating rate of 60°C/min in the bonding test compared with the DSC analyses in Fig. 3. However, Ag nanoparticles also formed at the outer layer of the Ag_2O particle. The TEM image of Fig. 9(a) clearly shows that such formed Ag nanoparticles connected to the Au substrate. This indicates an active nature of the nanoparticles resulting from their high surface energy.

Figure 10 shows the TEM image of the interface of the joint bonded at 400°C. At this bonding temperature, fully sintered dense Ag layer was observed and the Ag layer successfully bonded to the Ag substrate. Thus, the metal-to-metal joint through the sintered Ag layer can be attained using the bonding process with Ag_2O paste. In this bonding experiment, the bonding specimen was heated to a bonding temperature followed by quenching without holding. The holding at a bonding temperature is thought to promote the sintering of the Ag nanoparticles, and thereby improve the joint strength. The strength of the joint bonded at 250°C is plotted against the holding time in Fig. 8(b). The joint strength was improved with holding time. The joint strength of approximately 60MPa was obtained after 5min holding. This joint strength is considered to be sufficient for an electronics assembly compared with conventional soldering using a Pb-rich high temperature sonlder [3].

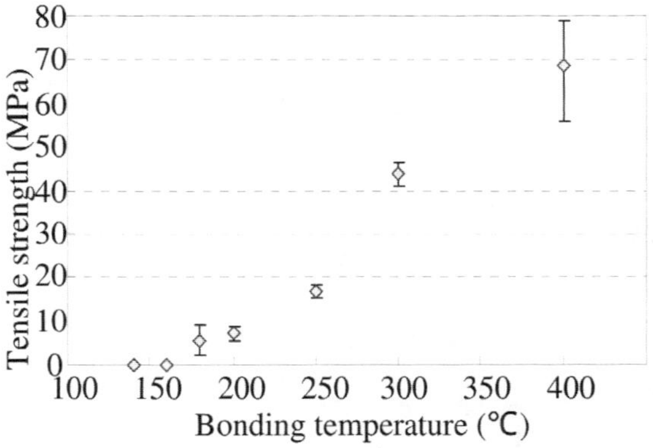

(a) Bonding temperature vs. tensile strength.

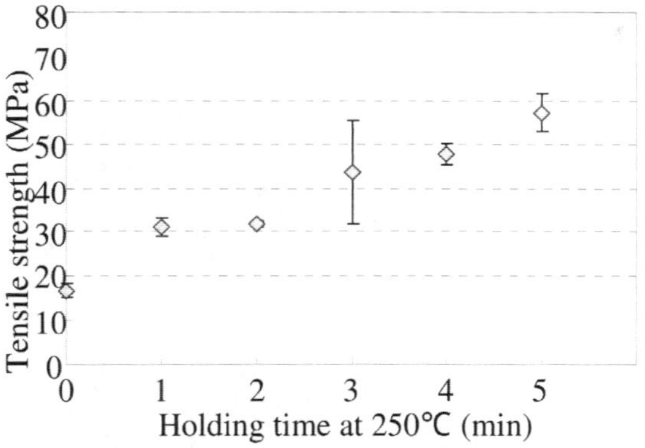

(b) Holding time at 250℃ vs. tensile strength.

Fig. 8 Relationship between bonding parameters and tensile strength of the Cu/Ni/Au-to-Cu/Ni/Au joint bonded using the Ag_2O paste at various bonding temperatures with a pressure of 5MPa.

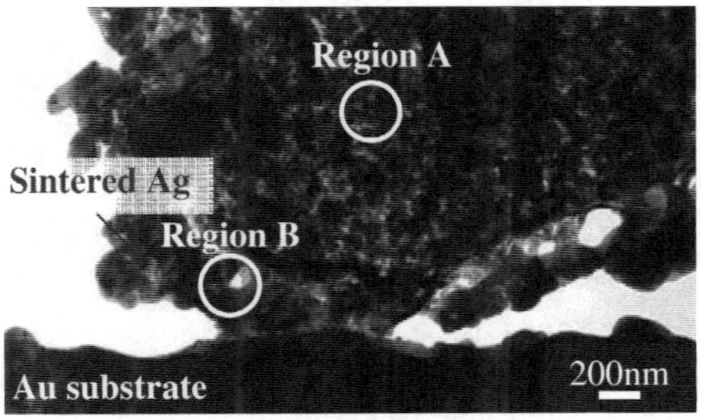

(a) Bright field image.

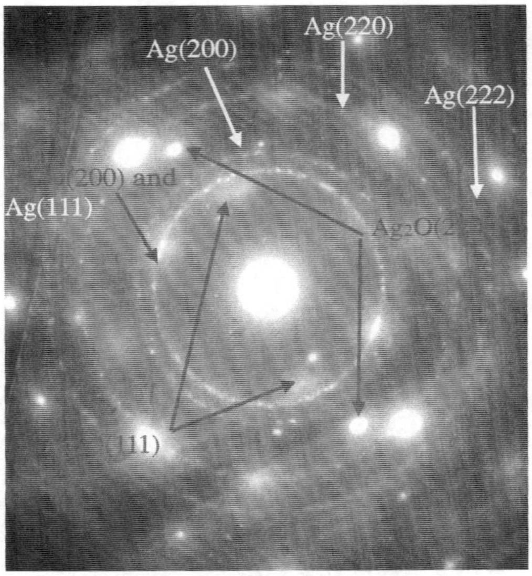

(b) Diffraction pattern of region A.

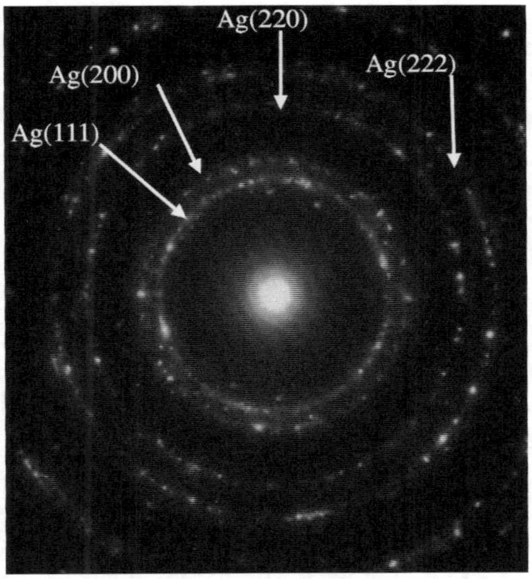

(c) Diffraction pattern of region B.

Fig. 9 TEM image and diffraction patterns of the Ag/Au interface bonded using the Ag_2O paste at 160℃ with a pressure of 5MPa and quenched.

Fig. 10 TEM image of the Ag/Au interface bonded using the Ag_2O paste at 400℃ with a pressure of 5MPa and quenched.

4. Conclusions

In the present study, a new metal-to-metal bonding process through in-situ formation of Ag nanoparticles by reduction of Ag_2O particles. The main results obtained are as follows.

1) Ag_2O particles were reduced by triethylene glycol and Ag nanoparticles formed at temperature ranging from130℃ to 160℃.
   Such formed Ag nanoparticles were sintered rapidly.
2) A certain joint strength was attained at a bonding temperature above 160℃ with a bonding pressure of 5MPa. The joint strength increased with bonding temperature.
3) Interfacial bonding was achieved by the Ag nanoparticles reduced from Ag_2O particles at 160℃. A dense bonded layer formed by sintering of the Ag nanoparticles at 400℃.
4) The tensile strength of the joint bonded at 250℃ with a holding time of 5min under a bonding pressure of 5MPa was around 60MPa.

Acknowledgement

This study was supported by Priority Assistance for the Formation of Worldwide Renowned Centers of Research - The Global COE Program (Project: Center of Excellence for Advanced Structural and Functional Materials Design) from the Ministry of Education, Culture, Sports, Science and Technology (MEXT), Japan.

References

[1] E. Ide, S. Angata, A. Hirose, and K. F. Kobayashi, *Acta Mater.*, **53**, 2385 (2005).
[2] Y. Akada, H. Tatsumi, T. Yamaguchi, A. Hirose, T. Morita and E. Ide, *Mater. Trans.*, **49**, 1537 (2008).
[3] E. Ide, A. Hirose and K. F. Kobayashi, *Mater. Trans.*, **47**, 211 (2006)

Amorphization and subsequent crystallization in $Zr_{66.7}Ni_{33.3}$ alloy under MeV electron irradiation

T Nagase[1,2], K Takizawa[2], M Nakamura[2], H Mori[1] and Y Umakoshi[3]
[1] Research Center for Ultra-High Voltage Electron Microscopy, Osaka University, 7-1, Mihogaoka, Ibaraki, 567-0047, Japan.
[2] Division of Materials and Manufacturing Science, Graduate School of Engineering, Osaka University, 2-1 Yamadaoka, Suita, 565-0871, Japan.
[3] National Institute for Materials Science, 1-2-1, Sengen, Tsukuba, 305-0047, Japan.
E-mail: t-nagase @uhvem.osaka-u.ac.jp

Abstract. MeV electron irradiation introduces Frenkel pairs in a crystalline phase and free volume in a glassy phase, resulting in solid-state amorphization of metallic crystals and crystallization of metallic glasses, respectively, in some alloy systems. C16-Zr_2Ni intermetallic compound transformed to a f.c.c.-type nano-crystalline phase through an amorphous state under electron irradiation. Namely, a metastable nano-crystalline structure was formed by a Crystal-to-Amorphous-to-Crystal (C-A-C) transition in the C16-Zr_2Ni thermal equilibrium phase. An overlap of amorphization and crystallization was observed at 298K, while such an overlap could not be observed at 103K. This behavior can be explained by the difference in the temperature dependence of electron dose required for amorphization and crystallization.

1. Introduction

MeV electron irradiation introduces Frenkel pairs in a crystalline phase and free volume in a glassy phase. Figure 1 shows a schematic illustration of Frenkel pair and free volume formation by the electron knock-on effect. It is reported that in some alloy systems, solid-state amorphization of metallic crystals [1-3] and crystallization of metallic glasses [4-6] occurs by the accumulation of atomic defects under the irradiation. Various unfavourable factors such as temperature rise, contamination, oxidation, change in chemical composition and cascade effect can be neglected in the case of electron irradiation, while these factors must be taken into account in other mechanical processes such as mechanical milling, severe plastic deformation, shot peening, ion and neutron irradiation. Furthermore, in-situ observation during MeV electron irradiation by a high voltage electron microscope (HVEM) makes it possible to carry out a real-time analysis of changes in microstructure. HVEM offers therefore a unique opportunity to investigate the phase stability of a crystalline phase against excess Frenkel pairs and of a glass phase against excess free volume.

Recently unique disordering-ordering phase transitions in metallic alloys were induced by mechanical milling and electron irradiation techniques: namely, cyclic crystalline-amorphous (Cyclic-CA) transformation during mechanical milling [7,8], electron irradiation induced crystal-to-amorphous-to-crystal (C-A-C) transitions [9,10], and electron irradiation induced quasicrystal-to-amorphous-to-crystal (Q-A-C) transition [11]. In the present study, a C-A-C transition in $Zr_{66.7}Ni_{33.3}$ alloy was investigated by high voltage electron microscopy, with emphasis placed on the temperature dependence of the transition.

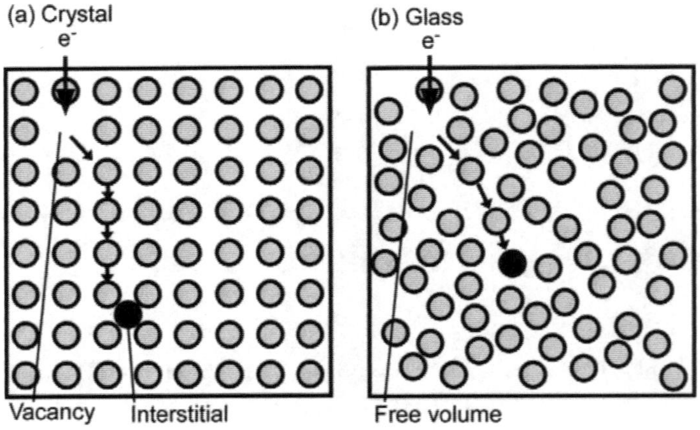

Figure 1. Schematic illustration of Frenkel pair introduction in a crystal (a) and free volume introduction in a glass phase (b) by the electron knock-on effect.

2. Experimental procedures

A master ingot of $Zr_{66.7}Ni_{33.3}$ alloy was prepared by arc melting of raw materials of pure Zr (purity, >99.9 at.%) and Ni (purity, >99.9 at.%) metals in a purified Ar atmosphere. Rapidly quenched ribbons with thickness between 0.02mm and 0.04mm were produced from the ingot at a rotation speed of 42 ms^{-1} by a single roller melt-spinning method in the same atmosphere. Thermal properties of the melt-spun amorphous phase were evaluated by differential scanning calorimetry (DSC) measurements in Ar atmosphere. C16-Zr_2Ni intermetallic compound was obtained by isothermal annealing at 1273 K for $1.8x10^3$ s in Ar atmosphere and vacuum. Thin foils for electron irradiation were prepared from the ribbons by twin-jet polishing in a solution of 30 % nitric acid and 70 % methanol kept at about 243 K. Electron irradiation was performed with the H-3000 ultrahigh voltage electron microscope (UHVEM) at Osaka University. The acceleration voltage was 2.0 MV and the dose rate was in the range between $3.3x10^{24}$ and $9.2x10^{25}$ $m^{-2}s^{-1}$. Changes in the bright field (BF) images and selected area diffraction (SAD) patterns during electron irradiation were monitored in situ by UHVEM.

3. Results

3.1. Thermal crystallization and electron irradiation induced crystallization of amorphous phase

Figure 2(a) shows a DSC curve of melt-spun specimens measured at a heating rate of 0.67 Ks^{-1}. An exothermic peak corresponding to the thermal crystallization of the melt-spun amorphous phase is seen at T_x. The temperature of T_x is evaluated to be 673K. Fig. 2(b) shows isothermal DSC curves of melt-spun amorphous specimens. Isothermal annealing in a DSC furnace began just after the rapid heating of the alloy sample up to the required temperature at a constant heating rate of 0.83 Ks^{-1}. Time-temperature transformation (TTT) diagram of thermal crystallization of melt-spun amorphous can be constructed based upon isothermal DSC measurements shown in Fig.2 (b). The TTT diagram obtained is shown in Fig. 2 (c). The onset time of an exothermic peak increases with decreasing annealing temperature. The solid and dotted lines, corresponding to onset and offset time of thermal crystallization, respectively, can be fitted to the following equation [12],

$$\tau = \tau_0 \cdot \exp(-E/RT) \qquad (1)$$

where τ is the onset time or offset time, E is the activation energy, R is the gas constant and T is temperature, respectively. Based on equation (1), the onset time of thermal crystallization at 103K and 298K is estimated to be of the order of 10^{128} and 10^{30} s, respectively. It seems then reasonable to consider that the metastable amorphous phase maintains the glassy state stably and devitrification of this phase through thermal crystallization can be ignored at room temperature.

Figure 3 shows a typical example of electron irradiation induced crystallization of the amorphous phase in a $Zr_{66.7}Ni_{33.3}$ alloy [13]. Changes in BF images and corresponding SAD patterns of the melt-spun amorphous phase in a $Zr_{66.7}Ni_{33.3}$ alloy during 2.0 MV electron irradiation at 298K, are illustrated. Before irradiation (a), the amorphous phase shows only featureless contrast over the entire BF image and only broad halo rings in SAD pattern. Grains with nano-granular contrast appear in the BF image and the Debye rings appear together with decrease in the intensity of the halo ring after irradiation for 600s (b). It is evident that the amorphous phase is not stable under electron irradiation at 298K in spite of the high phase stability against thermal crystallization, resulting in the formation of a nanocrystalline structure. The phase selection of irradiation induced crystallization was investigated in the previous study; not the thermal equilibrium C16-Zr_2Ni but a metastable f.c.c.-type nanocrystalline phase appeared from the amorphous phase under electron irradiation [13].

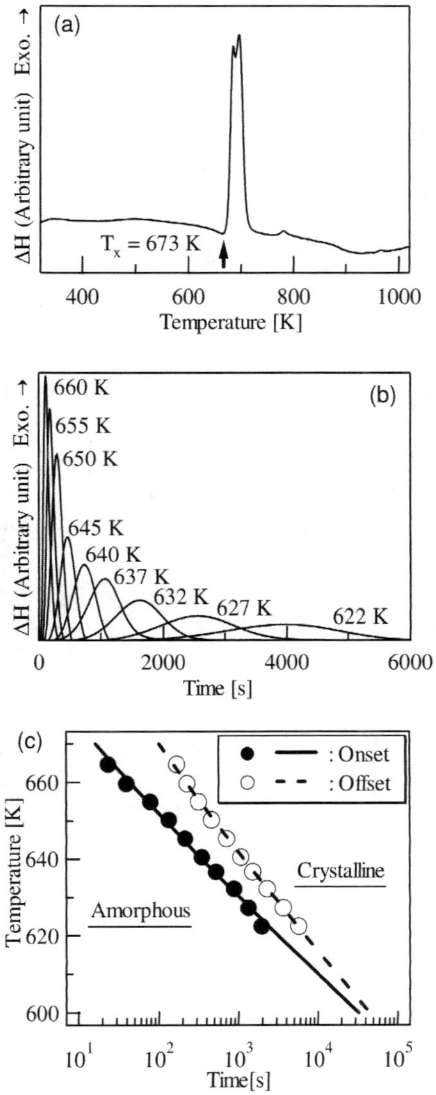

Figure 2. Conventional DSC analysis at a heating rate of 0.67 Ks^{-1} (a), isothermal DSC measurements (b) and TTT-diagram of thermal crystallization of the melt-spun amorphous phase (c) in $Zr_{66.7}Ni_{33.3}$ alloy.

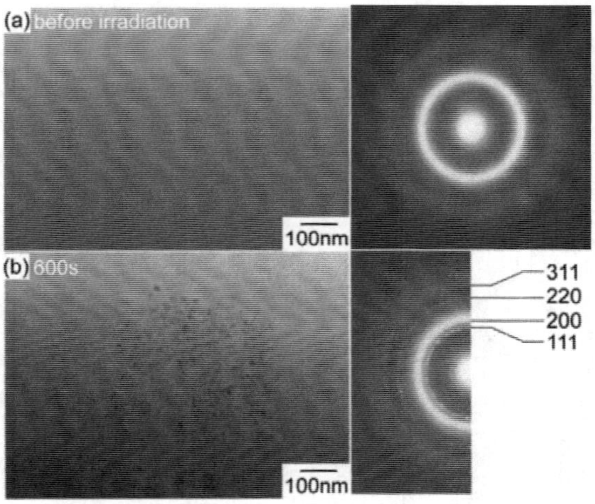

Figure 3. Typical example of electron irradiation induced crystallization under MeV electron irradiation. Changes in BF images and SAD patterns of melt-spun amorphous phase under 2.0MV electron irradiation at 298K are illustrated. (a) before irradiation, (b) after irradiation for 600s. The dose rate is $9.2 \times 10^{24}\ m^{-2}s^{-1}$.

3.2. Electron irradiation induced C-A-C transition

Figure 4 shows changes in BF images and corresponding SAD patterns of C16-Zr_2Ni intermetallic compound during 2.0MV electron irradiation at 103K. In situ TEM observation was performed by UHVEM at 2.0 MV. The specimen before electron irradiation shows bend contours in the BF image and electron diffraction spots in the SAD pattern (a). After 30s electron irradiation (b), the contrast of bend contours disappeared at the central area of the irradiation in the

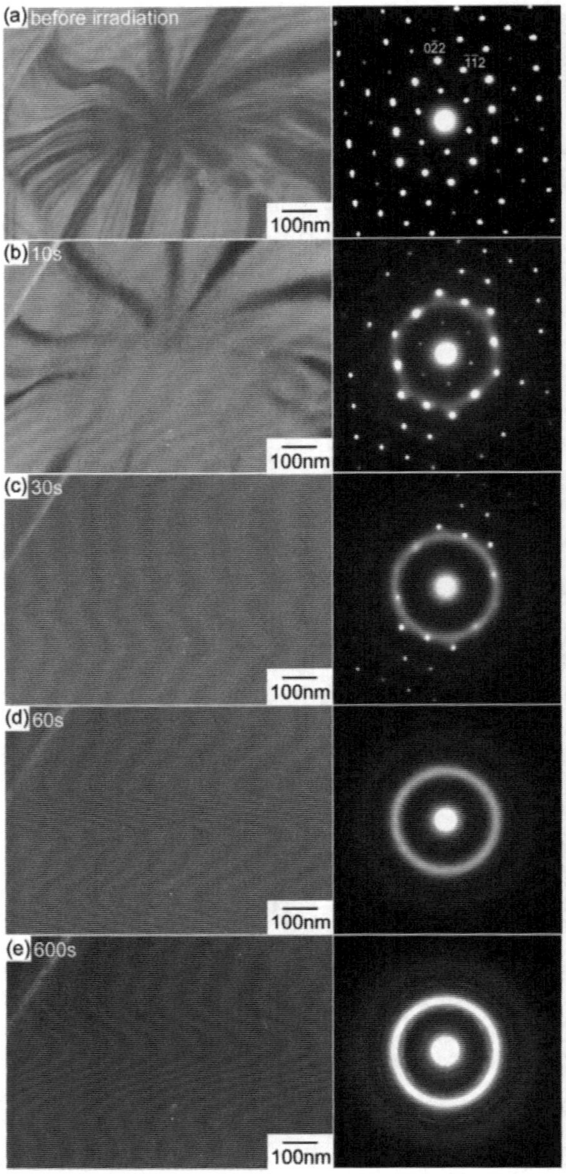

Figure 4. Changes in BF images and SAD patterns of C16-Zr_2Ni intermetallic compound under 2.0MV electron irradiation at 103K. The dose rate is $4.0x10^{24}$ $m^{-2}s^{-1}$.

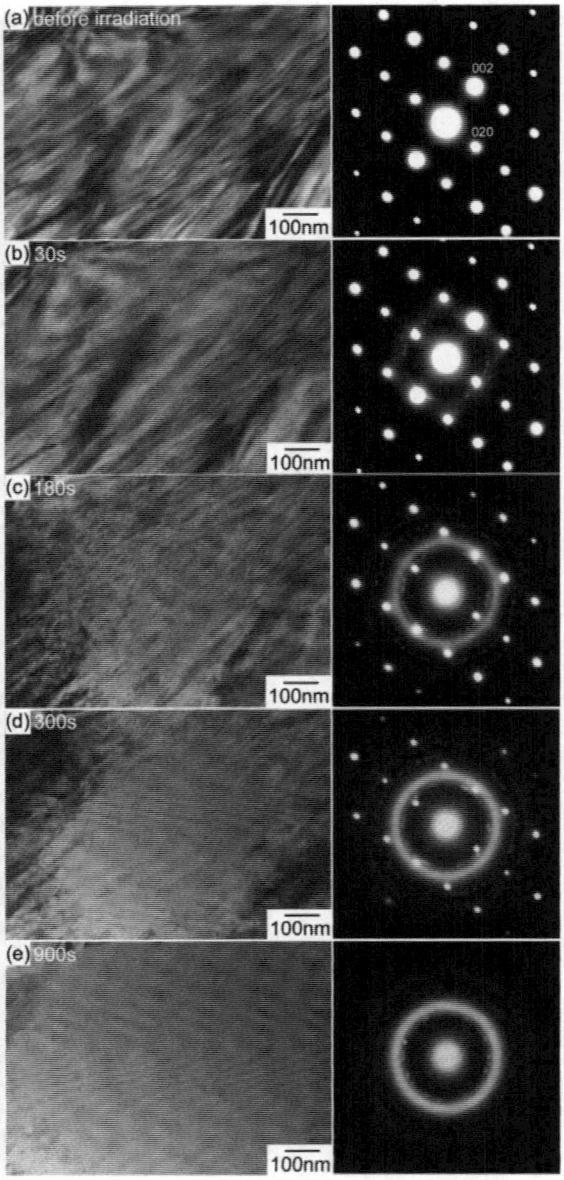

Figure 5. Changes in BF images and SAD patterns of C16-Zr_2Ni intermetallic compound under 2.0MV electron irradiation at 298K. The dose rate is $3.3x10^{24}$ $m^{-2}s^{-1}$.

BF image. Concurrently, halo rings corresponding to an amorphous phase appeared in the SAD pattern. Further electron irradiation brought about an increase in the intensity of the halo rings at the expense of that of the diffraction spots. In the specimen irradiated for 60s (d), the crystalline contrast (i.e., contrast of bend contours) vanished in the BF image. In the SAD pattern, electron diffraction spots vanished and only halo rings can be seen. The phase stability of intermetallic compound under electron irradiation is not high enough to maintain the original structure against electron irradiation, and an amorphous single phase was obtained through a crystal-to-amorphous (C-A) transition. In the specimen irradiated for 600s (e), grains with dark and bright granular contrast, with about 10 nm in size, appeared at the central part of the irradiated area in the BF image. In the SAD pattern, Debye rings appeared in addition to the halo rings. Nano-crystalline grains were formed through electron irradiation induced crystallization of an amorphous phase that had been obtained through the C-A transition by the irradiation. Namely, an electron irradiation induced C-A-C transition was observed in C16-Zr_2Ni intermetallic compound. The nanocrystalline grains obtained by the C-A-C transition are considered to be a f.c.c.-type phase which is similar to that obtained by electron irradiation induced crystallization of melt-spun amorphous phase in $Zr_{66.7}Ni_{33.3}$ alloy.

Figure 5 shows the C-A-C transition in C16-Zr_2Ni intermetallic compound at 298K. Nanocrystalline f.c.c. phase was formed under electron irradiation. One can notice that the C-A-C transition behavior at 298K is different from that at 103K. An amorphous single state can not be observed during the C-A-C transition. After electron irradiation for 30s (b), not only halo rings but also Debye rings appeared in the SAD pattern. Further electron irradiation brought about an increase in the intensities of halo rings and Debye rings at the expense of that of the diffraction spots. In the specimen irradiated for 180s (c), not featureless contrast typical of an amorphous phase but nano granular contrast was observed in the BF image. After irradiation for 900s (e), both electron diffraction spots in the SAD pattern and typical crystalline contrast (i.e., bend contours) in the BF image vanished. Instead, nano granular contrast can be seen in the BF image, and a mixture of halo rings and Debye rings corresponding to the f.c.c.-type phase can be observed in the SAD pattern. The amorphous-to-crystal (A-C) transition in this amorphous phase seems to be overlapped with the C-A transition of the original C16-Zr_2Ni intermetallic compound. The authors believe such an overlap behavior of C-A and A-C transitions during the C-A-C transition in intermetallic compounds under MeV electron irradiation has been observed here for the first time.

4. Discussion

An overlap behavior was observed at 298K, while such a behavior could not be observed at 103K during electron irradiation induced C-A-C transition in C16-Zr_2Ni intermetallic compound. It is well known that the temperature dependence of electron irradiation induced crystallization shows the opposite tendency to that of amorphization [14]. Higher the irradiation temperature is, smaller the total dose required to induce the crystallization is. In contrast, the total dose for amorphization increases with increasing irradiation temperature. The overlap effect can be explained by the difference in the temperature dependence of required electron dose between amorphization and crystallization. Figure 6 shows a schematic illustration of the temperature dependence of the doses in a $Zr_{66.7}Ni_{33.3}$ alloy. At temperature indicated by A in Fig. 6, a C-A transition occurs under electron irradiation. At this temperature, the incubation time of the crystallization of an amorphous phase to the f.c.c.-type phase is long enough for amorphization to complete. Then, the crystallization starts after complete amorphization of the intermetallic compound. This case corresponds to the present experimental result at 103K. The incubation time of crystallization decreases with increasing irradiation temperature, while that of amorphization increases with increasing temperature. At temperature indicated by B, the amorphous phase obtained by the C-A transition of the intermetallic compound can immediately transform to the f.c.c.-type phase. The crystallization starts before complete amorphization under electron irradiation. Overlap effect during C-A-C transition at 298K in the present study may correspond to the condition indicated by B in Fig. 6.

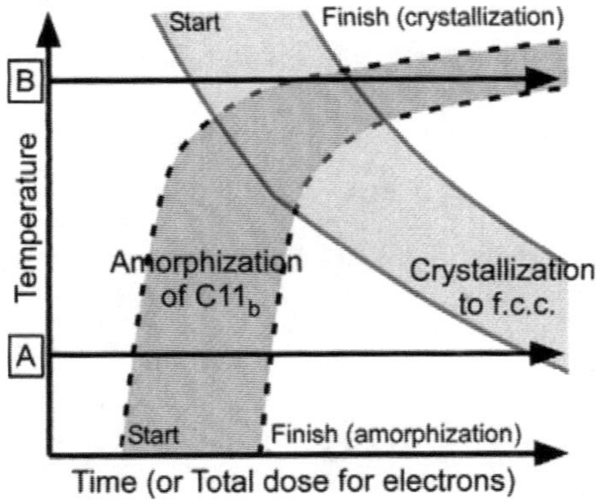

Figure 6. Schematic illustration of temperature dependence of electron dose required for amorphization of $C11_b$-Zr_2Ni and for crystallization of amorphous phase under MeV electron irradiation.

5. Conclusion

(1) Overlap of amorphization (C-A) and crystallization (A-C) was observed in the C-A-C transition of C16-Zr_2Ni intermetallic compound under 2.0MeV electron irradiation.
(2) The overlap behavior can be explained by the difference in the temperature dependence of electron dose required for amorphization and crystallization.

Acknowledgments

The authors are grateful to Dr. T. Sakata for operating the H-3000 UHVEM. This study was supported by Priority Assistance for the Formation of Worldwide Renowned Centers of Research - The Global COE Program (Project: Center of Excellence for Advanced Structural and Functional Materials Design) from the Ministry of Education, Culture, Sports, Science and Technology (MEXT), Japan.

References

[1] Thomas G, Mori H, Fujita H and Sinclair R 1982 *Scripta Mater.* **16** 589–592
[2] Mori H, Fujita H 1982 *Jpn. J. Appl. Phys.* **21** L494–496
[3] Mogro-Camprero A, Hall E L, Walter J L and Ratkowski A J 1982 *in Metastable Materials Formation by Ion Implantation*, eds. S T Picraux and W J Choyke (North-Holland: Amsterdam) pp 203–208
[4] Doi M and Imura T 1980 *J. Mater. Sci.* **15** 2867–2874
[5] Brimhall J L 1984 *J. Mater. Sci.* **19** 1818-1826
[6] Nagase T, Umakoshi Y and Sumida N 2002 *Mater. Sci. Eng.* A. **323** 218–225
[7] El-Eskandarany M S, Aoki K, Sumiyama K and Suzuki K 1997 *Appl. Phys. Lett.* **70** 1679–1681
[8] El-Eskandarany M S, Aoki K, Sumiyama K and Suzuki K 1997 *Scripta Mater.* **36** 1001–1009
[9] Nagase T and Umakoshi Y 2003 *Scripta Mater.* **48** 1237–1242
[10] Nagase T, Nino A, Hosokawa T and Umakoshi Y 2007 *Mater. Trans,* **48** 1651–1658
[11] Nagase T, Hosokawa T and Umakoshi Y 2006 *Intermetallics.* **14** 1027–1032
[12] Shingu P H 1988 *Mater. Sci. Eng.* **97** 137–141
[13] Nagase T, Nakamura M and Umakoshi Y 2007 *Intermetallics.* **15** 211–224
[14] Okamoto P R, Lam N Q and Rehn L E 1999 *Physics of crystal-to-glass transformations, in Solid State Physics* vol 52, eds H Ehrenreich and F Spaepen (Academic Press: San Diego) pp 1–135.
[15] Nagase T and Umakoshi Y 2003 *J. Appl. Phys.* **93** 912–918

Novel applications of ionic liquids in materials processing

Ramana G Reddy

ACIPCO Professor and Head, Department of Metallurgical and Materials Engineering, P.O. Box 870202, A129 Bevill Building, Tuscaloosa, AL 35487-0202, U.S.A.

rreddy@eng.ua.edu

Abstract. Ionic liquids are mixtures of organic and inorganic salts which are liquids at room temperature. Several potential applications of ionic liquids in the field of materials processing are electrowinning and electrodeposition of metals and alloys, electrolysis of active metals at low temperature, liquid-liquid extraction of metals. Results using 1-butyl-3-methylimidazolium chloride with $AlCl_3$ at low temperatures yielded high purity aluminium deposits (>99.9% pure) and current efficiencies >98%. Titanium and aluminium were co-deposited with/without the addition of $TiCl_4$ with up to 27 wt% Ti in the deposit with current efficiencies in the range of 78-85 %. Certain ionic liquids are potential replacements for thermal oils and molten salts as heat transfer fluids in solar energy applications due to high thermal stability, very low corrosivity and substantial sensible heat retentivity. The calculated storage densities for several chloride and fluoride ionic liquids are in the range of 160-210 MJ/m^3. A 3-D mathematical model was developed to simulate the large scale electrowinning of aluminium. Since ionic liquids processing results in their low energy consumption, low pollutant emissions many more materials processing applications are expected in future.

1. Introduction

Ionic liquids are mixtures of organic and inorganic salts and at room temperature they are in liquid state. They are comprised entirely of cations and anions and hence the name ionic liquids. A range of ionic liquids can be synthesized by mixing various salts containing different anions and cations. It is possible to obtain ionic liquids with a variety of physical and chemical properties for specific applications. Many room temperature ionic liquids based on imidazolium cations have been extensively explored in the recent years. Ionic liquids have a very low melting point, normally below room temperature. They have emerged as the next generation solvents which are capable of replacing traditionally used volatile organic compounds as industrial solvents. They have several unique properties such as wide electrochemical potential window for electrochemical processing, high chemical and thermal stability, very low vapor pressure, non-inflammability, high ionic conductivity, high solvating capability, and very low corrosivity. They are environmentally benign with no organic emissions. These characteristics make ionic liquids excellent candidates for a range of potential applications, including batteries, catalysis, supercapacitors, photovoltaics, solvents, and heat-transfer fluids. Due to their unique properties ionic liquids can be used as electrolytes in the extraction and refining of metals such as aluminum, titanium, copper, magnesium, at ambient temperature that are otherwise refined at high temperatures [1-3].

Ionic liquids are synthesized using the standard procedure [4]. Most popular and extensively studied ionic liquids are imidazolium cation based salts. Ionic liquids are characterized using nuclear magnetic resonance (NMR) spectrometer, differential thermal analysis (DTA) and differential Scanning Calorimeter (DSC). The properties of these liquids such as thermal stability, heat capacities, and thermodynamic properties are measured [5, 6]. We present here the results obtained in our research group on the electrochemical processing of aluminum, titanium and Al-Ti alloy, thermal storage and heat-transfer capabilities of some ionic liquids, and a 3-D mathematical model to simulate the electrowinning of aluminum in a batch reactor.

2. Electrochemical Processing of Metals and Alloys using Ionic Liquids

Recent advancements in the area of ionic liquids for electrochemical extraction and refining of metals had proved that electrolysis in ionic liquids electrolytes is the future direction of energy-efficient, environmentally benign metal extraction/ refining technology. Physicochemical properties of ionic liquids are critical for considering them as alternatives to organic solvents or molten salts. However, not all the physical properties are important for the metals extraction applications. Liquid temperature range, thermal stability, and electrolytic conductivity have the highest priority in determining the operating temperature range and electrical parameters. One of the most important and attractive features of ionic liquids is the extended temperature range of liquid state. The wide liquidus temperature range is manifested in their low melting points and high thermal decomposition temperatures that have been reported [7, 8]. Liquid state of the ionic liquids at or below room temperature is a unique feature that facilitates several chemical reactions that are otherwise impossible at these temperatures using molten salts. We will now look at the use of ionic liquids in the electrowinning and electrorefining of aluminum and titanium and also electrodeposition of Ti-Al alloys.

2.1. Extraction of Aluminum

The use of ionic liquids in the electrowinning of aluminum has several advantages such as low consumption of energy and electrode materials, no pollutant emissions, and low operating costs. The electrochemical processing of aluminum in ionic liquids media saves up to 85% of electrical energy, prevents any gaseous pollutant emission (CO, CF_4, etc.) and avoids any solid waste (e.g., aluminum dross and spent potliner) [7]. Electrowinning of aluminum from a solution containing 1-butyl-3-methyl imidazolium chloride and anhydrous $AlCl_3$ at ambient temperature were reported [3]. Experiments were performed using an electrolytic cell consisting of a graphite anode and copper cathode. The electrochemical reactions for the process are as follows:

$$\text{Anodic: } AlCl_3 + 7AlCl_4^- \rightarrow 4Al_2Cl_7^- + 1.5Cl_2 + 3e^-$$

$$\text{Cathodic: } 4Al_2Cl_7^- + 3e^- \rightarrow Al + 7AlCl_4^-$$

$$\text{Overall reaction: } AlCl_3 \rightarrow Al + 1.5Cl_2$$

During the electrolysis, chlorine was released on the anode and aluminum was deposited on the cathode. After the electrolysis, the anode and cathode were taken out from the cell, washed with water, and weight changes were determined. The morphologies of the deposited aluminum were dependant on experimental conditions. Dense deposition of aluminum on the cathode was obtained which was characterized by XRD.

The cathode current density and efficiency change with temperature, cell voltage, and molar ratio of $AlCl_3$ and C_4mimCl. The cathode efficiencies of greater than 99 pct were obtained with a cell voltage of 3.1 V, temperature of 105 °C and molar ratio of $AlCl_3$ to C_4mimCl above 1.7. The experimental conditions such as cell voltage and energy consumption are much lower than those of the current industrial method. Aluminum electrodeposition in an analogous system that comprised of 1-hexyl-3-methyl imidazolium chloride and anhydrous $AlCl_3$ was studied [9, 10].

Figure 1. Thickness of aluminum deposition on copper cathode with change of current density (a) Current density: 460A/m^2, 105°C; (b) Current density: 460A/m^2, 130°C (c) Current density: 160A/m^2, 105°C; (d) Current density: 160A/m^2, 130°C [2].

Process parameters influencing current density, current efficiency and morphology of aluminium deposition were investigated. Aluminum was electrowon from C_6mimCl-$AlCl_3$ melts with molar ratio from 1:1 to 1:2 at temperature range of 90 ° to 140 °C and with cell voltages from 2.5 to 3.4 V. It is established that the current density and current efficiency increase with temperature and applied cell voltage. The thickness of deposited aluminium increased with increase in current density and temperature, as shown in figure 1. The optimum electrowinning parameters under laboratory conditions were determined to be 110 °C, 3.5 V and electrolyte molar ratio (C_6mimCl-$AlCl_3$) of 1:1.7. Large scale aluminium electrowinning experiments in ionic liquid electrolyte were carried out in a batch recirculation type of electrolytic cell [11]. Recycling of aluminum scrap and refining aluminium alloys was successfully carried out using $AlCl_3$-C_4minCl and $AlCl_3$-C_6mimCl ionic liquids [12-14].

2.2 Modeling of the Electrowinning Process

A 3-D mathematical model was developed for the batch reactor of low temperature aluminium electrowinning using ionic liquid electrolytes [15]. This model solves the Poisson equation which describes the potential distribution within the bulk electrolyte. The electrical double layer at the electrode caused due to the charge of each species is also taken into consideration while solving the Poisson equation in order to calculate the cathode current density and current efficiency.

The model describes the deposition process taking into account the mass transport of participating ionic specie, homogeneous chemical reactions within the diffusion layer, and the associated electrochemical kinetics. Processing parameters for the optimal reactor performance have been assessed in this research. Aluminum electrowinning experiments were conducted in a batch recirculation cell using C_6mimCl-$AlCl_3$ as electrolyte with $AlCl_3$ concentration of 5ml/L, graphite as anode and copper plate as cathode. The conditions used in the cell were: temperature: 80 °C, electrolyte flow rate between 5 to 20 ml/min, and an applied cell voltage of 3 to 3.5 V. The results are shown in figure 2 where the model results are in good agreement with the experimental data.

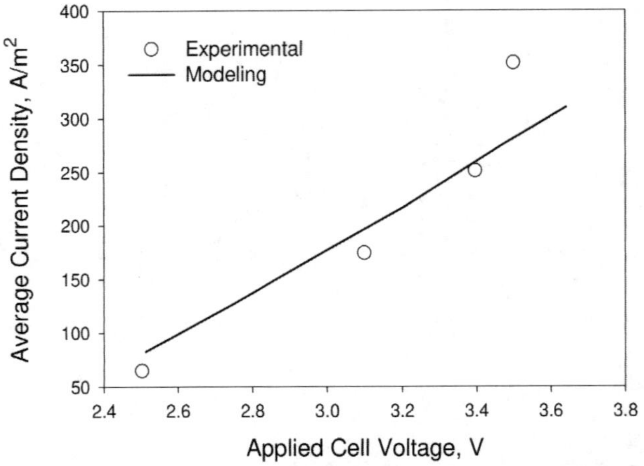

Figure 2. Effect of cell voltage on the current densities for large scale aluminium electrowinning. Comparison of modelling results with the experimental data [15].

2.3 Electrochemical Processing of Titanium-Aluminum Alloys

The electrochemical production of Ti and Ti-Al alloys is complicated due to the variable oxidation states of titanium (Ti^{+2}, Ti^{+3}, and Ti^{+4}). Electrochemical processing of Ti and Ti-Al alloys in ionic liquids was investigated by several authors [16-20]. In our research group we have studied the electrodeposition of Ti-Al alloy from the $AlCl_3$-1-butyl-3-methyl imidazolium chloride(BmimCl) mixture in the temperature range of 70 °C to 125 °C [21].

The bulk electrodeposition of Ti-Al alloy was carried out at different applied potentials in the range of 1.5 – 3.0 V with an electrolyte having a molar ratio of $AlCl_3$-BmimCl of 2:1 in the temperature range of 70 °C to 125 °C.

The cathodic reactions are:

$$[\mathrm{Ti}(\mathrm{Al_2Cl_7})_4]^{2-} + 2e^- = \mathrm{Ti(Cathode)} + 4(\mathrm{Al_2Cl_7})^-$$

$$4\mathrm{Al_2Cl_7^-} + 3e^- = \mathrm{Al(Cathode)} + 7\mathrm{AlCl_4^-}$$

The anodic reactions are:

$$\mathrm{Ti(Anode)} + 4(\mathrm{Al_2Cl_7})^- = [\mathrm{Ti}(\mathrm{Al_2Cl_7})_4]^- + 2e^-$$

$$AlCl_3 + 7AlCl_4^- = 4Al_2Cl_7^- + 1.5Cl_2 + 3e^-$$

The effect of applied cell voltage on the cathodic current density, current efficiency and deposit morphology of Ti-Al alloy was investigated in the range of 1.5-3.0V at constant experimental temperature of 100°C. The average compositions of the deposits obtained during the experiments as a function of applied voltage were in the range of 17-27 wt% Ti and 72-82 wt% Al. The cathodic current density increased linearly with applied voltage in the range of 36-187 A/m^2 [21].

The current densities varied linearly with applied cell voltage. Increase in cell voltage also enhances the cathodic reduction reaction, hence increase the current density. Current efficiencies did not vary much as a function of applied cell voltage and were within the range of 78-85%. Figure 3 shows the SEM micrographs of the deposits obtained at 100 °C with applied cell voltages of 1.5V and 3.0V. At low voltage (1.5V) a uniform spherical microstructure of the deposit was observed and the deposit tends to preferentially agglomerate at various sites with increase in applied cell voltage. A uniform spherical microstructure of Ti-Al particles in a very narrow size range between 5-20 μm was obtained. A similar trend was observed with the deposits with increase in temperature at a fixed applied cell voltage of 2.5V. At low temperatures the microstructure had even distribution of uniform spherical particles with a narrow size range between (50-70 μm). Aluminum and titanium were co-deposited and the compositional analysis confirms the formation of Al_3Ti. The deposit tends to agglomerate with increase in temperature and applied voltage.

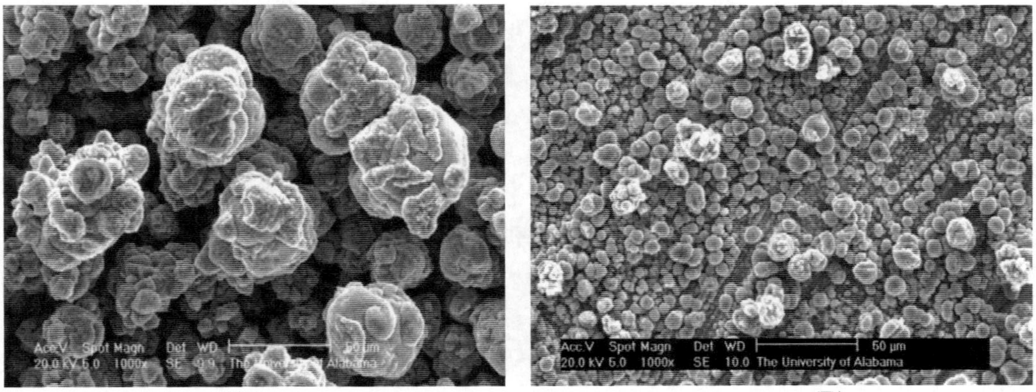

(A) (B)

Figure 3. SEM micrographs of the Ti-Al deposits obtained at 100 °C (A) 3.0V, (B) 1.5V [21]

The feasibility of obtaining a Ti-Al alloy deposit by electrochemical method using an $AlCl_3$-BmimCl electrolyte (molar ratio 2:1) at low temperatures (75-125°C) and low applied cell voltages (1.5-2.5V) was successfully demonstrated. The electrodeposited alloy containing up to 27 wt% of Ti was produced. High current efficiency of up to 85% was achieved [21].

3. Ionic Liquids as Thermal Storage Fluids

Ionic liquids have been proposed as thermal storage and heat transfer media due to their several advantages over thermal oils, liquid metals and molten salts traditionally used in thermal power plants. These advantages include wide liquid temperature range, high thermal/chemical stability, and low vapor pressure. Several researchers have evaluated certain ionic liquids as thermal storage fluids and their general aspects of application in terms of physic-chemical properties. Reddy et al. [6] have studied the thermal stability and corrosivity of a series of ionic liquids and compared them with thermal storage molten salts. Thermal stability and thermodynamic properties of [C_4mim][Tf_2N] ionic liquids were studied [22]. It is found that [C_4mim][Tf_2N] has several advantages over other ILs such as wider liquid temperature range, lower viscosity, higher chemical stability, moderate heat capacity and storage density. The calculated sensible heat storage density for [C_4min][Tf_2N] was more than 170 MJ/m^3 when the inlet and outlet field temperature were 210°C and 310°C. It is concluded that the evaluated ionic liquids are suitable for use as heat-transfer fluids and they are superior to present commercial heat-transfer fluids in terms of wide liquidus temperature range, low vapor pressure and ability to store substantial sensible heat [23].

In our research group we evaluated the thermal stability and thermodynamic properties of several ionic liquids and then assessed their viability as thermal storage and heat transfer media. The ionic liquids chosen for this study are widely used imidazolium-based ionic liquids due to their relatively low viscosities and high thermal stabilities. To achieve high thermal conversion efficiency, sensible thermal storage systems require the heat transfer fluid possess several important characteristics, such as wide liquid temperature range, high thermal stability and storage density. In particular the operating temperature higher than 400°C in a parabolic trough system would greatly enhance the thermal cycle efficiency.

4. Conclusions

Some of the novel applications of ionic liquids in materials processing have been presented in this paper. Aluminum was electrowon, electrorefined and recycled using $C_4mimCl-AlCl_3$ ionic liquid. High purity deposits (>99.9%) were obtained with very high current efficiency (>98%). Ti-Al alloys were deposited with or without $TiCl_4$ in the electrolyte with Ti content up to 27 wt% and current efficiency in the range of 78-85%. A uniform spherical well dispersed microstructure was obtained. This technique is being extended to other metals such as titanium, magnesium, etc whose industrial processing involves consumption of enormous amount of energy with the emission of pollutants. Heat capacities and densities for several ionic liquids were measured. Sensible heat storage capacities for ionic liquids were determined and compared with those of the existing heat transfer fluids. Sensible heat storage densities of ionic liquids were in the range of 160-210 MJ/m^3. The 3-D mathematical model developed to simulate the large scale electrowinning of aluminum showed good agreement with the experimental data. The novel electrochemical technique of extraction, processing and refining of metals such as aluminum and titanium is fast realized as an alternate route in the field of metals processing technology. Ionic liquids are realized as alternatives to heat transfer fluids in solar energy applications.

Acknowledgments

The author gratefully acknowledges the financial support from The U.S. Department of Energy, National Science Foundation, ACIPCO, and The University of Alabama.

References

[1] Reddy R G 2003 *Metall. Mater. Trans.* B **34** 137-52
[2] Zhang M, Kamavaram V and Reddy R G 2006 *Min. Metall. Process.* **23** 177-86
[3] Kamavaram V, Mantha D, Reddy R G 2003 J. Min. Metall. B **39** 43-58
[4] Wasserscheid P and Welton T 2003 *Ionic Liquids in Synthesis* Wiley-Vch Verlag, Weinheim, Germany p 7
[5] Zhang M, Kamavaram V and Reddy R G 2005 *J. Phase Equilibria and Diffusion* **26** 124-30
[6] Reddy R G, Zhang Z, Arena M F and Blake D M 2003 *High Temp. Mater. Process.* **22** 87
[7] Kim S K and Reddy R G 2001 *J Korean Inst. Surf. Engg.* **34** 553
[8] Zhang M, Kamavaram V and Reddy R G 2003 *JOM* **55** 54
[9] Zhang M, Kamavaram V and Reddy R G 2004 *JOM* **56** 334
[10] Zhang M, Kamavaram V and Reddy R G 2005 *Light Metals* 583
[11] Zhang M and Reddy R G 2006 Cell Development Part III and Emerging Technology – Aluminum Reduction Technology, vol 2 ed S Lindsay, T Pederson and T Galloway p 451
[12] Wu B, Reddy R G and Rogers R D 2000 *The 4th International Symposium on Aluminum Recycling via Room Temperature Electrolysis in Ionic Liquids, Recycling of Metals and Engineered Materials* ed D L Stewart Jr., J C Daley and R L Stephens p 845
[13] Kamavaram V, Reddy R G 2002 *Molten Salts XII* ed H C Delong, R W Bradshaw, M Matsunaga, G R Stafford and P C Trulove (Electrochemical Society) p 840
[14] Kamavaram V, Mantha D, Reddy R G 2005 *Electrochimica Acta* **50** 3286-95
[15] Zhang M, Reddy R G 2007 *ECS Transactions* **2** 27
[16] Janowski G M and Stafford G R 1992 *Metall. Trans. A* **23** 2715
[17] Ali M R, Nishikata A and Tsuru T 2003 *Ind. J. Chem. Tech.* **10** 21
[18] Tsuda T, Hussey C L, Stafford G R and Bonevich J E 2003 *J. Electrochem. Soc.* **150** C234
[19] Dent A, Seddon K and Welton T 1990 *J. Chem. Soc, Chem. Comm.* **4** 315
[20] Stafford G R 1994 *J. Chem. Soc.* **141** 945
[21] Pradhan D and Reddy R G 2008 *Electrochimica Acta* (accepted)
[22] Zhang M and Reddy R G 2007 *Materails Science and Technology:ENERGY:Energy Materials, Solar Energy Materials II* Sept 16-20, Detroit, Michigan 151-160
[23] Valkenburg M 2005 *Thermochimica Acta* **425** 181

Hydrothermal metallurgy for recycling of slag and glass

Toshihiro Tanaka, Takeshi Yoshikawa, Nobumitsu Hirai and Shigeru Katsuyama

Division of Materials and Manufacturing Science, Graduate School of Engineering, Osaka University, 2-1 Yamadaoka, Suita, 565-0871 Osaka, Japan.

E-mail: tanaka@mat.eng.osaka-u.ac.jp

Abstract. The authors have applied hydrothermal reactions to develop recycling processing of slag or glass. As an example, under hydrothermal conditions such as 200 ~ 300°C and 30 ~ 40MPa with H_2O, powders made of glass can be sintered to become solidified glass materials containing about 10mass% H_2O. When the glass containing H_2O is heated again under normal pressure, the glass expands releasing H_2O to make porous microstructure. H_2O starts to emit just above the glass transition temperature. Therefore, when we have a glass with low glass transition temperature, we can make low temperature foaming glass. The SiO_2-Na_2O-B_2O_3 glass is a candidate to be such a foaming glass. In this paper, we describe our recent trial on the fabrication of the low temperature foaming glass by using hydrothermal reaction.

1. Introduction

A large amount of slag is discharged from iron & steelmaking industries and mainly recycled for construction materials such as road bed materials. In addition, waste glass is also required to promote recycling by developing a reasonable recycling processes. Since slag and glass mainly consists of SiO_2, CaO, Al_2O_3, MgO, FeO etc, which are general components of ceramics materials, general sintering processes can be applied to produce some solidified materials from slag or glass. We have to, however, keep in mind that heating processes accompany energy consumption and sometimes CO_2 emission. Therefore, the authors have focused on the application of hydrothermal reactions to solidify the slag or glass powders to produce functional ceramic materials such as architecture materials etc. [1,2]. Hydrothermal reaction occurs with liquid H_2O under high pressure, that is to say, water at 120 ~ 350 °C as shown in Fig.1. This temperature can be obtained from exhausted heat coming out of iron & steelmaking processes or waste melting furnace etc. The application of the hydrothermal reactions is one of ideal environmental-friendly processes to cope with recycling issues of slag as Jung et al. have pointed out [3,4] and they have already applied the hydrothermal reactions to solidify slag powder with some additives.

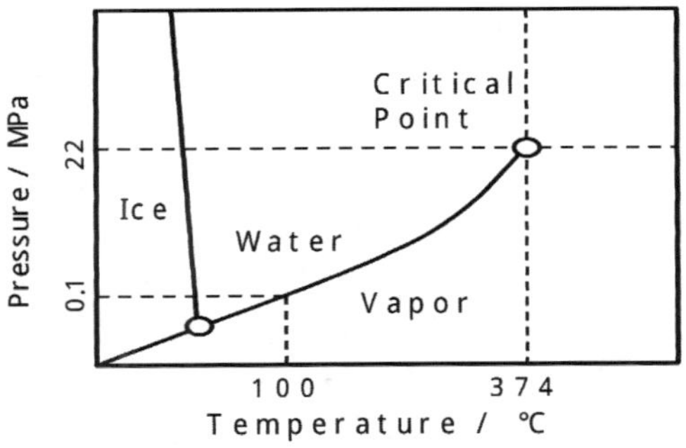

Figure 1. Phase diagram of H_2O.

The microstructure of molten slag or glass, which mainly consists of SiO_2 based network structure, is controlled by the addition of alkaline or alkaline-earth oxides to change chemical and physical properties such as basicity, viscosity etc. at high temperature. Under hydrothermal conditions, the microstructure of slag or glass might be controlled by H_2O, and this approach can be called "Hydrothermal Metallurgy".

Here, even when we can solidify glass or slag powder to make some materials, we are still required to add some additional functions to those solidified glass or slag materials. In order to equip more useful functions to those hydrothermally solidified materials, we have tried to make porous materials from slag or waste glass by applying hydrothermal reactions as described below. When we have porous glass or slag materials, those materials can be used as an insulator to control heat transfer, as a filter to remove impurities from polluted air or water, and furthermore as water reserve materials which can be applied to decrease temperature of wall or sidewalk by vaporizing H_2O absorbed in the porous structure of the materials in sunny days after H_2O is absorbed and kept during rainy days. Thus, if porous materials can be made from waste glass or slag by hydrothermal reactions, we may propose an environmental-friendly processing to create value-added materials to contribute largely to eco-society. The authors have been attempting to make various kinds of porous materials based on slag and glass. In this paper, we describe an example on fabrication of low temperature foaming glass by using "Hydrothermal Metallurgy"

2. Fabrication of low temperature foaming glass

Matamoros-Veloza et al.[5,6)] dissolved H_2O in waste glass mainly composed of SiO_2, Na_2O and CaO by hydrothermal synthesis and obtained a hydrated glass compact. When the prepared glass was heated, it started to expand and foam around 923 K with the release of water and became a porous material. The expansion of the glass is one of the reasonable approaches to make porous materials, but a foaming at a lower temperature would be beneficial. Since glass containing H_2O can be made at around 473K by hydrothermal reactions, it is to be desired that the foaming or expansion occurs at around the same temperature.

When the soda-silicate based glass was hydrothermally reacted, the constituents of the glass considerably influenced the water content of the glass under an identical hydrothermal condition. Moreover, it was found [7,8,9)] that an increase in water content lowers the glass transition temperature as shown in Fig.2. For the fabrication of low temperature foaming glass, 63mass%SiO_2- 27mass%Na_2O- 10mass% B_2O_3 glass, which exhibited adequate low glass transition temperature around 423K after the hydrothermal treatment as reported in the previous study [7,8,9)], was selected in the present work. The glass was subjected to the hydrothermal treatment at 523 K and its water releasing and foaming behavior with a heat treatment at 423 – 673 K was investigated [10)].

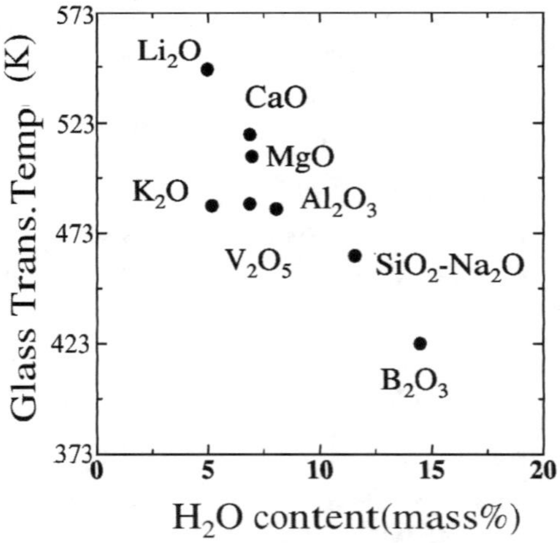

Figure 2. Relationship between glass transition temperature and H_2O content in 63mass% SiO_2-27mass%Na_2O-10mass%MO glass (MO= Li_2O, CaO, MgO, K_2O, Al_2O_3, V_2O_5, B_2O_3).

3. Experimental

The original glass sample was prepared from reagent grade quartz, Na_2CO_3 and H_3BO_3. These mixed powders corresponding to a composition of 63 mass% SiO_2 – 27 mass% Na_2O – 10 mass% B_2O_3 were melted in a Pt-20%Rh crucible at 1473 K for 3 h in air, and then rapidly cooled on a copper cooling block. The glass was subjected to a hydrothermal hot-pressing (HHP) technique. The experimental apparatus of HHP is shown in Fig.3. Pre-melted glass was ground and sieved at 63 μm. Glass powders were well-mixed with small amount of purified water and charged into the autoclave. After the mixture was pressed at 40 MPa, the autoclave was heated to 523 K within 20 min. and then immediately cooled to room temperature within the same time.

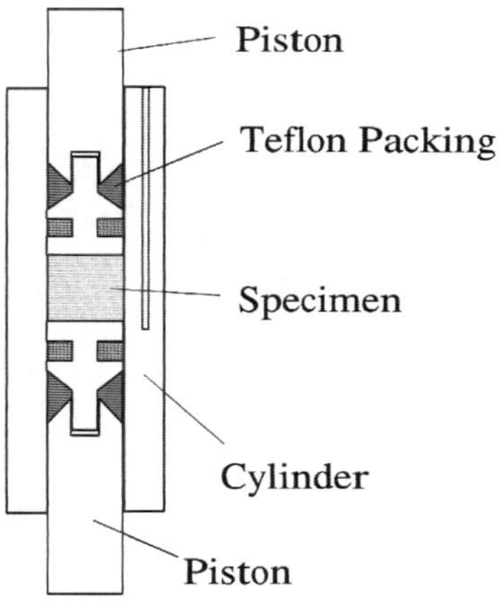

Figure 3. Hydrothermal hot pressing machine.

The HHP treated SiO_2-Na_2O-B_2O_3 glass prepared above was cut into cubic blocks of almost 6 mm. Each block was placed on a platinum pan and heated in a horizontal electric resistance furnace controlled to 423 – 673 K in air for 5 min. The weight change in the sample was measured before and after the heat treatment. The apparent density of the sample after firing was determined by measuring its shape and weight.

4. Results and Discussion

To investigate the macroscopic change in the HHP treated SiO_2-Na_2O-B_2O_3 glass, a block of the glass was heat-treated for 5 min. at 423 – 623 K in air. The samples after the heat treatment are shown in Fig.4.

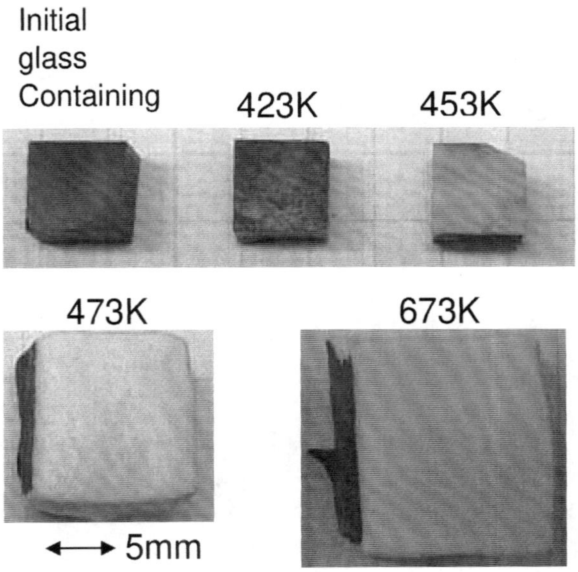

Figure 4. Change in macroscopic shapes of the HHP treated 63mass%SiO_2 - 27mass%Na_2O - 10mass%B_2O_3 glass after firing for 5 min. at 423 – 673 K.

Although no apparent change was observed when heated at 423 K, the sample heated at 453 K appeared to be whitened. Macroscopic expansion, namely foaming, of hydrated glass was observed for the samples heated over 473 K, which corresponds to the starting temperature for water release observed by the TG-DTA analysis, in which Al_2O_3 powder was used as a reference. A higher heating temperature resulted in a larger expansion of the glass materials. Here, a low temperature foaming was successfully obtained [10)] compared with the previously reported temperature of around 923 K by Matamoros-Veloza et al. [5,6)] with soda-lime silicate glass, of which specimen size was 20mm diameter × 10 mm height. It was confirmed by XRD analysis that the foamed glass at any temperature possesses the glass structure.

Apparent densities of the foamed glasses were determined from the sample shapes under the assumption of isotopic expansion. The apparent density decreases drastically at 473 K due to foaming, and continues to gradually decrease as the heating temperature increases. The lowest apparent density of 0.25g /cm^3 was obtained when the heat treatment was conducted at 673 K although the density of the initial specimen was 2.6g /cm^3. The weight loss was larger at a higher heating temperature, and the value of 11 % weight loss at 673 K was almost in accordance with the water content of the HHP treated glass. Those porous materials can be used for a filter to remove impurities in water or air.

5. Conclusion

Fabrication of low temperature foaming glass by using hydrothermal reaction was described as an example of the application of Hydrothermal Metallurgy. The HHP treated SiO_2-Na_2O-B_2O_3 glass was heated at 423 – 673 K in air to determine its macroscopic change by heating. A foaming behavior was observed even at 473 K and this foaming temperature was much lower than that of hydrated glass reported previously.

Acknowledgement

This study was partially supported by Priority Assistance for the Formation of Worldwide Renowned Centers of Research - The Global COE Program (Project: Center of Excellence for Advanced Structural and Functional Materials Design) from the Ministry of Education, Culture, Sports, Science and Technology (MEXT), Japan.

References

[1] T. Tanaka, N. Hirai, S. Maeda, M. Nakamoto and M. Hosokawa : *Proc. ICSTI2006, ISIJ, Osaka, Japan,* 704 (2006).
[2] T. Tanaka, S. Maeda, N. Takahira, N. Hirai and J. Lee, *Materials Science Forum*, **512**, 305 (2006).
[3] Z. Jing, F. Jin, T. Hashida, N. Yamasaki and E.H. Ishida, *J. Mater. Sci.*, **42** 8236 (2007).
[4] Z. Jing, E.H. Ishida, F. Jin, T. Hashida and N. Yamasaki, *Ind. Eng. Chem. Res.*, **45**, 7470 (2006).
[5] Z. Matamoros-Veloza, K. Yanagisawa, J.C. Rendon-Angeles and N. Yamasaki, *J. Mater. Sci. Lett.*, **21**, 1855 (2002).
[6] Z. Matamoros-Veloza, K. Yanagisawa, J.C. Rendon-Angeles and S. Oishi, *J. Phys. Condensed Matter*, **16**, S1361 (2004).
[7] M. Nakamoto, J. Lee, T. Tanaka, J. Ikeda and S. Inagaki, *ISIJ International*, **45**, 1567 (2005).
[8] M. Nakamoto, J. Lee and T. Tanaka : *Materials Science Forum*, **512,** 319 (2006).
[9] S. Sato, T. Yoshikawa, M. Nakamoto, T, Tanaka, J. Ikeda, *ISIJ International*, **48**, 245 (2008).
[10] T. Yoshikawa, S. Sato and T. Tanaka, *ISIJ International*, **48**, 130 (2008).

Journal of Physics: Conference Series **165** (2009) 012078 doi:10.1088/1742-6596/165/1/012078

Prediction of phase separation in multi-component oxide glass for the fabrication of porous glass materials from waste slag

Masanori Suzuki[1,2] and Toshihiro Tanaka[1]

[1] Division of Materials and Manufacturing Science, Graduate School of Engineering, Osaka University, 2-1 Yamadaoka, Suita, 565-0871 Osaka, Japan.
[2] Research fellow of the Japan Society for the Promotion of Science.

E-mail: masanori.suzuki@mat.eng.osaka-u.ac.jp

Abstract. Thermodynamic analyses were performed to evaluate the composition ranges for metastable immiscibility including spinodal decomposition in multi-component silicate systems, where glass was regarded as a super-cooled liquid phase. Experimental studies on two-phase separation in glass were carried out, and the phase separation was observed in those silicate glasses corresponding to calculated metastable miscibility gaps.

The authors also attempted the fabrication of porous glass using phase separation in oxide glasses containing B_2O_3. A multi-component borosilicate glass composition was prepared based on the silicate glass composition where spinodal decomposition had been confirmed. One of the glass phases formed by the spinodal decomposition in the boro-silicate glass was selectively removed by acid leaching and porous glass was successfully obtained.

1. Introduction

Phase separation in glass represents a phenomenon in which a single glass phase separates into two glass phases with different compositions during the heat-treatment of the glass at a given temperature.

To create functional glass materials using waste glass and slag from metallurgical processes or waste melting processes, the authors have investigated phase separation in multi-component oxide glass. For instance, spinodal decomposition as one of phenomena of the phase separation forms three-dimensionally interconnected microstructure, and porous glass is obtained by leaching one of those phases with acid solution. This porous glass can be expected in many applications such as a filter to remove impurities from polluted water or air. Thus, slag can be transformed into value-added functional glass materials using phase separation in oxide glass.

To obtain phase separation in glass from slag, the composition ranges for metastable immiscibility for multi-component oxide systems should be evaluated. In this study, the metastable immiscibility regions including spinodal decomposition were evaluated in several multi-component silicate systems by thermodynamic analyses, where glass was regarded as super-cooled liquid phase. Experimental studies were performed to confirm the phase separation in heat-treated glasses in the predicted miscibility gap. The fabrication of porous glass has been attempted using the phase separation in a multi-component oxide glass containing B_2O_3.

2. Estimation of the Composition Ranges for Phase Separation and Spinodal Decomposition [1,2]

Liquid-liquid phase separation can be generally predicted from the Gibbs energy curve of the liquid phase. For instance, the miscibility gap in a binary system is obtained from two points of contact on the Gibbs energy curve with a common tangent. The composition range for spinodal decomposition, referred to as the "spinodal region," is determined from two inflexion points in the Gibbs energy curve.

In our study, the composition ranges for phase separation and spinodal decomposition were estimated by calculating Gibbs energies in the metastable liquid phase, where glass in silicate systems was regarded as a super-cooled liquid phase. The Gibbs energies were calculated with the FactSage thermodynamic computing program and the F*A*C*T oxide databases for molten oxide systems [3].

To definitively determine the spinodal region, the activity of SiO_2 was calculated from the Gibbs energy in the super-cooled liquid phase. In an A-B binary system, the spinodal region is evaluated with the derivative of the activity with respect to composition derived by the following Equation (1):

$$\frac{\partial a_B}{\partial X_B} = \frac{\exp(\mu_B / RT)}{RT} \cdot \frac{\partial \mu_B}{\partial X_B} = (1 - X_B)\frac{a_B}{RT} \cdot \frac{\partial^2 G^{Mix}}{\partial X_B^2}, \tag{1}$$

where G^{Mix} denotes the Gibbs energy for mixing in the liquid phase. μ_B and a_B are the chemical potential and activity of component B and described by Eq. (2) under the pure liquid standard state:

$$\mu_B = RT \ln a_B = G^{Mix} + (1 - X_B)\frac{\partial G^{Mix}}{\partial X_B}. \tag{2}$$

Equation (1) indicates that the spinodal region is determined from the extremum points in the activity curve, because the second derivative of the Gibbs energy should be zero at its inflexion points.

The above estimation procedure for the spinodal region can be extended to ternary and multi-component systems as follows: if pure component A constitutes one of the separated phases, the spinodal region in an A-B-C ternary system is determined by calculating activity of component A across straight lines with constant composition ratios of components B and C that correspond to tie-lines of the miscibility gap. The spinodal region in a quaternary system is derived, similarly, by calculating activities of A in A-$B_{1-x}C_x$-D quasi-ternary systems (here x varies from 0 to 1).

3. Experimental Procedure

3.1 Preparation of Two-phase Glass Materials

Phase separation in glass was investigated experimentally to verify the calculated results of the metastable miscibility gaps. SiO_2-CaO-MgO-Na_2O and SiO_2-CaO-Al_2O_3-Na_2O glasses were prepared from silica, magnesia, alumina, sodium carbonate and calcium carbonate. These raw materials were melted in air for 5 h at 1873 K, and glass samples were obtained by quenching the melts in water. The glasses were then heat-treated at 948 K or 993 K. X-ray diffraction analysis was conducted on annealed glasses to confirm that they were glassy. Microstructures in these glasses were observed by a transmission electron microscopy.

3.2 Fabrication of Porous Glass Materials using Phase Separation in Glass

Porous glass was fabricated using two-phase glasses obtained from spinodal decomposition by leaching one of decomposed phases with acid solution. In the leaching process, glass sample was immersed in a HCl solution for 14 h, where bulk size of the sample was less than 2 mm in diameter. Microstructure on the surface of the porous glass was observed by field emission scanning electron microscopy. Composition of the glass dissolved in acid was examined with ICP emission spectrometry.

4. Results and Discussion

4.1 Prediction of Phase Separation in Multi-component Silicate Glasses [1,2]

Figure 1 shows the calculated metastable miscibility gap and the spinodal region of SiO_2-CaO-MgO-5 mol% Na_2O system, where molar fractions of each component are divided by the sum of the molar concentrations of all components except Na_2O.

The occurrence of phase separation was investigated for the glasses A-C, having compositions indicated by square marks in **Fig. 1**. The selected chemical compositions are shown in **Table 1**.

Electron micrographs of the heat-treated glasses A-C are shown in **Figure 2**, and they depict the occurrence of a phase separation. An interconnected microstructure is observed in glass A, and it corresponds to the spinodal decomposition. In glass B, a particle structure indicating the binodal decomposition was observed. As the composition of glass C is outside the miscibility gap, no specific microstructure was observed after annealing for 192 h. Since electron diffraction patterns for each specimen showed no evidence for the presence of crystalline phases, the heat-treated specimens were glassy in microscopic observation area. The above experimental results indicated that the phase separation occurs in SiO_2-CaO-MgO-Na_2O glasses according to the predicted miscibility gap.

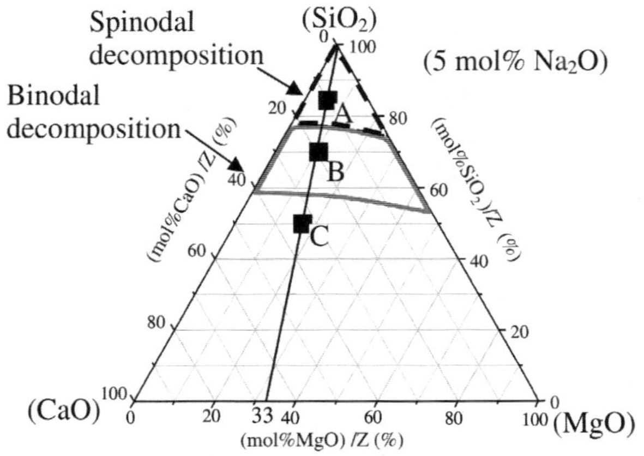

Figure 1. Calculated composition ranges for spinodal decomposition and binodal decom-position for the SiO_2-CaO-MgO-5 mol% Na_2O system at 948 K. (Z= mol% (SiO_2+CaO+MgO))

Table 1. Compositions of SiO_2-CaO-MgO-Na_2O glass samples used in the experiment.

	Composition [mol%]			
	SiO_2	CaO	MgO	Na_2O
A	80.8	9.5	4.7	5.0
B	66.5	19.0	9.5	5.0
C	47.5	31.7	15.8	5.0

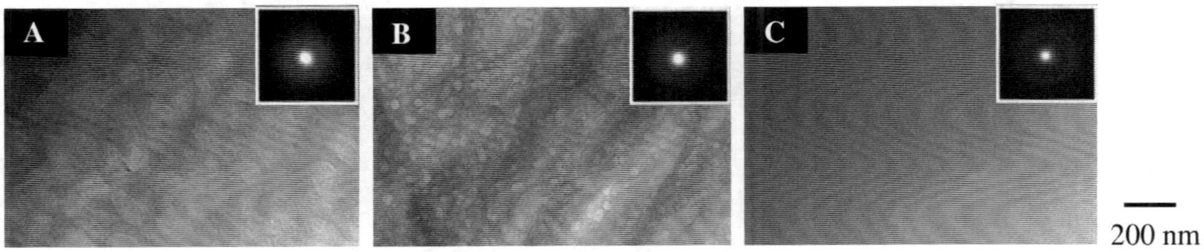

Figure 2. Electron micrographs of SiO_2-CaO-MgO-Na_2O glasses after annealing for 192 h at 948 K.

The metastable miscibility gap and the spinodal region in the SiO_2-CaO-Al_2O_3-5 mol% Na_2O system was calculated and is shown in **Figure 3**. Molar fractions of each component are divided by the sum of the molar concentrations of all components except Na_2O. The occurrence of spinodal de-composition in this system was investigated for glass D, whose composition is described by a square mark in **Fig. 3**. An interconnected microstructure indicating the spinodal decomposition was observed in the annealed glass D, as shown in **Figure 4**. This experimental result corresponds to the calculated miscibility gap as shown in **Fig. 3**.

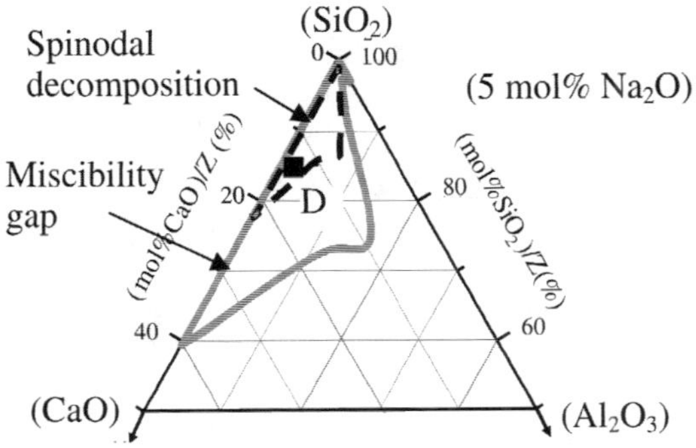

Figure 3. Calculated metastable immiscibility regions including spinodal decomposition in the SiO_2-CaO-Al_2O_3-5 mol% Na_2O system at 973 K. (Z= mol% (SiO_2+CaO+Al_2O_3))

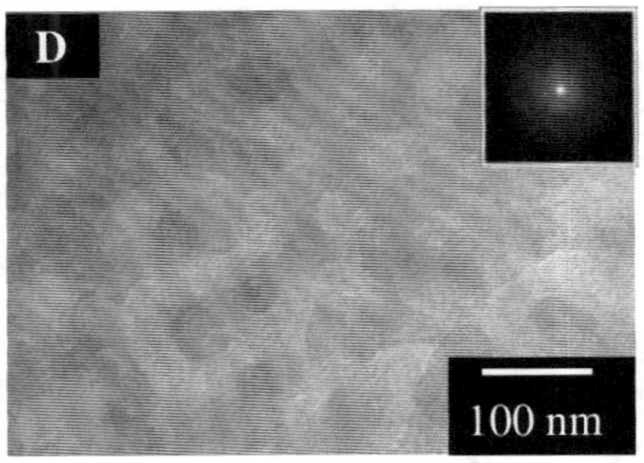

Figure 4. Electron micrograph of glass D (80.8SiO_2-11.7CaO-2.5Al_2O_3-5.0 mol% Na_2O) annealed for 96 h at 993 K.

4.2 Fabricating Porous Glass using Spinodal Decomposition in Multi-component Borosilicate Glass
The fabrication of porous glass was attempted using two-phase glasses with spinodal decomposition by leaching one of the decomposed glass phases with an acid solution.

It should be noted that in the spinodal decomposition in glasses containing B_2O_3, one of separated phases contains plenty of B_2O_3 and can be leached out with acid solutions [4,5]. In this study, a multi-component borosilicate glass composition D-2 was designed by partially substituting SiO_2 with B_2O_3 in the SiO_2-CaO-Al_2O_3-Na_2O glass D. Its composition is included in the spinodal region of the silicate system, and it was used to remove one of the separated phases in the leaching process.

Figure 5 shows microstructures of the borosilicate glass D-2 after heat-treatment and acid leaching. Morphology with interconnectivity was observed in the heat-treated glass, and an inter-connected porous structure was obtained after leaching. These microstructures indicate that one of the glass phases formed by spinodal decomposition was removed selectively with acid leaching. The chemical composition analysis for the glass phase dissolved into the acid solution revealed that a high proportion of B_2O_3 was dissolved in acid, and the porous glass contains a high concentration of SiO_2.

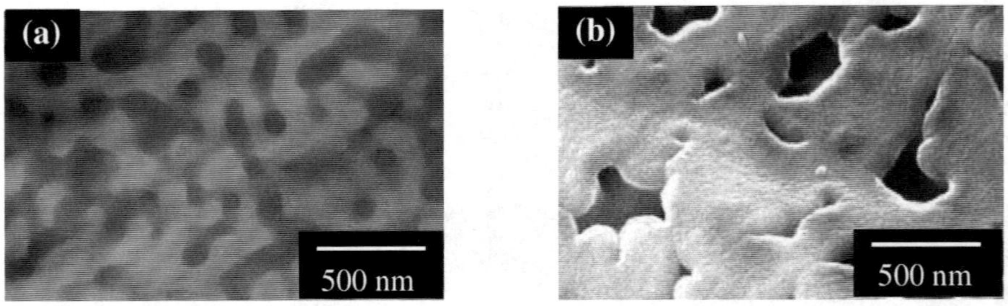

Figure 5. Microstructures in glass D-2 (59.7SiO_2-13.8CaO-2.9Al_2O_3-5.9Na_2O-17.7 mol% B_2O_3), (a) after annealing for 96 h at 993 K; (b) after leaching one of separated phases with acid solution.

5. Conclusions

Immiscibility regions in multi-component silicate glasses were determined by calculating Gibbs energies in the metastable liquid phase, where glass was regarded as a super-cooled liquid phase. The occurrence of phase separation in those silicate glasses was verified by observing microstructures in heat-treated glasses, and these experimental results corresponded to the predicted miscibility gaps.

For the fabrication of porous glass, a multi-component borosilicate glass was prepared by partially replacing SiO_2 with B_2O_3 in the silicate glass composition where spinodal decomposition had been confirmed. One of the glass phases formed by the spinodal decomposition in the borosilicate glass was leached out with an acid solution, and porous glass with a three-dimensional interconnected structure was obtained.

Acknowledgements

This study was supported by Priority Assistance for the Formation of Worldwide Renowned Centres of Research - The Global COE Program (Project: Center of Excellence for Advanced Structural and Functional Materials Design) from the Ministry of Education, Culture, Sports, Science and Technology (MEXT), Japan. Observation of the microstructures in the samples by transmission electron microscopy was supported by professor Hirotaro Mori and technical official Eiji Taguchi (Research Center for Ultra-High Voltage Electron Microscopy, Osaka University), and carried out in a facility in the Research Centre for Ultra-High Voltage Electron Microscopy, Osaka University.

References

[1] M.Suzuki and T.Tanaka, *ISIJ Int.,* **46** (10), 1391 (2006).
[2] M.Suzuki and T.Tanaka, *ISIJ Int.,* **48** (4), 405 (2008).
[3] A.D.Pelton and M.Blander, *Metall. Trans.*, **17B**, 805 (1986).
[4] T.H.Elmer, M.E.Nordberg, G.B.Carrier, E.J Korda: *J. Am. Ceram. Soc.,* **53** (4), 171 (1970).
[5] T.Nakashima and Y.Kuroki, *Nippon Kagaku-kaishi*, (10), 1231 (1981).

Journal of Physics: Conference Series **165** (2009) 012079 doi:10.1088/1742-6596/165/1/012079

A fundamental approach for the measurement of solid-liquid interfacial energy

Atsushi Fukuda, Takeshi Yoshikawa, Toshihiro Tanaka

Osaka University, 2-1 Yamadaoka, Suita, Osaka 565-0871 Japan

atsushi.fukuda@mat.eng.osaka-u.ac.jp

Abstract. Solid-liquid interfacial energies in alloys are important properties for the process design of materials production such as casting and crystal growth. Inadequate information exists on solid-liquid interfacial energies. Our aim for this study was thus to establish a method to measure solid-liquid interfacial energies in alloys from equilibrium interfacial shapes of solid-liquid-gas phases. Experiments were carried out on Cu-B and Ag-Bi systems where copper and silver were treated as solid phases, respectively. Since the determined values for solid-liquid interfacial tensions for both systems agreed with reported values or estimated values, it was clarified that solid-liquid interfacial energies can be measured by observing the interfacial shape.

1. Introduction

Solid-liquid interfacial energies are fundamental properties of materials and play a key role in material processing at the coexistence of solid and liquid phases such as for the processes of casting and crystal growth. Since little information exists on solid-liquid interfacial energies, further research is required for the simulation of the materials process including the solid-to-liquid transformation in various systems.

Solid-liquid interfacial energies have been measured based on the nucleation rate, maximum super-cooling, dihedral angles of grain boundary grooves and Young's contact angle but these techniques show considerable discrepancies in experimental results. These discrepancies are caused by experimental conditions being unsuitable for adaptation to nucleation theory. Additionally, inaccuracies in reference values of surface tensions and grain boundary energies are responsible for these discrepancies. Recently, the measurement of solid-liquid interfacial energies has been done by measuring the equilibrium interfacial shape of the solid-liquid-gas phases for Al-Si, Al-Al_2O_3 and Fe-C systems [1-3]. This method allows for the determination of the solid surface energy and the solid-interfacial energy if the liquid surface tension is available. No reports, however, exist where the measurement of solid-liquid interfacial energies of metal-metal or metal-alloy systems was undertaken.

In this work, to establish a method for the determination of both the solid-liquid interfacial energy and the solid surface energy in an alloy system, experiments measuring the equilibrium interfacial shape of solid-liquid-gas phases of Cu-B and Ag-Bi systems were carried out. The validity of this method was then verified through comparison of the experimental results with the reported or estimated values.

2. Experimental principle

2.1. Dupre's equation

When a liquid droplet is placed on a soluble solid substrate its equilibrium shape may be illustrated as in Figure 1. At physical equilibrium the three interfacial energies, the liquid surface tension (σ_L), the solid surface energy (σ_S) and the solid-liquid interfacial energy (σ_{SL}) are balanced at the triple point where the sine theorem gives Dupre's equation as expressed by equation (1).

$$\frac{\sigma_{SL}}{\sin\theta_V} = \frac{\sigma_L}{\sin\theta_S} = \frac{\sigma_S}{\sin(\theta_1 + \theta_2)} \tag{1}$$

If the value of the liquid surface tension is available we can determine the solid-liquid interfacial energy and the solid surface energy by measuring the equilibrium dihedral angles θ_V and θ_S. In this work, the liquid surface tensions of binary alloys were estimated using Butler's equation.

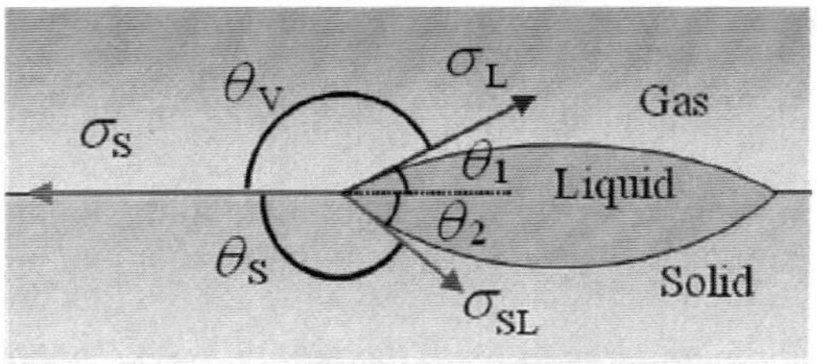

Figure 1. The equilibrium shape of a liquid droplet on a solid substrate.

2.2. Butler's equation

The surface tension of a liquid alloy σ_L can be estimated well by Butler's equation as expressed in equation (2) with thermodynamic properties of the alloy [4].

$$\begin{aligned}\sigma_L &= \sigma_A + \frac{RT}{A_A}\ln\frac{(1-N_B^S)}{(1-N_B^B)} + \frac{1}{A_A}\overline{G}_A^{E,S}(T,N_B^S) - \frac{1}{A_A}\overline{G}_A^{E,B}(T,N_B^B) \\ &= \sigma_B + \frac{RT}{A_B}\ln\frac{N_B^S}{N_B^B} + \frac{1}{A_B}\overline{G}_B^{E,S}(T,N_B^S) - \frac{1}{A_B}\overline{G}_B^{E,B}(T,N_B^B)\end{aligned} \tag{2}$$

where σ_X and A_X are the surface tension and the molar surface area of pure liquid X, respectively. $N_X^{B/S}$ is the mole fraction of component X of a bulk or a surface which is defined as the outermost monolayer. $\overline{G}_X^{E,B/S}(T,N_B^{B/S})$ is the partial excess Gibbs energy of component X in the surface as a function of T and $N_B^{B/S}$. The partial excess Gibbs energy of the surface is obtained as equation (3) assuming it has a comparable composition dependence to that in the bulk.

$$\overline{G}_X^{E,S}(T,N_B^S) = \frac{Z^S}{Z^B}\cdot\overline{G}_X^{E,B}(T,N_B^S) \tag{3}$$

where $Z^S/Z^B = 0.83$ which denotes the ratio of the coordination number in the surface to that in the bulk [5]. Liquid surface tensions of Cu-B and Ag-Bi systems were calculated from thermodynamic properties [6,7] at a given temperature and composition on the liquidus for each alloy system.

3. Experimental procedure

A Cu-rich Cu-B droplet was melted on a copper substrate in the Cu-B system. A copper substrate of 3 mm thickness (purity 99.96 %) into which a hole was bored to a depth of 1-2 mm in the centre was polished mechanically and chemically. Cu-B alloys for droplets were prepared by melting copper shots (purity 99.99 %) and a boron lump (purity 99.8 %) in an alumina crucible under an Ar-10%H_2 atmosphere at 1573 K.

The Cu-B alloy was placed in the hole of the copper substrate in a horizontal electric resistance furnace. It was heated to 1273 K within 120 min under an Ar-10%H_2 atmosphere. Water and oxygen in the atmospheric gas were removed by passing it through magnesium perchlorate and magnesium chips then heated to around 773 K. The temperature was then raised to 1348 K at a rate of 5 K/min and kept there for 5-30 min. The apparent contact angle (θ_1 in Figure 1) was measured using a CCD camera. The sample was cooled down at a rate of 6 K/min. The vertical cross-section was observed by optical microscope and the hidden angle (θ_2 in Figure 1) was determined.

Liquid bismuth was reacted with a solid silver substrate to form Ag-Bi melt for the Ag-Bi system. A silver substrate of 1-2 mm thickness (purity 99.99 %) with a hole in the centre, was polished mechanically and chemically. Bismuth shot (purity 99.999 %) was placed in the hole of the silver substrate in a horizontal infrared furnace. It was heated to 773 K at a rate of 2.5 K/min under vacuum and kept at that temperature for 5-60 min. The apparent contact angle and the hidden angle were determined as for the Cu-B system.

4. Result and discussion

4.1. Experimental results for the Cu-B system

A Cu- 10.0 mol% B alloy was melted on a copper substrate at 1348 K and kept there for 5 min. Figure 2(a) shows the liquid alloy on the copper substrate after being kept at that temperature for 5 min. The apparent contact angle θ_1 became constant within 5 min and was determined to be 6.8 ± 0.7°. The cross-section of the sample after quenching is shown in Figure 2(b). The liquid alloy dissolved the substrate and gave a hidden angle θ_2 of 62.4 ± 2.5°. Here, the volume of the liquid alloy is estimated to be 5 times the initial volume of the alloy that was placed on the substrate which indicates that the liquid composition became the Cu-liquidus (Cu- 1.9 mol% B) by taking into account the lever rule for the alloy composition. The liquid surface tension was thus estimated for the corresponding alloy composition to be 1366 mN/m using equation (2). The solid-liquid interfacial energy between solid copper and liquid Cu-B alloy and the solid surface energy of copper were determined to be 182 ± 23 mN/m and 1440 ± 17 mN/m using equation (1), respectively.

The determined interfacial energy between solid copper and liquid Cu-B alloy was in good agreement with reported values between solid and liquid copper of 177-232 mN/m [9-11]. In addition, the obtained solid surface energy of copper agrees fairly well with the reported value of 1478 mN/m [12] although it seems to be slightly smaller because of the effect of adsorption of boron on the copper surface. Hence, the possibility exists to determine the solid-liquid interfacial energy by observing the equilibrium shape of the three phase interface in the copper-based system.

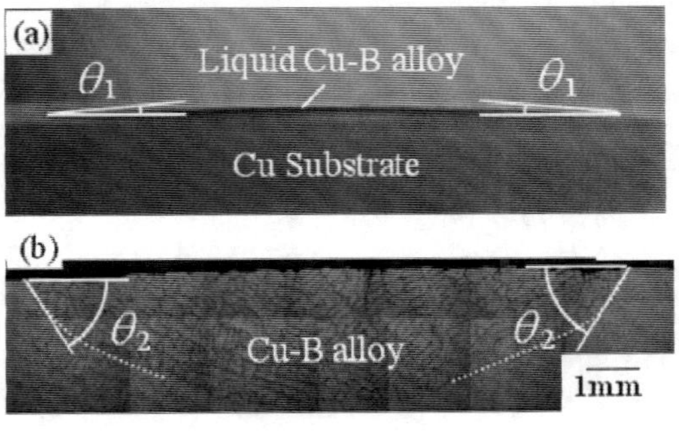

Figure 2. The liquid Cu-B droplet on the solid Cu substrate (a) at 1348 K and (b) the cross-section after the experiment.

4.2. Experimental results for the Ag-Bi system

A bismuth shot was melted on a silver substrate at 773 K and kept there for 60 min. Figure 3(a) shows the liquid alloy on the silver substrate after being kept at that temperature for 60 min. The apparent contact angle θ_1 became constant within 10 min and was determined to be 21.8 ± 0.3°. The cross-section of the sample after quenching is shown in Figure 3(b). The liquid alloy dissolved the substrate and gave a hidden angle θ_2 of 55.5 ± 1.7°. Here, the liquid size is estimated to be 3 times that of the initial size of the bismuth granule that was placed on the silver substrate which indicates that the liquid composition was the Ag-liquidus (Ag- 38.8 mol% Bi). The liquid surface tension was thus estimated for the corresponding alloy composition to be 416 mN/m using equation (2). The solid-liquid interfacial energy between solid silver and liquid Ag-Bi alloy, and the solid surface energy of silver were determined to be 187 ± 6 mN/m and 493 ± 8 mN/m using equation (1).

The determined interfacial energy between solid silver and liquid Ag-Bi alloy was in good agreement with the estimated value of the solid-liquid interfacial energy through the prediction proposed by Eustathopoulos [8,9] for the corresponding solid silver and liquid alloy of 178 mN/m. On the other hand, the solid surface energy of silver was much smaller than the reported value for pure silver of 1339 mN/m at 773K [12]. This might be influenced by the absorption of bismuth onto the silver surface although this should be confirmed by further investigation.

The applicability of this new method for the determination of the solid-liquid interfacial energy by observing the equilibrium interfacial shape of the three phases was also confirmed for the silver-based system.

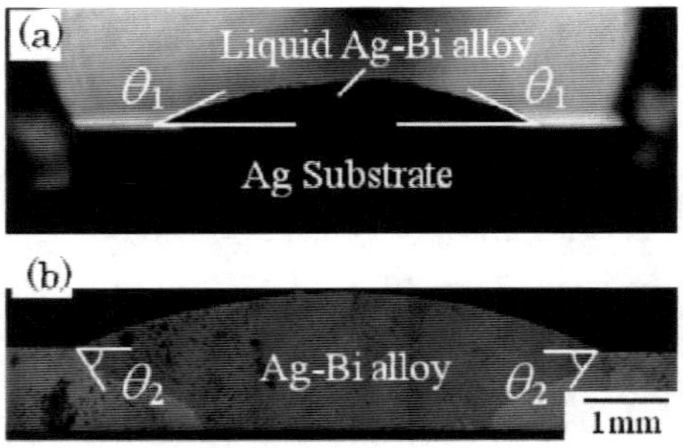

Figure 3. The liquid Ag-Bi droplet on the solid Ag substrate (a) at 773 K and (b) the cross-section after the experiment.

5. Conclusions

A measurement method was established for interfacial energies in alloy systems by observing the equilibrium shape of the solid-liquid-gas interface of an alloy system. Experiments were conducted on Cu-B and Ag-Bi systems. The determined interfacial energies from the observed interfacial shapes were in accordance with reported or estimated values. Hence, it is possible to measure the solid-liquid interfacial energies in alloy systems from the interfacial shape of the three phases.

References

[1] Nakae H and Katoh H 1999 *J. Japan Inst. Metals* **63** 1356
[2] Levi G and Kaplan W D 2003 *Acta Materialia* **51** 2793
[3] Shinozaki N, Satoh N, Shinozaki H , Wasai K and Era H 2006 *J. Japan Inst. Metals* **70** 950
[4] Butler J A V 1932 *Procedings of the Royal Society of London. Series A, Containing Papers of a Mathematical and Physical Character* **135** 348
[5] Tanaka T, Hack K, Iida T and Hara S 1996 *Z. Metallkunde* **87** 380
[6] Weber L and Tavangar R 2007 *Scripta Materialia* **57** 988
[7] Zoro E, Servant C and Legendre B 2007 *Calphad* **31** 89
[8] Eustathopoulos N 1983 *Int. Met. Rev.* **28** 189
[9] Turnbull D 1950 *J. Applied Physics* **21** 1022
[10] Vinet B, Magnusson L, Fredriksson H and Desre P J 2002 *J. Colloid Interf. Sci.* **255** 363
[11] Kaptay G 2001 *Trans. JWRI* **30** 245
[12] Alchagiro A B, Alchagirov B B, Taova T M and Khokonov K B 2001 *Trans. JWRI* **30** 287

Acknowledgement

This research is partly supported by a Grant-in-Aid for Scientific Research on Priority Area (B) and Priority Assistance for the Formation of Worldwide Renowned Centers of Research - The Global COE Program (Project: Centre of Excellence for Advanced Structural and Functional Materials Design) from the Ministry of Education, Culture, Sports, Science and Technology (MEXT), Japan.

Journal of Physics: Conference Series **165** (2009) 012080 doi:10.1088/1742-6596/165/1/012080

Numerical simulation of transformation-induced microscopic residual stress in ferrite-martensite lamellar steel

Y Mikami[1], **A Inao**[1], **M Mochizuki**[1] **and M Toyoda**[1]

Division of Materials and Manufacturing Science, Graduate School of Engineering, Osaka University, 2-1 Yamada-oka, Suita, Osaka, Japan

E-mail: mikami@mapse.eng.osaka-u.ac.jp

Abstract. The effect of transformation-induced microscopic residual stress on fatigue crack propagation behavior of ferrite-martensite lamellar steel was discussed. Fatigue tests of prestrained and non-prestrained specimens were performed. Inflections and branches at ferrite-martensite boundaries were observed in the non-prestrained specimens. On the other hand, less inflections and branches were found in the prestrained specimens. The experimental results showed that the transformation induced microscopic residual stress has influence on the fatigue crack propagation behavior. To estimate the microscopic residual, a numerical simulation method for the calculation of microscopic residual stress stress induced by martensitic transformation was performed. The simulation showed that compressive residual stress was generated in martensite layer, and the result agree with the experimental result that inflections and branches were observed at ferrite-martensite boundaries.

1. Introduction

Mechanical properties of steel plate are controlled by microstructure. Recently, an improvement in fatigue strength was achieved by introducing lamellar structure of ferrite and martensite by controlling rolling and cooling condition. Shown in figure 1 is the microstructure of ferrite-martensite lamellar

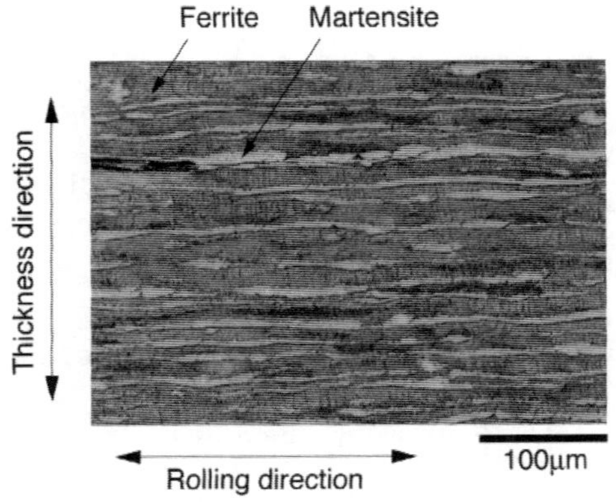

Figure 1. Microstructure of ferrite-maritensite lamellar steel.

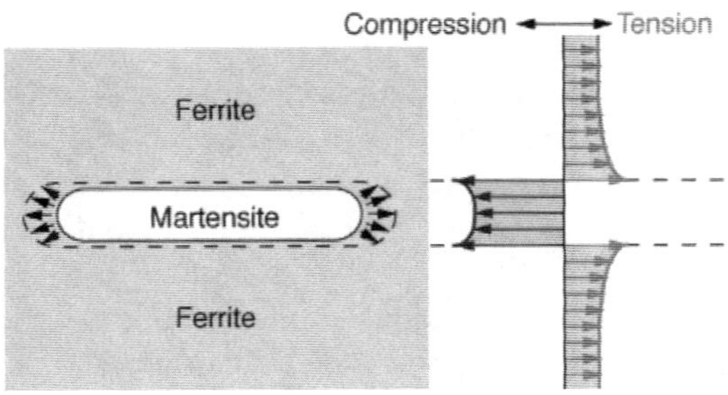

Figure 2. Schematic illustration of transformation induced microscopic residual stress in ferrite-martensite lamellar steel.

steel [1]. The ferrite-martensite lamellar steel has the yield stress of 520 MPa and the tensile strength of 581 MPa. Detailed observation of fatigue crack propagation paths showed that inflections and branches were observed at boundaries of ferrite and martensite. For the reason of the inflections and branches, the following two major points related to martensite phase might be considered: (1) the hardness of martensite as secondary phase, (2) microscopic compressive residual stress caused from transformation expansion of martensite. The schimatic illustration is shown in figure 2. In this paper, the effect of microscopic residual stress due to martensitic transformation expansion on fatigue strength is discussed.

2. Fatigue crack propagation behavior in ferrite-martensite lamellar steel

First, fatigue tests were performed in order to clarify the effect of microscopic residual stress. The steel plate with ferrite-martensite lamellar microstructure as shown in figure 1 was 1% prestrained along rolling direction. The prestrained and non-prestrained specimens were used in the fatigue tests. The fatigue tests were carried out with three-point bending under following conditions: ΔK increase, stress ratio $R = 0.1$, repetition frequency $f = 10$ Hz. Then the observation of fatigue crack propagation paths was performed.

Inflections and branches at ferrite-martensite boundaries were observed in the non-prestrained specimens. On the other hand, less inflections and branches were found in the prestrained specimens. As a result, fatigue crack growth rate da/dN was higher in 1% prestrained specimens than that in 0% prestrained specimens. It could be expected that microscopic compressive residual stress would be changed when specimens were prestrained, microscopic residual stress have some effect on the inflections and branches of fatigue crack propagation. Therefore a numerical simulation was performed to estimate the distribution of transformation-induced microscopic residual stresses.

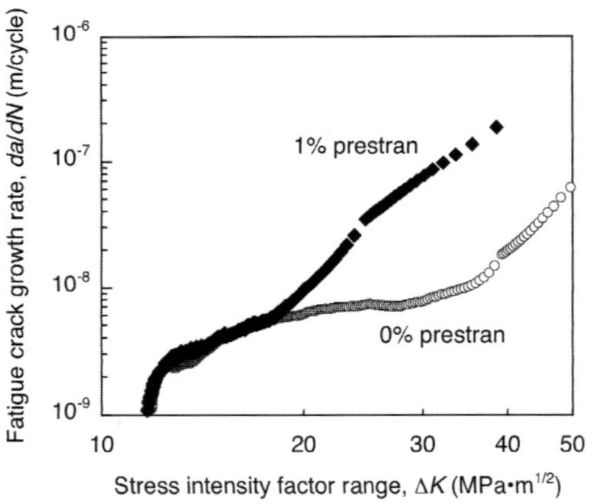

Figure 3. Fatigue crack growth rate of ferrite-martensite lamellar steels with and without prestrain.

3. Numerical simulation procedure of microscopic residual stress distribution

A numerical simulation was performed to calculate the microscopic residual stress in ferrite-martensite lamellar structure. A numerical simulation model of ferrite-martensite lamellar structure was generated as shown in figure 4. The geometrical parameters, such as the thickness and length of martensite, the distance between martensite phases, etc., were determined based on the experimental results of the observation of microstructure as shown in figure 1. The volumetric expansion and change in mechanical properties due to phase transformation was taken into consideration to calculate the transformation-induced microscopic stress distribution [2]. The transformation expansion curves during cooling process used in the numerical simulation are shown in figure 5. Figure 6 shows the temperature dependency of yield stress (σ_Y) and Young's modulus (E). The curves and transformation temperatures were selected from database. The data to use were determined based on the chemical composition measured by EPMA analysis of ferrite and martensite.

The simulation model was cooled from 900°C to 20°C: from the fully austenite condition to room temperature. The boundary conditions for the model was free to expansion and contraction. The microscopic residual stress is generated during the cooling process due to the difference in the transformation temperature and mechanical properties between ferrite and martensite.

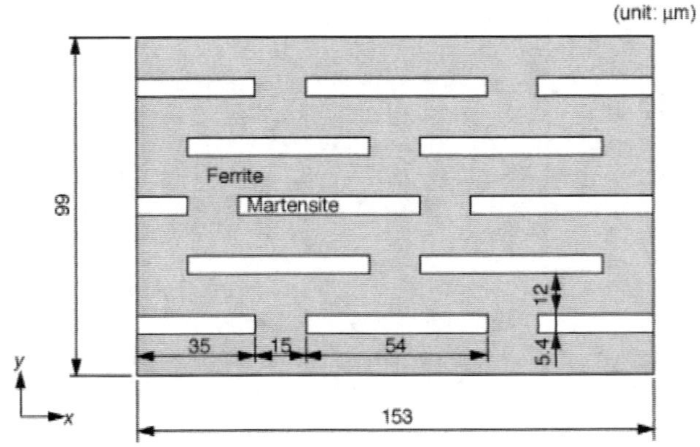

Figure 4. Numerical simulation model for transformation induced microscopic residual stress.

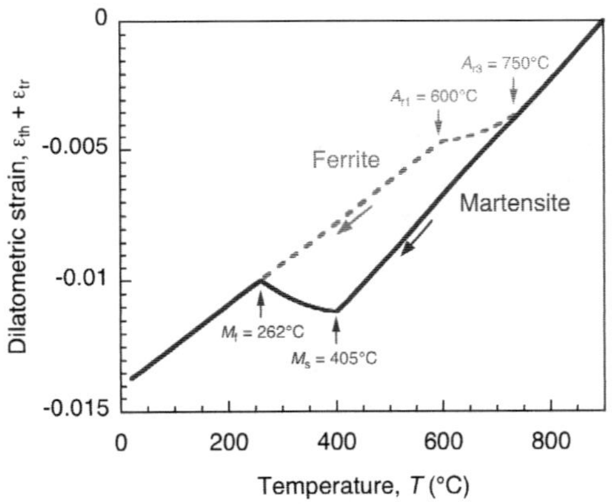

Figure 5. Transformation expansion curves used in the numerical simulation.

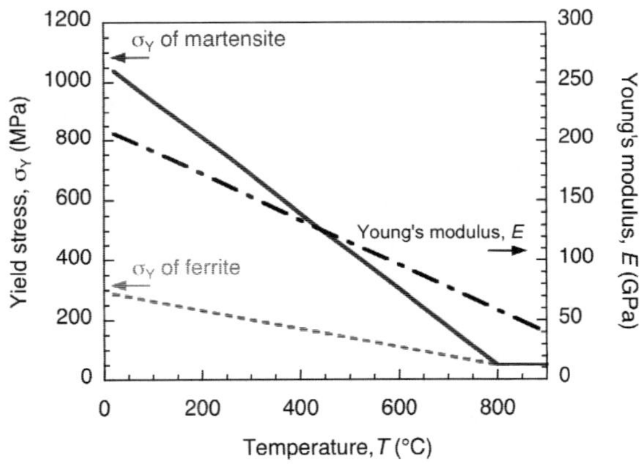

Figure 6. Temperature dependency of yield stress and Young's modulus used in the numerical simulation.

4. Microscopic residual stress distribution in ferrite-martensite lamellar steel

A result of the numerical simulation of transformation-induced microscopic residual stress is shown in figure 7. The distribution of residual stress along the direction of martensite layer σ_x over the simulation model is shown in figure 7(a). In addition, figure 7(b) shows the distribution of σ_x along the black line shown in figure 7(a). As seen from these figures, the microscopic stress distribution induced by transformation expansion of martensite is compressive in martensite and tensile in ferrite. From the viewpoint of the fatigue crack propagation behavior, when the residual stress in martensite was compressive, the penetration of the crack through martensite would be difficult. This agrees with the experimental result of the observation of fatigue crack path in non-prestrained ferrite-martensite lamellar steel.

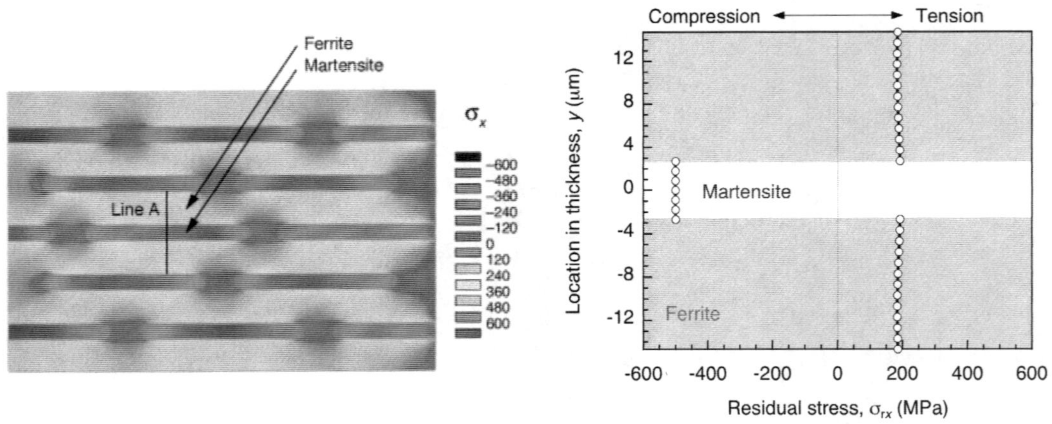

(a) contour figure (b) distribution on line A

Figure 7. Numerical simulation results of transformation-induced microscopic residual stress distribution.

5. Summary

In this study, the effect of microscopic residual stress on fatigue crack propagation behavior of ferrite-martensite lamellar steels was discussed. A numerical simulation method for transformation-induced microscopic residual stress was proposed. The following results were obtained.

- The fatigue test results of prestrained and non-prestrained specimens of ferrite-martensite lamellar steel showed that microscopic residual stress have some effect on the inflections and branches of fatigue crack propagation.
- The microscopic stress distribution induced by transformation expansion of martensite is compressive in martensite and tensile in ferrite.

Acknowledgement

This work was supported by the Global COE Program (Project: Center of Excellence for Advanced Structural and Functional Materials Design).

References

[1] Nakashima K, Nose T and Ishikawa T 2004 *Proc. Symp. on Welded Structure 2004 (Yousetsu-Kouzou-Symposium 2004)*, pp 335-342

[2] Mikami Y, Morikage Y, Mochizuki M and Toyoda M 2006 *Quarterly Journal of the Japan Welding Society* **24** 312-323

Nanocomposites synthesized using supercritical fluids

Zhimin Liu[1]
Beijing National Laboratory for Molecular Sciences, Institute of Chemistry, The Chinese Academy of Sciences, Beijing 100190, China

E-mail: liuzm@iccas.ac.cn

Abstract. Herein, our recent work on synthesis of nanocomposites including carbon nanotube-based, polymer-based and porous material-based composites via chemical reactions in supercritical fluids (SCFs) is introduced. The described examples highlight that SCFs allow facile synthesis of nanocomposites, leading to some new nanomaterials with special structures that are very difficult to achieve by conventional methods. The potential applications of these nanocomposites were also discussed.

1. Introduction

Nanocomposites have attracted great interest due to their unusual properties and promising applications in many fields. Up to date, various methods have been developed for the preparation of this kind of materials. In particular, clean synthesis of functional materials has been paid much attention with development of green chemistry in recent years. Green solvents including supercritical fluids (SCFs) and ionic liquids (ILs) provide new opportunity for the development of advanced functional materials. A supercritical fluid (SCF) can be defined as a substance, the temperature and pressure of which are higher than its critical temperature (T_c) and critical pressure (P_c), respectively, and which has a density close to or higher than its critical density (ρ_c) [1]. SCFs possess some unique properties such as low viscosity, high diffusivity, near zero surface tension, etc, and these properties can be tuned via changing pressure and temperature. SC CO_2 and SC H_2O are the widely used SCFs due to their unusual properties as well as green feature, and have been applied in many areas. Especially, they display promising applications in material science. Some new and clean methods involved with SCFs have been developed for the synthesis and processing of functional materials, and many materials with special structures and enhanced performances have been prepared.

In recent years, our group focuses on the synthesis of functional materials using green solvents. Taking advantage of SCFs and ILs, we proposed some novel methods for the synthesis of nanomaterials, especially for nanocomposites [2-6]. Herein the work on the synthesis of nanocomposites using SCFs is summarized, concentrating on the description of the methods and microstructures of the as-prepared nanocomposites.

2. Synthesis of nanocomposites using SCFs

Besides their common features, different SCFs including SC CO_2 solutions, SC H_2O, SC ethanol, SC diethyl amine have their inherent characteristics. By means of SCFs, several methods to prepare

[1] To whom any correspondence should be addressed.

nanocomposites have been developed, and a series of nanocomposites such as carbon nanotube-based, polymer-based and porous material-based nanocomposites have been fabricated in our previous work.

2.1 Synthesis of carbon nanotube-based composites

Functionalization of carbon nanotubes (CNTs) has attracted much attention since it can widen their applications in electronic devices, sensing area, catalysis, etc. Synthesis of CNT-based composites is an important way to functionalize CNTs. However, it is a challenging topic because the surface of CNTs is rather inert and it is not easy to be wetted by liquids with surface tension higher than 100-200 $mN{\cdot}m^{-1}$. SCFs have near zero surface tension, which show intriguing advantages for the synthesis of various CNT-based composites. In this paper, our recent advances in the synthesis of CNT composites using SC CO_2 and SC CO_2-based solutions, SC water, SC methanol will be summarized, focusing on the synthesis of metal/CNT, metal oxide/CNT and polymer/CNT composites [7,8].

2.1.1 Deposition of Ru nanocrystals on CNTs in SC H_2O

Different from normal and heated water, SC H_2O has dielectronic constant close to that of some organic solvents (e.g, acetone). Therefore, SC H_2O has been proven a good solvent for many organic reactions, and it also has displayed excellent advantages for synthesis of inorganic materials. Using SC H_2O as reaction medium, we developed a one-pot route to deposit metal nanoparticles (e.g, Ru) on CNTs [5]. In this method, CNTs were first dispersed in $RuCl_3$ aqueous solution at room temperature and placed in a high-pressure autoclave; then the autoclave was put into an air-bath with temperature of 400 °C or 450 °C; keeping at these conditions for certain time, Ru/CNT nanocomposites were obtained. It was demonstrated that Ru nanoparticles with size less than 5 nm were uniformly distributed on the surface of CNTs, and some particles seemed to be present in the inner channels of CNTs. This special structure may result from the property change of H_2O during the reaction process. At room temperature, the precursor was soluble in water. With increase in temperature, the solvent power of water for the precursor declined, which could render some precursor molecules to precipitate from the aqueous solution and further to be adsorbed on the surface of CNTs. At high temperature, these precursor molecules chemically converted into metallic form, and preferentially nucleated on the surface of CNTs and formed particles, producing metal/CNT composites finally. Since the surface tension of SC H_2O is zero, some precursor molecules could infuse into the inner cavity of CNTs to yield metal particles inside CNTs after chemical conversion. Changing the weight ratio of CNTs to $RuCl_3$, the loading content and particle size of Ru particles were readily tuned. Reaction temperature also influenced the particle size of Ru particles, and larger particles were obtained at higher temperature (e.g., 450 °C).

2.1.2 Synthesis of metal oxide/CNT composites using SC CO_2 solutions

SC CO_2 is the most widely used SCF because it is non-toxic, non-flammable, chemically inert and easily available. The SC CO_2-involved process is usually operated at mild conditions because the critical temperature of CO_2 (Tc=31.1°C) is close to room temperature and its critical pressure (Pc=7.38MPa) is not too high. However, the major shortcoming of SC CO_2 is that it cannot dissolve most of polar compounds, especially, ionic compounds (e.g, metal salts). SC CO_2 is miscible with some polar organic solvents to form solutions, which possess the similar properties to SCFs. Therefore, the SC CO_2-based solutions are intensively investigated, and have been applied in many processes. In our recent work, SC CO_2 solutions including SC CO_2-ethanol and SC CO_2-methanol have been successfully applied in the synthesis of CNT composites.

Metal oxide decorated CNTs are a kind of important materials, and have many potential applications. We developed an one-pot route to synthesize this kind of composites via decomposition of metal nitrates in SC CO_2 solutions. In this method, CNTs were first dispersed in the ethanol solution of metal nitrate, and CO_2 was then charged into the ethanol solution at relatively low temperature (e.g, 35°C) up to a desired pressure to form a CO_2-ethanol solution. Subsequently, temperature was increased to a higher value, at which the reaction media reached supercritical state and the metal nitrate decomposed into metal oxide as well, resulting in metal oxide/CNT composites. Using different metal nitrates as precursors, a series of metal oxides, such as Eu_2O_3 [9], Al_2O_3 [10], Co_3O_4 [11], Fe_2O_3 [6], ZrO_2 [12], CeO_2 [13], Mn_3O_4 [14], Cr_2O_3 [15], SnO_2 [16], etc, were decorated on CNTs in SC CO_2-ethanol solutions in the temperature range of 100-150°C.

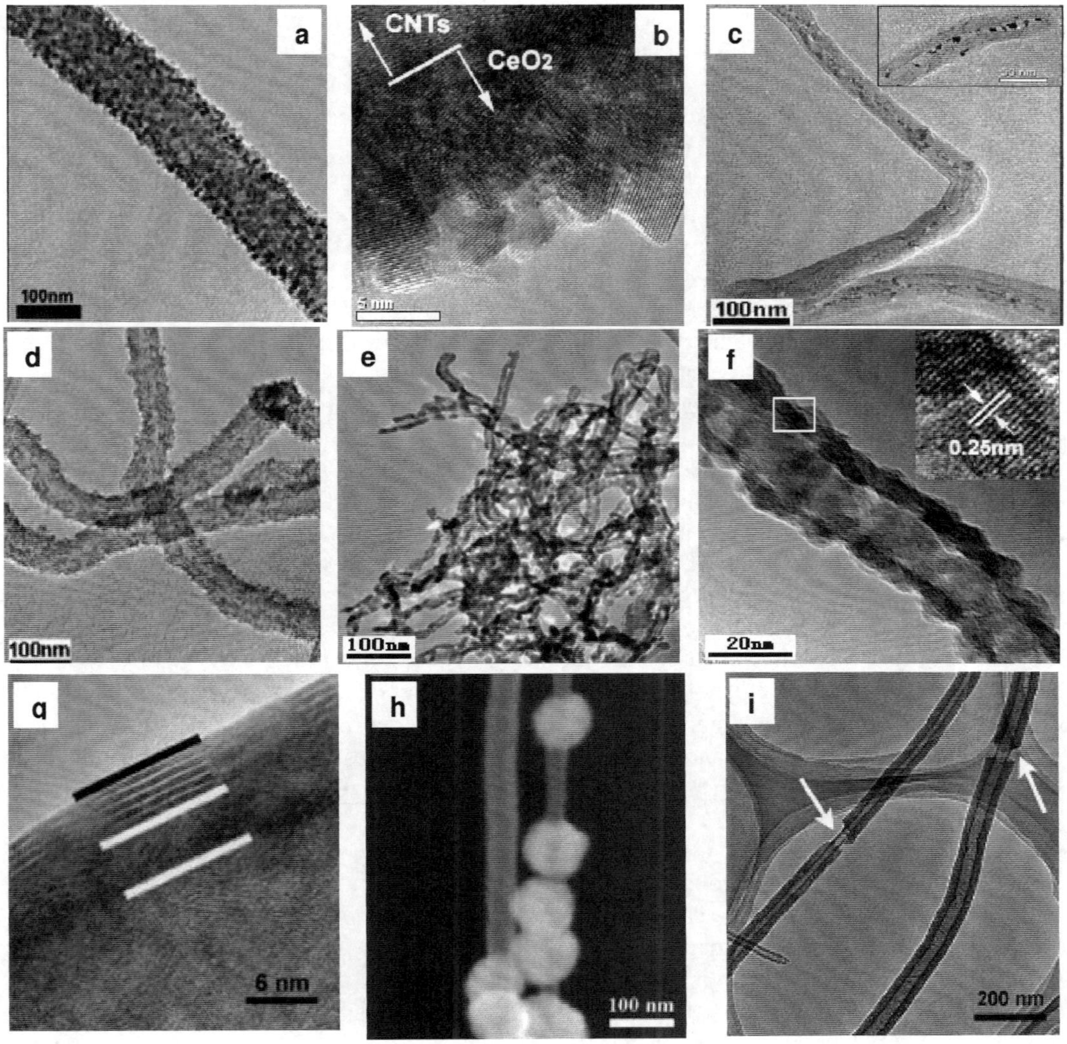

Figure 1 TEM images of the metal oxide/CNT composites prepared in SC CO_2-ethanol solutions: (a) Mn_3O_4/CNT, (b) CeO_2/CNT, (c) SnO_2/CNT, (d) Fe_2O_3/CNT, (e, f) α-Fe_2O_3 nanotubes, (g) Eu_2O_3/CNT, (h) Co_3O_4/CNT, (i) Al_2O_3/CNT. Reprinted from references [6], [9-14], and [16b]

The resultant composites displayed different morphology, as illustrated in Figure 1. In general, the metal oxides exist in the form of nanoparticles on CNTs (Figures 1a, b), and some particles could be intercalated into the inner cavity of CNTs under suitable conditions (Figure 1c). Several metal oxides with amorphous structure can be coated outside the CNTs to form core-shell structures (Figures 1d). Calcining the core-shell structures to remove the CNT at high temperature (e.g., 600 °C) , metal oxide nanotubes with structure similar to CNTs could be produced (Figures 1e, f) [6,15]. This means that CNTs can act as a template for the preparation of some metal oxide nanotubes. Some special structures of metal oxide/CNT composites were also achieved. For instance, crystalline Eu_2O_3 nanotubes were coated on CNTs for the first time (Figure 1g) [9]; Co_3O_4 spheres composing of nanocrystals were beaded on CNTs to form necklace-like structure (Figure 1h) [11]; discontinuous Al_2O_3 layer was wrapped outside CNTs (Figure 1i) [10]. These unique structures are difficult to prepare by conventional methods. For comparison, some control experiments were performed to prepare metal oxide/CNT composites in ethanol under the similar conditions. It was demonstrated that large particles of metal oxide formed out of CNTs. This suggests that the property of the reaction media plays a key role in the formation of metal oxide/CNT composites. In the process to synthesize metal oxide/CNT composites using SC CO_2-ethanol as reaction media, with the addition of CO_2 to the ethanol solution, the solvent power of the ethanol and CO_2 mixture to dissolve the precursor molecules reduced gradually, which caused some precursor molecules to precipitate out of the mixture and be adsorbed on the surface of CNTs. The dissolved precursor molecules could infuse into the inner cavity of CNTs due to the very low surface tension and high diffusivity of the solvent mixture. At the reaction temperature (e.g, 120 °C), the precursor on the CNT surface decomposed into metal oxide and preferentially nucleated on CNTs to form particles distributing on the CNTs or oxide layer wrapping the CNTs. It should be mentioned that most oxides wrapping outside the CNTs are amorphous, probably because the reaction temperature was too low to form crystalline phases. The differences in the microstructures of the resultant metal oxide-CNT composites may be caused by the differences in the interactions of the precursors and the CNTs used. The above method also indicates that SC CO_2-ethanol is a good medium for the synthesis of metal oxide-CNT composites under mild conditions.

2.1.3 Synthesis of CNT-based composites using other SCFs

Some organic solvents including ethanol, methanol and diethyl amine, which have critical temperatures around 250 °C, were also used as SCFs for the synthesis of CNT-based composites, and they displayed the similar advantages to the above described SCFs (e.g. SC CO_2 solutions and SC water). The procedures to prepare the CNT-based composites using these solvents are similar to that performed in SC water. Heating the precursor solution to reach supercritical state of the solvent, the CNT-based composites were in-situ obtained. In an example to synthesize TiO_2/CNT composites, the CNTs were dispersed in the ethanol solution of isopropanol titanium, and then heating the solution at 250 °C, a temperature which is higher than the critical temperature of ethanol, TiO_2/CNT composites were successfully obtained. The anatase nanoparticles with size less than 10 nm were homogeneously distributed on CNTs [17]. Similarly, RuO_2 nanocrystals with size around 3 nm were decorated on CNTs uniformly using SC diethyl amine as reaction medium [18].

2.2 Synthesis of polymer-based composites using SCFs

It is well known that SC CO_2 can swell most organic polymers (e.g, polystyrene) while it cannot dissolve them. Based on the particular interactions between SC CO_2 and polymers, several methods involved with SC CO_2 have been proposed to prepare polymer-based composites. A two-step method was developed to prepare silica-polypropylene (PP) nanocomposites [19]. In this method, tetraethyl orthosilicate (TEOS) was first impregnated into PP matrix using SC CO_2 as a swelling agent for PP and carrier for TEOS, and the TEOS intercalated PP was then treated in acidic water, resulting in hydrolysis/condensation reaction of TEOS confined in polymer network. The resulting composites showed unusual microstructures: fine silica nanoparticles distributed uniformly in the PP matrix without macroscopic phase segregation, and the polymer molecules entangled with the silica network. This unique structure led to a significant increase in Young's modulus and tensile strength relative to the virgin PP. This impregnation/hydrolysis method represents a straightforward and effective way to prepare silica/polymer hybrid materials. It is particularly advantageous for polymers insoluble in common organic solvents and for synthesis of the composites in which the compatibility between the polymer and silica is poor. Similarly, the mixture of TEOS and titanium isopropoxide (TIP) were simultaneously impregnated into a SC CO_2-swollen PP matrix in SC CO_2, and SiO_2-TiO_2/PP composites were produced via the hydrolysis and condensation reactions of the precursors confined in a polymer network [20]. The resultant composites showed similar structure to the above silica/PP composites, and also displayed enhanced mechanical performances.

Recently, we developed a method to deposit metal or metal oxide in polymer substrate using SC CO_2-ethanol solutions [21]. Hollow polymer spheres with size of about 400 nm, which cannot be dissolved and swollen by ethanol, were used as substrate. In a typical procedure to prepare Eu_2O_3/polymer composites, the polymer spheres were first dispersed in ethanol solution of europium nitrate; CO_2 was then charged into the solution up to a designed pressure at low temperature (e.g, 35°C), which mixed with ethanol to form homogeneous solution that can swell the polymer spheres; keeping at this condition for certain time, the temperature was increased to a high value (e.g, 120°C), at which europium nitrate decomposed into Eu_2O_3 and deposited onto polymer spheres to form Eu_2O_3/polymer composites. The resultant composites exhibited unique structure: the Eu_2O_3 nanoparticles were not only decorated on the outer surface, but also intercalated into the polymer shell and further into the inner cavity of the hollow polymer spheres. This special structure is also caused by the unusual properties of SC CO_2-ethanol solution. Similarly, Pd nanoparticles were embedded into polymer spheres using SC CO_2-ethanol solution as a reaction medium, and the as-prepared Pd/polymer spheres had high activity and stability for alkenen hydrogenations [22]. In another work, zinc acetate was used as Zn^{2+} precursor and thiourea as S^{2-} resource, and ZnS nanoparticles were decorated on the polymer spheres in SC CO_2-ethanol solution at 100 °C. The procedure to prepare this composite was similar to that for preparing Eu_2O_3/polymer composite. However, the ZnS nanoparticles had much smaller particle size of around 3nm. Since the ZnS/polymer composite was dispersible in water and could form stable suspension, it was used as a photocatalyst to degrade some organic dyes, which was very effective to degrade dyes under the visible light irradiation [23].

2.3 Synthesis of porous material-based composites

Porous materials including molecular sieves (e.g, zeolite, SBA-15, MCM-41, etc), natural clays usually have high surface areas, which are good supports for catalysts and have promising applications in catalysis. Deposition of chemically active components on the porous materials to prepare the supported catalysts is an interesting topic. SCFs, especially SC CO_2 and its solutions, provide new methods for the synthesis of this kind of materials due to their low viscosity, high diffusivity, zero surface tension as well as green features.

SBA-15 possesses a highly ordered mesoporous hexagonal structure with pore diameters varying from 5 to 30 nm. Functionalization of SBA-15 provides it more properties, and can widen its applications. Recently, we presented a monolayer and double-layer coating of TiO_2 on the inner surfaces of SBA-15 via surface sol–gel process in SC CO_2 [24]. In this work, SBA-15 with high level of silanol groups was adopted as substrate. In a typical experiment, the TIP dissolved in SC CO_2 passed through SBA-15, and grafted onto SBA-15 via the reaction of TIP with the surface silanol groups on SBA-15, forming a coating layer onto the inner surface of SBA-15. The unreacted TIP was subsequently extracted by SC CO_2. Repeating the impregnating step can result in the multilayer formation of TiO_2 on the silica surfaces. Calcining the impregnated SBA-15 at 550°C, TiO_2 coated SBA-15 composites were formed. Raman spectroscopy and XPS analyses confirm the formation of Si-O-Si and Ti-O-Ti. N_2 sorption analysis indicated that the TiO_2 loaded samples retained the similar structure to the SBA-15 used, and the pore diameter was reduced. In this study, the silanol groups on SBA-15 serve as active sites for titanium grafting and were consumed during the reaction process. The high concentration of hydroxyls on the silica surface was required to prepare high surface coverage of molecularly dispersed titanium species. Similarly, coating europium oxide on the walls of SBA-15 was achieved in a SC CO_2-ethane solution [25]. Europium nitrate hexahydrate as oxide precursor was first dissolved in ethanol, and mixed with SBA-15. Then ethane was charged into the mixture at 35°C to reach the supercritical state of ethane-ethanol solution, which facilitated the precursor to infuse into the channels of SBA-15. Following in situ thermal decomposition of europium nitrate at 120 °C in the supercritical solution, SBA-15/Eu_2O_3 composites were fabricated. In another work [26], an effective route was described to prepare TiO_2-montmorillonite (MMT) composites via impregnating TTBO into the interlayers of MMT in supercritical ethanol, followed by hydrolysis in water and calcination at 500 °C. The layered structure of MMT in the as-prepared samples was destroyed to some extent based on the XRD analysis and TEM observation. Anatase particles with size less than 5 nm were deposited in the partially exfoliated MMT layers. These samples showed blue-shifted UV absorption edge compared to TiO_2 particles. The MMT treated with supercritical ethanol has excellent ability to adsorb MB, and the TiO_2–MMT composites can combine the adsorptive ability of MMT and the catalytic degradation ability of TiO_2 to remove MB from its aqueous solution efficiently.

3. Summary

In summary, SCFs have provided new routes and displayed many unique advantages for the synthesis of nanocomposites though the processes have to be operated at high pressure. Remarkable advantages of the SCF techniques lie in their flexibility, simplicity, and large-scale production of the composites. The resultant materials including CNT-based, polymer-based and porous material-based composites usually possess special structures that are difficult to prepare by the conventional methods, and they also have enhanced performances and promising applications in many areas. The utilization of SCFs in material science has opened a new way for the development of advanced materials, and the related technologies may have bright future.

References

[1] J. A. Darr, and M. Poliakoff, Chem. Rev. **99**, 495 (1999).
[2] S. D. Miao, Z. M. Liu, B. X. Han, X. Yu, and Z. J. Miao, Angew. Chem. Int. Ed. **45**, 266 (2006).
[3] K. L. Ding, Z. J. Miao, Z. M. Liu, Z. F. Zhang, B. X. Han, S. D. Miao, and Y. Xie, J. Am. Chem. Soc. **129**, 6362 (2007).
[4] S. D. Miao, Z. M. Liu, B. X. Han, and G. Y. Yang, Micropor. Mesopor. Mater. **95**, 26 (2006).
[5] Z. Y. Sun, Z. M. Liu, B. X. Han, Y. Wang, J. M. Du, and Z. L. Xie, Adv. Mater. **17**, 928 (2005).
[6] Z. Y. Sun, H. Q. Yuan, Z. M. Liu, B. X. Han, and X. R. Zhang, Adv. Mater. **17**, 2993 (2005).
[7] Z. Liu, X. Dai, J. Xu, B. Han, J. Zhang, Y. Wang, Y. Huang, and G. Yang, Carbon **42**. 458 (2004).
[8] X. H. Dai, Z. M. Liu, B. X. Han, Z. Y. Sun, Y. Wang, J. Xu, X. L. Guo, and N. Zhao, J. Chen, Chem. Commun. 2190 (2004).
[9] L. Fu, Z. M. Liu, Y. Q. Liu, B. X. Han, J. Q. Wang, and L. C. Cao, Adv. Mater. **16**, 350 (2004).
[10] L. Fu, Z. M. Liu, Y. Q. Liu, L. C. Cao, D. C. Wei, and D. B. Zhu, Adv. Mater. **18**, 181 (2006).
[11] L Fu, Y. Q. Liu, Z. M. Liu, B. X. Han, P. Hu, L. C. Cao, D. B. Zhu, Adv. Mater. **17**, 217 (2005).
[12] Z. Y. Sun, X. R. Zhang, Z. M. Liu, B. X. Han, S. D. Miao, and G. Y. Yang, J. Phys. Chem. B **110**, 13410 (2006).
[13] Z. Y. Sun, X. R. Zhang, B. X. Han, Z. M. Liu, and S. D. Miao, Carbon **45**, 2589 (2007).
[14] G. M. An, P. Yu, M. J. Xiao, Z. M. Liu, and Z. J. Miao, Nanotechnology **19**, 275709 (2008).
[15] G. M. An, Y. Zhang, Z. M. Liu, Z. J. Miao, and K. L. Ding, Nanotechnology **19**, 035504 (2008).
[16] (a) G. M. An, N. Na, X. R. Zhang, Z. J. Miao, S. D. Miao, and Z. M. Liu, Nanotechnology **18**, 435707 (2007); (b) Z. Y. Sun, Z. M. Liu, B. X. Han, G. M. An Mater. Lett. **61**, 4565 (2007).
[17] G. M. An, W. H. Ma, Z. Y. Sun, and Z. M. Liu, Carbon **45**, 1795 (2007).
[18] Z. Y. Sun, Z. M. Liu, B. X. Han, S. D. Miao, J. M. Du, and Z. J. Miao, Carbon **44**, 888 (2006).
[19] D. H. Sun, R. Zhang, Z. M. Liu, Y. Huang, Y. Wang, J. He, B. X. Han, and G. Y. Yang, Macromolecules 38, 5617 (2005).
[20] D. H. Sun, Y. Huang, B. X. Han, and G. Y. Yang, Langmuir **22**, 4793 (2006).
[21] J. Q. Wang, C. L. Zhang, Z. M. Liu, and Z. Z. Yang, Macromol. Rapid Commun. **27**, 787 (2006).
[22] S. D. Miao, C. L. Zhang, Z. M. Liu, B. X. Han, Z. Z. Yang, J. Phys. Chem. C. 1**12**, 774 (2008).
[23] Y. Xie, C. L. Zhang, S. D. Miao, Z. M. Liu, K. L. Ding, Z. J. Miao, G. M. An, and Z. Z. Yang, J. Colloid & interf. Sci. **318**, 110 (2008).
[24] D. H. Sun, Z. M. Liu, B. X. Han, and Y. Huang, Micropor. Mesopor. Mater. **80**, 165 (2005).
[25] Z. M. Liu, J. Q. Wang, B. X. Han, and Z. Y. Sun, Micropor. Mesopor. Mater. **75**, 101 (2004).
[26] S. D. Miao, Z. M. Liu, B. X. Han, J. L. Zhang, and Z. Y. Sun, J. Mater. Chem. **16**, 579 (2006).

Journal of Physics: Conference Series **165** (2009) 012082 doi:10.1088/1742-6596/165/1/012082

Terahertz wave behaviours in ceramic and metal structures fabricated by spatial joining of micro-stereolithography

S Kirihara[1], T Niki[1] and M Kaneko[1]
Joining and Welding Research Institute, Osaka University
11-1 Mihogaoka Ibaraki, 567-0047 Osaka, JAPAN

E-mail: kirihara@jwri.osaka-u.ac.jp

Abstract. We have newly developed micro-stereolithography system to realize freeform fabrication of micrometer order 3D metal structures. In this process, the photo-sensitive resin paste mixed with nanometer sized ceramic and metal particles was spread on a glass substrate with 10 μm in layer thickness by using a mechanical knife edge, and two-dimensional images of UV ray were exposed using DMD (Digital Micro-mirror Device) with 2 μm in part accuracy. Through the layer by layer stacking process, micrometer order three-dimensional objects were formed. Dense metal structures could be obtained by dewaxing and successive sintering of the formed objects. In our recent investigation, micro photonic crystals with lattice structures of alumina or pure copper were fabricated in order to control electromagnetic wave propagation in a terahertz (THz) frequency range. The micro photonic crystals with a diamond structure perfectly reflected the THz wave by Bragg diffraction

1. Introduction

We have newly developed micro-sterelolithography system to realize spatial joining of micrometer order ceramic or metal structures with three-dimensional distributions of dielectric or conductive materials. The final goal of our investigation is to control electromagnetic waves in a terahertz frequency range with micrometer wavelength effectively. In near future, the terahertz wave will be expected to apply to various types of novel sensors which can detect gun powders, drugs, bacteria in foods, micro cracks in electric devices, cancer cells in human skin and other physical, chemical and living events. To control terahertz waves effectively, micrometer sized electromagnetic devices composed of dielectric ceramics applying for cavities, filters and antennas will be necessary [1,2].

Photonic crystals composed of dielectric lattices form band gaps for electromagnetic waves [3-6]. These artificial crystals can totally reflect light or electromagnetic wave at wavelengths comparable to the lattice spacings by Bragg deflection. The two different standing waves vibrating in the air and dielectric matrix form higher and lower frequency bands in the first and second Brillouin zones, respectively. The band gap width can be controlled by varying the structure, filling ratio, and dielectric constant of the lattice. Structural modifications by introducing defects can control the transmission of electromagnetic wave as well [7-9]. The introduced structural defects in the periodic arrangement can localize the electromagnetic wave energy and form the transmission mode in the band gap frequency range according to the size and dielectric constant of a defect region. The photonic crystal with a diamond structure can form the perfect band gap which opens for all crystal directions [10,11]. We

[1] Joining and Welding Research Institute, Osaka University, 11-1 Mihogaoka Ibaraki, 567-0047 Osaka, Japan

have fabricated the millimeter sized dielectric photonic crystal with a diamond structure to control the microwave by using a structural joining method of stereolithography [12,13]. In our recent study, micrometer sized alumina and pure copper lattices with a diamond structure were fabricated by using a newly developed micro-stereolithography system [14]. These photonic crystals showed the perfect band gap, which prohibited the terahertz wave propagation in all directions. Recently, we successfully fabricated a twinned diamond structure with a plane defect between mirror symmetric lattice patterns. A localized mode to transmit terahertz wave selectively was formed in the photonic band gap. In this study, transmission spectra of terahertz wave through alumina and pure copper diamond structures were measured. The selective transmission mode in the band gaps were observed for the twinned diamond structure and discussed in relation to the terahertz wave localization at the plane defect.

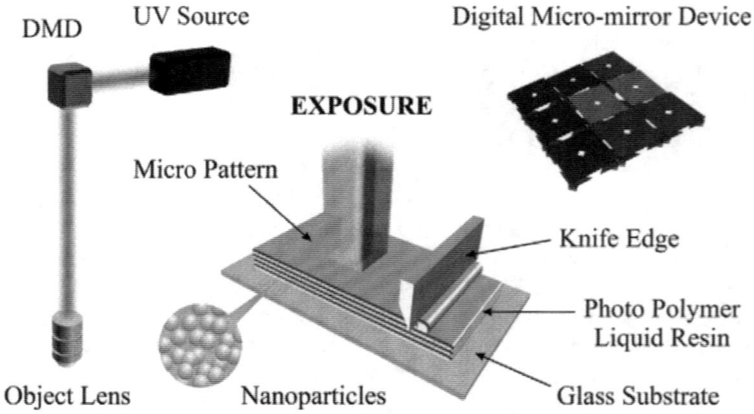

Figure 1. A schematic illustration of micro-stereolithography.

2. Experimental Procedure

The three-dimensional diamond structures were designed by using a 3D-CAD program (Toyota Caelum Co. Ltd., thinkdesign ver. 7.0). In the photonic crystal with diamond structure, the lattice constant was 1 mm. The whole structure was 6×6×2 mm^3 in size, consisting of 6×6×2 unit cells. The aspect ratio of the metal lattices was designed to be 1.5. The CAD data was converted into a STL file of a rapid prototyping format. After the slicing process of 3D model into a series of two-dimensional cross-sectional data into thin sections, this data was transferred to micro-stereolithographic equipment (D-MEC Co. Ltd., Japan, SI-C10008). In our system, photo sensitive acrylic resins dispersed with alumina or pure copper particles of 170nm or 5 μm in diameter at 40 or 54 vol. %, respectively, were fed over substrates from dispenser nozzles. The highly viscous resin paste was fed with controlled air pressure. It was spread uniformly by moving a knife edge. The thickness of each layer was controlled to 10 μm. A two-dimensional pattern was formed by illuminating visible laser of 405 nm in wavelength on the resin surface. The high resolution has been achieved by using a Digital Micromirror Device (DMD) and an objective lens. Figure 1 shows a schematic of the micro-stereolitographic system. The DMD is an optical element assembled by mirrors of 14 μm in edge length. The tilting of each tiny mirror can be controlled according to the two-dimensional cross sectional data by a computer. The three-dimensional structures were built by stacking these patterns layer by layer. In order to avoid deformation and cracking during dewaxing, careful investigation for the dewaxing process was required. The precursors with diamond structures were heated at various temperatures from 100°C to 600°C while the heating rate was 1.0 °C /min. The dewaxing process was observed in respect to the weight and color changes. The alumina or pure copper particles could be sintered at 1500

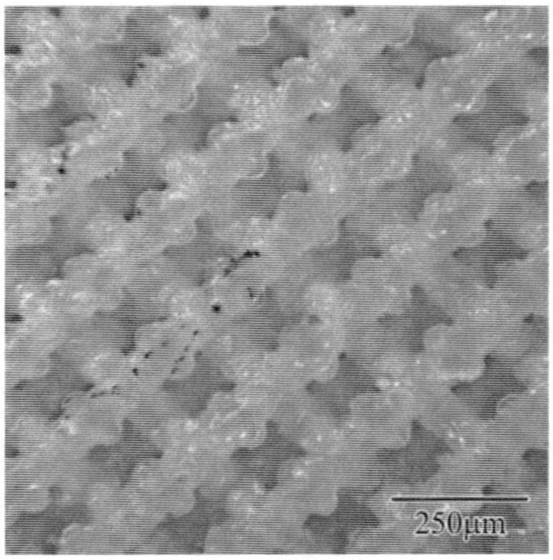

Figure 2. A dielectric photonic crystal composed of micro alumina lattices with diamond structure fabricated by using stereolithography.

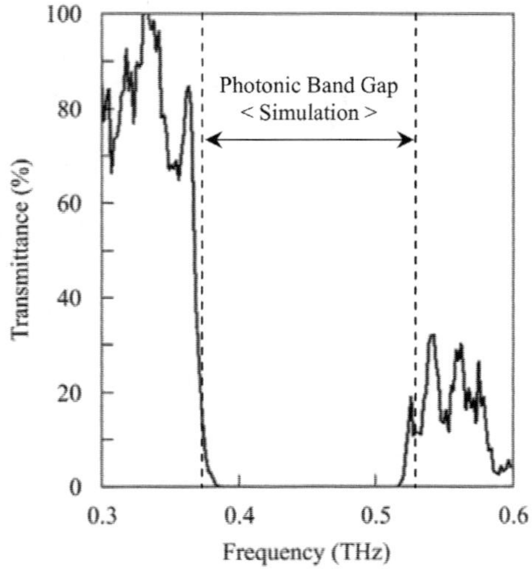

Figure 3. A transmission spectrum of terahertz wave for Γ-X <100> direction in the alumina photonic crystal with the diamond structure.

Figure 4. A twinned diamond photonic crystal with a plane defect between mirror symmetric alumina lattices parallel to (100) layers.

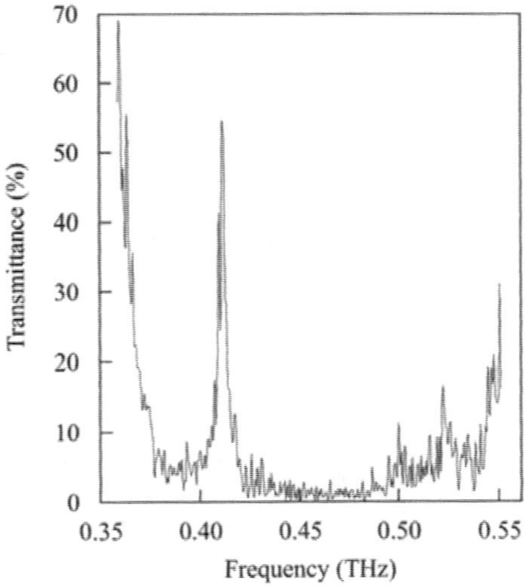

Figure 5. A localized mode of a transmission peak in a photonic band gap formed through the twinned diamond structure of the alumina lattice.

or 1000°C, respectively. The heating rate was 8.0°C/min. The density of the sintered sample was measured by the Archimedes method. The microstructures were observed by optical microscope and scanning electron microscopy (SEM). The transmittance and the phase shift of incident terahertz waves were measured by using terahertz time-domain spectroscopy (Advanced Infrared Spectroscopy Co. Ltd., Japan, J-Spec2001 spc/ou). Measured terahertz properties were compared with simulation by using a Transmission Line Modeling (TLM) simulation program (Flomerics: Micro-Stripes Ver. 7.0).

3. Results and Discussion

Three-dimensional lattice structures composed of the alumina dispersed acrylic resin were processed exactly by using micro-stereolithography. The spatial resolution was approximately 0.5 %. Figure 2 shows a (100) plane of the sintered diamond structure composed of the micrometer order alumina lattice. The deformation and cracking were not observed. The linear shrinkage on the horizontal axis was 23.8 % and that on the vertical axis was 24.6 %. The relative density reached 97.5 %. Dense alumina microstructure was formed, and the average grain size was approximately 2 μm. The measured dielectric constant of the lattice was about 10. The terahertz wave attenuation of the transmission amplitude through the alumina diamond structure for Γ-X <100> crystal direction is shown in figure 3. The forbidden gap is formed at the frequency rage of from 0.37 to 0.52 THz. The dotted lines show the higher and lower band gap edges calculated by using the TLM method of a finite element method. The similar transmission spectra for Γ-K <110> and Γ-L <111> directions were obtained. These measured results of the photonic band gap frequencies were verified to have good agreements with the simulated ones within a tolerance of 5 %. A common band gap was observed in every direction at the frequency range form 0.40 to 0.47 THz, where the electromagnetic wave cannot transmit through the crystal and is totally reflected in all directions. This common band was included in the calculated perfect photonic band gap by using a plane wave expansion (PWE) method. In this theoretical calculation, the plane wave propagations were simulated for all directions in the periodic arrangements of the dielectric materials by solving Maxwell's equations. And the photonic band distributions were drown along the symmetry lines in the Brillouin zone. Figure 4 shows the twinned diamond structure composed of the mirror symmetric alumina lattices. The plane defect forms parallel to the (100) crystal layer. The transmission spectrum for the Γ-X <100> crystal direction of the twinned diamond structure is shown in figure 5. The localized mode forms in the photonic band gap. At the peak frequency, the incident terahertz wave localized in the plane defect, and the amplified wave propagated to the opposite side. The three-dimensional photonic band gap structure to form the localized mode can be applied to the terahertz wave filters.

Figure 6 shows a precursor of the metal photonic crystal consisting of copper particles dispersed resin fabricated by micro-stereolithography. The spatial resolution was approximately 0.5 %. By the dewaxing and sintering process, full-metal diamond structures were successfully obtained. Figure 7 shows the sintered samples. The deformation and cracking were not observed. The lattice constant of the diamond structure is 900 μm. The linear shrinkage was 10%. It is possible to obtain the uniform shrinkage by designing an appropriate elongated structure in the vertical direction for compensation to the gravity effect. The relative density reached 89.7 %. Figure 8 shows the transmittance intensity and the phase shift spectra along Γ-X <100> directions. Opaque regions were formed and phase shifts became discontinuous, indicating the formation of photonic band gaps, at the frequency range from 0.39 to 0.48 THz. The band gap frequency was compared with calculations by using the TLM method as shown in figure 9. The measured opaque region corresponded to the calculation. In the photonic band gap, the incident wave are diffracted by multiple reflections in the periodic structure composed of the three dimensional pure copper lattices. As shown in figure 8 and figure 3, the transmission spectra obtained through the pure copper metal structures and the alumina ceramic ones exhibit different profiles. The electromagnetic wave can transmit with resonances and absorptions into the alumina ceramic lattices. However, the pure copper metal lattices reflect the wave on the surface without transmission. These different behaviours of the electromagnetic wave diffractions in the periodic lattices cause the variations of band gap profiles in the transmission spectra.

Figure 6. The diamond photonic crystal of acryl lattices with the pure copper dispersion formed by using micro-stereolithography.

Figure 7. The pure copper photonic crystal with the diamond lattice structure composed of the micrometer order conductive metal rods.

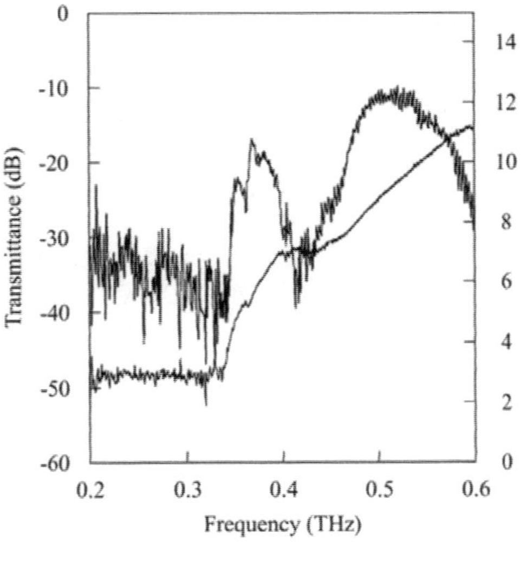

Figure 8. The terahertz wave transmittance and phase shift spectrum for the Γ-X <100> direction through the pure copper photonic crystal with the diamond lattice structure.

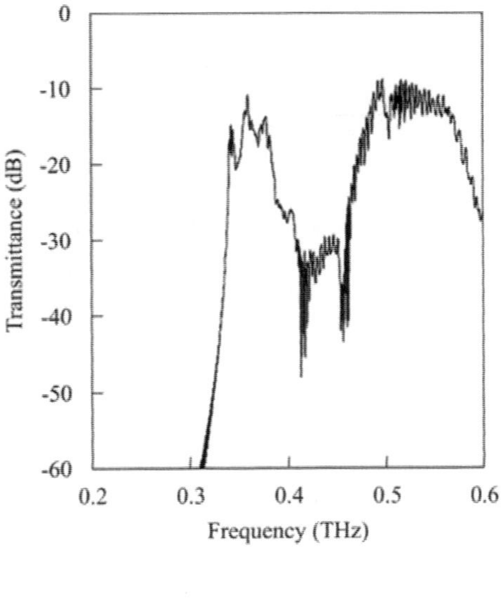

Figure 9. The simulated transmission spectrum in the terahertz wave frequency for the photonic crystal composed of the pure copper diamond lattices by using TLM method.

4. Conclusion

We have fabricated three-dimensional micro photonic crystals with a diamond structure composed of alumina or pure copper dispersed acrylic resin by using micro-stereolithography. By the careful optimization of process parameters regarding dewaxing and sintering, we have succeeded in fabricating dense alumina or pure copper micro diamond structures. The sintered photonic crystals formed complete photonic band gaps at the terahertz region. A twinned diamond photonic crystal composed of alumina lattices with a plane defect between mirror symmetric lattice structures was also fabricated. A localized mode of transmission peak was observed in the forbidden bands. In electromagnetic wave simulations, the localized mode with multiple reflections was formed in the plane defect between the twinned lattice patterns. These micro components of ceramic and metal photonic crystals have potentials to be used as cavities, filters and antennas in a THz range.

5. Acknowledgments

This study was supported by Priority Assistance for the Formation of Worldwide Renowned Centers of Research - The Global COE Program (Project: Center of Excellence for Advanced Structural and Functional Materials Design) from the Ministry of Education, Culture, Sports, Science and Technology (MEXT), Japan.

6. References

[1] Woodward M R, Wallacel P V, Arnonel D D and Linfield H E 2003 *J. Biological Phys.* **29** 257
[2] Exter V M, Fattinger C and Grischkowsky D 1989 *Optics Lett.* **14** 1128
[3] Yablonovitch E 1987 *Phys. Rev. Lett.* **58** 2059
[4] Ohtaka K 1979 *Phys. Rev. B* **19** 5057-5067.
[5] John S and Wang J 1990 *Phys. Rev. Lett.* **64** 2418
[6] Soukoulis C M 1996 Photonic Band Gap Materials (Netherlands: Kluwer Academic Publsher)
[7] Brown R E, Parker D C, Yablonovich E 1993 *J. Optical Society of America B*, **10** 404
[8] Noda S, Yamamoto N, Kobayashi H, Okano M and Tomoda K 1999 *Appl. Phys. Lett.* **75** 905
[9] Kawakami S 2002 Photonic Crystals (Tokyo: CMC)
[10] Ho H K, Chan T C and Soukoulis M C 1990 *Phys. Rev. Lett.* **65** 3152
[11] Haus W J 1994 *J. Modern Optics* **41** 195
[12] Kirihara S, Takeda W M, Sakoda K, Miyamoto Y 2002 *Solid State Communications* **124** 135
[13] Miyamoto Y, Kirihara S and Takeda W M 2006 *Chem. Lett.* **35** 342
[14] Kirihara S, Miyamoto Y 2008 Ceram. *Interconnect and Ceram. Microsystems Technol.* **5** 254

Nano-hydroxyapatite/poly ε-caprolactone composite 3D scaffolds for mastoid obliteration

S E Kim[1], H S Yun[1], Y T Hyun[1], J W Shin[2], J J Song[3]

[1] Center for Future Technology, Korea Institute of Materials Science, Changwon, Gyeongnam, 641-010, Korea
[2] Department of Biomedical Engineering, Inje University, Gimhae, Gyeongnam 621-749, Korea
[3] Department of Otorhinolaryngology, Dongguk University Hospital, Goyang-si, Gyeonggi-do, 410-773, Korea

E-mail: sek24@kims.re.kr

Abstract. The aim of this study is to evaluate the use of our nano-HA/PCL composite 3D scaffolds as graft materials for mastoid cavity obliteration in an animal model. Nano-HA particles were synthesized by chemical precipitation technique and mixed them with PCL solution to make composite paste. 3D scaffolds were fabricated by a paste extruding deposition process. The nano-HA/PCL 3D scaffolds showed good in vivo bone regeneration behaviour in a rabbit model after 4 and 8 week implantation. To characterize the 3D scaffolds as a grafting material for mastoid obliteration, mastoid cavities were introduced in rats and implanted the scaffolds. After two week implantation, histological examination showed good tissue in-growth and new bone formation behaviour. It can be argued that our nano-HA/PCL composite 3D scaffold is a promising alternative material for mastoid obliteration.

1. Introduction

Scaffolds for tissue engineering should be highly porous, tissue supportable and biodegradable. Furthermore, scaffolds for bone regeneration should be osteoinductive as well as osteoconductive. Hydroxyapatite (HA) and poly ε-caprolactone (PCL) composites have been paid much attention as one of the prospective bone scaffold materials, due to their good combination of biodegradability, osteoconductivity and osteoinductivity [1]. In addition to materials, pore structure of the scaffolds is very important to lead cell or tissue in-growth. There are several conventional methods for fabricating porous scaffolds such as particulate leaching, freeze drying, electrospinning and so on [2-4]. However, if the scaffold size is large, it is difficult to obtain three dimensionally interconnected pores for cell or

[1] Corresponding author; S E Kim, sek24@kims.re.kr

tissue in-growth by the conventional scaffolding processes. Further, the conventional techniques have some limitations in reproducibility and three dimensionally complex shaping.

Recently layer manufacturing technology has been paid much attention to fabricate porous scaffolds for tissue engineering, because it can provide different pore structure designs with three dimensionally interconnected pores, and it makes on-demand shaping possible by computer aided processes. There have been suggested several methods for fabricating scaffolds, such as fused deposition modeling, selective laser sintering, 3D printing™, multiphase jet solidification, and 3D plotting [5–7]. Among those methods, a modified fused deposition modeling is a preferred process today. We have also recently developed a paste extruding deposition process to fabricate three dimensional (3D) scaffolds. Using this process, we have developed nano-crystalline HA/PCL composite 3D scaffolds for bone regeneration [8].

By the way, the surgical treatment of serious otitis media often requires canal wall down mastoidectomy to ensure complete removal of decease. This otologic surgery yields mastoid cavity which should be eventually obliterated. Different types of biomaterials have been so far used for mastoid obliteration. However, no single graft material has been proved to be ideal [9-11]. The aim of this study is to characterize the in-vivo bone regeneration behaviour of nano-HA/PCL composite 3D scaffold as a candidate graft material for mastoid cavity obliteration. This study discusses the structure of nano-HA/PCL composite 3D scaffolds and the results of bony implantation in a rabbit and rat model.

2. Materials and method

2.1. Synthesis of nano-HA

Nano-hydroxyapatite (nano-HA) was synthesized by a precipitation method using calcium nitrate (Ca(NO3)2·4H2O, Junsei Chemical) and ammonium phosphate ((NH4)2HPO4, Junsei Chemical). Calcium nitrate and ammonium phosphate were dissolved separately in de-ionized water. Calcium nitrate solution was dropped very slowly into the ammonium phosphate solution while stirring and heating at 70°C. The pH of the solution was maintained between 10 and 12 by additions of ammonium hydroxide according to the following equation (1):

$$10Ca(NO_3)_2 + 6(NH_4)_3PO_4 + 2NH_4OH = Ca(PO_4)_6(OH)_2 + 20NH_4NO_3 \quad (1)$$

After titration and reaction were finished, the mixture was aged at room temperature for 24 h to precipitate apatite particles. The apatite precipitates were rinsed more than five times with de-ionized water and treated in de-ionized water for 4 h at 90–100°C. After treatment, the nano-crystalline apatite precipitates were obtained in slurry. This slurry and dimethylformamide (DMF) were mixed in a beaker with stirring, and the mixture was gradually heated up to 100–120°C. After complete evaporation of water, nano-HA/DMF slurry was obtained. The slurry was finally vacuum-dried at room temperature to obtain nano-HA powders.

2.2. Fabrication of nano-HA/PCL 3D scaffolds

Poly ε-caprolactone (PCL) (Sigma Aldrich, Mw 65,000) was dissolved in chloroform at 40 °C to prepare PCL solution. The nano-HA was mixed with PCL solution to make nano-HA/PCL composites. The composite paste extruded out of a syringe and stacked layer by layer to fabricate 3D scaffolds with a rectangular lattice shape using a robotic layer manufacturing machine. The layer manufactured scaffolds were rinsed in distilled water for at least 48 hours and freeze-dried for 72 hours to remove residual solvent completely. The structure of nano-HA/PCL scaffolds were examined by scanning electron microscopy. The porosity of 3D scaffolds was measured using mercury intrusion porosimetry (Auto-pore IV 9500, Micromeritics Instruments, Norcross, USA).

2.3. In-vivo evaluation

New Zealand white rabbits were anesthetized with ketamine hydrochloride and 2% xylazine hydrochloride. A 15 mm-defect was introduced in the tibia of each rabbit. The n-HA/PCL 3D scaffold was implanted in the defect and fixed with a titanium bone plate. The rabbits were sacrificed at 4 and 8 weeks after operation. Female rats were anesthetized with 50 mg/ml of ketamine and 25 mg/ml of rompun. A 3 mm-defect was introduced in the right mastoid of each rat and mastoid mucosa was removed with 2% trichloroacetic acid. Mastoid obliteration was conducted by implantation of n-HA/PCL 3D scaffolds. The rats were sacrificed at 2 weeks after operation. Retrieved samples were fixed in 10% neutral formalin followed by decalcification and paraffin- embedding. The samples were examined radiologically using a high resolution micro-computed tomography system (CT 40, Scanco Medical, Bassersdorf, Switzerland). The scanner was set to a voltage of 55 kV and a current of 145 mA to allow sufficient energy. The samples were scanned at 8-μm voxel (3D pixel) resolution with an integration time of 120 ms to produce reconstructed 3D images. Samples for histological examination were prepared by sectioning the paraffin-embedded samples and staining with hematoxylin and eosin.

3. Results and discussion

3.1. Structure of nano-HA/PCL 3D scaffolds

Fig. 1 shows the morphology, pore and surface structure of a nano-HA/PCL 3D scaffold fabricated by a paste extruding deposition process. The morphology of the scaffold was rectangular lattice shape with a regular array of extruded wires having a diameter of 0.4 mm and a road gap of 1 mm. The views of the scaffold in the x, y, z directions were three dimensionally interconnected pore channels with good regularity. Fig. 1(c) shows that nano-HA particles are exposed and dispersed well on the surface of scaffold. Concerning that the size of synthesized nano-HA particles is ranging from 50 to 100 nm, no big agglomeration was found. We have confirmed that these well dispersed nano-HA particles are very effective to enhance cell attachment and osteogenetic behavior owing to those hydrophilic characteristics and chemical stimulus, respectively [8]. The macro porosity of the scaffold was about 73 vol.%.

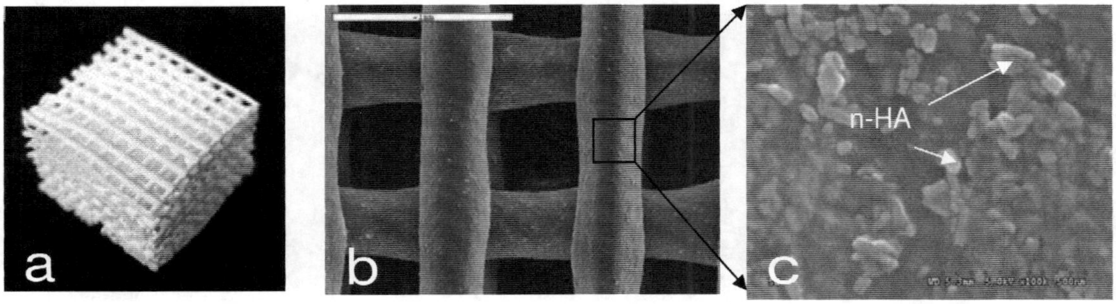

Fig. 1. The morphology of nano-HA/PCL 3D scaffold(a), and its pore structure(b) and surface(c).

3.2. Micro-CT images

Micro-CT images were taken from the bony tissue formed in the nano-HA/PCL 3D scaffolds after 4 and 8 week implantation in rabbit tibias, as shown in Fig. 2. After 4 week implantation, micro-CT images showed that all pores in 3D scaffolds are uniformly filled with bone-like tissues (brown color). This means that 3D scaffold structure is good for tissue in-growth. After 8 week implantation, the images were somehow dissimilar to 4 week. Brown colored region was decreased, while bright colored region was increased. The bright contrast in micro-CT images may be caused by higher density of materials. Therefore, increase in bright region is believed that mineralization was progressed in the scaffold. The bright contrast was clearer in the vicinity of host bone than in the center part of scaffolds. However, these micro-CT images can not provide us what kinds of tissue correspond to brown color and bright image in detail.

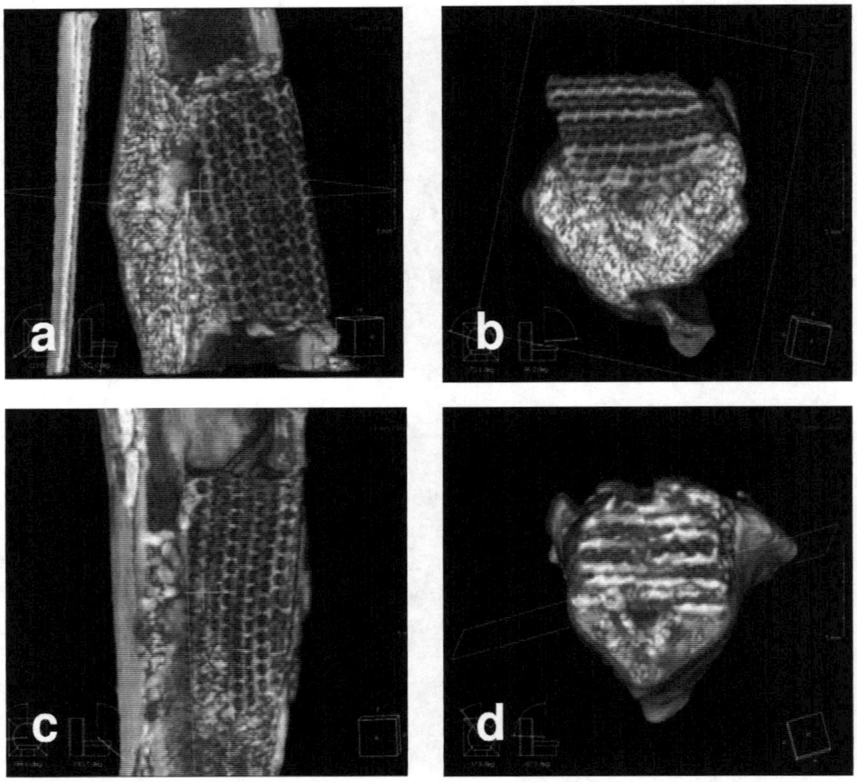

Fig. 2. Micro-CT images of 3D scaffolds and tissues taken from the rabbit tibias after 4 week (a, b) and 8 week (c, d) implantation. (a, c) and (b, d) correspond to longitudinal and transverse section, respectively.

3.3. Histological views

Histological examination provided us more clear information about tissue regeneration in vivo. Fig. 3 shows histological views of H&E stained tissues with the nano-HA/PCL 3D scaffolds after 4 and 8 week implantation in rabbit tibias. Likewise with micro-CT observation, all pore spaces in 3D scaffolds are filled with bone tissues. This indicates that three dimensionally interconnected pore structure is very important to lead in vivo tissue in-growth. There was no severe inflammatory reaction or adverse tissue reaction in any of the implanted scaffolds. At 4 week implantation (Fig. 3 (a, b)), immature bone tissue (osteoid) was found in the center of scaffold as well as in the vicinity of host bone. A large number of osteoblasts were evident around the newly formed osteoid matrix.

At 8 week implantation (Fig. 3 (c, d)), on the contrary, mineralized new bone tissue was found in the center of scaffolds as well as in the vicinity of host bone, and a lot of osteocytes and osteoblasts were found in the newly formed bone matrix. This means that three dimensionally interconnected pore structure facilitates micro- or macro-vascularization into the scaffold, followed by callus and new bone formation. Besides pore structure, of course, it is believed that nano-HA material must have also contributed to osteogenetic procedure. Therefore, it can be argued that nano-HA/PCL 3D scaffolds are really effective for in vivo bone regeneration, owing to three dimensionally interconnected pore structure and osteoinductive biomaterials. Some sporadic fatty marrow cells and fat-like tissues were observed around and inside the scaffolds. This is interpreted as a large part of scaffold was placed in the bone marrow cavity of rabbit tibia.

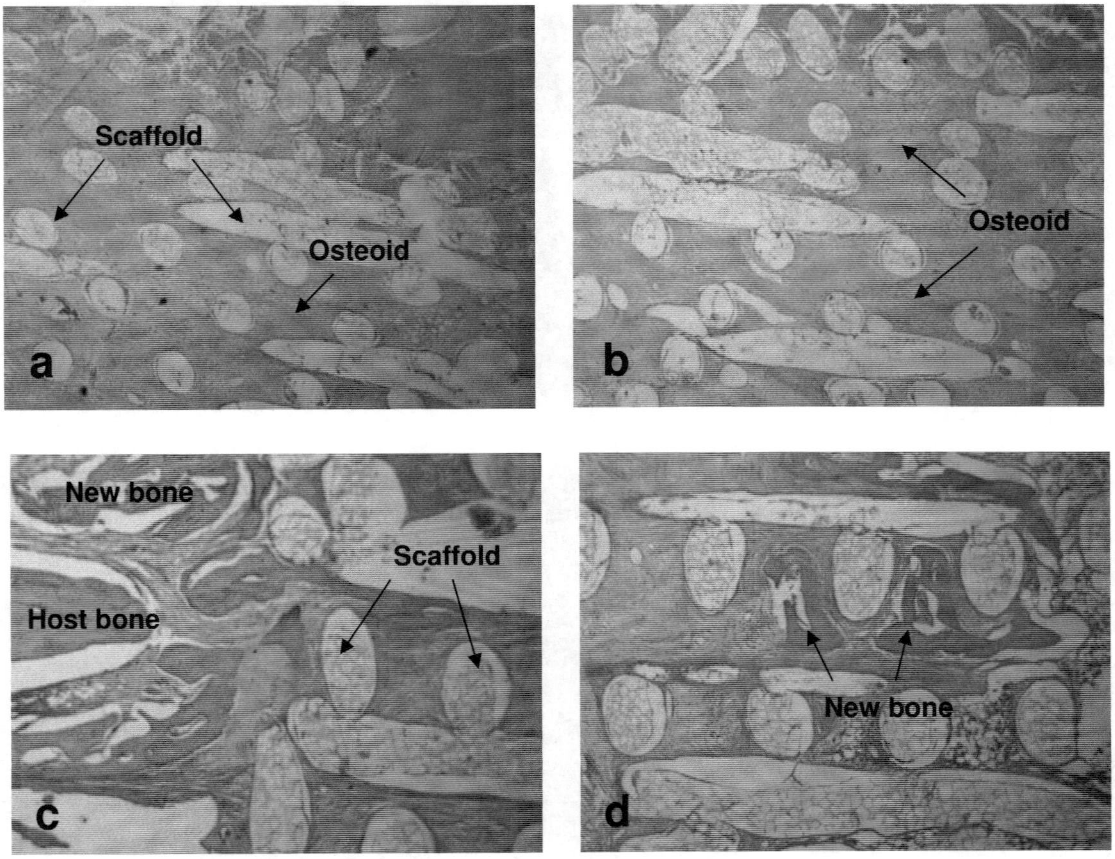

Fig. 3. Histological views of H&E stained tissues with 3D scaffolds after 4 week (a, b) and 8 week (c, d) implantation in rabbit tibia. (a, c) and (b, d) are the vicinity of host bone and the interior of scaffold, respectively.

3.4. Use of 3D scaffolds for mastoid obliteration

As aforementioned, mastoidectomy surgery yields large cavity corresponding to a critical size defect, which can not be regenerated naturally in the entire life. Therefore, the cavity should be filled with suitable filler materials for bone regeneration and obliteration. We tried to use our nano-HA/PCL 3D scaffolds for mastoid obliteration with an animal model. Fig. 4 shows histological views of H&E stained tissues with nano-HA/PCL 3D scaffolds after 2 week implantation in the mastoid cavity of rats. It is noted that immature bone tissue (osteoid) was already found in the center of cavity as well as in the vicinity of host bone even after 2 week implantation. It can be therefore argued that the nano-HA/PCL 3D scaffold could be a promising filler material for mastoid obliteration.

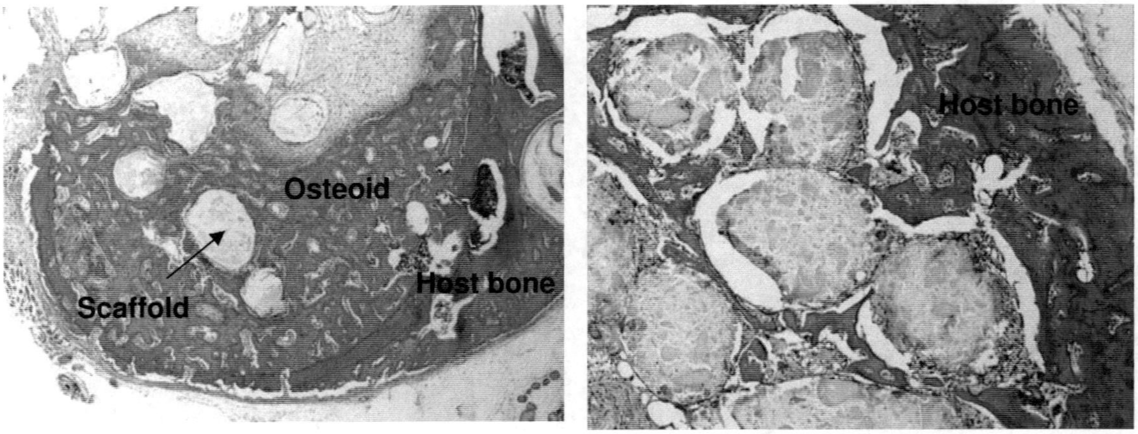

Fig. 4. Histological views of H&E stained tissues with 3D scaffolds after 2 week implantation in rats.

4. Conclusion

Nano-HA/PCL composite 3D scaffolds showed good bone regeneration behavior in animal models, owing to three dimensionally interconnected pore structure and osteoinductive biomaterials. The nano-HA/PCL 3D scaffold could be a promising candidate material for mastoid obliteration in otologic surgery.

Acknowledgements

The authors would like to acknowledge Mr. S. J. Heo and Dr. T. R. Yoon for technical assistance and discussion. The authors also would like to thank Dr. S. A. Park for help with micro-CT analysis.

References

[1] Shor L, Guceri S, Wen X, Gandhi M and Sun W 2007 *Biomaterials* **28** 5291
[2] Mikos K F, 1994 *Polymer* **35** 1068
[3] Cai Q, Yang J, Bei J, Wang S 2002 *Biomaterials* **23** 4483
[4] Mooney D J 1996 *Biomaterials* **17** 1417
[5] Tan K H, Chua C K, Leong K F 2003 *Biomaterials* **24** 3115
[6] Yeong W Y, Chua C K 2004 *Trends Biotechnology* **22** 643
[7] Dietmar W H, Thorsten S, Iwan Z 2001 *J Biomed Mater Res* **55** 203
[8] Heo S J, Kim S E, Wei J, Hyun Y T, Yun H S, Kim D H, Shin J W 2008 *J Biomed Mater Res* Part A DOI:10.1002/jbm.a.31726.
[9] Leatherman B D, Dornhoffer J L, Fan C Y, Mukunyadzi P 2001 *Otology & Neurotology* **22** 731
[10] Leatherman B D, Dornhoffer J L 2002 *Otology & Neurotology* **23** 657
[11] Minoda R, Hayashida M, Masuda M, Yumoto E 2007 *Otology & Neurotology* **28** 1

Adaptation of BAp crystal orientation to stress distribution in rat mandible during bone growth

T Nakano[1], W Fujitani[1], T Ishimoto[1] and Y Umakoshi[2]
[1]Division of Materials and Manufacturing Science, Graduate School of Engineering, Osaka University, 2-1, Yamada-oka, Suita, Osaka 565-0871, Japan
[2]National Institute for Materials Science, 1-2-1, Sengen, Tsukuba, Ibaragi, 305-0471, Japan

E-mail: nakano@mat.eng.osaka-u.ac.jp (T. Nakano)

Abstract. Biological apatite (BAp) *c*-axis orientation strongly depends on stress distribution *in vivo* and tends to align along the principal stress direction in bones. Dentulous mandible is subjected to a complicated stress condition *in vivo* during chewing but few studies have been carried out on the BAp *c*-axis orientation; so the adaptation of BAp crystal orientation to stress distribution was examined in rat dentulous mandible during bone growth and mastication. Female SD rats 4 to 14 weeks old were prepared, and the bone mineral density (BMD) and BAp crystal orientation were analyzed in a cross-section of mandible across the first molar focusing on two positions: separated from and just under the tooth root on the same cross-section perpendicular to the mesiodistal axis. The degree of BAp orientation was analyzed by a microbeam X-ray diffractometer using Cu-Kα radiation equipped with a detector of curved one-dimensional PSPC and two-dimensional PSPC in the reflection and transmission optics, respectively. BMD quickly increased during bone growth up to 14 weeks, although it was independent of the position from the tooth root. In contrast, BAp crystal orientation strongly depended on the age and the position from the tooth root, even in the same cross-section and direction, especially along the mesiodistal and the biting axes. With increased biting stress during bone growth, the degree of BAp orientation increased along the mesiodistal axis in a position separated from the tooth root more than that near the tooth root. In contrast, BAp preferential alignment clearly appeared along the biting axis near the tooth root. We conclude that BAp orientation rather than BMD sensitively adapts to local stress distribution, especially from the chewing stress *in vivo* in the mandible.

1. Introduction

Bone is a well organized nano-structure dominantly composed of collagen fibers and biological apatite (BAp) crystals, as schematically described in Fig. 1 [1]. Since BAp crystal structure has an anisotropic hexagonal lattice [2], its crystal orientation and the related alignment of the collagen fibers, which are important parameters for controlling bone quality, are closely related to the mechanical properties of bone [3]. Bone quality denotes the important parameters except bone mineral density (BMD) and bone volume to explain the bone strength [4].

The preferential alignment of the BAp *c*-axis depends strongly on the bone portion [3, 5]. The long bone exhibits a one-dimensional alignment along its axis [6], but the rabbit skull has a two-dimensional alignment along the flat bone surface [5]. In the case of the mandible, the cortical portion separated from the tooth basically exhibits one-dimensional alignment along the mesiodistal axis [5, 7].

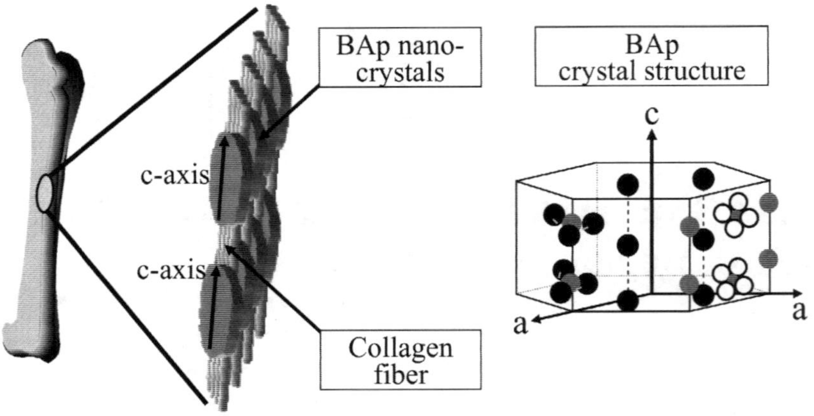

Figure 1. Schematic illustration of bone and bone microstructure composed of collagen fibers and biological apatite (BAp) crystals. BAp crystal shows an anisotropic hexagonal lattice.

Preferential BAp alignment along the mesiodistal direction also appears in the osteonal direction in human edentulous mandible [8]. Thus, BAp crystal orientation is closely related to stress distribution *in vivo* and the mechanical property of bone, but few studies have focused on BAp *c*-axis orientation in dentulous mandible in spite of its importance for understanding complicated stress distribution *in vivo*, the mandibular properties, the optimal design of dental implants, etc.

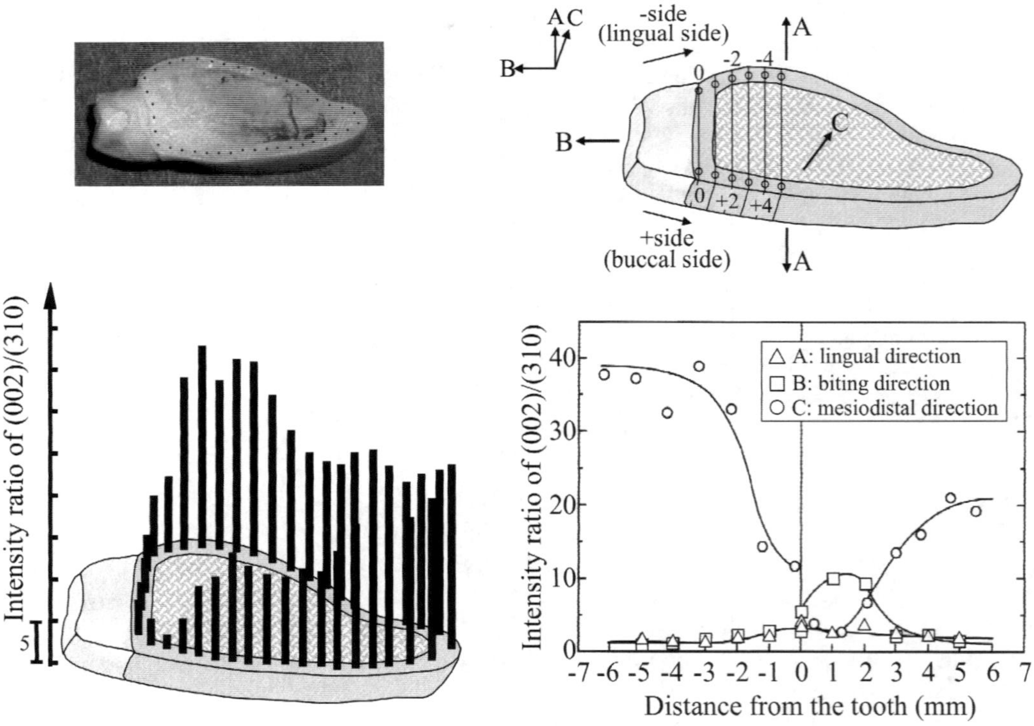

Figure 2. Variations in BAp crystal orientation in mature monkey mandible [5]. Degree of BAp *c*-axis corresponding to diffraction intensity ratio of (002)/(310) along the mesiodistal axis decreases near the tooth and that in the biting direction shows a local maximum on buccal side.

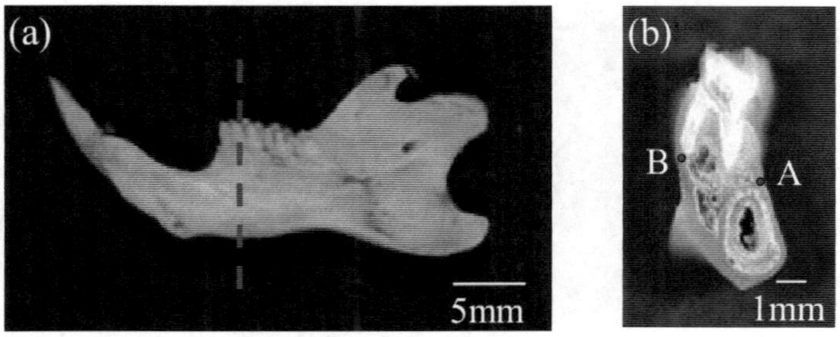

Figure 3. (a) View of mandible and (b) its cross-section showing cut and analyzed positions A and B separated from and near tooth root, respectively, in an 8-week-old rat.

Based on three-dimensional analysis by microbeam X-ray diffraction, a unique change was found near the tooth crown, as shown in Fig. 2 [5]. The preferential degree of BAp crystal orientation quickly decreases in the mesiodistal axis when it approaches the tooth crown. On the buccal side, the local maximum in the BAp orientation degree corresponding to the diffracted intensity ratio of (002)/(310) especially appears in the biting direction. This suggests that local stress distribution *in vivo* is closely related to the BAp crystal orientation in spite of the complicated stress field in the mandible.

This study's objective is to clarify the adaptation to the stress distribution during bone growth in the mandible by focusing on an increase in mastication force and/or number. In addition to bone shape and volume, two important parameters of bone were analyzed: BMD and BAp orientation. Finally, the changes in these two parameters were clarified to understand the variation in bone microstructure and to clarify the more important parameter during mandible growth.

2. Materials and Methods

Fifteen, 4-, 8-, and 14-week-old female Sprague-Dawley (SD) rats (Japan SLC Inc., Japan) (five per group) were used. Rats continuously grow due to their bone modeling and reach maturity around 20 weeks. The specimen cross-section contains the first molar tooth cut perpendicular to the lower border of the mandible (Fig. 3). The two positions, shown in Fig. 3(b) and labeled A and B, were analyzed because positions A and B are separated from and near the tooth root, respectively. Thus, position B must be directly influenced by the biting force.

A microbeam X-ray diffractometer (M18XHF^{22}-SR, Mac Science; D8 Discover with GADDS, Bruker AXS) using Cu-Kα radiation was applied to analyze the preferential orientation of the BAp *c*-axis in two optics, reflection and transmission. Incident microbeams were focused into 50 and 100 μmϕ in reflection and transmission optical systems, respectively. The diffracted beams were detected by curved one-dimensional position sensitive proportional counter (PSPC) and flat two-dimensional PSPC in the reflection and transmission optics, respectively [5, 9]. The transmission optical system was applied to the two-dimensional analysis of BAp crystal orientation using thin specimens polished to a thickness of about 150 μm. The degree of BAp *c*-axis orientation was evaluated by a diffracted intensity ratio of (002)/(310). See our previous article for details [9]. The bone morphology and volume were analyzed by micro CT, and BMD was evaluated by peripheral quantitative computed tomography (pQCT) (XCT Research SA+, Stratec Medizintechnik). Data were represented as a mean value ± standard deviation (SD). Statistical analysis was performed by one-way ANOVA and a two-tailed t-test. A *P*-value less than 0.05 was considered statistically significant.

3. Results and Discussion

It is common knowledge that the mandible grows under intermembranous ossification, and micro CT analysis showed that the bone area and the cortical bone thickness on the cross-section remarkably

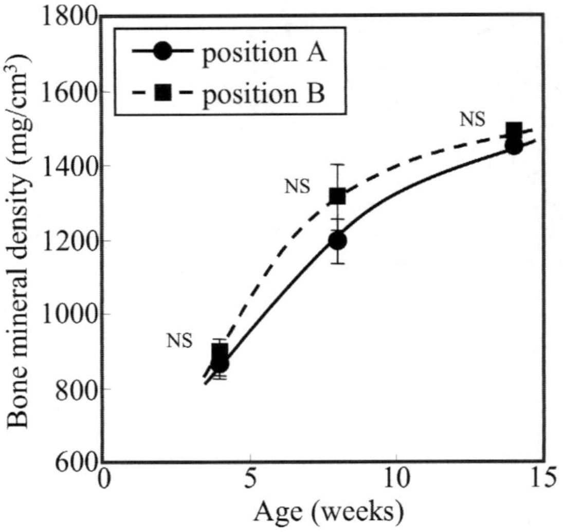

Figure 4. Variations in BMDs during bone growth in positions A and B separated from and near the tooth root, respectively. NS: not significant vs. position A.

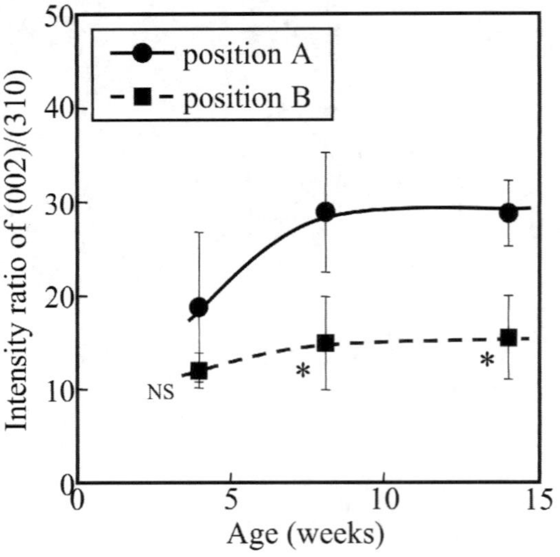

Figure 5. Effect of mastication on degree of BAp *c*-axis orientation along the mesiodistal axis during bone growth. NS: not significant and *: $P<0.05$ vs. position A.

increased by 14 weeks during bone growth. The alveolar bone portion of the mandible seems to grow rapidly with mastication.

Figure 4 shows that the BMD variations during bone growth in positions A and B. BMDs increase very rapidly in the early stage of bone growth until eight weeks and subsequently gradually up to 14 weeks. There is, however, no statistically difference in BMD between positions A and B despite different *in vivo* local stress distribution from the tooth root during mastication. This strongly suggests that BMDs in both positions are not so sensitive for adapting the local stress distribution that originated from the biting.

On the other hand, the variation in degree of BAp *c*-axis orientation along the mesiodistal axis appeared during bone growth depending on positions A and B. The mesiodistal axis is roughly perpendicular to the biting direction. Figure 5 shows the variations in the BAp *c*-axis orientation in the two positions separated from the tooth root (position A) and near the tooth root (position B). It is quite clear that the preferential alignment in position B where the biting force was efficiently applied is quite lower than that in position A. This strongly demonstrates no BMD differences but a big difference in BAp orientation exists between positions A and B. The biting stress seems very sensitive to the change in BAp orientation, which resembles the case of monkey mandible [5]. The analytical direction parallel to the mesiodistal axis is, however, perpendicular to the biting direction. Thus, the preferential degree of the BAp *c*-axis along the biting direction should be directly evaluated by the transmission optical system [9].

BAp orientation in all directions within the cross-section of the mandible perpendicular to the mesiodistal axis was detected using the two-dimensional PSPC. The quite large difference in intensity distribution in the diffraction rings between positions A and B separated from and near the tooth root, respectively. In position B near the tooth root, a crescent-shaped ring of the (002) diffraction appeared roughly along the biting and elongated tooth root directions, but no remarkable distribution of intensity along the (002) diffraction existed in position A. This different distribution of BAp *c*-axis alignment was found quantitatively as a function of the angle around the mesiodistal axis.

Thus, the integrated area of the (002)/(310) intensity ratio as a function of the above angle corresponds to the aligned degree of the BAp *c*-axis orientation on the cross-section based on biting.

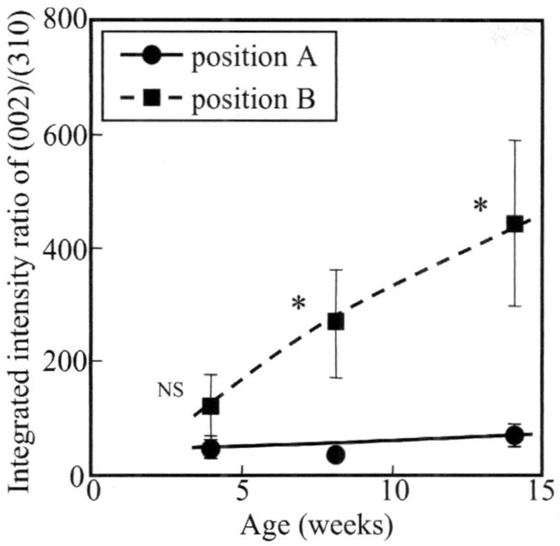

Figure 6. Effect of mastication on aligned degree of BAp *c*-axis orientation inside cross-section containing biting direction during bone growth. Integrated intensity ratio of (002)/(310) was adopted to evaluate preferential degree of BAp c-axis parallel to cross-section of mandible, as described in Fig. 3. NS: not significant and *: $P<0.05$ vs. position A.

Figure 6 shows the variations in the aligned degree of the BAp orientation inside the cross-section of the mandible containing the biting direction. The degree corresponds to the mastication effect as a function of age. In position B near the tooth root, the integrated intensity ratio roughly corresponds to the preferential aligned degree along the biting direction that increases quickly during bone growth up to 14 weeks. This good agreement with weight change may be closely related to the mass of the masticated food and may suggest that the aligned degree of the BAp crystal orientation along the biting direction corresponds to the mastication force and/or number due to the functional adaptation from local stress distribution *in vivo*.

A relationship among the biting, BAp preferential alignment, and cortical bone surface directions seems to exist. Detailed quantitative analysis will be described in a separate paper, but the preferred direction of the BAp *c*-axis seems to be located between the biting and bone surface directions, suggesting that the formation of BAp preferential alignment is strongly caused by biting and the morphology of the growing mandible.

4. Conclusions

The changes in two important parameters of BMD and BAp orientation were quantitatively evaluated in rat mandible during bone growth and the resultant increase in biting force and/or number; the following conclusions were reached:

(1) In the growth period up to 14 weeks, bone volume, BMD, and BAp orientation degree remarkably increase.

(2) The preferred BAp *c*-axis orientation is basically parallel to the mesiodistal direction to support the jaw, but mastication induces local alignment of the BAp *c*-axis roughly parallel to the biting direction. As a result, the BAp *c*-axis near the tooth root in position B exhibits two-dimensional alignment along the mesiodistal axis and biting direction as a function of age until 14 weeks.

(3) The change in the BAp crystal orientation is more sensitive to local stress distribution *in vivo* caused by mastication than that in BMD.

Acknowledgments
This study was supported by Priority Assistance for the Formation of Worldwide Renowned Centers of Research-The Global COE Program (Project: Center of Excellence for Advanced Structural and Functional Materials Design) and a Grant-in-Aid for Scientific Research and Development from the Ministry of Education, Culture, Sports, Science and Technology (MEXT), Japan.

References
[1] Landis WJ 1995 *Bone* **16** 533
[2] Elliot JC 1994 *Structure and Chemistry of the Apatites and Other Calcium Orthophosphates*. Amsterdam, Elsevier, 1
[3] Nakano T, Tabata and Y, Umakoshi Y 2005 Texture and Bone Reinforcement. In: *Encyclopedia of Materials, Science and Technology Updates*, Elsevier, Oxford, MS2061-1-8
[4] 1991 *Am. J. Med.* **90** 107
[5] Nakano T, Kaibara K, Tabata Y, Nagata N, Enomoto S, Marukawa E, and Umakoshi Y 2002 *Bone* **31** 479
[6] Sasaki N, Matsushima N, Ikawa T, Yamamura H, and Fukuda A 1989 *J. Biomech.* **22** 157
[7] Bacon GE, Bacon PJ, and Griffiths RK 1980 *J. Biomech.* **13** 725
[8] Momura T, Gold E, Powers MP, Shingaki S, and Katz JL 2003 *Dent. Mater.* **19** 167
[9] Miyabe S, Nakano T, Ishimoto T, Takano N, Adachi T, Iwaki H, Kobayashi A, Takaoka K, and Umakoshi Y 2007 *Mater. Trans.* **48** 343

Changes in bone microstructure and toughness during the healing process of long bones

T Ishimoto[1], T Nakano[1], Y Umakoshi[1], and Y Tabata[2]

[1] Division of Materials and Manufacturing Science, Graduate School of Engineering, Osaka University, 2-1 Yamadaoka, Suita, Osaka 565-0871, Japan
[2] Institute for Frontier Medical Sciences, Kyoto University, 53 Kawahara-cho, Shogoin, Sakyo-ku, Kyoto 606-8507, Japan

E-mail: nakano@mat.eng.osaka-u.ac.jp

Abstract. It is of great importance to understand how bone defects regain the microstructure and mechanical function of bone and how the microstructure affects the mechanical function during the bone healing process. In the present study on long bone defects, we investigated the relationship between the recovery process of fracture toughness and biological apatite (BAp)/collagen (Col) alignment as an index of the bone microstructure to clarify the bone toughening mechanisms. A 5-mm defect introduced in the rabbit ulna was allowed to heal naturally and a three-point bending test was conducted on the regenerated site to assess bone toughness. The bone toughness was quite low at the early stage of bone regeneration but increased during the postoperative period. The change in toughness agreed well with the characteristics of the fracture surface morphology, which reflected the history of the crack propagation. SEM and microbeam X-ray diffraction analyses indicated that the toughness was dominated by the degree and orientation of the preferred BAp/Col alignment, i.e. bundles aligned perpendicular to the crack propagation clearly contributed to the bone toughening owing to extra energy consumption for resistance to crack propagation. In conclusion, regenerated bone improves fracture toughness by reconstructing the preferred BAp/Col alignment along the bone longitudinal axis during the healing process of long bones.

1. Introduction

Many techniques have been successfully developed to reconstruct bone defects. However, it remains less understood how bone defects regain the microstructure and mechanical function of bone. Bone mechanical functions vary depending on the microstructure of the bone tissue constituents, which are mainly collagen (Col) fibers and biological apatite (BAp) crystals. The microstructural characteristics, such as the BAp density, BAp crystallinity, Col cross-linking and BAp/Col texture, have been reported to affect some mechanical properties of bone [1, 2]. Among these parameters, the preferred BAp/Col alignment, as well as the BAp density, dominantly contributes to the material properties of regenerated bone tissue, such as the ultimate stress [3].

In the present study, we focused on bone toughness, which is one of the important mechanical aspects, to understand the fracture healing mechanism of bone. The purposes of the study were to analyze the recovery process of the preferred BAp/Col alignment and toughness of bone, and to investigate how the BAp/Col alignment affects the bone toughness.

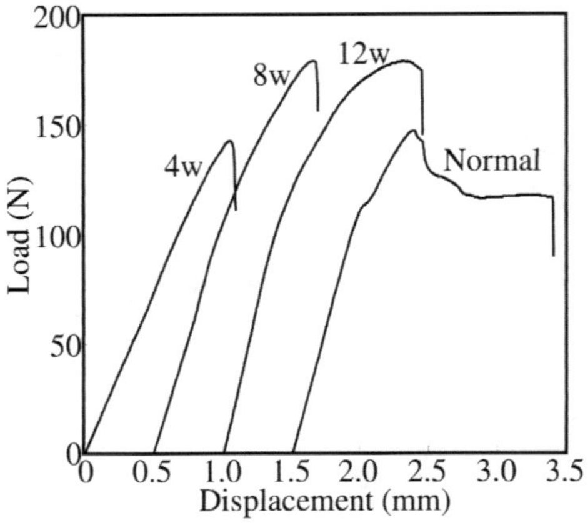

Figure 1. Typical load-displacement curves in three-point bending tests conducted on regenerated ulnae.

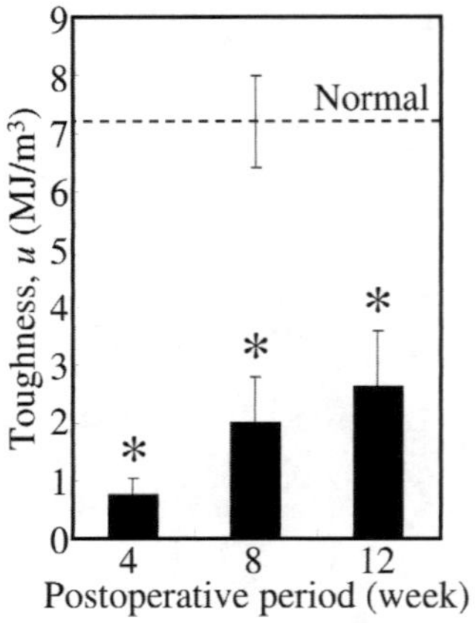

Figure 2. Variations in bone toughness evaluated by load-displacement curves. *P<0.05 vs. the normal value.

2. Materials and Methods

A rabbit ulnar osteotomy model was prepared by introducing 5-mm long segmental defects in the right ulnar middiaphysis of mature male New Zealand White Rabbits. The defects were allowed to naturally regenerate for 4, 8 and 12 weeks (N=5 each), before the rabbits were sacrificed by an overdose of pentobarbital. The right ulnae were removed, wrapped in saline-soaked gauze and stored at -80°C until mechanical testing. Immediately after the ulna was thawed in saline at room temperature and separated from the neighboring intact radius, a three-point bending test (Model 5565; Instron) was conducted on the regenerated portion until fracture occurred. The bending force was applied on the medial surface at the center of the regenerated site at room temperature. The distance between the two supports (L) and the crosshead speed were 30 mm and 10 mm/min, respectively. An absorption energy to failure (U) corresponding to the area under the load-displacement curve was evaluated. Finally, toughness (u) was calculated using the following equation [4]:

$$u = U \frac{3}{L} \frac{y_{\max}^2}{I_x} \quad (1)$$

where I_x is the cross-sectional moment of inertia and $y_{\max}$ is the maximum distance from the neutral axis for bending. I_x and $y_{\max}$ were measured in saline by peripheral quantitative computed tomography (pQCT) (XCT Research SA+; Stratec Medizintechnik) with a resolution of $0.08 \times 0.08 \times 0.46$ mm^3. Bone tissue was detected as pQCT values of >267 mg/cm^3 [5]. Despite the irregular morphology of the regenerated portion, I_x and $y_{\max}$ were calculated at the center of the regenerated portion because the bending moment was largest at the loading point during the three-point bending test. The fracture surface was observed by a SEM (JSM-5600; JEOL). The BAp orientation, which is closely related to the arrangement of Col fibers [6], was analyzed on the fracture surface by microbeam X-ray diffractometry (μXRD) (D8 Discover with GADDS; Bruker AXS) with an incident beam of 20 μm in diameter. Data were represented as the mean ± standard deviation (SD). Statistical analyses were performed by one-way ANOVA and a two-tailed t-test. Values of P<0.05 were considered to be statistically significant.

3. Results and Discussion

At 4 weeks after surgery, the defects were already filled with newly formed bone tissue [3], and a bending force could be applied to the regenerated portion. Figure 1 shows typical load-displacement curves obtained by three-point bending tests. Normal bone showed significant deformation after the

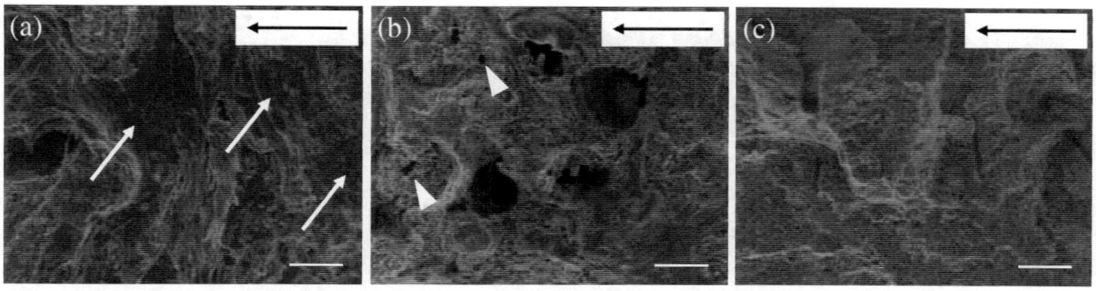

Figure 3. SEM images (×100) of the fracture surfaces of the regenerated bone at (a) 4 and (b) 8 weeks after the osteotomy, and (c) the normal bone. The top-right arrows represent the direction of the main crack propagation. Scale bars: 100 μm.

peak load was reached. In contrast, the regenerated bone at 4 weeks fractured immediately after the peak load was reached. After 8 weeks, the fracture strain tended to become larger as the postoperative period increased. However, the absorption energy to failure, which was expressed as the area under the load-displacement curve, was significantly lower in the regenerated bone than in the normal bone throughout the experimental period (data not shown). Figure 2 shows the variations in toughness of the regenerated bone as a function of the postoperative period. The toughness continuously increased over time as evaluated by one-way ANOVA, but was still quite low compared with that of normal bone even at 12 weeks after the operation.

The toughness was closely related to both the initiation and propagation behaviors of cracks. Figure 3 shows SEM images of fracture surfaces. At 4 weeks, the fracture surface was flat and smooth, indicating low consumption of fracture energy during crack propagation. A cleavage-like surface pattern was observed in the early stage of bone regeneration (Figure 3(a), arrows). At 8 weeks, the fracture surface became rougher and a cleavage-like surface pattern was not seen. Osteons, concentric lamellar structures surrounding a blood vessel, were observable on the surface at 8 weeks (Figure 3(b), arrowheads), and have been reported to play a role in bone toughening [7]. The fracture surface at 12 weeks was very similar to that at 8 weeks (data not shown). Normal bone exhibited a rugged surface (Figure 3(c)), clearly indicating strong resistance to crack propagation and the related high energy consumption during crack propagation.

These macroscopic crack propagation behaviors reflect the structural characteristics of bone tissue at various scale levels [7, 8]. We focused on the correlation between the resistance to crack propagation and the orientation of the preferred BAp/Col alignment during bone regeneration. Figure 4 shows high-magnification SEM images of fracture surfaces of regenerated bone at 8 weeks. The microscopic arrangements of the Col bundles were clearly observed. The Col bundles in Figure 4(a) were parallel to the fracture surface, whereas the Col bundles in Figure 4(b) were somewhat perpendicular to the surface. These observations were confirmed by μXRD analysis using the diffraction peaks of inorganic BAp crystals because the c-axis of BAp crystals is nearly parallel to the Col direction [6]. Figure 5 shows the integrated intensity ratios of the (002) diffraction peak to the (310) peak taken on the surfaces corresponding to the images in Figures 4(a) and 4(b), and the intensity ratio for randomly oriented hydroxyapatite (HAp) powders. The μXRD measurements were obtained using a reflecting optical system, such that the BAp c-axis alignment along the normal direction to the fracture surface, i.e. almost parallel to the ulnar long axis, was analyzed. The parallel BAp/Col bundles (//) showed a significantly lower intensity ratio than the randomly oriented HAp powders, whereas the somewhat perpendicularly aligned BAp/Col bundles (⊥) showed a significantly higher intensity ratio than the randomly oriented HAp powders. These μXRD results are consistent with the surface patterns observed in Figure 4.

The SEM and μXRD analyses showed that the orientations of the BAp/Col alignment in the regenerated bone were not uniform on the fracture surface, although clusters of well-arranged

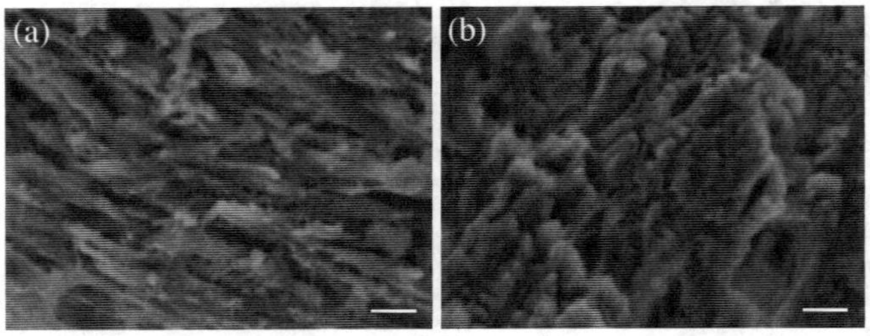

Figure 4. SEM images (×1000) of fracture surfaces of regenerated bone at 8 weeks. Col bundles seem to be aligned (a) parallel and (b) somewhat perpendicular to the crack propagation direction. Scale bars: 10 μm.

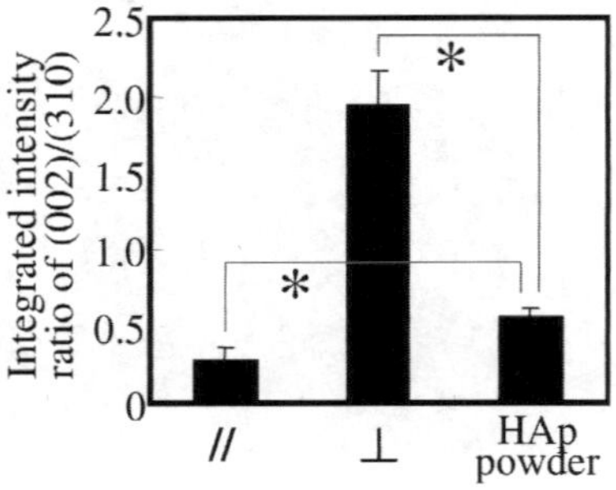

Figure 5. BAp orientations to the normal direction to the fracture surface as well as randomly oriented HAp powders. "//" and "⊥" correspond to Figures 4(a) and 4(b), respectively. $*P < 0.05$.

BAp/Col bundles could be seen locally. When Col bundles are perpendicular to the direction of crack propagation, the crack must propagate by cutting the Col bundles. The resultant crack bridged by the Col bundles causes extra energy consumption [8] and surely contributes to toughening of the regenerated bone. As bone healing proceeds, the bone microstructure is reconstructed toward that of normal bone by enhancing the degree of preferred alignment along the long bone axis [9]. This suggests that the regenerated bone improves the fracture toughness by reconstructing the preferred BAp/Col alignment along the bone longitudinal axis during the long bone healing process.

4. Conclusions

The toughness and preferred BAp/Col alignment of regenerated bone were investigated, and the following conclusions were reached:

1. The fracture surface was flat at the early stage of bone healing at 4 weeks, and became rougher during the postoperative period. As a result, the toughness of the regenerated bone gradually increased. However, it remained significantly lower than that of normal bone throughout the bone healing period.
2. The toughness was closely related to resistance to crack propagation and clearly seemed to reflect the degree and orientation of BAp/Col alignment, which can be quantitatively analyzed by μXRD.

Acknowledgments

This work was supported by funds from the "Priority Assistance of the Formation of Worldwide Renowned Centers of Research – The Global COE Program (Project: Center of Excellence for Advanced Structural and Functional Materials Design)" and a Grant-in-Aid for Scientific Research and Development from the Ministry of Education, Culture, Sports, Science and Technology of Japan.

References

[1] Nakano T, Kaibara K, Tabata Y, Enomoto S, Marukawa E and Umakoshi Y 2002 *Bone* **31** 479
[2] Saito M, Fujii K, Mori Y and Marumo K 2006 *Osteoporos. Int.* **17** 1514
[3] Ishimoto T, Nakano T, Umakoshi Y, Tabata Y 2007 *Mater. Sci. Forum* **561-565** 1451
[4] Komatsubara S et al. 2005 *Bone* **36** 678
[5] Tanaka M, Sakai A, Uchida S, Tanaka S, Yamaguchi K and Nakamura T 2004 *Bone* **34** 940
[6] Sasaki N, Matsushima N, Ikawa T, Yamamura H and Fukuda A 1989 *J. Biomech.* **22** 157
[7] Hiller LP, Stover SM, Gibson VA, Gibeling JC and Martin RB 2003 *J. Orthop. Res.* **21** 481
[8] Nalla RK, Kinney JH and Ritchie RO 2003 *Nat. Mater.* **2** 164
[9] Ishimoto T, Nakano T, Umakoshi Y, Tabata Y and Yamamoto M 2006 *Mater. Sci. Forum* **512** 261

Single-crystal growth and plastic deformation behaviour of a Ti-15Mo-5Zr-3Al alloy for biomedical application

S-H Lee[1], K Hagihara[1], M-H Oh[2] and T Nakano[1]

[1]Division of Materials and Manufacturing Science, Graduate School of Engineering, Osaka University, 2-1 Yamada-oka, Suita, Osaka 565-0871, Japan
[2]Department of Information & Nano Materials Engineering, Kumoh National Institute of Technology, 1 Yangho-dong, Gumi, Gyeongbuk 730-701, Korea
E-mail: nakano@mat.eng.osaka-u.ac.jp

Abstract. A Ti-15Mo-5Zr-3Al alloy with a bcc structure are promising materials for biomedical application and were examined. The focus of this study was on the effect of heat treatment on microstructure and plastic deformation behaviour using a single crystal. The single crystal was successfully obtained by a floating zone method at a crystal growth rate of 2.5 mm/h. A slip at the <111> dislocation was present irrespective of heat treatment at 573K or 673K for 1.2 ks or 300 ks. The yield stress at the $[\bar{1}49]$ loading axis varied significantly depending on microstructure, especially from the precipitation of the α phase. Al addition suppresses generation of the ω phase and increases the yield stress at the same time.

1. Introduction

Recently, β-type titanium alloys with a bcc structure have received a great deal of attention as structural and biomedical materials because of their excellent mechanical properties, good cold formability, low Young's modulus etc. [1]. Various types of β-Ti alloys have been developed for engineering parts and also biomedical applications [2, 3]. Ti-15Mo-5Zr-3Al (wt.%) alloy is one of the most promising materials for biomedical applications with outstanding bio-compatibility as well as mechanical characteristics [4]. The alloy was accepted for commercial use by being registered in ISO in 2007. Many studies on the effect of heat treatment on microstructure and deformation behaviour for this alloy have been reported. However, to obtain more comprehensive knowledge such as the crystal orientation dependence of plastic and elastic behaviour, further studies using single crystals are required. Hanada et al. previously examined plastic deformation behaviour of Ti-15Mo-5Zr (wt.%) using a single crystal [5, 6, 7] that had a similar composition to Ti-15Mo-5Zr-3Al. Concerning the effect of Al-addition to the microstructure in β-Ti alloys, Ikeda et al. reported a Ti-V system [8] wherein Al-addition enhances precipitation of the α-phase during annealing since Al is an α-phase stabilizing element. In addition, they determined that Al-addition hinders the formation of the athermal and the isothermal ω-phase. They also investigated the effect of Al-addition in Ti-15Mo-5Zr-3Al using a polycrystal [9] and made similar conclusions. Their experiment was not done by the direct observation by TEM but through measuring the variation in electric resistivity during annealing.

In this study, the influence of heat-treatment on the microstructure and the plastic deformation behaviour in a Ti-15Mo-5Zr-3Al alloy was examined using a single crystal. The results will be compared with those for the Ti-15Mo-5Zr reported by Hanada et al. and the effect of Al-addition to this alloy system will also be discussed.

2. Experimental procedure

The mother alloy with a composition of Ti-15Mo-5Zr-3Al (wt.%) was supplied by Kobe Steel Co. Ltd. The single crystal was grown by an optical floating zone melting method in high-purity Ar gas flow at a crystal growth rate of 2.5 mm/h [10]. The obtained single crystal was solution treated (ST) at 1108K for 3.6 ks in an Ar atmosphere, followed by water quenching. In some of the ST specimens, heat-treatment was further carried out at 573K and 673K for 1.2 ks and 300 ks, which is the same as that reported by Hanada et al. for Ti-15Mo-5Zr [5, 6]. The constituent phases in specimens were examined by transmission electron microscopy (TEM). For compression tests, rectangular specimens with dimensions of 2×2 $mm^2\times5$ mm were cut out. The loading axis was selected to be parallel to $[\bar{1}49]$, where the Schmid Factor (SF) for the $(\bar{1}01)$ [111] primary slip had a maximum value of 0.5. The tests were carried out at room temperature at a nominal strain rate of 1.67×10^{-4} s^{-1} in vacuum.

3. Results

Figure 1 shows electron diffraction patterns (DPs) along the [113] direction and corresponding dark-field images in ST and annealed specimens. In the ST specimen, only the spots derived from the β-phase was observed and no other extra spot was detected, although very weak streaks were observed along the $<12\bar{1}>$ direction as shown in Fig. 1(a). This indicates that the ST specimen is mainly composed of the β-phase. No significant difference was observed in the DPs for specimens from short-time annealing (1.2 ks both at 573K and 673K) as seen in Fig. 1(c), (e). No clear precipitates were observed from the dark-field observation (Fig. 1(d), (f)) although a slight increase in the intensity of the streaks was detected, especially at 673K. On the other hand, a large amount of ω-precipitates were formed and clear spots, derived from the four ω-variants, appeared after long-time annealing for 300 ks at 573K as shown in Fig. 1(g). For the specimen annealed at 673K, the observed zone DP was strongly distorted because of the large amount of precipitates. Many extra spots appeared between the fundamental spots of the β-matrix phase and these were identified as spots that derived from the α-phase and not from the ω-phase. Indeed, in the dark-field image the large lath-like shapes of α-precipitates were abundant as shown in Fig. 1(j).

Figure 2 shows the change in room temperature yield stress according to different heat-treatment conditions and compressed at the $[\bar{1}49]$ loading axis. The yield stress in the as-ST specimen was about 670 MPa. The yield stress increased as the annealing time and temperature increased but the increment was small for the short-time annealing at 1.2 ks both at 573 and 673K. The increment in stress became relatively large and was about 800 MPa for the long-time annealing specimen at 300 ks and 573K. In the specimen annealed at 673K for 300 ks the yield stress showed a significant increment and was about 1285 MPa.

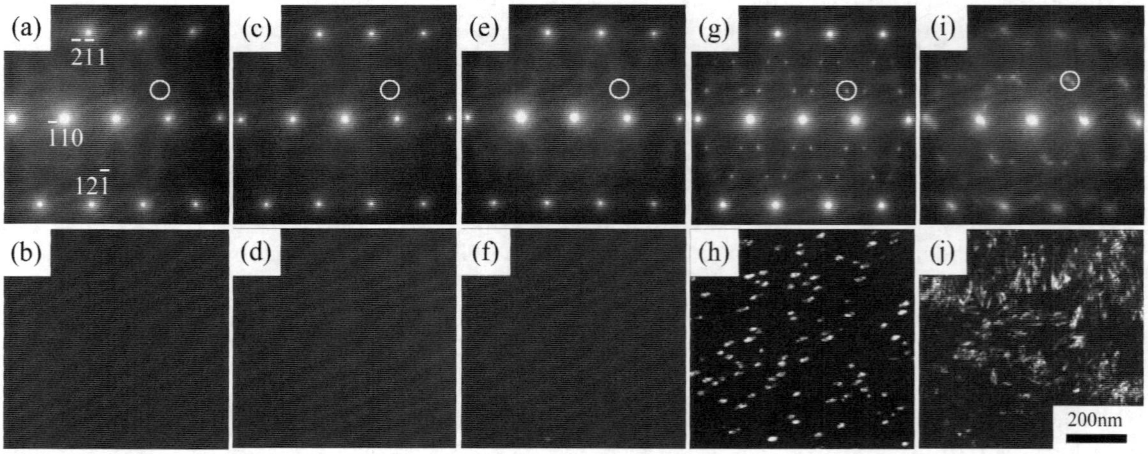

Figure 1. Electron diffraction patterns along the [113] direction (a, c, e, g, i) and corresponding dark-field images (b, d, f, h, j) in ST (a, b) and annealed specimens at 573K for 1.2 ks (c, d), 673K for 1.2 ks (e, f), 573K for 300 ks (g, h) and 673K for 300 ks (i, j). Spots used for the dark-field observation are indicated by circles in the DPs.

With regard to ductility of the specimens, at least 20% of the compressive strain could be obtained in all specimens except for the specimen annealed at 673K for 300 ks, which showed very high stress but it broke at a strain of less than 10%.

Figure 3 shows the slip traces of the specimens deformed to 1% plastic strain as observed on their $(\bar{1}2\bar{1})$ and $(\overline{11}\,\bar{5}\,1)$ side surfaces. The morphology of deformation markings indicates that the deformation occurred by slips in all specimens, irrespective of heat-treatment. The contrast of slip traces is strong on the $(\overline{11}\,\bar{5}\,1)$ but is very faint on the $(\bar{1}2\bar{1})$ face, indicating that the Burgers vector of dislocations is parallel to [111]. By a two-face slip analysis the slip plane of the [111] dislocation was determined to be nearly $(\bar{3}12)$, which is a slight deviation from the $(\bar{1}01)$ maximum resolved shear stress plane at the $[\bar{1}49]$ loading axis. Such a deviation of the slip plane from $(\bar{1}01)$ toward $(\bar{2}11)$ was also reported for Ti-15Mo-5Zr single crystals [5] and it is due to a dislocation core structure effect in the bcc-structured crystal.

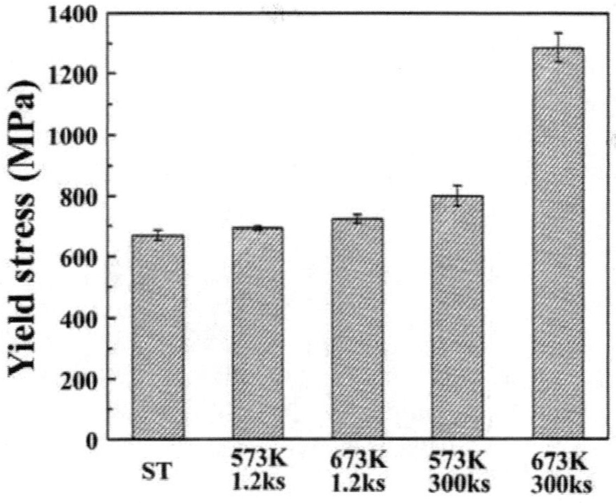

Figure 2. The change in yield stress after compression at room temperature after various heat-treatments.

Although the same [111] slip system was operative, independent of annealing conditions, the morphology of the traces was slightly different for each specimen. Slip traces were introduced relatively homogeneously to the ST specimen and no large difference in features was observed for specimens of short-time annealing at 1.2 ks and 573K or 673K, as shown in Fig. 3 (c, d and e, f). However, in the specimen annealed at 573K for 300 ks, in which a large amount of ω-phases precipitated, the distribution of slip traces became very coarse and localized. In the specimen aged at 673K for 300 ks, in which formation of α-precipitates occurred, the contrast of slip traces became very faint and difficult to identify.

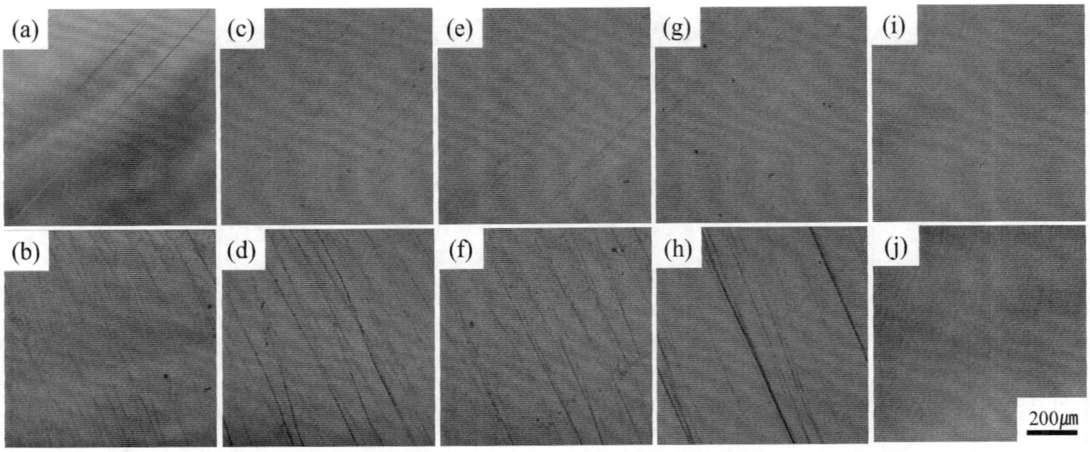

Figure 3. Optical microscope images of slip traces after compression tests to 1% plastic strain as observed on $(\bar{1}2\bar{1})$ (a to i) and $(\overline{11}\,\bar{5}\,1)$ side surfaces (b to i) after ST (a, b). Annealed specimens at 573K for 1.2 ks (c, d), 673K for 1.2 ks (e, f), 573K for 300 ks (g, h) and 673K for 300 ks (i , j). The loading axis was chosen as the horizontal direction in the figure.

4. Discussion

We clearly determined that the microstructure and plastic deformation behaviour of a β-type Ti-15Mo-5Zr-3Al single crystal depends on annealing conditions. In this section, our result is discussed by comparison to that of Ti-15Mo-5Zr as reported by Hanada et al. and the effect of Al addition on this alloy system is also considered. Our first focus was on the difference between microstructures. The observed microstructure showed a significant difference between our Al-addition alloy and the Ti-15Mo-5Zr alloy, even though heat-treatments were conducted under using equivalent conditions for both single crystals.

The large difference was due to a suppression of ω-phase precipitation in Ti-15Mo-5Zr-3Al. For Ti-15Mo-5Zr it was reported that precipitation of the ω-phase occurred even at short-time annealing for 1.2 ks and 573K. The ω-phase precipitation was clearly recognizable at 673K. For Ti-15Mo-5Zr-3Al, the ω-phase precipitation was not clear after the short-time annealing, even at 673K, and the alloy was only composed of the β-phase although a slight increase in intensity of streaks along $<12\bar{1}>$ was accompanied by annealing. These results are in good agreement with previous reports by Komatsu et al. [9] and Hanada et al. [11]. Al acts as a strong inhibitor for the formation of the ω-phase.

These changes in microstructure gave a strong influence on the plastic deformation behaviour. In Ti-15Mo-5Zr-3Al, the yield stress did not show a significant change after the short-time annealing compared with that of the ST specimen. This obviously indicates that the suppression of ω-phase precipitation in Ti-15Mo-5Zr-3Al reduces an increase in yield stress by annealing. However, in case of annealing to 300 ks at 573K the annealing led to ω-phase precipitation even in Ti-15Mo-5Zr-3Al resulting in a relatively large increase in yield stress. On the other hand, in the specimen annealed for a long-time at 300 ks and 673K large amounts of the α-phase precipitation occurred. This caused a significant increase in yield stress, although α-phase precipitation was accompanied by a strong reduction in ductility.

5. Conclusions

A single crystal of Ti-15Mo-5Zr-3Al (wt.%) was successfully grown by a floating zone method at a growth rate of 2.5 mm/h in a high purity Ar gas atmosphere. The microstructure and plastic deformation behaviour of the Ti-15Mo-5Zr-3Al single crystal varied significantly depending on the heat-treatment. For Ti-15Mo-5Zr-3Al the suppression of ω-phase precipitates compared with that of Ti-15Mo-5Zr and the effect of Al was clearly confirmed by TEM observations. The variation in microstructure leads to a large change in yield stress and this depends on the annealing condition.

Acknowledgement

This work was supported by funds from the "Priority Assistance of the Formation of Worldwide Renowned Centres of Research – The Global COE Program (Project: Centre of Excellence for Advanced Structural and Functional Materials Design)" and a Grant-in-Aid for Scientific Research and Development from the Ministry of Education, Culture, Sports, Science and Technology of Japan.

References

[1] Niinomi M 2003 *Biomaterials* **24** 2673
[2] Inamura T, Hosoda H, Wakashima K, and Miyazaki S 2005 *Mater. Trans.* **46** 1597
[3] Matsumoto H, Watanabe S, and Hanada S 2005 *Mater. Trans.* **46** 1070
[4] Okazaki Y, Ito Y, Kyo K, and Tateishi T 1996 *Mater. Sci. Eng. A* **213** 138
[5] Hanada S, and Izumi O 1980 *Titanium' 80 Science and Technology* 691
[6] Hanada S, and Izumi O 1980 *Metall. Trans. A* **11** 1447
[7] Hanada S, and Izumi O 1982 *Trans. Jpn. Inst. Metals* **23** 85
[8] Ikeda M, Komatsu S, Sugimoto T, and Kamei K 1994 *J. Jpn. Inst. Light Met.* **44** 35
[9] Komatsu S, Ikeda M, Sugimoto T, Kamei K, Maesaki O, and Kojima M 1996 *Mater. Sci. Eng. A* **213** 61
[10] Nakano T, Hagihara K et al 2007 *Ti-2007 Science and Technology* 1437
[11] Hanada S, and Izumi O 1980 *Trans. Jpn. Inst. Metals* **21** 201

Preferential orientation of biological apatite in normal and osteoporotic human vertebral trabeculae

S Miyabe, T Ishimoto and T Nakano

Division of Materials and Manufacturing Science, Graduate School of Engineering, Osaka University, 2-1 Yamadaoka, Suita, Osaka 565-0871, Japan

E-mail: nakano@mat.eng.osaka-u.ac.jp

Abstract. The preferential orientation of biological apatite (BAp) is a possible bone quality parameter for the comparison of the bone mechanical property. The preferential BAp orientation undergoes sensitive changes according to the change in the *in vivo* stress distribution, bone turnover rate etc., resulting in a variation of bone function. Osteoporosis is a metabolic bone disease characterized by reduced bone mass and deterioration of bone microstructure. The effect of osteoporosis on the preferential BAp orientation is however unknown. In this study, a microbeam-X-ray diffraction (μXRD) study was carried out on a trabecula extracted from osteoporotic and normal human vertebral bones and the degree of orientation for the BAp c-axis along its craniocaudal axis was analysed based on our previous report. A micro-computed tomography (μCT) measurement was also performed to analyze trabecular density and structure. In osteoporotic human vertebra, the trabecular number is markedly lower than that in normal vertebra. To sustain increased stress because of bone loss, the primary trabeculae, which are aligned parallel to the craniocaudal axis, tend to selectively remain while the secondary trabeculae, which are perpendicular to the craniocaudal axis, mostly disappear. Moreover, the primary trabecula from osteoporotic vertebra showed a significantly higher degree of BAp preferential orientation than the normal bone. This suggests that the remaining primary trabecula in osteoporotic vertebra is further reinforced by an increase in applied stress *in vivo* by enhancing the preferred BAp c-axis orientation along the trabecular direction.

1. Introduction

With the arrival of an aging society, osteoporosis is becoming a serious problem in the world. Osteoporosis is characterized by reduced bone mass and deterioration of bone microstructure, which increases the risk of fractures [1]. Vertebral bones are frequently evaluated in the diagnosis of osteoporosis and much attention has been paid to the trabecular bone in the vertebral bone because trabecular bones are believed to sustain 30-50% of vertebral strength [2]. Many studies have focused on the relationship between trabecular architecture and mechanical properties, while trabecular architecture has been reported to be a contributing factor to the mechanical properties of trabecular bone [3, 4]. Recently, our group reported that the preferential orientation of the biological apatite (BAp) c-axis as analyzed by microbeam X-ray diffraction (μXRD) is crucial in terms of bone reinforcement and is a possible candidate in the determination of bone quality parameters because of the crystallographic anisotropy of the hexagonal structure of the BAp crystallite [5]. However, little is understood about how osteoporosis changes the microstructural features and the related mechanical

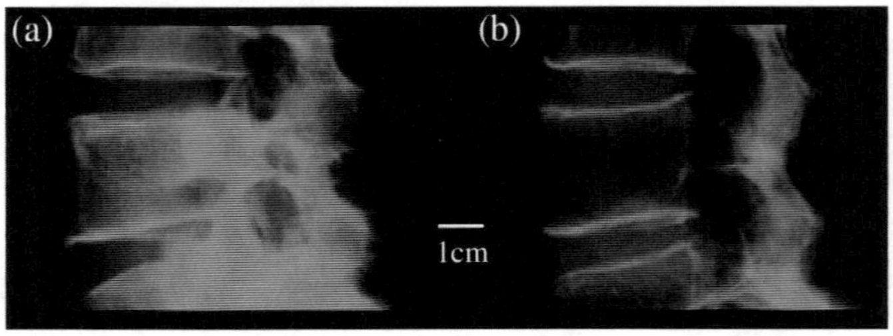

Figure 1. Mediolateral projection of the fourth lumbar vertebrae of (a) a 69 year old male and (b) a 86 year old female. Male and female bones were judged as normal and osteoporotic, respectively.

function of bone. In this study, μXRD analysis was conducted on isolated osteoporotic and normal trabecular bones from human vertebral bones to investigate the pathological condition from a crystallographic point of view, in addition to the number and structure of the trabeculae.

2. Materials and Methods

Fourth lumbar vertebrae were extracted from male (69 year old) and female (86 year old) cadavers. Figure 1 shows soft X-ray photographs of the two specimens. Radiopacity of the male vertebra (a) is obviously higher than that of the female bone (b). According to the criteria for the diagnosis of osteoporosis as based on a two-dimensional X-ray projection and shown in Figure 1, the female and male vertebra were judged as osteoporotic and normal, respectively. These bone specimens were fixed in a 10% formalin neutral buffered solution to avoid infections and prevent denaturation of the organic matrix.

The three-dimensional trabecular structure was reconstructed by micro-computed tomography (μCT) (SMX-100CT-SV; Shimadzu). The volume fraction of trabecular bone (BV/TV, %) was calculated in a cube of 10 mm × 10 mm × 10 mm at the central part of the fourth lumbar vertebra by Tri 3D-BON software (Ratoc). Based on the 3D image, the rod-like primary trabeculae aligned within 5 degrees from the craniocaudal axis were removed from the central part of the fourth lumbar vertebral body to analyze BAp orientation (N=3).

Two-dimensional quantitative analysis of preferential BAp orientation was performed by a microbeam X-ray diffractometer system (D8 Discover with GADDS, Bruker AXS) equipped with a two-dimensional position sensitive proportional X-ray counter (2D PSPC) (Hi-STAR, Bruker AXS) using a transmission optical system. Cu-Kα radiation was generated at a tube voltage of 45 kV and a tube current of 110 mA and the incident beam was focused onto a beam spot of 20 μm in diameter by a mono capillary collimator. The degree of BAp orientation was evaluated as the integrated intensity ratio of the (002) diffraction peak to the (310) peak. The two peaks were measured separately based on their Bragg's angles. Hydroxyapatite powder (NIST calcium hydroxyapatite #2910) with random orientation shows 0.7 of the intensity ratio in this system. Detail conditions for μXRD measurements were reported our previous paper [6].

Data is expressed as the mean ± standard deviation (SD). Statistical analysis was performed using the student's *t*-test. A *P*-value of <0.05 was considered to be significant.

3. Results and Discussion

Figure 2 shows μCT images of the central part of human fourth lumbar vertebral bodies. There is abundant trabecular volume and a three-dimensional trabecular network in the normal vertebra. In the osteoporotic vertebra, the trabecular number was markedly less, and the trabecular network was lost. The bone volume fractions (BV/TV) of the normal and the osteoporotic vertebrae are 16.3% and 5.5%, respectively. This difference in the bone volume between the osteoporotic and normal vertebrae suggests that *in vivo* stress applied to individual trabecula is greater in the osteoporotic bone than that in the normal bone. To sustain increased *in vivo* stress due to the loss of trabeculae, the primary trabeculae parallel to the craniocaudal axis tend to selectively remain while the secondary trabeculae perpendicular to the craniocaudal axis predominantly disappear.

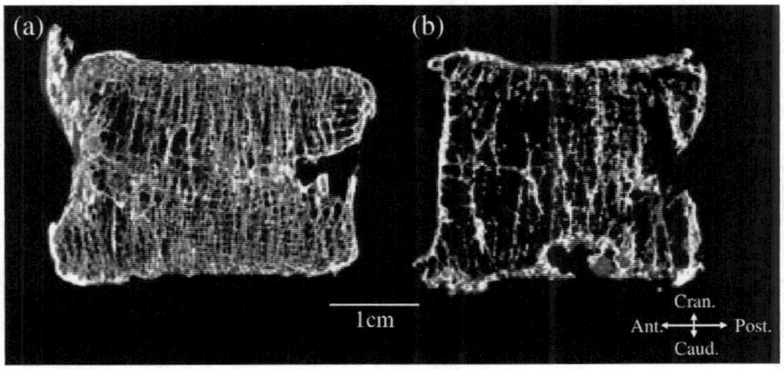

Figure 2. μCT images of the central cross section of human fourth lumbar vertebrae; (a) Normal and (b) osteoporotic bone.

Primary trabeculae were removed from the central part of the vertebra and μXRD measurements were conducted on them to analyze the BAp orientation in each trabecula. Figure 3 shows a typical isolated trabecula and a two-dimensional diagram of the degree of BAp orientation along all axes within a plane including trabecular direction. The higher value in the integrated intensity ratio of (002)/(310) indicates a more significant BAp c-axis orientation. A homogeneous distribution of the BAp c-axis within the plane results in a concentric diagram. Two-dimensional BAp distribution in the trabecula obviously indicates that the BAp c-axis distribution is highly anisotropic. The direction showing the maximum degree of BAp orientation corresponds precisely to the trabecular direction and the minimum BAp orientation is found in a direction perpendicular to the trabecular direction. Thus rod-like trabeculae have a uniaxial preferred orientation of the BAp c-axis, i.e. fibre-texture, along the trabecular direction. The maximum degree of uniaxial BAp c-axis orientation appears along the direction of the trabecular axis and is suitable to characterize the fibre-like BAp texture in the normal and osteoporotic trabeculae.

Figure 4 shows a comparison of the BAp preferred orientation between the normal and the osteoporotic trabeculae. The degree of BAp preferential orientation of the trabecula from the osteoporotic vertebra is significantly higher than that from the normal vertebra. It has been reported that *in vivo* stress is one of the dominant controlling factors of BAp orientation [5]. Figure 5 shows the correlation between the integrated intensity ratio of (002)/(310) and a type of *in vivo* stress component, which is represented by the reciprocal value of the trabecular volume fraction BV/TV. The degree of BAp orientation has a positive correlation to the *in vivo* stress, which indicates that even under the osteoporotic condition the trabecular bone can develop an anisotropic crystal texture and sustain increased stress craniocaudally when applied *in vivo* and based on the functional adaptation mechanism of bone.

Bone functional adaptation is known to be expressed as the change in bone mass and bone mineral density when the applied stress distribution changes [7]. We conclude that if a decrease in bone mass is inevitable because of metabolic bone diseases like osteoporosis a functional adaptation can be

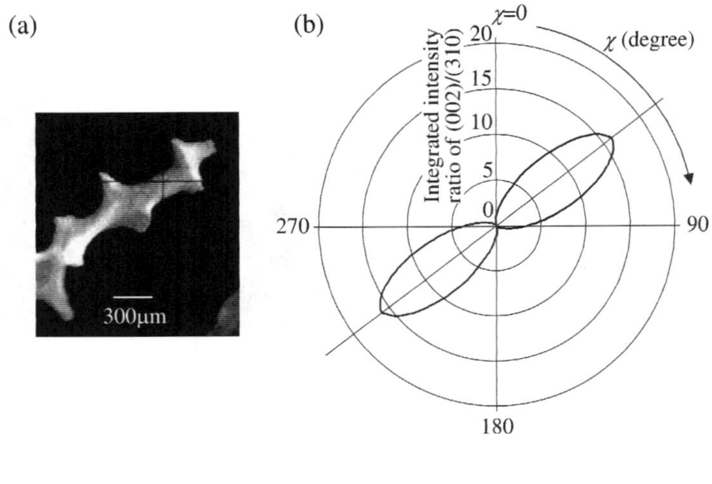

Figure 3. (a) Structure of the analyzed rod-like trabecula and (b) the distribution of the integrated intensity ratio of (002)/(310) along all directions in a plane including the trabecula. The μXRD measurement was done on the center of the crosshair in (a).

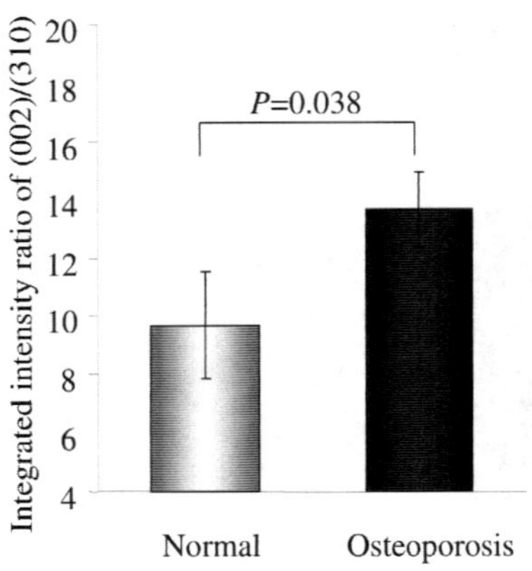

Figure 4. The maximum degree of preferential BAp orientation along the trabecular direction.

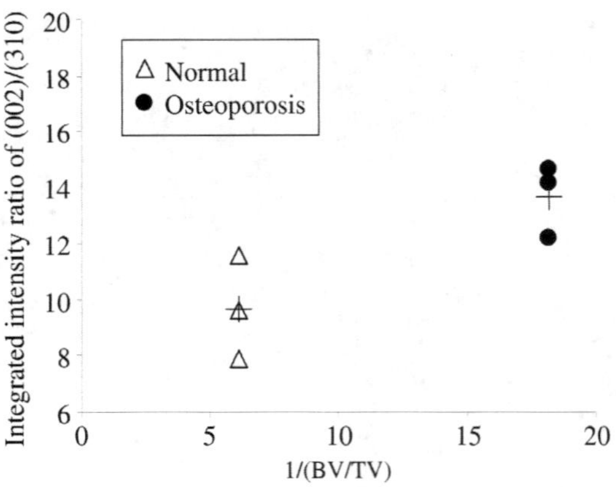

Figure 5. Relationship between the degree of preferential BAp orientation and the reciprocal value of the bone volume fraction (BV/TV) as an indicator of *in vivo* stress applied along the trabeculae.

achieved by changing the degree of BAp orientation. Investigating the BAp preferential alignment mechanism is important for the understanding of changes in the mechanical function of bones when they are subjected diseases such as osteoporosis.

4. Conclusions

The crystal orientation of BAp in osteoporotic and normal human vertebral trabeculae was investigated by means of the microbeam X-ray diffraction technique and the following conclusions are made:

1. The preferred BAp c-axis orientation was increased significantly in the osteoporotic trabecula compared with that in the normal trabecula.
2. In the osteoporotic trabecula, increased *in vivo* stress, due to bone loss, enhances the degree of BAp preferred orientation in the trabecular direction and is based on the functional adaptation of bone.

Acknowledgements

This study was supported by the Priority Assistance for the Formation of Worldwide Renowned Centres of Research - The Global COE Program (Project: Centre of Excellence for Advanced Structural and Functional Materials Design) and a Grant-in-Aid for Scientific Research and Development from the Ministry of Education, Culture, Sports, Science and Technology (MEXT), Japan. The work was partly supported by a Grant-in-Aid for JSPS Fellows (20 ·799). The authors would like to thank Prof. Y Nakajima and Prof. H Kiyama, Osaka City University for providing the bone specimen.

References

[1] 1991 *Am. J. Med.* **90** 107
[2] Mazess R B 1990 *Calcif. Tissue Int.* **47** 191
[3] Parfitt A M 1987 *Am. J. Med.* **82** 68
[4] Ito M, Nishida A, Koga A, Ikeda S, Shiraishi A, Hayashi K and Nakamura T 2002 *Bone* **31** 351
[5] Nakano T, Kaibara K, Tabata Y, Nagata N, Enomoto S and Umakoshi Y 2002 *Bone* **31** 479
[6] Miyabe S, Nakano T, Ishimoto T, Takano N, Adachi T, Iwaki H, Kobayashi A, Takaoka K and Umakoshi Y 2007 *Mater. Trans.* **48** 343
[7] Umemura Y, Ishiko T, Yamauchi T, Kurono M, Mashiko S 1997 *J. Bone Miner. Res.* **12** 1480

The optimal design of an implant to improve bone quality of implant surroundings based on stress analysis

Yoshihiro Noyama[1], Noriyuki Nagayama[2], Koichi Kuramoto[1] and Takayoshi Nakano[3]

[1]Nakashima Medical Division, Nakashima Propeller Co., Ltd., 688-1 Joto-Kitakata, Okayama-shi, Okayama 700-8691, Japan

[2]Department of Mechanical Engineering, Applied Electronics Group Industrial Technology Centre of Okayama Prefecture, 5301 Haga, Okayama 701-1296, Japan

[3]Division of Materials and Manufacturing Science, Graduate School of Engineering, Osaka University, 2-1 Yamadaoka, Suita, Osaka 565-0871, Japan

E-mail: nakano@mat.eng.osaka-u.ac.jp

Abstract. Research on how implant surface shape contributes to long-term stability after implantation is important in the field of orthopaedics. In particular, technology that controls various bone quality parameters and voluntary bone inducement in implant surroundings should be developed for the next generation of implants and this will improve the patient's quality of life (QOL). For this research, we focused on the inducement of the appropriate alignment for biological apatite (BAp) crystallites and related collagen (Col.) fibres as a bone quality parameter. In this study, we predicted that when stress is applied to bone, the BAp/Col. preferential alignment can be formed if osteocytes are in an environment that is aligned with the principle stress vector. We tested this idea by introducing grooves in the principal stress direction on the surface of an implant. This work thus analyzes the effect of stress transmission by a load at the proximal femur on the bone inside and near the grooves by using mechanical simulation in which groove angles can be changed on the implant surface. Coordinate data from the mechanical simulation of the combined bone/implant environment was verified against the coordinate data obtained by CT scans of actual canine bone. Results suggest that the tendency of stress transmission differs depending on the position and angle of the grooves and based on a vector diagram of the maximum and minimum principal stresses. The simulation was able to predict bone dynamics *in vivo* and enabled a best design of an implant to control the BAp/Col. alignment as an index of bone quality.

1. Introduction

In the field of orthopaedics, research on implant surface profiles, which influences long-term stability after implantation, is extremely important. In particular, technology that controls the maintenance of bone quality and voluntary bone induction in implant surroundings has attracted attention for the development of next-generation implants that are anticipated to improve the patient's QOL. We thus focused on the specific alignment of BAp crystallites and the related Col. fibres as a possible bone quality parameter candidate and attempted to clarify the mechanism of bone tissue formation *in vivo* by correlating BAp alignment, stress distribution *in vivo* and bone dynamic functions [1]. The term "bone quality" is a concept proposed by the US National Institute of Health (NIH) and represents a controlling factor for bone strength, other than the bone density. Our group reported that appropriate design of pores or grooves to promote the formation of new bone under a stress environment should be considered based on the BAp/Col. preferential alignment, which is closely related to the function of an osteocyte [2]. This is due to formation of the BAp/Col. preferential alignment which improves the dynamic characteristics, especially the bone strength between bone and implant. In this study, we investigated the optimal design of surface grooves for hip implants by calculating the stress distribution in implant surroundings such as the implant itself and the newly formed bone. Principal stresses which may control the BAp/Col. preferential alignment was used as a bone quality parameter.

2. Materials and Methods

To examine a stress environment *in vivo* of implant surroundings, we performed a dynamic simulation with canine bone as an example. As regards the stress transmission by a load at the proximal femur on the bone to the implant surroundings, grooves introduced on the implant surface were changed and this influence on the bone inside the grooves was investigated with finite element analysis (FEA). The analysis procedure was as follows: First, implant modelling was performed with three-dimensional (3-D) CAD software Unigraphics NX4.0 (SIEMENS) (Figure 1). Next, from the dog's medical images (DICOM data) the bone contour was extracted with 3-D image processing software Mimics 11 (Materialise) and a 3-D bone model was prepared with 3-D CAD software Imageware 10 (SIEMENS) (Figure 2). From this 3-D bone model various sizes were measured by reference to the X-ray picture (Figure 3). The 3-D model of the implant was matched to reproduce the bone/implant composite model in a state close to the clinical state. Using this composite model, the stress environment formed in the implant surroundings was investigated using non-linear FEA software MSC Marc 2005 (MSC Software). The unipolar-type artificial femoral head and the femoral stem, making up the implant, were made from SUS316L and Titanium-6wt.%Aluminum-4wt.%Vanadium (Ti-6Al-4V) alloy, respectively. The femur was regarded as a double-layer structure of cortical bone and cancellous bone. Table 1 shows material properties for each element. The loading condition was 700 N (5 times the body weight of 140 N [3]) toward the distal direction along the femoral axis (Figure 4). As for the boundary condition, the coefficient of friction was assumed to be 0 between the implant and the bone and also between the implant and the implant. The edge face after cutting the bone at a position 20 mm apart from the stem distal end was restrained completely. Furthermore, it was assumed that cancellous bone was induced inside all the grooves introduced to the implant surface. In the analysis model, the number of elements was 48,170 and the number of nodal points was 592,088 as shown in Figure 5.

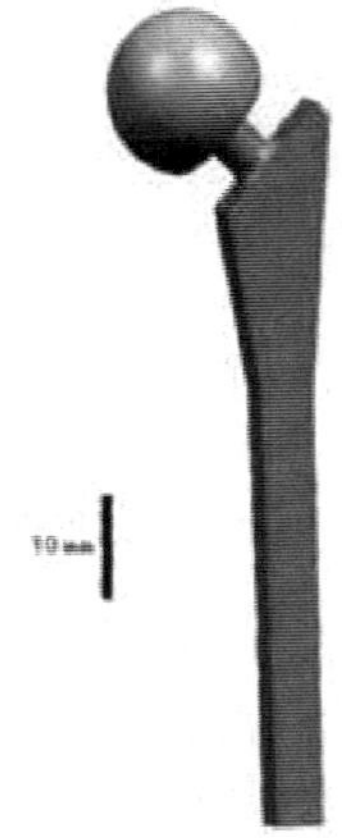

Figure 1. A virtual model of the canine implant.

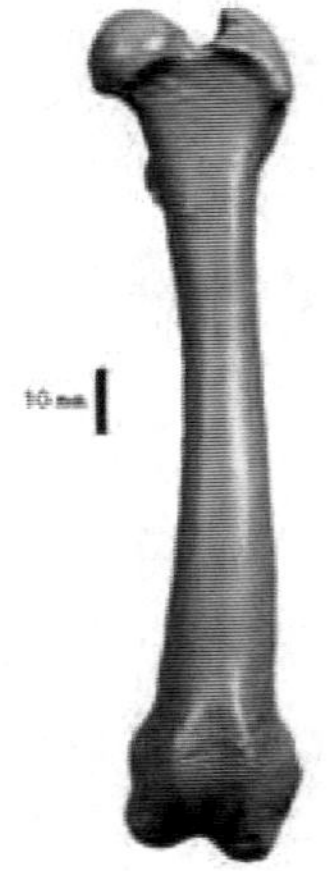

Figure 2. A virtual model of the canine femur.

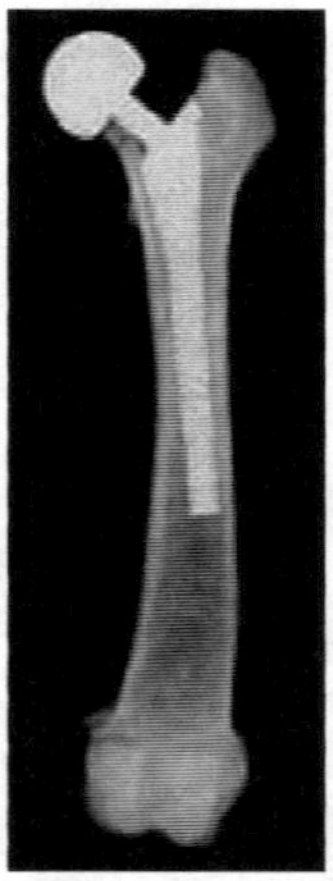

Figure 3. A radiograph of the canine femur after implantation.

Table 1. Material properties used for analysis [4,5].

	Material	Young's modulus (GPa)	Poisson's ratio (-)	Yield Strength (MPa)
Femoral stem	Ti-6Al-4V alloy	110	0.34	795
Femoral head	SUS316L	190	0.30	310
Canine bone	Cortical bone	13	0.30	-
	Cancellous bone	1	0.30	-

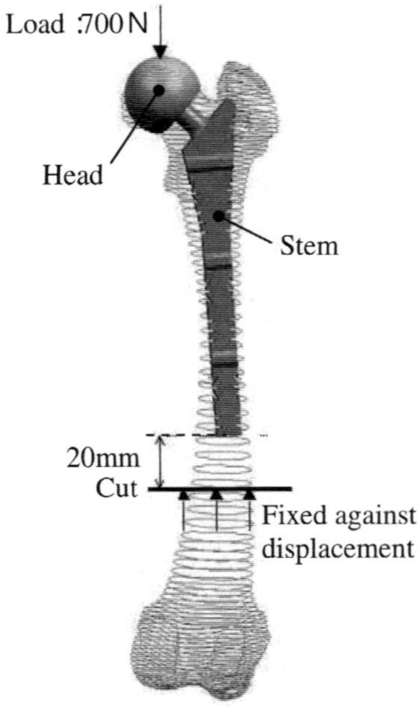

Figure 4. Boundary conditions used for the analysis.

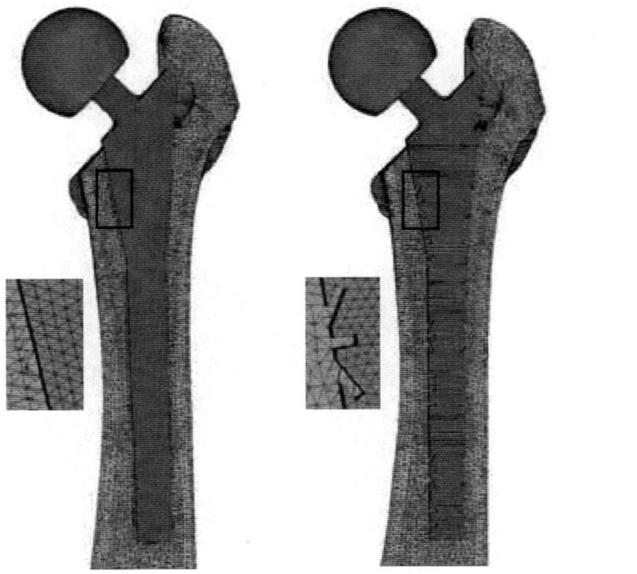

(A) control (B) with grooves

Figure 5. 3-D finite element models; (A) control, (B) with grooves.

3. Results and Discussion

Figure 6 shows the distribution of principal stress vectors in implant surroundings under a stress environment. This figure is a cross sectional image of the femur from an anterior view and displays the maximum and minimum principal stress vectors on setting the upper limit of range to 150 MPa and the lower limit to -150 MPa. Looking at the proximal inside part, the principal stress vectors inside the grooves introduced on the implant surface tended to be different depending on groove angle and groove position. The principal stress vector length tended to increase when the grove position got close to the proximal end. Furthermore, in some grooves, the principal stress vector aligned along the groove suggesting that bone formation possibly occurred inside the groove, i.e., the BAp/Col. preferential alignment depends on the direction of groove extension. It was, therefore, shown feasible to estimate dynamic characteristics between bone and implant by FEA which mimics the clinical status. The interaction between the external load and the stress acting on the bone inside the groove was, however, not determined.

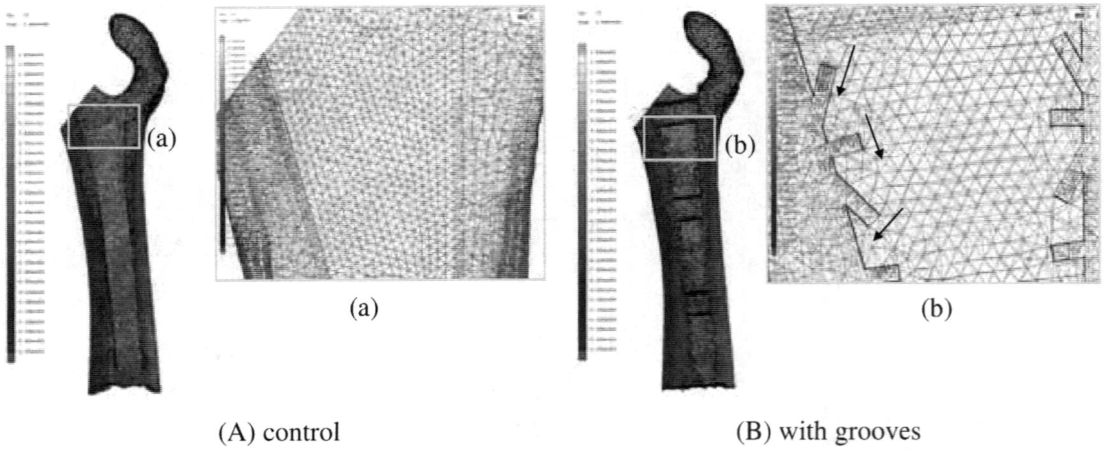

Figure 6. A vector plot of the maximum and minimum principal stresses; (A) control, (B) with grooves

4. Conclusions

In this study, to qualitatively estimate the BAp/Col. preferential alignment and the dynamic behaviour of osteocytes in implant surroundings under a stress environment *in vivo*, a dynamic simulation was performed with canine bone. The results of this study are summarized as follows:

1) As a method to evaluate the bone in the implant surroundings under a stress environment *in vivo*, a 3-D FEA was established.
2) The figure of maximum and minimum principal stress vectors on the bone cross sectional surface suggested that principal stress vectors inside the grooves tended to be different depending on groove angle and groove position.
3) Although the interaction between the external load and the stress acting on the bone inside the groove is not clear, this technique is capable of a qualitative estimation of dynamic behaviour *in vivo* and can be used to estimate dynamic characteristics between bone and implant. It is also useful for the optimal design of an implant when the bone quality of the implant surroundings can be evaluated or controlled in advance before a surgical operation.

Acknowledgements

This work was supported by funds from the "Priority Assistance of the Formation of Worldwide Renowned Centres of Research – The Global COE Program (Project: Centre of Excellence for Advanced Structural and Functional Materials Design)" and a Grant-in-Aid for Scientific Research and Development from the Ministry of Education, Culture, Sports, Science and Technology of Japan.

References

[1] Nakano T, Kaibara K, Tabata Y, Nagata N, Enomoto S, Marukawa E and Umakoshi Y 2002 *Bone* **31** 479
[2] Nakano T, Kan T, Ishimoto T et al. 2006 *Materials Transactions* **47** 2233
[3] Kanai A et al. 2005 *The Journal of Japanese Physical Therapy Association* **32** 513
[4] ASTM, American Standards for Testing and Materials, F136-02a, 2007 *Annual Book of ASTM Standards*, **13** 70
[5] H. Qian, J. Chen and T.R. Katona INTERNATIONAL SOCIETY OF BIOMECHANICS TECHNICAL GROUP COMPUTER SIMULATION, August 5-7, 1999

A model for nucleation of tin whisker through dislocation behavior

K Nakai[1,4], T Sakamoto [1], S Kobayashi[1], M Takamizawa[2], K Murakami[3] and M Hino[3]
[1]Course of Materials Science and Engineering, Department of Materials Science and Biotechnology, Graduate School of Science and Engineering, Ehime University, 3 Bunkyo-cho, Matsuyama 790-8577, Japan
[2]OM Sangyo Co. Ltd., 3-18-48 Noda, Okayama 700-0971, Japan
[3]Industrial Technology Center of Okayama Prefecture, 5301 Haga, Okayama 701-1296, Japan

E-mail: k-nakai@eng.ehime-u.ac.jp

Abstract. A model for the nucleation and growth processes of Sn whisker is offered. High density of localized screw dislocations by deformation form the dense spiral steps of atomic scale on Sn surface. The spiral steps would induce the nucleation of Sn whisker. Edge dislocations localized at the same region where dense screw dislocations exist supply Sn atoms to the Sn whisker through pipe diffusion. Both screw and edge dislocations would bend along almost one direction, namely, to relax the external shear stress. The image force also helps to bend the dislocations perpendicular to the whisker side-surface. The bending of dislocations at root of whisker leads the bend of whisker. The pipe diffusion of Sn atoms through edge dislocations from bulk Sn toward whisker is suppressed at the bent part of edge dislocation, resulting in release of Sn atoms inside whisker and leading to the growth of whisker near its root.

1. Introduction

Whisker formation on an electroplate of Sn is an important problem in electronic packages because a whisker induces electric short-circuit in electronic devices. In order to prevent whisker formation, fundamental researches for mechanism of the whisker formation have been carried out until now [1]. Addition of Pb to Sn is effective to reduce whisker formation [2]. However, the use of Pb is strictly prohibited according to RoHS. An alternative Pb-free Sn electroplate is required.

Sn whisker formation is promoted in the region where stress is applied, such as electronic joint parts [3]. This fact suggests that whisker formation is in association with dislocation behavior.

In this study, the formation of Sn whisker in bulk Sn under tensile stress has been investigated.

2. Experimental procedures

Two specimens of pure bulk Sn were cut into the parallelepiped shape of about 15mm×50mm×3mm, and they were kept with and without tensile stress for 7 to 20 days, respectively. The tensile stress was applied as shown in figure 1. The specimen was deformed around 10%, and the cross-head of tensile

[4] The corresponding author

test equipment (Shimadzu Autograph AG-10kNIS) was stopped. Stress relaxation then started. After about two thirds of yield stress is relaxed, the tensile stress was applied again. Such operation was repeated for 7 to 20 days. The microstructures of specimens were observed by a scanning electron microscope (SEM).

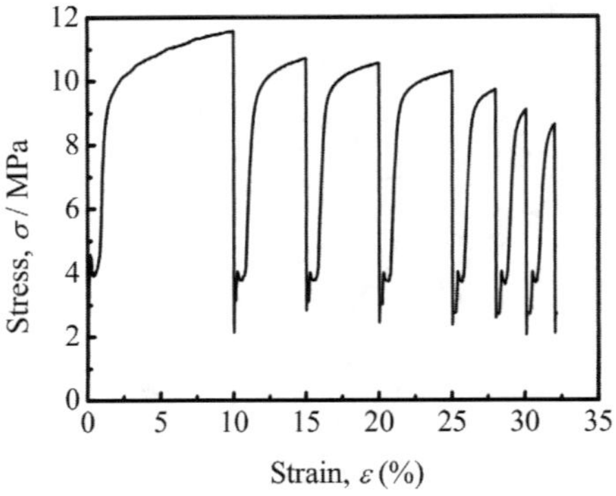

Figure 1. Stress-strain curve during tensile test for whisker formation.

3. Results and discussion

Figure 2 shows that nodular whiskers are observed on the specimen with tensile stress while few whiskers are observed in the specimen without tensile stress. The diameter of whisker is much smaller than grain size (average grain size is 35μm) unlike whiskers on a tin electroplate has nearly the same diameter with the grain size [1]. Another feature of whiskers on bulk Sn is that the whiskers are mainly formed at and/or near grain boundary.

Figure 3 shows a SEM micrograph of a whisker near grain boundary. A lot of slip bands are observed as indicated by arrows in the figure. The whisker is formed at the area where slip bands intersect.

The whisker formation at the area where the intersection of slip bands occurs suggests that the whisker formation is caused by the accumulation of dislocations. According to the results of figures 2 and 3, nucleation of whiskers would be affected by the inhomogeneity of the deformation. The

Figure 2. SEM micrograph taken from the bulk Sn kept under tensile stress for 20 days at ambient temperature. Whiskers (encircled) are formed at and/or near grain boundary.

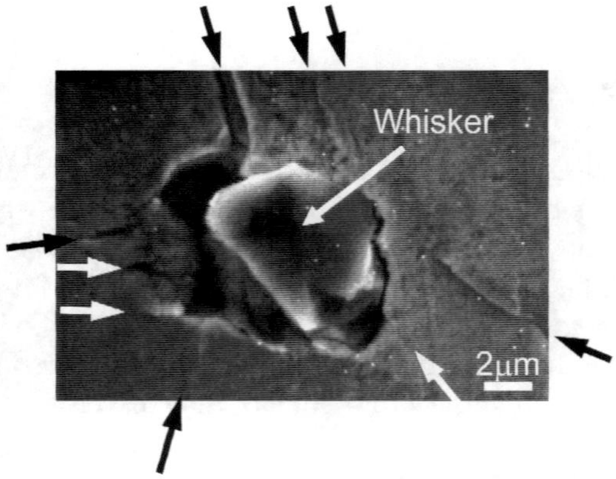

Figure 3. SEM micrograph taken from the bulk Sn under tensile stress for 7.5 days. A whisker was formed at the intersection of slip bands indicated by arrows.

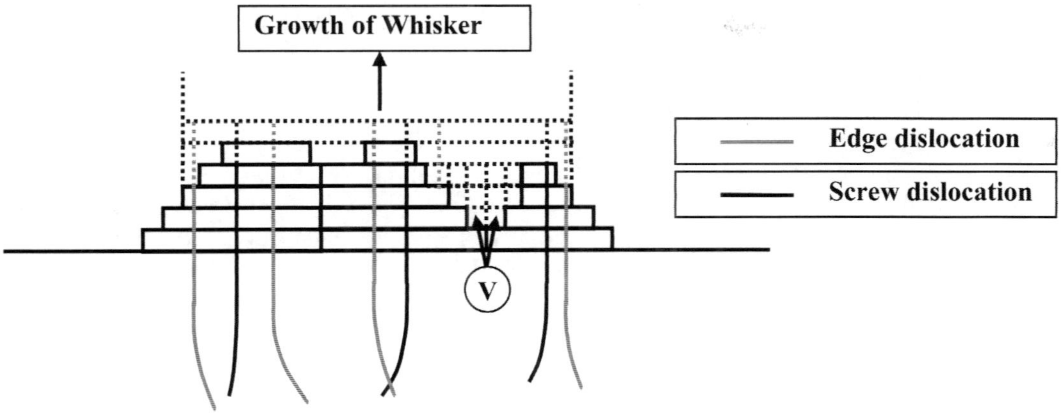

Figure 4. Growth of whisker in association with screw and edge dislocations.

deformation induces multiplication of dislocations and their conservative motions, resulting in their local accumulation around grain boundary. High density of dislocations are localized around the ends of slip bands at grain boundary. It could be suggested that the dislocations consist of both screw and edge dislocations, because various directions of slip bands due to the operations of various slip systems accumulate around grain boundary as shown in figure 3.

Based on above results, a model of whisker formation is considered as follows.

The motion of screw dislocation causes a step on Sn surface in atomic scale. The step tends to trap Sn atoms and/or their aggregates migrating on Sn surface because the step has combination sites with Sn atoms. Such a process would form a spiral step around a screw dislocation. The height of a step is one atomic distance, and the spiral step continues to increase with increasing trapped Sn atoms [4]. However, the spiral steps combine with each other when screw dislocations are densely localized by deformation [4]. When Sn atom migrates into the valley between spiral steps (designated as "V" in figure 4), the Sn atom is easily combined with spiral steps, resulting in reduction of surface energy of spiral steps. The regions where screw dislocations are densely localized act as nucleation sites for whiskers because diffusing Sn atoms on crystal surface are preferably captured in the region.

For the growth of whisker, it is indispensable that Sn atoms are supplied from the bulk Sn into whisker. The dense localization of edge dislocations at the same place where screw dislocations exist is also indispensable because pipe diffusion of Sn atoms should occur along edge dislocations. The pipe diffusion of Sn atoms along edge dislocations facilitates the supply of Sn atoms from bulk Sn into whisker.

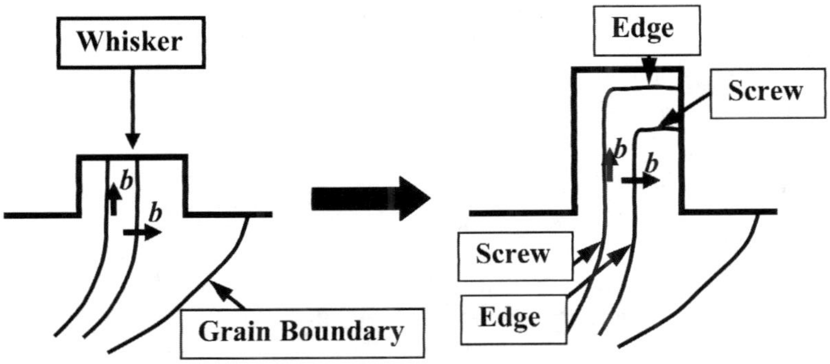

Figure 5. Dislocations inside whisker during growth process of whisker. $\boldsymbol{b}$ is the Burgers vector.

Figure 5(a) shows a whisker just after its nucleation. An edge dislocation and a screw dislocation are shown for simplicity. In the course of growth of whisker, both dislocations would bend towards side-surface of whisker in almost the same direction because external stress and image force are applied to the dislocations.

The bending of dislocations changes their natures, namely, screw and edge dislocations bent near the whisker side-surface change into edge and screw dislocations, respectively (see figure 5(b)). This phenomenon could induce the whisker growth from its root [1]. Migrating Sn atoms in bulk Sn are supplied into whisker through the pipe diffusion only along edge dislocation, but Sn atoms scarcely migrate along screw dislocation which is formed by the bending of edge dislocation. Sn atoms are accumulated around the area where edge dislocation changes to screw one. Jogs are formed here. In order to lower the edge dislocation energy which increases by the jogs of Sn atoms, namely, the increase in dislocation length, the jogs of Sn atoms would easily dissociate from the edge dislocation, resulting in diffusion of Sn atoms into the root of whisker.

Concentration of vacancies in whisker would be lower than that in bulk Sn, because a unit volume of whisker has more amount of surface where vacancies disappear than a unit volume of bulk Sn has. Then, vacancy migrates into whisker along the concentration gradient of vacancies. Sn atoms dissociated from edge dislocation easily diffuse into whisker, and the growth of whisker occurs remarkably at its root.

Trace analysis of a slip band near a whisker shows that the slip plane is {101}. It is suggested that the slip system is $\{101\}<10\bar{1}]$ according to the literature [5]. The magnitude of the Burgers vector of $<10\bar{1}]$ is larger than that of <001], along which the atomic spacing between Sn atoms is the smallest in β-Sn. Edge dislocation with the Burgers vector of $<10\bar{1}]$ has sufficiently broad space under the extra half plane so that Sn atoms should move easily through pipe diffusion, while other metals usually has the Burgers vectors with the smallest atomic spacing. This fact enables whisker to be formed on Sn unlike on other metals.

4. Conclusions

A model of whisker formation on bulk Sn under tensile stress has been offered. Whiskers, whose diameters are much smaller than that of grain in matrix, are nucleated at the area where intersection of slip bands occurs near grain boundary of the matrix, because high density of screw dislocations accumulated by deformation make atomic steps and valleys on the surface of bulk Sn. Edge dislocations accumulated in the same region supply Sn atoms for growth of whisker through pipe diffusion. The diffusion of Sn atom along the edge dislocation is retarded at the point where the edge dislocation changes into screw one because the edge dislocation bends towards the side-surface of whisker by external stress and image force. The accumulated Sn atoms are dissociated from the edge dislocation in order to lower the dislocation energy, resulting in the growth of whisker near its root.

5. References

[1] Tu K N and Li J C M 2005 *Mater. Sci. Eng.* A **409** 131-9
[2] Boettinger W J, Johnson C E, Bendersky L A, Moon K -W, Williams M E and Stafford G R 2005 *Acta Mater.* **53** 5033-50
[3] Shibutani T, Wu J, Yu Q and Pecht M 2008 *Microelectronics Reliability* **48** 1613-27
[4] Min Y K, Munir Z A and Schmaltzried H 1995 *Philos. Mag.* A **71** 815-29
[5] Yomogita K 1972 *Jpn. J. Appl. Phys.* **11** 1-5

Acknowledgments

Authors would like to express sincere thanks to the Small and Medium Enterprise Agency, the Ministry of Economy, Trade and Industry, Japan for providing the Research Grant-in-Aid for the promotion of this research. Authors also thank graduate students, T. Nishikawa and K. Tsuji, for the cooperation in the experiments.

International Conference on Advanced Structural and Functional Materials Design 2008 IOP Publishing
Journal of Physics: Conference Series **165** (2009) 012090 doi:10.1088/1742-6596/165/1/012090

A Cu-added Zr-Ce-Sn-Pr-O mixed oxide phase with a high oxygen storage capacity

Tadatoshi Murota, Kosuke Sogawa, Shinya Otsuka-Yao-Matsuo, Takahisa Omata

Division of Materials and Manufacturing Science, Graduate School of Engineering, Osaka University, 2-1 Yamada-oka, Suita, Osaka, 565-0871, Japan

E-mail: shinya@mat.eng.osaka-u.ac.jp

Abstract. Phases, oxygen release and absorption behaviour was investigated for Zr-Ce-Sn-Pr-O mixed oxides $(Zr_{2/8}Ce_{x/8}Sn_{y/8}Pr_{(6-x-y)/8})O_{2-z}$ (x=1.5, 4.5, y=0.5) together with the effect of Cu addition. Major phases of Sn-added samples were similar to those of Sn-free $(Zr_{2/8}Ce_{x/8}Pr_{(6-x)/8})O_{2-z}$ phases, that is, CaF_2-like λ and C phases for Pr-rich and Ce-rich compounds, respectively. All samples contained a SnO_2 or a Sn phase. When Cu was added, repetitive oxidation and reduction appeared to promote the fine dissolution of Sn and Cu through bulk phases. OSC values of both $(Zr_{2/8}Ce_{1.5/8}Sn_{0.5/8}Pr_{4/8})O_{2-z}$ and $(Zr_{2/8}Ce_{4.5/8}Sn_{0.5/8}Pr_{1/8})O_{2-z}$ were larger than those of the corresponding Sn-free samples. For the Ce-rich sample, Cu addition improved both the OSC value and its on-set temperature.

1. Introduction

Oxide materials that have high oxygen storage capacity (OSC) at low temperatures and that possess high thermal stability at high temperatures are indispensable for use as advanced automotive exhaust catalysts. Zr-Ce-O mixed oxide is currently used as an OSC material [1-2]. Candidates for high efficiency OSC materials should have compositions of $(Zr_{2/8}Ce_{x/8}Pr_{(6-x)/8})O_{2-y}$ (x=0~6) because small amounts of Ce stabilize single phases of λ-$(Zr_{0.25}Pr_{0.75})O_{2-y}$ that have a broad oxygen nonstoichiometry. One may expect that the addition of cations of three valence states: 0, +2 and +4 rather than Ce ions with +3 and +4 under a conventional atmosphere may be effective. We thus investigated different phases and their oxygen release behaviour for Zr-Ce-Sn-Pr-O mixed oxides $(Zr_{2/8}Ce_{x/8}Sn_{y/8}Pr_{(6-x-y)/8})O_{2-z}$ (x=1.5, 4.5, y=0.5). Another approach to improve the OSC property may be the addition of noble catalytic metals such as Cu but not rare-elements such as Pt. We have recently discovered that Cu addition greatly improved OSC behaviour at temperatures lower than 573 K for Zr-Ce-Pr-O mixed oxides [3]. An interesting feature was that Cu addition induced a striking change in the phase structure of the bulk oxide. In this study, the phases of Cu-added Zr-Ce-Sn-Pr-O mixed oxides $(Zr_{2/8}Ce_{x/8}Sn_{y/8}Pr_{(6-x-y)/8})Cu_{0.5/8}O_{2-z}$ (x=1.5, 4.5, y=0.5) as well as their oxygen release and absorption behaviour were investigated. Samples were subjected to XRD and the release and absorption properties were determined by OSC analysis and TPD oxygen gas analysis by means of TG/DTA in Ar+5% H_2 gas.

2. Experimental

2.1. Preparation of sample powders and OSC measurements

$ZrO(NO_3)_2$, $Ce(NO_3)_3$, $Pr(NO_3)_3$, and $SnCl_4$ were dissolved in water so that the solution concentration of the composition corresponding to $(Zr_{2/8}Ce_xSn_yPr_{(6-x-y)})O_{2-z}$ was 100 g/L. To obtain hydroxides of Zr^{4+}, Ce^{4+} and Pr^{4+} the same molar quantity of H_2O_2 as the molar quantity of the metal elements was added to the aqueous solution. The aqueous mixture was added dropwise to ammonia-water to enable uniform precipitation. The hydroxide mixture was calcined at 673 K after filtration and washing. Cu was then added as a $Cu(NO_3)$ solution by an impregnation technique and calcined at 673 K. Powder samples were annealed at 1473 K in air for 1 h. These powders were then reduced at 1373 K in H_2 gas. The powders were oxidized again at 973 K in air. Sample phases obtained through these reduction and oxidation processes were examined using the powder XRD analysis technique. As described below for TPD analysis it has been previously found that oxygen release was greatly improved by repetitive reduction and oxidation [3]. Oxide powders that were re-oxidized at 973 K in air were thus used as samples for OSC measurements. OSC was measured using a gas adsorption device (Ohkura Riken Inc.:BP-1). Powder samples of 0.5 g were loaded into a special measurement cell and heated to preset temperatures (373 K to 973 K). Samples were reduced by passing H_2 gas for 0.9 ks. The gas was then substituted by passing He gas for 0.3 ks. Finally, 1 cm^3 of O_2 gas was introduced into the cell several times using a pulse and the total amount of oxygen absorbed by the samples was measured by TCD (Thermal Conductivity Detector).

2.2. TPD oxygen gas analysis

Oxygen release behaviour was examined using TG/DTA equipment (Seiko Instruments Inc.: TG/DTA 6300). The hydroxide was calcined at 673 K and then annealed in O_2 gas at 1373 K. Powders of 100 mg were loaded into a platinum crucible and heated from 303 K to 773 K at a rate of 5 $K(60\ s)^{-1}$ in an O_2 gas atmosphere. The sample powders were then cooled to 303 K at a rate of 5 $K(60\ s)^{-1}$ in O_2 gas. The O_2 gas was then replaced by Ar for 0.9 ks. An Ar+5%H_2 atmosphere mixture was then used. After 7.2 ks, samples were heated from 303 K to 1373 K at a rate of 5 $K(60\ s)^{-1}$ (1st TPD oxygen analysis). The samples were then maintained at 1373 K for 10.8 ks and cooled to 303 K over 13.2 ks. The samples were then heated to 773 K in O_2 gas at a rate of 5 $K(60\ s)^{-1}$ and then cooled to 303 K followed by the 2nd TPD oxygen analysis in a flowing Ar+5%H_2 gas mixture. This operation was repeated three times. The oxygen discharge rate and reproducibility were improved greatly by repeating this operation. In this research, the result of the 3rd oxygen release experiment was adopted as the discharge characteristic of oxygen.

3. Results and Discussion

Figure 1 and Figure 2 show XRD patterns for Pr-rich and Ce-rich samples subjected to repetitive oxidation and reduction, respectively. As seen in those figures, when Cu was not added, major phases for the samples containing Sn resembled those for the Sn-free sample [3]. The main diffraction peaks for $(Zr_{2/8}Ce_{1.5/8}Sn_{0.5/8}Pr_{4/8})O_{2-z}$ and $(Zr_{2/8}Ce_{4.5/8}Sn_{0.5/8}Pr_{1/8})O_{2-z}$ could be identified based on CaF_2 structures and were therefore attributed to the λ and C phases for Pr-rich and Ce-rich compounds, respectively. In the case of the Pr-rich compound, the A-Pr_2O_3 phase appeared after reduction. The appearance of a λ phase for the Pr-rich composition, λ_{Pr}, after re-oxidation at 973 K resulted from the uneven dissolution of A-Pr_2O_3 into the λ phase [3]. The reduction process in H_2 gas promoted the precipitation and growth of the Sn metal phase. Thus, the XRD peaks of SnO_2 that were observed in the oxidation atmosphere increased during repetitive reduction and oxidation.

For all samples prepared in this work, when Cu was added, the pyrochlore-like phase appeared together with λ and C phases for Pr-rich and Ce-rich compositions, respectively. We have recently investigated the effect of Cu addition on the phase states for $(Zr_{2/8}Ce_{x/8}Pr_{(6-x)/8})Cu_{0.5/8}O_{2-y}$ with various Pr and Ce composition ratios and we found that the pyrochlore-like phase appears together with a CaF_2-type phase in the composition range of $0<x<5$ [3]. The lattice constant of the latter phase was almost unaffected by the appearance of the pyrochlore-like phase. Our current results are consistent with this previous report. A small change in the lattice constant of the λ phase by Cu addition may be attributed to the effect of Sn. For both Pr-rich and Ce-rich samples, the diffraction peaks due to SnO_2 showed an obvious decrease with Cu-addition. The appearance of Cu-Sn compounds for the Ce-rich sample may be an important feature. We found that the Cu_6Sn_5 phase decreased faster than the Cu_3Sn phase by repeating the successive reduction and oxidation several times. This result confirms that Cu and Sn ions gradually disperse into the bulk oxide by repetitive oxidation and reduction.

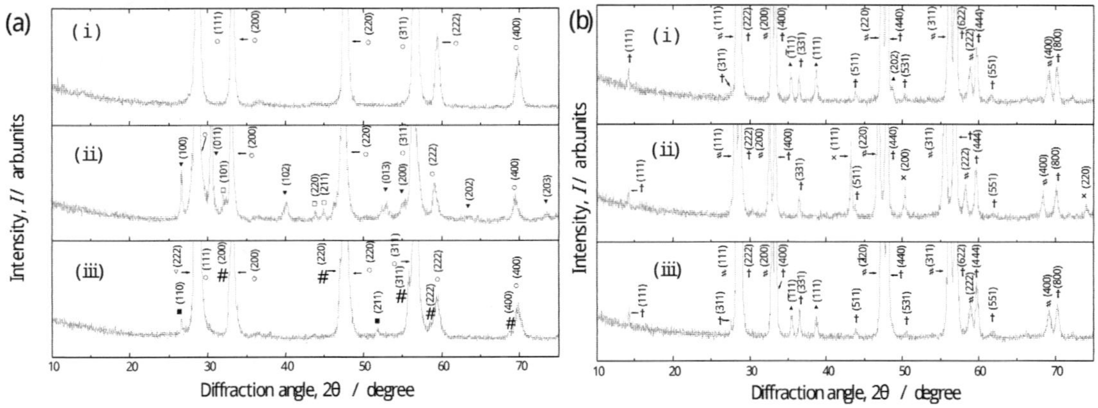

Figure 1 XRD patterns of Pr-rich Zr-Ce-Sn-Pr-O mixed oxide powders obtained by repetitive oxidation and reduction and compared with those of Cu-added Zr-Ce-Sn-Pr-O mixed oxide powders. (a): $(Zr_{2/8}Ce_{1.5/8}Sn_{0.5/8}Pr_{4/8})O_{2-y}$, (b): $(Zr_{2/8}Ce_{1.5/8}Sn_{0.5/8}Pr_{4/8})Cu_{0.5/8}O_{2-y}$. (i): annealed at 1473 K in air, (ii): reduced at 1373 K in H_2 gas after annealing at 1473 K in air, (iii): re-oxidized at 973 K in air. ○: λ-phase, #:λ_{Pr}, †:pyrochlore-like ,■:SnO_2, □:Sn, ▼:A-Pr_2O_3, ▲:CuO, ×:Cu.

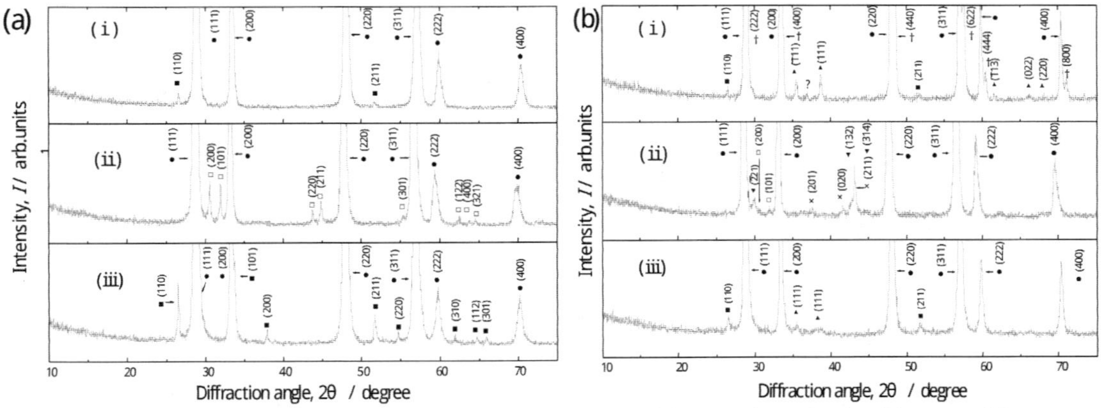

Figure 2 XRD patterns of Ce-rich Zr-Ce-Sn-Pr-O mixed oxide powders obtained by repetitive oxidation and reduction and compared to those of Cu added Zr-Ce-Sn-Pr-O mixed oxide powders. (a): $(Zr_{2/8}Ce_{4.5/8}Sn_{0.5/8}Pr_{1/8})O_{2-y}$, (b): $(Zr_{2/8}Ce_{4.5/8}Sn_{0.5/8}Pr_{1/8})Cu_{0.5/8}O_{2-y}$. (i): annealed at 1473 K in air. (ii): reduced at1373 K in H_2 gas after annealing at 1473 K in air, (iii): re-oxidation at 973 K in air. ●: C-phase, †:pyrochlore-like, ■:SnO_2, □:Sn, ▲:CuO, ▼:Cu_6Sn_5, ×:Cu_3Sn.

Figure 3 shows OSC data for Pr-rich and Ce-rich samples. OSC values for samples containing Sn were larger than those for the corresponding Sn-free samples. The on-set temperature shifted to a lower temperature. For the Ce-rich sample, Cu-addition further improved the OSC value and its on-set temperature and this result is in line with the fact that SnO_2 and CuO particles are distributed throughout the bulk phase after repetitive reduction and oxidation. Figure 4 shows TPD results used to obtain oxygen release properties as measured by TG/DTA. The weight loss of samples and the slopes of their graphs with time correspond to the total amount of oxygen released and the release rate, respectively. As seen in the figure for the Ce-rich composition both the total amount and rate of oxygen release increased according to the following order: Sn-free < Sn-addition < (Sn+Cu)-addition. This result is consistent with the OSC data shown in Figure 3(b). However, the on-set temperature for the TPD data did not increase in this order. Further investigation of the chemical properties of oxide samples with fine precipitation is needed.

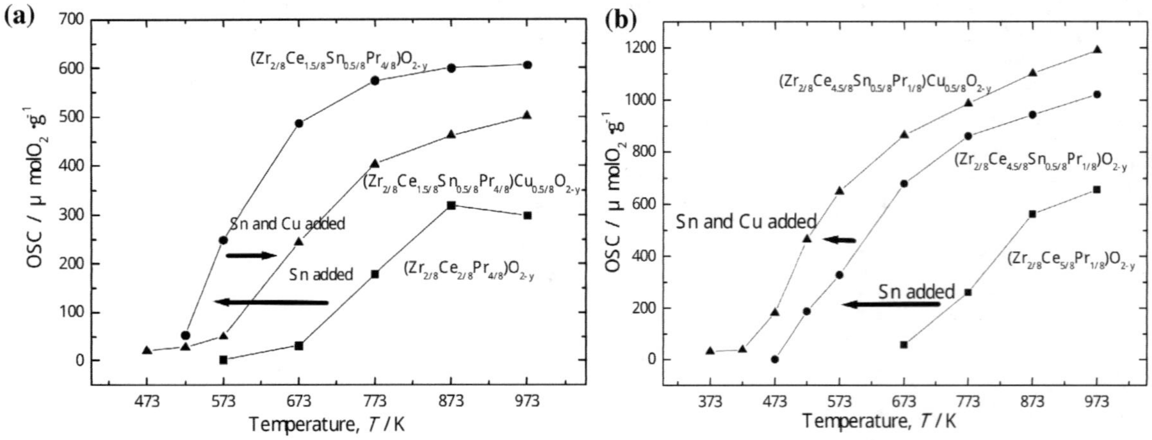

Figure 3 OSC data plotted against temperature. (a): Pr-rich composition, (b): Ce-rich composition.

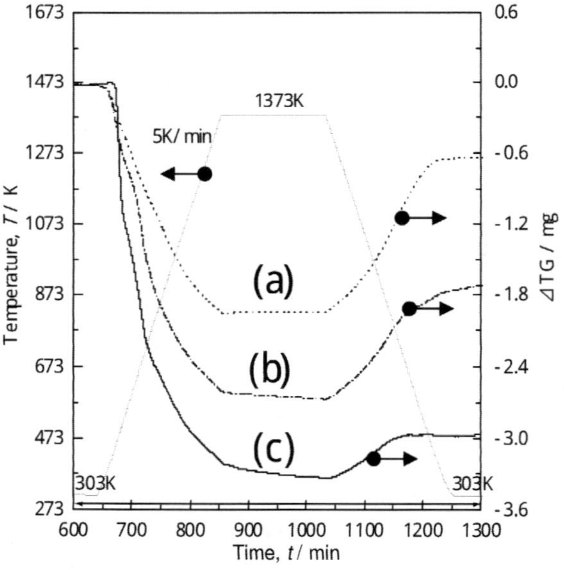

Figure 4 TPD oxygen gas analysis from TG/DTA in Ar+5%H_2 gas for various Ce-rich samples.
(a):$(Zr_{2/8}Ce_{5/8}Pr_{1/8})O_{2-y}$,
(b):$(Zr_{2/8}Ce_{4.5/8}Sn_{0.5/8}Pr_{1/8})O_{2-y}$,
(c):$(Zr_{2/8}Ce_{4.5/8}Sn_{0.5/8}Pr_{1/8})Cu_{0.5/8}O_{2-y}$,

4. Conclusion

For Cu-added samples, Cu_6Sn_5 and Cu_3Sn phases were formed during reduction at 1373 K in H_2 and successive oxidation appeared to promote a fine distribution of Sn and Cu ions throughout the bulk phases. Improvement of OSC values and on-set temperatures may be attributed to a change in the phase-state. We expect that a fine distribution of oxide phases would lead to highly durable automobile catalysts and a reduction in the amount of noble metals used in them.

Acknowledgements

This work was supported by a Grant-in-Aid for Scientific Research (S) (19106014), Japan Society.

References

[1] Otsuka-Yao S, Morioka H, Izu N and Okuda K 1995 *J.* Jpn. Ins. Met. **59** 1237-1246

[2] Ozawa M, Kimura M and Isogai A 1993 *J. Alloys Compd.* **193** 73-75

[3] Murota T, Sogawa K, Otsuka-Yao-Matsuo S and Omata T 2008 Collect abstract of the 2008 Autum Meeting of the Japan Inst.Metals pp. 260

Transition of rate-controlling process for reactive diffusion between Ta and bronze in superconductor

M Kajihara and Y Tejima

Department of Materials Science and Engineering, Tokyo Institute of Technology, Nagatsuta 4259-J2-59, Yokohama 226-8502, Japan

E-mail: kajihara@materia.titech.ac.jp

Abstract. The reactive diffusion between Ta and a bronze was experimentally examined using sandwich diffusion couples consisting of Ta and a Cu–9.3Sn–0.3Ti alloy. The diffusion couples were isothermally annealed at a temperature of T = 1053 K for various times up to t = 1200 h. During annealing, a layer of Ta_9Sn is formed at the interface in the diffusion couple. The mean thickness of the Ta_9Sn layer increases in proportion to a power function of the annealing time. The exponent of the power function is equal to unity at t < 167 h but smaller than 0.5 at t > 167 h. This means that the transition of rate-controlling process occurs at t = 167 h. For the reactive diffusion between Ta and the bronze at T = 1053 K, the growth of Ta_9Sn is controlled by the interface reaction at t < 167 h but by the volume and boundary diffusion at t > 167 h.

1. Introduction

In the electronics industry, a bronze method has been conveniently used to manufacture superconducting Nb_3Sn composite-wires with multifilamentary structure [1–12]. Here, Nb_3Sn is an intermetallic compound with the A-15 type crystal structure. In the bronze method, a ductile rod of Nb is embedded in a Cu–Sn alloy matrix and then extruded into a filament form. A bundle of filaments is inserted into a Ta tube covered with a Cu tube and then protruded in a wire. The resultant multifilamentary wire is annealed at an adequate temperature. Owing to reactive diffusion during annealing, a layer of Nb_3Sn is formed at the interface between the Cu–Sn alloy and Nb. The Cu–Sn alloy is called bronze, and thus this technique is denominated the bronze method. The growth behavior of the Nb_3Sn layer in the bronze method was investigated by many researchers [5–12]. However, monofilamentary and multifilamentary diffusion couples were adopted in most of the investigations. In such diffusion couples, interdiffusion takes place in rather complicated manners, and hence the Nb_3Sn layer with a uniform thickness cannot be formed easily. As a consequence, the filamentary diffusion couples are not appropriate to observe the growth behavior of the Nb_3Sn layer.

Instead of the filamentary diffusion couples, sandwich (Cu–Sn)/Nb/(Cu–Sn) diffusion couples were used to study experimentally the growth behavior of the Nb_3Sn layer in the bronze method [13, 14]. In the sandwich diffusion couple, interdiffusion occurs along the direction perpendicular to the interface, and hence the Nb_3Sn layer with a rather uniform thickness is formed during the interdiffusion. In the experiment by Osamura et al. [13], sandwich diffusion couples were prepared from Nb and a binary Cu–Sn alloy containing 7.4 at% of Sn, and then isothermally annealed in the temperature range of T = 973–1073 K. Different sandwich diffusion couples were prepared from Nb and a binary Cu–8.3Sn alloy and then isothermally annealed at T = 923–1053 K in a previous study [14]. To improve the superconducting properties of Nb_3Sn, a small amount of Ti is added into the Cu–Sn alloy. The influence of Ti on the growth of Nb_3Sn was experimentally studied by Osamura et al. at T = 973–1073 K using (Cu–7.5Sn–1Ti)/Nb/(Cu–7.5Sn–1Ti) diffusion couples [13]. The influence of Ti was also experimentally examined at T = 923–1053 K using sandwich diffusion couples consisting of Nb and a Cu–8.1Sn–0.3Ti alloy in a previous study [15]. According to these experiments [13, 15], the addition of Ti into the Cu–Sn alloy considerably accelerates the growth of Nb_3Sn at higher annealing temperatures but hardly at lower annealing temperatures.

Recently, Cu–Sn alloys with the $\alpha + \beta$ two-phase microstructure are preferentially used to manufacture the superconducting Nb_3Sn composite-wire. Here, α is the primary solid-solution phase of Cu with the face-centered cubic structure, and β is the intermediate phase with the body-centered cubic structure. Since the concentration of Sn is higher in the β phase than in the α phase, the amount of Sn supplied to Nb_3Sn is greater for the Cu–Sn alloy with the $\alpha + \beta$ two-phase microstructure than for that with the α single-phase microstructure. Hence, the influence of the β phase on the growth of Nb_3Sn was experimentally examined at T = 923–1053 K using sandwich diffusion couples composed of Nb and a Cu–9.3Sn–0.3Ti alloy in a previous study [16]. In these experiments [13–16], the mean thickness of the Nb_3Sn layer is expressed as a power function of the annealing time, and the exponent of the power function is typically close to unity. This means that the growth of Nb_3Sn is predominantly controlled by the interface reaction.

In the bronze method, Ta is used as a barrier against the diffusion of Sn from the bronze to the Cu tube. However, Ta may be gradually consumed by the reactive diffusion with the bronze during annealing. Unfortunately, no reliable information is available for the consumption of Ta during annealing. In the present study, the kinetics of the reactive diffusion between Ta and a bronze with the $\alpha + \beta$ two-phase microstructure was experimentally examined at T = 1053 K using sandwich diffusion couples consisting of Ta and a Cu–9.3Sn–0.3Ti alloy.

2. Experimental

Plate specimens with a dimension of 2 mm × 5 mm × 12 mm were prepared by spark erosion from a hot-rolled rectangular ingot of a ternary Cu–9.3Sn–0.3Ti alloy in a manner similar to a previous study [16]. Parallel surfaces with an area of 5 mm × 12 mm of each Cu–Sn–Ti plate specimen were mechanically polished on 120–800 emery papers. One of the polished parallel-surfaces was again mechanically polished on 1500–4000 emery papers, and then finished using diamond with a size of 1 μm. Pure Ta sheet specimens with a dimension of 0.1 mm × 5 mm × 12 mm were degreased in acetone with an ultrasonic cleaner. After degreasing, a Ta sheet specimen was immediately sandwiched between the finished surfaces of two freshly prepared Cu–Sn–Ti plate specimens in ethanol by a technique adopted in a previous study [16]. Each sandwich couple was completely dried, and then isothermally heat-treated for diffusion bonding in an evacuated silica tube, followed by air-cooling. The diffusion bonding was carried out at a temperature of T = 1053 K for 24 h. The diffusion couples were separately encapsulated in evacuated silica capsules, and then isothermally annealed at T = 1053 K for various times up to 1176 h. Hereafter, the summation of the heat-treating and annealing times is merely called the annealing time t. Cross-sections of the annealed diffusion couple were mechanically polished using 1500–4000 emery papers and diamond with a size of 1 μm, and then finished with an OP-S liquid by Struers Ltd. The microstructure of the cross-section was observed with a backscattered electron image (BEI) by scanning electron microscopy (SEM). The observation was also conducted by optical microscopy (OM). Concentration profiles of Cu, Sn, Ti and Ta for various phases on the cross-section were measured by electron probe microanalysis (EPMA).

3. Results and discussion

Under the present annealing conditions, a layer of Ta_9Sn with a uniform thickness is formed at the interface in the diffusion couple. Concentration profiles of the constituent components across the

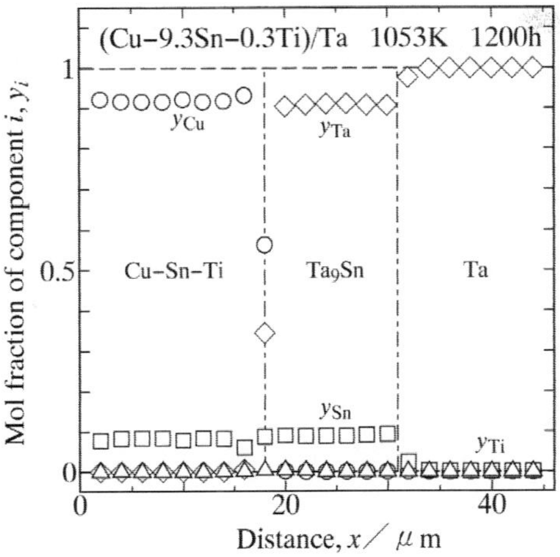

Figure 1. Concentration profiles in diffusion couple annealed at $T = 1053$ K for $t = 1200$ h.

Ta_9Sn layer and the α phase of the Cu–Sn–Ti alloy were determined by EPMA along the direction normal to the interface. A typical result for the diffusion couple annealed at $T = 1053$ K for $t = 1200$ h is shown in Fig. 1. In this figure, the ordinate and the abscissa show the mol fraction y_i of component i and the distance x, respectively, and open circles, squares, triangles and rhombuses indicate the values of y_{Cu}, y_{Sn}, y_{Ti} and y_{Ta}, respectively. As can be seen, y_{Cu} monotonically decreases from the α phase to Ta, and y_{Ta} from Ta to the α phase. On the other hand, y_{Sn} is similar in Ta_9Sn and the α phase. However, y_{Ti} is negligible in Ta_9Sn and Ta. Such a result was observed for all the annealed diffusion couples.

The mean thickness l of the Ta_9Sn layer is plotted against the annealing time t as open circles in Fig. 2. In this figure, the ordinate and the abscissa show the logarithms of l and t, respectively. As can be seen, the thickness l monotonically increases with increasing annealing time t. The open circles for shorter annealing times are located well on a straight line, and those for longer annealing times lie well on another straight line. Hence, in each annealing time range, l is expressed as a power function of t by the following equation:

$$l = k\left(\frac{t}{t_0}\right)^n . \qquad (1)$$

Here, t_0 is unit time, 1 s. It is adopted to make the argument t/t_0 of the power function dimensionless. The proportionality coefficient k has the same dimension as the thickness l, but the exponent n is dimensionless. From the open circles in Fig. 2, k and n were estimated by the least-squares method in each annealing time range. The estimated values are shown in Fig. 2. Using these values of k and n, l was calculated as a function of t from equation (1). The results are indicated as two solid lines in Fig. 2. These two lines are intersected each other at $t = 167$ h. The exponent n is equal to unity at $t < 167$ h but smaller than 0.5 at $t > 167$ h. When the interdiffusion across the Ta_9Sn layer is the rate-controlling process for the growth of Ta_9Sn and volume diffusion governs the interdiffusion, n is equivalent to 0.5. If boundary diffusion contributes to the interdiffusion, however, n becomes smaller than 0.5 [17]. On the other hand, n is equal to unity for the growth of Ta_9Sn controlled by the interface reaction [18]. Consequently, the growth of Ta_9Sn is controlled by the interface reaction at $t < 167$ h but by the volume and boundary diffusion at $t > 167$ h. The corresponding result for Nb_3Sn at $T = 1053$ K in a previous study [16] is indicated as open squares in Fig. 2. As can be seen, n is rather close to unity for Nb_3Sn. Therefore, the growth of Nb_3Sn is predominantly controlled by the interface reaction.

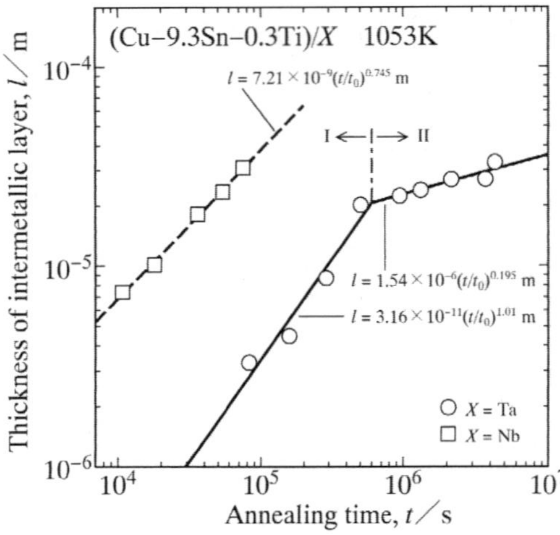

Figure 2. The thickness l of Ta_9Sn versus the annealing time t at T = 1053K shown as open circles. The corresponding result for Nb_3Sn at T = 1053 K in a previous study [16] is indicated as open squares.

However, the growth rate is one order of magnitude smaller for Ta_9Sn than for Nb_3Sn. This means that Ta works as an effective diffusion barrier in the bronze method.

4. Conclusions

The kinetics of the reactive diffusion between Ta and the Cu–9.3Sn–0.3Ti alloy was experimentally examined at a temperature of T = 1053 K. During isothermal annealing at T = 1053 K, Ta_9Sn is produced as a layer at the flat interface between Ta and the Cu–Sn–Ti alloy. The thickness l of Ta_9Sn is proportional to a power function of the annealing time t. The exponent of the power function is equal to unity at $t < 167$ h but smaller than 0.5 at $t > 167$ h. Hence, the growth of Ta_9Sn is controlled by the interface reaction at $t < 167$ h but by the volume and boundary diffusion at $t > 167$ h.

References

[1] Kaufmann AR and Pickett JJ 1970 *Bull. Am. Phys. Soc.* **15** 838
[2] Suenaga M and Sampson WB 1971 *Appl. Phys. Lett.* **18** 584
[3] Tachikawa K, Yoshida Y and Rinderer L 1972 *J. Mater. Sci.* **7** 1154
[4] Suenaga M and Sampson WB 1972 *Appl. Phys. Lett.* **20** 443
[5] Farrell HH, Gilmer GH and Suenaga M 1974 *J. Appl. Phys.* **45** 4025
[6] Farrell HH, Gilmer GH and Suenaga M 1975 *Thin Solid Films* **25** 253
[7] Dew-Hughes H, Luhman TS and Suenaga M 1976 *Nucl. Tech.* **29** 268
[8] Murase S, Koike Y and Shiraki H 1978 *J. Appl. Phys.* **49** 6020
[9] Kwasnitza K, Narlikar AV, Nissen HU and Salathé D 1980 *Cryogenics* 715
[10] Upadhyay P, Samanta SB and Narlikar AV 1981 *Mat. Res. Bull.* **16** 741
[11] Reddi BV, Raghavan V, Ray S and Narlikar AV 1983 *J. Mater. Sci.* **18** 1165
[12] Cheng CC and Verhoeven JD 1988 *J. Less-Com. Met.* **139** 15
[13] Osamura K, Ochiai S, Kondo S, Namatame M and Nosaki M 1986 *J. Mater. Sci.* **21** 1509
[14] Muranishi Y and Kajihara M 2005 *Mater. Sci. Eng. A* **404** 33
[15] Hayase T and Kajihara M 2006 *Mater. Sci. Eng. A* **433** 83
[16] Mikami K and Kajihara M 2007 *J. Mater. Sci.* **42** 8178
[17] Furuto A and Kajihara M 2008 *Mater. Trans.* **49** 294
[18] Sutton AP and Balluffi RW 1995 *Interfaces in Crystalline Materials* (Oxford Science Publications, Clarendon Press, Oxford) p. 603

AUTHOR INDEX

AUTHOR INDEX

AUTHOR INDEX

Institute of Physics
Dirac House, Temple Back
Bristol BS1 6BE UK